Mathematical Logic

with special reference to the natural numbers

# Mathematical Logic

## with special reference to the natural numbers

S. W. P. STEEN

*Sometime Cayley Lecturer in pure mathematics in the University of Cambridge*

Cambridge
at the University Press 1972

CAMBRIDGE UNIVERSITY PRESS
Cambridge, New York, Melbourne, Madrid, Cape Town, Singapore, São Paulo, Delhi

Cambridge University Press
The Edinburgh Building, Cambridge CB2 8RU, UK

Published in the United States of America by Cambridge University Press, New York

www.cambridge.org
Information on this title: www.cambridge.org/9780521080538



First published 1972
This digitally printed version 2008

*A catalogue record for this publication is available from the British Library*

*Library of Congress Catalogue Card Number: 77-152636*

ISBN 978-0-521-08053-8 hardback
ISBN 978-0-521-09058-2 paperback

*To my Wife*

# Contents

# Preface

About ten years ago I conceived the idea of writing a book on the natural numbers because I thought that what had appeared up till then seemed to have reached a point where there was a certain amount of completeness – of course there never will be absolute completeness – and this is one of the attractions of the subject. Anyway it was not until I had retired that I had the time to get down to the task properly. The result is a book which begins with an account of formal languages including the two most basic, namely, the propositional calculus and the predicate calculus, and then goes on to arithmetic; beginning with a very simple arithmetic; finding this inadequate; extending it to overcome this inadequacy; finding the resulting system, though richer in modes of expression, still, but for a different reason, inadequate; extending this in turn to remedy this inadequacy; finding the resulting system has lost some of the 'nice' qualities of its predecessor, but is again, for a new reason, inadequate; extending this and so on.

Before I come to develop arithmetic formally, it is convenient to have a primitive notation for the natural numbers (mainly to avoid lengthy circumlocutions) from which the concept of order and the operations of addition and multiplication can easily be obtained. I use sequences of tally marks, this is sufficient for our purposes. The real difficulty with arithmetic, as with other things, enters with the universal quantifier, when we want to make statements about all natural numbers. This use of tally marks is mentioned in the text but in the main is left to the reader to fill in.

There are several topics absent from the book which might have been included, these are partly off the main line of development, partly applications of the general theory developed, partly sidelines, etc. Among these topics are: recursive analysis, constructive ordinals, recursive equivalence types, recursive probability theory, the word problem, algorithms, finite automata, $\lambda$-conversion, combinations, productions, intuitionism, various forms of propositional calculus, many-valued logics, and so on. Of these the constructive ordinals are mentioned several times because now and again we come across a process which can be continued into the constructible transfinite, but we do not go into it further.

The matter developed in this book was developed over the years in a course of lectures delivered at Cambridge, except that very little was said about the contents of Chs. 10, 11, 12, so these three chapters have not come under the fire of criticism of young scholars, and I feel that in consequence that they are not of the same quality as the earlier chapters, particularly the account of cut elimination in Ch. 10. The remaining chapters have been fairly well thrashed out in lecture and I am very appreciative of the comments of my classes and of the elegant demonstrations they gave me from time to time. I hope that I have acknowledged them all.

With regard to the language in which the book is written, this is meant to consist of instructions and descriptions and occasionally of pointing out that such and such a procedure would lead to an impossible situation. Later in the book, when treating with ultra products I have transgressed and used Zorn's lemma, but a purist can tear that piece out of the book.

Each chapter is followed by a short historical account of the matter treated in that chapter, it is this way that I make acknowledgement to those who first invented the matter, if I have made omissions then I apologize. After the historical account there follow a few examples. Many more examples can be found in books by Rogers (1967), Shoenfeld (1967) and Church (1956).

I must thank Professor R. Harrop and Dr N. Routledge for comments on a former, now completely discarded draft which developed a much more complicated system. The present system owes its simplicity to the iterator symbol. I must also thank Dr G. T. Kneebone for reading the draft of Chs. 1–7 inclusive and providing valuable comments, and Drs T. J. Smiley and L. Drake for reading the draft of the remaining chapters and again providing valuable comments; also to the University Press for courtesy and consideration during the production of the book, and finally to my wife for help with the tedious business of making an index.

S. W. P. S.

*Christ's College*
*Cambridge*
*June 1971*

# Introduction

Mathematics is the art of making vague intuitive ideas precise and then studying the result. Many examples can be given of the wealth of interesting matter that has arisen when a vague intuitive idea has been made precise. Half the solution to a problem is to state it precisely. Among these vague intuitive ideas is that of *natural number* and that of *preciseness* itself, there is also the vague intuitive idea of *correctness*. In this book we are mainly concerned with making these three vague intuitive ideas precise and with inventing a method whereby our thoughts can be either communicated to others or stored for our own memory.

It may be that our concept of natural number arises from our perception of our own heart beats; this gives us a linear (and as far as we can perceive) unending *progression*, without conscious beginning or ending. The concept of an unending progression of distinct things with a definite starting entity and never returning to any entity previously encountered is the essence of the concept of natural number; this concept is given to us by our perception of our own heart beats. May-be our perception of time arises from our sensing of the circulation of the blood in the brain. This gives a linear background to our thoughts and sense data. It is amusing to imagine a creature with a two-dimensional flow of fluid through its body, such a creature might have a two-dimensional conception of time and be quite unable to conceive of natural numbers as we do.

Apart from making the intuitive idea of natural number precise we want to construct a method for communicating our thoughts on natural numbers to other persons and a method of storing them so that they don't get lost and forgotten. This we do by constructing a written *language*; this consists of a linear series of signs or shapes and will be the main medium in which our thoughts will be made precise. The language will be constructed with the utmost precision so that there can be no dispute as to what we wish to convey. We then translate our thoughts into this language, write them down as we shall say; the reverse process of converting something written in such a language into a thought is called understanding what has been written. Whether another human being gets a similar thought to the writer is a philosophical question which we do not discuss.

Our first task then will be to construct a language suitable for our purposes; but we also want to teach it to other persons. There are two methods of doing this.

The first method is to take advantage of the fact that we all already know a language of sorts, namely the imprecise language of daily use. This language has been invented over the ages for the purposes of descriptions, commands, instructions, explanations, excuses, deceits, lies, warnings, songs, etc., it is imprecise in that it is impossible to give a precise definition of 'sentence', 'word', 'noun' and many other syntactic terms. For instance, try and give a definition of 'word in the English language' and then test this definition against the works of Shakespeare and see if it is satisfactory. Such a definition to be of any use must be applicable to all writers at all times and not to a particular writer. For instance a definition such as 'any consecutive set of letters (i.e. without a blank between them) written by Shakespeare and printed in the first folio edition is a word and these are all the words' though a precise definition would be quite unacceptable. Since we know the English language, or at any rate some part of it, then we can use it to describe what we are constructing and this will be quite satisfactory; we are quite accustomed to doing this sort of thing. It is unsatisfactory in that it makes our development of arithmetic depend on the English language or at any rate on part of the English language, and we wish our development to be independent of imprecise concepts. This method is the usual method used for teaching a language.

But there is another method which is quite satisfactory because if we use it then our development does not depend on any thing outside itself. This method could be used in a modified form for inter-planetary communication, and is a modification of the method by which we learnt English in the first place. It consists of teaching the language by repeatedly writing down the signs of the language (its alphabet) until the pupil has understood which are the correct signs and which are foreign, for instance foreign shapes could be put down from time to time and immediately obliterated. Then repeatedly writing down correctly formed sequences of correct signs and others which are incorrectly formed and immediately erasing these latter; continue them until the pupil has understood which sequences of signs are correctly formed (meaningful sentences). This process would be possible, but lengthy and boring in the extreme. We shall employ the first method in this book.

It would be quite possible to use part of the English language instead of inventing a new language to deal with the matter treated in this book. But the result would be impossibly lengthy and much of the matter treated here would hardly have been thought of in the first place if we restricted ourselves entirely to the English language. For example in the English language there is (except for idiomatic variations such as gender, plurals, etc.) only one pronoun namely 'it'. In our language variables correspond to pronouns and we require an unending list of them. In the English language with its one pronoun we have to introduce lengthy circumlocutions such as the 'first', 'the second', etc. to obviate the lack of different pronouns. This has the obvious disadvantage of introducing natural numbers before one has defined them. But the main advantage in a symbolic language is that it makes the metamathematical investigations far simpler and the whole subject far easier to handle. Without it very little headway would ever have been made.

The language we shall construct will only be suitable for expressing our thoughts on natural numbers. If we wanted to construct a language for some other purpose, say chemistry, then we should want many more new signs standing for further undefined concepts.

The main difficulty in constructing a language for extramathematical purposes is that we are unable to give precise definitions of the concepts used. For instance the solution of the problem 'which came first the egg or the hen?' is that we are as yet unable to give a precise definition of egg or of hen which would be applicable to all times and places; thus the ancestors of a given present day hen if pursued far enough back into geological times would ultimately contain creatures that no one would now call hen, in between such a creature and the given live hen of today would be a series of creatures changing by imperceptible degrees. Maybe at some future date we shall be able to define 'hen' as a creature with such and such a protein molecule in its genetic code.

In other words all extramathematical concepts have furry edges. Thus to count the grains of sand on a beach (i) it is not clear where the boundaries of the beach are (ii) it is not clear what objects are grains of sand, they range from minute pebbles to impalpable dust. We do not want any vagueness of this sort which is inherent in all colloquial languages. Again, words in a colloquial language are used differently as time goes on; in fact it has been said that a word changes its meaning each time it is used. The sentence 'it is artful, aweful and painful' would be used now-a-days in quite different contexts to its use 300 years ago, then it was praise now it would be ridicule. We want to avoid anything like this. Also a phrase like 'a rose red city half as old as time' is utter nonsense but extremely pleasing and conjures up all sorts of images, but we want to avoid this sort of thing too, we want our language to be permanent, definite and absolutely precise.

We shall find it very useful to have a primitive form of notation for natural numbers so that when describing our language we can use expressions like 'formula A has more signs than formula B'. Since our language is designed to talk about the natural numbers then the properties of natural numbers should be avoided when talking about it. But expressions such as the above can easily be eliminated as follows: replace each sign in the formulae A and B by a tally mark, then the expression is an abbreviation for 'the sequence of tally marks obtained from formula B is a proper initial segment of the sequence of tally marks arising from formula A'. Here we have a primitive form of notation for the natural numbers namely: a tally mark standing alone is a natural number, if $\mathcal{N}$ is a natural number then so is $\mathcal{N}^\frown |$, namely, the result of adding a tally at the end of the sequence of tallies $\mathcal{N}$; these are all the natural numbers.

These sequences of tallies form an unending progression of distinct things with a definite starting place and so have the fundamental property of natural numbers. With them we can do addition by juxtaposition, multiplication by replacing each tally of one set of tallies by replicas of the sequence of tallies of another set of tallies. In a similar manner we can deal with other simple situations, thus avoiding the use of the term 'natural number'.

The language that we are going to construct is called the *object language* and the language that we use to describe it is called the *metalanguage*. The metalanguage was learnt by a method somewhat like the second of the two methods of teaching languages mentioned above. We have just shown how to avoid the use of natural numbers in the metalanguage.

Thoughts primarily are languageless and are only put into language for purposes of communication with others or with oneself (memory). We are so accustomed to put thoughts straight into words that one fails to realize that they are languageless and that a language though useful is unnecessary in order to be able to think. If one required a language to think in then a child would be without thoughts at all until it had learnt a language; as it began to learn a language at first it would have only one or two words to think with. Everyone is familiar with the situation when one is momentarily without the right word to express a thought; again when asked to describe some situation an image of the situation arises in the mind and one reads off from it, one puts what one perceives in the mind image into words. When one is working, thoughts not sentences, are coming into being and are perceived first as languageless thoughts, some of them are discarded before being put into words.

To construct our language we must first make our tools; these will consist of a *tape* divided into squares and such that we can always add more tape at the end when desired, so that we are always with spare tape at the end. Also we shall want a pen or other instrument to make marks in the squares: the marks are called *signs*.

At this point we have to rely on the goodwill of the reader to read our writing; by this we mean that he must realize that such and such a mark is intended to represent such and such a sign. If anyone takes the trouble to examine all the letters printed in this, or any other, book he will find that no two letters are exactly alike even if they are supposed to represent the same letter. This fact is sometimes useful in crime detection, in comparing letters typed on different typewriters even if of the same make. One would like this point about goodwill not to occur and to say something like this 'the signs are: a simple closed curve, a simple segment', etc.; but the definition of these things is away beyond the end of the matter treated in this book. The list of signs that we are going to use will first be displayed and we rely on the good sense of the reader to identify any other marks we make with one of them or to realize that it is intended to be a *new* mark, i.e. one which has as yet not appeared. If a reader is unable to do this then he will be unable to read what has been written, he may as well do something else. The signs will be written in consecutive squares of the tape; the object of the tape is so that the signs are of reasonable size (a sign occupying a square mile would be impracticable; we shall get on with squares in which the letters in this book could be printed), and so that we know when we have come to the end of a sequence of signs by coming to a blank square. Without some such tacit or expressed restriction reading would be impossible, the next symb o l might be miles away. If we had an unending sequence of distinct signs then we should be unable to distinguish between every pair and reading would be impossible, in this case there would be two signs which differed so little from each other that even an electron microscope would be unable to detect any difference. For instance if the symbols were circles $\frac{1}{20}$th inch in diameter lying in a one inch square and with centres at the rational points.

Certain sequences of signs will be called *well-formed*, the rest are called *ill-formed*.

Our language must be *constructive*, that is to say it must be possible to decide by a terminating process whether a given formula is well-formed or is ill-formed. It is plain that if this was otherwise then our language would be unreadable. The well-formed formulae in our language correspond to the words and sentences of a conversational language. If we were teaching a conversational language then we should stop at this place. Having listed and explained a certain set of words and told how they can be formed into sentences we would then use the language for descriptions, instructions, etc. But in our case our motive is different, we want to obtain the *true* statements about natural numbers.

*Truth* is a property of statements. The statement 'it snows' is true if and only if it snows. The inverted commas round a statement give us a name of that statement. We wish to give a truth definition for our language which will make precise our vague intuitive conception of arithmetic truth. This we do by noting that our statements are built up from *atomic statements* by *connectives* (often called logical connectives, because they can be used in any language). Our atomic statements are equations and inequations. For these we can give a simple and intuitively satisfying truth-definition. The truth of *compound statements* is then determined in a perfectly precise manner from the truth-values of the statements from which it is compounded. Unfortunately we are sometimes unable to find the truth-value of a statement because we are referred to an unbounded set of more elementary statements. But perhaps this is fortunate because if we had a method which would tell us the truth-value of any statement then all the interest would have gone from mathematics, we could turn the whole thing over to a computer which would tell us the answer. The theory of truth for arithmetic is a very difficult and complicated business. We shudder to think what difficulties we should encounter if we attempted to deal with truth in such disciplines as theology, law or politics, even if we had a suitable language for them, these disciplines make considerable use of the concept of truth; judging by history both ancient and modern it appears that the only way the human race has found of settling the question of truth in any of these disciplines is by the use of force. Even a truth-definition for physics would be extremely difficult to handle.

Since our motive is to find true arithmetic statements and since we are without a terminating method of testing a given arithmetic statement for truth then we proceed by what is called the *deductive method*. We first display a class of statements which we accept as true, these we call *axioms*, they are statements that anyone studying arithmetic would be bound to accept as true. Then we list certain methods whereby from true statements we can obtain other true statements. These methods are usually expressed as figures, the given true statements are called the *premisses* and the one that is obtained from them is called the *conclusion*. They and the axioms virtually play the part of implicit definitions of our primitive signs and logical connectives. Thus starting with the axioms and applying the figures repeatedly we continue to produce more and more true statements. The statements produced in this way are called *theorems*. But we shall see that in all except the simplest languages some true statement will always elude us, i.e. theorems will be a proper subset of true statements.

Our intuition makes us believe that a given meaningful statement is either true or false, that we are without a third possibility. This belief comes from the examination of testable cases – usually reducible to bounded sets. But when we come to deal with unbounded sets then a third possibility arises namely when we are unable to decide between truth and falsity. It seems rather useless to believe that each sentence is either true or false when we are unable to decide which is the case. Anyway we desire our language to be free from all beliefs.

The theory of truth belongs to *semantics*, this is the theory of *meaning*. The structure of a language belongs to *syntax*, this includes the signs and the rules which tell us which sequences of signs are well-formed and which figures are deductions. From syntax alone we obtain a language without meaning.

Meaning will be given to the languages which we construct, in that, for the arithmetic languages (without going into details) each statement will give rise to a structure consisting of pairs of sequences of tallies and according as these pairs of sequences of tallies are the same or are different then the statement will be true or false. A true statement will then correspond to a possible situation as regards these sets of pairs of tallies and a false statement will correspond to an impossible situation as regards these sets of pairs of sequences of tallies. In this way our language will have meaning and the true statements will correspond to possible constructions involving sequences of tallies. The full details of this sketch will become apparent as the book proceeds, but we omit them here.

We have just said that our motive is to obtain true arithmetic statements. But we shall only obtain a few of the most elementary such statements. Our interest is in what our languages can do and what their limitations are. That is to say we are interested in the *metatheory* of our languages. If we were interested in obtaining true arithmetic statements then we would be writing a book on arithmetic, it would differ from an ordinary book on arithmetic because we should begin with a very full syntactic introduction.

Frequently we want to show that each formula which has one property also has another property. We do this by showing outright that the shortest formulae with the first property also have the second property, then we show that if each formula having the first property and shorter than a given formula having the first property also has the second property then the given formula also has the second property. In this way we give sufficient instructions to obtain for a given formula having the first property a detailed demonstration that it also has the second property. This method we call *formula induction*, it is distinct from mathematical induction.

It has often been said that a formal system is just a meaningless game with symbols. In our case this is not so, we have something definite to say and have invented a language in which to say it. It is of course easy to invent a formal system void of meaning, sometimes such systems are useful to investigate, some like Post's productions are obtained by formalizing the essential properties of formal systems such as ours.

## Historical remarks

Brouwer (1947) suggested that our concept of number derives from our perception of our own heartbeats. Progressions are discussed in *Principia Mathematica*, Whitehead & Russell. For remarks on formal and informal languages see Carnap (1957). The method suggested for planetary intercourse is due to Freudental (1960). The terms 'object language' and 'metalanguage' are due to Carnap (1957). The idea of using a tape divided into squares with only one symbol on a square is due to Turing (1936, 1967), who used it in his machines. Russell once said 'no two x's are alike'. The term 'new' in the sense in which we have used it is due to Quine (1951). Turing makes remarks (1936, 1967) about the unfeasibility of using an unbounded set of distinct symbols by making a definition of 'neighbourhood' for a symbol. Tarski (1933–56) exhaustively discusses the concept of truth in formalized languages. Brouwer in his intuitionism – ably expounded by Heyting (1956) – discards the hypothesis that a statement is either true or false. This is commonly called T.N.D. (Tertium non datur). Lorenzen (1955) has a useful discussion about formula induction and various other kinds of induction, which are used in the metalanguage.

# Chapter 1

# Formal systems

## 1.1 *Nature of a formal system*

A *formal system* is constructed by choosing a set of *signs* and laying down *rules* for their manipulation. We have a *tape* marked into consecutive *squares* which can always be lengthened, so that we always have vacant squares at the right. The signs are placed on consecutive squares of the tape from left to right, at most one sign to a square. We use capital ell in script type ($\mathscr{L}$) with or without superscripts or subscripts to denote an undetermined formal system.

## 1.2 *The signs and symbols*

The signs of a formal system $\mathscr{L}$, called $\mathscr{L}$-signs, must be displayed by representative figures and it must be possible to distinguish between them and to recognise them on different occasions and to decide of an object whether it is intended to represent an $\mathscr{L}$-sign or whether it is irrelevant. These conditions are required in order that reading the system be possible. We lack means of displaying an unending list of distinct signs and if we make a list of distinct signs we shall have to stop at some place. Moreover at any stage in our work we shall only require a set of signs that can be displayed, so that the restriction to an initial displayed list of signs is without restriction on our work.

The judgement whether a given mark or figure is intended to represent an $\mathscr{L}$-sign must be left to the reader. The $\mathscr{L}$-signs in the initial displayed list are called *primitive $\mathscr{L}$-signs.*

In some formal systems we require a method whereby we can always obtain a *new sign* of a certain kind, that is a sign which is distinct from any sign of that kind, used up to that place. In order to be able to do this, starting with an initial displayed list, we use what we call *compound $\mathscr{L}$-signs*, that is certain sequences of $\mathscr{L}$-signs which can be generated according to some fixed plan. If the *scheme of generation* can be carried as far as we wish we can generate a new compound $\mathscr{L}$-sign according to the scheme at any place we desire. The method we shall adopt is to select a primitive $\mathscr{L}$-sign and obtain compound $\mathscr{L}$-signs by continually adding superscript primes to the right.

For example, in this book we shall only need one scheme of generation, we shall use the sequence which begins as follows:

$$x \quad x' \quad x'' \quad x''' \quad x''''$$

to obtain an unlimited supply of compound $\mathscr{L}$-signs which we shall call *variables*.

We can describe this scheme of generation thus:

(i) '$x$' is a variable,

(ii) if $\Sigma$ is a variable then so is $\Sigma^{\frown\prime}$,

(iii) a sequence of primitive $\mathscr{L}$-signs is a variable if and only if it is obtained from (i) and (ii).

Here $\Sigma$ stands for an undetermined primitive or compound $\mathscr{L}$-sign and $\Sigma^{\frown\prime}$ stands for the result of superscripting a prime at the end (in the next square on the tape) of that undetermined primitive or compound $\mathscr{L}$-sign. The clause (ii) could have been worded 'the result of attaching a superscript prime at the end of a variable is a variable'. The sign between capital sigma and the prime is called the *concatenation sign*. It is to be distinct from the $\mathscr{L}$-signs, it is used to augment the English language when we talk about a formal system. A primitive $\mathscr{L}$-sign used like the prime above is called a *generating sign*.

The *symbols* of a formal system, called $\mathscr{L}$*-symbols*, are the primitive $\mathscr{L}$-signs other than any generating signs together with the compound $\mathscr{L}$-signs which are obtained by a scheme of generation. Thus the set of primitive $\mathscr{L}$-signs is limited and can be displayed, but if there are any schemes of generation then the $\mathscr{L}$-symbols can only be generated as far as desired. We shall use capital sigma with or without superscripted primes or subscripts to denote an undetermined $\mathscr{L}$-symbol. Thus $\Sigma'$ will stand for an undetermined $\mathscr{L}$-symbol, $\Sigma^{\frown\prime}$ will stand for the result of superscripting a prime at the right of an undetermined $\mathscr{L}$-symbol. We use the notation $(\nu)$ to stand for an undetermined succession of generating signs and $(S\nu)$ for the result of adding another generating sign at the end. Similarly for $(\kappa)$, $(\theta)$, $(\lambda)$, and $(\mu)$. Thus $\Sigma', \Sigma'', \Sigma''', \ldots, \Sigma^{(\nu)}$, will stand for a sequence of undetermined $\mathscr{L}$-symbols. We shall frequently use Greek letters and the concatenation sign in this way, it will make our descriptions and instructions easier to follow.

There is one further requirement: any linear sequence of $\mathscr{L}$-symbols must be uniquely constructed from $\mathscr{L}$-symbols. For instance if 1 00 10 01 were $\mathscr{L}$-symbols obtained from the primitive $\mathscr{L}$-signs 0 and 1 then the linear sequence 1001 could be constructed from $\mathscr{L}$-symbols in two different ways. Similarly a linear sequence of generating signs must be uniquely constructed. For instance if $'$ and $''$ were different generating signs then $x'''$ would be ambiguous. Again if we allowed the same generating sign to be used fore and aft so that $'''x$, $''x$, and $'x$ were $\mathscr{L}$-symbols as well as $x'$, $x''$ and $x'''$ then $x'''x$ could be constructed in several ways.

## 1.3 *The formulae*

An $\mathscr{L}$-formula is a terminating sequence of $\mathscr{L}$-symbols. Thus an $\mathscr{L}$-symbol standing alone is an $\mathscr{L}$-formula. The *null $\mathscr{L}$-formula* is the empty terminating sequence of $\mathscr{L}$-symbols. A non-null terminating sequence of generating signs (if $\mathscr{L}$ contains any such signs) fails to be an $\mathscr{L}$-formula. Given a terminating sequence of signs (whether $\mathscr{L}$-signs or other signs) it is possible to decide whether it is an $\mathscr{L}$-formula or contains signs foreign to the system $\mathscr{L}$ or contains generating signs incorrectly placed. It is possible to decide of an $\mathscr{L}$-symbol whether it is the end symbol of a given $\mathscr{L}$-formula or is the initial $\mathscr{L}$-symbol of that $\mathscr{L}$-formula and if neither is the case then it is possible to find the $\mathscr{L}$-symbol which precedes it and the $\mathscr{L}$-symbol which follows it, our device of the tape divided into consecutive squares does this.

We shall use capital phi, psi, chi and omega with or without superscripts or subscripts to denote undetermined $\mathscr{L}$-formulae. If $\Phi$ and $\Psi$ are $\mathscr{L}$-formulae then $\Phi^\frown\Psi$ shall denote that $\mathscr{L}$-formula which is obtained by extending the $\mathscr{L}$-formula $\Phi$ by the addition of the $\mathscr{L}$-symbols of the $\mathscr{L}$-formula $\Psi$ to the right of $\Phi$ in the order in which they occur in $\Psi$. Similarly if $\Xi$ is an $\mathscr{L}$-formula then $\Phi^\frown\Psi^\frown\Xi$ shall denote that $\mathscr{L}$-formula which is obtained from the $\mathscr{L}$-formula $\Phi^\frown\Psi$ by addition of the $\mathscr{L}$-symbols of $\Xi$ to the right of the $\mathscr{L}$-formula $\Phi^\frown\Psi$ in the order in which they occur in $\Xi$. Clearly the same $\mathscr{L}$-formula is obtained if we attach $\Psi^\frown\Xi$ to the right of the $\mathscr{L}$-formula $\Phi$. $\Phi$ is called an *initial segment* of $\Phi^\frown\Psi^\frown\Xi$, $\Xi$ is called an *end segment* of $\Phi^\frown\Psi^\frown\Xi$, $\Psi$ is called a *consecutive part* of $\Phi^\frown\Psi^\frown\Xi$, here $\Phi$ or $\Psi$ or $\Xi$ may be null. $\Phi\Psi$ without concatenation sign denotes the sequence of separate formulae $\Phi$ and $\Upsilon$ in that order, there is then at least one blank square between them on the tape. The absence of concatenation sign denotes that we are dealing with a sequence of separate formulae rather than with a single formula. Usually written $\Phi\ \Upsilon$ or $\Phi, \Upsilon$.

## 1.4 *Occurrences*

An *occurrence of an $\mathscr{L}$-symbol* $\Sigma$ in an $\mathscr{L}$-formula $\Phi$ is an initial segment of $\Phi$ which ends with the $\mathscr{L}$-symbol $\Sigma$. Note that by definition of initial segment the part of $\Phi$ which is left when the initial segment is removed is an $\mathscr{L}$-formula, the end segment may be null. Thus if $\Phi'$ is an $\mathscr{L}$-formula and $\Sigma$ is an $\mathscr{L}$-symbol belonging to a scheme of generation with the prime as generating sign then $\Phi'{}^\frown\Sigma$ fails to be an occurrence of the $\mathscr{L}$-symbol $\Sigma$ in the $\mathscr{L}$-formula $\Phi'{}^\frown\Sigma^\frown{}'$ because by definition of initial segment $\Phi'{}^\frown\Sigma$ fails to be an initial segment of the $\mathscr{L}$-formula $\Phi'{}^\frown\Sigma^\frown{}'$, the part left when $\Phi'{}^\frown\Sigma$ is removed is only the prime which fails to be an $\mathscr{L}$-formula. A given $\mathscr{L}$-symbol may have several distinct occurrences in a given $\mathscr{L}$-formula. If the last $\mathscr{L}$-symbol of the $\mathscr{L}$-formula $\Phi$ is $\Sigma$ then $\Phi$ itself is an occurrence of $\Sigma$ in $\Phi$. The null formula fails to be an occurrence of any $\mathscr{L}$-symbol in any $\mathscr{L}$-formula. Similarly an *occurrence of a non-null $\mathscr{L}$-formula* $\Upsilon$ in an $\mathscr{L}$-formula $\Phi$ is an initial segment $\Phi'$ of $\Phi$ which has the end segment $\Upsilon$. Note that by definition the remainder of $\Phi$ when $\Phi'$ is removed is an $\mathscr{L}$-formula. The remainder may be null.

## 1.5 *Rules of formation*

Any formal system that has been constructed till the present time can be modified so that it uses the rules of formation that we are now going to give. Thus we say that the rules of formation are the same for all formal systems.

The symbols of a formal system are of two kinds: *proper symbols* and *improper symbols*. The improper symbols are:

$\boldsymbol{\lambda}$ *abstraction symbol*,
( *left parenthesis*,
) *right parenthesis*.

The abstraction symbol may be absent. There must be some proper symbols. The proper symbols are of two species at most; *variables* and *constants*. Each proper symbol has a *type* associated with it, the improper symbols are type-less. If $\Sigma$ is a variable and $'$ is a generating symbol then $\Sigma^{\frown\prime}$ has the same type as $\Sigma$. We use small alpha, beta with or without superscript primes for undetermined *type symbols*. The type symbols are generated according to the following scheme:

(i) omicron is a type symbol,

(ii) iota with or without superscript primes is a type symbol,

(iii) if $\alpha$, $\beta$ are type symbols then so is $({}^\frown\alpha^\frown\beta^\frown)$,

(iv) these are the only type symbols.

Some of these type symbols may be absent from a given formal system. Thus a formal system might have type symbols only of the types $o$, $(oo)$, $((oo)o)$. In writing type symbols we frequently omit the *outer pair of parenthesis* and omit other parentheses by *association to the left*. The omitted parentheses can then be replaced uniquely so that the result is a correctly formed type symbol. Thus we sometimes write:

$$ooo \quad \text{for} \quad ((oo)o),$$

$$o(oo) \quad \text{for} \quad (o(oo)),$$

and so on.

The *rules of formation* uniquely associate a type to certain $\mathscr{L}$-formulae and enable us to find it or to decide that the $\mathscr{L}$-formula is without type. The $\mathscr{L}$-formulae which have a type according to the rules of formation are called *well-formed* formulae (w.f.f.). The rules of formation also define a *status* (*free* or *bound*) for each occurrence of each variable in a well-formed formula and enable us to find it. In this way each occurrence of each variable in a well-formed formula is classified as a *free occurrence* or as a *bound occurrence*. When a type symbol is a suffix to a symbol then that symbol stands for an object of that type.

The rules of formation are:

(i) A symbol $\Sigma_\alpha$ of type $\alpha$ standing alone is a formula of type $\alpha$.

(ii) If $\Sigma$ is a variable of type $\alpha$ then its occurrence in the formula $\Sigma$ is free.

(iii) If $\Phi_{({}^\frown\alpha^\frown\beta^\frown)}$ is a formula of type $({}^\frown\alpha^\frown\beta^\frown)$ and $\Psi'_\beta$ is a formula of type $\beta$ then $({}^\frown\Phi_{({}^\frown\alpha^\frown\beta^\frown)}{}^\frown\Psi'^{\frown}_{\beta})$ is a formula of type $\alpha$. If $\Phi'$ is a $\binom{\text{free}}{\text{bound}}$ occurrence of a variable $\Sigma$ in the formula $\Phi_{({}^\frown\alpha^\frown\beta^\frown)}$ then $({}^\frown\Phi'$ is a $\binom{\text{free}}{\text{bound}}$

occurrence of the variable $\Sigma$ in the formula $(\ulcorner\Phi_{(\ulcorner\alpha\ulcorner\beta\urcorner)}\ulcorner\Psi'^{\urcorner}_{\beta})$, if $\Psi'$ is a $\binom{\text{free}}{\text{bound}}$ occurrence of the variable $\Sigma$ in the formula $\Psi_\beta$ then $(\ulcorner\Phi_{(\ulcorner\alpha\ulcorner\beta\urcorner)}\ulcorner\Psi'$ is a $\binom{\text{free}}{\text{bound}}$ occurrence of the variable $\Sigma$ in the formula $(\ulcorner\Phi_{(\ulcorner\alpha\ulcorner\beta\urcorner)}\ulcorner\Psi'^{\urcorner})$.

(iv) If $\Sigma_\beta$ is a variable of type $\beta$ and $\Phi_\alpha$ is a formula of type $\alpha$ then $(\ulcorner\boldsymbol{\lambda}\ulcorner\Sigma^{\urcorner}_{\beta}\Phi^{\urcorner}_{\alpha})$ is a formula of type $(\ulcorner\alpha\ulcorner\beta\urcorner)$. Each occurrence of the variable $\Sigma_\beta$ in the formula $(\ulcorner\boldsymbol{\lambda}\ulcorner\Sigma^{\urcorner}_{\beta}\Phi^{\urcorner}_{\alpha})$ is a bound occurrence of the variable $\Sigma_\beta$ in the formula $(\ulcorner\boldsymbol{\lambda}\ulcorner\Sigma^{\urcorner}_{\beta}\Phi^{\urcorner}_{\alpha})$. If $\Phi'$ is a $\binom{\text{free}}{\text{bound}}$ occurrence of a variable $\Sigma$, distinct from the variable $\Sigma_\beta$, in the formula $\Phi_\alpha$ then $(\ulcorner\boldsymbol{\lambda}\ulcorner\Sigma^{\urcorner}_{\beta}\Phi'$ is a $\binom{\text{free}}{\text{bound}}$ occurrence of the variable $\Sigma$ in the formula $(\ulcorner\boldsymbol{\lambda}\ulcorner\Sigma^{\urcorner}_{\beta}\Phi^{\urcorner}_{\alpha})$.

(v) A formula has a type if and only if a type is given to it by (i), (iii) and (iv). An occurrence of a variable is $\binom{\text{free}}{\text{bound}}$ if and only if it is $\binom{\text{free}}{\text{bound}}$ according to (i), (ii), (iii) and (iv).

An $\mathscr{L}$-formula which fails to be well-formed is said to be *ill-formed*.

Given an $\mathscr{L}$-formula we can discover whether it is well-formed or ill-formed and if it is well-formed then we can find its associated type, the details will be given shortly. According to these rules the null formula is without type. The formation of the formula $(\ulcorner\Phi_{(\ulcorner\alpha\ulcorner\beta\urcorner)}\ulcorner\Psi^{\urcorner}_{\beta})$ from the formulae $\Phi_{(\ulcorner\alpha\ulcorner\beta\urcorner)}$ and $\Psi_\beta$ is called *application*. A formula $\Phi_{(\ulcorner\alpha\ulcorner\beta\urcorner)}$ of the type $(\ulcorner\alpha\ulcorner\beta\urcorner)$ is called a *functor* of type $(\ulcorner\alpha\ulcorner\beta\urcorner)$. In the application process a functor of type $(\ulcorner\alpha\ulcorner\beta\urcorner)$ is given an *argument* of type $\beta$ and the resulting formula of type $\alpha$ is called the *value* of the functor for that argument. The formula $(\ulcorner\boldsymbol{\lambda}\ulcorner\Sigma^{\urcorner}_{\beta}\Phi^{\urcorner}_{\alpha})$ formed from the formula $\Phi_\alpha$ and the variable $\Sigma_\beta$ is called the *abstract of* $\Phi_\alpha$ *with respect to the variable* $\Sigma_\beta$. The abstract is a functor, by application it can be given an argument of type $\beta$ and the resulting formula is of type $\alpha$, thus $(\ulcorner(\ulcorner\boldsymbol{\lambda}\ulcorner\Sigma^{\urcorner}_{\beta}\Phi^{\urcorner}_{\alpha})\ulcorner\Psi^{\urcorner}_{\beta})$ is of type $\alpha$.

If the formula $\Phi'$ is a free occurrence of the variable $\Sigma_\beta$ in the well-formed formula $\Phi_\alpha$ then the occurrence $(\ulcorner\boldsymbol{\lambda}\ulcorner\Sigma^{\urcorner}_{\beta}\,\Phi'$ of the variable $\Sigma_\beta$ in the well-formed formula $(\ulcorner\boldsymbol{\lambda}\ulcorner\Sigma^{\urcorner}_{\beta}\,\Phi^{\urcorner}_{\alpha})$, which by (iv) is a bound occurrence, is said to be bound by the occurrence of $(\ulcorner\boldsymbol{\lambda}\ulcorner\Sigma_\beta$ in $(\ulcorner\boldsymbol{\lambda}\ulcorner\Sigma^{\urcorner}_{\beta}\,\Phi^{\urcorner}_{\alpha})$. The well-formed part $\Phi_\alpha$ of the well-formed formula $({}^\frown\boldsymbol{\lambda}{}^\frown\Sigma^\frown_\beta\Phi^\frown_\alpha)$ is called the *scope of the abstraction of the variable* $\Sigma_\beta$. If $\Phi'^\frown({}^\frown\boldsymbol{\lambda}{}^\frown\Sigma^\frown\Xi^\frown)$ is an occurrence of $({}^\frown\boldsymbol{\lambda}{}^\frown\Sigma^\frown\Xi^\frown)$ in $\Phi_\alpha$ then each free occurrence of the variable $\Sigma$ in $\Xi$ is bound by the occurrence $\Phi'^\frown({}^\frown\boldsymbol{\lambda}{}^\frown\Sigma$ of $({}^\frown\boldsymbol{\lambda}{}^\frown\Sigma$ in $\Phi_\alpha$. A bound occurrence $\Phi'$ of a variable $\Sigma$ in a well-formed formula $\Phi$ is bound by an occurrence $\Phi''$ of $({}^\frown\boldsymbol{\lambda}{}^\frown\Sigma$ in $\Phi'$, and $\Phi''$ is such that the scope of the occurrence $\Phi''$ of $({}^\frown\boldsymbol{\lambda}{}^\frown\Sigma$ in $\Phi'$ is the shortest well-formed formula $\Psi$ such that $\Phi'$ is an occurrence of $\Sigma$ in $\Phi''^\frown\Psi$. We also say that any well-formed part of $\Xi$ that contains a free occurrence of the variable $\Sigma$ is bound by the occurrence $\Phi'^\frown({}^\frown\boldsymbol{\lambda}{}^\frown\Sigma$ of $({}^\frown\boldsymbol{\lambda}{}^\frown\Sigma$ in $\Phi_\alpha$. We say that a well-formed part $\Xi$ of $\Phi_\alpha$ is *free for* $\Sigma$ in $\Phi_\alpha$ if it fails to be bound by any occurrence of $({}^\frown\boldsymbol{\lambda}{}^\frown\Sigma$ in $\Phi_\alpha$. Thus the occurrence $({}^\frown\boldsymbol{\lambda}{}^\frown\Sigma^\frown_\beta({}^\frown\boldsymbol{\lambda}{}^\frown\Sigma^\frown_\beta({}^\frown\Xi_{({}^\frown\alpha{}^\frown\beta{}^\frown)}{}^\frown\Sigma_\beta$ of the variable $\Sigma_\beta$ in the well-formed formula $({}^\frown\boldsymbol{\lambda}{}^\frown\Sigma^\frown_\beta({}^\frown\boldsymbol{\lambda}{}^\frown\Sigma^\frown_\beta({}^\frown\Xi_{({}^\frown\alpha{}^\frown\beta{}^\frown)}{}^\frown\Sigma^\frown_\beta)^\frown)^\frown)$ is bound by the occurrence $({}^\frown\boldsymbol{\lambda}{}^\frown\Sigma^\frown_\beta({}^\frown\boldsymbol{\lambda}{}^\frown\Sigma_\beta$ of $({}^\frown\boldsymbol{\lambda}{}^\frown\Sigma_\beta$ in that formula.

## 1.6 *Parentheses*

Suppose that $\Xi^\frown({}^\frown\Phi_{({}^\frown\alpha{}^\frown\beta{}^\frown)}{}^\frown\Psi^\frown_\beta)^\frown\Xi'$ is a well-formed formula where the parts $\Xi$ or $\Xi'$ or both may be null. Then the occurrence $\Xi^\frown($ of the left parenthesis is said to *correspond* to the occurrence $\Xi^\frown({}^\frown\Phi_{({}^\frown\alpha{}^\frown\beta{}^\frown)}{}^\frown\Psi^\frown_\beta)$ of the right parenthesis, and vice versa. These occurrences of parentheses are called *mates*. Similarly, if $\Xi^\frown({}^\frown\boldsymbol{\lambda}{}^\frown\Sigma^\frown_\beta\Phi^\frown_\alpha)^\frown\Xi'$ is a well-formed formula in which the parts $\Xi$ or $\Xi'$ or both may be null, then the occurrences $\Xi^\frown($ and $\Xi^\frown({}^\frown\boldsymbol{\lambda}{}^\frown\Sigma^\frown_\beta\Phi^\frown_\alpha)$ of parentheses are called mates. Each occurrence of a parenthesis in a well-formed formula has a unique mate. If $\Phi'^\frown($ is an occurrence of a left parenthesis in a well-formed formula $\Phi$ then the occurrence of its mate is $\Phi'^\frown({}^\frown\Phi''^\frown)$ where $({}^\frown\Phi''^\frown)$ is a well-formed formula and $\Phi'^\frown({}^\frown\Phi''^\frown)$ is an occurrence of a right parenthesis in $\Phi$. Similarly for the mate of a right parenthesis.

LEMMA (i). *If $\Phi$ is well-formed $\mathscr{L}$-formula then each occurrence of a left parenthesis in $\Phi$ has a unique mate, similarly for right parenthesis. The mate of a mate of the occurrence of a parenthesis is that occurrence of that parenthesis itself. Any proper initial segment $\Phi'$ of $\Phi$ contains an excess of occurrences of left parentheses, each occurrence of a right parenthesis in $\Phi'$ has its mate in $\Phi'$ but some occurrences of left parenthesis in $\Phi'$ lack mates in $\Phi'$. Similarly a proper end segment $\Phi''$ of $\Phi$ contains an excess of occurrences of right parentheses, each occurrence of a left parenthesis in $\Phi''$ has* *its mate in $\Phi''$ but some occurrences of right parentheses in $\Phi''$ lack mates in $\Phi''$. The mates of a consecutive well-formed part of $\Phi$ are mates in $\Phi$.*

If $\Phi$ is a single $\mathscr{L}$-symbol then $\Phi$ is well-formed and is without parentheses and the lemma follows trivially, otherwise $\Phi$ is of one of the forms:

$$({}^\frown\Psi_{({}^\frown\alpha{}^\frown\beta{}^\frown)}{}^\frown\Xi_\beta^\frown) \quad \text{or} \quad ({}^\frown\boldsymbol{\lambda}{}^\frown\Xi_\beta^\frown\Psi_\alpha^\frown).$$

Suppose the lemma has been demonstrated for well-formed formulae of shorter length than $\Phi$, then the lemma holds for $\Psi_{({}^\frown\alpha{}^\frown\beta{}^\frown)}$, $\Xi_\beta$ and $\Psi_\alpha$. Clearly the lemma then holds for $\Phi$.

Given an $\mathscr{L}$-formula $\Phi$ we can discover whether it is well-formed or is ill-formed and find its type if it is well-formed as follows: if $\Phi$ is a single symbol then it is well-formed and its type is known. Otherwise $\Phi$, if well-formed, must be of one of the forms: $({}^\frown\Psi_{({}^\frown\alpha{}^\frown\beta{}^\frown)}{}^\frown\Xi_\beta^\frown)$ or $({}^\frown\boldsymbol{\lambda}{}^\frown\Sigma_\beta^\frown\Psi_\alpha^\frown)$, the associated types are $\alpha$, $({}^\frown\alpha{}^\frown\beta{}^\frown)$ respectively. In either case $\Phi$ must begin with a left parenthesis and end with a right parenthesis, otherwise it is ill-formed. Suppose the former, then remove the outer pair of parentheses obtaining an $\mathscr{L}$-formula $\Phi'$, if $\Phi'$ is null then $\Phi$ is ill-formed. If $\Phi$ is well-formed then $\Phi'$ must begin with $\boldsymbol{\lambda}$, (or a proper $\mathscr{L}$-symbol of type $({}^\frown\alpha{}^\frown\beta{}^\frown)$) for some $\alpha, \beta$, otherwise $\Phi$ is ill-formed. If $\Phi'$ begins with $\boldsymbol{\lambda}$ then the next symbol must be a variable if $\Phi$ is well-formed, otherwise $\Phi$ is ill-formed. Suppose $\Phi'$ begins with $\boldsymbol{\lambda}{}^\frown\Sigma_\beta$, where $\Sigma_\beta$ is a variable, remove this, we are left with an $\mathscr{L}$-formula $\Phi''$, $\Phi$ is well-formed if and only if $\Phi''$ is well-formed, hence $\Phi''$ fails to be null if $\Phi$ is well-formed. Thus in this case we are referred to an $\mathscr{L}$-formula shorter than $\Phi$. The next case is when $\Phi'$ begins with a left parenthesis. Find the least initial segment of $\Phi'$ which fails to contain an excess of either kind of parenthesis. If we fail to find such a proper initial segment then, by the lemma, $\Phi$ is ill-formed. Suppose we find this initial segment $\Psi$ then $\Phi'$ is of the form $\Psi{}^\frown\Xi$. Now $\Phi$ is well-formed if and only if $\Psi$ and $\Xi$ are both well-formed and of respective types $({}^\frown\alpha{}^\frown\beta{}^\frown)$, $\beta$. The $\mathscr{L}$-formulae $\Psi$ and $\Xi$ are shorter than $\Phi$. The last case is when $\Phi'$ begins with the $\mathscr{L}$-symbol $\Psi_{({}^\frown\alpha{}^\frown\beta{}^\frown)}$, remove it and we are left with an $\mathscr{L}$-formula $\Phi''$. Now $\Phi$ is well-formed if and only if $\Phi''$ is a well-formed $\mathscr{L}$-formula of type $\beta$ also $\Phi''$ is shorter than $\Phi$. Thus again we are referred to $\mathscr{L}$-formulae shorter than $\Phi$. Clearly the process will terminate and we shall discover whether $\Phi$ is well-formed or is ill-formed and if it is well-formed then we shall find its type.

We say that the formulae $\Phi$ and $\Psi$ *overlap* if some non-null proper end segment of $\Phi$ is a non-null proper initial segment of $\Psi$, similarly with $\Phi$ and $\Psi$ interchanged.

LEMMA (ii). *If $\Phi$ and $\Psi$ are well-formed $\mathscr{L}$-formulae then either $\Phi$ is a consecutive part of $\Psi$ or $\Psi$ is a consecutive part of $\Phi$ or $\Phi$ and $\Psi$ fail to overlap.*

The first two alternatives arise in application and in abstraction. Suppose $\Phi$ is $\Phi'^{\frown}\Xi$ and $\Psi$ is $\Xi'^{\frown}\Psi''$, where $\Xi$, $\Xi'$, $\Phi'$ and $\Psi''$ are non-null. Since $\Xi$ is a proper end segment of $\Phi$ it will contain occurrences of right parentheses which lack their mates, but every occurrence of a left parenthesis in $\Xi$ will have its mate in $\Xi$, since $\Xi'$ is a proper initial segment of $\Psi$ it will contain occurrences of left parentheses which lack their mates, but every occurrence of a right parenthesis in $\Xi'$ will have its mate in $\Xi'$. Hence $\Xi$ is distinct from $\Xi'$ and so $\Phi$ and $\Psi$ fail to overlap.

COROLLARY. *If $\Phi$ is a well-formed $\mathscr{L}$-formula then the scopes of the various occurrences of $\boldsymbol{\lambda}$ in $\Phi$, if any, fail to overlap, but the scope of an occurrence may be contained in the scope of another occurrence.*

## 1.7 *Abstracts*

Suppose that $\phi\{\xi\}$ is of type $\beta$, where $\xi$ is of type $\alpha$, then $\boldsymbol{\lambda}\xi\,.\,\phi\{\xi\}$ is of type $(\beta\alpha)$. Given an argument $\delta$ of the type $\alpha$, the result, by the $\boldsymbol{\lambda}$-rules becomes $\phi\{\delta\}$, which depends on the formula $\delta$ of type $\alpha$. Thus the abstract $\boldsymbol{\lambda}\xi\,.\,\phi\{\xi\}$ depends on all formulae of type $\alpha$.

Now consider the following situation:

Let $\psi\{\eta, \xi\}$ be of type $\beta$, where $\eta$ is of type $(\beta\alpha)$, and $\xi$ is of type $\alpha$, then $\boldsymbol{\lambda}\eta\,.\,\psi\{\eta, \xi\}$, call this K, is of type $\beta(\beta\alpha)$ and depends on all formulae of type $(\beta\alpha)$. Let $\Delta$ be of type $\beta(\beta(\beta\alpha))$, then $\Delta\boldsymbol{\lambda}\eta\,.\,\psi\{\eta, \xi\}$ is of type $\beta$. Let $\phi\{\zeta, \xi\}$ be of type $\beta$, where $\zeta$ is of type $\beta$, then $\boldsymbol{\lambda}\xi\,.\,\phi\{\Delta\boldsymbol{\lambda}\eta\,.\,\psi\{\eta, \xi\}, \xi\}$, call this H, is of type $(\beta\alpha)$, and since it contains K then it depends on all formulae of type $(\beta\alpha)$ including itself! This form of circularity is known as *predicativity*, and H is called a *predicative formula*. This circularity seems unsatisfactory, to avoid it we have to make complicated changes in type theory. The situation will be reopened later.

## 1.8 *The rules of consequence*

Well-formed formulae of a formal system $\mathscr{L}$ of type $o$ (for some systems further effectively testable structural conditions are required) will be called $\mathscr{L}$*-statements*, if there are any further conditions then it must be possible to decide if they are fulfilled or if they are violated. An $\mathscr{L}$-statement will be called *closed* if every occurrence of each variable in the $\mathscr{L}$-statement is a bound occurrence, otherwise the $\mathscr{L}$-statement will be called *open*. In some formal systems it is required that an $\mathscr{L}$-statement be closed.

Certain $\mathscr{L}$-statements may be called $\mathscr{L}$*-axioms*, and if there are $\mathscr{L}$-axioms then there must be a method whereby we can decide of an $\mathscr{L}$-statement whether it is an $\mathscr{L}$-axiom or is distinct from every $\mathscr{L}$-axiom. Thus we could display the $\mathscr{L}$-axioms or we could give a description of them by laying down that any $\mathscr{L}$-statement satisfying such and such structural conditions is an $\mathscr{L}$-axiom (provided we have a method whereby we can decide of an $\mathscr{L}$-statement whether it satisfies the conditions or fails to do so). Such a description is called an $\mathscr{L}$*-axiom scheme.*

There may also be given some *rules of procedure* called $\mathscr{L}$*-rules*, these are relations between $\mathscr{L}$-statements whereby given a set of $\mathscr{L}$-statements satisfying certain structural conditions we may by applying one of the $\mathscr{L}$-rules produce another $\mathscr{L}$-statement whose structure depends in some definite manner on the structure of the members of the given set. The given set of $\mathscr{L}$-statements is called the *premisses* (or premiss if there is only one in the given set) and the $\mathscr{L}$-statement produced by application of the rule is called the *conclusion* of that rule. It must be possible to decide whether a given $\mathscr{L}$-statement is the result of an application of an $\mathscr{L}$-rule to given premisses or whether this fails to be the case. The $\mathscr{L}$-rules are depicted as follows:

$$\frac{\Phi}{\Xi}, \quad \frac{\Phi\ \Psi}{\Xi}$$

and so on.

There may be conditions on $\Phi$, $\Psi$ or on $\Phi$, $\Psi$, $\Xi$, if so then it must be possible to decide if these conditions are satisfied or are violated. The conditions must be checked before using the rule. The $\mathscr{L}$-statements above the line are called the premisses or *upper formulae*, the $\mathscr{L}$-statement below the line is called the conclusion or *lower formula*. The $\mathscr{L}$-axioms and the $\mathscr{L}$-statements which result from applications of the $\mathscr{L}$-rules are called $\mathscr{L}$*-theorems*. A partially ordered set of $\mathscr{L}$-statements which contains a unique last $\mathscr{L}$-statement and is such that each $\mathscr{L}$-statement in the set is either an $\mathscr{L}$-axiom or results from previous

$\mathscr{L}$-statements in the set by application of an $\mathscr{L}$-rule is called an *$\mathscr{L}$-proof* of the last $\mathscr{L}$-statement in the set. In particular cases the partially ordered set might be linearly ordered. A connected tree-like figure consisting of columns of $\mathscr{L}$-statements with an $\mathscr{L}$-axiom at the head of each column and an application of an $\mathscr{L}$-rule between each consecutive vertical pair of members of a column or at a place where two or more columns terminate and are replaced by a single column and which finally ends in a single $\mathscr{L}$-statement $\Phi$ is called an *$\mathscr{L}$-proof of $\Phi$ in tree-form* or an *$\mathscr{L}$-proof-tree of* $\Phi$. The $\mathscr{L}$-statement $\Phi$ is called the *base* of the tree. The portion of the tree which can be reached by proceeding upwards from a given $\mathscr{L}$-statement $\Psi$ in the tree is an $\mathscr{L}$-proof of $\Psi$ in tree-form and is called the *branch of the tree ending in* $\Psi$. An *$\mathscr{L}$-proof thread* is a linear column of $\mathscr{L}$-statements which forms a consecutive part of an $\mathscr{L}$-proof-tree. An $\mathscr{L}$-proof can be checked since it is possible to decide if a sequence of signs is an $\mathscr{L}$-formula and to decide the type of a well-formed $\mathscr{L}$-formula and so decide whether it is an $\mathscr{L}$-statement and to decide whether an $\mathscr{L}$-statement is an $\mathscr{L}$-axiom and to decide applications of $\mathscr{L}$-rules. The set of $\mathscr{L}$-axioms and $\mathscr{L}$-rules are collectively known as the *$\mathscr{L}$-rules of consequence.*

*An $\mathscr{L}$-theorem-scheme* is a description of a set of $\mathscr{L}$-theorems together with instructions for obtaining the $\mathscr{L}$-proof of any member of the set. $\mathscr{L}$-theorem-schemes are frequently stated when a set of $\mathscr{L}$-theorems have $\mathscr{L}$-proofs on the same general pattern, we then describe the pattern. In a formal system the $\mathscr{L}$-rules are normally *$\mathscr{L}$-rule-schemes*, that is to say they are descriptions applicable to a variety of cases.

We use the notation $\frac{\Phi}{\Psi}*$, $\frac{\Phi\ \Phi'}{\Psi}*$, etc. to denote that from an $\mathscr{L}$-proof of $\Phi$ (of $\Phi$ and $\Phi'$, etc.) we can find an $\mathscr{L}$-proof of $\Psi$. We use the notation $\genfrac{}{}{}{}{\Phi}{\overline{\Psi}}$, $\genfrac{}{}{}{}{\Phi\ \Phi'}{\overline{\Psi}}$, etc. to denote that $\Psi$ can be obtained from $\Phi$ (from $\Phi$ and $\Phi'$, etc.) by the $\mathscr{L}$-rules. In this case if the upper formula (formulae) are $\mathscr{L}$-theorems then so is the lower formula. $\frac{\Phi}{\Psi}*$, $\frac{\Phi\ \Phi'}{\Psi}*$, $\genfrac{}{}{}{}{\Phi}{\overline{\Psi}}$, $\genfrac{}{}{}{}{\Phi\ \Phi'}{\overline{\Psi}}$, etc. are called *derived $\mathscr{L}$-rules*. The demonstration of $\frac{\Phi}{\Psi}*$ consists in giving instructions to find an $\mathscr{L}$-proof of $\Psi$ from an $\mathscr{L}$-proof of $\Phi$. For $\genfrac{}{}{}{}{\Phi}{\overline{\Psi}}$ we must give the $\mathscr{L}$-rules used to transform $\Phi$ to $\Psi$. We also use the notation

$\frac{\Phi' \ldots \Phi^{(\theta)}}{\overline{\Psi' \ldots \Psi^{(\kappa)}}}$ to denote that each of $\Psi', \ldots, \Psi^{(\kappa)}$ may be obtained from $\Phi', \ldots, \Phi^{(\theta)}$ by the $\mathscr{L}$-rules. Similarly we use the notation $\frac{\Phi' \ldots \Phi^{(\theta)}}{\Psi' \ldots \Psi^{(\kappa)}}*$ to denote that from the $\mathscr{L}$-proofs of $\Phi', \ldots, \Phi^{(\theta)}$ we can effectively find $\mathscr{L}$-proofs for each of $\Psi', \ldots, \Psi^{(\kappa)}$. If $\theta = 0$ the above notations reduce to $\overline{\overline{\Psi' \ldots \Psi^{(\kappa)}}}$ and $\overline{\Psi' \ldots \Psi^{(\kappa)}}^{\,*}$ respectively. These mean that $\Psi', \ldots, \Psi^{(\kappa)}$ are $\mathscr{L}$-theorems. Note that $\frac{\Phi}{\Xi}$ whenever $\Xi$ is an $\mathscr{L}$-theorem, in this case $\Phi$ is unused. Also if $\frac{\Phi}{\Xi}$ then $\frac{\Phi\ \Psi}{\overline{\Xi}}$ here again $\Psi$ is unused.

## 1.9 *Corresponding and related occurrences*

Let capital gamma with or without superscripts or subscripts be signs foreign to a formal system $\mathscr{L}$. Then $\Phi\{\Gamma\}$ shall denote a terminating linear succession of $\mathscr{L}$-symbols and $\Gamma$'s, $\Phi\{\Gamma, \Gamma'\}$ shall denote a terminating linear succession of $\mathscr{L}$-symbols $\Gamma$'s and $\Gamma'$'s, here $\Phi\{\Gamma'\}$ is to be without occurrences of $\Gamma$, $\Phi\{\Gamma'', \Gamma'''\}$ without occurrences of $\Gamma$, $\Gamma'$, and so on. The terminating linear successions $\Phi\{\Gamma\}$, $\Phi\{\Gamma, \Gamma'\}$, and so on, are called *$\mathscr{L}$-formula-forms*. If $\Psi$, $\Psi'$ are non-null $\mathscr{L}$-formulae then $\Phi\{\Psi\}$, $\Phi\{\Psi, \Psi'\}$ shall denote the results of everywhere replacing $\Gamma$, $\Gamma'$ by $\Psi$, $\Psi'$ respectively in the $\mathscr{L}$-formula-form $\Phi\{\Gamma\}$, and so on. This is done by inserting a piece of tape which exactly contains the $\mathscr{L}$-formula $\Psi$ in place of the square which contains $\Gamma$, this is to be done for each occurrence of $\Gamma$, in $\Phi\{\Gamma\}$. Similarly each square containing $\Gamma'$ is replaced by a piece of tape which exactly contains $\Psi'$. Thus if $\Phi\{\Gamma, \Gamma'\}$ is $\Phi'^\frown\Gamma^\frown\Phi''^\frown\Gamma'^\frown\Phi'''^\frown\Gamma^\frown\Phi''''$ where $\Phi'$, $\Phi''$, $\Phi'''$ and $\Phi''''$ are $\mathscr{L}$-formulae, hence without occurrences of $\Gamma$, $\Gamma'$, then $\Phi\{\Psi, \Psi'\}$ is $\Phi'^\frown\Psi^\frown\Phi''^\frown\Psi'^\frown\Phi'''^\frown\Psi^\frown\Phi''''$. The $\mathscr{L}$-formulae $\Phi\{\Psi\}$, $\Phi\{\Psi, \Psi'\}$ are without occurrences of $\Gamma$ or $\Gamma'$. The $\mathscr{L}$-formulae $\Psi$, $\Psi'$ might be single $\mathscr{L}$-symbols, the $\mathscr{L}$-formulae $\Psi$, $\Psi'$ might be identical. $\Phi\{\mathfrak{G}\}$ denotes one of $\Phi\{\Gamma\}$, $\Phi\{\Gamma, \Gamma'\}$, etc. when it occurs the context will show which is intended.

Let $\Phi\{\Gamma\}$ be an $\mathscr{L}$-formula-form and let $\Psi'^\frown\Xi$ be an occurrence of the $\mathscr{L}$-formula $\Xi$ in the $\mathscr{L}$-formula $\Psi$, and let $\Phi'\{\Gamma\}^\frown\Gamma$ be an occurrence of $\Gamma$ in the $\mathscr{L}$-formula-form $\Phi\{\Gamma\}$. Then $\Phi'\{\Psi\}^\frown\Psi'^\frown\Xi$ is called an occurrence of $\Xi$ in $\Phi\{\Psi\}$ which corresponds to the occurrence of $\Psi'^\frown\Xi$ of $\Xi$ in $\Psi$ by substitution of $\Psi$ for $\Gamma$ in $\Phi\{\Gamma\}$. The $\mathscr{L}$-formula-form must be specified because there might be occurrences of $\Xi$ or of $\Psi$ in $\Phi\{\Gamma\}$. Suppose $\Phi''\{\Gamma\}^\frown\Xi$ is an occurrence of $\Xi$ in $\Phi\{\Gamma\}$ then $\Phi''\{\Psi\}^\frown\Xi$ is distinct from any occurrence of $\Xi$ in $\Phi\{\Psi\}$ which corresponds to the occurrence of $\Psi'^\frown\Xi$ of $\Xi$ in $\Psi$ by substitution of $\Psi$ for $\Gamma$ in $\Phi\{\Gamma\}$. We require the notion of corresponding occurrences because we shall build up $\mathscr{L}$-formula by certain rules and shall require to trace $\mathscr{L}$-symbols or $\mathscr{L}$-formulae through the build up. Sometimes we build up an $\mathscr{L}$-formula by certain rules and in the process transform some consecutive parts by other rules, and we may wish to trace these changing parts through the build up and transformation process. In this case we shall speak of *related occurrences*, which we now define. Let $\Phi\{\Gamma, \mathfrak{G}'\}$ be an $\mathscr{L}$-formula-form, let $\Psi'\{\mathfrak{G}'\}^\frown\Xi\{\mathfrak{G}'\}$ be an occurrence of $\Xi\{\mathfrak{G}'\}$ in $\Psi\{\mathfrak{G}'\}$. Let $\Phi'\{\Psi\{\mathfrak{G}'\}, \mathfrak{G}'\}^\frown\Psi'\{\mathfrak{G}'\}^\frown\Xi\{\mathfrak{G}'\}$ be an occurrence of $\Xi\{\mathfrak{G}'\}$ in $\Phi\{\Psi\{\mathfrak{G}'\}, \mathfrak{G}'\}$ corresponding to the occurrence $\Psi'\{\mathfrak{G}'\}^\frown\Xi\{\mathfrak{G}'\}$ of $\Xi\{\mathfrak{G}'\}$ in $\Psi\{\mathfrak{G}'\}$ by substitution of $\Psi\{\Gamma'\}$ for $\Gamma$ in $\Phi\{\Gamma, \mathfrak{G}'\}$. Then $\Phi'\{\Psi\{\mathfrak{W}'\}, \mathfrak{W}'\}^\frown\Psi'\{\mathfrak{W}'\}^\frown\Xi\{\mathfrak{W}'\}$ is called an occurrence of $\Xi\{\mathfrak{W}'\}$ in $\Phi\{\Psi\{\mathfrak{W}'\}, \mathfrak{W}'\}$ related to the occurrence $\Psi'\{\mathfrak{W}\}^\frown\Xi\{\mathfrak{W}\}$ of $\Xi\{\mathfrak{W}\}$ in $\Psi\{\mathfrak{W}\}$ by substitution of $\Psi\{\mathfrak{G}'\}$ for $\Gamma$ in $\Phi\{\Gamma, \mathfrak{G}'\}$ and the substitution of $\mathfrak{W}'$ for $\mathfrak{W}$ in $\Phi\{\Psi\{\mathfrak{G}'\}, \mathfrak{G}'\}$.

If $\Phi_\alpha\{\Gamma_\beta\}$ is an $\mathscr{L}$-formula form and if $\Phi_\alpha\{\Psi_\beta\}$ is a well-formed $\mathscr{L}$-formula of type $\alpha$ whenever $\Psi_\beta$ is of type $\beta$ then we shall say that $\Phi_\alpha\{\Gamma_\beta\}$ is an *$\mathscr{L}$-formula-form of type $\alpha$*. Similarly if $\Phi\{\Gamma_\beta, \Gamma'_{\beta'}\}$ is an $\mathscr{L}$-formula-form and if $\Phi_\alpha\{\Psi_\beta, \Psi'_{\beta'}\}$ is an $\mathscr{L}$-formula of type $\alpha$ whenever $\Psi_\beta$ is of type $\beta$ and $\Psi'_{\beta'}$ is of type $\beta'$, then $\Phi\{\Gamma_\beta, \Gamma'_{\beta'}\}$ is called an $\mathscr{L}$-formula-form of type $\alpha$, and so on. If $\alpha$ is omicron we sometimes omit the suffix $\alpha$ and speak of an *$\mathscr{L}$-statement-form*, similarly we use the term '*functor-form*', etc. We define 'free', 'bound', 'open', 'closed', etc. for formula-forms as for formulae. Thus if $\Phi\{\Gamma_\beta\}$ is a closed $\mathscr{L}$-statement-form and $\Sigma_\beta$ is a variable new to $\Phi\{\Gamma_\beta\}$ then $\Phi\{\Sigma_\beta\}$ has $\Sigma_\beta$ as sole free variable provided that some occurrence of $\Gamma_\beta$ in $\Phi\{\Gamma_\beta\}$ is outside the scope of any occurrence of $\boldsymbol{\lambda}\Gamma_\beta$ in $\Phi\{\Gamma_\beta\}$.

## 1.10 *The* $\boldsymbol{\lambda}$*-rules*

If lambda is an $\mathscr{L}$-symbol then $\mathscr{L}$ may contain the $\boldsymbol{\lambda}$*-rules* applicable to $\mathscr{L}$-variables and $\mathscr{L}$-formulae of certain types.

The $\boldsymbol{\lambda}$*-rules* for $\mathscr{L}$-variables of type $\beta$ and $\mathscr{L}$-formulae of type $\alpha$ are:

(i) If $\Sigma_\beta$ is an $\mathscr{L}$-variable of type $\beta$ and $\Psi_\beta$ is an $\mathscr{L}$-formula of type $\beta$

and $\Phi_\alpha\{\Psi_\beta\}$ is an $\mathscr{L}$-formula of type $\alpha$ whose $\mathscr{L}$-formula-form $\Phi_\alpha\{\Gamma_\beta\}$ lacks occurrences of the $\mathscr{L}$-variable $\Sigma_\beta$ then

$$\frac{\Xi\{(^\frown(^\frown\boldsymbol{\lambda}^\frown\Sigma^\frown_\beta\,\Phi_\alpha\{\Sigma_\beta\}^\frown)^\frown\Psi^\frown_\beta)\}}{\Xi\{\Phi_\alpha\{\Psi_\beta\}\}}$$

is a rule of procedure provided that each occurrence of an $\mathscr{L}$-variable $\Sigma$ in $\Phi_\alpha\{\Psi_\beta\}$ which corresponds to a free occurrence of that variable in $\Psi_\beta$ is a free occurrence of $\Sigma$ in $\Phi_\alpha\{\Psi_\beta\}$.

(ii) Conversely with the same proviso and the same notation

$$\frac{\Xi\{\Phi_\alpha\{\Psi_\beta\}\}}{\Xi\{(^\frown(^\frown\boldsymbol{\lambda}^\frown\Sigma^\frown_\beta\,\Phi_\alpha\{\Sigma_\beta\}^\frown)^\frown\Psi^\frown_\beta)\}}$$

is a rule of procedure.

(iii) Let $\Phi\{\Gamma\}$ be an $\mathscr{L}$-formula-form, $\Sigma$ an $\mathscr{L}$-variable. Let $\Phi\{\Sigma\}$ be the scope of an occurrence of $(^\frown\boldsymbol{\lambda}^\frown\Sigma$ in an $\mathscr{L}$-formula $\Psi\{(^\frown\boldsymbol{\lambda}^\frown\Sigma^\frown\Phi\{\Sigma\}^\frown)\}$. Let $\Sigma'$ be an $\mathscr{L}$-variable distinct from and of the same type as the $\mathscr{L}$-variable $\Sigma$. Let $\Phi\{\Gamma\}$ lack free occurrences of the $\mathscr{L}$-variables $\Sigma$, $\Sigma'$. Let each occurrence of $\Gamma$ in $\Phi\{\Gamma\}$ be outside the scope of any occurrence of $(^\frown\boldsymbol{\lambda}^\frown\Sigma'$ or of $(^\frown\boldsymbol{\lambda}^\frown\Sigma$ that there may be in $\Phi\{\Gamma\}$ then

$$\frac{\Psi\{(^\frown\boldsymbol{\lambda}^\frown\Sigma^\frown\Phi\{\Sigma\}^\frown)\}}{\Psi\{(^\frown\boldsymbol{\lambda}^\frown\Sigma'^\frown\Phi\{\Sigma'\}^\frown)\}}$$

is a rule of procedure. It is called *change of bound variable.*

If each occurrence of $\Gamma$ in $\Phi\{\Gamma\}$ lies outside any scope of $\boldsymbol{\lambda}^\frown\Sigma$ then we say that $\Gamma$ *is free for* $\Sigma$ *in* $\Phi\{\Gamma\}$.

## 1.11 *Definitions and abbreviations*

We shall frequently introduce new symbols to *abbreviate* closed formulae of a given formal system $\mathscr{L}$. The new symbol is to be considered as everywhere replaced by the $\mathscr{L}$-formula it abbreviates. This device is adopted merely to prevent $\mathscr{L}$-formulae becoming of unmanageable length; it also facilitates reading, and enables one more clearly to see certain aspects of the structure of an $\mathscr{L}$-formula. An $\mathscr{L}$-formula occupying a few pages of print would be difficult to assess; such a formula might begin with several lines consisting entirely of left parentheses; but if broken up into parts and new symbols used as abbreviations for the different parts we might arrive at a formula which could be printed on a line and certain features of its construction would be apparent at a glance. If the new symbol abbreviates a well-formed $\mathscr{L}$-formula then it can be given a type. Sometimes we shall introduce formulae containing new symbols to stand for certain other structurally related $\mathscr{L}$-formulae. When this is done the new symbol itself fails to stand for an $\mathscr{L}$-formula but certain formulae containing the new symbol stand for certain $\mathscr{L}$-formulae. This device is usually adopted if $\mathscr{L}$ fails to contain the abstraction symbol and its rules. Thus '$N$' could be introduced by

$$({}^\frown N^\frown \Phi^\frown) \quad \text{for} \quad ({}^\frown({}^\frown S^\frown \Phi^\frown)^\frown \Phi^\frown),$$

where $\Phi$ is of type $o$ and '$S$' is of type $ooo$, so that '$N$' is of type $oo$. But if abstraction is present together with the $\boldsymbol{\lambda}$-rules then we could define

$$N \quad \text{for} \quad ({}^\frown \boldsymbol{\lambda}^\frown \Sigma^\frown({}^\frown({}^\frown S^\frown \Sigma^\frown)^\frown \Sigma^\frown)^\frown),$$

where $\Sigma$ is a variable of type $o$.

Another case of this type of definition which occurs in Ch. 4 is: $N^\frown[^{\frown``\frown}\Phi^{\frown''\frown}]$, for the formula which results when we everywhere replace

$$\begin{array}{ccc} \neq & \text{by} & = \\ = & \text{by} & \neq \\ \vee & \text{by} & \& \\ \& & \text{by} & \vee \end{array}$$

Here $N$, [, ], " and " are new symbols.

Sometimes we replace an $\mathscr{L}$-formula by another one containing the same proper symbols but in a different order. This rearrangement of the order of $\mathscr{L}$-symbols in a well-formed $\mathscr{L}$-formula will sometimes be used when the order required by the $\mathscr{L}$-rules of formation is different from that to which we have been accustomed. This again makes for easier reading.

Thus $(x = y)$ will often be written instead of $((= x)y)$, $=$ is of type $o\iota\iota$.

## 1.12 *Omission of parentheses*

Lastly we shall frequently omit parentheses according to the following:

(i) The outer pair of parentheses may be omitted.

(ii) Parentheses may be omitted by association to the left.

Thus $\alpha\beta\gamma\delta$ will stand for $(((\alpha\beta)\gamma)\delta)$.

$\alpha\beta(\gamma\delta)$ will stand for $((\alpha\beta)(\gamma\delta))$.

This device is adopted because it soon becomes difficult to see the structure of a formula on account of a multitude of parentheses.

Parentheses round applications could be entirely omitted without affecting the use of a formal system.

LEMMA (iii). *If all parentheses round applications in a well-formed $\mathscr{L}$-formula are struck out then there is a unique method of restoring them so that the result is a well-formed $\mathscr{L}$-formula.*

Let $\Phi$ be a well-formed $\mathscr{L}$-formula and let $\Phi'$ be the result of removing parentheses round those applications which are outside abstracts. The only parentheses left in $\Phi'$ will then be round or inside abstracts; each abstract is a well-formed $\mathscr{L}$-formula, replace it by a new symbol of the same type. Let this transform $\Phi'$ into $\Phi''$. Then $\Phi''$ is a formula void of parentheses, it is a linear sequence of symbols each having a type. We know that it is possible to replace the parentheses so that $\Phi''$ is converted into a well-formed $\mathscr{L}$-formula $\Phi'''$ ($\Phi'''$ is obtained from $\Phi$ by replacing abstracts by new symbols of the same type). We want to show that there is a unique way of replacing the parentheses in $\Phi''$ so that the resulting formula is well-formed. The procedure we give is effective and will also tell us of any linear sequence of proper symbols whether it derives from a well-formed formula by omission of parentheses or whether this fails to be the case. For brevity we omit concatenation signs in type symbols.

If $\Phi''$ consists of a single symbol then it is well-formed as it stands and we have finished. If $\Phi''$ consists of several symbols let the left-most symbol be $\Sigma_{(\alpha\beta)}$ of type $(\alpha\beta)$. If the type of the left-most symbol fails to be compound then it is impossible to replace parentheses so that the result is well-formed. If the next symbol is $\Sigma_\beta$ of type $\beta$ (case ($a$)), then parentheses are put in as $({}^\frown\Sigma_{(\alpha\beta)}{}^\frown\Sigma_\beta^\frown)$. This is a well-formed formula of type $\alpha$, replace it by a new symbol of type $\alpha$. This converts $\Phi''$ into a shorter formula $\Phi^{\text{iv}}$. We then proceed with $\Phi^{\text{iv}}$. If the symbol after $\Sigma_{(\alpha\beta)}$ in $\Phi''$ is $\Sigma_\gamma$, where $\gamma$ is a type symbol which must be $(\ldots(\beta\delta')\ldots)\,\delta^{(s\theta)})$, where $\beta, \delta', \ldots, \delta^{(S\theta)}$ are type symbols, (case ($b$)), because it must be possible to insert parentheses so that $\Sigma_{(\alpha\beta)}$ is followed by a well-formed formula of type $\beta$. If $\gamma$ fails to be the above type symbol then it is impossible to insert parentheses in $\Phi''$ so that the result is well-formed. If $\gamma$ is the above type symbol then we insert parentheses as follows:

$$({}^\frown\Sigma_{(\alpha\beta)}{}^\frown\underbrace{(^\frown\ldots{}^\frown(}_{(S\theta)\text{-times}}{}^\frown\Sigma_\gamma.$$

The symbol next to $\Sigma_\gamma$ must be of type $\delta^{(S\theta)}$ or $(\ldots(\delta^{(S\theta)}\eta')\ldots\eta^{(S\pi)}$, where $\eta',\ldots,\eta^{(S\pi)}$ are type symbols, otherwise it is impossible to replace parentheses in $\Phi''$ so that the result is well-formed. If it is $\Sigma_{\delta(S\theta)}$ of type $\delta^{(S\theta)}$ then parentheses go back as

$$\underbrace{({}^\frown\Sigma_{(\alpha\beta)}{}^\frown({}^\frown\ldots{}^\frown(}_{(S\theta)\text{-times}}{}^\frown\Sigma_\gamma^\frown\,\Sigma_{\delta(S\theta)}^\frown \quad (\text{case } (a)).$$

Here $(\Sigma_\gamma^\frown\,\Sigma_{\delta(S\theta)}^\frown$ is a well-formed formula of type $(\ldots(\beta\delta')\ldots\delta^{(\theta)})$. Replace it by a new symbol of this type. This converts $\Phi''$ into a shorter formula $\Phi^{\text{iv}}$; we then proceed with $\Phi^{\text{iv}}$. If the next symbol is of type $(\ldots(\delta^{(S\theta)}\eta')\ldots\eta^{(S\pi)})$, call this $\gamma'$, then parentheses are inserted as follows:

$$\underbrace{({}^\frown\Sigma_{(\alpha\beta)}^\frown({}^\frown\ldots{}^\frown(}_{(S\theta)\text{-times}}{}^\frown\underbrace{\Sigma_\gamma^\frown({}^\frown\ldots{}^\frown(}_{(S\pi)\text{-times}}{}^\frown\Sigma_{\gamma'} \quad (\text{case } (b)).$$

So we continue, but the process will terminate with case $(a)$ occurring, we can then put in a right parenthesis and replace a part $({}^\frown\Sigma_{(\alpha'\beta')}^\frown\Sigma_{\beta'}^\frown)$ by a new symbol of type $\alpha'$. This converts $\Phi''$ into a shorter formula $\Phi^{\text{v}}$, we then proceed with $\Phi^{\text{v}}$. The whole process terminates in at most as many steps as there are proper symbols in $\Phi$. Thus given a linear sequence of proper symbols we can either insert parentheses uniquely so that the result is a well-formed formula or we can discover that it is impossible to do so. In a similar manner we can deal with parentheses inside abstracts. But if we omit parentheses round abstracts in a well-formed formula then there may be several ways of replacing them so that the result is a well-formed formula. Consider the formula

$$\boldsymbol{\lambda}{}^\frown\Sigma_\beta^\frown\boldsymbol{\lambda}{}^\frown\Sigma_\beta'^\frown\,\Phi_\alpha\{\Sigma_\beta,\Sigma_\beta'\}{}^\frown\Psi_\beta,$$

where $\Sigma_\beta$ and $\Sigma'_\beta$ are distinct variables of type $\beta$. This arises from either of the well-formed formulae, both of type $(\alpha\beta)$:

$$\underset{3}{(}{}^\frown\underset{2}{(}{}^\frown\boldsymbol{\lambda}{}^\frown\Sigma_\beta^\frown\underset{1}{(}{}^\frown\boldsymbol{\lambda}{}^\frown\Sigma_\beta'^\frown\,\Phi_\alpha\{\Sigma_\beta,\Sigma_\beta'\}{}^\frown\underset{1}{)}{}^\frown\underset{2}{)}{}^\frown\Psi_\beta^\frown\underset{3}{)},$$

$$\underset{2}{(}{}^\frown\boldsymbol{\lambda}{}^\frown\Sigma_\beta^\frown\underset{3}{(}{}^\frown\underset{1}{(}{}^\frown\boldsymbol{\lambda}{}^\frown\Sigma_\beta'^\frown\,\Phi_\alpha\{\Sigma_\beta,\Sigma_\beta'\}{}^\frown\underset{1}{)}{}^\frown\Psi_\beta^\frown\underset{3}{)}{}^\frown\underset{2}{)}$$

by omission of parentheses round abstractions, and omission of the outer pair of parentheses. Mates are shown by subscript signs. Using the $\boldsymbol{\lambda}$-rules these two formulae may be replaced respectively by:

$$({}^\frown\boldsymbol{\lambda}{}^\frown\Sigma_\beta'^\frown\,\Phi_\alpha\{\Psi_\beta,\Sigma_\beta'\}{}^\frown) \quad\text{and}\quad ({}^\frown\boldsymbol{\lambda}{}^\frown\Sigma_\beta^\frown\,\Phi_\alpha\{\Sigma_\beta,\Psi_\beta\}{}^\frown),$$

provided that the proviso of the $\boldsymbol{\lambda}$-rules is satisfied.

After this chapter we shall usually omit the concatenation sign. We shall usually write $(\boldsymbol{\lambda}\Sigma_\beta . \Phi_\alpha)$ for $(\boldsymbol{\lambda}\Sigma_\beta \Phi_\alpha)$, the dot before the scope of $\boldsymbol{\lambda}\Sigma_\beta$ makes for easier reading. Also we shall usually write

$$(\boldsymbol{\lambda}\Sigma_\beta \Sigma_{\beta'} . \Phi_\alpha) \quad \text{for} \quad (\boldsymbol{\lambda}\Sigma_\beta . (\boldsymbol{\lambda}\Sigma_{\beta'} . \Phi_\alpha)),$$

a formula of type $((\alpha\beta')\beta)$.

## 1.13 *Formal systems*

To sum up, a formal system is an ordered quartet $(\mathscr{S}, \mathscr{F}, \mathscr{A}, \mathscr{P})$, $\mathscr{S}$ is a display of signs, some of which may be designated as generating signs, $\mathscr{F}$ is a description of rules of formation, $\mathscr{A}$ is a display of axioms or a description of axiom schemes, $\mathscr{P}$ is a description of rules of procedure. It must be possible to decide of an object whether it comes under one of these cases or is foreign to them, only then is it possible to read the formal system and to check proofs. Thus we say that a formal system is *constructive*.

A formal system $\mathscr{L}$ may be without rules of procedure, in this case the $\mathscr{L}$-theorems are just the $\mathscr{L}$-axioms. A formal system $\mathscr{L}$ may lack axioms, in this case the formal system $\mathscr{L}$ is without theorems and we are then only interested in transforming $\mathscr{L}$-statements into other $\mathscr{L}$-statements by the $\mathscr{L}$-rules. Usually we then speak of transforming $\mathscr{L}$-formulae of a certain type into other $\mathscr{L}$-formulae, and the formal system $\mathscr{L}$ is then often used as a system of calculations, say of the value of functors.

If we know a procedure which will decide whether an $\mathscr{L}$-statement is an $\mathscr{L}$-theorem, we can omit the $\mathscr{L}$-rules and take as $\mathscr{L}$-axioms the $\mathscr{L}$-theorems, because the requirement that it be possible to decide whether an $\mathscr{L}$-statement is an $\mathscr{L}$-axiom remains satisfied. Thus $\mathscr{L}$-rules are only required when we lack a procedure to decide whether an $\mathscr{L}$-statement is an $\mathscr{L}$-theorem. A formal system $\mathscr{L}$ is called *decidable* if we have a procedure to decide if an $\mathscr{L}$-statement is an $\mathscr{L}$-theorem. But we can write down the $\mathscr{L}$-theorems one after the other, so that if we continue long enough any $\mathscr{L}$-theorem will appear in the list. To do this we select a new symbol, say $\Box$. We then denote a sequence of $\mathscr{L}$-formulae $\Phi'\ \Phi'' \ldots \Psi$ by the formula $\Phi'{}^\frown\Box^\frown\Phi''{}^\frown\Box^\frown \ldots {}^\frown\Box^\frown\Psi$ of a system $\mathscr{L}'$ obtained from the system $\mathscr{L}$ by adding the typeless symbol $\Box$. We then give an order of preference, called the *alphabetical order*, to the symbols, we then order the $\mathscr{L}'$-formulae first by length and lexicographically for those of equal length. In this manner we can generate the $\mathscr{L}'$-formulae one after the other. When an $\mathscr{L}'$-formula has been generated we test it whether it is of the form $\Phi'^\frown\Box^\frown\Phi''^\frown\Box^\frown\ldots^\frown\Box^\frown\Psi$ where the sequence $\Phi'\ \Phi''\ldots\Psi$ is an $\mathscr{L}$-proof of $\Psi$. If the test is affirmative we write $\Psi$ down in a list. In this manner we generate the $\mathscr{L}$-theorems one after the other without omissions but with repetitions.

## 1.14 *Extensions of formal systems*

A formal system $\mathscr{L}'$ is called a *primary extension* of a formal system $\mathscr{L}$ if the $\mathscr{L}$-symbols are $\mathscr{L}'$-symbols of the same type and if the $\mathscr{L}$-axioms and $\mathscr{L}$-rules are $\mathscr{L}'$-axioms and $\mathscr{L}'$-rules respectively. Thus a primary extension of a formal system $\mathscr{L}$ is obtained by doing some of the following operations: adding new symbols to $\mathscr{L}$, adding new axioms, adding new rules of procedure. If $\mathscr{L}'$ is a primary extension of $\mathscr{L}$ then $\mathscr{L}$ is called a *sub-system* of $\mathscr{L}'$. $\mathscr{L}$ is an *improper primary extension* and an *improper subsystem* of itself.

Two formal systems $\mathscr{L}$ and $\mathscr{L}'$ are *equivalent* when their variables are of the same type (by a trivial adjustment we can then use the same symbols for variables in both systems) and when the constants of the one system can respectively be replaced by suitable formulae of the other system of the same respective types in such a way that the theorems of the one system translate via the replacements into theorems of the other system.

If the formal system $\mathscr{L}$ is equivalent to the formal system $\mathscr{L}'$ and if $\mathscr{L}''$ is a primary extension of $\mathscr{L}'$ then $\mathscr{L}''$ is a *secondary extension* of $\mathscr{L}$.

Two formal systems, $\mathscr{L}$, $\mathscr{L}'$, can be equivalent merely because of different shapes in the choice of primitive symbols, or because though they both have exactly the same primitive symbols and each symbol has the same type in both systems yet the $\mathscr{L}$-axioms are $\mathscr{L}'$-theorems and the $\mathscr{L}$-rules are derived $\mathscr{L}'$-rules and vice versa. $\mathscr{L}$, $\mathscr{L}'$ can be equivalent when the variables are the same and of the same types in the two systems yet the proper constants are different and perhaps of different types. For instance some symbols introduced into $\mathscr{L}$ by definitional abbreviation might be primitive $\mathscr{L}'$-symbols of the same type as the $\mathscr{L}$-defined symbol. This, in general, would require that the $\mathscr{L}$-axioms and $\mathscr{L}$-rules be distinct from the $\mathscr{L}'$-axioms and $\mathscr{L}'$-rules respectively.

## 1.15 *Truth definitions*

Let $\mathscr{L}$ be a formal system and let an $\mathscr{L}$-statement be of type $o$. A *truth-definition* for $\mathscr{L}$ is a set of conditions $\mathscr{T}_{\mathscr{L}}$ applicable to closed $\mathscr{L}$-statements. We say that a closed $\mathscr{L}$-statement $\Phi$ is $\mathscr{T}_{\mathscr{L}}$-*true* if the conditions $\mathscr{T}_{\mathscr{L}}$ tested on $\Phi$ lead to an affirmative result. It may happen that it is impossible to test the conditions $\mathscr{T}_{\mathscr{L}}$ on some closed $\mathscr{L}$-statements. (For instance consider the intuitive definition of truth for arithmetic statements and then ask if Goldbach's conjecture is true.) We shall say that a closed $\mathscr{L}$-statement $\Phi$ is $\mathscr{T}_{\mathscr{L}}$-*false* if the conditions $\mathscr{T}_{\mathscr{L}}$ applied to $\Phi$ lead to a result which fails to be affirmative. It may happen that for a formal system $\mathscr{L}$ the conditions $\mathscr{T}_{\mathscr{L}}$ can be applied to each closed $\mathscr{L}$-statement and yield a result. In such a case we have an *effective truth-definition.*

We say that a formal system $\mathscr{L}$ is *consistent with respect to a truth-definition* $\mathscr{T}_{\mathscr{L}}$ if each closed $\mathscr{L}$-theorem is $\mathscr{T}_{\mathscr{L}}$-true.

We say that a formal system $\mathscr{L}$ is *complete with respect to a truth definition* $\mathscr{T}_{\mathscr{L}}$ if each $\mathscr{T}_{\mathscr{L}}$-true $\mathscr{L}$-statement is an $\mathscr{L}$-theorem.

It may be possible to give several distinct truth-definitions for a given formal system $\mathscr{L}$. A trivial truth-definition for a formal system $\mathscr{L}$ is to say that a closed $\mathscr{L}$-statement is true if and only if it is an $\mathscr{L}$-theorem. The concepts of truth and provability then coincide. A useless truth-definition for a formal system $\mathscr{L}$ containing conjunction is to say that an $\mathscr{L}$-statement is true if and only if it is of the form

$$(^\frown(^\frown\&^\frown(^\frown(^\frown\&^\frown\Psi'^\frown)^\frown\Psi'^\frown)^\frown)^\frown\Psi'^\frown)$$

(c.f. 'The hunting of the shark', 'what I say three times is true').

In a similar manner we can define a *falsity-definition* (denoted by $\mathscr{F}_{\mathscr{L}}$) for a formal system $\mathscr{L}$. If we define $\mathscr{T}_{\mathscr{L}}$ and $\mathscr{F}_{\mathscr{L}}$ for a formal system $\mathscr{L}$ then we require that they be exclusive.

## 1.16 *Negation*

Let $\mathscr{L}$ be a formal system and let $\mathscr{L}$-statements be of type $o$. Let $N$ be an $\mathscr{L}$-symbol of type $oo$ and let $\Phi$ be a closed $\mathscr{L}$-statement then $(^\frown N^\frown\Phi^\frown)$ is an $\mathscr{L}$-statement. Let $\mathscr{T}_{\mathscr{L}}$ be a truth-definition for $\mathscr{L}$ and let $\mathscr{F}_{\mathscr{L}}$ be a falsity-definition for $\mathscr{L}$ and let $(^\frown N^\frown\Phi^\frown)$ satisfy $\mathscr{F}_{\mathscr{L}}$ when $\Phi$ satisfies $\mathscr{T}_{\mathscr{L}}$ and $(^\frown N^\frown\Phi^\frown)$ satisfy $\mathscr{T}_{\mathscr{L}}$ when $\Phi$ satisfies $\mathscr{F}_{\mathscr{L}}$. Then we say that $N$ is a *two-valued* $\mathscr{T}_{\mathscr{L}}$-*negation symbol.* If a formal system $\mathscr{L}$ contains a two-valued negation symbol then we say that $\mathscr{L}$ is *consistent with respect to negation* if one of $\Phi$ and $(^\ulcorner N^\urcorner \Phi^\urcorner)$ fails to be an $\mathscr{L}$-theorem whenever $\Phi$ is a closed $\mathscr{L}$-statement, and we say that $\mathscr{L}$ is *complete with respect to negation* if exactly one of $\Phi$ and $(^\ulcorner N^\urcorner \Phi^\urcorner)$ is an $\mathscr{L}$-theorem whenever $\Phi$ is a closed $\mathscr{L}$-statement.

## Historical remarks to Chapter 1

The invention of language, written or otherwise, is lost in the mists of antiquity, but the invention of formal systems is of recent date, though the use of special symbols to augment a conversational language is of much older origin. Aristotle used capital letters, in the way we now use variables, to stand for undetermined propositions. Mathematicians used various symbols to denote mathematical terms, operators and concepts. But the first idea of a formal system goes back to Leibniz, Bibliography of Symbolic Logic in *J.S.L.* **I**. He wished to invent a formal system (characteristica universalis) which would suffice for all science. As yet we have only got as far as inventing formal systems for logic and mathematics, though one for some sciences such as Newtonian Dynamics or Thermodynamics would be possible. Leibniz also wished to invent a method (calculus ratiocinator) for manipulating statements in his projected formal system. But no school resulted and only fragments, though significant ones, were left.

The next investigators who are of interest to us are G. Boole (1847–54), A. de Morgan (1847), E. Schröder (1890–5), G. Frege (1879–1903), G. Peano (1889–1908), and C. S. Pierce (1933). Much of our modern logical and mathematical notation derives from Peano and Schröder. Modern symbolic logic really starts with Boole, his work was greatly enhanced by Schröder. Frege's 'concept writing' is very complicated and has been avoided by all subsequent writers. Boole noticed the similarity in use of the logical constants of conjunction and of disjunction with the mathematical operations of multiplication and addition. Thereby he could bring to bear mathematical methods into logic, this proved very fruitful, and his treatment gave great impetus to the development of formal systems.

The *Principia* of Whitehead and Russell (1910–13) built on the work of Boole, Schröder, Frege and Peano, was the first attempt at a completely formalized language and was worked out in great detail. Even so the authors said very little about the construction of such systems, so that they had little to correspond to our Chapter 1. In fact in the first edition they failed to mention a rule of proof which they used on practically every page, namely the rule of substitution. This omission was corrected in the second edition. This monumental work had tremendous influence in the development of formal systems, symbolic logic, etc. The λ-symbol occurs there for the first time. The theory of types given is very complicated, it is called the ramified theory of types, this was invented by Russell (1908) to obviate the paradoxes which were cropping up in set theory and in the theory of infinite cardinals. These paradoxes are of a kind known as syntactic because they can be eliminated by change in the syntactic rules, i.e. by change in the rules of formation in the formal system.

Variables really stem from Newton and the definition of a variable which we have given is a product of modern symbolic logic and recursive function theory. The concatenation sign is due to Tarski (1933). The idea of using a tape divided into squares is due to Turing (1936). We use it right at the start so as to be in no doubt as to when a formula begins or ends or whether we are dealing with a formula or with a sequence of formulae. The idea of using Greek (or Gothic) letters to augment the *syntax language* (the language in which we talk about the object language (formal system) we are inventing) is a product of modern symbolic logic, see for instance Carnap (1937). The definition of 'occurrence' we have given is due to Quine (1951). Carnap had great influence in defining a formal system as an ordered quartet. He and Church (1932) established the use of the abstraction symbol. Hilbert and Bernays (1934) introduced the use of proof-threads and formulae-forms in their dissection of proofs. The λ-rules were first correctly stated by Church (1941). Definitions as abbreviations are due to Russell and emphasized by Quine (1951), (*P.M.* vol. 1, p. 11), see also Markov (1954). Lemmas of the type of lemmas (i) and (ii) are due to Church (1941) and lemma (iii) is due to Łukasiewicz who wrote most of his work on the Propositional Calculus without using parentheses. The concept of a decidable system is due to Hilbert, he wanted to find a decision procedure for each formal system, this is now known to be impossible.

The concept of truth goes back to the ancient Greeks at least. Epimenides was the first to find something really interesting about it, viz. the antinomy of the liar, this we will come across later in the book. Truth definitions for formal systems stem from Tarski (1933). The concept of a consistent system is due to Hilbert (1904, 1922, 1930); he wanted to show that formal systems of arithmetic and analysis are consistent. We shall have much to say about this as the book proceeds.

Earlier writers like Frege, Peano, and Russell were mainly concerned with proving theorems in their systems. Hilbert was mainly concerned with finding out whether a given formal system was consistent, complete, decidable, etc. In other words he was mainly interested in *metamathematics*, theorems about theorems, and in the methods used in metamathematics (this term comes from Hilbert), in the methods of proof allowed in metamathematics, here he touched on effectiveness, finiteness, etc.

We shall be mainly concerned with what we can do with a given formal system. What concepts we can express in it, whether it is consistent, complete, decidable, etc. Whether it has a truth definition, and so on. We shall usually find that our systems have limitations and that in extending the system to remove such a limitation we get another system which fails to have some 'nice' property which the un-extended system had.

Our main interest is to construct a language in which we can talk about the natural numbers, to formalize our intuitive concept of natural number. But we are only interested in the properties of these systems, we are not much interested in proving theorems in the system itself.

## Examples 1

1. Give a method for enumerating the well-formed formulae (w.f.f.) of a formal system.

2. A formal system is given as follows:
signs: $x$ ′ ( ) $\lambda$, ′ is a generating sign, $x$ is a variable, if $\xi$ is a variable then so is $\xi^{n\prime}$, these are all the variables.

Rules of formation: a variable standing alone is w.f.f., if $\phi$ and $\psi$ are w.f.f. then so is $(\phi\psi)$, if $\xi$ is a variable and $\phi$ is w.f.f. then so is $(\lambda\xi.\,\phi)$. By adding a new symbol put this system into another form so that its rules of formation are as described in this chapter.

3. Put parentheses back in:

$$CCsCpqCCspCsq,$$
$$CCCpqCrsCCpCtsCrCts,$$
$$SSSpSqrStSttSSsqSSpsSps,$$

where $C$ and $S$ are of type $ooo$ and the other letters are symbols of type $o$.

4. Apply the $\boldsymbol{\lambda}$-rule (i) repeatedly to:

$$(\boldsymbol{\lambda} a(\boldsymbol{\lambda} b((\chi a)((\psi ab)))),$$

where $\chi$ for $\boldsymbol{\lambda} ab.(ab)$, $\psi$ for $\boldsymbol{\lambda} ab.a(ab)$

and obtain $\boldsymbol{\lambda} ab.a(a(a(a(ab))))$.

Similarly for $(\boldsymbol{\lambda} a(\chi(\psi a)))$

and obtain $\boldsymbol{\lambda} ab.a(a(a(a(a(ab)))))$.

5. Define:

$J$ for $\boldsymbol{\lambda} abcd.ab(adc)$

$B$ for $\boldsymbol{\lambda} abc.a(bc)$

$C$ for $\boldsymbol{\lambda} abc.acb$

$W$ for $\boldsymbol{\lambda} ab.abb$

show that $B(BC(BC))(BW(BBB))C)$ reduces to $J$ by applying the $\boldsymbol{\lambda}$-rule (i). And that if $T$ for $\boldsymbol{\lambda} ab.ba$ then

$$B(B(T(BD(B(TT)(B(BBB)T))))(BBT))(B(T(B(TI)(TI)))B)$$

reduces to $W$ by $\boldsymbol{\lambda}$-rule (i) where $D$ for $\boldsymbol{\lambda} a.aa$ and $I$ for $\boldsymbol{\lambda} a.a$.

6. Apply $\boldsymbol{\lambda}$-rule (i) to $(\boldsymbol{\lambda} x.xxx)(\boldsymbol{\lambda} x.xxx)$.

7. If $\Phi$ can be obtained from $\Psi$ by the $\boldsymbol{\lambda}$-rules then it is possible to do so when all applications of $\boldsymbol{\lambda}$-rule (ii) come after applications of $\boldsymbol{\lambda}$-rule (i).

8. A formula is said to be in normal form if it is impossible to apply $\boldsymbol{\lambda}$-rule (i) to it. Show that if a formula can be reduced to one in normal form by the $\boldsymbol{\lambda}$-rules then this normal form is unique to within change of bound variable.

9. A system has signs $S\ C\ a\ b\ x\ '$ the variables are $x\ x'\ x''\ \ldots$.

The axioms are $Sa$, $Sb$, $Sab$, $Sba$, $CSxaSxab$, $CSxbSxba$.

The rules are $\dfrac{S\alpha\{\xi\}}{S\alpha\{\beta\}}$, $\dfrac{S\alpha\, CS\alpha\beta}{S\beta}$, where $\xi$ is a variable and $\alpha$, $\beta$ are strings in $a$ and $b$ and variables, $\xi$ fails to occur in $\alpha\{\Gamma\}$.

Put this system into the type notation of a formal system as defined in this chapter.

# Chapter 2

# Propositional calculi

## 2.1 *Definition of a propositional calculus*

A *propositional calculus* is a formal system in which the types of the symbols are among those given by the scheme:

(i) $o$ is a type symbol,

(ii) if $\alpha$ is a type symbol then so is $(\alpha o)$,

(iii) these are the only type symbols.

A propositional calculus must have at least one symbol of type $o$. Symbols of type $(\alpha o)$ are called *connectives*. The only variables, if any, are of type $o$. These are called *propositional variables*. The abstraction symbol is absent.

A *pure propositional calculus* is a propositional calculus without constants of type $o$. An *applied propositional calculus* is without variables. A *mixed propositional calculus* contains both constants and variables of type $o$.

If $\mathscr{P}'$ and $\mathscr{P}''$ are propositional calculi and if both contain propositional variables then by a trivial change of notation we may use the same symbols for variables in the two systems. Thus we shall take the propositional variables in a propositional calculus to be: $p, p', p'', \ldots$ where the prime is a generating symbol. In this chapter we use $\pi$ with or without superscripts or subscripts to denote an undetermined propositional variable.

Suppose we have a propositional calculus $\mathscr{P}$ which contains constants $N$ and $D$ of types $oo$ and $ooo$ respectively. In terms of these constants by adjoining the abstraction symbol to $\mathscr{P}$ we can define a constant $K$ of type $ooo$ thus:

$$K \quad \text{for} \quad \lambda pp'.N(D(Np)(Np')).$$

This is the *normal form for definition of a new symbol*. The new symbol stands for a formula. We could have avoided the use of the abstraction symbol as follows:

$$K\Phi\Phi' \quad \text{for} \quad N(D(N\Phi)(N\Phi')),$$

where $\Phi$, $\Phi'$ are formula of type $o$. In this kind of definition the new symbol $K$ is without definition by itself but is only defined in contexts in which it will be used. By adjoining the abstraction symbol and variables of suitable types the second kind of definition can always be reduced to the first kind. If a propositional calculus is without propositional variables then we can adjoin them and the abstraction symbol solely to use the first kind of definition. Thus using the first kind of definition we see that $K$ is of type $ooo$ and that $K\Phi\Phi'$ is an abbreviation for

$$(\boldsymbol{\lambda} pp' . N(D(Np)(Np')))\,\Phi\Phi',$$

which by $\boldsymbol{\lambda}$-rule (i) becomes $N(D(N\Phi)(N\Phi'))$.

## 2.2 *Equivalence of propositional calculi*

A propositional calculus $\mathscr{P}'$ is *weaker* than a propositional calculus $\mathscr{P}$ under the following circumstances: First add a suffix 1 to each symbol of $\mathscr{P}$ and a suffix 2 to each symbol of $\mathscr{P}'$, except that if there are variables then they are to be the same in both systems. This makes the symbols of $\mathscr{P}$ and $\mathscr{P}'$ distinct, except possibly variables, if any, which are to be the same symbols in both systems.

(i) $\mathscr{P}'$ is without variables if $\mathscr{P}$ is without variables.

(ii) Definitions are given of the constants of $\mathscr{P}'$ in terms of the constants of $\mathscr{P}$; for this purpose the abstraction symbol and variables may be adjoined.

(iii) A $\mathscr{P}'$-theorem $\Phi'$ becomes a $\mathscr{P}$-theorem $\Phi$ when the constants in $\Phi'$ are replaced by their definitions in terms of the constants of $\mathscr{P}$ and the $\boldsymbol{\lambda}$-rules are used to eliminate the $\boldsymbol{\lambda}$-symbol.

Thus if $\mathscr{P}'$ is weaker than $\mathscr{P}$ then $\mathscr{P}$ can express anything that $\mathscr{P}'$ expresses and $\mathscr{P}'$-theorems translate into $\mathscr{P}$-theorems. If a propositional calculus $\mathscr{P}'$ is weaker than a propositional calculus $\mathscr{P}$ and if $\mathscr{P}$ is weaker than $\mathscr{P}'$ then $\mathscr{P}$ and $\mathscr{P}'$ are *equivalent*. If a propositional calculus $\mathscr{P}'$ is weaker than a propositional $\mathscr{P}$ and if they fail to be equivalent then $\mathscr{P}'$ is *strictly weaker* than $\mathscr{P}$. It can happen that $\mathscr{P}$ and $\mathscr{P}'$ are equivalent according to the above definition but some $\mathscr{P}$-theorem fails to be the translation of any $\mathscr{P}'$-theorem and vice versa. If some constant of $\mathscr{P}$ is the same symbol as some constant of $\mathscr{P}'$ before we put on the suffixes 1 and 2 and if all such constants correspond in the translation then we say that $\mathscr{P}$ and $\mathscr{P}'$ are equivalent by *natural translation*.

## 2.3 *Dependence and independence*

Suppose that $\mathscr{P}'$ has the same constants as $\mathscr{P}$ except that $\mathscr{P}$ has a further constant $\Gamma$ and that $\mathscr{P}$ is equivalent to $\mathscr{P}'$ by natural translation, then the *constant* $\Gamma$ *is dependent.* In this case the constant $\Gamma$ can be defined in terms of the other constants of $\mathscr{P}$ (possibly using the abstraction symbol). We can express this as follows: if a propositional calculus $\mathscr{P}$ contains a constant $\Gamma$ and if there is a definition of $\Gamma$ in terms of the other symbols of $\mathscr{P}$ such that the system which results when $\Gamma$ is everywhere replaced by its definition is equivalent to $\mathscr{P}$ then $\Gamma$ is dependent. Similarly if several symbols of $\mathscr{P}$ can be defined in terms of the remaining symbols of $\mathscr{P}$ and if the resulting system is equivalent to the original system then these symbols are dependent. If the propositional calculus $\mathscr{P}'$ is the same as the propositional calculus $\mathscr{P}$ except that $\mathscr{P}'$ lacks the axiom $\Phi$ of $\mathscr{P}$ and if $\mathscr{P}$ and $\mathscr{P}'$ are equivalent by natural translation then the axiom $\Phi$ *is dependent*, similarly for several axioms and for axiom schemes. Similarly if $\mathscr{P}'$ is the same as $\mathscr{P}$ except that $\mathscr{P}'$ lacks the rule $\mathscr{R}$ of $\mathscr{P}$ and if $\mathscr{P}$ and $\mathscr{P}'$ are equivalent by natural translation then the *rule* $\mathscr{R}$ *is dependent.* If in any of the above cases $\mathscr{P}'$ is strictly weaker than $\mathscr{P}$ then the *symbol, axiom or rule is independent.*

## 2.4 *Models of propositional calculi*

A *model* of a propositional calculus $\mathscr{P}$ is an applied propositional calculus $\mathscr{M}$ which satisfies the following conditions:

(i) $\mathscr{M}$ and $\mathscr{P}$ have exactly the same constants of compound type.

(ii) $\mathscr{M}$ has constants of type $o$, called *elements*, some of these are *designated* the remainder are *undesignated*, at least one element is designated and at least one is undesignated.

(iii) $\mathscr{M}$ is without axioms.

(iv) The rules of $\mathscr{M}$ enable one to replace any compound $\mathscr{M}$-statement $\Phi$ by a unique element, called the *value of the* $\mathscr{M}$*-statement* $\Phi$.

(v) If we replace $\mathscr{P}$-constants of type $o$ (if any) by suitably chosen fixed $\mathscr{M}$-elements and propositional variables (if any) by arbitrary $\mathscr{M}$-elements then a $\mathscr{P}$-statement $\Phi$ translates into an $\mathscr{M}$-statement, whose $\mathscr{M}$-value is called an $\mathscr{M}$-value of $\Phi$. Every $\mathscr{M}$-value of a $\mathscr{P}$-theorem is designated.

A $\mathscr{P}$-statement is called *$\mathscr{M}$-valid* if each of its $\mathscr{M}$-values is designated. Since a model $\mathscr{M}$ is without axioms then it is without theorems. We can only find the $\mathscr{M}$-value of $\mathscr{M}$-statements. We could have taken the designated $\mathscr{M}$-elements as axioms and reversed the rules, the $\mathscr{M}$-theorems would then be the $\mathscr{M}$-statements with designated $\mathscr{M}$-values. But we wish to use the model to find $\mathscr{M}$-values of $\mathscr{M}$-statements so the formulation given above is more natural for our purposes.

A model $\mathscr{M}$ of a propositional calculus $\mathscr{P}$ will be called *trivial* if the $\mathscr{M}$-value of every compound $\mathscr{M}$-statement is designated. Every propositional calculus possesses a trivial model.

Let $\Gamma$ be a $\mathscr{P}$-constant of type $\underbrace{o \ldots o}_{(S\theta)\text{-times}}$, to obtain a model $\mathscr{M}$ we have to fix the $\mathscr{M}$-elements and give the $\mathscr{M}$-values of $\Gamma\tau' \ldots \tau^{(\theta)}$, where $\tau', \ldots, \tau^{(\theta)}$ are $\mathscr{M}$-elements, the $\mathscr{M}$-value is to be an $\mathscr{M}$-element. Thus to obtain a model $\mathscr{M}$ of a propositional calculus $\mathscr{P}$ we require to give *$\mathscr{M}$-tables* for each $\mathscr{P}$-connective. Having obtained a set of tables for these connectives we then have to test if condition (v) is satisfied, that is that $\mathscr{P}$-theorems always take designated $\mathscr{M}$-values. It suffices to show that the $\mathscr{P}$-axioms are $\mathscr{M}$-valid and that the $\mathscr{P}$-rules preserve $\mathscr{M}$-validity.

Let $\mathscr{M}$ be a model of a propositional calculus $\mathscr{P}$ and let $\Gamma$, $\Gamma'$ be $\mathscr{P}$-connectives of the same type. If the tables of $\Gamma$, $\Gamma'$ are the same then $\Gamma$ and $\Gamma'$ are *$\mathscr{M}$-identifiable.* If $\tau'$, $\tau''$ are both designated $\mathscr{M}$-elements or both undesignated $\mathscr{M}$-elements and if the model $\mathscr{M}$ becomes a model $\mathscr{M}'$ when $\tau'$ and $\tau''$ are both replaced by a new element $\tau$ of the same designation as $\tau'$ and $\tau''$ then $\tau'$ and $\tau''$ are *indistinguishable.*

A model $\mathscr{M}$ of a propositional calculus $\mathscr{P}$ is *basic* if each pair of designated elements is distinguishable and similarly for each pair of undesignated elements. A propositional calculus is *$\nu$-valued* if it has a basic model with exactly $\nu$ elements. We must have $\nu \geqslant 2$. (| an initial segment of $(\nu)$).

A model $\mathscr{M}$ of a propositional calculus $\mathscr{P}$ has been defined as a formal system of a special kind, it must then be constructive. This will be the case if the designated and undesignated elements are displayed.

Two $\mathscr{P}$-statements $\omega$ and $\omega'$ are *equivalent* if for each $\Phi\{\Gamma\}$, $\Phi\{\omega\}$ is a $\mathscr{P}$-theorem if and only if $\Phi\{\omega'\}$ is a $\mathscr{P}$-theorem. Let $\Omega_\omega$ be the class of $\mathscr{P}$-statements equivalent to the $\mathscr{P}$-statement $\omega$. Clearly equivalence is a reflexive, symmetric and transitive relation. Hence $\Omega_{\omega'}$ is the same as $\Omega_\omega$ if $\omega'$ is equivalent to $\omega$. An equivalence class $\Omega$ will be called designated if and only if a member of $\Omega$ is a $\mathscr{P}$-theorem, then each member of $\Omega$ is a $\mathscr{P}$-theorem. This fixes the designated and the undesignated elements and the association of classes to $\mathscr{P}$-constants of type $o$. Let $\Gamma$ be a $\mathscr{P}$-connective of type $o \ldots o$, let $\Gamma\Omega' \ldots \Omega^{(\theta)}$ have value $\Omega$ if and only if $\Gamma\omega' \ldots \omega^{(\theta)}$ is in $\Omega$ where $\omega^{(\lambda)}$ is in $\Omega^{(\lambda)}$ $1 \leqslant \lambda \leqslant \theta$. This gives us tables. These classes and tables form a model of $\mathscr{P}$ provided the result is constructive and that there is an undesignated class. In this model we take as elements the equivalence classes, and as connectives we take the $\mathscr{P}$-connectives. The tables we have found provide the rules of the model. It remains to show that the $\mathscr{P}$-theorems are valid in the model. Suppose that $\phi\{\pi', \ldots, \pi^{(\theta)}\}$ containing exactly the variables $\pi', \ldots, \pi^{(\theta)}$ is a $\mathscr{P}$-theorem, then so is $\phi\{\omega', \ldots, \omega^{(\theta)}\}$, where $\pi'$ is equivalent to $\omega', \ldots, \pi^{(\theta)}$ is equivalent to $\omega^{(\theta)}$. Thus $\phi\{\pi', \ldots, \pi^{(\theta)}\}$ is valid in the model.

A propositional calculus $\mathscr{P}$ is *model-consistent* if it has a non-trivial model, it is *model-complete* if it has a model $\mathscr{M}$ such that every $\mathscr{M}$-valid $\mathscr{P}$-statement is a $\mathscr{P}$-theorem. Sometimes we say *consistent with respect to the model $\mathscr{M}$, complete with respect to the model $\mathscr{M}$*. A propositional calculus $\mathscr{P}$ is *functionally complete with respect to a model $\mathscr{M}$* under the following circumstances:

(i) $\mathscr{M}$ is a model of $\mathscr{P}$,

(ii) $\mathscr{M}$ is displayed,

(iii) $\mathscr{M}'$ is an extension of $\mathscr{M}$ which contains connectives of all types $\underbrace{(o \ldots o)}_{(SS\theta)\text{-times}}$ and every possible table for each type, $\mathscr{M}'$ has the same elements as $\mathscr{M}$,

(iv) $\mathscr{M}'$ is a model for a propositional calculus $\mathscr{P}'$,

(v) $\mathscr{P}'$ is equivalent to $\mathscr{P}$ by natural translation.

Let $\mathscr{M}$ be an applied propositional calculus which satisfies conditions (ii), (iii), (iv) of the conditions for being a model. We seek a pure propositional calculus $\mathscr{P}$ for which $\mathscr{M}$ is a model. $\mathscr{P}$ then contains variables $p, p', \ldots$ and the same connectives as $\mathscr{M}$. We could take as $\mathscr{P}$-axioms all $\mathscr{M}$-valid $\mathscr{P}$-statements provided that this is constructive. Rules would then be unnecessary. If the set of $\mathscr{M}$-valid $\mathscr{P}$-statements fails to be constructive then we have to try to discover axioms and rules by trial an error. The problem of finding $\mathscr{P}$ is the *problem of formalizing $\mathscr{M}$*. Given a formalization of $\mathscr{M}$ then we might try to find another formalization with the least possible set of axioms or a set of axioms containing the least possible set of symbols. We might try to find a formalization which is complete or which is functionally complete or an equivalent propositional calculus with a single connective. A formalization $\mathscr{P}$ of $\mathscr{M}$ is model-consistent because it has $\mathscr{M}$ for model.

## 2.5 *Deductions*

Deduction is represented in a formal system by application of the rules. The rules are of the form $\frac{\phi}{\psi}, \frac{\phi' \; \phi''}{\psi}, \ldots$, where the lower formula can be constructively obtained from the upper formulae. An example is $\frac{\phi \; C\phi\psi}{\psi}$, known as *Modus Ponens*, here $C$ is a connective of type *ooo*. A proof in a formal system is a tree-like figure with axioms at the tops of the branches and the statement proved at the base and we proceed from the axioms at the tops of the branches downwards to the base by applications of the rules. Thus in a proof the upper formulae of each application of a rule are theorems of the system, the part of the tree ending at an upper formula is a proof of that formula. Thus each statement in the proof is a theorem.

A *deduction* in a formal system is a tree-like figure such that we proceed from the tops of the branches to the base by applications of the rules. Thus a deduction differs from a proof only in that we can have statements other than axioms at the tops of the branches. Thus a deduction in a propositional calculus $\mathscr{P}$ amounts to a proof in another propositional calculus $\mathscr{P}'$ obtained from $\mathscr{P}$ by adding certain $\mathscr{P}$-statements as additional axioms, they are called *hypotheses*. We have already used the figure:

$$\begin{array}{c} \phi', \ldots, \phi^{(S\theta)} \\ \hline\hline \psi \end{array}$$

to denote that $\psi$ may be obtained from $\phi', \ldots, \phi^{(S\theta)}$ by use of the rules. This amounts to there being a deduction of $\psi$ from $\phi', \ldots, \phi^{(S\theta)}$ as hypotheses. This figure and the figure $\frac{\phi', \ldots, \phi^{(S\theta)}}{\psi}$ must be distinguished from the figure $\frac{\phi', \ldots, \phi^{(S\theta)}}{\psi}*$ which means that from the proofs of $\phi', \ldots, \phi^{(S\theta)}$ we can constructively obtain a proof of $\psi$. In this case it may happen that $\begin{array}{c} \phi', \ldots, \phi^{(S\theta)} \\ \hline\hline \psi \end{array}$ is impossible. Thus we have three different kinds of figures, namely:

(*a*) $\frac{\phi', \ldots, \phi^{(S\theta)}}{\psi}$ rules,

(b) $\dfrac{\phi', \ldots, \phi^{(S\theta)}}{\psi}$ (double line) deductions,

(c) $\dfrac{\phi', \ldots, \phi^{(S\theta)}}{\psi}*$ derivations.

In a formal system it usually happens that we are without means to express by a statement of the system that the statements $\phi', \ldots, \phi^{(S\theta)}, \psi$ are related in either of these ways. The fact that the statements $\phi', \ldots, \phi^{(S\theta)}, \psi$ are related in either of these ways must normally be expressed in the meta-language. The horizontal stroke, and the above modifications of it, are distinct from any symbol in the formal system, so that each of the three kinds of figure fail to be statements of the formal system.

However in certain formal systems there is a constant $C$ or constants $C', \ldots, C^{(S\pi)}$ all of type *ooo* which partly express derivations. Suppose that in a formal system we have:

$$\frac{\phi \;\; C\phi\psi}{\psi}*,$$

then a theorem $C\phi\psi$ can be taken to express the fact that from a proof of $\phi$ we can obtain one of $\psi$.

Similarly if in a formal system we have

$$\frac{\phi \;\; C\phi\psi}{\psi} \quad \text{or} \quad \frac{\phi \;\; C\phi\psi}{\psi} \text{ (double line)}$$

then $C\phi\psi$ can be taken to express a figure. Note the following relations between (*a*), (*b*), (*c*): if (*a*) then (*b*). If (*b*) and if $\phi', \ldots, \phi^{(\theta)}$ are theorems then (*c*).

If in a formal system we have a primitive or defined symbol $C_a^{(\theta)}$ or $C_b^{(\theta)}$ or $C_c^{(\theta)}$ such that:

(*d*) if $\dfrac{\phi', \ldots, \phi^{(S\theta)}}{\psi}$ then $C_a^{(\theta)}\phi' \ldots \phi^{(S\theta)}\psi$ is a theorem and conversely,

similarly in cases (*b*), (*c*), then the appropriate figure would be fully expressed in the system. Usually we can only satisfy the direct part of (*d*). Systems which contain a primitive or derived symbol $C_b^{(\theta)}$ satisfying the direct part of (*d*) are important. In such a system we have the *meta-theorem*:

(*e*) if $\dfrac{\phi', \ldots, \phi^{(S\theta)}}{\psi}$ (double line) then $C_b^{(\theta)}\phi' \ldots \phi^{(S\theta)}\psi$ is a theorem. Use of this meta-theorem frequently enables us to shorten proofs, the full proof can be found from the meta-theorem. The condition will hold in a propositional calculus if the following condition:

($f$) if $\frac{\phi', \ldots, \phi^{(S\theta)}}{\psi}$ then $\frac{\phi', \ldots, \phi^{(\theta)}}{C\phi^{(S\theta)}\psi}$ for some primitive or defined symbol $C$ of type $ooo$. For suppose ($f$) then: if $\frac{\phi', \ldots, \phi^{(S\theta)}}{\psi}$ then $C\phi'(C\ldots(C\phi^{(S\theta)}\psi)\ldots)$ is a theorem. We can then define $C_b^{(\theta)}$ by

$$C_b^{(\theta)}\phi' \ldots \phi^{(S\theta)}\psi \quad \text{for} \quad C\phi'(C\ldots(C\phi^{(S\theta)}\psi)\ldots).$$

Thus we obtain ($e$).

In a propositional calculus in which (iv) below holds a necessary and sufficient condition that ($f$) hold is that the conditions (i), (ii), (iii) below should hold,

(i) $C\phi\phi$ be a theorem.

(ii) $C\phi\chi$ be a theorem whenever $\chi$ is an axiom.

(iii) $\frac{\chi}{C\phi\chi}$.

(iv) If $\frac{\omega', \ldots, \omega^{(S\pi)}}{\chi}$ is a rule then $\frac{C\phi\omega', \ldots, C\phi\omega^{(S\pi)}}{C\phi\chi}$.

For suppose (i), (ii), (iii), (iv) and $\frac{\phi', \ldots, \phi^{(S\theta)}}{\psi}$.

In the tree obtained by writing out $\frac{\phi', \ldots, \phi^{(S\theta)}}{\psi}$ in full replace each premiss and conclusion $\chi$ by $C\phi^{(S\theta)}\chi$. The tops of the branches become $C\phi^{(S\theta)}\phi^{(S\theta)}$ or $C\phi^{(S\theta)}\phi', \ldots, C\phi^{(S\theta)}\phi^{(\theta)}$, or $C\phi^{(S\theta)}\chi$, where $\chi$ is an axiom. In the first and last cases we add the proofs of the theorems given in (i) and (ii), in the other cases we add the deductions given in (iii). In the rest of the tree we replace use of a rule by the full proof of the derived rule given in (iv). This gives us $\frac{\phi', \ldots, \phi^{(\theta)}}{C\phi^{(S\theta)}\psi}$. Conversely suppose ($f$): we have $\frac{\phi}{\phi}$ whence $\overline{\overline{C\phi\phi}}$ by ($f$), that is $C\phi\phi$ is a theorem hence (i). Again we have $\frac{\phi}{\chi}$ where $\chi$ is an axiom hence $C\phi\chi$ by ($f$) and so (iii). Hence the result.

## 2.6 *The classical propositional calculus*

A pure propositional calculus is called *classical* if it is equivalent to the following propositional calculus, $\mathscr{P}_C$. Symbols:

| | |
|---|---|
| $N_{oo}$ | *negation* symbol, |
| $D_{ooo}$ | *disjunction* symbol, |
| $p_o$ | variable, |
| $'$ | generating sign, |
| ( | left parenthesis, |
| ) | right parenthesis. |

The parentheses can be omitted because we only use application, so that if a formula is well-formed the parentheses can be omitted because there is a unique method of replacing them so that the result is well-formed. This was shown in Ch. 1, lemma (iii).

Axiom scheme. $D\pi N\pi$, where $\pi$ is a variable. Rules:

*Remodelling*

I$a$ $\dfrac{DDD\omega\phi\psi\omega'}{DDD\omega\psi\phi\omega}$,
*permutation.*

*Building*

II$a$ $\dfrac{\chi}{D\phi\chi}$, *dilution*

II$b$ $\dfrac{DN\phi\omega \quad DN\psi\omega}{DND\phi\psi\omega}$, *composition*

II$c$ $\dfrac{D\phi\omega}{DNN\phi\omega}$. *double negation*

$\omega$, $\omega'$ are called *subsidiary formulae* and may be omitted, $\chi$ is a *secondary formula* and must be present. $\phi$, $\psi$ are called the *main formulae.* If we omit a subsidiary formula then one occurrence of $D$ is struck out.

The premiss I$a$ written in full without omission of parentheses is $((D((D((D\omega)\phi))\psi))\omega')$, which is the only way of replacing the parentheses so that the result is well-formed. Notice that any symbol introduced into a $\mathscr{P}_C$-proof remains in that $\mathscr{P}_C$-proof from that place onwards. Hence any $\mathscr{P}_C$-theorem must contain the variables which occur in the axioms used in that proof. A formal system $\mathscr{L}$ is called *direct* if any symbol introduced into an $\mathscr{L}$-proof remains in the $\mathscr{L}$-proof from that place onwards. Thus the system $\mathscr{P}_C$ is direct.

## 2.7 *Some properties of the remodelling and building schemes*

PROP. 1 *Disjunction is communtative and associative.*

Commutativity of disjunction:

$$\frac{D\phi\psi}{D\psi\phi} \quad \text{I}a \text{ with both subsidiary formulae absent.}$$

Associativity of disjunction:

$$\begin{array}{ll} DD\phi\psi\chi & \\ \hline DD\psi\phi\chi & \text{I}a \text{ with left subsidiary formula absent,} \\ \hline DD\psi\chi\phi & \text{I}a \text{ with right subsidiary formula absent,} \\ \hline D\phi D\psi\chi & \text{I}a \text{ with both subsidiary formulae absent.} \end{array}$$

Again

$$\begin{array}{ll} D\phi D\psi\chi & \\ \hline DD\psi\chi\phi & \text{I}a \text{ with both subsidiary formulae absent,} \\ \hline DD\psi\phi\chi & \text{I}a \text{ with right subsidiary formula absent,} \\ \hline DD\phi\psi\chi & \text{I}a \text{ with left subsidiary formula absent.} \end{array}$$

Hence disjunction is associative.

PROP. 2. *The schemes* I$a$, II$b$, $c$ *are reversible.*

That is to say, if we have a $\mathscr{P}_C$-proof of the lower formula of one of these schemes then from that $\mathscr{P}_C$-proof we can obtain a $\mathscr{P}_C$-proof of the upper formula (upper formulae) by carrying out the procedure that we are about to describe. Thus in our notation

$$\frac{DDD\omega'\psi\phi\omega}{DDD\omega'\phi\psi\omega}*, \quad \frac{DND\phi\psi\omega}{DN\phi\omega \quad DN\psi\omega}* \quad \text{and} \quad \frac{DNN\phi\omega}{D\phi\omega}*.$$

The scheme I$a$ is reversible. This is clear.

The scheme II$b$ is reversible. Suppose we have a $\mathscr{P}_C$-proof of $DND\phi\psi\chi$ then we have a tree whose base is $DND\phi\psi\omega$. Follow corresponding occurrences of $ND\phi\psi$ up the tree. If $ND\phi\psi$ occurs in the subsidiary or secondary formula of the lower part of a rule then there will be a corresponding occurrence or occurrences of $ND\phi\psi$ in the subsidiary or secondary formula or formulae of the upper part of that rule. If there is an occurrence of $ND\phi\psi$ in a main formula of the lower part of I$a$ then there is a corresponding occurrence in the upper part of I$a$. If there is an occurrence of $ND\phi\psi$ in the main formula of the lower part of II$a$

there fails to be a corresponding occurrence of $ND\phi\psi$ in the upper part of II$a$, the same holds for II$b$ if $\phi$ and $\psi$ are the main formulae. But there will be an occurrence of $ND\phi\psi$ in a main formula of the upper part of II$b$ if there is an occurrence of $ND\phi\psi$ in a main formula of the lower part of II$b$ (see lemma (ii) Ch. 1). Similarly for II$c$. If the scheme $\dfrac{DD\phi\psi\omega'}{DNND\phi\psi\omega'}$, II$c$ occurs in a tree whose base is $DND\phi\psi\omega$ then the occurrence $DNND\phi\psi$ of $ND\phi\psi$ in $DNND\phi\psi\omega'$ will fail to correspond to the occurrence $DND\phi\psi$ of $ND\phi\psi$ in $DND\phi\psi\omega$ because once $N$ is introduced by a rule it remains in the $\mathscr{P}_C$-proof from that place, so that if there are two occurrences of $N$ before an occurrence of $\phi$ then this will persist. Lastly the axioms are without occurrences of $ND\phi\psi$.

Thus as we follow corresponding occurrences of $ND\phi\psi$ up the tree we may have branches at subsidiary formulae of II$b$ and the only places where we shall stop will be at main formulae of II$a$ or II$b$. In any case we shall stop before we reach the top of any branch. In the portion of the tree containing these occurrences of $ND\phi\psi$ replace $ND\phi\psi$ by $N\phi$ and if an occurrence of $ND\phi\psi$ disappears at the main formula of II$b$ then delete the right-hand upper formula of that application of II$b$ and the branch above it. The resulting tree, when repetitions are discarded, is a $\mathscr{P}_C$-proof of $DN\phi\omega$, for at the top stand $\mathscr{P}$-axioms and each step downwards is an application of one of the $\mathscr{P}_C$-rules. Thus we are left with a $\mathscr{P}_C$-proof of $DN\phi\omega$. Similarly we get a $\mathscr{P}_C$-proof of $DN\psi\omega$. Similarly from a $\mathscr{P}_C$-proof of $ND\phi\psi$ we get $\mathscr{P}_C$-proofs of $N\phi$, and $N\psi$.

The scheme II$c$ is reversible. We proceed in a similar manner. Given a $\mathscr{P}_C$-proof of $DNN\phi\omega$ we trace corresponding occurrences of $NN\phi$ up the $\mathscr{P}_C$-proof-tree and note that $NN\phi$ can only be introduced at applications of II$a$, $c$. Replace corresponding occurrences of $NN\phi$ by $\phi$ and we are left with a $\mathscr{P}_C$-proof-tree (apart from repetitions which can be deleted) of $D\phi\omega$. Similarly if the subsidiary formula $\omega$ is absent.

PROP. 3. *We have the derived rule*

$$\text{I}b \quad \frac{DD\phi\phi\omega}{D\phi\omega}* \quad \textit{cancellation.}$$

Suppose we have a $\mathscr{P}_C$-proof of $D\phi\phi$. Using I$a$ repeatedly we obtain $D\ldots D\Phi\Phi$, where $\Phi$ is a formula of the form $\psi'{}^\frown \ldots {}^\frown\psi^{(S\theta)}$, parentheses are omitted as in I$a$, each $\psi^{(\lambda)}$, $1 \leqslant \lambda \leqslant S\theta$ is either a variable or the negation of a variable or is $NN\chi^{(\lambda)}$ where $\chi^{(\lambda)}$ is a $\mathscr{P}_C$-statement or is $ND\chi^{(\lambda)}\omega^{(\lambda)}$, where $\chi^{(\lambda)}$ and $\omega^{(\lambda)}$ are $\mathscr{P}_C$-statements. By repeatedly using I$a$ and the reversibility of II$c$, then I$a$ repeatedly we may replace $NN\chi^{(\lambda)}$ by $\chi^{(\lambda)}$. (I$a$ is used to bring $NN\chi^{(\lambda)}$ to the left.) By repeatedly using the reversibility of II$b$ then I$a$ repeatedly we may replace $ND\chi^{(\lambda)}\omega^{(\lambda)}$ by $N\chi^{(\lambda)}$ or $N\omega^{(\lambda)}$. Continuing these three operations as long as possible (this amounts to moving occurrences of $N$ as far to the right as possible) we obtain $\mathscr{P}_C$-proofs of formulae

$$DD\ldots DD\ldots D\Psi'^{(\mu)}\Psi'^{(\mu)}, \quad 1 \leqslant \mu \leqslant \nu, \tag{1}$$

where $\Psi'^{(\mu)}$, $1 \leqslant \mu \leqslant \nu$ is a formula formed from a sequence of variables and negated variables. The process must cease because at each of the last two steps we decrease the length of the formula dealt with, and applications of I$a$ are limited. The statements (1) are $\mathscr{P}_C$-theorems and their $\mathscr{P}_C$-proofs proceed from $\mathscr{P}_C$-axioms using $\mathscr{P}_C$-rules I$a$, II$a$ only, because (1) is without occurrences of $ND$ or $NN$. Thus each $\Psi'^{(\mu)}$ must contain a variable and the same variable negated. Thus $D\ldots D\Psi'^{(\mu)}$ is a $\mathscr{P}_C$-theorem. By repeated use of II$b$, $c$, and of course I$a$, following the inverse order of the applications we made of their reversibilities, we obtain a $\mathscr{P}_C$-proof of $\phi$ as desired. From a $\mathscr{P}_C$-proof of $DD\phi\phi\omega$ by I$a$, II$a$ we can obtain a $\mathscr{P}_C$-proof of $DD\phi\omega\, D\phi\omega$ and hence one of $D\phi\omega$, as above shown. Thus I$b$ is a derived rule.

We define

D 1 $\qquad K$ for $\lambda pp'.NDNpNp'$.

$K$ is called the *conjunction sign*. It is of type $ooo$, $K\phi\psi$ for $NDN\phi N\psi$.
We have

II$b'$ $$\frac{D\phi\omega \quad D\psi\omega}{DK\phi\psi\omega}.$$

We have

$$\cfrac{\cfrac{\cfrac{D\phi\omega}{DNN\phi\omega}\ \text{II}c\ \cfrac{D\psi\omega}{DNN\psi\omega}}{DNDN\phi N\psi\omega}\ \text{II}b}{DK\phi\psi\omega}\ \text{by definition of } K.$$

PROP. 4. (i) *Rule* II$b'$ *is reversible.*

(ii) *Conjunction is commutative and associative.*

(iii) *Disjunction is distributive with conjunction.*

(iv) *Conjunction is distributive with disjunction.*

(i) From the reversibility of II$b$ we can obtain $\mathscr{P}_C$-proofs of $DNN\phi\omega$ and $DNN\psi\omega$ from a $\mathscr{P}_C$-proof of $DK\phi\psi\omega$. From the reversibility of II$c$ we then get $\mathscr{P}_C$-proofs of $D\phi\omega$ and $D\psi\omega$.

(ii) We have $\dfrac{K\phi\psi}{\phi\quad\psi}*$ by the reversibility of II$b'$, hence by II$b'$ we get $K\psi\phi$, so conjunction is commutative.

We have: to show $\dfrac{KK\phi\psi\chi}{K\phi K\psi\chi}*$ and $\dfrac{K\phi K\psi\chi}{KK\phi\psi\chi}*$.

We have
$$\begin{array}{lll}
KK\phi\psi\chi\;* & K\phi K\psi\chi\;* & \\
\cline{1-1}\cline{2-2}
K\phi\psi\ \chi\;* & \phi\ K\psi\chi\;* & \text{by reversibility of II}b' \\
\cline{1-1}\cline{2-2}
\phi\ \psi\ \chi & \phi\ \psi\ \chi & \text{ditto} \\
\cline{1-1}\cline{2-2}
\phi\ K\psi\chi & K\phi\psi\ \chi & \text{by II}b' \\
\cline{1-1}\cline{2-2}
K\phi K\psi\chi & KK\phi\psi\chi & \text{by II}b'.
\end{array}$$

Hence conjunction is associative.

We have to show
$$\frac{DK\phi\chi\psi}{KD\phi\chi D\psi\chi}*\quad\text{and}\quad\frac{KD\phi\chi D\psi\chi}{DK\phi\psi\chi}*.$$

(iii) We have
$$\begin{array}{lll}
DK\phi\psi\chi\;* & \text{reversibility of II}b & KD\phi\chi D\psi\chi\;* \\
\cline{1-1}\cline{3-3}
D\phi\chi\ \ D\psi\chi & \text{II}b' & D\phi\chi\ \ D\psi\chi\quad \text{II}b'. \\
\cline{1-1}\cline{3-3}
KD\phi\chi D\psi\chi & & DK\phi\psi\chi.
\end{array}$$

Hence disjunction is distributive with conjunction.

(iv) We have to show
$$\frac{KD\phi\psi\chi}{DK\phi\chi K\psi\chi}*\quad\text{and}\quad\frac{DK\phi\chi K\psi\chi}{KD\phi\psi\chi}*.$$

We have
$$\begin{array}{ll}
KD\phi\psi\chi\;* & \\
\hline
D\phi\psi\ \ \chi & \text{reversibility of II}b' \\
\hline
D\psi\phi\ \ D\chi\phi\ \ D\chi K\psi\chi & \text{I}a,\ \text{II}a \\
\hline
DK\psi\chi\phi\ \ D\chi K\psi\chi & \text{I}a,\ \text{II}b' \\
\hline
DK\phi\chi K\psi\chi & \text{I}a,\ \text{II}b'
\end{array}
\qquad
\begin{array}{ll}
DK\phi\chi K\psi\chi\;* & \\
\hline
D\phi K\psi\chi\ \ D\chi K\psi\chi & \text{I}a \\
\hline
DK\psi\chi\phi\ \ DK\psi\chi\chi\;* & \text{reversibility of II}b \\
\hline
D\psi\phi\ \ D\chi\chi\;* & \text{I}a, b \\
\hline
D\phi\psi\ \ \chi & \text{II}b. \\
\hline
KD\phi\psi\chi &
\end{array}$$

Hence conjunction is distributive with disjunction.

We use the notation $\phi\equiv\psi$ to denote that $\dfrac{\chi\{\phi\}}{\chi\{\psi\}}*$ and $\dfrac{\chi\{\psi\}}{\chi\{\phi\}}*$, for any $\chi$, we then say that $\phi$ and $\psi$ are *equivalent*.

PROP. 5

| | | | |
|---|---|---|---|
| 1. $D\phi\phi \equiv \phi$ | *idem-potency* | 1*. $K\phi\phi \equiv \phi$, |
| 2. $D\phi\phi' \equiv D\phi'\phi$ | *commu-tativity* | 2*. $K\phi\phi' \equiv K\phi'\phi$, |
| 3. $DD\phi\phi'\phi'' \equiv D\phi D\phi'\phi''$ | *associ-ativity* | 3*. $KK\phi\phi'\phi'' \equiv K\phi K\phi'\phi''$, |
| 4. $DK\phi\phi'\phi'' \equiv KD\phi\phi''D\phi'\phi''$ | *distri-butivity* | 4*. $KD\phi\phi'\phi'' \equiv DK\phi\phi''K\phi'\phi''$, |
| 5. $ND\phi\phi' \equiv KN\phi N\phi'$ | *de Mor-gan laws* | 5*. $NK\phi\phi' \equiv DN\phi N\phi'$, |

6. $NN\phi \equiv \phi$ *double negation.*

We have to show: $\dfrac{\chi\{\phi\}}{\chi\{\psi\}}*$ and $\dfrac{\chi\{\phi\}}{\chi\{\psi\}}*$, for any $\chi$, and any pairs $\phi$ and $\psi$ listed. $\chi\{\phi\}$ is $\phi$: 1 is I$b$ and II$a$, 1* is II$b'$ and its reversibility, 2 and 2*, 3 and 3*, 4 and 4* have already been demonstrated. 5*, 6 come from II$c$ and its reversibility. And 5, if $KN\phi\, N\phi'$ is a $\mathscr{P}_C$-theorem then so are $N\phi$ and $N\phi'$, whence by II$b$ so is $ND\phi\phi'$. If $ND\phi\phi'$ is a $\mathscr{P}_C$-theorem then so is $KN\phi\, N\phi'$.

Otherwise we note the place where $\phi$ is introduced into the $\mathscr{P}_C$-proof of $\chi\{\phi\}$ and then introduce $\psi$ instead. If $\phi$ is introduced at II$a$ then introduce $\psi$ instead, otherwise $\phi$ is introduced by different rules according to which of 1–6 we are considering. Take, for instance 4* with $KD\phi\phi'\phi''$ for $\phi$ and $DK\phi\phi''\,K\phi'\phi''$ for $\psi$. $KD\phi\phi'\phi''$ can be introduced at II$b'$

$$\frac{DD\phi\phi'\omega \quad D\phi''\omega}{DKD\phi\phi'\phi''\omega}\ \text{II}b'.$$

Replace this by

$$\frac{\text{II}b'\ \dfrac{\text{I}a\ \dfrac{DD\phi\phi'\omega}{D\phi\, D\phi'\omega} \qquad \dfrac{\dfrac{D\phi''\omega}{D\phi'\, D\phi''\omega}\,\text{II}a}{D\phi''\, D\phi'\omega}\,\text{I}a}{\text{I}a\ \dfrac{DK\phi\phi''\, D\phi'\omega}{D\phi'\, DK\phi\phi''\omega}} \qquad \dfrac{\dfrac{D\phi''\omega}{DK\phi\phi''\, D\phi''\omega}\,\text{II}a}{D\phi''\, DK\phi\phi''\omega}\,\text{I}a}{\dfrac{DK\phi'\phi''\, DK\phi\phi''\omega}{DDK\phi'\phi''\, K\phi\phi''\omega}\,\text{I}a}\ \text{II}b'$$

between this and the base of the $\mathscr{P}_C$-proof-tree replace $KD\phi\phi'\phi''$ by $DK\phi\phi''K\phi'\phi''$ and the $\mathscr{P}_C$-proof-tree of $\chi\{KD\phi\phi'\phi''\}$ is converted into a $\mathscr{P}_C$-proof-tree of $\chi\{DK\phi\phi''K\phi'\phi''\}$. The other cases are dealt with in a similar manner.

## 2.8 *Deduction theorem*

PROP. 6. *The deduction theorem holds in* $\mathscr{P}_C$.

That is to say (*f*) holds: if

$$\frac{\phi', \ldots, \phi^{(S\theta)}}{\psi} \quad \text{then} \quad \frac{\phi', \ldots, \phi^{(\theta)}}{C\phi^{(S\theta)}\psi},$$

where

D 2. $C$ for $\lambda pp'.DNpp'$. Then $C$ is of type *ooo*, it is called the *conditional symbol*; in our parenthesesless notation we can define $C$ for $DN$. We have to show (i), (ii), (iii) and (iv). $C$ expresses *material implication*.

(i) $DN\phi\phi$ is a $\mathscr{P}_C$-theorem, this is known as *tertium non datur*. We proceed by induction on the construction of $\phi$, this is known as *forumula induction*. If $\phi$ is atomic, that is if $\phi$ is a variable, say $p$, then the result follows at once from the axiom $DpNp$ by I$a$. If it is of the form $N\psi$ and the result holds for $\psi$ then we have

$$\cfrac{\cfrac{DN\psi\psi}{D\psi N\psi}\ \text{I}a}{DNN\psi N\psi}\ \text{II}c$$

as desired. If $\phi$ is a disjunction $D\psi'\psi''$ and the result holds for $\psi'$ and for $\psi''$ then we have

$$\cfrac{\cfrac{DN\psi'\psi'}{DN\psi' D\psi'\psi''}\ \text{II}a, \qquad \text{I}a\ \cfrac{DN\psi''\psi''}{DN\psi'' D\psi'\psi''}}{DND\psi'\psi'' D\psi'\psi''}\ \text{II}b$$

thus $DN\phi\phi$ is a $\mathscr{P}_C$-theorem for any $\mathscr{P}_C$-statement $\phi$.

(ii) $DN\phi\chi$ is a $\mathscr{P}_C$-theorem whenever $\chi$ is a $\mathscr{P}_C$-axiom. This follows at once from II$a$.

(iii) $\dfrac{\chi}{C\phi\chi}$, this again follows at once from II$a$.

(iv) If $\frac{\chi}{\psi}$ and $\frac{\chi' \ \chi''}{\psi}$ are $\mathscr{P}_C$-rules then

$$\frac{DN\phi\chi}{\overline{\overline{DN\phi\psi}}} \quad \frac{DN\phi\chi' \quad DN\phi\chi''}{\overline{\overline{DN\phi\psi}}},$$

these follow at once from II$a$, I$a$ and the rule in question by putting $N\phi$ into the subsidiary formulae. Thus

$$\begin{array}{ll} \underline{\quad\chi\quad} & \text{II}a \\ \underline{DN\phi\chi} & \text{I}a \\ \underline{D\chi N\phi} & \\ D\psi N\phi & \end{array}$$

using the rule $\frac{\chi}{\psi}$ with $N\phi$ in the subsidiary formula. The other cases follow similarly. Thus the deduction theorem holds in $\mathscr{P}_C$.

## 2.9 *Modus Ponens*

The rule
$$\frac{D\omega\phi \quad DN\phi\chi}{D\omega\chi},$$

where $\omega$ is subsidiary and can be absent, $\chi$ is secondary and must be present, is known as *Modus Ponens*, the *rule of detachment* or the *cut*. The formula $\phi$ is known as the *cut formula*.

PROP. 7 *Modus Ponens is a derived rule in* $\mathscr{P}_C$.

We have to show:
$$\frac{D\omega\phi \quad DN\phi\chi}{D\omega\chi}*$$

$D\omega$ may be absent but $\chi$ must be present. We have to show how we can obtain a $\mathscr{P}_C$-proof of $D\omega\chi$ when we are given $\mathscr{P}_C$-proofs of $D\omega\phi$ and $DN\phi\chi$. The demonstration is by formula induction on the cut formula $\phi$.

($a$) $\phi$ is atomic, that is $\phi$ is a variable, $\pi$. We use *theorem induction* on $DN\phi\chi$, that is we suppose the result holds for the formula or formulae in the $\mathscr{P}_C$-proof-tree immediately above $DN\phi\chi$. If $DN\phi\chi$ is a remodelling of a $\mathscr{P}_C$-axiom then $\chi$ must be $\phi$ and we have done. Otherwise if $N\phi$ is in the subsidiary formula or secondary formula of the building rule immediately above $DN\phi\chi$ in the $\mathscr{P}_C$-proof of $DN\phi\chi$ so that we have for a one premiss rule

$$\frac{DN\phi\chi'}{\overline{\overline{DN\phi\chi}}} \quad \text{I}a, \text{II}a \text{ or } c$$

thus we have by our induction hypothesis:

$$\cfrac{\cfrac{D\omega\phi \quad DN\phi\chi'}{D\omega\chi'}\,*}{D\omega\chi}\quad \mathrm{I}a,\ \mathrm{II}a \text{ or } c, \text{ exactly as before.}$$

Similarly for a two premiss rule II$b$. If $N\phi$ is in the main formula of the building rule immediately above $DN\phi\chi$ in the $\mathscr{P}_C$-proof-tree of $DN\phi\chi$ then this rule can only be II$a$ since $\phi$ is atomic ($DND\psi'\chi'\omega$ fails to become $DN\pi\chi$ by I$a$ if $\pi$ is in $D\psi'\chi'$ and $DNN\psi'\chi'$ fails to become $DN\pi\chi$ by I$a$ if $\pi$ is in $\psi'$, where $\pi$ is a variable). Thus the part of the tree above $DN\phi\chi$ is

$$\frac{\chi'}{DN\phi\chi}\quad \mathrm{II}a,\ \mathrm{I}a,$$

where $\chi$ is $D\chi'\chi''$ or is just $\chi'$. Hence we obtain

$$\frac{\chi'}{D\omega\chi}\quad \mathrm{II}a,\ \mathrm{I}a$$

as desired.

($b$) $\phi$ is $D\phi'\phi''$ and the result holds for $\phi'$ and $\phi''$. We have $\mathscr{P}_C$-proofs of $D\omega D\phi'\phi''$ and of $DND\phi'\phi''\chi$. From the reversibility of II$b$ we can obtain $\mathscr{P}_C$-proofs of $DN\phi'\chi$ and of $DN\phi''\chi$. Hence by formula induction:

$$\cfrac{\cfrac{\cfrac{\cfrac{\cfrac{\cfrac{D\omega D\phi'\phi''}{DD\omega\phi'\phi''}\ \mathrm{I}a \quad DN\phi''\chi}{DD\omega\phi'\chi}\ * \text{ by formula induction}}{DD\omega\chi\phi'}\ \mathrm{I}a \quad DN\phi'\chi}{DD\omega\chi\chi}\ * \text{ by formula induction}}{D\omega\chi}\ * \text{ Prop. 3.}}{}$$

($c$) $\phi$ is $N\phi'$ and the result holds for $\phi'$. We have $\mathscr{P}_C$-proofs of $D\omega N\phi'$ and of $DNN\phi'\chi$. From the reversibility of II$c$ we can obtain a $\mathscr{P}_C$-proof of $D\phi'\chi$. Thus if $\omega$ is present we have by formula induction

$$\cfrac{\cfrac{D\chi\phi' \quad DN\phi'\omega}{D\chi\omega}\ *}{D\omega\chi}\quad \mathrm{I}a.$$

If $\omega$ is absent we have

$$\cfrac{\cfrac{D\chi\phi' \qquad \cfrac{N\phi'}{DN\phi'\chi}\ \mathrm{II}a,\ \mathrm{I}a}{D\chi\chi}\ *}{\chi}\ * \quad \text{Prop. 3.}$$

## 2.10 *Regularity*

D 3 $\qquad B$ for $\lambda pp'.KCpp'Cp'p$.

Thus $B\phi\psi$ for $KC\phi\psi C\psi\phi$. $B$ is called the *biconditional symbol*. If $B\phi\psi$ is a $\mathscr{P}_C$-theorem then $\phi$ and $\psi$ are called *equivalent*. A propositional calculus $\mathscr{P}$ is called *regular* if from the $\mathscr{P}$-proofs of $B\phi\psi$ and of $\chi\{\phi\}$ we can obtain a $\mathscr{P}$-proof of $\chi\{\psi\}$. $B$ expresses *material equivalence*.

PROP. 8. *$\mathscr{P}_C$ is regular.*

We have to show
$$\frac{B\phi\psi \quad \chi\{\phi\}}{\chi\{\psi\}}*,$$
i.e. from the $\mathscr{P}_C$-proofs of $B\phi\psi$ and $\chi\{\phi\}$ we can find a $\mathscr{P}_C$-proof of $\chi\{\psi\}$. We proceed by formula induction on $\chi\{\pi\}$. $\chi\{\pi\}$ is $\pi$, we have:
$$\cfrac{\cfrac{B\phi\psi}{DN\phi\psi}* \quad \phi}{\psi}* \qquad \begin{array}{l}\text{reversibility of II}b'\\ \text{Modus Ponens.}\end{array}$$
$\chi\{\pi\}$ is $N\pi$ we have
$$\cfrac{\cfrac{\cfrac{B\phi\psi}{DN\psi\phi}* \ \text{reversibility of II}b'}{DNN\phi N\psi}\ \text{II}c,\text{I}a \quad N\phi}{N\psi}* \quad \text{Modus Ponens.}$$
$\chi\{\pi\}$ is $D\pi\chi$ we have
$$\cfrac{\cfrac{B\phi\psi}{DN\phi\psi}* \quad D\phi\chi}{D\psi\chi}* \qquad \begin{array}{l}\text{reversibility of II}b'\\ \text{Modus Ponens.}\end{array}$$
Thus $\mathscr{P}_C$- is regular.

COR. (i)
$$\frac{C\phi\psi \qquad C\psi\phi}{\overline{C\chi\{\phi\}\chi\{\psi\} \quad C\chi\{\psi\}\chi\{\phi\}}} \quad \textit{and} \quad \frac{B\phi\psi}{B\chi\{\phi\}\chi\{\psi\}}*.$$

We demonstrate the result by formula induction on $\chi$.

If $\chi\{\phi\}$ is $\phi$ then the result is trivial.

If $\chi\{\phi\}$ is $D\chi'\{\phi\}\chi''\{\phi\}$ and the result holds for $\chi'\{\phi\}$ and $\chi''\{\phi\}$, then we have:
$$\frac{C\phi\psi \qquad\qquad C\psi\phi}{\overline{C\chi'\{\phi\}\chi'\{\psi\} \quad C\chi''\{\phi\}\chi''\{\psi\} \quad C\chi'\{\psi\}\chi'\{\phi\} \quad C\chi''\{\psi\}\chi''\{\phi\}}},$$

whence by dilution and permutation

$$\frac{DN\chi'\{\phi\}D\chi'\{\psi\}\chi''\{\psi\} \quad DN\chi''\{\phi\}D\chi'\{\psi\}\chi''\{\psi\}}{DND\chi'\{\phi\}\chi''\{\phi\}D\chi'\{\psi\}\chi''\{\psi\}}\ \text{II}b$$

similarly with $\phi$ and $\psi$ interchanged, and the result follows.

If $\chi\{\phi\}$ is $N\chi'\{\phi\}$ and the result holds for $\chi'\{\phi\}$ then we have

$$\frac{C\phi\psi \qquad\qquad C\psi\phi}{C\chi'\{\phi\}\chi'\{\psi\} \quad C\chi'\{\psi\}\chi'\{\phi\}}$$

whence by II$c$, I$a$

$$DN\chi'\{\phi\}NN\chi'\{\psi\} \quad DN\chi'\{\psi\}NN\chi'\{\phi\}$$

which is $\quad CN\chi'\{\psi\}N\chi'\{\phi\} \quad CN\chi'\{\phi\}N\chi'\{\psi\}\quad$ as required.

Cor. (ii)

$$\frac{\phi \ \ \chi\{B\phi\psi\}}{\chi\{\psi\}}*\quad \frac{\phi \ \ \chi\{K\phi\psi\}}{\chi\{\psi\}}*\quad \frac{\phi \ \ \chi\{C\phi\psi\}}{\chi\{\psi\}}*.$$

This follows from Prop. 8 since

$$\frac{\phi}{B\psi B\phi\psi}, \quad \frac{\phi}{B\psi K\phi\psi}, \quad \frac{\phi}{B\psi C\phi\psi}.$$

These are easily established, consider the second one,

$$\cfrac{\cfrac{\cfrac{\phi}{D\phi N\psi}\ \text{II}a,\text{I}a \quad D\psi N\psi}{DK\phi\psi N\psi}\ \text{II}b'}{DN\psi K\phi\psi}\ \text{I}a \qquad \cfrac{\cfrac{\cfrac{D\psi N\psi}{DD\psi N\psi N\phi}\ \text{II}a,\text{I}a}{DDN\phi N\psi\psi}\ \text{I}a}{DNK\phi\psi\psi}\ \text{II}c \text{ and definition of } K.$$

Similarly for the other cases.

## 2.11 *Duality*

Prop. 9

| | | |
|---|---|---|
| 1. $B\phi D\phi\phi$ | *idempotency* | 1*. $B\phi K\phi\phi$ |
| 2. $BD\phi\phi'D\phi'\phi$ | *commutativity* | 2*. $BK\phi\phi'K\phi'\phi$ |
| 3. $BDD\phi\phi'\phi''D\phi D\phi'\phi''$ | *associativity* | 3*. $BKK\phi\phi'\phi''K\phi K\phi'\phi''$ |
| 4. $BDK\phi\phi'\phi''KD\phi\phi''D\phi'\phi''$ | *distributivity* | 4*. $BKD\phi\phi'\phi''DK\phi\phi''K\phi'\phi''$ |
| 5. $BND\phi\phi'KN\phi N\phi'$ | *de Morgan laws* | 5*. $BNK\phi\phi'DN\phi N\phi'$ |

6. $B\phi NN\phi$ *double negation.*

Using the Deduction Theorem we have

$$\dfrac{\dfrac{\dfrac{\dfrac{D\phi\phi'' \quad D\phi'\phi''}{DK\phi\phi'\phi''}\,\mathrm{II}b'}{CD\phi\phi''CD\phi'\phi''DK\phi\phi'\phi''}\,\text{Prop. 6, twice}}{DNNDND\phi\phi''ND\phi'\phi''DK\phi\phi'\phi''}\,\mathrm{I}a, \mathrm{II}c}{CKD\phi\phi''D\phi'\phi''DK\phi\phi'\phi''}\,\text{Def. of } C, K. \qquad \text{(i)}$$

Again we have

$$\dfrac{\dfrac{\dfrac{\dfrac{\dfrac{\dfrac{\dfrac{\phi}{D\phi\phi''}\ \mathrm{II}a, \mathrm{I}a\ \dfrac{\phi'}{D\phi'\phi''}}{KD\phi\phi''D\phi'\phi''}\,\mathrm{II}b'}{C\phi C\phi'KD\phi\phi''D\phi'\phi''}\,\text{Prop 6.}}{DNNDN\phi N\phi'KD\phi\phi''D\phi'\phi''}\,\mathrm{I}a, \mathrm{II}c.}{DNK\phi\phi'KD\phi\phi''D\phi'\phi''}\,\text{Def. of } K \quad \dfrac{\dfrac{\dfrac{\dfrac{\phi''}{D\phi\phi'' \quad D\phi'\phi''}\,\mathrm{II}a}{KD\phi\phi''D\phi'\phi''}\,\mathrm{II}b'}{C\phi''KD\phi\phi''D\phi'\phi''}\,\text{Prop. 6.}}{DN\phi''KD\phi\phi''D\phi'\phi''}}{\dfrac{DNDK\phi\phi'\phi''KD\phi\phi''D\phi'\phi''}{CDK\phi\phi'\phi''KD\phi\phi''D\phi'\phi''}\,\text{Def. of } C.}\,\mathrm{II}b \qquad \text{(ii)}$$

4 now follows from (i) and (ii) by II$b'$ and definition of $K$. The rest are dealt with similarly and are left as exercises to the reader.

The *dual* of a $\mathscr{P}_C$-statement $\phi$ is the result of replacing $D$ by $K$ and $K$ by $D$ throughout $\phi$.

PROP. 10. *If $\phi$ is a $\mathscr{P}_C$-theorem then so is $N\phi'$, where $\phi'$ is the dual of $\phi$.* In the $\mathscr{P}_C$-theorem $\phi$ replace each variable $\pi$ by $N\pi$, the result is a $\mathscr{P}_C$-theorem $\phi''$, say. The original $\mathscr{P}_C$-proof-tree becomes a $\mathscr{P}_C$-deduction from $\mathscr{P}_C$-statements of the form $DN\pi NN\pi$ which are $\mathscr{P}_C$-theorems hence $\phi''$ is a $\mathscr{P}_C$-theorem. Now use the de Morgan laws Prop. 5 (5) repeatedly and the result follows.

## 2.12 *Independence of symbols, axioms and rules*

PROP. 11. *The symbols, axioms and rules of $\mathscr{P}_C$ are independent*

Suppose that $N$ can be defined in terms of $D$

$$N \quad \text{for} \quad \lambda p.(D,p), \quad \text{of type } oo,$$

where $(D,p)$ is some $\mathscr{P}_C$-statement built up from $D$ and $p$ alone. Then $N\phi$ may be replaced by $(D,\phi)$ in any $\mathscr{P}_C$-theorem and the result is another $\mathscr{P}_C$-theorem. In particular the axiom $DpNp$ becomes $Dp(D,p)$. By the commutative and associative laws for disjunction this becomes $D \dots Dp \dots p$. By the derived rule I$b$ we obtain $p$ as a $\mathscr{P}_C$-theorem. This is absurd because a $\mathscr{P}_C$-theorem must contain at least four symbols. Statements fail to decrease as we go down a $\mathscr{P}_C$-proof.

Suppose that $D$ can be defined in terms of $N$

$$D \quad \text{for} \quad \lambda pp'.(N,p,p'), \quad \text{of type } ooo,$$

where $(N,p,p')$ is some $\mathscr{P}_C$-statement built up from $N$, $p,p'$ alone. The only such $\mathscr{P}_C$-statements are:

$$p, Np, NNp, \dots \quad \text{and} \quad p', Np', NNp', \dots.$$

Thus $D\phi\psi$ would be independent of one of its arguments. The axiom $DpNp$ would become one of $p, Np, NNp, \dots$, one of these would be a $\mathscr{P}_C$-theorem. By the reversibility of II$c$ repeatedly one of $p$, $Np$ would be a $\mathscr{P}_C$-theorem. This is absurd because a $\mathscr{P}_C$-theorem must contain at least four symbols.

The $\mathscr{P}_C$-axiom $DpNp$ is independent because it is impossible to obtain it from the other axioms. Once a variable is introduced into a $\mathscr{P}_C$-proof it remains in that $\mathscr{P}_C$-proof from that place onwards. Hence if we are denied use of the axiom $DpNp$ then the resulting $\mathscr{P}_C$-theorems will all contain some variable distinct from the variable $p$.

Lastly the $\mathscr{P}_C$-rules are independent. Suppose we omit the rule I$a$ then we are unable to obtain the $\mathscr{P}_C$-theorem $DNpp$ because the other $\mathscr{P}_C$-rules increase the length of a $\mathscr{P}_C$-statement. If we omit rule II$a$ then we are unable to obtain the $\mathscr{P}_C$-theorem $DpDpNp$ because the lower formula of I$a$ with a subsidiary formula, II$b$, $c$ begin in a different way. In the case of I$a$ without subsidiary formulae we could only obtain $DpDpNp$ from $DDpNpp$ and this is distinct from the lower formula of I$a$ with $DpDpNp$ or $DDNppp$ or $DDppNp$ as upper formula and these fail to be lower formulae of II$b$, $c$. If we omit II$b$ we are unable to obtain $DNDppp$, this fails to be the lower formula of II$a$ in a $\mathscr{P}_C$-proof, because the upper formula would then be $p$ which is impossible. It also fails to be the lower formula of II$c$, if it were the lower formula of I$a$ then the upper formula would be $DpNDpp$ this fails to be the lower formula of II$c$, if it is the lower formula of II$a$ then the upper formula would be $NDpp$ from which by idempotency we could obtain a $\mathscr{P}_C$-proof of $Np$, which is impossible. If we omit II$c$ then we are unable to obtain the $\mathscr{P}_C$-theorem $DNNpNp$ by similar considerations.

We have just shown that the symbols of $\mathscr{P}_C$ are independent, yet it is possible to define $N$ and $D$ in terms of a single symbol, $S$ or $S'$. This can be done as follows:

D 4. $\quad S$ for $\lambda pp'.DNpNp'$.

D 5. $\quad S'$ for $\lambda pp'.KNpNp'$.

then $S$ and $S'$ are both of type $ooo$. We can then define

$$N \text{ for } \lambda p.Spp.$$

$$D \text{ for } \lambda pp'.SSppSp'p'.$$

And

$$N \text{ for } \lambda p.S'pp.$$

$$D \text{ for } \lambda pp'.S'S'pp'S'pp'.$$

The axiom scheme would become

$$SS\pi\pi SS\pi\pi S\pi\pi,$$

$$S'S'\pi S'\pi\pi S'\pi S'\pi\pi.$$

Thus there can be *syzygies* between independent symbols. Independence of axioms can also be shown by finding a model $\mathscr{M}$ for the formal system less one axiom and showing that this axiom fails to be $\mathscr{M}$-valid.

## 2.13 *Consistency and completeness of* $\mathscr{P}_C$

PROP. 12. *$\mathscr{P}_C$ is model-consistent*

Consider the applied propositional calculus $\mathscr{M}_C$ with the elements $\boldsymbol{t}$, $\boldsymbol{f}$ of type $o$ called *truth-values*, $\boldsymbol{t}$ is designated and $\boldsymbol{f}$ is undesignated, and constants $N_{oo}$, $D_{ooo}$ of the types shown in the subscripts. The rules are (type symbols are omitted)

$$\frac{\phi\{N\boldsymbol{t}\}}{\phi\{\boldsymbol{f}\}} \quad \frac{\phi\{N\boldsymbol{f}\}}{\phi\{\boldsymbol{t}\}} \quad \frac{\phi\{D\boldsymbol{tt}\}}{\phi\{\boldsymbol{t}\}} \quad \frac{\phi\{D\boldsymbol{tf}\}}{\phi\{\boldsymbol{t}\}} \quad \frac{\phi\{D\boldsymbol{ft}\}}{\phi\{\boldsymbol{t}\}} \quad \frac{\phi\{D\boldsymbol{ff}\}}{\phi\{\boldsymbol{f}\}}$$

and conversely. It is easily verified that $\mathscr{M}_C$ is a model for $\mathscr{P}_C$. Thus $\mathscr{P}_C$ is a two-valued propositional calculus. An $\mathscr{M}_C$-valid $\mathscr{P}_C$-statement is called a *tautology*.

PROP. 13. *$\mathscr{P}_C$ is complete with respect to $\mathscr{M}_C$.*

Let $\phi$ be a $\mathscr{P}_C$-statement, by repeated use of the de Morgan laws and double negation move negations to the right as far as possible until they act on variables only, this introduces the connective $K$. Then using the distributive and commutative laws reduce the resulting $\mathscr{P}_C$-statement to a conjunction of disjunctions (*conjunctive normal form*).

$$K \dots K(D \dots D\psi' \dots \psi^{(\pi)}) \dots (D \dots D\chi' \dots \chi^{(\theta)}),$$

where $\psi', \dots, \psi^{(\pi)}, \dots, \chi', \dots, \chi^{(\theta)}$ are variables or negated variables. A *disjunctive normal form* is obtained by interchanging the roles of $K$ and $D$. A conjunctive normal form is $\mathscr{M}_C$-valid if and only if each conjunctand contains as disjunctands a variable $\pi$ and the same variable negated, otherwise by suitable choice of $\boldsymbol{t}$ and $\boldsymbol{f}$ to replace the variables we can make that conjunctand take the value $\boldsymbol{f}$. Starting from the axiom $D\pi N\pi$ and using II$a$, I$a$ we can easily $\mathscr{P}_C$-prove any disjunctand of an $\mathscr{M}_C$-valid conjunctive normal form, and hence the conjunction of these disjunctands. If $\psi$ is the conjunctive normal form of $\phi$ then from a $\mathscr{P}_C$-proof of $\psi$ we can find one of $\phi$ by using the distributive laws and the law of double negation and the definition of $K$. Thus an $\mathscr{M}_C$-valid $\mathscr{P}_C$-statement is a $\mathscr{P}_C$-theorem. Thus $\mathscr{P}_C$ is complete with respect to $\mathscr{M}_C$.

PROP. 14. *$\mathscr{P}_C$ is functionally complete with respect to $\mathscr{M}_C$.*

Let $\mathscr{M}'_C$ be an extension of $\mathscr{M}_C$ which contains every possible table for every connective of types $oo, ooo, \dots$. Let $\Gamma$ be one such connective and suppose it has $\nu$ arguments. If $H', \dots, H^{(\nu)}$ are all either $\boldsymbol{t}$ or $\boldsymbol{f}$ then let the $\mathscr{M}_C$-value of $\Gamma$ with these arguments be denoted by $\Gamma[H', \dots, H^{(\nu)}]$. We wish to show that there is a $\mathscr{P}_C$-statement containing exactly the distinct variables $p', \dots, p^{(\nu)}$ whose $\mathscr{M}_C$-value when the variables are replaced by $H', \dots, H^{(\nu)}$ is $\Gamma[H', \dots, H^{(\nu)}]$. Consider $K \dots Kq' \dots q^{(\nu)}$, where $q'$ is $p'$ if $H'$ is $\boldsymbol{t}$ otherwise $q'$ is $Np', \dots, q^{(\nu)}$ is $p^{(\nu)}$ if $H^{(\nu)}$ is $\boldsymbol{t}$ otherwise $q^{(\nu)}$ is $Np^{(\nu)}$. Thus if we replace $p'$ by $H', \dots, p^{(\nu)}$ by $H^{(\nu)}$ then $K \dots Kq' \dots q^{(\nu)}$ has $\boldsymbol{t}$ for $\mathscr{M}_C$-value while $K \dots Kr' \dots r^{(\nu)}$, where $r', \dots, r^{(\nu)}$ are respectively $p'$ or $Np', \dots, p^{(\nu)}$ or $Np^{(\nu)}$ but the set $r', \dots, r^{(\nu)}$ is different from the set $q', \dots, q^{(\nu)}$, has $\boldsymbol{f}$ for $\mathscr{M}_C$-value, on the same replacement. Now form the disjunction of all $K \dots Kq' \dots q^{(\nu)}$ for just those sets $q', \dots, q^{(\nu)}$ for which

$\Gamma[H', ..., H^{(\nu)}]$ is $\boldsymbol{t}$. This is a $\mathscr{P}_C$-statement which has $\mathscr{M}_C$-value $\boldsymbol{t}$ just in case $\Gamma$ reduces to $\boldsymbol{t}$. Thus $\mathscr{P}_C$ is functionally complete with respect to $\mathscr{M}_C$.

## 2.14 *Decidability*

PROP. 15. *$\mathscr{P}_C$ is decidable*

It is an effective process to decide whether a $\mathscr{P}_C$-statement is $\mathscr{M}_C$-valid, hence it is an effective process to decide whether a $\mathscr{P}_C$-statement is an $\mathscr{M}_C$-theorem. Furthermore if a $\mathscr{P}_C$-statement is found to be a $\mathscr{P}_C$-theorem by the test of $\mathscr{M}_C$-validity then it is an effective process to supply the $\mathscr{P}_C$-proof. For all we need do is to put the $\mathscr{P}_C$-statement into conjunctive normal form, this is easily tested for $\mathscr{M}_C$-validity and if the test is affirmative it is a routine matter to supply a $\mathscr{P}_C$-proof.

We can obtain this result in another way. We note that apart from I$a$ the length of a $\mathscr{P}_C$-statement increases as we proceed down a $\mathscr{P}_C$-proof-tree and I$a$ leaves the length unaltered, repeated use of I$a$ will reproduce a previous formula so use of I$a$ is limited. Thus given a $\mathscr{P}_C$-statement $\phi$ we can construct all possible deduction-trees with base $\phi$. We can then decide if any of these trees are $\mathscr{P}_C$-proof-trees of $\phi$ or whether each fails to be a $\mathscr{P}_C$-proof-tree of $\phi$. Thus we can decide if $\phi$ is a $\mathscr{P}_C$-theorem and if so we can find a $\mathscr{P}_C$-proof for it.

For instance $DNDDNpp'Npp'$, call it $\chi$, fails to be a $\mathscr{P}_C$-theorem. It is of the form $DND\phi\psi\omega$ with $DNpp'$ for $\phi$ and $Np$ for $\psi$ and $p'$ for $\omega$. This can arise (i) from I$a \dfrac{D\omega ND\phi\psi}{DND\phi\psi\omega}$ or (ii) from II$a \dfrac{\omega}{DND\phi\psi\omega}$ or (iii) from II$b \dfrac{DN\phi\omega \quad DN\psi\omega}{DND\phi\psi\omega}$ only. Case (i) $D\omega ND\phi\psi$ can arise from $\chi$ by I$a$ which brings us back to where we started or from $\dfrac{ND\phi\psi}{D\omega ND\phi\psi}$ II$a$ only. $ND\phi\psi$ can arise from $\dfrac{N\phi \quad N\psi}{ND\phi\psi}$ II$b$ only. Now $N\phi$ is $NDNpp'$ which can only arise from $\dfrac{NNp \quad Np'}{NDNpp'}$ II$b$. The second upper formula fails to arise from any other $\mathscr{P}_C$-statement by the $\mathscr{P}_C$-rules and fails to be a $\mathscr{P}_C$-axiom. Thus case (i) fails to provide a $\mathscr{P}_C$-proof of $\chi$. Case (ii) $\omega$ is $p'$ and so fails to provide a $\mathscr{P}_C$-proof of $\chi$. Case (iii) $DN\phi\omega$ can only arise from $\dfrac{D\omega N\phi}{DN\phi\omega}$ I$a$ or $\dfrac{\omega}{DN\phi\omega}$ II$a$ this we can reject as before since $\omega$ is $p'$.

$D\omega N\phi$ is $Dp'NDNpp'$ and this can only arise from $DN\phi\omega$ by I$a$, which we started with, or from $\dfrac{NDNpp'}{Dp'NDNpp'}$ II$a$, $NDNpp'$ can only arise from $\dfrac{NNp \quad Np'}{NDNpp'}$ II$b$ and this we can reject as before. Lastly $DN\phi\omega$ is $DNDNpp'p'$ and, apart from cases already considered, can only arise from

$$\frac{DNNpp' \quad DNp'p'}{DNDNpp'p'} \text{ II}b.$$

$DNp'p'$ can only arise from an axiom by I$a$, but $DNNpp'$ can only arise from $Dpp'$ or $Dp'p$ each of which fails to be a $\mathscr{P}_C$-axiom. Hence case (iii) fails to produce a $\mathscr{P}_C$-proof of $\chi$. Case (iii) is more easily dealt with if we consider $DN\psi\omega$ which is $DNNpp'$.

A system with I$b$ as an independent rule would fail to be decidable in this way.

We can add any $\mathscr{P}_C$-statement as an extra axiom without causing every $\mathscr{P}_C$-statement to be a $\mathscr{P}_C$-theorem, provided the statement contains a connective. A variable would then fail to be a $\mathscr{P}_C$-theorem. The resulting system would however fail to have many of the properties of $\mathscr{P}_C$.

## 2.15 *Truth-tables*

The simplest way of testing a $\mathscr{P}_C$-statement for being a tautology is by the use of truth-tables. Suppose that a $\mathscr{P}_C$-statement $\phi$ is written in terms of $D$, $N$, $K$, $C$ and $B$; we replace $D$, $K$, $C$ and $B$ by two place functions, $d$, $k$, $c$ and $b$ respectively, and replace $N$ by a one-place function $n$. We then replace the propositional variables by elements $\boldsymbol{t}$ and $\boldsymbol{f}$ in any manner, always replacing different occurrences of the same propositional variable by the same element. We then evaluate the resulting composition of functions by using the following *truth-tables*:

| $p$ | $p'$ | $dpp'$ | $p$ | $p'$ | $kpp'$ | $p$ | $p'$ | $cpp'$ | $p$ | $p'$ | $bpp'$ | $p$ | $np$ |
|---|---|---|---|---|---|---|---|---|---|---|---|---|---|
| $\boldsymbol{t}$ | $\boldsymbol{t}$ | $\boldsymbol{t}$ | $\boldsymbol{t}$ | $\boldsymbol{t}$ | $\boldsymbol{t}$ | $\boldsymbol{t}$ | $\boldsymbol{t}$ | $\boldsymbol{t}$ | $\boldsymbol{t}$ | $\boldsymbol{t}$ | $\boldsymbol{t}$ | $\boldsymbol{t}$ | $\boldsymbol{f}$ |
| $\boldsymbol{t}$ | $\boldsymbol{f}$ | $\boldsymbol{t}$ | $\boldsymbol{t}$ | $\boldsymbol{f}$ | $\boldsymbol{f}$ | $\boldsymbol{t}$ | $\boldsymbol{f}$ | $\boldsymbol{f}$ | $\boldsymbol{t}$ | $\boldsymbol{f}$ | $\boldsymbol{f}$ | $\boldsymbol{f}$ | $\boldsymbol{t}$ |
| $\boldsymbol{f}$ | $\boldsymbol{t}$ | $\boldsymbol{t}$ | $\boldsymbol{f}$ | $\boldsymbol{t}$ | $\boldsymbol{f}$ | $\boldsymbol{f}$ | $\boldsymbol{t}$ | $\boldsymbol{t}$ | $\boldsymbol{f}$ | $\boldsymbol{t}$ | $\boldsymbol{f}$ | | |
| $\boldsymbol{f}$ | $\boldsymbol{f}$ | $\boldsymbol{f}$ | $\boldsymbol{f}$ | $\boldsymbol{f}$ | $\boldsymbol{f}$ | $\boldsymbol{f}$ | $\boldsymbol{f}$ | $\boldsymbol{t}$ | $\boldsymbol{f}$ | $\boldsymbol{f}$ | $\boldsymbol{t}$ | | |

These tables give the values of the five functions for values of the arguments on the same line.

For example let us test *CCKpp'p''CpCp'p''* for tautology. This $\mathscr{P}_C$-statement is a conditional, let us see if it is possible to make it take the value ***f*** by suitable values, ***t*** or ***f***, given to *p*, *p'* and *p''*. The only way of making a conditional take the value ***f*** is to make the first component ***t*** and the second component ***f***. Thus we want to make *CKpp'p''* take the value ***t*** and *CpCp'p''* take the value ***f***. Both of these are again conditionals so we want *p* to have the value ***t*** and *Cp'p''* have the value ***f***, this requires *p'* to have the value ***t*** and *p''* to have the value ***f***. This then fixes the values of *p*, *p'* and *p''* in order that our original statement take the value ***f***. Putting these values for *p*, *p'* and *p''* in the original statement we easily calculate that it is ***t***, thus it must always be ***t***, and so is a tautology. We put the working down as follows:

```
CCKpp'p"CpCp'p"
f
 t      f
         tf
           t f
   tt f
  t
 f
```

The first line with ***f*** under the first occurrence of *C* indicates that we want to make the whole statement take the value ***f***. To do this we must make the first component take the value ***t*** and the second component take the value ***f***, this is indicated by placing in the second line ***t*** under the second occurrence of *C* and ***f*** under the third occurrence of *C*. Since the third occurrence of *C* is to take the value ***f*** then its first component must take the value ***t*** and its second component must take the value ***f***; this is indicated by placing ***t*** in the third line under the second occurrence of *p*, and ***f*** in the third line under the fourth occurrence of *C*. We have now found values that *p*, *p'* and *p''* must have if our statement is to take the value ***f***. Because *p'* must have the value ***t*** and *p''* must have the value ***f*** since *Cp'p''* is to have the value ***f***, this is centred in the fourth line. In the fifth line we enter the values of *p*, *p'* and *p''* under the first occurrence of these symbols. In the sixth line we evaluate *Kpp'* and find that it is ***t***, in the seventh line we evaluate the first component of our original conditional and find that it is $f$, but we had found in the second line that it should be $t$, hence it is impossible to make the original conditional take the value $f$, hence it always takes the value $t$ and so is a tautology.

The working can be put down on one line, since except for the last entry we have always used different columns. Thus:

$$\begin{array}{ccccccccccc} C & C & K & p & p' & p'' & C & p & C & p' & p'' \\ f & t & t & t & t & f & f & t & f & t & f \\ & & f & & & & & & & & \\ 1 & 7 & 6 & 5 & 5 & 5 & 2 & 3 & 3 & 4 & 4 \end{array}$$

The numerals indicate the order in which the letters $t$ and $f$ are put in. The final line puts both $t$ and $f$ under $K$ indicating an impossibility.

Since $\mathscr{P}_C$ is decidable we could have another formulation for it, namely: the axiom-scheme 'a tautology is an axiom' and dispense with rules. The conditions for being a formal system are satisfied because we have a test for being an axiom.

Sometimes it is simpler to evaluate directly for all possible argument values, we would then put down the work as follows:

| $p$ | $p'$ | $p''$ | $BBpp'BBpp''Bp'p''$ |
|---|---|---|---|
| $t$ | $t$ | $t$ | $t$ |
| $t$ | $t$ | $f$ | $t$ |
| $t$ | $f$ | $t$ | $t$ |
| $t$ | $f$ | $f$ | $t$ |
| $f$ | $t$ | $t$ | $t$ |
| $f$ | $t$ | $f$ | $t$ |
| $f$ | $f$ | $t$ | $t$ |
| $f$ | $f$ | $f$ | $t$ |

Thus we have a tautology.

Independence of axioms and rules may also be shown by means of models. We find a model $\mathscr{M}'$ for the formal system $\mathscr{L}'$ obtained from the formal system $\mathscr{L}$ by omitting one axiom, axiom-scheme or rule and such that the omitted axiom or axiom-scheme fails to be valid in the model $\mathscr{M}'$ or the omitted rule fails to preserve $\mathscr{L}'$-validity.

For instance the following three axiom-schemes and Modus Ponens give a Propositional Calculus $\mathscr{P}_1$ equivalent to $\mathscr{P}_C$, $f$ is a constant.

(i) $C\phi C\phi'\phi$,
(ii) $CC\phi C\phi'\phi''CC\phi\phi'C\phi\phi''$,
(iii) $CCC\phi ff\phi$.

Consider the following truth-tables for the connective $C$.

| $p$ | $p'$ | M.P. $Cpp'$ | (i) $Cpp'$ | (ii) $Cpp'$ | (iii) $Cpp'$ |
|---|---|---|---|---|---|
| **t** | **t** | **t** | **t** | **t** | **t** |
| **t** | **g** | **t** | **f** | **g** | **g** |
| **t** | **f** | **f** | **f** | **f** | **f** |
| **g** | **t** | **t** | **f** | **t** | **t** |
| **g** | **g** | **t** | **g** | **t** | **t** |
| **g** | **f** | **f** | **t** | **g** | **f** |
| **f** | **t** | **t** | **t** | **t** | **t** |
| **f** | **g** | **g** | **t** | **t** | **t** |
| **f** | **f** | **t** | **t** | **t** | **t** |

The constant **f** is undesignated. The heading of the various columns denotes that the corresponding rule or axiom-scheme fails for that truth-table, but all the other rules and axiom-schemes hold for that truth-table. In this model we have three elements, **t** is designated, **f** and **g** are undesignated.

We leave the checking of this table as an exercise for the reader.

## 2.16 *Boolean Algebra*

A *Boolean Algebra* is a formal system with the symbols:

| symbol | type | name |
|---|---|---|
| $a$ | $\iota$ | variable for an element |
| $=$ | $o\iota\iota$ | equality |
| $\neq$ | $o\iota\iota$ | inequality |
| 0 | $\iota$ | null element |
| 1 | $\iota$ | unit element |
| $\cup$ | $\iota\iota\iota$ | union |
| $\cap$ | $\iota\iota\iota$ | intersection |
| $-$ | $\iota\iota$ | complement |
| $'$ | | generating symbol |
| ( | | left parenthesis |
| ) | | right parenthesis |

The axioms are given by the following schemes:

We have written $(\alpha \cup \beta)$ for $((\cup\, \alpha)\beta)$, etc.

$$\alpha = \alpha$$

$$0 \neq 1$$

$$\bar{0} = 1 \qquad \bar{1} = 0$$

$$\alpha \cap 0 = 0 \qquad \alpha \cup 1 = 1$$

$$\alpha \cap 1 = \alpha \qquad \alpha \cup 0 = \alpha$$

$$\alpha \cap \bar{\alpha} = 0 \qquad \alpha \cup \bar{\alpha} = 1$$

$$\bar{\bar{\alpha}} = \alpha$$

$$\alpha \cap \alpha = \alpha \qquad \alpha \cup \alpha = \alpha$$

$$\overline{(\alpha \cap \beta)} = \bar{\alpha} \cup \bar{\beta} \qquad \overline{(\alpha \cup \beta)} = \bar{\alpha} \cap \bar{\beta}$$

$$\alpha \cap \beta = \beta \cap \alpha \qquad \alpha \cup \beta = \beta \cup \alpha$$

$$\alpha \cap (\beta \cap \gamma) = (\alpha \cap \beta) \cap \gamma \qquad \alpha \cup (\beta \cup \gamma) = (\alpha \cup \beta) \cup \gamma$$

$$\alpha \cap (\beta \cup \gamma) = (\alpha \cap \beta) \cup (\alpha \cap \gamma) \qquad \alpha \cup (\beta \cap \gamma) = (\alpha \cup \beta) \cap (\alpha \cup \gamma)$$

Note the duality. We have neglected independence. We have written equalities in the customary manner. $\alpha$, $\beta$, $\gamma$ stand for arbitrary elements. We have omitted parentheses wholesale.

The rules are:

$$\frac{\alpha = \beta \quad \gamma\{\alpha\} = \delta\{\alpha\}}{\gamma\{\beta\} = \delta\{\beta\}}.$$

We could add the symbols and rules of the Propositional Calculus and so get compound statements, but this is unnecessary. Note that if we replace

| | | |
|---|---|---|
| $\cap$ | by | $K$ |
| $\cup$ | by | $D$ |
| $-$ | by | $N$ |
| $0$ | by | $p \,\&\, Np$ |
| $1$ | by | $p \vee Np$ |
| $=$ | by | $B$ |
| $\neq$ | by | $NB$ |

and the variables for elements by propositional variables, then the axioms become tautologies of $\mathscr{P}_C$. Again if we replace the elements by variables for subsets of a given set $X$, then the axioms become axioms for elementary set theory, as regards union, intersection and complementation, 0 becomes the null set and 1 the given set $X$. The simplest Boolean

Algebra consists of the two elements $\{0, 1\}$. This is called the two-valued Boolean Algebra and corresponds to the pair $\{\boldsymbol{f}, \boldsymbol{t}\}$.

**Prop. 16.** *We have*

(i) $\alpha \cup (\alpha \cap \beta) = \alpha \qquad \alpha \cap (\alpha \cup \beta) = \alpha$

(ii) $(\alpha \cap \bar{\beta}) \cup \beta = \alpha \cup \beta \qquad (\alpha \cup \bar{\beta}) \cap \beta = \alpha \cap \beta$

(iii) $\alpha \cup \beta = \alpha$ *if and only if* $\alpha \cap \beta = \beta$.

We have
$$\begin{aligned}\alpha \cup (\alpha \cap \beta) &= (\alpha \cap 1) \cup (\alpha \cap \beta)\\ &= \alpha \cap (1 \cup \beta)\\ &= \alpha \cap 1\\ &= \alpha.\end{aligned}$$

Again
$$\begin{aligned}(\alpha \cap \bar{\beta}) \cup \beta &= (\alpha \cup \beta) \cap (\bar{\beta} \cup \beta)\\ &= (\alpha \cup \beta) \cap 1\\ &= \alpha \cup \beta.\end{aligned}$$

Lastly if $\alpha \cup \beta = \alpha$ then
$$\begin{aligned}\alpha \cap \beta &= (\alpha \cup \beta) \cap \beta\\ &= (\alpha \cap \beta) \cup (\beta \cap \beta)\\ &= (\alpha \cap \beta) \cup \beta\\ &= \beta \quad \text{by} \quad \text{(i)}.\end{aligned}$$

D 6.
$$\begin{aligned}&\alpha \rightarrow \beta \quad \text{for} \quad \bar{\alpha} \cup \beta\\ &\alpha \leftrightarrow \beta \quad \text{for} \quad (\alpha \rightarrow \beta) \cap (\beta \rightarrow \alpha)\\ &\alpha \leqslant \beta \quad \text{for} \quad \alpha \cap \beta = \alpha.\end{aligned}$$

Notice the correspondence with $\mathscr{P}_C$ on replacing $\rightarrow$ by $C$ and $\leftrightarrow$ by $B$.

**Prop. 17.** *We have*

*If* $\alpha \leqslant \beta$ *and* $\beta \leqslant \alpha$ *then* $\alpha = \beta$.

*If* $\alpha \leqslant \beta$ *and* $\beta \leqslant \gamma$ *then* $\alpha \leqslant \gamma$.

$0 \leqslant \alpha \qquad \alpha \leqslant 1.$

*If* $\alpha \leqslant \beta$ *then* $\alpha \cap \gamma \leqslant \beta \cap \gamma$.

*If* $\alpha \leqslant \beta$ *then* $\alpha \cup \gamma \leqslant \beta \cup \gamma$.

$$\alpha \cap \beta \leqslant \alpha \qquad \alpha \leqslant \alpha \cup \beta$$

$$\alpha \leqslant \beta \text{ if and only if } \bar{\beta} \leqslant \bar{\alpha}$$

$$\alpha \leqslant \beta \text{ if and only if } \alpha \rightarrow \beta = 1$$

$$\alpha = \beta \text{ if and only if } \alpha \leftrightarrow \beta = 1.$$

These are very easy and are left as exercises for the reader.

## 2.17 *Normal forms*

Using the commutative, associative, distributive, de Morgan laws and double negation we can express any $\mathscr{F}_C$-statement in an equivalent form, either as a conjunction of disjunctions of propositional variables and negated propositional variables or as a disjunction of conjunctions of propositional variables and negated propositional variables. The first case is called the conjunctive normal form (c.n.f.), and the second case is called the disjunctive normal form (d.n.f.). We push the negations to the right as far as they will go so that they act only on propositional variables or negated propositional variables then use double negation as long as possible and lastly the distributive laws. It is like multiplying out an algebraic formula.

We can further ensure that in the first case each propositional variable which occurs in the original $\mathscr{F}_C$-statement occurs negated or unnegated in each disjunctand in the c.n.f. and dually in the second case in each conjunctand in the d.n.f. This is achieved by disjuncting $KpNp$ in the first case and conjuncting $DpNp$ in the second case, if $p$ is a relevant variable. Clearly we are left with an equivalent $\mathscr{F}_C$-statement in each case.

The N. and S.C. that an $\mathscr{F}_C$-statement be a tautology is that each disjunctand of its c.n.f. contain a variable and the same variable negated. Dually the N. and S.C. that an $\mathscr{F}_C$-statement be refutable is that each conjunction of its d.n.f. contain a variable and the same variable negated.

We can further ensure that in the c.n.f. each disjunction contains each variable exactly once either negated or unnegated, unless the c.n.f. is a tautology. For we can omit any disjunction which contains a variable and the same variable negated and obtain an equivalent $\mathscr{F}_C$-statement. If the original $\mathscr{F}_C$-statement was a tautology by this removal we would finally remove all the disjunctands. Dually for the d.n.f.

## Historical remarks to Chapter 2

Propositional calculi have a long history. Aristotle's syllogistic, when expressed in modern symbology, amounts to a form of singularly predicate calculus or class theory. This is discussed by Łukasiewicz (1951) in detail. The early history of the classical propositional calculus is given in great detail by Bocheński (1951, 1961), his account covers all the middle ages up to Frege and then on to the present time. Church (1956) gives the history of propositional calculi since Boole. Lewis (1918) gives the history from Leibniz to Schröder in great detail. Other works on the early history of logic are Bocheński (1951) which deals with pre-Aristotlean times, the old Peripatetics, the Stoic–Megara school and the last period after Chrysippus. Another work is Moody (1953) which deals with the Medieval period, and lastly Dürr (1951) which deals with Boethius.

The definition we have given for a propositional calculus seems to cover all systems which are usually called propositional calculi.

The problem of the independence of symbols, rules, axioms, etc. is of quite modern origin. Huntingdon (1904) and Bernays (1926) were the earliest to discuss these matters.

Models for propositional calculi came in with truth-tables. Łukasiewicz (1920, 1941) was the first to formalize a three-valued propositional calculus. Post (1921) also considered many-valued propositional calculi. Independence of axioms, etc. was demonstrated by Bernays (1926) using many-valued models, this is called the *matrix method*. Extensions of formal systems were studied by Łukasiewicz and Tarski (1930). Modus Ponens first appears in *Scholastic Logic*.

In treating the classical propositional calculus we use the notation of Łukasiewicz (1920). This is easily read in conversational English if one reads as follows:

read $D\phi\psi$ as either $\phi$ or $\psi$,
read $N\phi$ as not $\phi$,
read $K\phi\psi$ as both $\phi$ and $\psi$,
read $C\phi\psi$ as if $\phi$ then $\psi$,
read $B\phi\psi$ as $\phi$ if and only if $\psi$.

so that $DND\phi\psi\omega$ is read:

either not either $\phi$ or $\psi$ or $\omega$. The symbolic form is, however, easier to understand.

The form we have taken for the classical propositional calculus is due to Gentzen (1934, 1955) (see also Anderson & Johnson (1962) and J. Dorp (1962)), who used the terms 'remodelling scheme' and 'building scheme'. The type of proof used in Prop. 2 is due to Gentzen (1934). Prop. 3 was pointed out to me by A. H. Lachlan. The direct method has been further developed by Schütte (1950, 1960).

The de Morgan laws were stated by de Morgan (1867), but were known long before that, and Prop. 5 probably first appeared in connection with Boolean Algebras.

The deduction theorem first appears in Herbrand (1930) and has been much used since, notably by Hilbert–Bernays (1934–6) and Church (1956).

Tertium non datur or the law of the excluded middle goes back a long way and was first called in question by Brouwer (1908) in the case of infinite classes. He maintained that rules that apply to finite classes might fail for infinite classes. Thus he invented Intuitionism, a form of mathematics which does not accept T.N.D. It gives rise to a propositional calculus. But it is far more complicated than the classical propositional calculus, and so is his form of analysis, much more complicated than classical analysis (see Heyting (1934, 1955, 1956) and Kleene–Vesley (1965)). But his methods go some way to clarifying ones ideas about effectiveness, finiteness, constructiveness, etc. which are conditions that Hilbert (1904) insisted should apply to metamathematical demonstrations.

Prop. 7, the elimination of Modus Ponens, is *Gentzen's Hauptsatz* (1955). We thought that since it can be eliminated then why have it at all. The word 'syzygy' is due to the algebraist Sylvester who used it in connection with non-linear relations between algebraic invariants, it means a yoking together. Many proofs of the *decision problem* for the classical propositional calculus have been given, notably by Church (1956), Kalmár (1935), etc.

A correct truth-table for implication was given by Philo of Megara about 300 B.C. Truth-tables were informally used by Frege in special cases, six years later Peirce stated them as a general *decision method* for the classical propositional calculus. Much of the recent development is due to Łukasiewicz and Post. The term 'tautology' is due to Wittgenstein (1922). D4, 5 are due to Sheffer (1913).

The modern treatment of propositional calculi stems from Boole and de Morgan in 1847. This was the algebra of logic. MacColl (1877) was probably the first to deal with a true propositional calculus. Frege gave the first formulation of the classical propositional calculus as a formal system in its own right. But his work was for long neglected and so the propositional calculus developed in the older form as in the work of Peirce, Schröder and Peano. Whitehead and Russell appreciated the work of Frege and gave the classical propositional calculus a formulation with negation and implication as primitives, Modus Ponens and substitution as rules. But substitution was not explicitly mentioned, though this omission was noted later. One of their axioms was found to be redundant by Bernays (1926). Nicod (1916) found a formulation of the classical propositional calculus with only one connective and only one axiom and only one rule apart from substitution. Since then a variety of formulations have been discovered. One by Hilbert (*Hilbert–Bernays* (1934–6) vol. 1) has 12 axioms in 4 groups of 3 axioms each, this was designed to separate out the roles of the connectives $N$, $C$, $K$, $D$. If in this formulation we omit one axiom then we get a formulation of the intuitional propositional calculus. Between the two world wars the Poles were very active in reasearch on the propositional calculus, see Jordan (1945), Storrs McCall (1969) and H. Skolimowski (1969). Another study is to formalize a partial system of the classical propositional calculus which has only the implication sign, and the object is to find axioms so that exactly all tautologies which only contain the implication sign and propositional variables are theorems of the system. The chief interest of such studies is to find a formulation of the classical propositional calculus which is an extension of this partial system. For instance

$$CCCpp'p''CCp''pCp'''p,$$

with Modus Ponens and substitution is an elegant formulation of the implicational propositional calculus. If to this we add the axiom $Cfp$, where $f$ is a constant, we get a formulation of the classical propositional calculus.

Implication in the classical propositional calculus allows as true an implication with a false antecedent, this seems in some sense unnatural. This has given rise at the hands of Lewis (1918, 1920, 1932) of various other propositional calculi designed to rectify this. He also considered other connectives such as '*possible*' and '*necessary*'. These give rise to what are called *modal logics*. We do not discuss them in this book. Much work has been done on these systems, finding decision procedures, formulations, etc.

Gentzen (1955) used *Sequenzen*, that is figures of the form

$$p', p'', \ldots, p^{(\nu)} \to q', q'', \ldots, q^{(\mu)}$$

where we have used $p$ and $q$ as variables. This behaves in the same way as

$$CK \ldots Kp'p'' \ldots p^{(\nu)}D \ldots Dq'q'' \ldots q^{(\mu)}.$$

We have written the sequenzen as

$$\frac{p'\,p'' \ldots p^{(\nu)}}{q'\,q'' \ldots q^{(\mu)}},$$

but have only used it with one lower formula. Gentzen allows the cases when either upper or lower formula may be void.

The idea of using axiom schemes is due to v. Neumann (1925), a substitution rule is then unnecessary. Prop. 8, the substitutivity of equivalent statements, is due to Post (1921). Conjunctive and disjunctive normal forms derive from Boolean Algebra.

A large number of examples on the propositional calculus are found in Church (1956).

Boolean Algebra was invented by George Boole (1847, 1854, 1916). He noticed the resemblance between the behaviour of $K$ and $D$ and $+$ and $\times$ in arithmetic. The history of modern symbolic logic can be traced back to Boole. More will be said about Boolean Algebra at the end of Chapters 3 and 12, there we shall consider Boolean-valued set-theory as an extension of the classical two-valued set-theory. Sheffer (1913) gave a set of independent axioms for Boolean Algebra.

## Examples 2

1. Complete the demonstration of Prop. 5 for the cases other than 4*.
2. Complete the demonstration of Prop. 8, Cor. (ii) for the first and third cases.
3. Obtain $\mathscr{P}_C$-proofs for $BBpp'Bp'p$, $CCpp'CNp'Np$.
4. Give a $\mathscr{P}_1$-proof for $Cpp$.
5. State and prove the Deduction Theorem in $\mathscr{P}_1$.

6. Obtain a $\mathscr{P}_C$-statement $\phi$ which has the following table:

| $p$ | $p'$ | $p''$ | $\phi$ |
|---|---|---|---|
| $t$ | $t$ | $t$ | $f$ |
| $t$ | $t$ | $f$ | $t$ |
| $t$ | $f$ | $t$ | $t$ |
| $t$ | $f$ | $f$ | $f$ |
| $f$ | $t$ | $t$ | $t$ |
| $f$ | $t$ | $f$ | $f$ |
| $f$ | $f$ | $t$ | $t$ |
| $f$ | $f$ | $f$ | $f$ |

7. Define $D$, $B$, $K$ in terms of $C$, $N$.
Define $C$, $D$, $B$ in terms of $K$, $N$.
Define $C$, $D$, $B$, $K$, $N$ in terms of $S$ and in terms of $S'$.

8. Show that $SSpSqrSpSSrpSSsqSSpsSps$ is a tautology, where $p$, $q$, $r$, and $s$ are of type $o$.

9. A Propositional calculus $\mathscr{P}_2$ has the axiom schemes:

(i) $C\phi C\psi\phi$,
(ii) $CC\chi C\phi\psi CC\chi\phi C\chi\psi$,
(iii) $CCN\phi N\psi C\psi\phi$.

And the rule Modus Ponens.

State and prove the deduction Theorem for $\mathscr{P}_2$.

10. Prove the following theorems of $\mathscr{P}_2$:

$$CNpCpp', \quad CNNpp, \quad CpNNp, \quad CCpp'CNpNp'.$$

11. Check the table giving the independences of the axioms of $\mathscr{P}_1$.

12. Give definitions of $C$ and $f$ of $\mathscr{P}_1$ in terms of $D$ and $N$ of $\mathscr{P}_C$, and give definitions of $D$ and $N$ of $\mathscr{P}_C$ in terms of $C$ and $f$ of $\mathscr{P}_1$.

13. Show that theorems of $\mathscr{P}_1$ translate into theorems of $\mathscr{P}_C$ by the definitions found in Ex. 12 but that there are theorems of $\mathscr{P}_C$ which are unobtainable in this way, show also that there are theorems of $\mathscr{P}_1$ which fail to be translations of theorems of $\mathscr{P}_C$.

14. Show that the axioms of $\mathscr{P}_C$ are theorems of $\mathscr{P}_1$ and that the rules of $\mathscr{P}_C$ are derived rules of $\mathscr{P}_1$.

15. Prove $BNBNpp'Bpp'$ in $\mathscr{P}_C$.

16. Show that if $\dfrac{\phi}{\overline{\psi}}$ and $\dfrac{\psi}{\overline{\phi}}$ then $\dfrac{\chi\{\phi\}}{\overline{\chi\{\psi\}}}$ and $\dfrac{\chi\{\psi\}}{\overline{\chi\{\phi\}}}$.

17. Show that $CKCK\phi\psi\chi CKN\phi\psi'\chi CK\psi\psi'\chi$ is a $\mathscr{P}_C$-theorem.

18. Define $t$ for $Cff$ in the system $\mathscr{P}_1$. Define $K$ of $\mathscr{P}_C$ in terms of $C$ and $f$ of $\mathscr{P}_1$. Write out $DKptNp$ without definitional abbreviation in terms of $C$ and $f$ of $\mathscr{P}_1$.

19. Write $\phi = \psi$ if and only if $B\phi\psi$ is a $\mathscr{P}_C$-theorem. Show that the result is a Boolean Algebra when $\cup$ $\cap$ are suitably defined for $\mathscr{P}_C$-statements.

20. Suppose that a Boolean Algebra $\mathscr{B}$ has an additional functor * of type $u$ with the axiom schemes:

$$\alpha \leqslant \alpha^*,$$

$$\alpha^{**} = \alpha^*,$$

$$(\alpha \cup \beta)^* = \alpha^* \cup \beta^*,$$

$$0^* = 0.$$

Show that the elements of $\mathscr{B}$ which satisfy $\alpha^{**} = \alpha$ form a Boolean sub-algebra $\mathscr{B}^*$ of $\mathscr{B}$. When the functions in $\mathscr{B}^*$ are defined as follows:

$$\text{write} \quad \alpha^{\circledast} \quad \text{for} \quad \alpha^{**}, \quad \text{then}$$

$$\alpha \mathbin{\overset{\frown}{*}} \beta \quad \text{for} \quad \alpha \cap \beta,$$

$$\alpha \mathbin{\underset{\smile}{*}} \beta \quad \text{for} \quad (\alpha \cup \beta)^{\circledast},$$

$$\overset{*}{\overline{\alpha}} \quad \text{for} \quad \overline{\alpha}^{\circledast}.$$

(If the elements of $\mathscr{B}$ are sets of points in a topological space then the specified sets are the regular open sets–open sets which have no pin holes or cracks.)

21. Define

$$\alpha + \beta \quad \text{for} \quad (\alpha \cap \overline{\beta}) \cup (\beta \cap \overline{\alpha}),$$

$$\alpha \times \beta \quad \text{for} \quad \alpha \cap \beta.$$

Show that with these definitions every Boolean Algebra becomes a Boolean Ring, i.e. a ring in which

$$\alpha + \alpha = 0,$$

if

$$\alpha + \beta = 0 \quad \text{then} \quad \alpha = \beta,$$

$$\alpha \times \beta = \beta \times \alpha.$$

Conversely show that with the definitions

$$\alpha \cup \beta \quad \text{for} \quad \alpha + \beta + (\alpha \times \beta),$$
$$\alpha \cap \beta \quad \text{for} \quad \alpha \times \beta,$$
$$\bar{\alpha} \quad \text{for} \quad 1 + \alpha,$$

every Boolean Ring becomes a Boolean Algebra. In both cases the algebraic zero and unit coincide with the Boolean zero and unit respectively.

# Chapter 3

# Predicate calculi

## 3.1 *Definition of a predicate calculus*

A *predicate* or *functional calculus* of the *first order* is a formal system $\mathscr{F}$ which is a primary extension of a propositional calculus $\mathscr{P}$, it is obtained from a propositional calculus by adding additional symbols for variables or constants of type $\iota$, called *individual variables* or constants respectively, and adding variables or constants of some or all the types $o\iota, o\iota\iota, o\iota\iota\iota, \ldots$, called *predicate variables* or *constant predicates* respectively; there may also be variables or constants of types $\iota\iota, \iota\iota, \ldots$, called *variables for functions* or *constant functions* respectively, and there may also be constants of some of the types $o(o\iota)$, $o(o\iota\iota)$, ..., $(o\iota)\,o(o\iota)\,(o\iota\iota)$, ... called *quantifiers* There must be individual variables given by a scheme of generation, and there must be some predicates. If quantifiers are present then the abstraction symbol is required in order to provide arguments of types $o\iota, o\iota\iota, \ldots$ for the quantifiers. If propositional variables and constants are excluded the resulting system (which fails to be a primary extension of a propositional calculus) is also called a predictate calculus.

If quantifiers are absent the resulting system is called a *free variable predicate calculus of the first order* with or without functions as the case may be. If individual variables are discarded so that all well-formed formulae of type $\iota$ are constants then quantifiers are useless and the resulting system reduces to a propositional calculus when propositional variables and constants are written in the form $\phi$, $\psi\{\alpha\}$, $\chi\{\alpha, \beta\}$, ..., where $\alpha, \beta$ are constant individuals and $\phi$ is a propositional variable or constant, $\psi, \chi, \ldots$ are predicates of types $o\iota, o\iota\iota, \ldots$. If individual variables are discarded and function variables present then in the resulting system there are variable well-formed formulae of type $\iota$ and the system becomes more manageable if these are replaced by individual variables. For these reasons we require individual variables to be present. If the only predicates present are of type $o\iota$ then we have a *monadic predicate calculus* of *the first order*, if the only predicates present are of types $o\iota$ or $o\iota\iota$ then we have a *diadic predicate calculus of the first order*, and so on. Quantifiers of type $o(o\iota)$ are called *simple quantifiers*, quantifiers of other types are called *compound quantifiers*.

A *predicate or functional calculus of the second order*, $\mathscr{F}^2$, is a primary extension of a predicate calculus of the first order obtained by adding *quantifiers (simple or compound) for predicates.* Simple predicate quantifiers are of types: read 'for at least one', 'for an unbounded set', 'for all' etc.

$$o(oo), \quad o(o(o\iota)), \quad o(o(o\iota\iota)), \ldots$$

generally $o(o\alpha)$, where $\alpha$ is a type of a predicate or of a propositional variable, and compound quantifiers are of types: read 'for all pairs', 'for all except a bounded set of pairs', etc.

$$o(ooo), \quad o(oo(o\iota)), \quad o(o(o\iota)\,o), \quad o(o(o\iota)\,(o\iota)), \quad o(oo(o\iota\iota)), \quad \ldots$$

generally $o(o\alpha\beta), o(o\alpha\beta\gamma), \ldots$, where $\alpha, \beta, \gamma, \ldots$ are types of predicates or of propositional variables.

*A predicate or functional calculus of the third order*, $\mathscr{F}^3$, is a primary extension of a predicate calculus of the second order obtained by adding variables or constants of types

$$oo, \quad o(o\iota), \quad o(o\iota\iota), \quad \ldots$$

generally $o\alpha$ where $\alpha$ is the type of a predicate or of a propositional variable, and $o\alpha\beta$ where $\alpha, \beta$ are types of predicates or of propositional variables, ..., these are called *predicates of predicates*, we could also add some mixed predicates requiring for some arguments predicates and for others requiring individuals, these would have types:

$$oo\iota, \quad o\iota o, \quad o(o\iota)\,\iota, \quad o\iota(o\iota), \quad \ldots$$

and generally $o\alpha\iota, o\iota\alpha, o\beta\gamma\iota, \ldots$, where $\alpha, \beta, \gamma$ are types of predicates or propositional variables.

*A predicate or functional calculus of the fourth order*, $\mathscr{F}^4$, is a primary extension of a predicate calculus of the third order obtained by adding quantifiers of various kinds over predicates of predicates or over mixed predicates. And so on.

A predicate calculus of the first order without constant individuals or constant functions or constant predicates or constant propositions is called a *pure predicate or functional calculus of the first order* with or without functions as the case may be. A pure predicate calculus of the second order is defined similarly. A predicate calculus of the third or higher order is called pure if it is an extension of a pure predicate calculus of one lower order and if only variables (for calculi of odd order) or quantifiers (for calculi of even order) are adjoined. For instance a pure calculus of the third order may have variables and constants of type $oo$, $o(o\iota)$, because there may be constants of these types in the pure predicate calculus of the second order of which it is a primary extension, but in the extension only variables of these types are adjoined.

A predicate calculus of the first order which is without propositional variables and predicate variables is an *applied predicate or functional calculus of the first order*, with or without functions as the case may be. Similarly a predicate calculus of odd order is applied if it is obtained from a predicate calculus of one order less by adding only constants and is mixed if both constants and variables are added. A predicate calculus of even order with some constant predicates is mixed, it must have variable predicates of appropriate types.

A formal system $\mathscr{S}$ is *based on a pure predicate calculus of the first order* $\mathscr{F}$ if the constants of $\mathscr{F}$ are constants of $\mathscr{S}$ and if $\mathscr{S}$ has symbols of each type for which there are variables in $\mathscr{F}$ and has individual variables (we may use the same symbols for individual variables in $\mathscr{S}$ and in $\mathscr{F}$) and if whenever $\phi$ is an $\mathscr{F}$-theorem and the free variables in $\phi$ are replaced in any manner by $\mathscr{S}$-formulae of the same type, where a given variable is always replaced by the same $\mathscr{S}$-formula, the result is an $\mathscr{S}$-theorem. Thus if a formal system $\mathscr{S}$ is based on a pure predicate calculus $\mathscr{F}$ then in setting up the system $\mathscr{S}$ we need only state that it is based on $\mathscr{F}$ and then list the special symbols, axioms and rules of $\mathscr{S}$. Alternately if a formal system $\mathscr{S}$ is based on a pure predicate calculus of the first order $\mathscr{F}$ then we could put down the symbols of $\mathscr{F}$, the axioms of $\mathscr{F}$ as axiom schemes and the rules of $\mathscr{F}$ as part of the construction of $\mathscr{S}$. We similarly define a formal system $\mathscr{S}$ to be based on a pure predicate calculus of higher order.

Note that a predicate calculus of the second order might contain a negation of type $oo$ and a quantifier $Q$ of type $o(oo)$ so that $QN$ would be a well-formed formula of type $o$. If $Q$ were an existential quantifier this could be read 'there is a statement whose negation is true'.

The predicate calculi discussed so far are *one-sorted predicate calculi* because they contain only one sort of individual, namely those of type $\iota$. A *many-sorted predicate calculus* contains individual variables having types from among, $\iota'$, $\iota''$, ... and possibly individual constants of some of these types, and possibly functions with values and arguments from these types. The order of a many-sorted predicate calculus is defined as for a one-sorted predicate calculus. A many-sorted predicate calculus $\mathscr{F}'$ is based on a pure one-sorted predicate calculus $\mathscr{F}$ when the constants of $\mathscr{F}$ are constants of $\mathscr{F}'$ and when the axioms and rules of $\mathscr{F}$ apply to each sort of individual variable in $\mathscr{F}'$. We shall show later that it is possible to reduce a many-sorted predicate calculus $\mathscr{F}'$ of the first order to a one-sorted predicate calculus $\mathscr{F}$ of the first order by adjoining to $\mathscr{F}$ some constant one-argument predicates. Each such predicate plays the part of saying that its argument is of a certain sort, distinct predicates referring to distinct sorts. Similarly we could consider many sorts of predicates in predicate calculi of higher orders.

A predicate calculus is a formal system hence it must be constructive in accordance with the definition of a formal system which was given in Ch. 1. If two predicate calculi have variables of the same types then by a trivial adjustment of notation we may use the same symbols for these variables in both systems.

A predicate calculus $\mathscr{F}'$ is *weaker* than a predicate calculus $\mathscr{F}$ under the following circumstances:

(i) $\mathscr{F}'$ is without variables of type $\alpha$ if $\mathscr{F}$ is without variables of type $\alpha$.

(ii) The constants of $\mathscr{F}'$ can be defined in terms of the constants of $\mathscr{F}$.

(iii) An $\mathscr{F}'$-theorem $\phi'$ becomes an $\mathscr{F}$-theorem $\phi$ when the $\mathscr{F}'$-constants in $\phi'$ are replaced by their definitions in terms of $\mathscr{F}$-constants and the abstraction rule is used if necessary to eliminate the abstraction symbol, and the same symbols are used for individual variables in both systems.

Note that (i) follows from (ii) and (iii). If $\mathscr{F}'$ is weaker than $\mathscr{F}$ and $\mathscr{F}$ is weaker than $\mathscr{F}'$ then $\mathscr{F}$ and $\mathscr{F}'$ are *equivalent*. If $\mathscr{F}'$ is weaker than $\mathscr{F}$ and fails to be equivalent to $\mathscr{F}$, then *$\mathscr{F}'$ is strictly weaker than $\mathscr{F}$*. 'Equivalent' is an equivalence relation.

'Dependent' and 'independent' applied to symbols, axioms and rules are defined as for a propositional calculus.

## 3.2 *Models*

A *model* $\mathscr{M}$ of a predicate calculus of the first order $\mathscr{F}$ with only simple quantifiers is a formal system which satisfies the following conditions:

(i) $\mathscr{M}$ has exactly the same constants as $\mathscr{F}$ of types other than

$$\iota, \iota\iota, \iota\iota\iota, \ldots, o, o\iota, o\iota\iota, \ldots.$$

(ii) $\mathscr{M}$ is without variables but has constants of types $\iota$ and $o$ and of any of the types $\iota\iota, \iota\iota\iota, \ldots, o\iota, o\iota\iota, \ldots$ which occur in $\mathscr{F}$. The $\mathscr{M}$-constants of type $o$ are *designated* or *undesignated*, at least one is designated and at least one is undesignated, the $\mathscr{M}$-constants of type $o$ are called *elements*, the constants of type $\iota$ are called *individuals*, the constants of types $\iota\iota, \iota\iota\iota, \ldots$ are called *functions* and the constants of types $o\iota, o\iota\iota, \ldots$ are called *predicates*.

(iii) $\mathscr{M}$ is without axioms.

(iv) The rules of $\mathscr{M}$ enable us to replace any $\mathscr{M}$-statement by a unique element, called the $\mathscr{M}$-value of the $\mathscr{M}$-statement.

(v) If $\Phi\{\alpha\}$ is an $\mathscr{M}$-statement and $\alpha$ an $\mathscr{M}$-individual then there is an $\mathscr{M}$-term $\Phi_\mu$ of type $o\iota$ whose $\mathscr{M}$-value for the argument $\alpha$ is the same as that of $\Phi\{\alpha\}$, for each individual $\alpha$.

(vi) $\mathscr{F}$-theorems reduce to designated elements of $\mathscr{M}$ however the following procedure is carried out:

(*a*) the $\mathscr{F}$-constants of types $o, \iota, \iota\iota, \ldots, o\iota, o\iota\iota, \ldots$ are replaced by fixed suitably chosen $\mathscr{M}$-terms of the same respective type,

(*b*) free $\mathscr{F}$-variables are replaced by arbitrary $\mathscr{M}$-terms of same type,

(*c*) if $\boldsymbol{\lambda}\xi.\Phi\{\xi\}$ is a closed $\mathscr{M}'$-formula of type $o\iota$, where $\mathscr{M}'$ is obtained from $\mathscr{M}$ by adjoining individual variables and the abstraction symbol, then it is replaced by the $\mathscr{M}$-term $\Phi_\mu$ of type $o\iota$ which when applied to an $\mathscr{M}$-individual $\alpha$ always has the same $\mathscr{M}$-value as $\Phi\{\alpha\}$, by (v) there is such an $\mathscr{M}$-term Thus $\Phi\{\alpha\}$ is replaced by $(\Phi_\mu\alpha)$.

The element of $\mathscr{M}$ which results when we apply the procedure (*a*), (*b*), (*c*) to an $\mathscr{F}$-statement $\phi$ is called an *$\mathscr{M}$-value of $\phi$*.

If we extend $\mathscr{M}$ to a system $\mathscr{M}'$ by adjoining individual variables then by (iv) $\mathscr{M}$ is to contain a constant of type $o\iota$ whose $\mathscr{M}$-value for an $\mathscr{M}$-individual $\alpha$ as argument is the same as that of $\boldsymbol{\lambda}\xi.\Phi\{\xi\}$ for the same argument, here $\Phi\{\xi\}$ is an $\mathscr{M}'$-statement whose sole free variable is $\xi$.

For instance if $\Phi\{\xi, \boldsymbol{\lambda}\eta . \Psi\{\xi, \eta\}\}$ is an $\mathscr{M}'$-statement whose sole free variable is $\xi$, then $\mathscr{M}$ is to contain a constant of type $o\iota$, $\Xi$ such that $\Xi\alpha$ has the same $\mathscr{M}$-value as $\Phi\{\alpha, \lambda\eta . \Psi\{\alpha, \eta\}\}$ for each $\mathscr{M}$-individual $\alpha$, where again there is to be an $\mathscr{M}$-constant $\Xi'$ of type $o\iota$ such that $\Psi\{\alpha, \beta\}$ has the same $\mathscr{M}$-value as $\Xi'\beta$ for each $\mathscr{M}$-individual $\beta$. $\Xi'$ will in general vary as $\alpha$ varies.

If $\mathscr{F}$ contains compound quantifiers then clauses (vi) $c$, (v) require amendment. Suppose $\mathscr{F}$ contains a compound quantifier of type $o(o\iota\iota)$ then if $\boldsymbol{\lambda}\xi\xi' . \Phi\{\xi, \xi'\}$ is a closed formula of type $o\iota\iota$ of the system $\mathscr{M}'$ obtained from $\mathscr{M}$ by adjoining individual variables and the abstraction symbol, it is replaced by that $\mathscr{M}$-term of type $o\iota\iota$ which when applied to $\mathscr{M}$-individuals $\alpha, \beta$ always has the same $\mathscr{M}$-value as that of $\Phi\{\alpha, \beta\}$.

A model of a predicate calculus of the first order is a formal system and as such it must be constructive. If a model of a predicate calculus of the first order is displayed then it is constructive. If the elements can be taken as $\boldsymbol{t}, \boldsymbol{f}$ of which $\boldsymbol{t}$ is designated and $\boldsymbol{f}$ is undesignated then the model is called *2-valued*, otherwise it is *many-valued*. A *semi-model* satisfies (i), (ii), (iii), (v) and so may fail to be constructive.

## 3·3 *Predicative and impredicative predicate calculi*

The $\boldsymbol{\lambda}$ symbol allows us to form predicates from statements, thus if $\phi\{\xi\}$ is an $\mathscr{F}^2$-statement then $\boldsymbol{\lambda}\xi . \phi\{\xi\}$ is an $\mathscr{F}^2$-property of type $o\iota$. In particular $\boldsymbol{\lambda}\xi . \phi\{(Qp)\, \psi\{p, \xi\}, \xi\}$ where $Q$ is a quantifier of type $o(o(o\iota))$ is an $\mathscr{F}^2$-property, call it $P$, of type $o\iota$. Then the property $P$ is defined in terms of all $\mathscr{F}^2$-properties of type $o\iota$, including itself. This seems an objectionable circular property, and it seems desirable to avoid it if possible. This is done by classifying properties of a given type into orders, we thus obtain a *ramified type theory*. It is more complicated but avoids the unpleasant circularity. Predicates of type $\alpha$ are given an *order* as follows. The order of a predicate variable is a numeral attached to that variable. The order of a predicate $\boldsymbol{\lambda}\xi . \phi\{\xi\}$ is the maximum of the orders of the free predicates which occur in $\phi\{\xi\}$ and the successors of the orders of the bound predicate variables which occur in $\phi\{\xi\}$. Thus the order of the predicate $P$ above is at least one greater than the order of the bound variable $p$. Thus $P$ fails to be defined in terms of itself, and the circularity has disappeared. It is called a *predicative predicate*, the others are called *impredicative predicates*.

## 3·4 *The classical predicate calculus of the first order*

A pure predicate calculus of the first order is *classical* if it is equivalent to the following predicate calculus of the first order, $\mathscr{F}_C$.

*Symbols*: those of $\mathscr{P}_C$ together with:

| | type | name |
|---|---|---|
| $x$ | $\iota$ | *individual variable* |
| $p_{o\iota}, p_{o\iota\iota}, \ldots$ | $o\iota, o\iota\iota, \ldots$ | *predicate variables* |
| $E$ | $o(o\iota)$ | *existential quantifier* |
| $\boldsymbol{\lambda}$ | – | *abstraction symbol* |

further variables are obtained by superscripting primes, as for propositional variables.

*Axioms*

$D\pi N\pi$, where $\pi$ is a propositional variable,
$D\pi\xi N\pi\xi$, where $\pi$ is a one-place predicate and $\xi$ is an individual variable, etc., for two or more place predicates.

*Rules*

Those for $\mathscr{P}_C$ together with:

*Remodelling scheme*

I$b$ $\dfrac{DD\phi\phi\omega}{D\phi\omega}$ *cancellation*,

*Building schemes*

II$d$ $\dfrac{D\phi\{\eta\}\,\omega}{DE(\boldsymbol{\lambda}\xi.\phi\{\xi\})\,\omega}$ *existential dilution*, $\xi$ fails to occur free in $\phi\{\Gamma_\iota\}$, $\eta$ free in $\phi\{\eta\}$

II$e$ $\dfrac{DN\phi\{\eta\}\,\omega}{DNE(\boldsymbol{\lambda}\xi.\phi\{\xi\})\,\omega,}$ *generalization*, where $\eta$ fails to occur free in $\omega$, $\phi\{\Gamma_\iota\}$ and $\xi$ fails to occur free in $\phi\{\Gamma_\iota\}$.

Here $\xi$, $\eta$ denote individual variables, $\phi\{\eta\}, \omega$ denote $\mathscr{F}_C$-statements, $\omega$ is a subsidiary formula and may be absent, $\phi$, $\phi(\eta)$, $N\phi(\eta)$ are main formulae of the upper formulae and $\phi$, $E(\boldsymbol{\lambda}\xi.\phi\{\xi\})$, $NE(\boldsymbol{\lambda}\xi.\phi\{\xi\})$ are the main formulae of the lower formulae. We write

D 7. $(E\xi)\,\phi\{\xi\}$ for $E(\boldsymbol{\lambda}\xi.\phi\{\xi\})$.

$(E\xi,\xi')$ for $E(\boldsymbol{\lambda}\xi.E(\boldsymbol{\lambda}\xi'.\phi\{\xi,\xi'\}))$, etc.

Note that an $\mathscr{F}_C$-proof is *direct*, the only formulae that can be omitted are duplicates as in I$b$. This rule causes the undecidability of $\mathscr{F}_C$, as we shall see later.

## 3·5 *Properties of the system* $\mathscr{F}_C$

PROP. 1. *The schemes* I$a$, $b$, II$b$, $c$, $e$ *are reversible.*

In Prop. 2, Ch. 2 we showed the reversibility of I$a$, II$b$, $c$ for the system $\mathscr{P}_C$. Similar demonstrations hold for $\mathscr{F}_C$.

I$b$ is reversible

$$\frac{D\phi\omega}{\overline{DD\,\phi\phi\omega}}\quad \text{II}a, \text{I}a.$$

II$e$ is reversible. Suppose we have an $\mathscr{F}_C$-proof of $DN(E\xi)\,\phi\{\xi\}\,\omega$. In this $\mathscr{F}_C$-proof corresponding occurrences of $N(E\xi)\,\phi\{\xi\}$ can only be introduced at II$a$, $e$. If a corresponding occurrence of $N(E\xi)\,\phi\{\xi\}$ is introduced at II$a$ then introduce $N\phi\{\zeta\}$ instead and apply II$e$, where $\zeta$ is *new* (is absent from the proof); then $N(E\xi)\,\phi\{\xi\}$ is introduced only at II$e$. If $N(E\xi)\,\phi\{\xi\}$ is introduced at II$e$ then omit the lower formula and replace all *descendants* of this occurrence by $N\phi(\zeta)$, where $\zeta$ is the same new variable as before, in all cases we use the same new variable, replace $N\phi\{\eta\}$ in the upper formula of this application of II$e$ by $N\phi\{\zeta\}$ and similarly replace all free occurrences of $\eta$ higher up the tree by $\zeta$. The result is an $\mathscr{F}_C$-proof of $DN\phi\{\zeta\}\,\omega$. The replacement of a variable by a new one, that is one that fails to occur in the tree, clearly fails to violate the condition on variables in any applications of II$d$, $e$ that may occur in the $\mathscr{F}_C$-proof of $DN(E\xi)\,\phi\{\xi\}\,\omega$. We must replace all free occurrences of the variable $\eta$ higher up the tree by $\zeta$ and not merely all corresponding occurrences of $\eta$, because we might have $\psi\{\eta,\eta\}$ higher up the tree and only the first $\eta$ in $\psi\{\eta,\eta\}$ might correspond to our $\eta$ in $N\phi\{\eta\}$. Since our $\eta$ is going to be *generalized*, that is II$e$ is applied to it, then to comply with the conditions on variables in the applications of II$e$ the second $\eta$ in $\psi\{\eta,\eta\}$ must be *restricted*, that is II$d$ applied to it, before we generalize the first $\eta$. If we now change both these $\eta$'s to $\zeta$'s where $\zeta$ is new, then we can still apply these rules in the same order and places as before. If we only changed corresponding occurrences of our $\eta$ then $\psi\{\eta,\eta\}$ would become $\psi\{\zeta,\eta\}$. Now $\psi\{\eta,\eta\}$ might be $D\pi\eta N\pi\eta$ in which case $\psi\{\zeta,\eta\}$ would become $D\pi\zeta N\pi\eta$ which fails to be an $\mathscr{F}_C$-theorem. This completes the demonstration of the proposition.

PROP. 2. *$D\phi N\phi$ is an $\mathscr{F}_C$-theorem.*

We proceed as in Prop 6 (i), Ch. 2 by formula induction on $\phi$. We have the extra case:

$\phi$ is $(E\xi)\,\psi\{\xi\}$ and the result holds for $\psi\{\xi\}$, then:

$$\cfrac{\cfrac{\cfrac{\cfrac{D\psi\{\xi\}N\psi\{\xi\}}{D(E\xi)\,\psi\{\xi\}N\psi\{\xi\}}\ \text{II}d,\ \ \xi\text{ fails to occur free in }\psi\{\Gamma_\iota\},}{DN\psi\{\xi\}(E\xi)\,\psi\{\xi\}}\ \text{I}a,}{DN(E\xi)\,\psi\{\xi\}(E\xi)\,\psi\{\xi\}}\ \text{II}e,\text{ conditions on variables are satisfied,}}{D(E\xi)\,\psi\{\xi\}N(E\xi)\,\psi\{\xi\}}\ \text{I}a.$$

Thus the result holds for $(E\xi)\,\psi\{\xi\}$ if it holds for $\psi\{\xi\}$. This completes the demonstration of the proposition. Thus *Tertium non datur* holds in the system $\mathscr{F}_C$.

PROP. 3. *The Deduction Theorem holds in $\mathscr{F}_C$ with the following restriction. If $\phi', \ldots, \phi^{(S\theta)} \vdash_{\mathscr{F}_C} \psi$, where* II*e fails to be applied to any variable which occurs in $\phi^{(S\theta)}$, then $\phi', \ldots, \phi^{(S\theta)} \vdash_{\mathscr{F}_C} DN\phi^{(S\theta)}\psi$.*

When II*e* fails to be applied to any variable in $\phi^{(S\theta)}$ we say that *the variables in $\phi^{(S\theta)}$ are held constant*. They could be restricted. We use the notation:

$$\phi', \ldots, \phi^{(S\theta)} \vdash_{\mathscr{F}_C} \psi \quad \text{for} \quad \frac{\phi', \ldots, \phi^{(S\theta)}}{\psi} \quad \text{in } \mathscr{F}_C.$$

The demonstration is similar to that of Prop. 6, Ch. 2. We have just shown that (i) (tertium non datur) holds in $\mathscr{F}_C$, (ii), (iii) follows exactly as in Prop. 6, Ch. 2 and similarly for (iv) except that if the rule used is II*e*,

$$\frac{\chi}{\psi}\ \text{II}e \quad \text{then} \quad \frac{DN\phi\chi}{DN\phi\psi}\ \text{II}e, \text{I}a$$

will only be valid if the variable generalized fails to occur free in $\phi$, hence the restriction. This completes the demonstration of the proposition. If we apply generalization to each free variable in an $\mathscr{F}_C$-statement $\phi$ the result is called the *closure* of $\phi$.

COR. (i). *If $D\phi'\omega, \ldots, D\phi^{(S\kappa)}\omega \vdash_{\mathscr{F}_C} D\psi\omega$ then*

$$D\phi'\omega, \ldots, D\phi^{(\kappa)}\omega \vdash_{\mathscr{F}_C} DDN\phi^{(S\kappa)}\psi\omega,$$

*provided that the variables in $D\phi^{(S\kappa)}\omega$ are held constant.*

We have $DND\phi^{(S\kappa)}\omega D\psi\omega$ from Prop. 3 whence the result by permutation and cancellation.

D8.
$$\begin{cases} \prod_{\theta=1}^{1} \chi^{(\theta)} \quad \text{for} \quad \chi', \\ \prod_{\theta=1}^{S\kappa} \chi^{(\theta)} \quad \text{for} \quad K \prod_{\theta=1}^{\kappa} \chi^{(\theta)} \chi^{(S\kappa)}. \end{cases}$$

Cor. (ii). *If* $\phi', \ldots, \phi^{(\kappa)} \vdash_{\mathscr{F}_C} \psi$ *then* $DN \prod_{\theta=1}^{\kappa} \overline{\phi}^{(\theta)} \psi$, *where* $\overline{\phi}$ *is the closure of* $\phi$.

We proceed as in Prop. 3 except that if II$e$ is used, say $\frac{\psi}{\chi}$ II$e$, where a variable $\xi$ free in an hypothesis $\phi$ is generalized then we replace it by

$$\frac{\dfrac{\dfrac{DN\phi\psi}{D(E\xi)N\phi\psi}\ \text{II}d}{D(E\xi)N\phi\chi}\ \text{II}e\text{, condition on variables is satisfied}}{DNN(E\xi)N\phi\chi}\ \text{II}c.$$

In this way from $\phi', \ldots, \phi^{(S\kappa)} \vdash_{\mathscr{F}_C} \psi$

we obtain $\phi', \ldots, \phi^{(\kappa)} \vdash_{\mathscr{F}_C} DN\overline{\phi}^{(S\kappa)}\psi$,

where $\overline{\phi}^{(S\kappa)}$ is obtained by generalizing all the variables in $\phi^{(S\kappa)}$ to which II$e$ was applied in the original deduction of $\psi$. The result now follows if we generalize all the free variables still left in $\overline{\phi}^{(S\kappa)}$, i.e. if we apply II$d$

$$\frac{\dfrac{DN\overline{\phi}^{(S\kappa)}\psi}{D(E\xi)N\overline{\phi}^{(S\kappa)}\chi}\ \text{II}d}{DNN(E\xi)N\overline{\phi}^{(S\kappa)}\psi.}$$

Now repeat for $\phi^{(\kappa)}, \ldots, \phi'$ and the result follows.

D9.
$$\begin{cases} \sum_{\theta=1}^{1} \chi^{(\theta)} \quad \text{for} \quad \chi', \\ \sum_{\theta=1}^{S\kappa} \chi^{(\theta)} \quad \text{for} \quad D \sum_{\theta=1}^{\kappa} \chi^{(\theta)} \chi^{(S\kappa)}. \end{cases}$$

Cor. (iii). *If* $\phi$ *is quantifier-free and is in conjunctive normal form, so that* $\phi$ *is of the form* $\prod_{\theta=1}^{\kappa} \phi^{(\theta)}$ *where* $\phi^{(\theta)}$ *is a disjunction of atomic statements or negations of atomic statements and if* $C\phi\psi$ *is an* $\mathscr{F}_C$*-theorem then*

$$\phi', \ldots, \phi^{(\kappa)} \vdash_{\mathscr{F}_C} \psi.$$

We may suppose that in the $\mathscr{F}_C$-proof of $C\phi\psi$ the free variables in $\phi$

are held constant. Because if one of them, say $\xi$, was generalized then this generalization must occur in the $\mathscr{F}_C$-proof of $C\phi\psi$ before the occurrences of $\xi$ in $\phi$ enter the proof. Now in the earlier part of the $\mathscr{F}_C$-proof we can replace $\xi$ by a new variable at all its occurrences, we are left with an $\mathscr{F}_C$-proof of $C\phi\psi$, in which the variables in $\phi$ are held constant.

Let $\phi$ be $\prod_{\theta=1}^{\kappa}\sum_{\theta'=1}^{\sigma\{\theta\}}\phi^{(\theta,\theta')}$, this distributed becomes $\sum_{\tau}\prod_{\theta=1}^{\kappa}\phi^{(\theta,\tau\{\theta\})}$, call it $\phi^*$, where $1 \leqslant \tau\{\theta\} \leqslant \sigma\{\theta\}$, and the summation is over all such $\tau$. From 'if $\dfrac{N\phi}{N\phi^*}$ then $\dfrac{DN\phi\psi}{DN\phi^*\psi}$' and $DN\phi\psi$ we obtain $DN\sum_{\tau}\prod_{\theta=1}^{\kappa}\phi^{(\theta,\tau\{\theta\})}\psi$, whence by the reversibility of II$b$ repeatedly we obtain $DN\prod_{\theta=1}^{\kappa}\phi^{(\theta,\tau\{\theta\})}\psi$, for each $\tau$. All these $\mathscr{F}_C$-proofs are obtained from one tree by replacing part disjunctions of $\sum_{\tau}\prod_{\theta=1}^{\kappa}\phi^{(\theta,\tau\{\theta\})}$ by $\prod_{\theta=1}^{\kappa}\phi^{(\theta,\tau\{\theta\})}$ and omitting certain branches. Thus we have $D\sum_{\theta=1}^{\kappa}N\phi^{(\theta,\tau\{\theta\})}\psi$ for each $\tau$. Now consider the places where $N\phi^{(\theta,\tau\{\theta\})}$ enters the $\mathscr{F}_C$-proof of $D\sum_{\theta=1}^{\kappa}N\phi^{(\theta,\tau\{\theta\})}\psi$. It will do so either at an $\mathscr{F}_C$-axiom (T.N.D.) or at II$a$ or at II$c$ only. If it enters by an $\mathscr{F}_C$-axiom $DN\phi^{(\theta,\tau\{\theta\})}\phi^{(\theta,\tau\{\theta\})}$ replace this by $\dfrac{\phi^{(\theta,\tau\{\theta\})}}{DN\phi^{(\theta,\tau\{\theta\})}\phi^{(\theta,\tau\{\theta\})}}$ II$a$. This converts the $\mathscr{F}_C$-proof of $D\sum_{\theta=1}^{\kappa}N\phi^{(\theta,\tau\{\theta\})}\psi$ into an $\mathscr{F}_C$-deduction of $D\sum_{\theta=1}^{\kappa}N\phi^{(\theta,\tau\{\theta\})}\psi$ from certain hypotheses $\phi^{(\theta,\tau\{\theta\})}$. Once $N\phi^{(\theta,\tau\{\theta\})}$ has entered this deduction it thereafter remains in the subsidiary formulae of building rules because it fails to be governed by a quantifier or by $ND$ or by $N$ in the theorem. We are only considering occurrences of $N\phi^{(\theta,\tau\{\theta\})}$ which correspond to the occurrence of $N\phi^{(\theta,\tau\{\theta\})}$ in $\sum_{\theta=1}^{\kappa}N\phi^{(\theta,\tau\{\theta\})}$. If $N\phi^{(\theta,\tau\{\theta\})}$ enters by II$c$ then $N\phi^{(\theta,\tau\{\theta\})}$ is $NN\bar{\phi}^{(\theta,\tau\{\theta\})}$, where $\bar{\phi}^{(\theta,\tau\{\theta\})}$ is atomic, and we have

$$\frac{D\bar{\phi}^{(\theta,\tau\{\theta\})}\psi'}{DNN\bar{\phi}^{(\theta,\tau\{\theta\})}\psi'} \quad \text{II}c,$$

consider the places where corresponding occurrences of $\bar{\phi}^{(\theta,\tau\{\theta\})}$ enter the deduction, $\bar{\phi}^{(\theta,\tau\{\theta\})}$ is an atomic statement so it enters by II$a$ or by (T.N.D.) only. If it enters by (T.N.D.) replace this by

$$\frac{N\bar{\phi}^{(\theta,\tau\{\theta\})}}{DN\bar{\phi}^{(\theta,\tau\{\theta\})}\bar{\phi}^{(\theta,\tau(\{\theta\})}} \quad \text{II}a.$$

These two cases altogether leave us with an $\mathscr{F}_C$-deduction of $D\sum_{\theta=1}^{\kappa} N\phi^{(\theta,\tau(\theta))}\psi$ from hypotheses $\phi^{(\theta,\tau(\theta))}$, $1 \leqslant \theta \leqslant \kappa$. In this deduction omit all occurrences of $N\phi^{(\theta,\tau(\theta))}$ which correspond to the occurrences of $N\phi^{(\theta,\tau(\theta))}$ we have been considering, so that $N\phi^{(\theta,\tau(\theta))}$ fails to be introduced by II$a$. We are left with an $\mathscr{F}_C$-deduction of $\psi$ from hypotheses $\phi^{(\theta,\tau(\theta))}$, $1 \leqslant \theta \leqslant \kappa$, because we have only altered the subsidiary formulae of building rules by omitting from the upper and lower formulae of a rule the same disjunctands and this converts an application of a rule into another application of the same rule, also the parts of upper and lower formulae of remodelling rules have been altered in the same way. These cancellations fail to effect $\psi$ because an occurrence of $N\phi^{(\theta,\tau(\theta))}$ in $\sum_{\theta=1}^{\kappa} N\phi^{(\theta,\tau(\theta))}$ fails to correspond to any occurrence of $N\phi^{(\theta,\tau(\theta))}$ in $\psi$. The final result when repetitions have been removed, is an $\mathscr{F}_C$-deduction of $\psi$ from hypotheses $\phi^{(\theta,\tau(\theta))}$, $1 \leqslant \theta \leqslant \kappa$, this holds for each $\tau$ we require.

LEMMA. *If $\omega, \chi', \ldots, \chi^{(\kappa)} \vdash_{\mathscr{F}_C} \psi$ and $\omega', \chi' \ldots, \chi^{(\kappa)} \vdash_{\mathscr{F}_C} \psi$ where the variables in $\omega, \omega'$ are held constant, then $D\omega\omega', \chi, \ldots, \chi^{(\kappa)} \vdash_{\mathscr{F}_C} \psi$.*

We have $D\omega\omega', \chi', \ldots, \chi^{(\kappa)} \vdash_{\mathscr{F}_C} D\psi\omega'$, by replacing all occurrences of $\omega$ which correspond to the occurrence of $\omega$ in the first hypothesis by $D\omega\omega'$ and then putting $\omega'$ into the subsidiary formulae, by remodelling, of any rules used: the conditions on variables are satisfied because the variables in $\omega'$ are held constant. Again we have $D\psi\omega', \chi', \ldots, \chi^{(\kappa)} \vdash_{\mathscr{F}_C} D\psi\psi$, by replacing $\omega'$ by $D\psi\omega'$ as before in the second hypothesis and then putting $\psi$ into the subsidiary formulae, the condition on variables is satisfied because the free variables in $\psi$ fail to be quantified. Thus we have: $D\omega\omega', \chi', \ldots, \chi^{(\kappa)} \vdash_{\mathscr{F}_C} D\psi\omega'$; $D\psi\omega', \chi', \ldots, \chi^{(\kappa)} \vdash_{\mathscr{F}_C} D\psi\psi$, using I$b$ we get the required result.

Returning to Cor. (iii) we have $\phi^{(1,\tau(1))}, \ldots, \phi^{(\kappa,\tau(\kappa))} \vdash_{\mathscr{F}_C} \psi$ for each $\tau$. Now the hypotheses arise from distributing $\prod_{\theta=1}^{\kappa} \sum_{\theta'=1}^{\sigma(\theta)} \phi^{(\theta,\theta)}$ so that we shall have sets of hypotheses which differ only in their first members; we have already noticed that we may assume that the variables in the hypotheses are kept constant so that we may apply the lemma repeatedly and obtain $\sum_{\theta'=1}^{\sigma(1)} \phi^{(1,\theta')}$, $\phi^{(2,\tau(2))}$, $\phi^{(\kappa,\tau(\kappa))} \vdash_{\mathscr{F}_C} \psi$. In a similar way we have this result when the hypotheses $\phi^{(2,\tau(2))}$ for all possible $\tau$ are the same,

whence by the lemma repeatedly we get $\sum_{\theta'=1}^{\sigma(1)} \phi^{(1,\theta')}, \sum_{\theta'=1}^{\sigma(2)} \phi^{(2,\theta')}, \phi^{(3,\tau(3))}, \ldots, \phi^{(\kappa,\tau(\kappa))}$.
Continuing in this manner we finally arrive at $\phi', \ldots, \phi^{(\kappa)} \vdash_{\mathscr{F}_C} \psi$, as desired.

Cor. (iv). *If $\bar{\phi}$ is the closure of $\phi$, where $\phi$ is of the form $\prod_{\theta=1}^{\kappa} \phi^{(\theta)}$ and where $\phi^{(\theta)}$ is a disjunction of atomic statements or negations of atomic statements, and if $C\bar{\phi}\psi$ is an $\mathscr{F}_C$-theorem then $\phi', \ldots, \phi^{(\kappa)} \vdash_{\mathscr{F}_C} \psi$.*

Consider the proof-tree of $C\bar{\phi}\psi$, if II$d$ is applied to a variable in $\phi$ and if this variable is held constant in $\psi$ or is absent from $\psi$ then omit that application of II$d$, the result is an $\mathscr{F}_C$-proof-tree of $C\bar{\phi}^*\psi$ where $\bar{\phi}^*$ differs from $\bar{\phi}$ by omission of generalizations. Every other variable in $\phi$ is first restricted in $\phi$ and later on generalized in $\psi$. Variables in $\phi$ fail to get generalized. Now consider the portion of the $\mathscr{F}_C$-proof-tree of $C\bar{\phi}^*\psi$ up to the place where the first variable in $\phi$ is restricted. We have

$$\cfrac{\cfrac{\cfrac{\cfrac{DN\phi\psi'}{D(E\xi)\,N\phi\psi'}\,\text{II}d}{D(E\mathfrak{y})\,(E\xi)\,N\phi D\psi''\omega}}{D(E\mathfrak{y})\,(E\xi)\,N\phi DN(E\eta)\,N\psi''\omega}\,\text{II}c,e \;\; \mathfrak{y} \text{ for } \eta',\ldots,\eta^{(\nu)}}{C\phi\psi,}$$

replace the first piece by $\phi', \ldots, \phi^{(\kappa)} \vdash_{\mathscr{F}_C} \psi'$ using Cor. (iii), then continue thus:

$$\cfrac{\cfrac{\cfrac{\psi'}{D\psi''\omega}\;\text{as before but without II}d\text{ on }N\phi,}{DN(E\xi)\,N\psi''\omega}\;\text{II}c,e,}{\psi}\;\text{as before but without II}d\text{ on }N\phi.$$

## 3.6 *Modus Ponens*

Prop. 4. *Modus Ponens is a derived $\mathscr{F}_C$-rule.*

We have to show

$$\frac{D\omega\phi \quad DN\phi\chi}{D\omega\chi}\,*$$

Given the $\mathscr{F}_C$-proofs of the upper formulae we have to show how to obtain an $\mathscr{F}_C$-proof of the lower formula. The demonstration is by formula induction on the cut formula $\phi$. With one of $\omega$, $\chi$ non-null.

(*a*) $\phi$ is atomic. We use theorem induction on $DN\phi\chi$. If $DN\phi\chi$ is a remodelling of an axiom then $\chi$ must be $\phi$ and the result follows at once. Otherwise we demonstrate the more general result

$$\frac{D\omega\phi \quad D\sum_{\theta=1}^{S\kappa} N\phi\chi}{D\omega\chi} *$$

(this is to account for all the cancellations of $N\phi$ by I*b* between $DN\phi\chi$ and the next building scheme above it). We use theorem induction on the right upper formula, that is we suppose that the result holds for the formula immediately above $D\sum_{\theta=1}^{S\kappa} N\phi\chi$ in the $\mathscr{F}_C$-proof-tree of this formula. We have already shown that the result holds if the right upper formula is an axiom. If $\sum_{\theta=1}^{S\kappa} N\phi$ is in the subsidiary formula of a rule and if the result holds for the upper formula or formulae of that rule then it follows at once that it holds for the lower formula of that rule.

Thus if we have

$$\frac{D\sum_{\theta=1}^{S\kappa} N\phi\chi'}{D\sum_{\theta=1}^{S\kappa} N\phi\chi}$$

then we have

$$\cfrac{\cfrac{D\omega\phi \quad D\sum_{\theta=1}^{S\kappa} N\phi\chi'}{D\omega\chi'} * \quad \text{by induction hypothesis}}{D\omega\chi} \quad \text{deduction as before,}$$

and similarly for a two-premiss rule.

If the whole or part of $\sum_{\theta=1}^{S\kappa} N\phi$ is in the main formula of a building rule then this building rule can only be II*a* since $\phi$ is atomic. By I*a* $D\sum_{\theta=1}^{S\kappa} N\phi\chi$ could become $DN\phi\psi$, which since $\phi$ is atomic is different from any of the forms of the lower formulae of building rules other than II*a*. Or, by I*a* $D\sum_{\theta=1}^{S\kappa} N\phi\chi$ could become $D\chi'\psi$, where $\chi$ is $D\chi'\chi''$ and $\sum_{\theta=1}^{S\kappa} N\phi$ occurs in $\psi$, this could be of the form of the lower formula of the building rules II$a, b, c, d, e$, but then $\sum_{\theta=1}^{S\kappa} N\phi$ would occur in the subsidiary formula contrary to the case considered. Thus the rule can only be II*a*. In this case the formula immediately above $D\sum_{\theta=1}^{S\kappa} N\phi\chi$ is either $\chi$ in which case the result is trivial or is of the form we are considering. Thus the only non-trivial case is when $\sum_{\theta=1}^{S\kappa} N\phi\chi$ is in the subsidiary formula. By our supposition I$b$ with $N\phi$ as main formula fails to be the rule immediately above $D\sum_{\theta=1}^{S\kappa} N\phi\chi$. If the rule is II$a$ with part or all of $\sum_{\theta=1}^{S\kappa} N\phi$ in the main formula then we have

$$\frac{D\sum_{\theta=1}^{\kappa'} N\phi\chi'}{D\sum_{\theta=1}^{S\kappa} N\phi\chi} \quad \text{II}a, \text{I}a,$$

where there may be more $N\phi$'s in the lower formula so that we have diluted with a formula of the form $D\sum_{\theta=1}^{\nu} N\phi\chi''$ ($\chi''$ may be null) or a permutation of this, and if the result holds for the upper formula then:

$$\frac{D\omega\phi \quad D\sum_{\theta=1}^{\kappa'} N\phi\chi'}{D\omega\chi'} *$$

then by dilution we easily get $D\omega\chi$. This completes this case.

It is impossible for $\chi'$ to be null, otherwise $\Sigma N\phi$ would arise from an axiom, which is absurd.

($b$) $\phi$ is $D\phi'\phi''$ and the result holds for $\phi'$ and $\phi''$. We have $\mathscr{F}_C$-proofs of $D\omega D\phi'\phi''$ and $DND\phi'\phi''\chi$, from the reversibility of II$b$ we then have $\mathscr{F}_C$-proofs of $DN\phi'\chi$ and $DN\phi''\chi$. Hence by formula induction:

$$\frac{D\omega D\phi'\phi''}{DD\omega\phi'\phi''} \text{I}a$$

$$\frac{DD\omega\phi'\phi'' \quad DN\phi''\chi}{DD\omega\phi'\chi} * \quad \text{by induction hypothesis,}$$

$$\frac{DD\omega\phi'\chi}{DD\omega\chi\phi'} \text{I}a$$

$$\frac{DD\omega\chi\phi' \quad DN\phi'\chi}{DD\omega\chi\chi} * \quad \text{by induction hypothesis,}$$

$$\frac{DD\omega\chi\chi}{D\omega\chi} \text{I}a, b.$$

($c$) $\phi$ is $N\phi'$ and the result holds for $\phi'$. We have $\mathscr{F}_C$-proofs of $D\omega N\phi'$ and $DNN\phi'\chi$, from the reversibility of II$c$ we then have an $\mathscr{F}_C$-proof of $D\phi\ \chi$. Thus we have, if $\omega$ is present

$$\frac{D\chi\phi' \quad DN\phi'\omega}{D\chi\omega} * \quad \text{by induction hypothesis,}$$

$$\frac{D\chi\omega}{D\omega\chi} \text{I}a.$$

If $\omega$ is absent we have

$$\cfrac{D\chi\phi' \quad \genfrac{}{}{1pt}{}{N\phi'}{\overline{DN\phi'\chi}}\;\text{II}a,\ \text{I}a,}{\cfrac{D\chi\chi}{\chi}\;\text{I}b.}\;*$$

(*d*) $\phi$ is of the form $(E\xi)\,\phi'\{\xi\}$ and the result holds for $\phi'\{\eta\}$; we have

$$\frac{D\omega\phi'\{\eta\} \quad DN\phi'\{\eta\}\chi}{D\omega\chi}\;* \quad \text{by induction hypothesis;}$$

we wish to demonstrate

$$\frac{D\omega(E\xi)\,\phi'\{\xi\} \qquad DN(E\xi)\,\phi'\{\xi\}\gamma}{D\omega\chi}\;*.$$

It suffices to demonstrate

$$\frac{D\omega \sum_{\theta=1}^{\kappa} (E\xi)\,\phi'\{\xi\} \quad DN\phi'\{\eta\}\chi}{D\omega\chi}\;*,$$

because from the reversibility of II*e* we can obtain an $\mathscr{F}_C$-proof of $DN\phi'\{\eta\}\chi$ for any variable $\eta$ which fails to occur free in $DN\phi'\{\Gamma_\iota\}\chi$, also immediately above $D\omega(E\xi)\,\phi'\{\xi\}$ there may have been several cancellations of $(E\xi)\,\phi'\{\xi\}$.

If $\sum_{\theta=1}^{\kappa} (E\xi)\,\phi'\{\xi\}$ fails to occur in the left upper formula then the result follows at once by dilution. Otherwise we use theorem induction on the left upper formula. If the left upper formula is an axiom then $(E\xi)\,\phi'\{\xi\}$ is absent. If the left upper formula is the lower formula of any building rule other than the introduction of $(E\xi)\,\phi'\{\xi\}$ by II*d* and if the result holds for the upper formula then we easily obtain $D\omega\chi$ on application of the same rule because $D \sum_{\theta=1}^{\kappa} (E\xi)\,\phi'\{\xi\}$ must be in the subsidiary formula of that rule. We had an almost similar situation under (*a*). But if the rule is an introduction of $(E\xi)\,\phi'\{\xi\}$ by II*d* and if the result holds for the upper formula we have

$$\cfrac{\cfrac{DD\omega\phi'\{\zeta\}\Sigma(E\xi)\,\phi'\{\xi\} \quad DN\phi'\{\zeta\}\chi}{DD\omega\phi'\{\zeta\}\chi}\;* \qquad DN\phi'\{\zeta\}\chi}{\genfrac{}{}{1pt}{}{DD\omega\chi\chi}{\overline{D\omega\chi}}\;\text{I}a, b.}\;* \quad \begin{array}{l}\text{by induction hypothesis,}\\ \text{by formula induction,}\end{array}$$

From the demonstration of Prop. 1 we see that the variable $\eta$ in $DN\phi'\{\eta\}\chi$ obtained from the reversibility of II*e* can be any new variable, the variable $\zeta$ which is restricted by II$d$ in the upper left formula might be any variable. If we change all free occurrences of $\zeta$ in the $\mathscr{F}_C$-proof of the upper left formula to a new variable then we obtain an $\mathscr{F}_C$-proof of a formula which differs from the upper left formula in that now $\zeta$ is a new variable, but $\omega$ and $(E\xi)\,\phi'\{\xi\}$ may have suffered a change of variable to this new one. In this case we should end up with $D\omega'\chi$, where $\omega'$ differs from $\omega$ in that one free variable in $\omega$ has been changed to a new variable (which fails to occur in $\omega$). By change of variable in the $\mathscr{F}_C$-proof of $D\omega\chi$ we can get back to $D\omega\chi$. This completes the demonstration of the proposition.

Note that we have given an effective method for eliminating the cut. This consists in taking a highest cut in the proof-tree and either eliminating it outright or replacing it by a cut higher up the proof-tree or by a cut or cuts with simpler cut formulae. Thus applying the process a highest cut ultimately gets replaced by cuts with atomic cut formulae and these can be made to disappear altogether.

## 3·7 *Regularity*

PROP. 5. *$\mathscr{F}_C$ is regular.*

We have to show
$$\frac{B\phi\psi}{B\chi\{\phi\}\,\chi\{\psi\}}*.$$

We show
$$\frac{C\phi\psi \quad C\psi\phi}{\overline{C\chi\{\phi\}\,\chi\{\psi\} \quad C\chi\{\psi\}\,\psi\{\phi\}}},$$

from this the result and
$$\frac{B\phi\psi \quad \chi\{\phi\}}{\chi\{\psi\}}*$$

follow, the first by the reversibility of II$b'$ and the last follows easily by Modus Ponens which can be eliminated. We proceed by formula induction on $\chi\{\pi\}$. The cases when $\chi\{\pi\}$ is $\pi$ or is $N\chi'\{\pi\}$ or is $D\chi'\{\pi\}\,\chi''\{\pi\}$ are dealt with as in Cor. (i), Prop. 8, Ch. 2.

If $\chi\{\pi\}$ is $(E\xi)\,\chi\{\pi,\xi\}$ and the result holds for $\chi\{\pi,\xi\}$ then we have

$$\frac{C\phi\psi \quad C\psi\phi}{\overline{C\chi\{\phi,\xi\}\,\chi\{\psi,\xi\} \quad C\chi\{\psi,\xi\}\,\chi\{\phi,\xi\}}}$$

whence by II$d$, I$a$ we get

$$DN\chi\{\phi,\xi\}\,(E\xi)\,\chi\{\psi,\xi\} \quad DN\chi\{\psi,\xi\}\,(E\xi)\,\chi\{\phi,\xi\}$$

and by II*e*

$$DN(E\xi)\chi\{\phi,\xi\}(E\xi)\chi\{\psi,\xi\} \quad DN(E\xi)\chi\{\psi,\xi\}(E\xi)\chi\{\phi,\xi\}$$

as desired. The variable $\xi$ can occur in $\phi$ or in $\psi$ or in both.

COR. (i).
$$\frac{DB\phi\psi\omega \quad D\chi\{\phi\}\omega}{D\chi\{\psi\}\omega}.$$

We proceed by formula induction on $\chi\{\pi\}$. The details are left to the reader.

D 10 $\qquad (A\xi)\,\phi\{\xi\}$ for $N(E\xi)\,N\phi\{\xi\}$.

$A$ is called the *universal quantifier*.

PROP. 6. 1–5 *and* 1*–5* *and* 6 *of Prop.* 5, *Ch.* 2 *hold in* $\mathscr{F}_C$, *also*

7. $B(E\xi)\,D\phi\{\xi\}\,\psi D(E\xi)\,\phi\{\xi\}\,\psi$; $\quad$ 7*. $B(A\xi)\,K\phi\{\xi\}\,\psi K(A\xi)\,\phi\{\xi\}\,\psi$;
8. $B(A\xi)\,D\phi\{\xi\}\,\psi D(A\xi)\,\phi\{\xi\}\,\psi$; $\quad$ 8.* $B(E\xi)\,K\phi\{\xi\}\,\psi K(E\xi)\,\phi\{\xi\}\,\psi$;

*distributivity of quantifiers.*

*In* 7–8* *incl. the variable* $\xi$ *fails to occur free in* $\psi$ *or* $\phi\{\Gamma_\iota\}$.

9. $BN(E\xi)\,\phi\{\xi\}\,(A\xi)\,N\phi\{\xi\}$; $\quad$ 9*. $BN(A\xi)\,\phi\{\xi\}\,(E\xi)\,N\phi\{\xi\}$;
10. $B\phi(E\xi)\,\phi$; $\quad$ 10*. $B\phi(A\xi)\,\phi$;

*in* 10, 10* $\xi$ *fails to occur free in* $\phi$.

II*d'* $\qquad \dfrac{D\phi\{\xi\}\,\omega}{D(A\xi)\,\phi\{\xi\}\,\omega}$, *where* $\xi$ *fails to occur free in* $\omega$ *or free in* $\phi\{\Gamma_\iota\}$.

II*d' is a derived rule in* $\mathscr{F}_C$. *The rule* II*d' is reversible.*

For 7–9* consider 7. We have from Prop. 2

$$\frac{\dfrac{DND\phi\{\xi\}\,\psi D\phi\{\xi\}\,\psi}{\overline{DND\phi\{\xi\}\,\psi D(E\xi)\,\phi\{\xi\}\,\psi}}\ \text{I}a,\ \text{II}d,}{DN(E\xi)\,D\phi\{\xi\}\,\psi D(E\xi)\,\phi\{\xi\}\,\chi}\ \text{II}e,\ \xi \text{ fails to occur free in } \psi \text{ or free in } \phi\{\Gamma_\iota\},$$

again

$$\frac{DND\phi\{\xi\}\,\psi D\phi\{\xi\}\,\psi}{DN\phi\{\xi\}\,D\phi\{\xi\}\,\psi \quad DN\psi D\phi\{\xi\}\,\psi}\ {*}\ \text{Tertium non datur reversibility of II}b,$$

$$\frac{DN\phi\{\xi\}\,D\phi\{\xi\}\,\psi \quad DN\psi D\phi\{\xi\}\,\psi}{\overline{DN\phi\{\xi\}\,(E\xi)\,D\phi\{\xi\}\,\psi \quad DN\psi(E\xi)\,D\phi\{\xi\}\,\psi}}\ \text{I}a,\ \text{II}d,$$

$$\frac{DN\phi\{\xi\}\,(E\xi)\,D\phi\{\xi\}\,\psi \quad DN\psi(E\xi)\,D\phi\{\xi\}\,\psi}{DN(E\xi)\,\phi\{\xi\}\,(E\xi)\,D\phi\{\xi\}\,\psi \qquad DN\psi(E\xi)\,D\phi\{\xi\}\,\psi}\ \text{II}e,\ \xi \text{ fails to occur free in } (E\xi)\,D\phi\{\xi\}\,\psi,$$

$$\frac{DN(E\xi)\,\phi\{\xi\}\,(E\xi)\,D\phi\{\xi\}\,\psi \qquad DN\psi(E\xi)\,D\phi\{\xi\}\,\psi}{DND(E\xi)\,\phi\{\xi\}\,\psi\,(E\xi)\,D\phi\{\xi\}\,\psi}\ \text{II}b$$

the result now follows by II*b* and definition of $B$.

7*, 8, 8* follow similarly and are left as exercises to the reader.

9. We have

$$\frac{\dfrac{DN(E\xi)N\phi\{\xi\}(E\xi)N\phi\{\xi\}}{DNNN(E\xi)N\phi\{\xi\}(E\xi)N\phi\{\xi\}}}{CN(A\xi)\phi\{\xi\}(E\xi)N\phi\{\xi\}} \quad \text{II}c \quad \begin{array}{l}\text{Tertium non datur,}\\ \text{by def. of } C, A,\end{array}$$

again

$$\frac{\dfrac{DN(E\xi)N\phi\{\xi\}(E\xi)N\phi\{\xi\}}{DN(E\xi)N\phi\{\xi\}NN(E\xi)N\phi\{\xi\}}}{C(E\xi)N\phi\{\xi\}N(A\xi)\phi\{\xi\}} \quad \text{I}a, \text{II}c \quad \begin{array}{l}\text{Tertium non datur,}\\ \text{by def. of } C, A\end{array}$$

the result now follows by II$b'$ and the definition of $B$.

9* follows similarly and is left to the reader. 10, 10* are trivial.

## 3.8 *The system* $\mathscr{F}''_C$

The system $\mathscr{F}''_C$ is like the system $\mathscr{F}_C$ except that we add the symbol $A$ of type $o(o\iota)$ instead of $E$ and replace the building rules II$d, e$ by

II$d'$ $\dfrac{D\phi\{\eta\}\omega}{D(A\xi)\phi\{\xi\}\omega}$ $\eta$ is absent from $\omega$ and $\phi\{\Gamma_\iota\}$, $\xi$ is absent from $\phi\{\Gamma_\iota\}$ or is everywhere bound in $\phi\{\Gamma_\iota\}$,

II$e'$ $\dfrac{DN\phi\{\eta\}\omega}{DN(A\xi)\phi\{\xi\}\omega.}$ $\xi$ is absent from $\phi\{\Gamma_\iota\}$, or is everywhere bound in $\phi\{\Gamma_\iota\}$,

we then replace D 10 by:

D 10′ $(E\xi)\phi\{\xi\}$ for $N(A\xi)N\phi\{\xi\}$.

The systems $\mathscr{F}_C$ and $\mathscr{F}''_C$ are equivalent via the definitions D 10 and D 10′.

A convenient system which is equivalent to $\mathscr{F}_C$ and to $\mathscr{F}''_C$ is the system $\mathscr{F}'_C$ where we use the building rules II$d$, II$d'$ and one of the definitions D 10, D 10′. All our results so far hold for $\mathscr{F}''_C$.

Prop. 7. *If $\phi$ is an $\mathscr{F}_C$-theorem then so is $N\bar{\phi}$, where $\bar{\phi}$ is obtained from $\phi$ by interchanging $D$ with $K$, $E$ with $A$, $\pi$ with $N\pi$ where $\pi$ is atomic and conversely.*

$\bar{\phi}$ is called the *dual* of $\phi$.

Proceed by induction on the construction of $\phi$. Use 5, 5*, 6, 9, 9* to obtain $B\phi N\bar{\phi}$. $\bar{\bar{\phi}}$ is $\phi$.

Cor. (i). *If* $\dfrac{\phi}{\psi}$ (double line) *without use of* II$e$ *then* $\dfrac{N\psi}{N\phi}$ (double line).

We verify that if $\dfrac{\phi}{\psi}$ is a rule other than II$e$ then $\dfrac{N\psi}{N\phi}$ is a derived rule. This fails for II$e$.

PROP. 8.

$$B(E\xi)\,D\phi\{\xi\}\,\psi\{\xi\}\,D(E\xi)\,\phi\{\xi\}\,(E\xi)\,\psi\{\xi\},$$
$$C(E\xi)\,K\phi\{\xi\}\,\psi\{\xi\}\,K\{E\xi\}\,\phi\{\xi\}\,(E\xi)\,\psi\{\xi\},$$
$$B(A\xi)\,K\phi\{\xi\}\,\psi\{\xi\}\,K(A\xi)\,\phi\{\xi\}\,(A\xi)\,\psi\{\xi\},$$
$$CD(A\xi)\,\phi\{\xi\}\,(A\xi)\,\psi\{\xi\}\,(A\xi)\,D\phi\{\xi\}\,\psi\{\xi\}.$$

Consider the first, we have

$$\dfrac{\genfrac{}{}{3pt}{}{\genfrac{}{}{3pt}{}{\genfrac{}{}{3pt}{}{DN\phi\{\xi\}\,\phi\{\xi\}\quad \text{Tertium non datur}}{DN\phi\{\xi\}\,D\phi\{\xi\}\,\psi\{\xi\}}\ \text{II}a,\text{I}a}{DN\phi\{\xi\}\,(E\xi)\,D\phi\{\xi\}\,\psi\{\xi\}}\ \text{II}d,\text{I}a}{DN(E\xi)\,\phi\{\xi\}\,(E\xi)\,D\phi\{\xi\}\,\psi\{\xi\}}\ \text{II}e \quad DN(E\xi)\,\psi\{\xi\}\,(E\xi)\,D\phi\{\xi\}\,\psi\{\xi\}\ \text{similarly}}{DND(E\xi)\,\phi\{\xi\}\,(E\xi)\,\psi\{\xi\}\,(E\xi)\,D\phi\{\xi\}\,\psi\{\xi\}.}\ \text{II}b$$

Again

$$\dfrac{\dfrac{\genfrac{}{}{3pt}{}{\genfrac{}{}{3pt}{}{DN\phi\{\xi\}\,\phi\{\xi\}\quad \text{Tertium non datur}}{DN\phi\{\xi\}\,D\phi\{\xi\}\,\psi\{\xi\}}\ \text{II}a,\text{I}a}{DN\phi\{\xi\}\,D(E\xi)\,\phi\{\xi\}\,(E\xi)\,\psi\{\xi\}}\ \text{II}d,\text{I}a \quad DN\psi\{\xi\}\,D(E\xi)\,\phi\{\xi\}\,(E\xi)\,\psi\{\xi\}\ \text{similarly}}{DND\phi\{\xi\}\,\psi\{\xi\}\,D(E\xi)\,\phi\{\xi\}\,(E\xi)\,\psi\{\xi\}}\ \text{II}b}{DN(E\xi)\,D\phi\{\xi\}\,\psi\{\xi\}\,D(E\xi)\,\phi\{\xi\}\,(E\xi)\,\psi\{\xi\},}\ \text{II}e$$

the result follows from these by II$b'$.

The remainder follow in a similar manner and are left to the reader.

## 3·9 *Prenex normal forms*

An $\mathscr{F}_C$-statement is said to be in *prenex normal form* if it is of the form $(Q\mathfrak{x})\,\chi\{\mathfrak{x}\}$, where $(Q\mathfrak{x})$ is a sequence of quantifiers, existential or universal or both, and $\chi\{\mathfrak{x}\}$ is an $\mathscr{F}_C$-statement void of quantifiers. $(Q\mathfrak{x})$ is called the *prefix* and $\chi\{\mathfrak{x}\}$ is called the *matrix*. By repeated application of 7–9* incl. of Prop. 6 and Cor. (i) of Prop. 5 we see that each $\mathscr{F}_C$-statement $\phi$ has an equivalent prenex normal form $\psi$, thus $\phi$ is an $\mathscr{F}_C$-theorem if and only if its prenex normal form $\psi$ is an $\mathscr{F}_C$-theorem. Note that the prenex normal form, in general, fails to be unique. Let $(Q\mathfrak{x})\,\chi\{\mathfrak{x}\}$ be a closed $\mathscr{F}_C$-statement in prenex normal form with prefix $(Q\mathfrak{x})$ and matrix $\chi\{\mathfrak{x}\}$. Let $\mathfrak{x}$ be the sequence of distinct variables $\xi', \ldots, \xi^{(\theta)}$ in that order and $(Q\mathfrak{x})$ a sequence of quantifiers on these variables in the same order. A variable $\xi^{(\kappa)}$, $1 \leqslant \kappa \leqslant \theta$ is called *general* if $(A\xi^{(\kappa)})$ occurs in $(Q\mathfrak{x})$ and is called *restricted* if $(E\xi^{(\kappa)})$ occurs in $(Q\mathfrak{x})$. If $1 \leqslant \nu \leqslant \kappa \leqslant \theta$ then $\xi^{(\nu)}$ is called *superior to* $\xi^{(\kappa)}$, and $\xi^{(\kappa)}$ *inferior to* $\xi^{(\nu)}$.

An $\mathscr{F}_C$-statement $\phi$ without bound variables is said to be *tautologous* under the following circumstances: $\phi$ is built up from other $\mathscr{F}_C$-statements joined together by $N$ and $D$; we replace each of these part statements by $\boldsymbol{f}$ or by $\boldsymbol{t}$, but make the same replacement at each occurrence of a variant, we then have a formula built up from $\boldsymbol{f}$ and $\boldsymbol{t}$ (regarded as of type $o$) by $N$ and $D$, we then calculate the value of the statement as follows: replace $N\boldsymbol{f}$ by $\boldsymbol{t}$, $N\boldsymbol{t}$ by $\boldsymbol{f}$; $D\boldsymbol{ff}$ by $\boldsymbol{f}$, $D\boldsymbol{tf}$, $D\boldsymbol{ft}$, $D\boldsymbol{tt}$ by $\boldsymbol{t}$. The final result is either $\boldsymbol{t}$ or $\boldsymbol{f}$. If the final result is always $\boldsymbol{t}$ however we make the initial replacements of the part statements then the $\mathscr{F}_C$-statement is said to be tautologous. It is easily seen from Ch. 2 that a $\mathscr{P}_C$-statement is a $\mathscr{P}_C$-theorem if and only if it is tautologous.

An $\mathscr{F}_C$-statement $\phi$ can be built up from other $\mathscr{F}_C$-statements in various ways by $N$ and $D$, e.g. $DDN\pi\pi\omega$ can be built up from itself alone, or from $DN\pi\pi$ and $\omega$ or from $N\pi$, $\pi$ and $\omega$ or just from $\pi$ and $\omega$. In the first case we replace the whole statement by $\boldsymbol{t}$ or $\boldsymbol{f}$ and the test for tautology fails, similarly in the second case we replace $DN\pi\pi$ and $\omega$ by $\boldsymbol{t}$ or $\boldsymbol{f}$ in all possible ways and again the test for tautology fails, similarly the test fails in the third case, but in the fourth case we have to replace $\pi$ and $\omega$ by $\boldsymbol{t}$ or $\boldsymbol{f}$ in all possible ways and the value is always $\boldsymbol{t}$, so we have a tautology.

PROP. 9. *An $\mathscr{F}'_C$-proof (using* II$d'$ *instead of* II$e$) *of an $\mathscr{F}'_C$-statement in prenex normal form can be modified so that applications of* I$b$, II$d$, $d'$ *come after applications of* II$a, b, c$.

An $\mathscr{F}'_C$-proof in which applications of I$b$, II$d$, $d'$ come after applications of II$a, b, c$ is called an *$\mathscr{F}'_C$-proof in normal form*. An $\mathscr{F}'_C$-proof in normal form starts with $\mathscr{F}_C$-axioms then applications of I$a$, II$a$, $b$, $c$ until we arrive at an $\mathscr{F}_C$-theorem $\phi$ without quantifiers, then applications of I$a$, $b$, II$b$, $d'$ until we finally arrive at an $\mathscr{F}'_C$-theorem in prenex normal form.

We shall find that the $\mathscr{F}_C$-statement $\phi$ is a disjunction $\sum_{\theta=1}^{\kappa} \psi^{(\theta)}$, where the disjunctands are of the same logical structure but merely differ by choice of individual variables, they are *variants* of a common form. I*a* is used to bring one of the disjunctands $\psi', \ldots, \psi^{(\kappa)}$ to the left when we apply II*d* or II*d'* to it. I*b* is used to discard duplicates as they occur. The final $\mathscr{F}'_C$-theorem is of the form $(Q'\xi')\ldots(Q^{(\pi)}\xi^{(\pi)})\,\psi\{\xi', \ldots, \xi^{(\pi)}\}$ where each $Q^{(\theta)}$ is either $E$ or $A$ and $\psi\{\xi', \ldots, \xi^{(\pi)}\}$ differs from any of the disjunctands $\psi', \ldots, \psi^{(\kappa)}$ merely by change of individual variables.

Suppose we have an $\mathscr{F}'_C$-proof of $\chi$ in prenex normal form. We first modify applications of II*a*, if necessary, so that they fail to introduce quantifiers. Suppose that $D\chi'\chi''$ is introduced by II*a* then introduce $\chi'\,\chi''$ one after the other, suppose $ND\chi'\chi''$ is introduced at II*a* then introduce $N\chi', N\chi''$ separately and apply II*b*, (this forms two branches), if $NN\chi$ is introduced then introduce $\chi$ and apply II*c*, suppose that $(E\xi)\,\chi'\{\xi\}$ is introduced at II*a* then introduce $\chi'\{\xi\}$ instead and apply II*d*, suppose that $(A\xi)\,\chi'\{\xi\}$ is introduced at II*a* then introduce $\chi'\{\eta\}$ instead, where $\eta$ is new, and apply II*d'*. Repeat this process as long as possible and we shall have used II*a* only with atomic statements or negations of atomic statements as main formulae. Thus we suppose that the $\mathscr{F}_C$-proof of $(Q\mathfrak{x})\,\psi\{\mathfrak{x}\}$ only uses II*a* with main formulae which are atomic formulae or negations of atomic formulae. Note that we do this without using I*b*.

Instead of using the system $\mathscr{F}'_C$ we shall use an equivalent system $\mathscr{F}''_C$ which has the rules II$d^*$, $d'^*$ in place of rules II*d*, *d'*, where

$$\text{II}d^* \quad \frac{D\sum_{\theta=1}^{\kappa}\phi\{\xi^{(\theta)}\}\,\omega}{D(E\xi)\,\phi\{\xi\}\,\omega},$$

$$\text{II}d'^* \quad \frac{D\sum_{\theta=1}^{\kappa}\phi\{\xi^{(\theta)}\}\,\omega}{D(A\xi)\,\phi\{\xi\}\,\omega},$$

in II$d'^*$ $\omega$ is free for $\xi', \ldots, \xi^{(\theta)}$ and so is $\phi\{\Gamma_\iota\}$, and $\xi', \ldots, \xi^{(\theta)}$ are distinct. An $\mathscr{F}''_C$-proof will use rules II$d^*$, $d'^*$ whenever possible, so it will be without a sequence of applications of II*d* followed by I*b*, etc.

We now want to show how to modify an $\mathscr{F}''_C$-proof of a prenex formula so that applications of I*b*, II$d^*$, $d'^*$ come after applications of II*a*, *b*, *c*. Clearly the systems $\mathscr{F}'_C$ and $\mathscr{F}''_C$ are equivalent.

We define the rank of an $\mathscr{F}''_C$-proof of an $\mathscr{F}''_C$-formula $(Q\mathfrak{x})\,\phi\{\mathfrak{x}\}$ in prenex normal form, where $\mathfrak{x}$ stands for $\xi', \ldots, \xi^{(\pi)}$, as the ordered $S\pi$-tuplet $\{\nu, \nu', \ldots, \nu^{(\pi)}\}$, where $\nu$ is the number of occurrences of applications of rules II$a$, $b$, $c$ beneath applications of rule I$b$, $\nu'$ is the number of applications of rules II$a$, $b$, $c$ beneath rules II$d^*$, $d'^*$ which bind a variable standing in the first argument place in $\phi\{\mathfrak{x}\}, \ldots, \nu^{(\pi)}$ is the number of applications of rules II$a$, $b$, $c$ beneath applications of rules II$d^*$, $d'^*$ which bind variables standing in the $\pi$th argument place in $\phi\{\mathfrak{x}\}$. These are calculated as follows: take for instance rule I$b$. Mark applications of I$b$ in the $\mathscr{F}''_C$-proof of $(Q\mathfrak{x})\,\phi\{\mathfrak{x}\}$, let these be denoted by $K', \ldots, K^{(\kappa)}$, let $\mu', \ldots, \mu^{(\kappa)}$ be respectively the number of applications of rules II$a$, $b$, $c$ beneath $K', \ldots, K^{(\pi)}$, then $\mu' + \mu'' + \ldots + \mu^{(\kappa)} = \nu$, similarly for the other cases. Ranks are ordered lexicographically.

We now show how to modify the $\mathscr{F}''_C$-proof of the prenex $\mathscr{F}_C$-formula $(Q\mathfrak{x})\,\phi\{\mathfrak{x}\}$ by steps so that all applications of rules I$b$, II$d^*$, $d'^*$ occur below applications of rules II$a$, $b$, $c$ in such a manner that the rank strictly decreases at each step. Thus the process will terminate when the rank is $\{0, \ldots, 0\}$, because we shall see that if the rank is greater we can always make a step. We can then find an $\mathscr{F}'_C$-proof in which all applications of rules I$b$, II$d$, $d'$ occur below all applications of rules II$a$, $b$, $c$.

Consider first rule I$b$. In what follows we omit mention of I$a$ in many places. Suppose we have, apart from I$a$

$$\cfrac{\cfrac{DD\phi\phi\omega}{D\phi\omega}\,\text{I}b,}{DD\psi\phi\omega,}\,\text{II}a,$$

where $\psi$ is atomic or the negation of an atomic formula. Replace this by

$$\cfrac{\cfrac{\cfrac{\cfrac{DD\phi\phi\omega}{D\psi DD\phi\phi\omega}\,\text{II}a,}{DD\psi\psi DD\phi\phi\omega}\,\text{II}a,}{DDD\psi\phi D\psi\phi\omega}\,\text{I}a,}{DD\psi\phi\omega}\,\text{I}b.$$

We are left with an $\mathscr{F}''_C$-proof of $(Q\mathfrak{x})\,\phi\{\mathfrak{x}\}$ of lower rank if we are dealing with a highest case of an application of I$b$ immediately above an application of II$a$.

Suppose we have

$$\cfrac{\cfrac{DDN\phi N\phi\omega}{DN\phi\omega}\,\text{I}b \quad DN\psi\omega}{DND\phi\psi\omega}\,\text{II}b.$$

Replace this by

$$\frac{\dfrac{\dfrac{DDN\phi N\phi\omega \quad DN\psi\omega}{DND\phi\psi DN\phi\omega \quad DN\psi\omega}\ \text{II}b,\ \text{II}a,\ \text{to introduce } N\phi \text{ in right upper formula,}}{DND\phi\psi ND\phi\psi\omega}\ \text{II}b,\ \text{II}a,\ \text{to introduce } ND\phi\psi \text{ in right upper formula,}}{DND\phi\psi\omega.}\ \text{I}b,$$

Call this case (i).

Case (ii) is

$$\frac{\dfrac{DDDN\phi\chi\chi\omega}{DDN\phi\chi\omega}\ \text{I}a,b, \quad DDN\psi\chi\omega}{DDND\phi\psi\chi\omega.}\ \text{II}b,$$

Replace by

$$\frac{\dfrac{DDDN\phi\chi\chi\omega \quad DDN\psi\chi\omega}{DDDND\phi\psi\chi\chi\omega}\ \text{II}b,\ \text{II}a \text{ to introduce another } \chi \text{ in right upper formula,}}{DDND\phi\psi\chi\omega.}\ \text{I}a,b,$$

Again we are left with an $\mathscr{F}''_C$-proof of $(Q\mathfrak{x})\,\phi\{\mathfrak{x}\}$ of lesser rank provided we are using a highest case of I$b$ above II$b$. The use of rule II$a$, as already observed, is without any use of I$b$, so that if we have a highest case of I$b$ above II$b$ then the rank has fallen.

Suppose we have

$$\frac{\dfrac{DD\phi\phi\psi}{D\phi\omega}\ \text{I}b,}{DNN\phi\omega.}\ \text{II}c,$$

Replace this by

$$\frac{\dfrac{\dfrac{DD\phi\phi\omega}{DDNN\phi\phi\omega}\ \text{II}c,}{DDNN\phi NN\phi\omega}\ \text{II}c,}{DNN\phi\omega.}\ \text{I}b,$$

Again we are left with an $\mathscr{F}''_C$-proof of $(Q\mathfrak{x})\,\phi\{\mathfrak{x}\}$ of lower rank if we are dealing with a highest case of I$b$ above II$c$. Call this case (i).

Case (ii) is

$$\frac{\dfrac{DD\phi\phi D\psi\omega}{D\phi D\psi\omega}\ \text{I}b,}{DD\phi NN\psi\omega.}\ \text{II}c,$$

Replace this by

$$\cfrac{\cfrac{DD\phi\phi D\psi\omega}{DD\phi\phi DNN\psi\omega}\ \mathrm{II}c,}{D\phi DNN\psi\omega'.}\ \mathrm{I}b,$$

Again we are left with an $\mathscr{F}''_C$-proof of $(Q\mathfrak{x})\,\phi\{\mathfrak{x}\}$ which is of lower rank.

Now consider rule II$d^*$. Suppose we have

$$\cfrac{\cfrac{D\Sigma\phi\{\eta\}\,\omega}{D(E\xi)\,\phi\{\xi\}\,\omega}\ \mathrm{II}d^*,}{D\chi D(E\xi)\,\phi\{\xi\}\,\omega.}\ \mathrm{II}a,$$

Replace this by

$$\cfrac{\cfrac{D\Sigma\phi\{\eta\}\,\omega}{D\chi D\Sigma\phi\{\eta\}\,\omega}\ \mathrm{II}a,}{D\chi D(E\xi)\,\phi\{\xi\}\,\omega.}\ \mathrm{II}d^*,$$

Again we are left with an $\mathscr{F}''_C$-proof of $(Q\mathfrak{x})\,\phi\{\mathfrak{x}\}$ of lower rank. Rule II$d'^*$ is dealt with similarly, but may require a change of variable.

Suppose we have

$$\cfrac{\cfrac{D\Sigma\phi\{\eta\}\,\omega}{D(E\xi)\,\phi\{\xi\}\,\omega}\ \mathrm{II}d^*,}{D(E\xi)\,\phi\{\xi\}\,\omega',}\ \mathrm{II}c,$$

where $\omega'$ differs from $\omega$ by having $NN$ placed over a disjunctand. The other case where $NN$ is placed over $(E\xi)\,\phi\{\xi\}$ is impossible because the theorem is in prenex normal form. Replace this by

$$\cfrac{\cfrac{D\Sigma\phi\{\eta\}\,\omega}{D\Sigma\phi\{\eta\}\,\omega'}\ \mathrm{II}c,}{D(E\xi)\,\phi\{\xi\}\,\omega'.}\ \mathrm{II}d^*,$$

Again we are left with an $\mathscr{F}''_C$-proof of $(Q\mathfrak{x})\,\phi\{\mathfrak{x}\}$ of lower rank. Rule II$d'^*$ is dealt with similarly.

Suppose we have

$$\cfrac{\cfrac{DD\Sigma\phi\{\eta'\}N\chi\omega}{DD(E\xi)\,\phi\{\xi\}\,N\chi\omega}\ \mathrm{II}d^*, \quad DD(E\xi)\,\phi\{\xi\}\,N\psi\omega}{D(E\xi)\,\phi\{\xi\}\,DND\chi\psi\omega.}\ \mathrm{II}b. \qquad (a)$$

Replace this by

$$\cfrac{\cfrac{DD\Sigma\phi\{\eta'\}N\chi\omega}{DD\Sigma\Sigma\phi\{\eta'\}\Sigma\phi\{\eta''\}\ldots\Sigma\phi\{\eta^{(\theta)}\}N\chi\omega}\,\text{II}a \qquad \cfrac{D\Sigma\phi\{\eta''\}\omega'' \quad D\Sigma\phi\{\eta^{(\theta)}\}\omega^{(\theta)}}{DD\Sigma\Sigma\phi\{\eta'\}\Sigma\phi\{\eta''\}\ldots\Sigma\phi\{\eta^{(\theta)}\}N\psi\omega}}{\cfrac{DNDχψD\Sigma\Sigma\psi\{\eta'\}\ldots\Sigma\phi\{\eta^{(\theta)}\}\omega}{DND\chi\psi(E\xi)\phi\{\xi\}\omega}\,\text{II}d^*,}\,\text{II}b \qquad (b)$$

where $(E\xi)\phi\{\xi\}$ is introduced into the branch above the right upper formula in $(a)$ at applications of II $d^*$:

$$\frac{D\Sigma\phi\{\eta''\}\omega''}{D(E\xi)\phi\{\xi\}\omega''}\cdots\frac{D\Sigma\phi\{\eta^{(\theta)}\}\omega^{(\theta)}}{D(E\xi)\phi\{\xi\}\omega^{(\theta)}}.$$

Now we have
$$\frac{D(E\xi)\phi\{\xi\}\omega''\ldots D(E\xi)\phi\{\xi\}\omega^{(\theta)}}{DD(E\xi)\phi\{\xi\}N\psi\omega}, \qquad (c)$$

where $(E\xi)\phi\{\xi\}$ is in the subsidiary formula. Hence we shall have the same figure with these occurrences of $(E\xi)\phi\{\xi\}$ everywhere replaced by

$$\Sigma\Sigma\phi\{\eta'\}\Sigma\phi\{\eta''\}\ldots\Sigma\phi\{\eta^{(\theta)}\}.$$

Now add applications of II $a$ as follows:

$$\frac{D\Sigma\phi\{\eta''\}\omega''}{D\Sigma\Sigma\phi\{\eta'\}\Sigma\phi\{\eta''\}\ldots\Sigma\phi\{\eta^{(\theta)}\}\omega''}\cdots\frac{D\Sigma\phi\{\eta^{(\theta)}\}\omega^{(\theta)}}{D\Sigma\Sigma\phi\{\eta'\}\Sigma\phi\{\eta''\}\ldots\Sigma\phi\{\eta^{(\theta)}\}\omega^{(\theta)}}. \qquad (d)$$

Now from $(b)$, $(c)$ and $(d)$ we obtain

$$\cfrac{\cfrac{D\Sigma\phi\{\eta''\}\omega''}{D\Sigma\Sigma\phi\{\eta'\}\ldots\Sigma\phi\{\eta^{(\theta)}\}\omega''}\,\text{II}a,\text{ etc.}\ldots\cfrac{D\Sigma\phi\{\eta^{(\theta)}\}\omega^{(\theta)}}{D\Sigma\Sigma\phi\{\eta'\}\ldots\Sigma\phi\{\eta^{(\theta)}\}\omega^{(\theta)}}\,\text{II}a,\text{ etc.}}{DD\Sigma\Sigma\phi\{\eta'\}\ldots\Sigma\phi\{\eta^{(\theta)}\}N\psi\omega}.$$

Use this as the right upper part of $(b)$, then finish up as in $(b)$ and we have placed the application of II $b$ above the application of II $d^*$. In doing this we have had to introduce various variants of $\phi\{\xi\}$, this is done without using I $b$ or any applications of II $d^*$, $d'^*$ that bind variables earlier in the list than the variables we are binding in $(a)$. The effect of this is that in the rank $\{\nu, \nu', \ldots, \nu^{(\pi)}\}$ the first component is unaltered because we have made our alteration without use of I $b$, and if the variables we are binding is $\xi^{(\theta)}$, then $\nu', \ldots, \nu^{(\theta \dot{-} 1)}$ are unaltered while $\nu^{(\theta)}$ is decreased by one, the other components may be increased. The total result is a reduction in rank.

Suppose we have

$$\dfrac{\dfrac{DD\Sigma\phi\{\eta\}N\chi\omega}{DD(A\xi)\,\phi\{\xi\}N\chi\omega}\ \text{II}d'^* \qquad DD(A\xi)\,\phi\{\xi\}N\psi\omega}{DND\chi\psi D(A\xi)\,\phi\{\xi\}\,\omega}\ \text{II}b.$$

Replace this by

$$\dfrac{\dfrac{\dfrac{DD\Sigma\phi\{\eta\}N\chi\omega}{DDD\Sigma\phi\{\eta\}\Sigma\phi\{\zeta\}N\chi\omega}\ \text{II}a,\text{etc.} \qquad \dfrac{\dfrac{DD\Sigma\phi\{\zeta\}N\psi\omega}{}\ \text{by the reversibility of II}d'^*,}{DDD\Sigma\phi\{\zeta\}\Sigma\phi\{\eta\}N\psi\omega}\ \text{II}a,\text{ etc.}}{DND\chi\psi DD\Sigma\phi\{\eta\}\Sigma\phi\{\zeta\}\,\omega}\ \text{II}b}{DND\chi\psi D(A\xi)\,\phi\{\xi\}\,\omega}\ \text{II}d'^*. \qquad (e)$$

In the reversibility of II$d'^*$ we may take the variables $\zeta$ to be new and distinct from the variables $\eta$. This allows us to apply II$d'^*$. As in the case of I$b$ below II$d^*$ we have decreased the rank. The reversibility of II$d'^*$ is performed without use of I$b$.

For completeness we add:

LEMMA. *Rule* II$d'^*$ *is reversible.*

By this we mean that if we have an $\mathscr{F}''_C$-proof of $D(A\xi)\,\phi\{\xi\}\,\omega$ then we can find an $\mathscr{F}''_C$-proof of $D\Sigma\phi\{\eta\}\,\omega$ for some $\Sigma\phi\{\eta\}$.

In the $\mathscr{F}''_C$-proof-tree of $D(A\xi)\,\phi\{\xi\}\,\omega$ note the places where corresponding occurrences of $(A\xi)\,\phi\{\xi\}$ are introduced by II$d'^*$. These will be of the form

$$\frac{D\Sigma\phi\{\eta'\}\,\omega'}{D(A\xi)\,\phi\{\xi\}\,\omega'}\cdots\frac{D\Sigma\phi\{\eta^{(\theta)}\}\,\omega^{(\theta)}}{D(A\xi)\,\phi\{\xi\}\,\omega^{(\theta)}}. \qquad (f)$$

In the $\mathscr{F}''_C$-proof from these places to $D(A\xi)\,\phi\{\xi\}\,\omega$ the part $(A\xi)\,\phi\{\xi\}$ will remain in the subsidiary formulae everywhere. Hence we may replace all these occurrences of $(A\xi)\,\phi\{\xi\}$ by the disjunction of

$$\Sigma\phi\{\eta'\}\ldots\Sigma\phi\{\eta^{(\theta)}\}.$$

These can enter by II$a$, etc., applied to the upper formulae of ($f$). In this way we obtain an $\mathscr{F}''_C$-proof of

$$D\Sigma\Sigma\phi\{\eta'\}\ldots\Sigma\phi\{\eta^{(\theta)}\}\,\omega,$$

instead of one of $D(A\xi)\,\phi\{\xi\}\,\omega$. II$a$, etc., as before observed, has been done without use of I$b$, and the only use of II$d^*$, $d'^*$ has been on variables later in the list $\xi', \ldots, \xi^{(\pi)}$ than the variable $\xi$, so that the rank of the $\mathscr{F}''_C$-proof of ($e$) has decreased.

To conplete the demonstration of Prop. 9 we make the alterations discussed above starting from the highest available places. Each time the rank is reduced, and as long as the rank is greater than the lowest rank we can always reduce it. This completes the demonstration of the proposition.

## 3.10 *H-disjunctions*

Let
$$(Q'\zeta')\dots(Q^{(\pi)}\zeta^{(\pi)})\,\psi\{\zeta',\dots,\zeta^{(\pi)}\} \tag{1}$$
be a closed $\mathscr{F}'_C$-statement in prenex normal form where the matrix $\psi\{\zeta',\dots,\zeta^{(\pi)}\}$ is quantifier-free. Here each $Q^{(\theta)}$ is either $A$ or $E$. If $Q^{(\theta)}$ is $E$ then $\zeta^{(\theta)}$ is called a *restricted variable*, if $Q^{(\theta)}$ is $A$ then $\zeta^{(\theta)}$ is called a *general variable*. Let there be $\pi'$ restricted variables in (1) and let there be $\pi''$ general variables in (1), then $\pi'+\pi''=\pi$. Form a list of all ordered $\pi'$-tuplets of natural numbers $\{\nu',\dots,\nu^{(\pi')}\}$ ordered by the sum $\nu'+\dots+\nu^{(\pi')}$ and lexicographically for those of equal sum. Take the initial segment consisting of the first $\kappa$ members. Now write down the list
$$\left.\begin{array}{l}\zeta'_1,\dots,\zeta^{(\pi)}_1,\\ \dots\dots\dots\dots\\ \zeta'_\kappa,\dots,\zeta^{(\pi)}_\kappa.\end{array}\right\} \tag{2}$$
where the restricted variables in the $\nu$th line are $x^{(\nu')},\dots,x^{(\nu^{(\pi')})}$, $\{\nu',\dots,\nu^{(\pi')}\}$ being the $\nu$th $\pi'$-tuplet in our list of $\pi'$-tuplets, and where the general variables in the first line are in order from left to right
$$x',\dots,x^{(\pi'')} \quad \text{if } \zeta' \text{ is general in (1)}$$
or
$$x'',\dots,x^{(S\pi'')} \quad \text{if } \zeta' \text{ is restricted in (1).}$$
Suppose that exactly the first $\lambda$ restricted variables in line $\theta'$, $\theta'<\theta$ are from left to right the same as in line $\theta$, then the general variables are the same from left to right in these two lines up to and including the general variable immediately following the $\lambda$th restricted variable. The remaining general variables in line $\theta$ are in order from left to right the next new variables in the alphabetical list $x, x', x'', \dots$ of variables. An example will make this clear. Let (1) be:
$$(Ex')\,(Ax'')\,(Ex''')\,(Ex^{\text{iv}})\,(Ax^{\text{v}})\,DNpx^{\text{iv}}x''x^{\text{v}}px'x'''x^{\text{iv}}, \tag{3}$$
where $p$ is a three-place predicate. Let $\kappa$ be 12. For greater clarity we write $x_\nu$ instead of $x'\dots{}'$ with $\nu$ superscript primes.

*H-scheme of order* 12

| r | g | r | r | g | line | triplet | sum | line | r | g | r | r | g |
|---|---|---|---|---|---|---|---|---|---|---|---|---|---|
| | | | | | | | | | variables | | | | |
| | | | $x_1$ | $x_3$ | 1 | [1, 1, 1] | 3 | 1 | $x_1$ | $x_2$ | $x_1$ | $x_1$ | $x_3$ |
| | | | $x_2$ | $x_4$ | 2 | [1, 1, 2] | | 2 | $x_1$ | $x_2$ | $x_1$ | $x_2$ | $x_4$ |
| | | $x_1$ | $x_3$ | $x_8$ | 5 | [1, 2, 1] | 4 | 3 | $x_1$ | $x_2$ | $x_2$ | $x_1$ | $x_5$ |
| | | | $x_4$ | $x_{15}$ | 11 | | | | | | | | |
| | | | $x_1$ | $x_5$ | 3 | [2, 1, 1] | | 4 | $x_2$ | $x_6$ | $x_1$ | $x_1$ | $x_7$ |
| | | | $x_2$ | $x_9$ | 6 | [1, 1, 3] | | 5 | $x_1$ | $x_2$ | $x_1$ | $x_3$ | $x_8$ |
| $x_1$ | $x_2$ | $x_2$ | $x_3$ | $x_{16}$ | 12 | [1, 2, 2] | | 6 | $x_1$ | $x_2$ | $x_2$ | $x_2$ | $x_9$ |
| | | $x_3$ | $x_1$ | $x_{10}$ | 7 | [1, 3, 1] | | 7 | $x_1$ | $x_2$ | $x_3$ | $x_1$ | $x_{10}$ |
| | | | $x_1$ | $x_7$ | 4 | [2, 1, 2] | 5 | 8 | $x_2$ | $x_6$ | $x_1$ | $x_2$ | $x_{11}$ |
| $x_2$ | $x_6$ | $x_1$ | $x_2$ | $x_{11}$ | 8 | [2, 2, 1] | | 9 | $x_2$ | $x_6$ | $x_2$ | $x_1$ | $x_{12}$ |
| | | $x_2$ | $x_1$ | $x_{12}$ | 9 | [3, 1, 1] | | 10 | $x_3$ | $x_{13}$ | $x_1$ | $x_1$ | $x_{14}$ |
| | | | | | | [1, 1, 4] | 6 | 11 | $x_1$ | $x_2$ | $x_1$ | $x_4$ | $x_{15}$ |
| $x_3$ | $x_{13}$ | $x_1$ | $x_1$ | $x_{14}$ | 10 | [1, 2, 3] | | 12 | $x_1$ | $x_2$ | $x_2$ | $x_3$ | $x_{16}$ |

The first twelve triplets have been written down in the prescribed order. In the column headed 'variables' the restricted variables occur in the first, third and fourth places and the general variables in the second and fifth places. The suffices of the restricted variables agree in order from left to right with the members of the ordered triplet in the same row. The general variables are then put in, $x_2$ and $x_3$ in the first line, $x_2$ and $x_4$ in the second line, since the first line begins with $x_1$ and is followed with $x_2$ and the second line begins with $x_1$ then the general variable in the second place in the second line is also $x_2$. Generally the second variable, which is a general variable, is $x_2$ whenever the first variable, which is a restricted variable, is $x_1$. In the fourth line the second variable (the first general variable in that line) is $x_6$ because this is the first available new variable in the alphabetical list of variables and this is the first time that $x_2$ has occurred in the first place. Generally whenever the first variable is $x_2$ then the second variable is $x_6$. The *H-scheme* is obtained by writing down line 1 followed by those lines whose initial segment is the same as in line 1 for as long as possible. Thus lines 1, 2, 5 and 11 agree in having initial segments $x_1 x_2 x_1$: These are followed by lines 3, 6 and 12 which agree in having initial segments $x_1 x_2 x_2$. This in turn is followed by line 7 which agrees with the above in having initial segments $x_1 x_2$. The agreement of initial segments is denoted by bracketing. The *H*-scheme of order 12 for (1) is the list (2) arranged by bracketing together lines with equal initial segments and ordering lexicographically within the brackets.

Write $\quad q\zeta_1\zeta_2\zeta_3\zeta_4\zeta_5 \quad$ for $\quad DNp\zeta_4\zeta_2\zeta_5 p\zeta_1\zeta_3\zeta_4$

$q_\theta \quad$ for $\quad q\zeta_1^{(\theta)}\zeta_2^{(\theta)}\zeta_3^{(\theta)}\zeta_4^{(\theta)}\zeta_5^{(\theta)}$

where $\zeta_1^{(\theta)}, \zeta_2^{(\theta)}, \zeta_3^{(\theta)}\zeta_4^{(\theta)}$ and $\zeta_5^{(\theta)}$ are variables in line $\theta$ of (2). We note that the *H-disjunction of order* 12 namely:

$$\sum_{\theta=1}^{12} q_\rho \tag{4}$$

is a tautology, in fact $Dq_1q_{12}$ is a tautology. $\sum_{\theta=1}^{11} q_\theta$ fails to be a tautology, because we can only have a tautology when $\zeta_4^{(\theta)}\zeta_2^{(\theta)}\zeta_5^{(\theta)}$ is the same as $\zeta_1^{(\theta')}\zeta_3^{(\theta')}\zeta_4^{(\theta')}$ and for $1 \leqslant \theta, \theta' \leqslant 12$ this only occurs when $\theta = 1$ and $\theta' = 12$. Thus the *H*-scheme of order 12 for the statement (3) makes the *H*-disjunction of order 12 a tautology.

Now consider

$$(Ex_1)(Ax_2)(Ex_3, x_4)(Ax_5)DNpx_5x_4x_1px_1x_2x_3. \tag{5}$$

Any *H*-scheme for the statement (5) fails to make an *H*-disjunction a tautology because $\zeta_5^{(\theta)}$ is alphabetically later than $\zeta_4^{(\theta)}$ while $\zeta_2^{(\theta')}$ is alphabetically later than $\zeta_1^{(\theta')}$, hence $\zeta_5^{(\theta)}\zeta_4^{(\theta)}\zeta_1^{(\theta)}$ fails to agree with $\zeta_1^{(\theta')}\zeta_2^{(\theta')}\zeta_3^{(\theta')}$ for any $\theta, \theta'$. Consider again the statement (5), an *H*-scheme of order $\kappa$ for (5) gives rise to an *H*-disjunction which fails to be a tautology for any numeral $\kappa$.

A disjunctand of the *H*-disjunction of (5) is

$$DNpx_{\rho[\nu_1,\nu_3,\nu_4]}x_{\nu_4}x_{\nu_1}px_{\nu_1}x_{\sigma[\nu_1]}x_{\nu_3}, \tag{6}$$

where $\sigma[\nu_1] > \nu_1$ and $\rho[\nu_1, \nu_3, \nu_4] > \nu_1, \nu_3, \nu_4$. Consider the 2-valued model $\mathscr{N}$ in which the individuals are the natural numbers. We can make (6) take the $\mathscr{N}$-value $\boldsymbol{f}$ by taking:

$$p\nu_1\nu_2\nu_3 = \boldsymbol{f} \quad \text{for} \quad \nu_1 < \nu_2 \quad \text{and} \quad p\nu_1\nu_2\nu_3 = \boldsymbol{t} \quad \text{for} \quad \nu_1 > \nu_2,$$

the values for $\nu_1 = \nu_2$ are immaterial. Consider the negation of (5)

$$(Ax_1)(Ex_2)(Ax_3, x_4)(Ex_5)Kpx_5x_4x_1Npx_1x_2x_3. \tag{7}$$

Thus we can obtain a *satisfaction* of (7) over the model $\mathscr{N}$.

We shall demonstrate later the general proposition that if the *H*-disjunctions of (1) all fail to be tautologies then there is a satisfaction of the negation of (1) over the 2-valued model in which the individuals are the natural numbers. Note that we lack a method for deciding whether there is a numeral $\kappa$ such that the $H$-disjunction of order $\kappa$ is a tautology. If we had such a method then the system $\mathscr{F}_C$ would be decidable, we show later on that the system $\mathscr{F}_C$ is undecidable.

From the tautology (4) we may obtain an $\mathscr{F}'_C$-proof in normal form of the statement (3). Apply universal quantification to the variables $x_5, x_7, x_8, x_9, x_{10}, x_{11}, x_{12}, x_{14}, x_{15}, x_{16}$ successively, these variables occur at one place only in (4) so the condition on variables in the rule for universal quantification is satisfied. Delete these variables from the $H$-scheme of order 12 this leaves:

| r | g | r | r | g | line |
|---|---|---|---|---|---|
| | | | $x_1$ | $x_3$ | 1 |
| | | $x_1$ | $x_2$ | $x_4$ | 2 |
| | | | $x_3$ | | 5 |
| | | | $x_4$ | | 11 |
| | | | $x_1$ | | 3 |
| $x_1$ | $x_2$ | $x_2$ | $x_2$ | | 6 |
| | | | $x_3$ | | 12 |
| | | $x_3$ | $x_1$ | | 7 |
| | | | $x_1$ | | 4 |
| | | $x_1$ | $x_2$ | | 8 |
| $x_2$ | $x_6$ | $x_2$ | $x_1$ | | 9 |
| $x_3$ | $x_{13}$ | $x_1$ | $x_1$ | | 10 |

The tautology (4) has become

$$\sum_{\theta=1}^{12} q'_\theta, \tag{4.1}$$

where $q'_\theta$ is $q_\theta$ if $\theta = 1, 2$ otherwise $q'_\theta$ is $(Ax_5)\, q\zeta_1^{(\theta)}\zeta_2^{(\theta)}\zeta_3^{(\theta)}\zeta_4^{(\theta)}x_5$, Now apply existential quantification to the fourth variable in every disjunction of (4.1) except $q_1$ and $q_2$ and delete those variables from the $H$-scheme of order 12. The disjunction (4.1) becomes

$$\sum_{\theta=1}^{12} q''_\theta, \tag{4.2}$$

where $q''_1 = q_1, q''_2 = q_2$ otherwise $q''$ is $(Ex_4)(Ax_5)\, q\zeta_1^{(\theta)}\zeta_2^{(\theta)}\zeta_3^{(\theta)}x_4x_5$. In (4.2) $q''_5$ is the same as $q''_{11}$ so cancel $q_{11}$, also $q''_3$, $q''_6$, $q''_{12}$ are the same so cancel $q''_6$ and $q''_{12}$, also $q''_8$ is the same as $q''_4$ so cancel $q''_8$. Thus (4.2) becomes

$$\underset{\text{7-times}}{D\ldots D}\, q_1 q_2 q''_5 q''_3 q''_7 q''_4 q''_9 q''_{10}. \tag{4.3}$$

The deleted $H$-scheme of order 12 has become:

| r | g | r | r | g | line |
|---|---|---|---|---|---|
| | | | $x_1$ | $x_3$ | 1 |
| | | $x_1$ | ~~$x_2$~~ | ~~$x_4$~~ | ~~5~~ |
| $x_1$ | $x_1$ | $x_2$ | | | 3 |
| | | $x_3$ | | | 7 |
| $x_2$ | $x_6$ | $x_1$ | | | 4 |
| | | $x_2$ | | | 9 |
| $x_3$ | $x_{13}$ | $x_1$ | | | 10 |

In the disjunction (4.3) the variable $x_4$ occurs only in $q_2$ hence we may apply universal quantification to it, we can then apply existential quantification to the fourth variable in $q_2$ this makes $q_2$ the same as $q_5''$ so cancel $q_5''$ and (4.3) becomes:

$$\underset{\text{6-times}}{D\ldots D} q_1 q_2'' q_3'' q_7'' q_4'' q_9'' q_{10}''. \tag{4.4}$$

Delete these variables from the $H$-scheme. This is indicated above by a stroke through them and through 5. Now apply existential quantifiers to the third variable in each disjunction of (4.4) except $q_1$ and cross these variables out of the $H$-scheme. The disjunction (4.4) becomes

$$\underset{\text{6-times}}{D\ldots D} q_1 q_2''' q_3''' q_7''' q_4''' q_9''' q_{10}''', \tag{4.5}$$

where $q_\lambda'''$ is

$$(Ex_3)(Ex_4)(Ax_5)\, q\zeta_1^{(\lambda)}\zeta_2^{(\lambda)}x_3x_4x_5, \quad \lambda = 2, 3, 4, 7, 9, 10.$$

In the disjunction (4.5) $q_2''', q_3''', q_7'''$ are the same, so are $q_4'''$ and $q_9'''$. Cancel duplicates and we obtain the disjunction

$$DDDq_1 q_2''' q_4''' q_{10}''' \tag{4.6}$$

and the deleted $H$-scheme:

| r | g | r | r | g | line |
|---|---|---|---|---|---|
| | | $x_1$ | $x_1$ | $x_3$ | 1 |
| $x_1$ | $x_2$ | | | | 2 |
| $x_2$ | $x_6$ | | | | 4 |
| $x_3$ | $x_{13}$ | | | | 10 |

In the disjunction (4.6) the variable $x_{13}$ occurs only in $q_{10}'''$ and the variable $x_6$ occurs only in $q_4'''$ hence we may apply universal quantification to them, we can then apply existential quantification to the first variables in $q_4'''$

and $q_{10}'''$. Cross out these variables from the deleted $H$-scheme. $q_4'''$ and $q_{10}'''$ have now become the same, so omit $q_{10}'''$. The disjunction (4.6) has become:

$$DDq_1 q_2'''(Ex_1)(Ax_2)(Ex_3, x_4)(Ax_5)\, qx_1x_2x_3x_4x_5,$$

and the deleted $H$-scheme is:

| r | g | r | r | g | line |
|---|---|---|---|---|---|
| $x_1$ | $x_2$ | [ $x_1$ | $x_1$ | $x_3$ | 1 |
| | | ⌊ | | | 2 |
| | | | | | 4 |

The variable $x_3$ occurs free only in $q_1$ so we may apply universal quantification to it, we can then apply existential quantification to the third and fourth variables in $q_1$. This makes $q_1$ the same as $q_2'''$, so cancel $q_2'''$. We then obtain the disjunction

$$D(Ex_3, x_4)(Ax_5)\, q\zeta_1^{(1)}\zeta_2^{(1)}x_3x_4x_5(Ex_1)(Ax_2)(Ex_3, x_4)(Ax_5)\, qx_1x_2x_3x_4x_5. \tag{4.7}$$

In the disjunction (4.7) the variable $x_2$ occurs free only in the first disjunctand, so we may apply universal quantification to it, we can then apply existential quantification to the first variable. This makes the disjunctands the same, cancel one of them, and we are left with (3).

The $H$-scheme for (1) of order $\kappa$ can be written down on a fixed plan for any numeral $\kappa$, hence the place number of a general variable is uniquely determined by the place number of the superior restricted variables. Thus for the statement (3) the place number of the second variable is uniquely determined by the place number of the first variable, and the place number of the fifth variable is uniquely determined by the triplet of the place numbers of the three superior restricted variables.

PROP. 10. *If for some numeral $\kappa$ the H-disjunction of order $\kappa$ of a closed $\mathscr{F}'_C$-statement $\phi$ in prenex normal form is a tautology then $\phi$ is an $\mathscr{F}'_C$-theorem.*

The method of demonstration is the same as that given in the worked example. We apply quantifications to the various disjunctands of the $H$-disjunction of order $\kappa$ and delete the corresponding variables from the $H$-scheme, and cancel duplicates as they occur. At any stage in the proceedings a variable in the $H$-scheme is *available* if it is at the end of its line in a deleted $H$-scheme and is a restricted variable or is a similarly situated general variable which fails to occur elsewhere in the deleted $H$-scheme. If there is always an available variable until all the variables are deleted from the $H$-scheme then we obtain an $\mathscr{F}'_C$-proof of $\phi$. Suppose that at some stage there fails to be an available variable, then in the deleted $H$-scheme at that stage the variable at the end of each line is a general variable and each such variable occurs elsewhere in the deleted $H$-scheme. The lines in the deleted $H$-scheme are always distinct because identical lines get deleted as soon as they arise by cancelling duplicates. A variable can only occur once as a general variable in a deleted $H$-scheme but it can occur again as a restricted variable. For instance the variable $x_2$ occurs once as a general variable and six times as a restricted variable in the complete $H$-scheme of order 12 for the statement (3). The restricted variables which precede a general variable in a line of an $H$-scheme are alphabetically earlier variables. Thus if there fails to be an available variable then each general variable $\xi$ at the end of a line occurs again as a restricted variable in another line which ends in an alphabetically later general variable $\eta$. In turn there is another general variable alphabetically later than $\eta$ and so on without end. This is absurd because the $H$-scheme of order $\kappa$ is displayed. Thus there is always an available variable and we may continue to quantify and remove duplicates until we obtain an $\mathscr{F}'_C$-proof of $\phi$. This demonstrates the proposition.

PROP. 11. *If $\phi$ is an $\mathscr{F}'_C$-theorem in prenex normal form then there is a numeral $\kappa$ such that the H-disjunction of order $\kappa$ is a tautology.*

According to Prop. 9 the $\mathscr{F}'_C$-proof of $\phi$ can be modified to one in normal form. We then have a tautology

$$\sum_{\theta=1}^{\nu} \psi^{(\theta)}, \tag{8}$$

where each $\psi^{(\lambda)}$ differs from $\psi\{\zeta', \ldots, \zeta^{(\pi)}\}$ by change of individual variables. From (8) we can obtain $\phi$ by I$a$, $b$, II$d$, $d'$. Let $\mathscr{F}'^*_C$ be the same as $\mathscr{F}'_C$ except that the individual variables are $x^*, x^{*\prime}, x^{*\prime\prime}, \ldots$, and let $\mathscr{F}'_C \cup \mathscr{F}'^*_C$ be the same as the system $\mathscr{F}'_C$ except that the variables are those of $\mathscr{F}'_C$ and those of $\mathscr{F}'^*_C$. We now change the individual variables in (8) to those of $\mathscr{F}'^*_C$ by superscripting an asterisk to each variable. In this way let $\psi^{(\lambda)}$ become $\psi^{(\lambda)*}$ and (8) become (8*). We will give a method of changing the individual variables in (8*) to $\mathscr{F}'_C$ variables in such a way that (8*) is changed into part of an $H$-disjunction. Let (8*) become (9) by this change.

Clearly if we change an individual variable at all its occurrences to another one then a tautology remains a tautology. Thus (9) will be a tautology and part of an $H$-disjunction. Thus there will be a numeral $\kappa$ such that the $H$-disjunction of order of $\kappa$ is a tautology. Let $\phi$ be

$$(A\zeta')\dots(A\zeta^{(\mu)})(E\zeta^{(S\mu)})(Q^{(SS\mu)}\zeta^{(SS\mu)})\dots(Q^{(\pi)}\zeta^{(\pi)})\,\psi\{\zeta',\dots,\zeta^{(\pi)}\},$$

where if $\mu$ is zero the initial set of universal quantifiers is absent. Let the variables in $\psi'^*,\dots,\psi^{(\nu)*}$ be

$$\left.\begin{array}{c}\zeta_1'^*,\dots,\zeta_1^{(\pi)*},\\ \dots\dots\dots\dots\\ \zeta_\nu'^*,\dots,\zeta_\nu^{(\pi)*}.\end{array}\right\} \qquad (10)$$

We now replace the $\mathscr{F}_C'^*$-variables by $\mathscr{F}'_C$-variables in (10) from left to right. We first replace through (8*), (10) the first $\mu$ variables in each line of (10) by $x',\dots,x^{(\mu)}$ respectively, that is $\zeta_1'^*,\dots,\zeta_\nu'^*$ are replaced by $x'$ at all their occurrences, …, $\zeta_1^{(\mu)*},\dots,\zeta_\nu^{(\mu)*}$ are all replaced by $x^{(\mu)}$ at all their occurrences. Let (8*) then become (8′), it is an $\mathscr{F}'_C\cup\mathscr{F}_C'^*$-statement, clearly it is a tautology and we can obtain (8) from it. This is because all the other general variables in (8*) are distinct from $x',\dots,x^{(\mu)}$. By this change (10) becomes (10′). Secondly if the lines $\nu'$ and $\nu''$ of (10) agree in having the same initial segment and if the next variable is different and is a general variable then we may alter this general variable in one of the lines so that they are both the same. Suppose these general variables are $\zeta_{\nu'}^{(S\lambda)*}$ and $\zeta_{\nu''}^{(S\lambda)*}$ so that the segments $\zeta_{\nu'}'^*,\dots,\zeta_{\nu'}^{(\lambda)*}$ and $\zeta_{\nu''}'^*,\dots,\zeta_{\nu''}^{(\lambda)*}$ are the same. In passing from (8′) to $\phi$ we shall at some stage generalize $\zeta_{\nu'}^{(S\lambda)*}$ and at another stage we shall generalize $\zeta_{\nu''}^{(S\lambda)*}$. Suppose that we generalize $\zeta_{\nu'}^{(S\lambda)*}$ before we generalize $\zeta_{\nu''}^{(S\lambda)*}$. We can modify the order of quantifying the individual variables in (8′) so that we generalize $\zeta_{\nu''}^{(S\lambda)*}$ immediately after (except for permutations) generalizing $\zeta_{\nu'}^{(S\lambda)*}$.

This follows because when we are about to generalize $\zeta_{\nu'}^{(S\lambda)*}$ every as yet unquantified variable will be distinct from $\zeta_{\nu'}^{(S\lambda)*}$ so that any variable in lines other than line $\nu'$ which is quantified between the generalizations of $\zeta_{\nu'}^{(S\lambda)*}$ and $\zeta_{\nu''}^{(S\lambda)*}$ could have been quantified before the generalization of $\zeta_{\nu'}^{(S\lambda)*}$. Any quantifications on variables in line $\nu'$ which occur between the generalizations of $\zeta_{\nu'}^{(S\lambda)*}$ and $\zeta_{\nu''}^{(S\lambda)*}$ (which must be restrictions, because the initial segments of the two lines are the same) can take place immediately after the generalization of $\zeta_{\nu''}^{(S\lambda)*}$, any such variable is distinct from $\zeta_{\nu''}^{(S\lambda)*}$, again because the initial segments are the same.

Having made these modifications so that we generalize $\zeta_{\nu''}^{(S\lambda)*}$ immediately after (except for permutations) generalizing $\zeta_{\nu'}^{(S\lambda)*}$ we now everywhere replace $\zeta_{\nu''}^{(S\lambda)*}$ by $\zeta_{\nu'}^{(S\lambda)*}$ and before generalizing $\zeta_{\nu'}^{(S\lambda)*}$ we cancel line $\nu''$ which is then the same as line $\nu'$. Similarly if several lines have the same initial segments. We then continue as before. The replacement of $\zeta_{\nu''}^{(S\lambda)*}$ everywhere by $\zeta_{\nu'}^{(S\lambda)*}$ might make a restricted variable, which was originally distinct from $\zeta_{\nu'}^{(S\lambda)*}$, the same as $\zeta_{\nu'}^{(S\lambda)*}$. If this occurs and if this variable was originally restricted before $\zeta_{\nu'}^{(S\lambda)*}$ was generalized then restrict it as before. If this variable is restricted between the generalizations of $\zeta_{\nu'}^{(S\lambda)*}$ and $\zeta_{\nu''}^{(S\lambda)*}$ (it is impossible for it to be restricted after the generalization of $\zeta_{\nu''}^{(S\lambda)*}$) then, as above, the restriction can take place before the generalization of $\zeta_{\nu'}^{(S\lambda)*}$, this variable fails to occur in line $\nu'$, for if it did then it would occur in line $\nu''$, since the initial segments up to $\lambda$ variables are the same in the two lines, and a general variable fails to be the same as a superior variable. Make these modifications as long as possible. Now suppose that we have replaced $\mathscr{F}'^{*}_{C}$-individual variables from left to right by $\mathscr{F}'_{C}$-individual variables up to a certain point. We call an $\mathscr{F}'^{*}_{C}$-individual variable *available* if its superior variables are $\mathscr{F}'_{C}$-individual variables. If an available variable is general then we replace it at all its occurrences by that $\mathscr{F}'_{C}$-individual variable $\zeta$ which agrees with the numbering in the $H$-scheme of $\phi$ where a general variable is determined by its superior restricted variables.

LEMMA. *If each available variable is restricted then one of them is different from each general variable.*

Suppose that for each line $\theta$ the available variable $\zeta_\theta$ is restricted and is the same as a general variable in line $\rho[\theta]$. Then $\rho[\theta] \neq \theta$ because this general variable is an $\mathscr{F}'^{*}_{C}$-variable (being the same as an available variable and all available variables are $\mathscr{F}'^{*}_{C}$-variables), if it is in line $\theta$ then it is inferior to $\zeta_\theta$, but a general variable is different from all its superior variables. (We quantify from right to left, but we are renaming the $\mathscr{F}'^{*}_{C}$-variables from left to right.) This general variable $\zeta_\theta$ in line $\rho[\theta]$ is inferior to the available variable in that line, this variable, say $\zeta_{\rho[\theta]}$, is restricted, by hypothesis, and can only be quantified after the general variable $\zeta_\theta$ in that line has been quantified. By hypothesis the restricted variable $\zeta_{\rho[\theta]}$ is the same as a general variable in line $\rho[\rho[\theta]]$, say $\rho^2[\theta]$, and so on.

Since the scheme (10) is displayed we must have

$$\alpha = \rho^{\mu'}[\theta] = \rho^{\mu''}[\theta],$$

for some $\mu' < \mu''$.

Thus we have a general variable in line $\rho^{\mu''}[\theta]$ is equal to a restricted variable (which is available) in line $\rho^{\mu''\dot{-}1}[\theta]$ and so it is impossible to generalize the former until the latter has been restricted, but this restricted variable is inferior to a general variable which must be generalized first, this in turn is the same as a restricted (available) variable in line $\rho^{\mu''\dot{-}2}[\theta]$ which must be restricted first, and so on until, a general variable in line $\rho^{S\mu'}[\theta]$ is the same as a restricted (available) variable in line $\rho^{\mu'}[\theta]$, but this variable is $\zeta_\alpha$ itself. Thus finally, we are to restrict a restricted variable $\zeta_\alpha$ in line $\alpha$ before we generalize an inferior general variable. This is absurd. Thus there is always an available restricted variable distinct from any general variable.

Replace the alphabetically earliest such restricted variable (which is an $\mathscr{F}'^*_C$-variable) at all its occurrences by the first as yet unused $\mathscr{F}'_C$-variable. This leaves the general variables unaffected. Thus an available variable (which is an $\mathscr{F}'^*_C$-variable) can always be replaced by an $\mathscr{F}'_C$-variable without upsetting our build-up of an $H$-scheme by renaming of general variables. Finally each $\mathscr{F}'^*_C$-variable is replaced by an $\mathscr{F}'_C$-variable in such a way that the resulting disjunction is a part disjunction of an $H$-disjunction, because we have chosen the general variables so that this should be so. This completes the demonstration of Prop. 11.

## 3.11 *Validity and satisfaction*

An $\mathscr{F}_C$-statement $\phi$ is called *generally valid over $\mathscr{N}$* when (vii) below has been demonstrated:

(i) We replace a part $(E\xi)\,\psi\{\xi\}$ of $\phi$ by $N(A\xi)\,N\phi\{\xi\}$.

(ii) (*a*) If $\pi$ is a propositional variable which occurs in $\phi$ then we replace it by ***t*** or by ***f***.

(*b*) If $\pi$ is a one-place predicate variable which occurs in $\phi$ then we replace each of $\pi\nu$ $\nu = 0, 1, 2, \ldots$ either by ***t*** or by ***f***.

(*c*) If $\pi$ is a two-place predicate variable which occurs in $\phi$ then we replace each of $\pi\nu\kappa$ $\nu, \kappa = 0, 1, 2, \ldots,$ either by ***t*** or by ***f***.

(*d*) similarly for many-place predicate variables.

(iii) We replace the free individual variables by numerals.

(iv) We replace a part $(A\xi)\,\psi\{\xi\}$ of $\phi$ by $\boldsymbol{t}$ if and only if $\psi\{\nu\}$ is replaced by $\boldsymbol{t}$ for $\nu = 0, 1, 2, \ldots$ otherwise we replace $(A\xi)\,\psi\{\xi\}$ by $\boldsymbol{f}$.

(v) We replace $\boldsymbol{Dff}$ by $\boldsymbol{f}$ and $\boldsymbol{Dft}$, $\boldsymbol{Dtf}$, $\boldsymbol{Dtt}$ by $\boldsymbol{t}$.

(vi) We replace $\boldsymbol{Nt}$ by $\boldsymbol{f}$ and $\boldsymbol{Nf}$ by $\boldsymbol{t}$.

(vii) $\phi$ reduces to $\boldsymbol{t}$ however the replacements (ii) (iii) are carried out.

We lack a test for general validity. This will be demonstrated in Ch. 7.

A closed $\mathscr{F}_C$-statement is said to be *satisfiable over* $\mathscr{N}$ if it can be shown to reduce to $\boldsymbol{t}$ for at least one replacement under (ii), (iii). For example consider:

$$DN(A\xi)\,DN\psi\{\xi\}\,\chi\{\xi\}\,DN(A\xi)\,\psi\{\xi\}\,(A\xi)\,\chi\{\xi\}. \tag{11}$$

We show that it is impossible for (11) to reduce to $\boldsymbol{f}$ when the above process is carried out. If (11) reduces to $\boldsymbol{f}$ then $N(A\xi)\,DN\psi\{\xi\}\,\chi\{\xi\}$ and $DN(A\xi)\,\psi\{\xi\}\,(A\xi)\,\chi\{\xi\}$ must both reduce to $\boldsymbol{f}$. In order that this happen $DN\psi\{\nu\}\,\chi\{\nu\}$ and $\psi\{\nu\}$ must both reduce to $\boldsymbol{t}$ for each numeral $\nu$, and $\chi\{\nu\}$ must reduce to $\boldsymbol{f}$ for at least one numeral, say $\kappa$. Then $DN\psi\{\kappa\}\,\chi\{\kappa\}$ reduces to $\boldsymbol{t}$ hence $\psi\{\kappa\}$ must reduce to $\boldsymbol{f}$, this is absurd. Thus (11) always reduces to $\boldsymbol{t}$ no matter how the replacements are carried out.

Consider (3), give $\boldsymbol{t}, \boldsymbol{f}$ to $p\lambda\mu\nu$ in any manner, if $p\nu'\nu''\nu'''$ is $\boldsymbol{t}$ for some set $\nu', \nu'', \nu'''$ of numerals then (3) is $\boldsymbol{t}$, if $p\nu'\nu''\nu'''$ is always $\boldsymbol{f}$ then (3) is $\boldsymbol{t}$.

Prop. 12. *A closed $\mathscr{F}_C$-statement is an $\mathscr{F}_C$-theorem if and only if it is generally valid over $\mathscr{N}$.*

The $\mathscr{F}_C$-axioms are generally valid over $\mathscr{N}$. The $\mathscr{F}_C$-rules preserve general validity over $\mathscr{N}$. Thus $\mathscr{F}_C$-theorems are generally valid over $\mathscr{N}$.

Now suppose that the $\mathscr{F}_C$-statement $\phi$ is generally valid over $\mathscr{N}$. We show that an $\mathscr{F}_C$-statement $\phi$ is either an $\mathscr{F}_C$-theorem or its negation is satisfiable over $\mathscr{N}$. Hence if $\phi$ is generally valid over $\mathscr{N}$ then its negation fails to be satisfied over $\mathscr{N}$ and so by the alternative $\phi$ is an $\mathscr{F}_C$-theorem.

Consider the $H$-disjunctions of $\phi$ of orders 1, 2, 3, ... if one of them is a tautology then $\phi$ is an $\mathscr{F}_C$-theorem. Suppose that each $H$-disjunction of $\phi$ fails to be a tautology, we will show that the negation of $\phi$ is satisfiable over $\mathscr{N}$.

In the $H$-disjunctions of $\phi$ replace $x^{(\nu)}$ by $\nu$. Each $H$-disjunction takes the value $f$ for some assignment of values $t, f$ to each $\pi\nu' \ldots \nu^{(\lambda)}$ for each predicate variable which occurs in $\phi$ and for each set of arguments which occurs in an $H$-disjunction. If a set of arguments fails to occur in any $H$-disjunction then we give $\pi\nu'\ldots\nu^{(\lambda)}$ the value $t$. Let $\mathscr{V}_\kappa$ be an assignment of values $t, f$ to each $\pi\nu'\ldots\nu^{(\lambda)}$ which occurs in $H_\kappa$, the $H$-disjunction of order $\kappa$. $\mathscr{V}_\kappa$ will only give values to $\pi\nu'\ldots\nu^{(\lambda)}$ for those argument sets $\nu'\ldots\nu^{(\lambda)}$ which occur in $H_\kappa$. Let $\mathscr{M}_\kappa$ be the set of assignments $\mathscr{V}_\kappa$. Now $H_\kappa$ is a part disjunction of $H_\theta$ for $\kappa < \theta$, hence one $\mathscr{V}_\theta$ will contain all the values given by at least one $\mathscr{V}_\kappa$. We can express this by saying that at least one $\mathscr{V}_\kappa$ can be extended to become a $\mathscr{V}_\theta$ or that a $\mathscr{V}_\theta$ with domain of definition restricted to that of $\mathscr{V}_\kappa$ becomes a $\mathscr{V}_\kappa$. Each $\mathscr{M}_\kappa$ contains at least one $\mathscr{V}_\kappa$. Hence there is a valuation $\mathscr{V}$ which defines $\pi\nu'\ldots\nu^{(\lambda)}$ for each argument set which occurs in some $H_\kappa$ and which gives the value $f$ to $\phi$. Now this valuation says that for any values given to the restricted variables there are values that can be given to the general variables, which values depend on the values given to the superior restricted variables, in such a manner that $\phi$ takes the value $f$. Then $N\phi$ takes the value $t$ and for any values given to the general variables in $N\phi$ there are values which can be given to the restricted variables of $N\phi$, which values depend on the values given to the superior general variables, in such a way that $N\phi$ takes the value $t$. But this is to say that $N\phi$ is satisfiable over $\mathscr{N}$, and we have finished. This is called the *denumerable model.*

COR (i). *An $\mathscr{F}_C$-theorem is valid over $\mathscr{N}_\kappa$* (the set of natural numbers $\leqslant \kappa$). The $\mathscr{F}_C$-axioms are valid over $\mathscr{N}_\kappa$ for any $\kappa$ and it is easily verified that the $\mathscr{F}_C$-rules preserve validity over $\mathscr{N}_\kappa$ so the result follows. Validity over $\mathscr{N}_\kappa$ is decidable, we need only replace $(E\xi)\,\phi\{\xi\}$ by $\sum_{\theta=0}^{\kappa} \phi\{\theta\}$ (and $(A\xi)\,\phi\{\xi\}$ by $\prod_{\theta=0}^{\kappa} \phi\{\theta\}$) replace $\phi\{\theta\}$ by a propositional variable $p^{(\theta)}$ and evaluate by truth-tables.

COR. (ii). *An $\mathscr{F}_C$-statement which is valid over $\mathscr{N}_\kappa$ for each natural number $\kappa$ but which fails to be valid over $\mathscr{N}$ can be found.*

Consider the conjunction $P$ of the following $\mathscr{F}_C$-statements:

$$(Ax)\,Npxx$$
$$(Ax, x', x'')\,CKpxx'px'x''pxx''$$
$$(Ax)\,(Ex')\,pxx'.$$

It is clear that $P$ fails to be satisfiable over any $\mathscr{N}_\kappa$. Hence $NP$ is valid over every $\mathscr{N}_\kappa$. But $P$ is satisfiable over $\mathscr{N}$ (let $pxx'$ be $x < x'$), hence $NP$ fails to be valid over $\mathscr{N}$.

Another example is the negation of the conjunction $Q$ of the following $\mathscr{F}_C$-statements:

$$(Ex)\,(Ax')\,Npx'x$$
$$(Ax, x', x'', x''')\,CKKpxx''px'x''px'''xpx'''x'$$
$$(Ax)\,(Ex')\,pxx'.$$

Again it is clear that $Q$ fails to be satisfiable over any $\mathscr{N}_\kappa$ but is satisfiable over $\mathscr{N}$ (let $pxx'$ be $Sx = x'$).

## 3.12 *Independence*

Prop. 13. *The symbols, axioms and rules of $\mathscr{F}_C$ are independent.*

We have to show that $N$, $D$, $E$, $p^{(\lambda)}$, $x^{(\lambda)}$ are independent. Clearly $p$ is independent otherwise $\mathscr{F}_C$-theorems would be without occurrences of $p$, similarly for $x$ and the other variables. But note that if we omit $p$ we get an equivalent system, similarly for $x$ and the other variables.

The only closed $\mathscr{F}_C$-formulae of type $oo$ formed from $D$, $E$, $\boldsymbol{\lambda}$ and variables are:

$$\boldsymbol{\lambda}p.p,\quad \boldsymbol{\lambda}p.E(\boldsymbol{\lambda}x.p),\quad \boldsymbol{\lambda}p.E(\boldsymbol{\lambda}x.DpE(\boldsymbol{\lambda}x'.p)),\quad \boldsymbol{\lambda}p.Dpp,\quad \boldsymbol{\lambda}p.DpE(\boldsymbol{\lambda}x.p),\quad \text{etc.}$$

but these all using $\lambda$-rule (i) give $B\Delta\phi\phi$, where $\Delta$ stands for any one of the above formula. Hence if we took $\Delta$ as a definition of $N$ then $BN\phi\phi$ whence

$$\cfrac{\cfrac{DpNp}{Dpp}\,*}{p}\ Ib$$

and this is absurd because an $\mathscr{F}_C$-theorem must contain $D$. Thus the symbol $N$ is independent.

The only closed $\mathscr{F}_C$-formulae of type $ooo$ without occurrences of $D$ are: $\boldsymbol{\lambda}pp'.\phi$ where $\phi$ can only contain the variables $p$ and $p'$ otherwise it fails to be closed. The only such $\phi$ are $p, p'$, $Np, Np'$ or equivalents such as $NNp$, $(Ex)p$, etc. Thus if $D$ were definable in terms of the other symbols of $\mathscr{F}_C$ then $Dpp'$ would be $p$, $p'$, $Np$ or $Np'$ whence $DNpp$ would be $Np$, $p$ or $NNp$, this is absurd.

$E$ is independent because the only $\mathscr{F}_C$-formula of type $o(o\iota)$ that we can construct from $N$, $D$ and variables is $\lambda p_{o\iota} . \phi$, where $\phi$ is of type $o$ and fails to contain $E$, but this fails to be closed and so violates the conditions for a definition.

The demonstration that the $\mathscr{F}_C$-axioms are independent is the same as for $\mathscr{P}_C$.

The $\mathscr{F}_C$-rules are independent. First, rule I$a$ is independent, because if we omit rule I$a$ then we are unable to obtain the $\mathscr{F}_C$-theorem $DNpp$. Any $\mathscr{F}_C$-proof of $DNpp$ fails to use II$d, e$, because once $E$ enters an $\mathscr{F}_C$-proof then it remains in that $\mathscr{F}_C$-proof from that place till the base. Thus any $\mathscr{F}_C$-proof of $DNpp$ will be a $\mathscr{P}_C$-proof possibly using I$b$. Any $\mathscr{F}_C$-proof of $DNpp$ will fail to use II$b, c$ because it is without occurrence of $NN$ or of $ND$, hence an $\mathscr{F}_C$-proof of $DNpp$ which omits I$a$ proceeds from the axiom $DpNp$ using I$b$, II$a$ only. We are unable to apply I$b$ to $DpNp$ so we can only apply II$a$ obtaining $D\phi DpNp$ where $\phi$ is an $\mathscr{F}_C$-statement built up from $D$, $N$ and $p$ only, other variables and $E$ must be absent because if they were introduced into the $\mathscr{P}_C$-proof by II$a$ then they would remain in the $\mathscr{P}_C$-proof from that place to the base. We can only use I$b$ on $D\phi DpNp$ if $\phi$ is $DpNp$ in which case we get back to where we started, or if $\phi$ is of the form $D\psi\psi$ in which case we get $D\psi DpNp$ so that we might have originally diluted with $\psi$. In any case we get something longer. And so on, starting with $DpNp$ and using only I$b$, II$a$ we get longer and longer $\mathscr{F}_C$-statements thus we never get $DNpp$.

Similar considerations apply to II$a$, $b$, $c$ using the examples chosen in the case of the independence of these rules in $\mathscr{P}_C$. Note that the demonstration that I$b$ is a derived rule in $\mathscr{P}_C$ requires use of all $\mathscr{P}_C$-rules whence if we omit a $\mathscr{P}_C$-rule we are denied use of the dependence of I$b$ on the other $\mathscr{P}_C$-rules. The $\mathscr{F}_C$-rules II$d$, $e$ are independent because they are the only means of obtaining $\mathscr{F}_C$-theorems starting with $DE$, $DNE$ respectively, or $E$, $NE$ respectively (when the rule is used without subsidiary formula). Clearly there are such $\mathscr{F}_C$-theorems, for example:

$$D(Ex)pxNpx \quad \text{and} \quad DN(Ex)px(Ex)px.$$

For the rule I$b$ we note that if we are denied the use of rule I$b$ then we are unable to obtain:

$$(Ex')(Ax'')(Ex''')(Ex^{\text{iv}})(Ax^{\text{v}})DNpx^{\text{iv}}x''x^{\text{v}}px'x'''x^{\text{iv}}.$$

Any $\mathscr{F}'_C$-proof of (3) which fails to use I$b$ also fails to use II$a$, $b$, $c$ (these lengthen the formula) and must start with the axiom $px'x''x'''Nx'x''x'''$ it must then use I$a$, II$e$, II$d$ twice II$e$ and lastly II$d$. But it is impossible to generalize only the first occurrence from the left of $x'''$.

## 3.13 *Consistency*

PROP. 14. *The system $\mathscr{F}_C$ is model consistent.*

We show that $\mathscr{F}_C$ has a model with a sole individual $\alpha$ and two elements $\boldsymbol{t}, \boldsymbol{f}$ of which $\boldsymbol{t}$ is designated and $\boldsymbol{f}$ is undesignated. It has the constants $N$, $D$, $E$. $N$ and $D$ obey the same rules as in the model $\mathscr{M}_C$ for $\mathscr{P}_C$. The rule for $E$ is

$$\frac{\phi\{E(\boldsymbol{\lambda}\xi.\psi\{\xi\})}{\phi\{\psi\{\alpha\}\}}.$$

It is easily verified that these rules give a model for $\mathscr{F}_C$.

PROP. 15. *The system $\mathscr{F}_C$ is consistent with respect to negation.*

We have to show that if $\phi$ is a closed $\mathscr{F}_C$-statement then at least one of $\phi$ or $N\phi$ fails to be an $\mathscr{F}_C$-theorem. Suppose that both $\phi$ and $N\phi$ are $\mathscr{F}_C$-theorems, then so are $\phi$ and $DN\phi\pi$ where $\pi$ is a variable whence by Modus Ponens (which may be eliminated) so is $\pi$. But this is impossible, because an $\mathscr{F}_C$-theorem must contain at least four symbols. Each $\mathscr{F}_C$-axiom contains four symbols and applications of the rules except I$b$ either increases the length of a formula or leaves it of the same length. But I$b$ only eliminates duplicates, we start off with $DN\pi\pi$ and by I$b$ we can only eliminate duplicates of $N\pi$ and $\pi$ thus we are certainly left with $D$, $N\pi$ and $\pi$, i.e. four symbols at least.

In Chs. 6, 7 we study a formal system of arithmetic which we call the system $\mathbf{A}_0$. We show that the system $\mathbf{A}_0$ has an unsolvable decision problem even though it is complete. We then show that $\mathscr{F}_C$ has an unsolvable decision problem by effectively showing how a decision procedure for $\mathscr{F}_C$ would yield one for $\mathbf{A}_0$.

## 3.14 *$\mathscr{F}_C$ with functors*

The system $\mathscr{F}_C$ with functors or constant individuals or both can be dealt with as the system $\mathscr{F}_C$. We have the additional rule:

II$f$ $\quad \dfrac{D\phi\{\eta\}\,\omega}{D\phi\{\alpha\}\,\omega}$ $\eta$ free in $\phi\{\eta\}$, absent from $\omega$, $\phi\{\Gamma_\iota\}$.

This is called the *rule of substitution*. Here $\alpha$ is a term of type $\iota$ and is free in $\phi\{\alpha\}$, i.e. if $\xi$ is a variable which is free in $\alpha$ then corresponding occurrences of $\xi$ are free in $\phi\{\alpha\}$. This rule could always be applied to the axioms before starting a proof, so the rule could be dispensed with because the altered axioms are again axioms.

An $\mathscr{F}_C$ with functors and/or constant individuals can be reduced to an $\mathscr{F}_C$ without functors or constant individuals merely by change of notation.

$$p_{f_0 g_{01} h_{f20}} \quad \text{for} \quad \lambda xx'x''.pfxgxx'hfx''x,$$

where $f$ is of type $\iota\iota$, $g$ is of type $\iota\iota\iota$ and $h$ is of type $\iota\iota\iota$ so that $p_{f_0 g_{01} h_{f20}}$ is of type $o\iota\iota\iota$. Similarly $p_{af_0 g_{0a}}$ for $\lambda x.pafxgxa$

where $f$ is of type $\iota\iota$ and $g$ is of type $\iota\iota\iota$ and $a$ is of type $\iota$, so that $p_{af_0 g_{0a}}$ is of type $o\iota$. The rule II$f$ would then amount to a rule for changing predicates, but this could always be done in the axioms before we began.

## 3.15 *Theories*

*A theory $\mathscr{T}$ based on $\mathscr{F}_C$* is an applied predicate calculus in which certain statements are specified as axioms, it may have some extra rules. Suppose that the $\mathscr{T}$-axioms can be displayed, let $\phi$ be their conjunction, and let $\bar{\phi}$ be the closure of $\phi$. If $\psi$ is a $\mathscr{T}$-theorem by the deduction theorem Prop. 3, Cor. (ii) we obtain the $\mathscr{F}_C$-theorem $D\bar{\phi}\psi$. Suppose that $D\omega\psi$ and $DN\psi\chi$ are $\mathscr{T}$-theorems then $C\bar{\phi}D\omega\psi$ and $C\bar{\phi}DN\psi\chi$ are $\mathscr{F}_C$-theorems whence so are $DDN\bar{\phi}\omega\psi$ and $DN\psi DN\bar{\phi}\chi$, whence by Modus Ponens $DDN\bar{\phi}\omega DN\bar{\phi}\chi$, but Modus Ponens can be eliminated from $\mathscr{F}_C$, thus $DDN\bar{\phi}\omega DN\bar{\phi}\chi$ is an $\mathscr{F}_C$-theorem, hence by I$b$ $DN\bar{\phi}D\omega\chi$, i.e. $C\bar{\phi}D\omega\chi$ is an $\mathscr{F}_C$-theorem.

Now suppose that $\phi$ is $\prod_{\theta=1}^{\kappa} \phi^{(\theta)}$ where $\phi^{(\theta)}, 1 \leqslant \theta \leqslant \kappa$, are disjunctions of atomic statements or negations of atomic statements, from the $\mathscr{F}_C$-theorem $C\bar{\phi}D\omega\chi$ we obtain by Prop. 3, Cor. (iv) $\phi', \ldots, \phi^{(\kappa)} \vdash_{\mathscr{F}'_C} D\omega\chi$, thus $D\omega\chi$ is a $\mathscr{T}$-theorem.

Prop. 3, Cor. (v). *Modus Ponens can be eliminated from a theory whose axioms are disjunctions of atomic statements or negations of atomic statements.*

Suppose that we have $\phi', \ldots, \phi^{(\kappa)} \vdash_{\mathscr{F}'_C} D\omega\psi$ and $\phi', \ldots, \phi^{(\kappa)} \vdash_{\mathscr{F}'_C} DN\psi\chi$ then we have $C\bar{\phi}D\omega\chi$ as above whence we have $\phi', \ldots, \phi^{(\kappa)} \vdash_{\mathscr{F}'_C} D\omega\chi$, by Prop. 3,

Cor. (iv). A theory whose axioms contain free variables would normally be based on $\mathscr{F}'_C$ rather than on $\mathscr{F}_C$. The rule for substitution of variables merely converts an axiom scheme (as used in $\mathscr{F}_C$) into a set of particular axioms. A theory whose axioms are disjunctions of atomic statements or negations of atomic statements is called a *theory in free disjunctive* form, free variables are allowed.

Cor. (vi). *If* $\dfrac{D\phi\omega}{D\psi\omega}$ *is a rule in a theory in free disjunctive form then* $\dfrac{DN\psi\omega}{DN\phi\omega}*$, *where* $\omega$ *is subsidiary.*

We have $\dfrac{D\phi N\phi}{D\psi N\phi}$ taking $N\phi$ for $\omega$. Thus $C\phi\psi$, now $CC\phi\psi CDN\psi\omega DN\phi\omega$ is a $\mathscr{P}_C$-theorem, whence by Modus Ponens twice we get $DN\phi\omega$ if we have $DN\psi\omega$. But Modus Ponens can be eliminated.

Cor. (vii). *The deduction theorem holds in a theory whose special rules are without restrictions on variables.*

The demonstration is the same as before, the extra rules of the theory behave just like the $\mathscr{F}_C$-rules other than II*e*.

Cor. (viii). *Modus Ponens can be eliminated from a theory without axioms and whose special rules are without restrictions on variables or iutroductions of E.*

We proceed as in Prop. 4. The case when $\phi$ is atomic and $DN\phi\chi$ is an axiom is the same as before because the only axiom it can be is, as before, an $\mathscr{F}_C$-axiom. Again in this case if $N\phi$ is introduced by a special rule then this acts just like a case of introduction by II*a*.

## 3.16 *Many-sorted predicate calculi*

*A $\kappa$-sorted classical predicate calculus of the first order* is formed from the symbols:

| | type | name |
|---|---|---|
| $x_{\iota'}$ | $\iota'$ | individual variable of the first sort |
| $\vdots$ | $\vdots$ | |
| $x_{\iota^{(\kappa)}}$ | $\iota^{(\kappa)}$ | individual variable of the $\kappa$th sort |
| $p_{o\iota'}, \ldots, p_{o\iota^{(\kappa)}}$ | $o\iota', \ldots, o\iota^{(\kappa)}$ | one-place predicate variable |

| | type | name |
|---|---|---|
| $p_{o\iota'\iota'}, p_{o\iota'\iota''}, \ldots, p_{o\iota^{(\kappa)}\iota^{(\kappa)}}$ generally | types as shown | two-place predicate variables |
| $p_{o\epsilon'\ldots\epsilon^{(\lambda)}}$, where $\epsilon^{(\theta)}$, for $1 \leqslant \theta \leqslant \lambda$, is one of $\iota', \ldots, \iota^{(\kappa)}$ | types as shown | $\lambda$-place predicate variables |
| $E', \ldots, E^{(\kappa)}$ | $o(o\iota'), \ldots, o(o\iota^{(\kappa)})$ | existential quantifiers of types shown |
| $\lambda$ | | abstraction symbol |
| ′ | | generating symbol |
| $N$ | $oo$ | negation symbol |
| $D$ | $ooo$ | disjunction symbol |
| ( ) | | parentheses |

The axioms are T.N.D. for all atomic statements. The rules are those of $\mathscr{F}_C$ with restriction and generalization for each type of individual variable, denoted by II$d'$, ..., II$d^{(\kappa)}$, II$e'$, ..., II$e^{(\kappa)}$. We denote the $\kappa$-sorted classical predicate calculus by $\mathscr{F}_{C\kappa}$. The situation is just as if in $\mathscr{F}_C$ we labelled the variables as $x^{(\mu\cdot\kappa+\theta)}, 1 \leqslant \theta \leqslant \kappa$, and stated II$d$, $e$ separately for each $\theta, 1 \leqslant \theta \leqslant \kappa$. But the main difference is in the argument places of the predicates.

Let $\mathscr{F}_C^{(\kappa)}$ be $\mathscr{F}_C$ plus constant one-place predicates $S', \ldots, S^{(\kappa)}$ and additional axioms $DNS'\xi S'\xi, \ldots, DNS^{(\kappa)}\xi S^{(\kappa)}\xi$. We give a method for translating $\mathscr{F}_{C\kappa}$-statements into $\mathscr{F}_C^{(\kappa)}$-statements in such a way that $\mathscr{F}_{C\kappa}$-theorems translate into $\mathscr{F}_C^{(\kappa)}$-theorems. The translation of an $\mathscr{F}_{C\kappa}$-statement is obtained as follows:

(*a*) $\phi$ is atomic, say $p_{o\iota^{(\theta')}\ldots\iota^{(\theta^{(\lambda)})}} x^{(\mu')}_{\iota^{(\theta')}} \ldots x^{(\mu^{(\lambda)})}_{\iota^{(\theta^{(\lambda)})}}$, the translation is

$$\text{(A)} \qquad K \prod_{\nu=1}^{\lambda} S^{(\theta^{(\nu)})} x^{(\kappa\cdot\mu^{(\nu)}+\theta^{(\nu)})} p_{o\iota^{(\theta')}\ldots\iota^{(\theta^{(\lambda)})}} x^{(\kappa\cdot\mu'+\theta')} \ldots x^{(\kappa\cdot\mu^{(\lambda)}+\theta^{(\lambda)})}.$$

If two of the individual variables are the same then we omit an occurrence of $S$ followed by that variable,

(*b*) $\phi$ is $N\psi$, its translation is $N\psi'$, where $\psi'$ is the translation of $\psi$,

(*c*) $\phi$ is $D\psi\chi$, its translation is $D\psi'\chi'$, where $\psi', \chi'$ are the translations of $\psi', \chi$ respectively.

(*d*) $\phi$ is $(E^\theta\xi_\theta)\,\psi\{\xi_\theta\}$, its translation is $(E\xi')\,\psi'\{\xi'\}$, where $\psi'\{\xi'\}$ is the translation of $\psi\{\xi_\theta\}$.

(e) $\phi'$ is the translation of $\phi$ if and only if it is obtained from $\phi$ by (a), ..., (d).

Some $\mathscr{F}_C^{(\kappa)}$-statements fail to be translations of $\mathscr{F}_{C\kappa}$-statements, for instance the $\mathscr{F}_C^{(\kappa)}$-axioms involving $S$.

We can demonstrate for $\mathscr{F}_{C\kappa}$ all the propositions we demonstrated for $\mathscr{F}_C$ and by similar methods. Thus we can demonstrate the Deduction Theorem and the substitutivity of the biconditional. For this purpose $\mathscr{F}_{C\kappa}$ appears as $\mathscr{F}_C$ in a different notation.

*A $\kappa$-sorted theory $\mathscr{T}_\kappa$* based upon $\mathscr{F}_{C\kappa}$ is obtained by taking an applied $\mathscr{F}_{C\kappa}$ and adding some of its statements as extra axioms, we could also add axiom-schemes or we could allow the rule of substitution, we might also have some extra rules. An example of a three-sorted theory is 3-dimensional geometry, the three sorts of individuals are: points, lines planes, Another example is set theory with various types of sets, or number theory with different variables for natural numbers and for classes of natural numbers. Some of these theories contain functors, i.e. symbols of types, $\iota'\iota'$, $\iota'\iota''$, $\iota'\iota''\iota''$, $\iota''\iota'$, etc. and constant individuals as well as constant predicates. We shall show later how functors and constant individuals can be eliminated. Thus we shall assume that our $\kappa$-sorted theories are without functors or constant individuals, but they may contain constant predicates and must contain at least one such. Corresponding to a $\kappa$-sorted theory $\mathscr{T}_\kappa$ we get a one-sorted theory $\mathscr{T}$ by adding the constant one-place predicates $S', \ldots, S^{(\kappa)}$ and the axioms $DNS'\xi S'\xi, \ldots, DNS^{(\kappa)}\xi S^{(\kappa)}\xi$, the remaining $\mathscr{T}$-axioms are the translations of the $\mathscr{T}_\kappa$-axioms, and T.N.D. for atomic statements. Sometimes

$$(E\xi)\,S'\xi, \ldots, (E\xi)\,S^{(\kappa)}\xi$$

are also taken as axioms, but seems unnecessary.

PROP. 16. (i) *A $\mathscr{T}_\kappa$- statement is a $\mathscr{T}_\kappa$-theorem if and only if its translation into $\mathscr{T}$ is a $\mathscr{T}$-theorem,*

(ii) *If $\mathscr{T}_\kappa$ is consistent with respect to negation then so is $\mathscr{T}$.*

(iii) *If $\mathscr{T}$ is consistent with respect to negation then so is $\mathscr{T}_\kappa$.*

(iv) *There is an effective method whereby given a $\mathscr{T}_\kappa$-proof of a $\mathscr{T}_\kappa$-statement $\phi$ we can find a $\mathscr{T}$-proof of the translation of $\phi$ into $\mathscr{T}$; and conversely, there is an effective method whereby given a $\mathscr{T}$-proof of a $\mathscr{T}$-statement $\psi$ which is the translation of a $\mathscr{T}_\kappa$-statement $\phi$ we can find a $\mathscr{T}_\kappa$-proof of $\phi$.*

(ii) follows from (i), so does (iii). Ad. (ii) if $\mathscr{T}$ is inconsistent with respect to negation then we can $\mathcal{T}$-prove $\phi$ and $N\phi$ for any $\mathcal{T}$-statement $\phi$, for the case when $\phi$ is the translation of a $\mathcal{T}_\kappa$-statement $\psi$ by (i) $\psi$ and $N\psi$ are $\mathcal{T}_\kappa$-theorems and so $\mathcal{T}_\kappa$ is inconsistent. Similarly for (iii). Again (i) follows from (iv). Thus we need only demonstrate (iv).

Given a $\mathcal{T}_\kappa$-proof of the $\mathcal{T}_\kappa$-statement $\phi$ we want to construct a $\mathcal{T}$-proof of the translation $\phi'$ of $\phi$. In the $\mathcal{T}_\kappa$-proof of $\phi$ replace every $\mathcal{T}_\kappa$-statement by its translation into $\mathcal{T}$. Since the translation of $N\phi$ is $N\phi'$ where $\phi'$ is the translation of $\phi$ and $D\phi'\psi'$ is the translation of $D\phi\psi$ where $\phi'$ and $\psi'$ are the translations of $\phi$ and $\psi$ respectively, then applications of rules I$a$, $b$ and II$a$, $b$, $c$ remain applications of the same rules. Rule II$d$ becomes $\dfrac{D\phi'\{\eta\}\,\omega'}{D(E\xi)\,\phi'\{\xi\}\,\omega'}$ $\xi$ fails to occur free in $\phi\,\{\Gamma_\theta\}$, where $\xi$ is a variable of sort $\theta$, and where $\phi'\{\Gamma\}$ is the translation of $\phi\{\Gamma\}$ and $\omega'$ is the translation of $\omega$. But this is still a case of II$d$. Similarly for II$e$. An axiom $D\pi N\pi$ translates into $D\pi' N\pi'$, where $\pi'$ is the translation of $\pi$, this is a case of T.N.D. and so is a $\mathcal{T}$-theorem, add its $\mathcal{T}$-proof. The other $\mathcal{T}_\kappa$-axioms translate into $\mathcal{T}$-axioms by definition of $\mathcal{T}$. Thus half of (iv) is demonstrated.

For the second half of (iv) first suppose that $\mathcal{T}_\kappa$ is $\mathcal{F}_{C\kappa}$; omit each occurrence of $S^{(\theta^{(\nu)})}x^{(\kappa\cdot\mu^{(\nu)}+\theta^{(\nu)})}$ and replace the remaining occurrences of $x^{(\kappa\cdot\mu^{(\nu)}+\theta^{(\nu)})}$ by $x^{(\mu^{(\nu)})}_{\iota\theta^{(\nu)}}$. Omit $K$ and $\Pi$ in $(A)$, also replace $(E\xi)$ by $(E^{(\theta)}\,\xi_\theta)$ whenever $S^{(\theta)}\xi$ occurs in the scope of $(E\xi)$. All this converts the $\mathcal{T}$-statement $\psi$ into a $\mathcal{T}_\kappa$-statement $\phi_\kappa$.

Now we may suppose that $\psi$ is in prenex normal form and that its $\mathcal{T}$-proof is also in normal form. Thus the $\mathcal{T}$-proof of $\psi$ begins with applications of rules I$a$, II$a$, $b$, $c$ and finishes with applications of rules I$a$, $b$, II$d$, $d'$. The first part of a normal proof produces a quantifier-free tautology. If in this we omit all occurrences of $S^{(\theta^{(\nu)})}x^{(\kappa\cdot\mu^{(\nu)}+\theta^{(\nu)})}$ and replace $x^{(\kappa\cdot\mu^{(\nu)}+\theta^{(\nu)})}$ by $x^{(\mu^{(\nu)})}_{\iota\theta^{(\lambda)}}$ then we are left with another tautology in $\mathcal{T}_\kappa$. This tautology is a $\mathcal{T}_\kappa$-theorem. Now put the quantifiers back in the same order as in the $\mathcal{T}$-proof of $\psi$ and we are left with a $\mathcal{T}_\kappa$-proof of $\phi$.

Note that in our translation we replace

$$(Ex_i)\,px_i \quad \text{by} \quad (Ex)\,(KS_i xpx).$$

Thus

$$\begin{aligned}(Ax_i)\,px_i \quad \text{becomes} \quad & N(Ex_i)\,Npx_i\\ & N(Ex)\,(KS_i xNpx)\\ & (Ax)\,NKpSxNpx\\ & (Ax)\,CSxpx.\end{aligned}$$

Let $\Phi$ be the conjunction of those $\mathcal{T}_\kappa$-axioms used in the $\mathcal{T}_\kappa$-proof of $\phi$. Let $\Psi$ be the translation of $\Phi$ into $\mathcal{T}$. By the Deduction Theorem we have $C\Phi\phi$ is an $\mathcal{F}_{C\kappa}$-theorem. Then we have just shown that $C\Psi\psi$ is an $\mathcal{F}_C$-theorem, whence $\psi$ is a $\mathcal{T}$-theorem. Note that the prenex normal form of $C\Phi\phi$ translates into the prenex normal form of $C\Psi\psi$. This completes the demonstration of Prop. 15.

## 3.17 *Equality*

Most theories based on $\mathcal{F}_C$ require the notion of equality. Hence we now set up an extension of $\mathcal{F}_C$ by adjoining a single constant binary predicate $I$ and a rule and an axiom involving $I$, the resulting system will be denoted by $I\mathcal{F}_C$. Then a theory based on $\mathcal{F}_C$ which requires the notion of equality becomes a theory based on $I\mathcal{F}_C$.

The system $I\mathcal{F}_C$ is the system $\mathcal{F}_C$ (with or without factors or constant individuals) plus the symbol $I$ of type $o\iota\iota$ and the axiom

$I(i)$ $$I\alpha\alpha$$

and the rule

$I(ii)$ $$\frac{D\phi\{\alpha\}\omega \quad DI\alpha\beta\omega}{D\phi\{\beta\}\omega},$$

$\omega$ is subsidiary and may be omitted. $\phi\{\beta\}$ is called *a variant of* $\phi\{\alpha\}$ *by* $I(ii)$. Naturally we require T.N.D. for the predicate $I$ to be an axiom.

The system $I\mathcal{F}_C$ is the system $\mathcal{F}_C$ with an extra constant binary predicate $I$, an extra axiom and an extra rule. Thus everything we have demonstrated so far for the systems $\mathcal{F}_C$, $\mathcal{F}_{C\kappa}$, etc. will still hold provided its demonstration can be modified to deal with the extra axiom and rule. We frequently write $(\alpha = \beta)$ for $I\alpha\beta$, $(\alpha \neq \beta)$ for $NI\alpha\beta$.

In detail we have

In Cor. (iii), Prop. 3 we may suppose that $\phi^{(\theta,\theta')}$ is distinct from axiom $I(i)$. For that matter we may suppose that for each $\theta$ $\phi^{(\theta,\theta'')}$ is different from $N\phi^{(\theta,\theta')}$ for all $\theta'$, $\theta''$, otherwise $\phi^\theta$ is useless, being a theorem.

We can dispense with rule $I(ii)$ if we push all uses of $I(ii)$ back into the axioms. That is to say: rule $I(ii)$ can be put above every other rule. Sometimes a change of variable may be required. We start with a highest application of $I(ii)$ and push it upwards until it disappears, then take a next highest and so on. When this is done Prop. 4 (the derivability of M.P.) goes through as before.

In Prop. 7 we interchange $\alpha = \beta$ with $\alpha \neq \beta$.

Prop. 9 goes through because we can push all applications of $I(ii)$ back into the axioms and so above rules I$a$, $b$, II$d$, $d'$.

In the definition of tautology $\alpha = \alpha$ must be given $\boldsymbol{t}$.

Prop. 9 becomes:

Prop. *I*9. *An $I\mathscr{F}_C$-proof can be modified so that all applications of rules* II$a$, $b$, $c$, $I(ii)$ *occur above all applications of rules* I$b$, II$d$, $d'$.

An $I\mathscr{F}_C$-proof then divides into two parts; the first part is a free variable $\mathscr{F}_C$-proof of a disjunction. The second part consists of applying quantifiers.

In Prop. 10, 11 we need to change 'tautology' to '$I\mathscr{P}_C$-theorem'.

In the definition of 'generally over $\mathscr{N}$' we add: $\nu = \nu$ is given $\boldsymbol{t}$ and $\nu = \mu$, where $\nu$ is different from $\mu$, is given $\boldsymbol{f}$.

Prop. 12 and 14 carry over without difficulty.

Prop. *I*13. *The rule $I(ii)$ is redundant otherwise the symbols, axioms and rules of $I\mathscr{F}_C$ are independent.*

We need only show that the symbol $I$ is independent of the other symbols and that $I(i)$ is independent of the other axioms. The symbol $I$ is independent of the other symbols of $I\mathscr{F}_C$ because if there was definition of $I$ in terms of the other symbols of $I\mathscr{F}_C$, i.e. in terms of the symbols of $\mathscr{F}_C$ then this will be without use of any free variables, because a free variable in the definiendum must be present as a free variable in the definiens, this leaves only $N$, $D$ and $E$ and bound variables and with these we are unable to construct a formula of type $o\iota\iota$. Similarly the axiom $I(i)$ is unobtainable from the axioms of $\mathscr{F}_C$.

Prop. *I*14. *$I\mathscr{F}_C$ is consistent with respect to negation.*

If we could $I\mathscr{F}_C$-prove $\phi$ and $N\phi$ for some $I\mathscr{F}_C$-statement $\phi$ then we could do so without use of $I(ii)$; by the deduction theorem we would then obtain an $\mathscr{F}_C$-proof of $D \sum_{\theta=1}^{\kappa} NI\alpha^{(\theta)}\alpha^{(\theta)}KN\phi\phi$ where $I\alpha^{(\theta)}\alpha^{(\theta)}$ were the cases of $I(i)$ used in the original proofs In this replace $I\alpha^{(\theta)}\alpha^{(\theta)}$ by $D\pi^{(\theta)}N\pi^{(\theta)}$ and we are left with an $\mathscr{F}_C$-proof of $D \sum_{\theta=1}^{\kappa} ND\pi^{(\theta)}N\pi^{(\theta)}KN\phi\phi$, but this is absurd because $D \sum_{\kappa=1}^{\theta} ND\pi^{(\theta)}N\pi^{(\theta)}KN\phi\phi$ fails to be generally valid.

D11. $(E!\,\xi)\,\phi\{\xi\}$ for $(E\xi)\,(\phi\{\xi\}\ \&\ (A\eta)\,C\phi\{\eta\}\,I\xi\eta)$

read 'there is exactly one thing with the property $\lambda\xi.\phi\{\xi\}$'.

Given a theory $\mathscr{T}$ there is another theory $\mathscr{T}'$ effectively obtainable from $\mathscr{T}$ such that $\mathscr{T}$-theorems are $\mathscr{T}'$-theorems and the $\mathscr{T}'$-axioms are disjunctions of atomic statements of their negations. To obtain $\mathscr{T}'$ we replace any $\mathscr{T}$-axiom of the form $K\phi\psi$ by the two axioms $\phi$ and $\psi$, from these we easily recover $K\phi\psi$. We replace any $\mathscr{T}$-axiom $(A\xi)\,\phi(\xi\}$ by $\phi\{\xi\}$, from which we easily recover the former one. We replace a $\mathscr{T}$-axiom of the form $(E\xi)\,\phi\{\xi\}$ by $\phi\{\alpha\}$ where $\alpha$ is a new individual constant of type $\iota$ again we easily recover the original $\mathscr{T}$-axiom. Start with the axioms in prenex normal form with matrix in conjunctive normal form, then doing this as long as possible we finally arrive at a theory $\mathscr{T}'$ in which each axiom is a disjunction of atomic statements or their negations. A theory of this type we call a *theory in free disjunctive form.*

PROP. 17. *If $\mathscr{T}$ is a theory in free disjunctive form and if $\mathscr{T}$ contains only a displayed list of predicates, then the rule $I(ii)$ can be replaced by a displayed list of special cases.*

Let $p', \ldots, p^{(\lambda)}$ be the displayed list of $\mathscr{T}$-predicates and suppose that they have $\theta', \ldots, \theta^{(\lambda)}$ argument places respectively. We replace $I(ii)$ by the displayed list of special cases:

$$\frac{Dp\alpha\omega \quad DI\alpha\beta\omega}{Dp\beta\omega}, \quad \text{for a one-place predicate,}$$

$$\frac{Dp\alpha\gamma\omega \quad DI\alpha\beta\omega}{Dp\beta\gamma\omega}, \quad \frac{Dp\gamma\alpha\omega \quad DI\alpha\beta\omega}{Dp\gamma\beta\omega}, \quad \text{for a two-place predicate, including } I \text{ itself,}$$

and so on, and similar rules for $Np\alpha, Np\alpha\gamma, Np\gamma\alpha$, etc.

We now show that rule $I(ii)$ is obtainable from these special cases. We proceed by formula induction.

(*a*) $\phi\{\alpha\}$ is atomic; the result holds by hypothesis.

(*b*) $\phi\{\alpha\}$ is $D\phi'\{\alpha\}\,\phi''\{\alpha\}$ and the result holds for $\phi'\{\alpha\}$ and for $\phi''\{\alpha\}$, we have

$$\cfrac{\cfrac{DD\phi'\{\alpha\}\,\phi''\{\alpha\}\,\omega \quad \cfrac{DI\alpha\beta\omega}{DDI\alpha\beta\omega\phi''\{\alpha\}}\,\text{II}a}{DD\phi'\{\beta\}\,\phi''\{\alpha\}\,\omega} \quad \cfrac{DI\alpha\beta\omega}{DDI\alpha\beta\omega\phi'\{\beta\}}\,\text{II}a}{DD\phi'\{\beta\}\,\phi''\{\beta\}\,\omega}, \quad \text{as desired.}$$

(c) $\phi\{\alpha\}$ is $N\phi'\{\alpha\}$ and the result holds for $\phi'\{\alpha\}$, we have by formula induction:

(c′) $\phi'\{\alpha\}$ is atomic, the result holds by hypothesis;

(c″) $\phi'\{\alpha\}$ is $N\phi''\{\alpha\}$ and the result holds for $\phi''\{\alpha\}$, we have $DNN\phi''\{\alpha\}\omega$ whence by the reversibility of II*c* we have $D\phi''\{\alpha\}\omega$, thus

$$\cfrac{\cfrac{D\phi''\{\alpha\}\omega \quad DI\alpha\beta\omega}{D\phi''\{\beta\}\omega}}{DNN\phi''\{\beta\}\omega}\,\text{II}c,$$

as described.

(c‴) $\phi'\{\alpha\}$ is $D\psi\{\alpha\}\psi'\{\alpha\}$ and the result holds for $\psi\{\alpha\}$ and for $\psi'\{\alpha\}$, we may assume that the result holds for $N\psi\{\alpha\}$ and for $N\psi'\{\alpha\}$ because these are shorter formulae, we have $DNDψ\{\alpha\}\psi\{\alpha\}\omega$, whence by the reversibility of II*b* we have $DN\psi\{\alpha\}\omega$ and $DN\psi'\{\alpha\}\omega$, with $DI\alpha\beta\omega$ these yield $DN\psi\{\beta\}\omega$ and $DN\psi'\{\beta\}\omega$ whence the result follows by II*b*, as desired.

(c⁗) $\phi'\{\alpha\}$ is $(E\xi)\psi\{\xi,\alpha\}$ and the result holds for $\psi\{\xi,\alpha\}$, $\psi\{\gamma,\alpha\}$. In the proof of $D(E\xi)\psi\{\xi,\alpha\}$ we may suppose that $E$ is introduced at II*d* only, because an introduction by II*a* can easily be replaced by an introduction by II*a* without $E$ followed by an application of II*d*. Suppose that $\dfrac{D\psi\{\gamma,\alpha\}\omega}{D(E\xi)\phi\{\xi,\alpha\}\omega}$ is an introduction of $E$, replace this by

$$\cfrac{\cfrac{D\psi\{\gamma,\alpha\}\omega \quad DI\alpha\beta\omega}{D\psi\{\gamma,\beta\}\omega}}{D(E\xi)\psi\{\xi,\beta\}\omega}$$

and the result follows.

This completes the demonstration of the proposition.

Prop. 18. (i) $\dfrac{\phi\{\alpha\}}{(A\xi)CI\xi\alpha\phi\{\xi\}}$; (i′) $\dfrac{(A\xi)CI\xi\alpha\phi\{\xi\}}{\phi\{\alpha\}}*$.

(ii) $\dfrac{\phi\{\alpha\}}{(E\xi)KI\xi\alpha\phi\{\xi\}}$; (ii′) $\dfrac{(E\xi)KI\xi\alpha\phi\{\xi\}}{\phi\{\alpha\}}*$.

Ad. (i)

$$\cfrac{\cfrac{\cfrac{\phi\{\alpha\}}{DNI\alpha\xi\phi\{\alpha\}}\,\text{II}a}{DNI\alpha\xi\phi\{\xi\}}\,I(ii)}{(A\xi)CI\alpha\xi\phi\{\xi\}.}\,\text{II}d'$$

using the axiom $DI\alpha\xi NI\alpha\xi$,

Ad. (i') $$\begin{array}{ll} (A\xi)\,CI\xi\alpha\phi\{\xi\} & * \quad \text{reversibility of II}d', \\ \cline{1-1} CI\xi\alpha\phi\{\xi\} & * \quad \text{substitution,} \\ \cline{1-1} CI\alpha\alpha\phi\{\alpha\} & \quad \text{M.P. using axiom } I(i), \\ \cline{1-1} \phi\{\alpha\}. & \end{array}$$

Ad. (ii) $$\begin{array}{ll} \phi\{\alpha\} & \quad \text{II}b \text{ and axiom } I(i), \\ \cline{1-1} KI\alpha\alpha\phi\{\alpha\} & \quad \text{II}d \\ \cline{1-1} (E\xi)\,KI\alpha\xi\phi\{\xi\}. & \end{array}$$

Ad. (ii') $$\begin{array}{ll} (E\xi)\,KI\alpha\xi\phi\{\xi\} & *, \\ \cline{1-1} \sum\limits_{\theta=1}^{\kappa} KI\alpha\beta^{(\theta)}\phi\{\beta^{(\theta)}\}, & \end{array}$$

one $\beta^{(\theta)}$, $1 \leqslant \theta \leqslant \kappa$, must be $\alpha$ otherwise lower formula is $\boldsymbol{f}$, this is absurd if upper formula is $\boldsymbol{t}$, hence $KI\alpha\alpha\phi\{\alpha\}$, hence by reversibility of II$b'$ $\phi\{\alpha\}$.

We easily obtain the derived rules:

$$\begin{array}{c} DI\alpha\beta\omega \\ \hline\hline DI\beta\alpha\omega \end{array} \quad \text{and} \quad \begin{array}{c} DI\alpha\beta\omega \quad DI\beta\gamma\omega \\ \hline\hline DI\alpha\gamma\omega \end{array}.$$

The first comes from

$$\begin{array}{c} I\alpha\alpha \\ \hline DI\alpha\alpha\omega \quad DI\alpha\beta\omega \\ \hline DI\beta\alpha\omega. \end{array}$$

The axiom expresses the *reflexiveness* of equality, the first of the derived rules expresses the *symmetry of* equality and the second derived rule expresses the *transitivity* of equality. A *relation* (a formula of type $o\iota\iota$) is called an *equivalence relation* if it is reflexive, symmetric and transitive.

## 3.18 *The predicate calculus with equality and functors*

We showed before that $\mathscr{F}_C$ with functors is virtually the same as $\mathscr{F}_C$ without functors in that we can easily translate from the one into the other. We now demonstrate a similar result for $I\mathscr{F}_C$ in a different way. Let $I\mathscr{F}_{Cf}$ be $I\mathscr{F}_C$ with functors and possibly with constant individuals (functors without argument places). We translate $I\mathscr{F}_{Cf}$ into a theory $\mathscr{T}$ without functors or constant individuals in such a way that $I\mathscr{F}_{Cf}$ theorems translate into $\mathscr{T}$-theorems and $\mathscr{T}$-theorems which are translations of $I\mathscr{F}_{Cf}$-statements are translations of $I\mathscr{F}_{Cf}$-theorems. If we had started with a theory based on $I\mathscr{F}_{Cf}$ then a similar translation is obtained. The translation is as follows. Replace

$I\alpha\xi$ by $I\alpha\xi$, where $\alpha$ is a constant and $\xi$ is a variable,
$I\xi\alpha$ by $I\alpha\xi$, where $\alpha$ is a constant and $\xi$ is a variable,
$I\xi\eta$ by $I\xi\eta$, where $\xi$ and $\eta$ are variables,
$I\alpha\beta$ by $(E\xi)KI\alpha\xi I\beta\xi$, where $\alpha$ and $\beta$ are constants,
$\phi\{\alpha\}$ by $(E\xi)K\overline{\phi}\{\xi\}I\alpha\xi$, where $\alpha$ is a constant, $\overline{\phi}$ the translation of $\phi$,
$\phi\{f\alpha',\ldots,\alpha^{(\kappa)}\}$ by $(E\xi)K\overline{\phi}\{\xi\}(E\eta',\ldots,\eta^{(\kappa)})KF\eta'\ldots\eta^{(\kappa)}\xi\prod_{\theta-1}^{\kappa}I\alpha^{(\theta)}\eta^{(\theta)}$,

where $\alpha',\ldots,\alpha^{(\kappa)}$ are constants, but if any of the $\alpha$'s are variables then omit the corresponding $I\alpha^{(\theta)}\eta^{(\theta)}$ and the corresponding quantifier, use different predicates $F$ for different functors $f$. $\overline{\phi}\{\xi\}$ is the translation of $\phi\{\xi\}$.

$N\phi$ by $N\overline{\phi}$, where $\overline{\phi}$ is the transform of $\phi$,
$D\phi\psi$ by $D\overline{\phi}\overline{\psi}$, where $\overline{\phi}$ and $\overline{\psi}$ are the transforms of $\phi$ and $\psi$ respectively,
$(E\xi)\phi\{\xi\}$ by $(E\xi)\overline{\phi}\{\xi\}$, where $\overline{\phi}\{\xi\}$ is the transform of $\phi\{\xi\}$.

Repeat until all the functors have been eliminated and all occurrences of constants $\alpha$ occur as $I\alpha\xi$, lastly, replace $I\alpha\xi$ by $a\{\xi\}$ using different one-place predicates $a$ for different constants $\alpha$.

We shall require some new axioms in the translated system, namely

$$\left.\begin{array}{l} CKF\eta'\ldots\eta^{(\kappa)}\xi F\eta'\ldots\eta^{(\kappa)}\xi' I\xi\xi', \\ (A\eta'\ldots\eta^{(\kappa)})(E\xi)F\eta'\ldots\eta^{(\kappa)}\xi, \\ (E\eta)a\{\eta\}, \\ CKa\{\eta\}a\{\eta'\}I\eta\eta', \end{array}\right\} \qquad (1)$$

for all $F$ and $a$ that we have introduced. If we had started with a theory $\mathscr{T}$ based on $I\mathscr{F}_{Cf}$ then we should also require the translations of the $\mathscr{T}$-axioms as axioms in the translated system, similarly for $\mathscr{T}$-rules.

PROP. 19. *An $\mathscr{F}_f$-statement is a $\mathscr{T}_f$-theorem if and only if its translation is a $\mathscr{T}$-theorem.*

Let $\phi$ be a $\mathscr{T}_f$-theorem, then its $\mathscr{T}_f$-proof proceeds from $\mathscr{T}_f$-axioms by the $\mathscr{T}_f$-rules and the $I\mathscr{F}_C$-rules and the rule of substitution, and the $I\mathscr{F}_C$-axioms.

Let $\bar{\phi}$ be the translation of $\phi$. The $\mathscr{T}_f$-proof-tree of $\phi$ translates into a tree at the tops of whose branches there stand $\mathscr{T}$-axioms or $I\mathscr{F}_C$-axioms, because the translations of $\mathscr{F}_f$-axioms are $\mathscr{T}$-axioms and the $I\mathscr{F}_C$-axioms remain unaltered if we take these to be without constant individuals or functors as we can if we use the rule of substitution. It remains to show that the $\mathscr{T}_f$-rules and the $I\mathscr{F}_C$-rules translate into derived $\mathscr{T}$-rules or derived $I\mathscr{F}_C$-rules. $\mathscr{F}_f$-rules translate into $\mathscr{T}$-rules by definition. Thus it remains to show that $I\mathscr{F}_C$-rules translate into derived $\mathscr{T}$-rules or derived $I\mathscr{F}_C$-rules, possibly using the new axioms (1). It is at once seen that I$a$, I$b$, II$a$, II$b$, II$c$, II$e'$ translate into cases of the same rules. For II$d$ we have

$$\frac{D\phi\{\alpha\}\,\omega}{D(E\xi)\,\phi\{\xi\}\,\omega};$$

the translation of this is $\dfrac{D(E\xi)\,Ka\{\xi\}\,\bar{\phi}\{\xi\}\,\bar{\omega}}{D(E\xi)\,\bar{\phi}\{\xi\}\,\bar{\omega}}$.

We have

$$\cfrac{\cfrac{D(E\xi)\,Ka\{\xi\}\,\bar{\phi}\{\xi\}\,\bar{\omega}}{DK(E\xi)\,a\{\xi\}\,(E\xi)\,\bar{\phi}\{\xi\}\,\bar{\omega}}\ \text{Prop., 8, M.P.}}{D(E\xi)\,\bar{\phi}\{\xi\}\,\bar{\omega}}\,*\ \text{reversibility of II}b'.$$

For the rule $I(ii)$ $\dfrac{D\phi\{\alpha\}\,\omega \quad DI\alpha\beta\omega}{D\phi\{\beta\}\,\omega}$,

the translation is

$$\frac{D(E\xi)\,Ka\{\xi\}\,\bar{\phi}\{\xi\}\,\bar{\omega} \quad D(E\xi)\,Ka\{\xi\}\,b\{\xi\}\,\bar{\omega}}{D(E\xi)\,Kb\{\xi\}\,\bar{\phi}\{\xi\}\,\bar{\omega}}.$$

We have

$$\cfrac{\cfrac{\cfrac{D(E\xi)\,Ka\{\xi\}\,\bar{\phi}\{\xi\}\,\bar{\omega} \quad D(E\xi)\,Ka\{\xi\}\,b\{\xi\}\,\bar{\omega}}{DK(E\xi)\,Ka\{\xi\}\,\bar{\phi}\{\xi\}\,(E\xi)\,Ka\{\xi\}\,b\{\xi\}\,\bar{\omega}}\ \text{II}b'}{D(E\xi,\xi')\,KKKa\{\xi\}\,a\{\xi'\}\,b\{\xi'\}\,\bar{\phi}\{\xi\}\,\bar{\omega}}\,*\ \text{Prop. 6}}{D(E\xi)\,Kb\,\{\xi\}\,\bar{\phi}\{\xi\}\,\omega}\,*\ \text{by } \mathscr{F}_C \text{ and (1)},$$

as desired.

For the substitution rule $\dfrac{D\phi\{\xi\}\,\omega}{D\phi\{\alpha\}\,\omega}$

this becomes $\dfrac{D\bar{\phi}\{\xi\}\,\bar{\omega}}{D(E\xi)\,Ka\{\xi\}\,\bar{\phi}\{\xi\}\,\bar{\omega}}$.

We have the tautology $C\chi CD\phi\psi DK\chi\phi\psi$, hence we have

$$\dfrac{Ca\{\xi\}CD\overline{\phi}\{\xi\}\varpi DKa\{\xi\}\overline{\phi}\{\xi\}\varpi}{Ca\{\xi\}CD\overline{\phi}\{\xi\}\varpi D(E\xi)Ka\{\xi\}\overline{\phi}\{\xi\}\varpi}\ \mathrm{II}d$$

$$\dfrac{Ca\{\xi\}CD\overline{\phi}\{\xi\}\varpi D(E\xi)Ka\{\xi\}\overline{\phi}\{\xi\}\varpi}{Ca\{\xi\}C(A\xi)D\overline{\phi}\{\xi\}\varpi D(E\xi)Ka\{\xi\}\overline{\phi}\{\xi\}\varpi}\ \mathrm{II}d$$

$$\dfrac{Ca\{\xi\}C(A\xi)D\overline{\phi}\{\xi\}\varpi D(E\xi)Ka\{\xi\}\overline{\phi}\{\xi\}\varpi}{C(E\xi)a\{\xi\}C(A\xi)D\overline{\phi}\{\xi\}\varpi D(E\xi)Ka\{\xi\}\overline{\phi}\{\xi\}\varpi}\ \mathrm{II}e'$$

$$\dfrac{\dfrac{D\overline{\phi}\{\xi\}\varpi}{(A\xi)D\overline{\phi}\{\xi\}\varpi}\ \mathrm{II}e \qquad \dfrac{C(E\xi)a\{\xi\}C(A\xi)D\overline{\phi}\{\xi\}\varpi D(E\xi)Ka\{\xi\}\overline{\phi}\{\xi\}\varpi}{C(A\xi)D\overline{\phi}\{\xi\}\varpi D(E\xi)Ka\{\xi\}\overline{\phi}\{\xi\}\varpi}\ (1),\ \mathrm{M.P.}}{D(E\xi)Ka\{\xi\}\overline{\phi}\{\xi\}\varpi}\ \mathrm{M.P.}$$

as desired. Thus the $\mathscr{T}_f$-rules and the $I\mathscr{F}_C$-rules translate into $\mathscr{T}$-rules or derived $\mathscr{T}$-rules, and the result follows.

## 3.19 *Elimination of axiom schemes*

PROP. 20. *A theory $\mathscr{T}$ without functors and based on $I\mathscr{F}_C$ and with axiom schemes can be replaced by an equivalent 2-sorted theory $\mathscr{S}$ with a terminating sequence of axioms.*

We are thinking of axiom schemes involving arbitrary statements; for instance Mathematical Induction. Axiom schemes which contain arbitrary terms but are without arbitrary statements can be replaced by axioms provided we add the rule of substitution II$f$.

To demonstrate the proposition we introduce two new symbols: $s$ and $\epsilon$ of types $\iota\iota\iota$ and $o(o\iota)\iota$ respectively, where $o\iota$ is the type of a second sort of variable. The construction will be clearer if the reader interprets

| | | |
|---|---|---|
| $Jx$ for | $N(Ex', x'')(x = sx'x'')$ | read '$x$ is an individual', |
| $sxx'$ | is the ordered pair | $xx'$, |
| $sxsx'x''$ | is the ordered triplet | $xx'x''$, etc. |

The first sort of variable is a variable for sequences of individuals.

We frequently write $(x\epsilon X)$ instead of $\epsilon xX$ and read it as '$x$ has the property $X$', where $X$ is a variable of the second sort, we call them properties.

We now translate the theory $\mathscr{T}$ into a two-sorted theory $\mathscr{S}$ as follows. Replace

$\pi\xi$ by $(\xi\epsilon\pi)$, where $\pi$ is atomic and of type $o\iota$,

$\pi\xi\xi'$ by $(s\xi\xi'\epsilon\pi)$, where $\pi$ is atomic and of type $o\iota\iota$,

$\pi\xi\xi\xi''$ by $(s\xi s\xi'\xi''\epsilon\pi)$, etc.

We shall frequently write $\mathfrak{x}_{(S\kappa)}$ for $s\xi' s\xi'' s \ldots s\xi^{(\kappa)}\xi^{(S\kappa)}$. $\mathfrak{x}^\frown\mathfrak{y}$ is the sequence made up from the members of the sequence $\mathfrak{x}$ in their proper order followed by the members of the sequence $\mathfrak{y}$ in their proper order. We frequently omit the suffix $S\kappa$ in $\mathfrak{x}_{(S\kappa)}$ and just write $\mathfrak{x}$.

We now give some definitions:

| | | | | |
|---|---|---|---|---|
| D 12 | $(\xi\epsilon\overline{\Xi})$ | for | $N(\xi\epsilon\Xi)$ | *complement* |
| D 13 | $(\xi\epsilon\Xi \cup \Xi')$ | for | $D(\xi\epsilon\Xi)(\xi\epsilon\Xi)$ | *union* |
| D 14 | $(\xi\epsilon\Xi \cap \Xi')$ | for | $K(\xi\epsilon\Xi)(\xi\epsilon\Xi)$ | *intersection* |
| D 15 | $KJ\xi'(\xi^\frown\xi'\epsilon\Xi \times V)$ | for | $(\xi\epsilon\Xi)$ | *direct product* |
| D 16 | $(\xi\epsilon\mathscr{D}\Xi)$ | for | $(E\xi')(KJ\xi'(\xi'^\frown\xi\epsilon\Xi))$ | *domain* |
| D 17 | $KJ\xi'(\xi^\frown\xi'\epsilon Cnv_1\Xi)$ | for | $(\xi'^\frown\xi\epsilon\Xi)$ | *inverse* |
| D 18 | $KKJ\xi' J\xi(\xi^\frown\xi'^\frown\xi''\epsilon Cnv_2\Xi)$ | for | $KKJ\xi' J\xi(\xi'^\frown\xi^\frown\xi''\epsilon\Xi)$ | *permutation* |
| D 19 | $KJ\xi(\xi^\frown\xi'\epsilon Id\Xi)$ | for | $KJ\xi(\xi^\frown\xi'^\frown\xi\epsilon\Xi)$ | *identification.* |

We now show that by means of these definitions a statement

$$\phi\{\xi', \ldots, \xi^{(S\kappa)}\} \quad \text{becomes} \quad (\mathfrak{x}_{(S\kappa)}\epsilon\Delta),$$

where $\Delta$ is constructed from atomic properties by means of complement, union, intersection, domain, direct product, inverse, permutation and identification.

Our second sort of variable is a variable for predicates or properties so we must ensure that we can perform operations on these variables corresponding to the operations we can perform on predicates by means of logical connectives. To do this we adopt the axioms:

(1) $B(\xi\epsilon\overline{\Xi})N(\xi\epsilon\Xi)$ (1′) $(EX)(A\xi)B(\xi\epsilon X)N(\xi\epsilon\Xi)$.

D 20 $X = X'$ for $(A\xi)B(\xi\epsilon X)(\xi\epsilon X')$

$(sxy\epsilon I)$ for $x = y$,

then (1′) becomes $(EX)(X = \overline{\Xi})$, i.e. $\overline{\Xi}$ is a thing of the second sort. We want similar axioms for (2)–(8) below.

(2) $B(\xi\epsilon\Xi \cup \Xi')D(\xi\epsilon\Xi)(\xi\epsilon\Xi')$;

(3) $B(\xi\epsilon\Xi \cap \Xi')K(\xi\epsilon\Xi)(\xi\epsilon\Xi')$;

(4) $BKJ\xi'(\xi^\frown\xi'\epsilon\Xi \times V)(\xi\epsilon\Xi)$;

(5) $B(\xi\epsilon\mathscr{D}\Xi)(E\xi')KJ\xi'(\xi^\frown\xi'\epsilon\Xi)$;

(6) $BKJ\xi'(\xi^\frown\xi'\epsilon Cnv_1\Xi)KJ\xi'(\xi'^\frown\xi\epsilon\Xi)$;

(7) $BKKJ\xi' J\xi''(\xi^\frown\xi'^\frown\xi''\epsilon Cnv_2\Xi)KKJ\xi' J\xi''(\xi'^\frown\xi^\frown\xi''\epsilon\Xi)$;

(8) $BKJ\xi'(\xi^\frown\xi'\epsilon Id\Xi)KJ\xi'(\xi^\frown\xi'^\frown\xi\epsilon\Xi)$.

We want $^- \cup \cap \times \mathscr{D}\ Cnv_1\ Cnv_2\ Id$ as operations on properties satisfying the above axioms, then we can show that we have $B(\xi\epsilon\Delta)\,\phi\{\xi\}$ for any statement $\phi$ and some property $\Delta$.

We proceed by formula induction on $\phi$.

(*a*) $\phi$ is atomic, say $\phi$ is $\pi\xi' \ldots \xi^{(\kappa)}$, replace as described above.

(*b*) $\phi$ is $N\phi'$ and the result holds for $\phi'$, so $\phi'\{\xi', \ldots, \xi^{(S\kappa)}\}$ has been replaced by $(\mathfrak{x}_{(S\kappa)}\epsilon\Delta')$, we replace $N\phi\{\xi', \ldots, \xi^{(S\kappa)}\}$ by $(\mathfrak{x}_{(S\kappa)}\epsilon\overline{\Lambda}')$.

(*c*) $\phi$ is $D\phi'\{\eta', \ldots, \eta^{(\lambda)}\}\phi''\{\zeta', \ldots, \zeta^{(\mu)}\}$ and the result holds for $\phi'$ and for $\phi''$. Thus we have replaced

$$\phi'\{\eta', \ldots, \eta^{(\lambda)}\} \quad \text{by} \quad (\mathfrak{y}_{(\lambda)}\epsilon\Delta'),$$
$$\phi''\{\zeta', \ldots, \zeta^{(\mu)}\} \quad \text{by} \quad (\mathfrak{z}_{(\mu)}\epsilon\Delta'').$$

We first replace $(\mathfrak{y}_{(\lambda)}\epsilon\Delta')$ by $(\mathfrak{x}_{(S\kappa)}\epsilon\Delta''')$ and $(\mathfrak{z}_{(\mu)}\epsilon\Delta'')$ by $(\mathfrak{x}_{(S\kappa)}\,\epsilon\Delta^{\mathrm{iv}})$ where $\xi', \ldots, \xi^{(S\kappa)}$ are the variables $\eta', \ldots, \eta^{(\lambda)}$ and $\zeta', \ldots, \zeta^{(\mu)}$ in some order, say in alphabetical order without repetitions. We have for individuals $\xi, \xi'$ and sequences $\eta$:

$$\xi^\frown\mathfrak{y}\epsilon X \leftrightarrow \mathfrak{y}^\frown\xi\epsilon Cnv_1 X,$$
$$\xi^\frown\xi'^\frown\mathfrak{y}\epsilon X \leftrightarrow \xi'^\frown\xi^\frown\mathfrak{y}\epsilon Cnv_2 X,$$
$$\mathfrak{y}\epsilon X \leftrightarrow \mathfrak{y}^\frown\xi\epsilon X \times V,$$
$$\xi^\frown\mathfrak{y}\,\xi\epsilon X \leftrightarrow \xi^\frown\mathfrak{y}\epsilon Id X.$$

Any permutation of a sequence can be brought about by repeatedly interchanging a pair of consecutive members. Thus a sequence of natural numbers can be brought into a sequence in order of magnitude by repeatedly interchanging consecutive members. Repeated application of $Cnv_1$ will bring any given member of a sequence to the front then applications of $Cnv_2$ will interchange the first two members then repeated applications of $Cnv_1$ will bring all the members back to their original places except that two consecutive members have been interchanged. Thus by repeated applications of $Cnv_1$ and $Cnv_2$ we can bring any sequence into a given permutation. Thus by applications of $Cnv_1$ and $Cnv_2$ we can bring $\eta' \ldots \eta^{(\mu)}$ into $\xi^{(\theta')} \ldots \xi^{(\theta^{(\mu)})}$ where $\theta' < \theta'' < \ldots < \theta^{(\mu)}$ if $S\theta^{(\pi)} < \theta^{(S\pi)}$ apply $Cnv_1$ until $\xi^{(\theta^{(\pi)})}$ is at the end then apply the direct product $\theta^{(S\pi)} - S\theta^{(\pi)}$-times. This will give us the sequence

$$\xi^{(S\theta^{(\pi)})} \ldots \xi^{(\theta^{(\mu)})}\xi^{(\theta')} \ldots \xi^{(\theta^{(\pi)})}\xi \ldots \xi^{(\theta^{(S\pi)} - S\theta^{(\pi)})}.$$

Now apply repeatedly until we have filled in all the gaps. Then apply $Cnv_1$ repeatedly until $\xi'$ is in front. The new variables introduced can be relabelled so that the whole sequence is $\xi' \dots \xi^{(\kappa)}$. If there were any duplicates in $\eta' \dots \eta^{(\mu)}$ then permute until one is in front and its duplicate at the rear then apply $Id$ to eliminate the rear one. Finally replace $\phi$ by $(\mathfrak{x}_{(S\kappa)}\epsilon\Delta''' \cup \Delta^{\mathrm{iv}})$. This completes case (c). (d) is $(E\xi)\,\phi\{\xi,\mathfrak{y}\}$ and we have replaced $\phi\{\xi,\mathfrak{H}\}$ by $\xi^\frown\mathfrak{y}\epsilon\Xi$, we replace $(E\xi)\,\phi'\{\xi,\mathfrak{y}\}$ by $\mathfrak{y}\epsilon\mathscr{D}\Xi$. This completes the description of the replacement.

For the full development we require some more axioms:

(9) $x\tilde{\epsilon}y \quad X\tilde{\epsilon}x \quad X\tilde{\epsilon}Y$

(10) $(Ax,y)\,(Ez)\,(z = sxy)$

(11) $CKJxJyLxy$

(12) $CKKLzu(x = szv)\,(y = suw)\,BLxyLvw$

(13) $CK(x = szu)\,(y = suz)\,Lxy$

(14) $(AX)\,(Ax,y)\,CK(x\epsilon X)\,(y\epsilon X)\,Lxy$,

where $Lxy$ expresses that the sequences $x$ and $y$ are of equal length. For instance the last of these axioms says that two sequences having the same property have the same length. This should suffice to complete the demonstration of Prop. 20.

In any system the rules are normally rule schemes. Our method allows us to replace rule schemes by single rules provided we have a substitution rule. For instance the $I\mathscr{F}_C$-rules become:

$$\mathrm{I}a \quad \frac{x\epsilon U' \cup X \cup Y \cup U}{x\epsilon U' \cup Y \cup X \cup U}; \qquad \mathrm{I}b \quad \frac{x\epsilon X \cup X \cup U}{x\epsilon X \cup U}.$$

$$\mathrm{II}a \quad \frac{x\epsilon X}{x\epsilon X \cup Y}; \qquad \mathrm{II}b \quad \frac{x\epsilon \overline{X} \cup U \quad x\epsilon \overline{Y} \cup U}{x\epsilon \overline{X \cup Y} \cup U}; \qquad \mathrm{II}c \quad \frac{x\epsilon X \cup U}{x\epsilon \overline{\overline{X}} \cup U};$$

$$\mathrm{II}d \quad \frac{yx\epsilon X \cup U}{x\epsilon\mathscr{D}(X \cup U)}; \qquad \mathrm{II}e \quad \frac{yx\epsilon \overline{X} \cup U}{x\epsilon\overline{\mathscr{D}X} \cup U}.$$

I$a$, ..., II$c$ give the rules of Boolean Algebra, if we add II$d$, $e$ we get a Boolean Algebra with a projection operation superimposed on it.

In the 2-sorted theory $\mathfrak{S}$ quantification is allowed over one sort of individual only. A predicate calculus of the second order is obtained from a predicate calculus of the first order by adding quantifiers over predicate variables. It can be shown that a theory $\mathscr{T}$ based on the pure predicate calculus of the second order with axiom schemes fails to be replaceable by a 2-sorted theory without axiom schemes.

### 3.20 *Special cases of the decision problem*

We are as yet unable to show that $\mathscr{F}_C$ has an unsolvable decision problem, namely the problem of finding a uniform method of deciding of any $\mathscr{F}_C$-statement whether it is an $\mathscr{F}_C$-theorem, because we are as yet without a precise definition of what we mean by 'solvable'. Later on we shall identify 'solvable' by 'calculable by a Turing machine'. We shall then show that there fails to be a Turing machine which will tell us whether an $\mathscr{F}_C$-statement is an $\mathscr{F}_C$-theorem. We have to postpone the undecidability of $\mathscr{F}_C$ (which is entirely due to I$b$) until we have given the theory of Turing machines. Meanwhile we can give some methods for deciding certain types of $\mathscr{F}_C$-statements. When we come to the theory of Turing machines it will be seen that these methods can be set up on a suitable Turing machine.

PROP. 21. *The monadic $\mathscr{F}_C$ is solvable. More explicitly an $\mathscr{F}_C$-statement containing only one-place predicates is an $\mathscr{F}_C$-theorem if and only if it is valid in a domain of $2^\kappa$ elements, where $\kappa$ is the number of one-place predicates in the statement.*

Let $\phi$ be a monadic $\mathscr{F}_C$-statement containing exactly $\kappa$ distinct one-place predicates $\phi', \ldots, \phi^{(\kappa)}$. Replace $\phi', \ldots, \phi^{(\kappa)}$ by functions $f', \ldots, f^{(\kappa)}$ over $\mathscr{N}$ (the natural numbers) whose values are $\boldsymbol{t}$ or $\boldsymbol{f}$. Two natural numbers $\lambda$, $\mu$ will be said to belong to the same class if $f^{(\theta)}\lambda = f^{(\theta)}\mu$ for $1 \leqslant \theta \leqslant \kappa$. There are $2^\kappa$ such classes. Let $\nu', \ldots, \nu^{(2^\kappa)}$ be the least natural numbers in these classes. Now replace the functions $f', \ldots, f^{(\kappa)}$ over $\mathscr{N}$ by functions $g', \ldots, g^{(\kappa)}$ over $\nu', \ldots, \nu^{(2^\kappa)}$, (call this set $\mathscr{N}_{2^\kappa}$), such that $g^{(\theta)}\pi = f^{(\theta)}\pi$ if and only if $\pi$ is the least natural number in the class to which it belongs.

Suppose that $\phi$ is generally valid over $\mathscr{N}$, clearly $\phi$ is then generally valid over $\mathscr{N}_{2^\kappa}$. Suppose $\phi$ is generally valid over $\mathscr{N}_{2^\kappa}$, replace $g^{(\theta)}\pi$ by $f^{(\theta)}\pi'$ in all possible ways, where $\pi'$ is in the same class as $\pi$, and we see that $\phi$ is generally valid over $\mathscr{N}$. The proposition now follows by Prop. 12

PROP. 22. *The $\mathscr{F}_C$-statement*

$$(A\xi', \ldots, \xi^{(\lambda)})\,(E\eta', \ldots, \eta^{(\mu)})\,\phi\{\xi', \ldots, \xi^{(\lambda)}, \eta', \ldots, \eta^{(\mu)}\}$$

*is an $\mathscr{F}_C$-theorem if and only if it is valid over a domain with $\lambda$ elements (one element if $\lambda = 0$). Thus it is solvable.*

First consider $(E\eta', \ldots, \eta^{(\mu)})\,\phi\{\eta', \ldots, \eta^{(\mu)}\}$. If $\phi\{\eta, \ldots, \eta\}$ is a tautology then it is an $\mathscr{F}_C$-theorem and hence by II$d$ repeatedly so is

$$(E\eta', \ldots, \eta^{(\mu)})\,\phi\{\eta', \ldots, \eta^{(\mu)}\}.$$

Thus if $(E\eta', \ldots, \eta^{(\mu)})\,\phi\{\eta', \ldots, \eta^{(\mu)}\}$ is valid over a one-element domain then it is an $\mathscr{F}_C$-theorem If $(E\eta', \ldots, \eta^{(\mu)})\,\phi\{\eta', \ldots, \eta^{(\mu)}\}$ is an $\mathscr{F}_C$-theorem then it is valid over a one-element domain, because it is valid over any domain.

Now consider $(A\xi', \ldots, \xi^{(\lambda)})\,(E\eta', \ldots, \eta^{(\mu)})\,\phi\{\xi', \ldots, \xi^{(\lambda)}, \eta', \ldots, \eta^{(\mu)}\}$ call this $\psi$. Let $\phi', \phi'', \ldots, \phi^{(\kappa)}$ be $\phi\{\xi', \ldots, \xi^{(\lambda)}, \zeta', \ldots, \zeta^{(\mu)}\}$ where $\zeta', \ldots, \zeta^{(\mu)}$ are the variables $\xi', \ldots, \xi^{(\lambda)}$ in some order possibly with repetitions, $\kappa = \lambda^{\mu}$. If the disjunction $\sum_{\theta=1}^{\kappa} \phi^{(\theta)}$ is a tautology then $\psi$ is an $\mathscr{F}_C$-theorem. We have:

$$\frac{D\phi^{(\theta)}\,\omega}{D(E\eta', \ldots, \eta^{(\mu)})\,\phi\{\xi', \ldots, \xi^{(\lambda)}, \eta', \ldots, \eta^{(\mu)}\}\,\omega.}\quad \text{II}d$$

By taking $\omega$ to be $\sum_{\lambda=1}^{\kappa} {}^{*(\theta)}\phi^{(\lambda)}$, where the asterisk indicates the omission of $\phi^{(\theta)}$, and repeating and then discarding duplicates by I$b$ we obtain

$$\frac{\sum_{\theta=1}^{\kappa} \phi^{(\theta)}}{(E\eta', \ldots, \eta^{(\mu)})\,\phi\{\xi', \ldots, \xi^{(\lambda)}, \eta', \ldots, \eta^{(\mu)}\}},$$

but the upper formula is an $\mathscr{F}_C$-theorem. Apply II$e'$ repeatedly to the lower formula and we obtain the $\mathscr{F}_C$-theorem $\psi$.

If $\psi$ is valid over a $\lambda$-element domain then the disjunction $\sum_{\theta=1}^{\kappa} \phi^{(\theta)}$ takes the value $t$ no matter how the values $t, f$ are given to its atomic formulae $\pi\xi^{(\theta')} \ldots \xi^{(\theta^{(\nu)})}$, $1 \leqslant \theta', \ldots, \theta^{(\nu)} \leqslant \lambda$ because one disjunctand will then take the value $t$, namely that disjunctand that gives the satisfying set $\zeta', \ldots, \zeta^{(\mu)}$. Thus the disjunctand is a tautology and so $\psi$ is an $\mathscr{F}_C$-theorem. Thus if $\psi$ is valid over a $\lambda$-element domain then $\psi$ is an $\mathscr{F}_C$-theorem. If $\psi$ is an $\mathscr{F}_C$-theorem then it is valid over any $\mathscr{N}_\kappa$ in particular it is valid over a $\lambda$-element domain. This completes the demonstration of the proposition.

The decision problem for $\mathscr{F}_C$ can be framed in several ways:

(i) To decide if an $\mathscr{F}_C$-statement is valid over $\mathscr{N}$, this is the same as to decide if an $\mathscr{F}_C$-statement is an $\mathscr{F}_C$-theorem.

(ii) To find the natural numbers $\kappa$ such that an $\mathscr{F}_C$-statement is valid over $\mathscr{N}_\kappa$ and to decide if it is valid over $\mathscr{N}$.

(iii) To decide if an $\mathscr{F}_C$-statement $\phi$ can be satisfied over $\mathscr{N}$. If $\phi$ fails to be satisfiable over $\mathscr{N}$ then $N\phi$ is valid over $\mathscr{N}$ and so $N\phi$ is an $\mathscr{F}_C$-theorem.

PROP. 23. *A closed $\mathscr{F}_C$-statement in prenex normal form with a prefix of the form* $(E\xi', \ldots, \xi^{(\mu)})(A\eta, \eta')(E\zeta', \ldots, \zeta^{(\lambda)})$ *can be decided as regards satisfiability.*

COR. (i) *A closed $\mathscr{F}_C$-statement in prenex normal form with a prefix of the form*

$$(A\xi', \ldots, \xi^{(\mu)})(E\eta, \eta')(A\zeta', \ldots, \zeta^{(\lambda)})$$

*can be decided as regards validity.*

We first reduce the problem to the case of a closed $\mathscr{F}_C$-statement in prenex normal form with a prefix of the form $(A\eta, \eta')(E\zeta', \ldots, \zeta^{(\lambda)})$. We give the demonstration for the case $\mu = \lambda = 1$ and where the matrix is built up from a single binary predicate variable $\pi$.

$$(E\xi)(A\eta, \eta')(E\zeta)\phi\{\pi; \xi, \eta, \eta', \zeta\}. \tag{1}$$

The case when $\lambda > 1$ is dealt with similarly, the case when $\mu > 1$ follows by repetition of the case $\mu = 1$. We show that (1) is satisfiable if and only if

$$(A\eta, \eta')(E\zeta, \zeta', \zeta'', \zeta''')KKK\phi\{\pi; '\xi', \eta, \eta', \zeta\}\phi\{\pi; '\xi', \eta, '\xi', \zeta'\}$$
$$\phi\{\pi; '\xi', '\xi', \eta, \zeta''\}\phi\{\pi; '\xi', '\xi', '\xi', \zeta'''\} \tag{2}$$

is satisfiable, where the variable $\xi$ has been eliminated by writing $\pi' v$ for $\pi\xi v$, $\pi'' v$ for $\pi v \xi$ and $a$ for $\pi\xi\xi$, $a$ is a propositional variable. This we have indicated by writing $'\xi'$ instead of $\xi$. We have, as it were, taken the resolved form as far as the first restricted variable is concerned. First suppose that (1) is satisfiable over $\mathscr{M}$, where $\mathscr{M}$ is $\mathscr{N}_\kappa$ or $\mathscr{N}$. Then we show that (2) is similarly satisfiable. Let $f$ be a function over $\mathscr{M}^2$ which satisfies (1), if $\xi$ is in $\mathscr{M}$ then for arbitrary $\eta, \eta'$ in $\mathscr{M}$ we have for suitable $\xi, \zeta, \zeta', \zeta'', \zeta'''$ in $\mathscr{M}$,

$$(E\zeta)\phi\{f; \xi, \eta, \eta', \zeta\}, \quad (E\zeta')\phi\{f; \xi, \eta, \xi, \zeta'\}$$

$$(E\zeta'')\phi\{f; \xi, \xi, \eta', \zeta''\} \quad \text{and} \quad (E\zeta''')\phi\{f; \xi, \xi, \xi, \zeta'''\}$$

all take the value $\boldsymbol{t}$. Hence the conjunction

$$(E\zeta,\zeta',\zeta'',\zeta'')\,KKK\phi\{f,\xi,\eta,\eta',\zeta\}\,\phi\{f,\xi,\eta,\xi,\zeta'\}\,\phi\{f,\xi,\xi,\eta,\zeta''\}$$
$$\{f,\xi,\xi,\xi,\zeta'''\} \quad (3)$$

also takes the value $\boldsymbol{t}$.

Now define $f'v$ for $f\xi v$, $f''v$ for $fv\xi$, $a$ for $f\xi\xi$ so $a$ is $\boldsymbol{t}$ or $\boldsymbol{f}$. Put these in (3) and we have a satisfaction of (2) over $\mathscr{M}$, satisfying functions being $f, f', f''$.

We now show that if (2) is satisfied over $\mathscr{M}$, where $\mathscr{M}$ is $\mathscr{N}_\kappa$ or $\mathscr{N}$, then (1) is satisfied over a domain obtained from $\mathscr{M}$ by adjoining a new element. Suppose then that the functions $f, f', f''$ and $a$ satisfy (2) over $\mathscr{M}$ and that $\zeta = g[\eta,\eta']$, $\zeta' = g'[\eta]$, $\zeta'' = g''[\eta']$, $\zeta''' = a$ give the values of $\zeta$, $\zeta'$, $\zeta''$, $\zeta'''$ in the model over $\mathscr{M}$. It is easily seen, by distribution of quantifiers, that $\zeta$, $\zeta'$, $\zeta''$ and $\zeta'''$ depend only on the variables shown. Now let $\alpha$ be a new element and define

$$g[\eta,\alpha] = f'\eta,$$
$$g[\alpha,\eta'] = f''\eta',$$
$$g[\alpha\alpha] = a.$$

Then $g$ is defined over $\mathscr{M} \cup \{\alpha\}$. The conjunction (3) takes the value $\boldsymbol{t}$ hence so do:

$$(E\zeta)\,\phi\{f,\alpha,\eta,\eta',\zeta\} = \boldsymbol{t} \quad \text{for arbitrary } \eta,\eta' \text{ in } \mathscr{M},$$
$$(E\zeta')\,\phi\{f,\alpha,\eta,\alpha,\zeta'\} = \boldsymbol{t} \quad \text{for arbitrary } \eta \text{ in } \mathscr{M},$$
$$(E\zeta'')\,\phi\{f,\alpha,\alpha,\eta',\zeta''\} = \boldsymbol{t} \quad \text{for arbitrary } \eta' \text{ in } \mathscr{M},$$
$$(E\zeta''')\,\phi\{f,\alpha,\alpha,\alpha,\zeta'''\} = \boldsymbol{t},$$

but this says that (1) is satisfied over $\mathscr{M} \cup \{\alpha\}$ as desired.

We now show that there is an effective method of deciding whether a closed $\mathscr{F}_C$-statement, $(A\xi,\xi')\,(E\eta',\ldots,\eta^{(\mu)})\,\phi\{\pi;\xi,\xi',\eta',\ldots,\eta^{(\mu)}\}$ (0) containing exactly one predicate variable and that binary, can be satisfied. The method with more predicates and with various place numbers is similar.

A table of order $\nu$ for a binary logical function $f$ is a $\nu\times\nu$ array of $\boldsymbol{t}$'s and $\boldsymbol{f}$'s giving the value of $f\lambda\mu$ for $1 \leqslant \lambda, \mu \leqslant \nu$. A table $T_\mu$ of order $(\mu+2)$ for a binary logical function $f$, is said to satisfy $\phi\{\pi;\xi,\xi',\eta',\ldots,\eta^{(\mu)}\}$ if

$$\phi\{f;1,2,3,\ldots,(\mu+2)\} = \boldsymbol{t}. \quad (4)$$

Denote by $[T_\mu/\kappa\kappa']$, $1 \leqslant \kappa,\kappa' \leqslant \mu$, that table of order 2 which is obtained from the table $T_\mu$ when we extract the $2 \times 2$ principal minor $(\kappa, \kappa')$, i.e. the intersection of the $\kappa$, $\kappa'$ rows and columns.

We show that N. and S.C. that (0) be satisfied is that there is a non-empty set $\Sigma$ of tables of order $(\mu+2)$ which satisfy (4) and which have the following properties:

(A) If $T_0$ is a table of $\Sigma$ and $1 \leqslant \kappa, \kappa' \leqslant \mu+2$, $\kappa \neq \kappa'$, then there is a table $T$ of $\Sigma$ such that $[T/12] = [T_0/\kappa\kappa']$.

(B) If $T_0$ is a table of $\Sigma$ then there is a table $T'$ of $\Sigma$ such that $[T'/1] = [T_0/1]$ and $T' = [T'/1134......(\mu+2)]$. i.e., the first two rows are columns are the same.

(C) If $T_1$, $T_2$ are tables of $\Sigma$ then there is a table $T$ of $\Sigma$ such that $[T/1] = [T_1/1]$ and $[T/2] = [T_2/1]$.

If (0) is satisfied then clearly there is a set $\Sigma$ of tables of order $(\mu+2)$ which satisfy (4) and which have the properties (A), (B) and (C). If (0) is satisfiable then it is satisfiable over $\mathscr{N}$ and there will be a table over $\mathscr{N} \times \mathscr{N}$ giving the value of $f\kappa\kappa'$ for all $1 \leqslant \kappa, \kappa'$. In this satisfaction we replace $\xi$, $\xi'$ by any two members of $\mathscr{N}$ and there will be $\mu$ other members of $\mathscr{N}$ for $\eta', \ldots, \eta^{(\mu)}$, call these $3, \ldots, (\mu+2)$, where $\xi$, $\xi'$ are 1, 2; there will be other $\mu$ members of $\mathscr{N}$, if $\xi$, $\xi'$ are $\kappa$, $\kappa'$, $1 \leqslant \kappa, \kappa' \leqslant \mu+2$ which gives (A). Similarly (B) if we replace $\xi$, $\xi'$ by the same element of $\mathscr{N}$, namely 1. If we replace $\xi$, $\xi'$ by $\theta$, $\theta'$ and again by $\pi$, $\pi'$ then there will be a satisfaction when we replace $\xi$, $\xi'$ by $\theta$, $\pi$ which gives us (C).

Let $\mathscr{T}$ be a table over $\mathscr{N}$ which satisfies (0) then we choose any two members of $\mathscr{N}$ for $\xi$, $\xi'$, say $\kappa$, $\kappa'$, and there will be $\mu$ other members of $\mathscr{N}$ for $\eta', \ldots, \eta^{(\mu)}$, say $\nu', \ldots, \nu^{(\mu)}$, then the $(\mu+2)$ minor of $\mathscr{T}$ formed from the $\kappa, \kappa', \nu', \ldots, \nu^{(\mu)}$ rows and columns is a table $T$ which satisfies (4). The set $\Sigma$ of tables which satisfy (4) and are of order $(\mu+2)$ is bounded because there are at most $2^{(\mu+2)^2}$ of them and these can be written down and tested to see if they satisfy (4).

We now show that if there is a non-empty set $\Sigma$ of tables of order $(\mu+2)$ which satisfy (4) and have properties (A), (B) and (C) then (0) is satisfiable over $\mathscr{N}$. It is an effective process to decide if a set $\Sigma$ of tables of order $(\mu+2)$ which satisfy (4) also satisfy (A), (B) and (C).

Let
$$\omega[\nu, \kappa] = \tfrac{1}{2}(\kappa^2 + 2\kappa\nu + \nu^2 - 3\kappa - \nu + 2),$$
this is the place number of the ordered pair $\langle \nu, \kappa \rangle$ in the list:
$$\langle 1, 1\rangle, \ \langle 1, 2\rangle, \ \langle 2, 1\rangle, \ \langle 1, 3\rangle, \ \langle 2, 2\rangle, \ \langle 3, 1\rangle, \ \langle 1, 4\rangle, \ldots$$
also $\kappa, \nu \leqslant \omega[\kappa, \nu]$.

Let $\Sigma$ be a set of tables of order $(\mu+2)$ which satisfy (4) and (A), (B) and (C). Denote by $\Xi(T_1, T_2)$ a table of $\Sigma$ such that

$$[\Xi(T_1, T_2)/1] = [T_1/1],$$

$$[\Xi(T_1, T_2)/2] = [T_2/1],$$

such a table exists by (C). By some numbering of all possible tables we can easily arrange that $\Xi(T_1, T_2)$ is uniquely defined.

Let $\Sigma_1$ be the set of all tables of order 1 which are minors of the tables of $\Sigma$. Let $\Sigma_2$ be the set of all tables of order 2 which are minors of tables of $\Sigma$. Then $\Sigma_1$ has at most 2 members and $\Sigma_2$ has at most 16 members.

We now define a set $Z_\theta, \theta = 0, 1, 2, \ldots,$ of tables of order $1+\theta.\mu$ as follows:

$Z_0$ is an arbitrary member of $\Sigma_1$. If $P_2$ is a member of $\Sigma_2$ we denote by $\Xi(P_2)$ a table of $\Sigma$ such that $[\Xi(P_2)/1, 2] = P_2$, by (A) there is such a table. When $Z_\theta$ has been determined in such a way that $[Z_\theta/\kappa, \kappa']$ is in $\Sigma_2$, $1 \leqslant \kappa, \kappa' \leqslant 1+\theta.\mu$, then we determine $Z_{\theta+1}$ as follows : (for $Z_0$ this condition becomes $[Z_0/1, 1]$ is in $\Sigma_2$, this is satisfied by (B)). Let $\theta = \omega[\kappa, \kappa']$, then $[Z_{\theta+1}/1, 2, \ldots, 1+\theta.\mu] = Z_\theta$, i.e. $Z_{\theta+1}$ is the same as $Z_\theta$ whenever $Z_\theta$ is defined. $[Z_{\theta+1}/\kappa, \kappa', 2+\theta.\mu, \ldots, 1+(\theta+1).\mu] = \Xi[Z_\theta/\kappa, \kappa']$, this follows from (A) and the hypothesis that $[Z_\theta/\kappa, \kappa']$ is in $\Sigma_2$. Altogether this gives the values for $Z_{\theta+1}$ at the points $\langle \nu, \nu' \rangle$ where

$$\nu, \nu' = \kappa, \kappa', 2+\theta.\mu, \ldots, 1+(\theta+1).\mu \quad \text{or} \quad \nu, \nu' = 1, 2, \ldots, 1+\theta.\mu.$$

We also want

$$[Z_{\theta+1}/\lambda, \lambda'] = [\Xi([Z_\theta/\lambda]), [\Xi([Z_\theta/\kappa, \kappa']/\lambda'])/1, 2],$$

this is possible by (C), here $\lambda = 0, 1, 2, \ldots, 1+\theta.\mu$ except $\kappa, \kappa'$,

$$\lambda' = 2+\theta.\mu, \ldots, 1+(\theta+1).\mu.$$

This gives the values of $Z_{\theta+1}$ at the points $\langle \lambda, \lambda' \rangle$, $\langle \lambda', \lambda \rangle$ where $\lambda, \lambda'$ are as just described. Thus altogether $Z_{\theta+1}$ is defined at the points $\langle \nu, \nu' \rangle$ for $1 \leqslant \nu, \nu' \leqslant 1+(\theta+1).\mu$.

It remains to show that $[Z_{\theta+1}/\nu, \nu']$ is in $\Sigma_2$ for $1 \leqslant \nu, \nu' \leqslant 1+(\theta+1).\mu$, so that we can proceed as above to determine $Z_{\theta+2}$ from $Z_{\theta+1}$. Clearly this already holds for $1 \leqslant \nu \leqslant 1+\theta.\mu$ and $2+\theta \leqslant \nu' \leqslant 1+(\theta+1).\mu$ by our construction. It also holds for $2+\theta.\mu \leqslant \nu \leqslant 1+(\theta+1).\mu$ because then $[Z_{\theta+1}/\nu, \nu'] = [\Xi([Z_\theta/\kappa, \kappa'])/\nu, \nu']$ which is in $\Sigma_2$. Thus it holds generally.

We now see that (0) is satisfied over $\mathcal{N}$, for let $\kappa, \kappa'$ be a pair of natural numbers, $\theta = \omega[\kappa, \kappa']$. Then $\phi\{f; \kappa, \kappa', 2+\theta.\mu, \ldots, 1+(\theta+1).\mu\} = \boldsymbol{t}$, where $f$ is the function such that

$$f[\nu, \nu'] \quad \text{is given by} \quad \Xi([Z_\theta/\kappa, \kappa']).$$

This completes the demonstration of the proposition.

The same procedure can be applied to

$$(A\xi)\,(E\eta', \ldots, \eta^{(\mu)})\,\phi\{\pi;\ \xi, \eta', \ldots, \eta^{(\mu)}\},$$

or we can treat this as

$$(A\xi, \xi')\,(E\eta', \ldots, \eta^{(\mu)})\,K\phi\{\pi;\ \xi, \eta', \ldots, \eta^{(\mu)})\,(D\pi\xi'\xi'N\pi\xi'\xi'),$$

to which it is equivalent. But if we apply the procedure to

$$(A\xi, \xi', \xi''))\,E(\eta', \ldots, \eta^{(\mu)})\,\phi\{\pi;\ \xi, \xi', \xi'', \eta', \ldots, \eta^{(\mu)}\}$$

then it fails, because though we can find conditions analogous to (A), (B) and (C) yet some are necessary and others are sufficient. We are unable to find necessary and sufficient conditions.

### 3.21 *The reduction problem*

Since we are unable to solve the decision problem for $\mathscr{F}_C$ it is of some interest to find types of $\mathscr{F}_C$-statements to which any $\mathscr{F}_C$-statement is equivalent as regards satisfiability or as regards validity.

We have just shown that any $\mathscr{F}_C$-statement of the form (0) can be decided as regards satisfiability, in this paragraph we show that any $\mathscr{F}_C$-statement is equivalent as regards satisfiability to an effectively constructible $\mathscr{F}_C$-statement of similar form to (0) but with three universal quantifiers at the head of the prefix instead of only two.

We thus have achieved some finality in the decision problem as regards satisfaction in the sense that statements of the form (0) are solvable as regards satisfaction, but those like (0) but with 3 universal quantifiers are unsolvable, (in fact equivalent to the general case). Again if the binary predicate is replaced by a unary predicate then we again have solvability. Thus we have gone as far as we can in this direction.

There are a variety of other directions in which we could try to get similar results. These come about by considering different types of prefix.

PROP. 24. *A closed $\mathscr{F}_C$ statement $\phi$ is satisfiable over $\mathscr{N}$ if and only if a certain binary $\mathscr{F}_C$-statement of the form*

$$(A\xi, \xi', \xi'')(E\eta', \ldots, \eta^{(\mu)})\,\psi\{\mathfrak{p}; \xi, \xi', \xi'', \eta', \ldots, \eta^{(\mu)}\} \tag{5}$$

*is satisfiable over $\mathscr{N}$, where $\mathfrak{p}$ stands for $\pi', \ldots, \pi^{(\kappa)}$, where $\pi', \ldots, \pi^{(\kappa)}$ are binary predicates and there is an effective method of finding $\psi$ given $\phi$.*

LEMMA (i). *If $\phi$ is a closed $\mathscr{F}_C$-statement then there is an effective process for finding a closed binary $\mathscr{F}_C$-statement $\psi$ such that $\phi$ is an $\mathscr{F}_C$-theorem if and only if $\psi$ is an $\mathscr{F}_C$-theorem.*

A binary $\mathscr{F}_C$-statement is an $\mathscr{F}_C$-statement which contains only binary and singularly predicates. Let $\pi', \ldots, \pi^{(\lambda)}$ be exactly all the predicate variables which occur in the $\mathscr{F}_C$-statement $\phi$. Let $\pi^{(\theta)}$ have $\mu^{(\theta)}$ argument places. Let $\mu = Max[\mu', \ldots, \mu^{(\lambda)}]$. Let $\pi_1', \ldots, \pi_1^{(\lambda)}$ be new and distinct one-place predicate variables, let $\pi_2', \ldots, \pi_2^{(\mu)}$ be new and distinct two-place predicate variables. Let $\upsilon$ be a new individual variable. We form $\psi$ from $\phi$ by replacing $\pi^{(\theta)}\xi' \ldots \xi^{(\mu^{(\theta)})}$ by

$$(A\upsilon)\,K \prod_{\theta'=1}^{\mu(\theta)} \pi_2^{(\theta')}\xi^{(\theta')}\upsilon\,\pi_1^{(\theta)}\upsilon \tag{6}$$

throughout $\phi$. Clearly $\psi$ is binary. If $\phi$ is an $\mathscr{F}_C$-theorem then $\phi$ is generally valid over $\mathscr{N}$. If we replace $\pi_2', \ldots, \pi_2^{(\mu^{(\theta)})}, \pi_1^{(\theta)}$ by logical functions over $\mathscr{N}$ then (6) gives a logical function over $\mathscr{N}$ but if in $\phi$ we replace $\pi', \ldots, \pi^{(\lambda)}$ by any logical functions over $\mathscr{N}$ the result is always $\boldsymbol{t}$, hence the result of replacing $\pi_2', \ldots, \pi_1^{(\lambda)}$ by arbitrary logical functions over $\mathscr{N}$ gives the value $\boldsymbol{t}$, i.e. $\psi$ is generally valid over $\mathscr{N}$, thus $\psi$ is an $\mathscr{F}_C$-theorem.

Now suppose that $\psi$ is generally valid over $\mathscr{N}$ (and hence is an $\mathscr{F}_C$-theorem). Let $p', \ldots, p^{(\lambda)}$ be any logical functions over $\mathscr{N}$ with $\mu', \ldots, \mu^{(\lambda)}$ argument places respectively. We have to show that $\phi$ is satisfied by these functions. Let $\phi$ become $\bar{\phi}$ when $p', \ldots, p^{(\lambda)}$ are substituted for $\pi', \ldots, \pi^{(\lambda)}$ respectively. Let $\mathscr{N}^\mu$ be the set of all $\mu$-tuplets of natural numbers arranged in some order (one such ordering will be displayed in Ch. 5), let $\nu$ be the place number of the $\mu$-tuplet $\langle \nu_1, \ldots, \nu_\mu \rangle$ in this ordering. We define one-place logical functions $q', \ldots, q^{(\lambda)}$ and two-place logical functions $r', \ldots, r^{(\mu)}$ over $\mathcal{N}^{\mu}$ as follows:

$$q^{(\theta)}\nu = p^{(\theta)}\nu_1 \ldots \nu_{\mu^{(\theta)}} \quad (1 \leqslant \theta \leqslant \lambda),$$

$$r^{(\theta)}\nu\nu' = t \quad \text{if and only if} \quad \nu_1 = \nu'_{\theta} \quad (1 \leqslant \theta \leqslant \lambda).$$

Since $\psi$ is generally valid over $\mathcal{N}^{\mu}$ (since $\mathcal{N}^{\mu}$ is enumerated) then it is satisfied over $\mathcal{N}^{\mu}$ by replacing $\pi_1^{(\theta)}$ by $q^{(\theta)}$ and $\pi_2^{(\theta)}$ by $r^{(\theta)}$. Let $\psi$ become $\overline{\psi}$ after this substitution. In particular (6) becomes:

$$(A\upsilon)\, C \prod_{\theta'=1}^{\mu(\theta)} r^{(\theta')}\xi^{(\theta')}\upsilon q^{(\theta)}\upsilon. \tag{7}$$

Let $\Delta\lambda\mu$ be that logical function over $\mathcal{N}$ which has the value $t$ if and only if $\lambda = \mu$. Then (7) becomes

$$(A\upsilon', \ldots, \upsilon^{(\mu^{(\theta)})})\, C \prod_{\theta'=1}^{\mu(\theta)} \Delta\xi^{(\theta')}\upsilon^{(\theta')}p^{(\theta)}\upsilon' \ldots \upsilon^{(\mu^{(\theta)})};$$

this has the same truth-value as $p^{(\theta)}\xi' \ldots \xi^{(\mu^{(\theta)})}$, hence $\overline{\phi}$ has the same truth-value as $\overline{\psi}$, thus $\phi$ is generally valid over $\mathcal{N}$ as we wished to show.

COR. *If $\phi$ is a closed $\mathscr{F}_C$-statement then there is an effective process for finding a closed binary $\mathscr{F}_C$-statement $\psi$ such that $\phi$ is satisfiable if and only if $\psi$ is satisfiable.*

An $\mathscr{F}_C$-statement is said to be in *Skolem V-normal form* if it is closed and in prenex normal form and has a prefix in which each existential quantifier preceeds each universal quantifier. i.e. if it is of the form

$$(E\mathfrak{x})\,(A\mathfrak{y})\,\phi\{\mathfrak{x}, \mathfrak{y}\},$$

where the matrix $\phi$ is quantifier-free. An $\mathscr{F}_C$-statement is said to be in *Skolem S-normal form* if it is closed and in prenex normal form and has a prefix in which each universal quantifier preceeds each existential quantifier, i.e. if it is of the form $(A\mathfrak{x})\,(E\mathfrak{y})\,\phi\{\mathfrak{x}, \mathfrak{y}\}$, where the matrix $\phi$ is quantifier-free. Here $\mathfrak{x}$ stands for $\xi', \ldots, \xi^{(\lambda)}$ and $\mathfrak{y}$ stands for $\eta', \ldots, \eta^{(\mu)}$.

LEMMA (ii). *If $\phi$ is a closed binary $\mathscr{F}_C$-statement then there is an effective method for finding a closed binary $\mathscr{F}_C$-statement $\psi$ in Skolem S-normal form such that $\phi$ is satisfiable if and only if $\psi$ is satisfiable.*

Let $\phi$ be a closed binary $\mathscr{F}_C$-statement, let $\phi'$ be a prenex normal form of it. Then $\phi$ is satisfiable if and only if $\phi'$ is satisfiable because $B\phi\phi'$ is an

$\mathscr{F}_C$-theorem. If $\phi'$ begins with an existential quantifier then conjunct to it $(A\zeta)DN\pi\zeta\pi\zeta$ and reduce the result to prenex normal form beginning with the universal quantifier $(A\zeta)$. Call the result $\phi''$. Then $B\phi'\phi''$ is an $\mathscr{F}_C$-theorem so $\phi'$ is satisfiable if and only if $\phi''$ is satisfiable. Thus $\phi$ is satisfiable if and only if $\phi''$ is satisfiable. Let $\phi''$ be $(Q)\chi$ where $(Q)$ is $(A\mathfrak{x}_1)(E\mathfrak{y}_1)(Q')$ and $(Q')$ is $(A\mathfrak{x}_2)(E\mathfrak{y}_2)(Q'')$, here $(Q)$, $(Q')$, $(Q'')$ denote sequences of quantifiers. Write:

$$\omega[\mathfrak{x},\mathfrak{y}] \quad \text{for} \quad (E\zeta)K\prod_{\theta=1}^{\lambda}\pi^{(\theta)}\zeta\xi^{(\theta)}\prod_{\theta'=1}^{\mu}\pi^{(\lambda+\theta')}\zeta\eta^{(\theta')}. \tag{8}$$

Also write $(E\zeta)\,\omega_0[\zeta,\mathfrak{x},\mathfrak{y}]$ for (8). Here $\pi',\ldots,\pi^{(\lambda+\mu)}$ are new. Let $\phi'''$ be

$$K(A\mathfrak{x}_1)(E\mathfrak{y}_1)\,\omega[\mathfrak{x}_1,\mathfrak{y}_1](A\mathfrak{x},\mathfrak{y})\,C\omega[\mathfrak{x},\mathfrak{y}](Q')\chi.$$

Let $\phi^{\text{iv}}$ be

$$(A\mathfrak{x}_1)(A\mathfrak{x})(A\mathfrak{y})(A\mathfrak{x}_2)(A\zeta)(E\mathfrak{y}_1)(E\mathfrak{y}_2)(E\zeta')(Q'')\,K\omega_0[\zeta',\mathfrak{x}_1,\mathfrak{y}_1]\;C\omega_0[\zeta,\mathfrak{x},\mathfrak{y}]\chi,$$

where the sets of variables $\mathfrak{x}_1,\mathfrak{y}_1,\mathfrak{x},\mathfrak{y},\mathfrak{x}_2,\mathfrak{y}_2$ are distinct.

If $\phi^{\text{iv}}$ is satisfiable then so is $(Q)\chi$, i.e. $\phi''$ is satisfiable. We have $B\phi'''\phi^{\text{iv}}$ is an $\mathscr{F}_C$-theorem, because $\phi^{\text{iv}}$ is a prenex normal form of $\phi'''$. Also

$$C\phi^{\text{iv}}(Q)\chi, \tag{9}$$

because we have the $\mathscr{F}_C$-theorems:

$$C(A\mathfrak{y})\,C\omega[\mathfrak{x},\mathfrak{y}](Q_1)\chi C(E\mathfrak{y})\,\omega[\mathfrak{x},\mathfrak{y}](E\mathfrak{y})(Q_1)\chi,$$
$$C(A\mathfrak{x})\,C(E\mathfrak{y})\,\omega[\mathfrak{x},\mathfrak{y}](E\mathfrak{y})(Q_1)\chi C(A\mathfrak{x})(E\mathfrak{y})\,\omega[\mathfrak{x},\mathfrak{y}](A\mathfrak{x})(E\mathfrak{y})(Q_1)\chi$$
$$C(A\mathfrak{x})(A\mathfrak{y})\,C\omega[\mathfrak{x},\mathfrak{y}](Q_1)\chi(A\mathfrak{x})\,C(E\mathfrak{y})\,\omega[\mathfrak{x},\mathfrak{y}](E\mathfrak{y})(Q_1)\chi,$$

whence by Modus Ponens:

$$C(A\mathfrak{x})(A\mathfrak{y})\,C\omega[\mathfrak{x},\mathfrak{y}](Q_1)\chi C(A\mathfrak{x})(E\mathfrak{y})\,\omega[\mathfrak{x},\mathfrak{y}](A\mathfrak{x})(E\mathfrak{y})(Q_1)\chi,$$

whence, using $(CC\phi C\psi\chi CK\phi\psi\chi)$,

$$CK(A\mathfrak{x})(E\mathfrak{y})\,\omega[\mathfrak{x},\mathfrak{y}](A\mathfrak{x})(A\mathfrak{y})\,C\omega[\mathfrak{x},\mathfrak{y}](Q')\chi(A\mathfrak{x})(E\mathfrak{y})(Q')\chi,$$

which is

$$C\phi'''(Q)\chi \tag{10}$$

as desired, hence $C\phi^{\text{iv}}(Q)\chi$, i.e. (9).

We now show that if $(Q)\chi$ is satisfiable over $\mathscr{N}$ then so is $\phi^{\text{iv}}$. Suppose then that $(Q)\chi$ is satisfiable over $\mathscr{N}$ by certain logical functions. Then for all natural numbers $\nu',\ldots,\nu^{(\lambda)}$ there are natural numbers $\kappa',\ldots,\kappa^{(\mu)}$

such that $(Q)\chi$ takes the value $\boldsymbol{t}$ when the predicates in $\chi$ are replaced by the logical functions just mentioned. Call the set of natural numbers $\nu', \ldots, \nu^{(\lambda)}, \kappa', \ldots, \kappa^{(\mu)}$ a $T$-set, since $(Q)\chi$ is satisfiable then there are $T$-sets. Enumerate the ordered $(\lambda+\mu)$-tuplets and let the $\theta$th be

$$\langle \theta_1, \ldots, \theta_{(\lambda+\mu)} \rangle.$$

Define logical functions $\Omega_1[\nu, \kappa], \ldots, \Omega_{(\lambda+\mu)}[\nu, \kappa]$ such that

$$\Omega_\pi[\theta, \theta_\pi] = \boldsymbol{t} \quad \text{for} \quad 1 \leqslant \pi \leqslant \lambda,$$
$$\Omega_{\lambda+\pi}[\theta, \theta_\pi] = \boldsymbol{t} \quad \text{for} \quad 1 \leqslant \pi \leqslant \mu,$$

if and only if $\theta_1, \ldots, \theta_{(\lambda+\mu)}$ is a $T$-set. Then $(E\zeta) \prod_{\pi=1}^{(\lambda+\mu)} \Omega_\pi[\zeta, \theta_\pi]$ takes the value $\boldsymbol{t}$ for $T$-sets otherwise the value $\boldsymbol{f}$. Thus for these logical functions $\phi^{\text{iv}}$ takes the value $\boldsymbol{t}$ because $(A\mathfrak{x})(E\mathfrak{y})(E\zeta) \prod_{\theta=1}^{\mu} \omega[\zeta, \xi^{(\theta)}]$ then takes the value taken by $(Q)\chi$ so that $C\omega[\mathfrak{x}, \mathfrak{y}](Q)\chi$ takes the value $\boldsymbol{t}$. Thus $K(A\mathfrak{x})(E\mathfrak{y})\,\omega[\mathfrak{x}, \mathfrak{y}](A\mathfrak{x})(A\mathfrak{y})\,C\omega[\mathfrak{x}, \mathfrak{y}](Q)\chi$ takes the value $\boldsymbol{t}$. Thus $\phi^{\text{iv}}$ is satisfiable as desired. Repeat the process until all the existential quantifiers succeed all the universal quantifiers. This completes the demonstration of the lemma.

Lemma. (iii). *If $\phi$ is a closed binary $\mathscr{F}_C$-statement then there is an effective method of finding a closed binary $\mathscr{F}_C$-statement $\psi$ in Skolem V-normal form such that $\phi$ is generally valid if and only if $\psi$ is generally valid.*

The demonstration is similar to that of lemma (ii). These are similar lemmas omitting the word 'binary'.

Lemma (iv). *If $\phi$ is a closed binary $\mathscr{F}_C$-statement in Skolem S-normal form then there is an effective method of finding a closed binary $\mathscr{F}_C$-statement $\psi$ in Skolem S-normal form and containing exactly three universal quantifiers such that $\phi$ is satisfiable if and only if $\psi$ is satisfiable.*

Let $\phi$ be $(A\xi', \ldots, \xi^{(\lambda)})(E\eta', \ldots, \eta^{(\mu)})\,\phi'\{\xi', \ldots, \xi^{(\lambda)}, \eta', \ldots, \eta^{(\mu)}\}$ where $\phi'$ is binary and is without quantifiers and $\phi$ is closed. Let $\phi''$ be the conjunction of the following three $I\mathscr{F}_C$-statements:

$$(A\xi''', \ldots, \xi^{(\lambda)})(A\zeta')(E\zeta'', \zeta''')(E\eta', \ldots, \eta^{(\mu)})\,K\pi'\zeta'\zeta''\pi''\zeta'\zeta'''$$
$$\phi\{\zeta'', \zeta''', \xi'''', \ldots, \xi^{(\lambda)}, \eta', \ldots, \eta^{(\mu)}\}, \quad (11)$$

$$(A\zeta'', \zeta''')(E\zeta')\,K\pi'\zeta'\zeta''\pi''\zeta'\zeta''', \quad (12)$$

$$(A\zeta', \zeta'', \zeta''')\,KCK\pi'\zeta'\zeta''\pi'\zeta'\zeta'''I\zeta''\zeta'''CK\pi''\zeta'\zeta''\pi''\zeta'\zeta'''I\zeta''\zeta''', \quad (13)$$

where $\zeta', \zeta''$ are new. Let the predicate variables which occur in $\phi'$ be $\pi''', \ldots, \pi^{(\kappa+2)}$, then these are binary. We show that if $\phi$ is satisfiable then so is $\phi''$. Let $p''', \ldots, p^{(\kappa+2)}$ be a set of logical functions over $\mathscr{N}$ which satisfy $\phi$, if $p', p''$ are logical functions over $\mathscr{N}$ such that $p'[\xi, \xi_1] = \boldsymbol{t}, p''[\xi, \xi_2] = \boldsymbol{t}$ if and only if $\xi$ is the place number of the ordered pair $\langle \xi_1, \xi_2 \rangle$ in our ordering of all ordered pairs of natural numbers, then $p', p'', \ldots, p^{(\kappa+2)}$ clearly satisfy $\phi''$. We now show that if $\phi''$ is satisfiable then so is $\phi$. If $p', p'', \ldots, p^{(\kappa+2)}$ is a set of logical functions which satisfy $\phi''$ over $\mathscr{N}$, then $p''', \ldots, p^{(\kappa+2)}$ is a set of logical functions over $\mathscr{N}$ which satisfy $\phi$, because if $\nu', \ldots, \nu^{(\lambda)}$ are natural numbers then by (12) there is a natural number $\pi'$ such that $p'[\pi', \nu']$ and $p''[\pi', \nu'']$ take the value $\boldsymbol{t}$. From (11) there are further natural numbers $\pi'', \pi'''$ and $\gamma', \ldots, \gamma^{(\mu)}$ such that $p'[\pi', \pi'']$ and $p''[\pi', \pi''']$ and $\overline{\phi}'\{\pi'', \pi''', \nu''', \ldots, \nu^{(\lambda)}, \gamma, \ldots, \gamma^{(\mu)}\}$ take the value $\boldsymbol{t}$ for some natural numbers $\gamma', \ldots, \gamma^{(\mu)}$, where $\overline{\phi}'$ is obtained from $\phi$ by replacing $\pi^{(\theta)}$ by $p^{(\theta)}$. From (13) we have $\pi'' = \nu', \pi''' = \nu''$ where

$$\overline{\phi}'\{\nu', \nu'', \nu''', \ldots, \nu^{(\lambda)}, \gamma', \ldots, \gamma^{(\mu)}\}$$

takes the value $\boldsymbol{t}$. Thus all together $\phi$ is satisfied if and only if $\phi''$ is satisfied. The prenex normal form of $\phi''$ contains $(\lambda - 1)$ initially placed universal quantifiers followed by existential quantifiers as long as $\lambda - 1 \geqslant 3$. Proceed until we get exactly three initially placed universal quantifiers. Call the result $\phi'''$, then $\phi$ is satisfiable if and only if $\phi'''$ is satisfiable.

We now eliminate the equality relation. Let $\phi^{\text{iv}}$ be the conjunction of $\phi'''$ and $(A\xi) I\xi\xi$, $(A\xi, \eta)(CI\xi\eta C\pi\xi\pi\eta$,

$$(A\xi, \eta, \zeta) CI\xi\eta C\pi\xi\zeta\pi\eta\zeta \quad \text{and} \quad (A\xi, \eta, \zeta) CI\xi\eta C\pi\zeta\xi\pi\zeta\eta,$$

for all singularity and binary predicates which occur in $\phi'''$ including the equality relation itself. The prenex normal form of $\phi^{\text{iv}}$ again contains exactly three universal quantifiers followed by existential quantifiers only. Now replace the equality relation by a new binary predicate and we have finished, because $\phi^{\text{iv}}$ is satisfied if and only if $\phi'''$ is satisfied.

PROP. 25.† *A binary $\mathscr{F}_C$-statement*

$$(A\xi', \xi'', \xi''')(E\eta', \ldots, \eta^{(\nu)})\, \phi\{\pi', \ldots, \pi^{(\lambda)}, \xi', \ldots, \eta^{(\mu)}\}$$

*in Skolem S-normal form is satisfiable if and only if a binary $\mathscr{F}_C$-statement $(A\xi', \xi'', \xi'')(E\eta', \ldots, \eta^{(\nu)})\, \psi\{\pi; \xi', \xi'', \xi'', \eta', \ldots, \eta^{(\nu)}$, also in Skolem S-normal*

† Prop. 25 follows closely the proof given by Kalmár and Surányi (1947), though the symbolism has been changed to that of the present author. It is reproduced by permission of the publisher, the Association for Symbolic Logic.

*form and containing only a single binary predicate, is satisfiable. Also there is an effective method for finding $\psi$ given $\phi$.*

Let $$(A\xi', \xi'', \xi''')(E\eta', \ldots, \eta^{(\mu)})\,\phi\{\pi', \ldots, \pi^{(\lambda)}; \xi', \xi'', \xi''', \eta', \ldots, \eta^{(\mu)}\} \quad (14)$$

be the given binary $\mathscr{F}_C$-statement in Skolem $S$-normal form, where $\phi$ is quantifier free. Suppose that (14) is satisfiable over $\mathscr{N}$. Then there are two-place logical functions $p', \ldots, p^{(\lambda)}$ defined over $\mathscr{N}$ such that for any natural numbers $\kappa', \kappa'', \kappa'''$ there exist natural numbers $\theta', \ldots, \theta^{(\mu)}$ such that

$$\bar{\phi}(p', \ldots, p^{(\lambda)}; \kappa', \kappa'', \kappa''', \theta', \ldots, \theta^{(\mu)}\} = t, \quad (15)$$

where $\bar{\phi}$ is what $\phi$ becomes when $\pi', \ldots, \pi^{(\lambda)}$ are replaced by $p', \ldots, p^{(\lambda)}$ respectively and the propositional connectives are replaced by their respective truth-functions. $\bar{\phi}$ is a $\lambda.(\mu+3)^2$-place logical function with arguments $p^{(\theta)}[\nu', \nu''], 1 \leqslant \theta \leqslant \lambda, 1 \leqslant \nu', \nu'' \leqslant \mu+3$. We want to construct

$$(A\xi', \xi'', \xi''')(E\eta', \ldots, \eta^{(\mu)})\,\psi\{\pi; \xi', \xi'', \xi''', \eta', \ldots, \eta^{(\mu)}\} \quad (16)$$

with a single binary predicate $\pi$ which is satisfiable if and only if (13) is satisfiable. (16) is then called a *reduction type*.

We can interpret $p^{(\mu)}\nu'\nu''$ occurring in (15) as the value $Qp^{(\mu)}\langle\nu', \nu''\rangle$ of a single binary predicate $Q$ defined over the set $J_0 = \mathscr{N}^2 \cup \{p', \ldots, p^{(\lambda)}\}$, where $\mathscr{N}^2$ is the set of ordered pairs of natural numbers $\langle\nu', \nu''\rangle$. Thus: there exist elements $\pi', \ldots, \pi^{(\lambda)}$ of $J_0$ and for arbitrary $\nu', \nu'', \nu'''$ of $\mathscr{N}$ there exists $\nu^{\text{iv}}, \ldots, \nu^{(\mu+3)}$ of $\mathscr{N}$ and ordered pairs $\langle\nu^{(\theta')}, \nu^{(\theta'')}\rangle$ of $J_0$ so that for $1 \leqslant \theta', \theta'' \leqslant \mu+3$ we have:

$$\Psi\{Q; \pi', \ldots, \pi^{(\lambda)}; \langle\nu', \nu'\rangle, \ldots, \langle\nu^{(\mu+3)}, \nu^{(\mu+3)}\rangle\} = t, \quad (17)$$

where $\Psi$ denotes the logical function which arises from $\phi\{p', \ldots, p^{(\lambda)}; x', \ldots, x^{(\mu+3)}\}$ when we replace $p^{(\theta)}x^{(\theta')}x^{(\theta'')}$ by $Q\pi^{(\theta)}\langle x^{(\theta')}, x^{(\theta'')}\rangle$.

The $\mathscr{F}_C$-statement (16) to be constructed has to formalize in $\mathscr{F}_C$ the clause (17). Instead of $J^0$ we shall use the set

$$J = \mathscr{N} \cup (\mathscr{N}^2 \times \{0, 1\}) \cup \{p', \ldots, p^{(\lambda)}\},$$

which consists of the natural numbers $\mathscr{N}$, triads of the forms $\langle\nu^{(\theta')}, \nu^{(\theta'')}, 0\rangle$ and $\langle\nu^{(\theta')}, \nu^{(\theta'')}, 1\rangle$ and $p', \ldots, p^{(\lambda)}$. The triads $\langle\nu^{(\theta')}, \nu^{(\theta'')}, 0\rangle$ will play the rôle of the pairs $\langle\nu^{(\theta')}, \nu^{(\theta'')}\rangle$ while the triads $\langle\nu^{(\theta')}, \nu^{(\theta'')}, 1\rangle$ serve merely to express the coincidence of the first and second components of pairs. Thus for

$$\tau = \langle\nu', \kappa', 0\rangle \quad \sigma = \langle\nu'', \kappa'', 1\rangle$$

we shall define

$$Q\tau\sigma = \boldsymbol{t} \quad \text{if and only if} \quad \nu' = \nu'',$$
$$Q\sigma\tau = \boldsymbol{t} \quad \text{if and only if} \quad \kappa' = \kappa''.$$

$\nu$ is the first component of $\tau$ is expressed by $Q\nu\tau$,

$\kappa$ is the second component of $\tau$ is expressed by $Q\tau\kappa$.

$p'$ is characterized as the only element of $J$ for which $Qp'p' = \boldsymbol{t}$,

$p''$ is characterized as the only element of $J$ for which $NQp''p' = \boldsymbol{t}$,

$p^{(\theta)}$ is characterized as the only element of $J$ for which $Qp^{(\theta)}p^{(\theta-1)} = \boldsymbol{t}$, $3 \leqslant \theta \leqslant \lambda$.

The triads $\tau = \langle \nu', \kappa', 0\rangle$ are characterized by $Q\tau p^{(\lambda)} = \boldsymbol{t}$, we distinguish between the natural numbers and the triads $\langle \nu'', \kappa'', 1\rangle = \sigma$ by $Qp'\nu'' = \boldsymbol{t}$ and $Qp'\sigma = \boldsymbol{f}$, then the triads $\sigma$ can be characterized by '$\sigma$ is different from $p', \ldots, p^{(\lambda)}$ and $\sigma$ is different from a triad $\langle \nu', \kappa', 0\rangle$ and $Qp'\sigma = \boldsymbol{f}$'. Now if (15) holds then, for numerical functions $\rho^{(\mathrm{iv})}, \ldots, \rho^{(\mu+3)}$,

$$\nu^{\mathrm{iv}} = \rho^{(\mathrm{iv})}\nu'\nu''\nu''', \ldots, \nu^{(\mu+3)} = \rho^{(\mu+3)}\nu'\nu''\nu'''.$$

Let $\mathscr{T}$ denote the set of triads $\langle \nu', \kappa', 0\rangle$,

$\mathscr{U}$ denote the set of triads $\langle \nu', \kappa', 1\rangle$,

$\mathscr{P}$ denote the set of $\{p', \ldots, p^{(\lambda)}\}$,

$\mathscr{N}$ denote the set of natural numbers,

$$\mathscr{J} = \mathscr{T} \cup \mathscr{U} \cup \mathscr{P} \cup \mathscr{N}.$$

We define a binary logical function $Q$ over $\mathscr{J}$ by the following table, in this $Q$ has the same truth-value at an entry in the table as the statement standing at that entry.

| | | $\mathscr{P}$ | $\mathscr{N}$ | $\mathscr{T}$ | $\mathscr{U}$ |
|---|---|---|---|---|---|
| | $Qxx'$ | $x' = p^{(\theta')}$<br>$1 \leqslant \theta' \leqslant \lambda$ | $x' = \kappa$ | $x' = \langle \kappa', \kappa'', 0\rangle$ | $x' = \langle \kappa', \kappa'', 1\rangle$ |
| $\mathscr{P}$ | $x = p^{(\theta)}$<br>$1 \leqslant \theta \leqslant \lambda$ | $D\begin{cases} K(\theta \neq 2)(\theta' = 1) \\ K(\theta = \theta'+1)(\theta' \neq 1) \end{cases}$ | $x = x$ | $p^{(\theta)}\kappa'\kappa''$ | $x \neq x$ |
| $\mathscr{N}$ | $x = \nu$ | $\theta' = 1$ | $x \neq x$ | $\nu = \kappa'$ | $\nu = \kappa'$ |
| $\mathscr{T}$ | $x = \langle \nu', \nu'', 0\rangle$ | $D(\theta' = 1)(\theta' = \lambda)$ | $\nu'' = \kappa$ | $x \neq x$ | $\nu' = \kappa'$ |
| $\mathscr{U}$ | $x = \langle \nu', \nu'', 1\rangle$ | $\theta' = 1$ | $\nu'' = \kappa$ | $\nu'' = \kappa''$ | $x \neq x$ |

Consulting the table we see that

$(a_1)$ $Qxx = t$ if and only if (iff) $x = p'$,
$(a_2)$ $NQxp' = t$ iff $x = p''$,
$(a_\theta)$ $Qxp^{(\theta-1)} = t$ iff $x = p^{(\theta)}$ $(3 \leqslant \theta \leqslant \lambda)$,
$(b)$ $Qxp^{(\lambda)} = t$ iff $x \in \mathscr{T}$,
$(c)$ $KKNQxxNQxp^{(\lambda)}Qp'x = t$ iff $x \in \mathscr{N}$,
$(d)$ $KQxp' \prod_{\theta=2}^{\lambda} NQxp^{(\theta)}NQp'x = t$ iff $x \in \mathscr{U}$.

Further supposing that $\tau = \langle \nu', \nu'', 0 \rangle$, $\sigma = \langle \kappa', \kappa'', 1 \rangle$ then

$(e)$ $Q\nu\tau = t$ iff $\nu = \nu'$,
$(f)$ $Q\nu\sigma = t$ iff $\nu = \kappa'$,
$(g)$ $Q\tau\nu = t$ iff $\nu'' = \nu$,
$(h)$ $Q\sigma\nu = t$ iff $\kappa'' = \nu$,
$(i)$ $Q\tau\sigma = t$ iff $\nu' = \kappa'$,
$(j)$ $Q\sigma\tau = t$ iff $\kappa'' = \nu''$.

Also supposing that $\tau_1 = \langle \nu_1', \nu_2'', 0 \rangle$, $\tau_2 = \langle \nu_2', \nu_2'', 0 \rangle$

$(k)$ if $KKKQ\tau_1\sigma Q\tau_2\sigma Q\sigma\tau_1 Q\sigma\tau_2 = t$ then $\tau_1 = \tau_2$,
$(l)$ for $1 \leqslant \theta \leqslant \lambda, Qp^{(\theta)}\tau = p^{(\theta)}\kappa'\kappa''$.

Write $I[x; y', y^{(\lambda)}]$ for $KKNQxxNQxy^{(\lambda)}Qy'x'$

$$\gamma[x; y', \ldots, y^{(\lambda)}] \text{ for } KQxy' \prod_{\theta=2}^{\lambda} KNQxy^{(\theta)}NQy'x.$$

Then for arbitrary elements $x'$, $x''$, $x'''$ of $\mathscr{J}$ and if $y' = p', \ldots, y^{(\lambda)} = p^{(\lambda)}$ we have

$(\bar{a})$ by $(a_\theta)$, $1 \leqslant \theta \leqslant \lambda$,

$$KKKQy'y'NQy''y' \prod_{\theta=3}^{\lambda} Qy^{(\theta)}y^{(\theta-1)}CDDKQx'x'Qx''x''KNQx'y'NQx''y'$$

$$\sum_{\theta=3}^{\lambda} KQx'y^{(\theta-1)}Qx''y^{(\theta-1)}KBQx'x'''Qx''x'''BQx'''x'Qx'''x''. \quad (18)$$

This says that if $x' = x'' = p^{(\theta)}$, $1 \leqslant \theta \leqslant \lambda$, then $BQx'x'''Qx''x'''$ and $BQx'''x'Qx'''x''$, and that $y^{(\theta)} = p^{(\theta)}$, $1 \leqslant \theta \leqslant \lambda$.

$(\bar{b})$ By $(c)$, $(d)$, $(f)$ and $(h)$,

$$CKI[x'; y', y^{(\lambda)}]\, I[x''; y', y^{(\lambda)}]\, (Eu)\, KK\gamma[u; y', \ldots, y^{(\lambda)}]\, Qx'uQux''. \quad (19)$$

This says that if $x', x'' \epsilon \mathscr{N}$ then there is a triad $\langle x', x'', 1\rangle$.

($\bar{c}$) By (c), (b) (d), (e), (f), (i), (g), (h) and (j)

$$CKKI[x'; y', y^{(\lambda)}]\, Qx''y^{(\lambda)}\gamma\, [x'''; y', \dots, y^{(\lambda)}]$$
$$CKCKQx'x''Qx'x'''Qx''x'''CKQx''x'Qx'''x'Qx'''x''. \quad (20)$$

This says that if $x' \epsilon \mathscr{N}$, $x'' \epsilon \mathscr{T}$, $x''' \epsilon \mathscr{U}$, $y' = p', \dots, y^{(\lambda)} = p^{(\lambda)}$ then if $x'$ is the first component of $x''$ and also is the first component of $x'''$ then $x''$ and $x'''$ have the same first components, and similarly for the second components.

($\bar{d}$) By (b), (d) and (k),

$$C\underbrace{K \dots K}_{6\text{ times}}Qx'y^{(\lambda)}Qx''y^{(\lambda)}\gamma[x'''; y', \dots, y^{(\lambda)}]$$
$$Qx'x'''Qx''x'''Qx'''x'Qx'''x'' \prod_{\theta=1}^{\lambda} BQy^{(\theta)}x'Qy^{(\theta)}x''. \quad (21)$$

This says that if $x', x'' \epsilon \mathscr{T}$ and $x''' \epsilon \mathscr{J}$ and if $x', x''$ have the same first and the same second components then $Bp^{(\theta)}x_1'x_2'p^{(\theta)}x_1''x_2''$, $1 \leqslant \theta \leqslant \lambda$, where $x' = \langle x_1', x_2', 0\rangle$, $x'' = \langle x_1'', x_2'', 0\rangle$.

Finally by (c), (b), (e), (g), (l) and the definition of $\Psi$ the fact that for $x', x'', x''' \epsilon \mathscr{N}$ and $x^{\text{iv}} = \rho^{\text{iv}}x'x''x''', \dots, x^{(\mu+3)} = p^{(\mu+3)}x'x''x'''$ (15) holds,

$$(Ex^{\text{iv}}, \dots, x^{(\mu+3)})\, K \prod_{\theta=4}^{\mu+3} I[x^{(\theta)}; y', y^{(\lambda)}]\, C \prod_{\theta=1}^{3} I[x^{(\theta)}; y', y^{(\lambda)}]\, (Eu_{1,1}, \dots, u_{\mu+3,\mu+3})$$
$$\prod_{\theta,\theta=1}^{\mu+3} KKQu_{\theta,\theta'}y^{(\lambda)}\, Qx^{(\theta)}u_{\theta,\theta'}Qu_{\theta,\theta'}x^{(\theta')}\Psi[Q; y', \dots, y^{(\lambda)}, u_{1,1}, \dots, u_{\mu+3,\mu+3}]. \quad (22)$$

This says that for any $x', x'', x''' \epsilon \mathscr{N}$ there are $x^{\text{iv}}, \dots, x^{(\mu+3)} \epsilon \mathscr{N}$ such that if $u_{\theta,\theta'} = \langle x^{(\theta)}, x^{(\theta')}, 0\rangle$ then (15) holds.

Now form the conjunction of (18), …, (22) inclusive and place $(Ax', x'', x''')\, (Ey', \dots, y^{(\lambda)})$ in front and replace $Q$ by the predicate variable $q$, we obtain an $\mathscr{F}_C$-statement of the form (14), viz.:

$$(Ax', x'', x''')\, (Ey', \dots, y^{(\lambda)})\, (Eu)\, (Ex^{\text{iv}}, \dots, x^{(\mu+3)})\, (Eu_{1,1}, \dots, u_{\mu+3,\mu+3})$$
$$\psi\{q; y', \dots, y^{(\lambda)}, u, x', \dots, x^{(\mu+3)}; u_{1,1}, \dots, u_{\mu+3\,\mu+3}\}. \quad (23)$$

We have just shown that if (14) can be satisfied then so can (23), namely by the logical function $Q$ given by the table. Now suppose that (23) can be satisfied over $\mathscr{N}$, we wish to show that (14) can also be satisfied over $\mathscr{N}$.

By hypothesis we can find a binary logical function over $\mathcal{N}$, say $Q$, and ternary numerical functions over $\mathcal{N}$

$$\chi', \ldots, \chi^{(\lambda)}, \quad \omega, \quad \rho^{\text{iv}}, \ldots, \rho^{(\mu+3)}, \quad \tau_{1,1}, \ldots, \tau_{\mu+3,\,\mu+3},$$

such that on replacing

$$\begin{array}{rll} q & \text{by} & Q, \\ y^{(\theta)} & \text{by} & \chi^{(\theta)}x'x''x''', \\ u & \text{by} & \omega x'x''x''', \\ x^{(\theta)} & \text{by} & \rho^{(\theta)}x'x''x''' \quad (4 \leqslant \theta \leqslant \mu+3), \\ u_{\theta,\theta'} & \text{by} & \tau_{\theta,\theta'}x'x''x''' \quad (1 \leqslant \theta, \theta' \leqslant \mu+3) \end{array}$$

in (23) and $x', x'', x'''$ by $\nu', \nu'', \nu'''$ respectively we obtain the value $\boldsymbol{t}$. Consequently we have

$$CKI[x'; y', y^{(\lambda)}]\, I[x''; y', y^{(\lambda)}]\, KK\gamma[\omega x'x''x'''; y', \ldots, y^{(\lambda)}]$$
$$Qx'\omega x'x''x'''Q\omega x'x''x'''x'', \quad (19')$$

(20′) and (21′) are obtained by writing $\rho^{(\theta)}x', x''x'''$ for $x^{(\theta)}$, $1 \leqslant \theta \leqslant 3$

$$K\prod_{\theta=4}^{\mu+3} I[\rho^{(\theta)}x'x''x'''; y', y^{(\lambda)}]\, C \prod_{\theta=1}^{3} I[x^{(\theta)}; y', y^{(\lambda)}]\, K \prod_{\theta,\theta'=1}^{\mu+3} KKQ\tau_{\theta,\theta'}x'x''x'''y^{(\lambda)}$$
$$Q\rho^{(\theta)}x'x''x'''\tau_{\theta,\theta'}x'x''x'''Q\tau_{\theta,\theta'}x'x''x'''\rho^{(\theta')}x'x''x'''\Psi\{Q; y', \ldots, y^{(\lambda)},$$
$$\tau_{1,1}x'x''x''', \ldots, \tau_{\mu+3,\,\mu+3}x'x''x'''\} \quad (22'')$$

for any $x', x'', x''' \in \mathcal{N}$ and for $y^{(\theta)} = \chi^{(\theta)}x'x''x'''$, $1 \leqslant \theta \leqslant \lambda$. Choose a fixed member of $\mathcal{N}$, say 1. Write $a^{(\theta)}$ for $\chi^{(\theta)}111$. Then by (18) for any

$$x', x'', x''' \in \mathcal{N},$$

$$Q\chi'x'x''x'''\chi'x'x''x''', \quad (23')$$

$$NQ\chi''x'x''x'''\chi'x'x''x''', \quad (23'')$$

$$Q\chi^{(\theta)}x'x''x'''\chi^{(\theta-1)}x'x''x''' \quad (3 \leqslant \theta \leqslant \lambda) \quad (23^{(\theta)})$$

and

$$CKQ\nu'\nu'Q\nu''\nu''KBQ\nu'\nu'''Q\nu''\nu'''BQ\nu'''\nu'Q\nu'''\nu'', \quad (24')$$

$$CKNQ\nu'\chi'\nu'\nu''\nu'''NQ\nu''\chi'\nu'\nu''\nu'''KBQ\nu'\nu'''Q\nu''\nu'''BQ\nu'''\nu'Q\nu'''\nu'', \quad (24'')$$

$$CKQ\nu'\chi^{(\theta-1)}\nu'\nu''\nu'''Q\nu''\chi^{(\theta-1)}\nu'\nu''\nu'''KBQ\nu'\nu'''Q\nu''\nu'''BQ\nu'''\nu'Q\nu'''\nu''. \quad (24^{(\theta)})$$

From $(23^{(\theta)})$, $1 \leqslant \theta \leqslant \lambda$, we obtain for $\nu' = \nu'' = \nu''' = 1$,

$$Qa'a', \tag{25'}$$

$$NQa''a', \tag{25''}$$

$$Qa^{(\theta)}a^{(\theta-1)}, \quad 3 \leqslant \theta \leqslant \lambda. \tag{25^{(\theta)}}$$

Now we show that for any $\nu', \nu'', \nu''' \epsilon \mathcal{N}$ and $1 \leqslant \theta \leqslant \lambda$,

$$BQ\chi^{(\theta)}\nu'\nu''\nu'''\nu Q a^{(\theta)}\nu, \tag{26_1^{(\theta)}}$$

$$BQ\nu\chi^{(\theta)}\nu'\nu''\nu'''Q\nu a^{(\theta)}. \tag{26_2^{(\theta)}}$$

*For* $\theta = 1$ in (24′) put $\chi'\nu'\nu''\nu'''$ for $\nu'$,

$a'$ for $\nu''$,

$\nu$ for $\nu'''$,

detach (23′) and (25′), and we get $(26_1')$ and $(26_2')$. For $\theta = 2$ in $(26_2')$ put $\chi''\nu'\nu''\nu'''$ for $\nu$ and use (23″) we get

$$NQ\chi''\nu'\nu''\nu'''a'. \tag{27''}$$

In (24″) put $\chi''\nu'\nu''\nu'''$ for $\nu'$,

$a''$ for $\nu''$,

$\nu$ for $\nu'''$,

and replace $X'\nu'\nu''\nu'''$ by $a'$ in virtue of $(26_2')$, then detach (27″) and we are left with $(26_1'')$ and $(26_2'')$.

Supposing that we have shown $(26_2^{(\theta-1)})$ for some $3 \leqslant \theta \leqslant \lambda$ then in $(26_2^{(\theta-1)})$ put $X^{(\theta)}\nu'\nu''\nu'''$ for $\nu$ use the resulting equivalence in $(23^{(\theta)})$ and we get

$$Q\chi^{(\theta)}\nu'\nu''\nu'''a^{(\theta-1)}, \tag{27^{(\theta)}}$$

in $(24^{(\theta)})$ put $\chi^{(\theta)}\nu'\nu''\nu'''$ for $\nu'$,

$a^{(\theta)}$ for $\nu''$,

$\nu$ for $\nu'''$,

replace $\chi^{(\theta-1)}\nu'\nu''\nu'''$ by $a^{(\theta-1)}$ in virtue of $(26_2^{(\theta(-1)})$ then detach $(25^{(\theta)})$ and $(27^{(\theta)})$ and we are left with $(26_1^{(\theta)})$ and $(26_2^{(\theta)})$.

Consequently we may replace $\chi^{(\theta)}\nu'\nu''\nu'''$ by $a^{(\theta)}$ for $1 \leqslant \theta \leqslant \lambda$ whenever it occurs as an argument of $Q$. In particular (18′), (20′), (21′), (22′), (23″) hold for any $\nu', \nu'', \nu''' \epsilon \mathcal{N}$ and $y^{(\theta)} = a^{(\theta)}$, $1 \leqslant \theta \leqslant \lambda$.

Let $\mathcal{N}'$ denote the set of elements of $\mathcal{N}$ for which $I[\nu; a', a^{(\lambda)}]$ holds. In virtue of (22′), $\mathcal{N}'$ has a member. Choose a fixed element $b$ of $\mathcal{N}'$, say $b = \rho^{\text{iv}}111$ (see (22′)). We define predicates $p', \ldots, p^{(\lambda)}$ over $\mathcal{N}'$ by

$$p^{(\theta)}\nu\kappa = Qa^{(\theta)}\tau_{1,2}\nu\kappa b \quad (1 \leqslant \theta \leqslant \lambda, \nu, \kappa\epsilon\mathcal{N}').$$

We now show that these satisfy (14) over $\mathcal{N}'$. Let $\nu$, $\kappa\epsilon\mathcal{N}'$ and $p = \tau_{1,2}\nu\kappa b$. By (22″) we have

$$Qpa^{(\lambda)} \quad (x''' = b), \tag{28}$$

$$Q\nu p \quad (\nu = \nu', \theta = 1), \tag{29}$$

$$Qp\kappa \quad (\kappa = \nu'', \theta' = 2). \tag{30}$$

LEMMA. *For any element $q\epsilon\mathcal{N}'$ for which*

$$Qqa, \tag{28′}$$

$$Q\nu q, \tag{29′}$$

$$Qq\kappa, \tag{30′}$$

*hold, we have $BQa^{(\lambda)}qp^{(\theta)}\nu\kappa$ for $\nu$, $\kappa\epsilon\mathcal{N}'$, $1 \leqslant \theta \leqslant \lambda$,* i.e. $BQa^{(\lambda)}qQa^{(\lambda)}p$. Indeed, for $r = \omega\nu\kappa l$ we have by (19′)

$$\gamma[r; a', \ldots, a^{(\lambda)}], \tag{28″}$$

$$Q\nu r, \tag{29″}$$

$$Qr\kappa. \tag{30″}$$

In (20′) put $\nu$ for $\nu'$, $p$ for $\nu''$, $r$ for $\nu'''$ detach $I[\nu; a', a^{(\lambda)}]$, (28) (28″), (29), (29″) and we obtain

$$Qpr. \tag{31}$$

In (20) put $\nu$ for $\nu'$, $q$ for $\nu''$, $r$ for $\nu'''$ detach $I[\nu; a', a^{(\lambda)}]$, (28′), (28″), (29′), (29″) and we obtain

$$Qqr. \tag{31′}$$

In (20) put $\kappa$ for $\nu'$, $p$ for $\nu''$, $r$ for $\nu'''$ detach $I[\nu; a', a^{(\lambda)}]$, (28), (28″), (30), (30″) and we obtain

$$Qrp. \tag{31″}$$

In (20) put $\kappa$ for $\nu'$, $q$ for $\nu''$, $r$ for $\nu'''$ detach $I[\kappa; a', a^{(\lambda)}]$, (28)′, (28″), (30′), (28″), (30′), (30″) and we obtain

$$Qrq. \tag{31‴}$$

Finally in (21′) put $p$ for $x'$, $q$ for $x''$, $r$ for $x'''$ detach (28), (28′), (28″), (31), (30′), (31″), (31‴) and we obtain

$$BQa^{(\theta)}pQa^{(\theta)}q,$$

which is the lemma.

Now let $\nu'$, $\nu''$, $\nu'''$ be elements of $\mathscr{N}'$, and let

$$\nu^{(\theta)} = \rho^{(\theta)}\nu'\nu''\nu''', \quad 4 \leqslant \theta \leqslant \mu+3$$

(this holds also for $1 \leqslant \theta \leqslant 3$ by definition). By (28′) $\nu^{\text{iv}}, \ldots, \nu^{(\mu+3)}$ also belong to $\mathscr{N}'$. Let $t_{\theta,\theta'} = \tau_{\theta,\theta'}\nu'\nu''\nu'''$, $1 \leqslant \theta, \theta' \leqslant \mu+3$. By (28″) we obtain, for $1 \leqslant \theta, \theta' \leqslant \mu+3$,

$$Qt_{\theta,\theta'}a^{(\lambda)}, \tag{32}$$

$$Q\nu^{(\theta)}t_{\theta,\theta'}, \tag{33}$$

$$Qt_{\theta,\theta'}\nu^{(\theta')}, \tag{34}$$

$$\Psi\{Q; a', \ldots, a^{(\lambda)}, t_{1,1}, \ldots, t_{\mu+3,\mu+3}\}. \tag{35}$$

The lemma now gives from (32), (33), (34)

$$BQa^{(\theta)}t_{\theta',\theta''}p^{(\theta)}\nu^{(\theta')}\nu^{(\theta'')}, \quad 1 \leqslant \theta \leqslant \lambda, 1 \leqslant \theta', \theta'' \leqslant \mu+3.$$

Thus by the definition of $\Psi$ we infer from (35)

$$\overline{\phi}\{p', \ldots, p^{(\lambda)}, \nu', \ldots, \nu^{(\mu+3)}\}$$

i.e. the binary logical functions $p', \ldots, p^{(\lambda)}$ over $\mathscr{N}'$ satisfy (14). This completes the demonstration of the proposition.

## 3.22 *Method of semantic tableaux*

We shall show in Ch. 7 that $\mathscr{F}_C$ is undecidable, that is to say that any proposed method of deciding whether an $\mathscr{F}_C$-statement is an $\mathscr{F}_C$-theorem or otherwise will fail to give a result in some cases. We now give a method for deciding whether an $\mathscr{F}_C$-statement is generally valid or whether its negation can be satisfied. Of course the method will fail to give a result in some cases. The method consists in trying to find a *counter example* by the use of *semantic tableaux*. We try to make a given $\mathscr{F}_C$-statement $\phi$ take the value $f$ over some individual domain $\mathscr{D}$. We suppose that $\phi$ is built up from atomic statements by negation, disjunction and generalization only, if other logical connectives are used the method is similar as can be seen from some of the examples we shall give.

Consider an $\mathscr{F}_C$-statement $\phi$. We form two columns and call one an ***f**-column* and the other one a ***t**-column*, each column will be called *opposite* to the other and they will be said to *correspond*. We place $\phi$ at the head of the **f**-column, this means that we wish to make $\phi$ take the value **f**. We then proceed as follows: If $\psi$ is $N\psi'$ and is in an **f**-column then we place $\psi'$ at the end of the corresponding **t**-column, if $\psi$ is $N\psi'$ and is in a **t**-column then we place $\psi'$ at the end of the corresponding **f**-column, if $\psi$ is $D\psi'\psi''$ and is in an **f**-column then we place $\psi'$ and $\psi''$ one after the other immediately below $\psi$ in that **f**-column. If $\psi$ is $D\psi'\psi''$ and is in a **t**-column then we divide the **t**-column and the corresponding **f**-column into two columns which correspond respectively with the two **t**-columns and we place $\psi'$ at the head of one of the two new **t**-columns and $\psi''$ at the head of the other new **t**-column, the previous **t**-column is to count as the earlier part of both the **t**-columns into which it has divided. The two columns denote alternative ways of making $\phi$ take the value **f**.

If $\psi$ is $(A\xi)\,\psi'\{\xi\}$ and is in an **f**-column then we place $\psi'\{\alpha\}$ at the end of that **f**-column, here $\alpha$ is a new element of $\mathscr{D}$, if $\psi$ is $(A\xi)\,\psi'\{\xi\}$ and is in a **t**-column then we place $\psi\{\xi\}$ below it at the end of that **t**-column we can also if we like place $\psi\{\alpha\}$ at any time at the end of that **t**-column. If $\psi\{\xi\}$ is a general statement and contains the free variables $\mathfrak{x}$, then in the **f**-column we must have $\alpha\{\mathfrak{x}\}$ instead of just $\alpha$, here $\mathfrak{x}$ stands for a set of free variables. Similarly for other connectives. The process will terminate, except for substitutions for variables, because at each step we get shorter and shorter statements. If at any time we come across $\chi$ in one column and in the corresponding opposite column then we close both these columns by a horizontal line, this means that in order to make $\phi$ take the value **f**, by the alternatives given by these columns, we have to give both **t** and **f** to $\chi$, since this is impossible we discard this alternative. If every column is closed then our attempt to make $\phi$ take the value **f** has failed and so $\phi$ is generally valid over any domain.

The process terminates, except for substitutions for variables, when we arrive at atomic formulae. Thus a semantic tableau will (i) terminate by all columns being closed in which case $\phi$ is an $\mathscr{F}_C$-theorem, or some columns fail to be closed either (ii) because we are unable to decide whether the end formula of one column also occurs in the corresponding opposite column, possibly after substitution for variables, or (iii) because we can see that it fails to do so for all the open columns. In this last case we can read off a counter-example, in the first case we can construct an

$\mathscr{F}_C$-proof. The method always comes to a definite conclusion, if the first or the last case holds, but fails if the second case arises. If all the atomic formulae are singulary we can obtain a counter example, if that case arises, over a bounded domain. But if there are binary predicates then we have the general case (see Prop. 25), and unbounded domains may be required and the functions introduced in dealing with a general statement in an ***f***-column may introduce such complications that we are unable to tell whether certain atomic formulae occur in both of two opposite corresponding columns.

Consider

$$CKN(Ex)\,Kpxp'x(Ax)\,Cp''xpxN(Ex)\,Kp'xp''x \quad (0)$$

| ***t*** | | ***f*** |
|---|---|---|
| | : | (0) |
| (i) $KN(Ex)\,Kpxp'x(Ax)\,Cp''xpx$ | : | (xi) $N(Ex)\,Kp'xp''x$ |
| (ii) $N(Ex)\,Kpxp'x$ | : | (xii) $(Ex)\,Kpxp'x$ |
| (iii) $(Ax)\,Cp''xpx$ | : | (xiii) $Kpap'a$ we have substituted for a variable |
| (iv) $(Ex)\,Kp'xp''x$ | : | |
| (v) $Cp''apa$ we have substituted for a variable | : | |
| (vi) $Kp'ap''a'$ | : | |
| (vii) $p'a$ | : | |
| (viii) $p''a$ | : | (xiv) $pa$ 1 : (xv) $p'a$ 2 |
| (ix) $Np''a$ 1 : (x) $pa$ 2 | : | (xvi) $p''a$ |

Below entry (xiii) the ***f***-column splits into two columns indicating the two ways in which (xiii) can be made ***f***. The second of these two columns can be closed because (xv) agrees with (vii). Now there is only one ***f***-column and the ***t***-column splits indicating the two ways in which (v) can be made ***t***. This gives us the pair of opposite columns labelled 1, 1 and the pair of opposite columns labelled 2, 1. Both these columns can be closed because the last member of ***t***-2, 1 (x) agrees with (xiv) and the last member of ***f***-1, 1 (xvi) agrees with (viii). Thus our attempt to make (0) ***f*** has failed. Hence (0) is generally valid and so is an $\mathscr{F}_C$-theorem. From the tableau we can find an $\mathscr{F}_C$-proof of (0), as follows:

We start with the hypotheses: (ii), (iii), (iv) and (vi)$_x$ from these we deduce $pa$ as shown in the tableau but we had already produced $p'a$ so that we get $Kpap'a$ and hence (xii), thus we have:

$$N(\text{xii}),\ (\text{iii}),\ (\text{iv}),\ (\text{vi}) \vdash_{\mathscr{F}_C} (\text{xii})$$

where $(\mathrm{vi})_x$ is (vi) with $x$ instead of $a$. By the Deduction Theorem we have

$$(\mathrm{iii}), (\mathrm{iv}) \vdash C(\mathrm{vi})_x CN(\mathrm{xii})(\mathrm{xii}),$$

$$(\mathrm{iii}), (\mathrm{iv}) \vdash C(\mathrm{vi})_x (\mathrm{xii}),$$

$$(\mathrm{iii}), (\mathrm{iv}) \vdash C(\mathrm{iv})(\mathrm{xii}), \quad \text{Prop. 6}$$

$$(\mathrm{iii}) \vdash C(\mathrm{iv})C(\mathrm{iv})(\mathrm{xii}),$$

$$(\mathrm{iii}) \vdash C(\mathrm{iv})(\mathrm{xii}),$$

$$(\mathrm{iii}) \vdash CN(\mathrm{xii})N(\mathrm{iv}),$$

$$\vdash C(\mathrm{iii})CN(\mathrm{xii})N(\mathrm{iv}),$$

which is equivalent to (0).

Now consider

$$CK(Ex)KpxNp'x(Ex)Kp'xNp''x(Ex)KpxNp''x \quad (0)$$

Our tableau is:

| *t* | | : | *f* | | | | : | | | |
|---|---|---|---|---|---|---|---|---|---|---|
| | | : | | (0) | | | | | | |
| (i) | $(Ex)KpxNp'x$ | : | (ix) | $(Ex)KpxNp''x$ | | | | | | |
| (ii) | $(Ex)Kp'xNp''x$ | : | (x) | $KpxNp''x$ | | | | | | |
| (iii) | $KpaNp'a$ | : | (xi) | $KpaNp''a$ | | | | | | |
| (iv) | $pa$ | : | (xii) | $pa$ | 1 | : | (xiii) | $Np''a$ | 2 |
| (v) | $Np'a$ | : | | | | : | (xiv) | $KpbNp''b$ | |
| (vi) | $Kp'bNp''b$ | : | | | | | | | |
| (vii) | $p'b$ | : | | | | | | | |
| (viii) | $Np''b$ | : | (xvi) | $pb$ | 21 | : | (xvii) | $Np''b$ | 22 |
| (xv) | $p''a$ | : | (xviii) | $p'a$ | | | | | |
| | | : | (xix) | $p''b$ | | | | | |

In this case the tableau terminates without the tableau being closed, and we are able to read off a counter example, namely the domain consists of two elements $a$ and $b$, $pa$, $p''a$ and $p'b$ are ***t*** while $pb$, $p'a$ and $p''b$ are ***f***. We take $pa$ to be ***t*** because as far as making the entry (ix) ***f*** we need only take (xiii) ***f*** independently of the value of (xii), but to make (i) and (ii) ***t*** we must have (xii), i.e. (iv) ***t***.

In the ***t***-column (i) gives rise to (iii), (iv) and (v) and (ii) gives rise to (vi), (vii) and (viii). In the ***f***-column (xii) and (xiii) are the two alternative ways in which (xi) can be made ***f*** of these (xii) closes with (iv). We still have (xiv) and this gives rise to two columns 21 and 22, of these 22 closes with (viii). This column consists of (ix), (x), (xiii), (xiv), (xvii).

But column 1 consisting of (ix), (x), (xiii), (xiv), (xvi) can be continued with (xviii) and (xix) from (v) and (viii) respectively, and this column is still open, we could go on with $KpcNp''c$ but it is pointless to do so. The final **t**-column consists of

(i), (ii), (iii), (iv), (v), (vi), (vii), (viii), (xv)

and the final **f**-column consists of

(0), (ix), (x), (xi), (xiii), (xiv), (xvi), (xviii), (xix).

We read off the counter example by the values given for the predicates $p, p', p''$ at $a, b$, in these two final columns.

Now consider:

$$C(Ax)(Ex')KDpxx'p'xx'Dpx'xp'xx(Ex)(Ax')DKpxxp'x'x'Kpxx'p'xx' \quad (0)$$

| **t** | (0) | **f** | | |
|---|---|---|---|---|
| *Dpxap'xa* | | *Kpxxp'bb* | | |
| *Dpaxp'xx* | | *Kpxbp'xb* | | |
| | 1 *pxb* | | 2 *p'xb* | |
| | 11 *pxx* | 12 *p'bb* | 21 *pxx* | 22 *p'bb* |
| | *paa* | *pab* | *paa* | *p'ab* |
| | *pab* | *pbb* | *pbb* | |
| | *pbb* | | *p'ab* | |
| | | | *p'bb* | |

This gives 4 **f**-columns each of which correspond to the single **t**-column at present consisting of only two entries. Now the **t**-column similarly splits up into 4 columns, thus

| | | | |
|---|---|---|---|
| 1 *pxa* | | 2 *p'xa* | |
| 11 *pax* | 12 *p'xx* | 21 *pax* | 22 *p'xx* |
| *paa* | *p'aa* | *paa* | *p'bb* |
| *pba* | *paa* | *p'ba* | *p'ba* |
| *pab* | *pba* | *pab* | *p'aa* |
| | *p'bb* | *p'aa* | |

Each of these four **t**-columns corresponds to each of the four **f**-columns. To make (0) take the value **f** we require that a pair of corresponding columns be open. We have entered in the columns all the results of substitution over a two-element domain. Thus we require to test 16 cases, viz. any **t**-column with any **f**-column, we tabulate the results:

| | | |
|---|---|---|
| *f*–11 | *t*–11 | closed by *paa* |
| *f*–12 | *t*–11 | closed by *pab* |
| *f*–21 | *t*–11 | closed by *paa* |
| *f*–22 | *t*–11 | open |
| *f*–11 | *t*–12 | closed by *paa* |
| *f*–12 | *t*–12 | closed by *p'bb* |
| *f*–21 | *t*–12 | closed by *paa* |
| *f*–22 | *t*–12 | closed by *p'bb* |
| *f*–11 | *t*–21 | closed by *paa* |
| *f*–12 | *t*–21 | closed by *pab* |
| *f*–21 | *t*–21 | closed by *paa* |
| *f*–22 | *t*–21 | open *p'bb* |
| *f*–11 | *t*–22 | open |
| *f*–12 | *t*–22 | closed by *p'bb* |
| *f*–21 | *t*–22 | closed by *p'bb* |
| *f*–22 | *t*–22 | closed by *p'bb* |

Thus we have found three open pairs Each of these gives a way of making (0) *f* over a two-element domain. Namely

| *f*–22 and | *t*–11 | *f*–22 and | *t*–21 | *f*–11 and | *t*–22 |
|---|---|---|---|---|---|
| *t* | *f* | *t* | *f* | *t* | *f* |
| *paa* | *p'bb* | *paa* | *p'bb* | *p'bb* | *paa* |
| *pab* | *p'ab* | *p'ba* | *p'ab* | *p'ab* | *pbb* |
| *pba* | | *pab* | | *p'ba* | *pab* |
| | | *p'aa* | | | |

the values omitted can be chosen arbitrarily.

## 3.23 *An application of the method of semantic tableaux*

As an application of the use of semantic tableaux we demonstrate the following proposition:

PROP. 26. *Let $C\phi\psi$ be a closed $\mathscr{F}_C$-theorem where neither $N\phi$ nor $\psi$ is an $\mathscr{F}_C$-theorem and where $\phi$ is built up from atomic statements $\alpha', \ldots, \alpha^{(\lambda)}$, $\gamma', \ldots, \gamma^{(\mu)}$ while $\psi$ is built up from atomic statements $\beta', \ldots, \beta^{(\nu)}, \gamma', \ldots, \gamma^{(\mu)}$, so that $\phi$ and $\psi$ have exactly the atomic statements $\gamma', \ldots, \gamma^{(\mu)}$ in common; then we can find a closed $\mathscr{F}_C$-statement $\chi$ built up from some of the atomic statements $\gamma', \ldots, \gamma^{(\mu)}$ only such that $C\phi\chi$ and $C\chi\psi$ are $\mathscr{F}_C$-theorems.*

Let then $C\phi\psi$ be a closed $\mathscr{F}_C$-theorem and suppose that $\phi$ and $\psi$ are in prenex normal form with matrixes $\phi_0$ and $\psi_0$ respectively, where $\phi_0$ is in conjunctive normal form while $\psi_0$ is in disjunctive normal form and that both are constructed from atomic statements as described in the enunciation of the proposition. Suppose also that $\phi_0$ is of the form $\prod_{\theta=1}^{\kappa} \phi_0^{(\theta)}$ while $\psi_0$ is of the form $\sum_{\theta=1}^{\lambda} \psi_0^{(\theta)}$ and that $\phi_0^{(\theta)}$ is of the form $\sum_{\pi=1}^{\theta^{(\nu)}} \phi_{0,\pi}^{(\theta)}$ while $\psi_0^{(\theta)}$ is of the form $\prod_{\pi=1}^{\theta^{(\mu)}} \psi_{0,\pi}^{(\theta)}$ where $\phi_{0,\pi}^{(\theta)}$ and $\psi_{0,\pi}^{(\theta)}$ are atomic statements or negations of atomic statements. Now form the semantic tableau for $C\phi\psi$; it will be closed. In forming the tableau we first treat the quantifiers introducing variables and functions and constants as required by the rules of forming a semantic tableau. When the quantifiers have been treated we have entries $\phi_0^*$ in the $\boldsymbol{t}$-column and $\psi_0^*$ in the $\boldsymbol{f}$-column, where $\phi_0^*$ and $\psi_0^*$ differ from $\phi_0$ and $\psi_0$ respectively by substitutions on the variables in the latter. We continue the tableau as follows: beneath $\phi_0^*$ in the $\boldsymbol{t}$-column there appear one below the other $\phi_0'^*, \ldots, \phi_0^{(\kappa)*}$, and beneath $\psi_0^*$ in the $\boldsymbol{f}$-column there appear one below the other $\psi_0'^*, \ldots, \psi_0^{(\lambda)*}$, these differ from $\phi_0', \ldots, \phi_0^{(\kappa)}$ and $\psi_0', \ldots, \psi_0^{(\lambda)}$ respectively by the substitutions we made above. Thereafter the single columns $\boldsymbol{t}$ and $\boldsymbol{f}$ both split; we will put it down as follows: the $\boldsymbol{f}$-column splits into $1^{(\mu)}$ columns corresponding to the $1^{(\mu)}$ ways in which $\psi_0'^*$ can be made $\boldsymbol{f}$, then each of these columns splits up into $2^{(\mu)}$ columns indicating the $2^{(\mu)}$ ways in which $\psi_0''^*$ can be made $\boldsymbol{f}$, and so on until we have the $\lambda$th split into $\lambda^{(\mu)}$ columns indicating the $\lambda^{(\mu)}$ ways in which $\psi_0^{(\lambda)*}$ can be made $\boldsymbol{f}$. Altogether we have $1^{(\mu)} \times 2^{(\mu)} \times \ldots \times \lambda^{(\mu)}$ columns on the $\boldsymbol{f}$-side. These columns contain one conjunctand from each of the $\lambda$ rows $\psi_0'^*, \ldots, \psi_0^{(\lambda)*}$, so each column indicates a way in which $\psi_0^*$ can be made $\boldsymbol{f}$ by making one conjunctand in each of the conjunctions $\psi_0'^*, \ldots, \psi_0^{(\lambda)*}$ $\boldsymbol{f}$. Each of these columns is to have a corresponding opposite column on the $\boldsymbol{t}$-side; we further transfer an atomic statement which is negated in the $\boldsymbol{f}$-side into the corresponding column on the $\boldsymbol{t}$-side. Now we go over to the $\boldsymbol{t}$-side.

Each of the columns on the $\boldsymbol{t}$-side will split up into $1^{(\nu)}$ columns indicating the various ways in which $\phi_0'^*$ can be made $\boldsymbol{t}$. Then each of these columns splits up into $2^{(\nu)}$ columns indicating the various ways in which $\phi_0''^*$ can be made $\boldsymbol{t}$, and so on until the $\kappa$th split into $\kappa^{(\nu)}$ columns indicating the various ways in which $\phi_0^{(\kappa)*}$, can be made $\boldsymbol{t}$. Altogether each of the $1^{(\mu)} \times 2^{(\mu)} \times \ldots \times \lambda^{(\mu)}$ columns which the $\boldsymbol{t}$-side acquired from the $\boldsymbol{f}$-side has split up into $1^{(\nu)} \times 2^{(\nu)} \times \ldots \times \kappa^{(\nu)}$ further columns. These require their opposite columns on the $\boldsymbol{f}$-side to which are transferred any negated atoms.

Thus each side of the final tableau has $1^{(\mu)} \times \ldots \times \lambda^{(\mu)} \times 1^{(\nu)} \times \ldots \times \kappa^{(\nu)}$ columns which correspond in pairs, we continue these columns with substitutions on the free variables, then if we make all the atoms in one $\boldsymbol{f}$- column $\boldsymbol{f}$ and all the atoms in the corresponding $\boldsymbol{t}$-column $\boldsymbol{t}$ then we shall make $C\phi\psi\boldsymbol{f}$; but this is impossible because $C\phi\psi$ is an $\mathscr{F}_C$-theorem. Thus each corresponding pair of columns is closed by the same atom appearing in both of them. Now we form an $\mathscr{F}_C$-statement built up from some of $\gamma', \ldots, \gamma^{(\mu)}$ only as follows: we may suppose that columns are closed by closed atoms, otherwise make a substitution.

There are four ways in which a pair of corresponding columns can be closed. Case (i), the common atom $\delta$ in the $\boldsymbol{t}$-side, can arise from $\phi$ and in the $\boldsymbol{f}$-side from $\psi$, or, case (ii), the other way about, the common atom $\delta$ in the $\boldsymbol{t}$-side can arise from $\psi$ and in the $\boldsymbol{f}$-side from $\phi$, in these two cases the common atom must be a $\gamma$, or case (iii) the common atom $\delta$ can arise in both sides from $\phi$, or case (iv) the common atom $\delta$ can arise in both sides from $\psi$. With each pair of corresponding columns we associate the $\mathscr{F}_C$-statement $\delta$, if $\delta$ is the common atom and we have case (i), $N\delta$ if we have case (ii), $\boldsymbol{f}$ if case (iii), and lastly $\boldsymbol{t}$ if we have case (iv). Now suppose that we have associated an $\mathscr{F}_C$-statement with each subtableau up to a certain point and that a set of these subtableaux join (on going up) or split (on going down), then to the subtableau formed from their union we associate the disjunction of the statements associated with each of them if the split is due to treatment of $\phi$, but their conjunction if the split is due to treatment of $\psi$. This gives us an $\mathscr{F}_C$-statement $\chi_0^*$, composed of $\gamma$'s, $\boldsymbol{f}$ and $\boldsymbol{t}$, by $N$, $D$ and $K$, associated with the whole tableau which we can effectively find when the closed tableau is given. Now the structure of both sides of the tableau is the same, because each column has its corresponding column and when one splits so does the opposite one. Now in the $\boldsymbol{t}$-side omit every entry except atoms which close columns and those which arise from $\psi$, and replace these atoms by $\delta$, $N\delta$, $\boldsymbol{f}$ or $\boldsymbol{t}$ according as which of the four cases occurred. We are left with a tableau for $C\chi_0^*\psi_0^*$, because if a pair of corresponding columns closed by case (i) then the situation is exactly the same as before, and similarly in case (ii), in case (iii) we have $\boldsymbol{f}$, but in order to get a counter example we ought to have $\boldsymbol{t}$, and in the last case we have $\boldsymbol{t}$, but to get a counter example we ought to have $\boldsymbol{f}$. Now the entry $\boldsymbol{f}$ can be replaced by $K\gamma'N\gamma'$, this in a $\boldsymbol{t}$-column becomes $\gamma'$ with $N\gamma'$ below it; this gives $\gamma'$ in that column and in the corresponding column, hence the column is closed as before.

Similarly the entry $\boldsymbol{t}$ can be replaced by $D\gamma' N\gamma'$, this in an $\boldsymbol{f}$-column becomes $\gamma'$ with $N\gamma'$ below it; this gives $\gamma'$ in that column and in the corresponding column, hence the column is closed as before. Thus columns which were originally closed by $\alpha$'s or $\beta$'s are now closed entirely by $\gamma$'s. We have the closed tableau for $C\chi_0^* \psi_0^*$ which is an $\mathscr{F}_C$-statement free of quantifiers. We call $\chi_0^*$ the *sentential power* of $\phi$. Similarly if we interchange the treatment of the two sides we get the tableau for $C\phi_0^* \chi_0^*$. $\chi_0^*$ is the same in both cases, the structure is the same in both cases, if one divides so does the other and for the same reason. We want to put the quantifiers back so as to get the *quantificational power* of $\phi$. Replace the argument places in $\chi_0^*$ by distinct variables, $\xi', \ldots, \xi^{(\theta)}$, say, which are distinct from the variables in $\psi_0^*$. Now put quantifiers on as follows: If $\xi^{(\lambda)}$ replaces a term which was introduced on account of treatment of $\phi$ or of a formula originating from $\phi$ then we generalize $\xi^{(\lambda)}$ otherwise we restrict $\xi^{(\lambda)}$. The order of putting back the quantifiers is the inverse order to that of the introduction of the terms which occupied the places of the variables $\xi', \ldots, \xi^{(\theta)}$, this gives us $\chi$. Then $C\chi\psi$ is an $\mathscr{F}_C$-theorem because its tableau is closed. Similarly we get $C\phi\chi$.

Note that if $\chi_0^*$ is $\boldsymbol{f}$ then $N\phi$ is an $\mathscr{F}_C$-theorem because the tableau for $N\phi$ is closed on its own, and if $\chi_0^*$ is $\boldsymbol{t}$ then $\psi$ is an $\mathscr{F}_C$-theorem because then the tableau for $\psi$ is closed on its own.

Note that $\boldsymbol{f}$ and $\boldsymbol{t}$ can be omitted from $\chi$ because in the formation of $\chi$ we can replace $K\delta\boldsymbol{f}$ and $K\boldsymbol{f}\delta$ by $\boldsymbol{f}$, $D\delta\boldsymbol{t}$ and $D\boldsymbol{t}\delta$ by $\boldsymbol{t}$, $D\delta\boldsymbol{f}$ and $D\boldsymbol{f}\delta$ by $\delta$, lastly $K\delta\boldsymbol{t}$ and $K\boldsymbol{t}\delta$ by $\delta$. So $\boldsymbol{f}$ and $\boldsymbol{t}$ will disappear unless the final result is $\boldsymbol{f}$ or $\boldsymbol{t}$, but we have discarded this case. Note also that only those $\gamma$'s are used in $\chi$ which occur positively in both $\phi$ and $\psi$ or occur negatively in both $\phi$ and $\psi$. Where an atom is said to *occur positively* in $\omega$ if it is on the same side of the tableau as $\omega$, otherwise it is said to *occur negatively* in $\omega$. Thus only some of the $\gamma$'s are used in $\chi$.

We restrict a variable which replaces a term which originated from treatment of $\phi$ because terms which at their first occurrence arise from $\phi$ do so by treatment of existential quantifiers. We generalize a variable which originated from treatment of $\psi$ because terms which at their first occurrence arise from $\psi$ do so by treatment of universal quantifiers.

Let us return to case (ii) where we were unable to decide whether the tableau closed or was open. A tableau might show that certain compositions of functions were the same, i.e. that $f \ldots gx = l \ldots mx$ for all $x$, we only need on the $\boldsymbol{t}$ side an $\mathscr{F}_C$-statement which has this as an interpretation. By substitutions for variables we can arrive at great complications as regards composition of functions. A composition of singulary functions can be written as a *word*, i.e. a string of letters or primed letters from a given set called the *alphabet*, then if we have $f...gx = l...mx$ for all $x$ a consecutive part $f...g$ of a word may be replaced by a corresponding part $l...m$. The *word problem*, which is undecidable, is to find a method which will tell us whether a given word in a given alphabet can be transformed into another given word by using a bounded set of replacements of the type; replace a consecutive part $f...g$ by $l...m$. In forming tableau we break down formulae until we come to atomic formulae we then use substitutions on the variables and try to close columns thereby. Thus supposing all functions are singulary the problem becomes: can

$$a'x', ..., a^{(\kappa)}x^{(\kappa)}$$

be made the same as $b'x'^*, ..., b^{(\kappa)}x^{(\kappa)*}$ respectively by substitutions on the variables using the replacements allowed? This is rather more complicated than the word problem itself.

As an example of Prop. 26 consider:

$$CKKDN\gamma'\gamma D\alpha\gamma' DK\gamma N\alpha\gamma' DDK\beta\gamma K\gamma\gamma' KN\gamma\beta.$$

The tableau is given on facing page.

There are 64 pairs of corresponding columns, viz. $\nu, \nu'$ where $\nu, \nu' = 1, ..., 8$. In the next table we show that the tableau is closed by giving the value of the associated $\mathscr{F}_C$-statement.

| | | | | | | | |
|---|---|---|---|---|---|---|---|
| $11'\boldsymbol{f}, \boldsymbol{t}, \gamma$ | $12'\boldsymbol{f}, \gamma$ | $13'\boldsymbol{f}$ | $14'\boldsymbol{f}$ | $15'\boldsymbol{f}, \boldsymbol{t}, \gamma$ | $16'\boldsymbol{f}, \gamma$ | $17'\boldsymbol{f}, \boldsymbol{t}, \gamma$ | $18'\boldsymbol{f}, \gamma$ |
| $21'\boldsymbol{f}, \boldsymbol{t}$ | $22'\boldsymbol{f}$ | $23'\boldsymbol{f}, \gamma'$ | $24'\boldsymbol{f}, \gamma'$ | $25'\boldsymbol{f}, \boldsymbol{t}$ | $26'\boldsymbol{f}$ | $27'\boldsymbol{f}, \boldsymbol{t}, \gamma'$ | $28'\boldsymbol{f}, \gamma'$ |
| $31'\boldsymbol{f}, \boldsymbol{t}, \gamma$ | $32'\boldsymbol{f}, \gamma$ | $33'\boldsymbol{f}, \gamma'$ | $34'\boldsymbol{f}, \gamma'$ | $35'\boldsymbol{f}, \boldsymbol{t}, \gamma$ | $36'\boldsymbol{f}, \gamma$ | $37'\boldsymbol{f}, \boldsymbol{t}, \gamma, \gamma'$ | $38'\boldsymbol{f}, \gamma, \gamma'$ |
| $41'\boldsymbol{f}, \boldsymbol{t}$ | $42'\boldsymbol{f}$ | $43'\boldsymbol{f}\gamma'$ | $44'\boldsymbol{f}, \gamma'$ | $45'\boldsymbol{f}, \boldsymbol{t}$ | $46'\boldsymbol{f}$ | $47'\boldsymbol{f}, \boldsymbol{t}, \gamma'$ | $48'\boldsymbol{f}, \gamma'$ |
| $51'\boldsymbol{f}, \boldsymbol{t}, \gamma$ | $52'\boldsymbol{f}, \gamma$ | $53'\boldsymbol{f}$ | $54'\boldsymbol{f}$ | $55'\boldsymbol{f}, \boldsymbol{t}, \gamma$ | $56'\boldsymbol{f}, \gamma$ | $57'\boldsymbol{f}, \boldsymbol{t}, \gamma$ | $58'\boldsymbol{f}, \gamma$ |
| $61'\boldsymbol{t}, \gamma$ | $62'\gamma$ | $63'\gamma'$ | $64'\gamma'$ | $65'\boldsymbol{t}, \gamma$ | $66'\gamma$ | $67'\boldsymbol{t}, \gamma, \gamma'$ | $68'\gamma, \gamma'$ |
| $71'\boldsymbol{t}, \gamma$ | $72'\gamma$ | $73'\gamma'$ | $74'\gamma'$ | $75'\boldsymbol{t}, \gamma$ | $76'\gamma$ | $77'\boldsymbol{t}, \gamma, \gamma'$ | $78'\gamma, \gamma'$ |
| $81'\boldsymbol{t}, \gamma$ | $82'\gamma$ | $83'\gamma'$ | $84'\gamma'$ | $85'\boldsymbol{t}, \gamma$ | $86'\gamma$ | $87'\boldsymbol{t}, \gamma, \gamma'$ | $88'\gamma, \gamma'$ |

In many cases corresponding columns can be closed in several different ways. Thus $55'$ can be closed in three different ways, viz. by $\alpha$ and $N\alpha$ being in column 5 which gives $\boldsymbol{f}$, by $\gamma$ and $N\gamma$ being in column $5'$ which gives us $\boldsymbol{t}$, and lastly by $\gamma$ being in both columns.

We have now associated an $\mathscr{F}_C$-statement with each column. As we go up the tableau these columns unite by treatment of the $\boldsymbol{t}$-side. Accord-

$\boldsymbol{t}$

$KKDN\gamma'\gamma D\alpha\gamma' DK\gamma N\alpha\gamma'$

$DN\gamma'\gamma$

$D\alpha\gamma'$

$DK\gamma N\alpha\gamma'$

$N\gamma'$ $\qquad\qquad$ $\gamma$

$\alpha$ $\quad$ $\gamma'$ $\qquad$ $\alpha$ $\quad$ $\gamma'$

| | | | | | | | |
|---|---|---|---|---|---|---|---|
| $K\gamma N\alpha$ | $\gamma'$ | $K\gamma N\alpha$ | $\gamma'$ | $K\gamma N\alpha$ | $\gamma'$ | $K\gamma N\alpha$ | $\gamma'$ |
| $\gamma$ | | $\gamma$ | | $\gamma$ | | $\gamma$ | |
| $N\alpha$ | | $N\alpha$ | | $N\alpha$ | | $N\alpha$ | |
| 1 | 2 | 3 | 4 | 5 | 6 | 7 | 8 |

$\boldsymbol{f}$

$DDK\beta\gamma K\gamma\gamma' KN\gamma\beta$

$K\beta\gamma$

$K\gamma\gamma'$

$KN\gamma\beta$

$\beta$ $\qquad\qquad$ $\gamma$

$\gamma$ $\quad$ $\gamma'$ $\qquad$ $\gamma$ $\quad$ $\gamma'$

| | | | | | | | |
|---|---|---|---|---|---|---|---|
| $N\gamma$ | $\beta$ | $N\gamma$ | $\beta$ | $N\gamma$ | $\beta$ | $N\gamma$ | $\beta$ |
| $1'$ | $2'$ | $3'$ | $4'$ | $5'$ | $6'$ | $7'$ | $8'$ |

ing to our rules for finding $\chi$ we take the disjunction of the $\mathscr{F}_C$-statements associated with lines:

$1\nu'$ and $2\nu'$, $\quad 3\nu'$ and $4\nu'$, $\quad 5\nu'$ and $6\nu'$, $\quad 7\nu'$ and $8\nu'$, $\quad 1 \leqslant \nu' \leqslant 8$.

The results we show in the following table:

| | | | | | | | |
|---|---|---|---|---|---|---|---|
| $11'\boldsymbol{f},\boldsymbol{t},\gamma$ | $12'\boldsymbol{f},\gamma$ | $13'\boldsymbol{f},\gamma'$ | $14'\boldsymbol{f},\gamma'$ | $15'\boldsymbol{f},\boldsymbol{t},\gamma$ | $16'\boldsymbol{f},\gamma$ | $17'\boldsymbol{f},\boldsymbol{t},\gamma,\gamma',D\gamma\gamma'$ | $18'\boldsymbol{f},\gamma,\gamma'$ |
| $21'\boldsymbol{f},\boldsymbol{t}$ | $22'\boldsymbol{f}$ | $23'\boldsymbol{f},\gamma'$ | $24'\boldsymbol{f},\gamma'$ | $25'\boldsymbol{f},\boldsymbol{t}$ | $26'\boldsymbol{f}$ | $27'\boldsymbol{f},\boldsymbol{t},\gamma'$ | $28'\boldsymbol{f},\gamma'$ |
| $31'\boldsymbol{t},\gamma$ | $32'\gamma$ | $33'\gamma'$ | $34'\gamma'$ | $35'\boldsymbol{f},\boldsymbol{t},\gamma$ | $36'\gamma$ | $37'\boldsymbol{f},\boldsymbol{t},\gamma,\gamma',D\gamma\gamma'$ | $38'\gamma,\gamma',D\gamma\gamma'$ |
| $41'\boldsymbol{t},\gamma$ | $42'\gamma$ | $43'\gamma'$ | $44'\gamma'$ | $45'\boldsymbol{t},\gamma$ | $46'\gamma$ | $47'\boldsymbol{t},\gamma,\gamma',D\gamma\gamma'$ | $48'\gamma,\gamma',D\gamma\gamma'$ |

Here $\nu\,\nu'$ indicates the disjunction of $(2\nu \dot{-} 1)\,\nu'$ and $(2\nu)\,\nu'$. In some cases there are several different ways in which the disjunction can be taken. Thus $47'$ can give either $\gamma$, $\gamma'$, $D\gamma\gamma'$ or $\boldsymbol{t}$ according as to which entry in $77'$ and $87'$ we use. The next stage is the disjunction of the pairs $(2\nu \dot{-} 1)\,\nu'$ and $(2\nu)\,\nu'$. The result is put down in the next table:

| | | | | | | | |
|---|---|---|---|---|---|---|---|
| $11'\boldsymbol{f},\boldsymbol{t},\gamma$ | $12'\boldsymbol{f},\gamma$ | $13'\boldsymbol{f},\gamma'$ | $14'\boldsymbol{f},\gamma'$ | $15'\boldsymbol{f},\boldsymbol{t},\gamma$ | $16'\boldsymbol{f},\gamma$ | $17'\boldsymbol{f},\boldsymbol{t},\gamma,\gamma',D\gamma\gamma'$ | $18'\boldsymbol{f},\gamma,\gamma',D\gamma\gamma'$ |
| $21'\boldsymbol{t},\gamma$ | $22'\gamma$ | $23'\gamma'$ | $24'\gamma'$ | $25'\boldsymbol{t},\gamma$ | $26'\gamma$ | $27'\boldsymbol{t},\gamma,\gamma',D\gamma\gamma'$ | $28'\gamma,\gamma',D\gamma\gamma'$ |

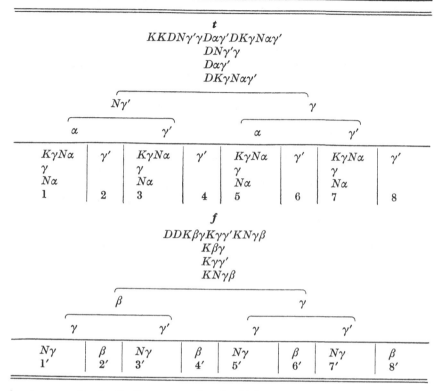

Again we take the disjunction of the two members in each column.

$$1'\boldsymbol{t},\gamma \quad 2'\gamma \quad 3'\gamma', \quad 4'\gamma' \quad 5'\boldsymbol{t},\gamma, \quad 6'\gamma \quad 7'\boldsymbol{t},\gamma,\gamma',D\gamma\gamma' \quad 8'\gamma,\gamma',D\gamma\gamma'.$$

We now form the conjunctions of consecutive pairs corresponding to the junction of columns on the $\boldsymbol{f}$-side. This gives

$$\gamma; \quad \gamma'; \quad K\gamma D\gamma\gamma',\gamma; \quad K\gamma D\gamma\gamma', K\gamma' D\gamma\gamma', K\gamma\gamma', \gamma, \gamma', D\gamma\gamma'.$$

We have to do this a second time, getting

$$K\gamma\gamma'; \quad \gamma, K\gamma D\gamma\gamma', KK\gamma\gamma' D\gamma\gamma', K\gamma' D\gamma\gamma', K\gamma\gamma'$$

and lastly we require the conjunction of these, however we do this we get something equivalent to $K\gamma\gamma'$; this then is the sentential power. We can easily verify that

$$CK\gamma\gamma' D^2 K\beta\gamma K\gamma\gamma' KN\gamma\beta \quad \text{and} \quad CK^2 DN\gamma'\gamma D\alpha\gamma' DK\gamma N\alpha\gamma' K\gamma\gamma'$$

are tautologies.

### 3.24 *Resolved* $\mathscr{F}_C$

The resolved $\mathscr{F}_C$ or the resolved $I\mathscr{F}_C$ is obtained as follows: $\eta$ is a new symbol of type $\iota(o\iota)$, the axioms and rules are the same as for $\mathscr{F}_C$ or $I\mathscr{F}_C$, which ever is the case, except that the symbol $E$ is omitted and the rules II $d, e$ are discarded and the following rule replaces them:

$$\text{H} \quad \frac{D\phi\{\alpha\}\,\omega}{D\phi\{\boldsymbol{\eta}_\xi\phi\{\xi\}\}\,\omega},$$

where we have written

D 21 $\qquad \boldsymbol{\eta}_\xi\phi\{\xi\}$ for $\boldsymbol{\eta}(\boldsymbol{\lambda}\xi\phi\{\xi\})$.

Here $\boldsymbol{\eta}_\xi\phi\{\xi\}$ plays the part of a thing which has the property $(\boldsymbol{\lambda}\xi\phi\{\xi\})$. It is a term of type $\iota$, it can only be used after $\phi\{\alpha\}$ has been proved for some term $\alpha$ of type $\iota$. The resulting system is called $H\mathscr{F}_C$ or $HI\mathscr{F}_C$ according as it is without or has the equality symbol. These systems fail to be formal systems because it is impossible to decide whether a term $\boldsymbol{\eta}_\xi\phi\{\xi\}$ can be used. The theorems of $H\mathscr{F}_C$ are the $\mathscr{F}_C$-theorems in resolved form. By this we mean that if $\phi$ is an $\mathscr{F}_C$-statement in prenex normal form then $\psi$ is obtained from $\phi$ by replacing each restricted variable by a new function of the superior general variables and omitting the quantifiers; the point of the new symbol is that it gives a uniform notation for these new functions. The system $H\mathscr{F}_C$ is without quantifiers but it is sometimes convenient to have a system using both II$d, e$ and rule H, we then have a system like $\mathscr{F}_C$ with functors and constant individuals.

A related system is the following system based on $I\mathscr{F}_C$: it is $I\mathscr{F}_C$ together with the symbol $\boldsymbol{\iota}$ of type $\iota(o\iota)$, we write as usual

D 22 $$\boldsymbol{\iota}_\xi \phi\{\xi\} \quad \text{for} \quad \boldsymbol{\iota}(\boldsymbol{\lambda}\xi\phi\{\xi\}).$$

We have the rule

$$\text{J} \quad \frac{(E\xi)\,\phi\{\xi\} \quad (A\xi,\eta)\,CK\phi\{\xi\}\phi\{\eta\}I\xi\eta}{\phi\{\boldsymbol{\iota}_\xi\phi\{\xi\}\}}.$$

$\boldsymbol{\iota}_\xi \phi\{\xi\}$ is read 'the thing with the property $(\boldsymbol{\lambda}\xi\phi\{\xi\})$'. The formula $\boldsymbol{\iota}_\xi\phi\{\xi\}$ is of type $\iota$ and can be used only after the premisses of J have been proved. The resulting system, which again fails to be a formal system, is denoted by $J\mathscr{F}_C$. From J we obtain

$$\text{J}' \quad \frac{(E\xi)\,\phi\{\xi\} \quad (E\xi)\,K\phi\{\xi\}\psi\{\xi\} \quad (A\xi,\eta)\,CK\phi\{\xi\}\,\phi\{\eta\}\,I\xi\eta}{\psi\{\boldsymbol{\iota}_\xi\phi\{\xi\}\}}.$$

From the hypotheses of J′ we obtain by $\mathscr{F}_C$

$$(A\xi,\eta)\,CKK\phi\{\xi\}\,\psi\{\xi\}\,K\phi\{\eta\}\,\psi\{\eta\}\,I\xi\eta,$$

whence from $J$ and $(E\xi)\,K\phi\{\xi\}\,\psi\{\xi\}$ we obtain

$$K\phi\{\boldsymbol{\iota}_\xi K\phi\{\xi\}\,\psi\{\xi\}\}\,\psi\{\boldsymbol{\iota}_\xi K\phi\{\xi\}\,\phi\{\xi\}\},$$

whence $\phi\{\boldsymbol{\iota}_\xi K\phi\{\xi\}\,\psi\{\xi\}\}$, but we have from the first and last premisses of J′ by J $\phi\{\boldsymbol{\iota}_\xi\phi\{\xi\}\}$, whence from the last premiss of J′

$$\boldsymbol{\iota}_\xi\phi\{\xi\} = \boldsymbol{\iota}_\xi K\phi\{\xi\}\,\psi\{\xi\}$$

(using the reversibility of II$e$). Thus we obtain $K\phi\{\boldsymbol{\iota}_\xi\phi\{\xi\}\}\,\psi\{\boldsymbol{\iota}_\xi\phi\{\xi\}\}$ whence $\psi\{\boldsymbol{\iota}_\xi\phi\{\xi\}\}$.

Since the systems $H\mathscr{F}_C$, $HI\mathscr{F}_C$, $J\mathscr{F}_C$ fail to be formal systems, because we are without means of deciding whether the terms $\boldsymbol{\eta}_\xi\phi\{\xi\}$ or $\boldsymbol{\iota}_\xi\phi\{\xi\}$ may be used, the following formal system is put forward. It is called $E\mathscr{F}_C$ or $EI\mathscr{F}_C$ according as it is without or with the equality symbol. It is based on $\mathscr{F}_C$ or on $I\mathscr{F}_C$ as the case may be. It has the additional symbol $\boldsymbol{\epsilon}$ again of type $\iota(o\iota)$. We write

D 23 $$\boldsymbol{\epsilon}_\xi\phi\{\xi\} \quad \text{for} \quad \boldsymbol{\epsilon}(\boldsymbol{\lambda}\xi\phi\{\xi\}),$$

this is called an $\epsilon$-term. $\xi$ is called the bound variable of the $\epsilon$-term $\epsilon_\xi \phi\{\xi\}$. The symbol $E$ and the rules II$d$, $e$ are omitted and replaced by

E $$\frac{D\phi\{\alpha\}\,\omega}{D\phi\{\epsilon_\xi \phi\{\xi\}\}\,\omega},$$

where $\omega$ is subsidiary and can be omitted, this is called the $\epsilon$-rule or the E-rule, we say that this application of the $\epsilon$-rule belongs to the $\epsilon$-term $\epsilon_\xi \phi\{\xi\}$,

E′ $$\frac{DN\phi\{\xi\}\,\omega}{DN\phi\{\epsilon_\xi \phi\{\xi\}\}\,\omega},$$

$\xi$ fails to occur free in $\omega$ and in $\phi\{\Gamma_\iota\}$.

If we allow the rule of substitution then rule E′ becomes a case of the rule of substitution and so can be dispensed with. If we write

D 24 $$E(\lambda\xi\phi\{\xi\}) \quad \text{for} \quad \phi\{\epsilon(\lambda\xi\phi\{\xi\})\}$$

then rules E and E′ translate respectively into rules II$d$, $e$. If we write

D 25 $$A(\lambda\xi\phi\{\xi\}) \quad \text{for} \quad \phi\{\epsilon(\lambda\xi N\phi\{\xi\})\}$$

then we have a definition for the universal quantifier. Thus in the systems $E\mathscr{F}_C$ and $EI\mathscr{F}_C$ we have definitions for both quantifiers, if we have the rule of substitution then we need only rule E. These two systems are formal because the $\epsilon$-term $\epsilon_\xi \phi\{\xi\}$ can be used without restriction. The $\epsilon$-term, $\epsilon_\xi \phi\{\xi\}$, is read 'the most $\phi$-like thing'; then $\phi\{\epsilon_\xi N\phi\{\xi\}\}$ reads 'the most un-$\phi$-like thing is $\phi$-like', if this is the case then everything is $\phi$-like. The term $\epsilon_\xi \phi\{\xi\}$ can be used when $(A\xi)\,N\phi\{\xi\}$ is an $\mathscr{F}_C$-theorem. For instance if $r$ is a variable for a rational number then $\epsilon_r(r^2 = 2)$ is read 'the rational number whose square is most nearly equal to 2' but we have $(\epsilon_r(r^2 = 2))^2 \neq 2$, so that $\epsilon(r^2 = 2)$ might be any rational number (see Ch. 12).

It is readily seen that if $\phi$ is an $E\mathscr{F}_C$-theorem without occurrences of any $\epsilon$-term then the $E\mathscr{F}_C$-proof of $\phi$ is without occurrences of $\epsilon$-terms because once an $\epsilon$-term enters an $E\mathscr{F}_C$-proof then an $\epsilon$-term will remain in the $E\mathscr{F}_C$-proof from that place to the base. Rule E rejects the term $\alpha$ which might contain occurrences of $\epsilon$-terms but it introduces another $\epsilon$-term, so an $\epsilon$-term remains in the $E\mathscr{F}_C$-proof, even if some are rejected.

The situation for $EI\mathscr{F}_C$ is different because an $\epsilon$-term can disappear at an application of $I(ii)$. The $\epsilon$-term acts as a *choice function*. If $\phi\{\alpha\}$ holds for some term $\alpha$ then $\epsilon_\xi \phi\{\xi\}$ could be such a term. So that $\epsilon$ picks out a particular term for which $\phi$ holds. If however $\phi\{\alpha\}$ fails for each term $\alpha$ then $\epsilon_\xi \phi\{\xi\}$ could be any term. This expresses the reversibility of E′.

LEMMA. *The rule of substitution can be eliminated from $EI\mathscr{F}_C$.*
First push all substitutions back into the axioms.

Suppose that we have $\dfrac{D\phi\{\xi\}\,\omega}{D\phi\{\alpha\}\,\omega}$, in an $EI\mathscr{F}_C$-proof-tree, where $\xi$ fails to occur free in $D\phi\{\Gamma_\iota\}\,\omega$. In this tree replace all corresponding occurrences of $\xi$ by $\alpha$. Axioms remain axioms, applications of rules remain applications of the same rules or repetitions and we are left with an $EI\mathscr{F}_C$-proof-tree of $D\phi\{\alpha\}\,\omega$. In other words, substitutions may be pushed back to the axioms. Note in particular that rule E is preserved.

PROP. 27. *Rules* II$b, c, e$ *are reversible in* $EI\mathscr{F}_C$.
II$b$ is reversible. We have to show

$$\frac{DND\phi'\phi''\omega}{DN\phi'\omega \quad DN\phi''\omega}{*}.$$

Consider the $EI\mathscr{F}_C$-proof-tree of $DND\phi'\phi''\omega$ and note the places at which a related occurrence of $ND\phi'\phi''$ enters this tree. This will occur at II$a$, $b$ only. If a related occurrence $ND\phi_1'\phi_1''$ enters the tree at II$a$ then replace it by $N\phi_1'$ and if at II$b$ then strike out the upper right formula and the branch above it and replace $ND\phi_1'\phi_1''$ by $N\phi_1'$, then II$b$ becomes a repetition. In each case in the lower parts of the tree replace related descendants of $ND\phi'\phi''$ by the result of omitting $N$ and $\phi''$. We are left with a tree with $DN\phi'\omega$ at its base and each node has an application of the rule previously used at that node if $ND\phi_1'\phi_1''$ was in the subsidiary or secondary formula of that rule or was in the main formula of $I(ii)$. $ND\phi_1'\phi_1''$ fails to be in the main formula of any rule other than II$a$, $b$, $I(ii)$ or E. II$a$, $b$ become repetitions $I(ii)$ remains but E can get upset if $ND\phi_1'\phi_1''$ was part of all of the main formula. Axioms remain axioms. If $ND\phi_1'\phi_1''$ occurs in the main formula of rule E, for instance if we have

$$\frac{DD\psi\{ND\phi'\{\alpha\}\,\phi''\}\}\,\omega}{DD\psi\{ND\phi'\{\epsilon_\xi D\psi\{ND\phi'\{\xi\}\,\phi''\}\}\,\phi''\omega.}\quad \text{E.}$$

If in this we strike out related occurrences of $D$—$\phi''$ then it becomes

$$\frac{D\psi\{N\phi'\{\alpha\}\}\,\omega}{D\psi\{\phi'\{\epsilon_\xi D\psi\{ND\phi'\{\xi\}\phi''\}\,\omega},$$

this fails to be a case of any rule. We shall have to call the occurrence of $ND\text{—}\phi''$ in the $\epsilon$-term related to the occurrence of $ND\text{—}\phi''$ in the upper formula. By this device we preserve the proof. This completes this case.

II*c* is reversible. We have to show $\dfrac{DNN\phi\omega}{D\phi\omega}*$.

In the $EI\mathscr{F}_C$-proof-tree of $DNN\phi\omega$ consider the corresponding or related occurrences of $NN\phi$. These will be introduced at II*a*, *c* only, omit $NN$ at each of these occurrences throughout the proof-tree and we are left with a tree with $D\phi\omega$ at its base. Applications of rules remain applications of the same rules or repetitions except that rule E can become upset if a related occurrence of $NN\phi$ occurs in the main formula of that rule axioms remain axioms. If $NN\phi$ occurs in the main formula of rule E, use the device as before. We are left with an $EI\mathscr{F}_C$-proof-tree of $D\phi\omega$.

II*e* is reversible. We have to show

$$\frac{DN(E\xi)\,\phi\{\xi\}\,\omega}{DN\phi\{\xi\}\,\omega}*.$$

Again in the $EI\mathscr{F}_C$-proof-tree of $DN(E\xi)\,\phi\{\xi\}\,\omega$ consider the corresponding or related occurrences of $N(E\xi)\,\phi\{\xi\}$ and in these omit $(E\xi)$ or the part related to it. These occurrences can only be introduced into the tree at II*a*, *e*, when we omit $(E\xi)$ applications of rules remain applications of the same rules except that II*a*, *e* can become repetitions, axioms remain axioms, but rule E can be upset if $(E\xi)$ occurs in the main formula of that rule. We use the same device as before, and the result follows.

PROP. 28. *Modus Ponens is a derived rule of* $EI\mathscr{F}_C$.

We have to show

$$\frac{D\omega\phi \quad DN\phi\chi}{D\omega\chi}*.$$

From the $EI\mathscr{F}_C$-proofs of the upper formulae we require to find $EI\mathscr{F}_C$-proof of the lower formula.

We proceed as in Prop. 4, by formula induction on the cut formula. There are only three cases, namely when the cut formula is atomic, a disjunction or a negation. The last two cases are dealt with exactly as before. The first case is also dealt with as before but the theorem induction requires the additional consideration of rule E. Thus suppose that in our theorem induction we have a case of rule E:

$$D\phi\{\epsilon_\xi N\phi\{\xi\}\}\,\omega \quad \frac{DN\phi\{\xi\}\,\chi}{DN\phi\{\epsilon_\xi N\phi\{\xi\}\}\,\chi}\ \mathrm{E}; \tag{1}$$

by the reversibility of II $e$ we obtain $D\phi\{\xi\}\,\omega$, thus we require:

$$\frac{D\phi\{\xi\}\,\omega \quad DN\phi\{\xi\}\,\chi}{D\omega\chi}\,* \tag{2}$$

Here $\phi\{\xi\}$ is atomic, say it is: $\pi\alpha'...\alpha^{(\kappa)}$ where some of the $\alpha$'s contain $\xi$ or are $\xi$. Then $\phi\{\epsilon_\xi N\phi\{\xi\}\}$ is $\pi\alpha'^*...\alpha^{(\kappa)*}$, where $\alpha^{(\lambda)*}$ is the result of replacing all free occurrences of $\xi$ in $\alpha^{(\lambda)}$ by $\alpha$, where $\alpha$ is $\epsilon_\xi N\pi\alpha'\ldots\alpha^{(\lambda)}$. Thus $\phi\{\xi\}$ contains fewer occurrences of $\epsilon$ than $\phi\{\epsilon_\xi N\phi\{\xi\}\}$. Now (2) holds when $\phi\{\xi\}$ is without the $\epsilon$-symbol, because then rule E is lacking and we can proceed as in Prop. 4. Thus if we assume as induction hypothesis that the result holds for fewer occurrences of $\epsilon$-symbols than there are in $\phi\{\epsilon_\xi N\phi\{\xi\}\}$ then it holds for $\phi\{\epsilon_\xi N\phi\{\xi\}\}$. The same applies to $\Sigma N\phi$ as to $N\phi$.

PROP. 29. *If a quantifier-free $I\mathscr{F}_C$-statement $\phi$ can be obtained by an $EI\mathscr{F}_C$-deduction from quantifier-free $I\mathscr{F}_C$-statements $\phi', \ldots, \phi^{(\kappa)}$, then it can be obtained by an $I\mathscr{F}_C$-deduction from the same statements without using quantifiers.*

$\phi$ is an $EI\mathscr{F}_C$-statement without the $\epsilon$-symbol or quantifiers, and so are the hypotheses $\phi', \ldots, \phi^{(\kappa)}$. In an $EI\mathscr{F}_C$-deduction if an $\epsilon$-symbol enters then it remains in the proof unless it disappears at an application of $I(ii)$, e.g.

$$\frac{D\phi\{\alpha\}\,\omega \quad DI\alpha\beta\omega}{D\phi\{\beta\}\,\omega}\ I(ii),$$

where $\alpha$ contains $\epsilon$-terms but $\beta$ is without $\epsilon$-terms. If $\omega$ and $\phi\{\Gamma_\iota\}$ are without $\epsilon$-terms then $D\phi\{\beta\}\,\omega$ is without $\epsilon$-terms.

Consider the highest such case of rule $I(ii)$, follow the related occurrences of $I\alpha\beta$ up the tree until we come to the highest case of $I\alpha'\beta'$ where $\alpha'$ contains $\epsilon$-terms but $\beta'$ is without them. Then in the tree above this each $I\gamma\delta$ must be such that $\gamma, \delta$ both contain $\epsilon$ or neither do.

Equations of the type of $I\gamma\delta$ just mentioned are incapable of producing equations of the type of $I\alpha\beta$ above. Hence the equation $I\gamma\delta$ must arise from an axiom T.N.D., but this introduces $NI\gamma\delta$ as well. The descendants of $NI\gamma\delta$ will be in the subsidiary formula of the application of I (ii) which eliminated the $\epsilon$-term This leaves an $\epsilon$-term in the lower formula of that application of $I(ii)$ which will have to be eliminated lower down the tree. This, in turn, in the same manner, will introduce another $\epsilon$-term, which in turn will have to be eliminated still further down the tree and so on without end. This is absurd. Hence the deduction must be without $\epsilon$-terms altogether.

Prop. 29 enables us to show the consistency with respect to negation of theories based on $EI\mathscr{F}_C$ whose axioms are without the $\epsilon$-symbol and which are verifiable. That is to say: a closed statement of the theory without $\epsilon$-terms can be decided using a suitable definition of truth for the theory. If the axioms $\phi', \ldots, \phi^{(\kappa)}$ of the theory are valid then so is every theorem $\phi$ of the theory which lacks $\epsilon$-terms because a proof of $\phi$ will fail to contain the $\epsilon$-symbol and will proceed from the axioms $I\alpha\alpha$, T.N.D. for atomic statements by the rules $I\,a, b$, $\text{II}\,a, b, c$, $I(ii)$ only, all of these preserve validity. Thus closed $\epsilon$-free theorems of the theory lacking the $\epsilon$-symbol are valid so that the theory fails to have as theorems both $\phi$ and $N\phi$. Thus the theory is consistent with respect to negation.

If $\mathscr{T}$ is a theory based on $EI\mathscr{F}_C$ whose axioms lack the $\epsilon$-symbol but may contain quantifiers then we first put the axioms in *resolved form* by introducing new function symbols, we then omit the universal quantifiers and we are left with quantifier-free statements. Call the resulting theory $\mathscr{T}^*$, this differs from $\mathscr{T}$ by having quantifier-free axioms but contains some new function symbols. If we can find a definition of $\mathscr{T}^*$-truth such that the $\mathscr{T}^*$-axioms are valid then $\mathscr{T}^*$ and hence $\mathscr{T}$ are consistent with respect to negation.

## 3.25 *The system $\mathscr{BF}_C$*

We have seen that an $\mathscr{F}_C$-statement $\phi$ is satisfiable if and only if an effectively constructible $\mathscr{F}_C$-statement $\psi$ containing exactly one predicate $\pi$, where $\pi$ is binary, is satisfiable. Similarly for validity. It is thus of some interest to consider the system $\mathscr{BF}_C$ which is $\mathscr{F}_C$ with a single binary predicate $p$. In $\mathscr{BF}_C$ we define equality

$$(\xi = \eta) \quad \text{for} \quad K(A\zeta)\,Bp\zeta\xi p\zeta\eta(A\zeta)\,Bp\xi\zeta p\xi\eta.$$

$=$ is a symbol of type $o\iota\iota$, we use the more familiar way of writing equalities. More generally equality can always be defined in a theory $\mathscr{T}$

which contains a terminating sequence of predicates, but is without functions. We define $(\xi = \eta)$ for the conjunction of

$$Bp_1^{(\theta)}\xi p_1^{(\theta)}\eta, \quad (A\zeta)\,Bp_2^{(\theta)}\zeta\xi p_2^{(\theta)}\zeta\eta, \quad (A\zeta)\,Bp_2^{(\theta)}\xi\zeta p_2^{(\theta)}\eta\zeta, \text{ etc.}$$

taken over all the predicates contained in the theory $\mathscr{T}$.

It is more usual to define

D 26 $\qquad (\xi = \eta)$ for $(A\zeta)\,Bp\zeta\xi p\zeta\eta$, $\quad (\xi \neq \eta)$ for $N(\xi = \eta)$,

and take $C(A\zeta)\,Bp\xi\zeta p\eta\zeta(\xi = \eta)$ as an axiom, or to define

D 26′ $\qquad (\xi = \eta)$ for $(A\zeta)\,Bp\xi\zeta p\eta\zeta$, $\quad (\xi \neq \eta)$ for $N(\xi = \eta)$,

and take $C(A\zeta)\,Bp\zeta\xi p\zeta\eta\ (\xi = \eta)$ as an axiom.

In either case if $(\xi = \eta)$ we have $(A\zeta)\,Bp\zeta\xi p\zeta\eta$ and $(A\zeta)\,Bp\xi\zeta p\eta\zeta$ so that by $\mathscr{F}_C$ $p\alpha\xi$ may be replaced by $p\alpha\eta$ and $p\xi\alpha$ by $p\eta\alpha$ in any $\mathscr{F}_C$-statement, in other words we have

$$(\alpha = \alpha) \quad \text{and} \quad \frac{D(\alpha = \beta)\,\omega \quad D\phi\{\alpha\}\,\omega}{D\phi\{\beta\}\,\omega}$$

by regularity $\phi$ being built up from the sole predicate $p$, this is the same axiom and rule as for the equality symbol $I$.

D 27 $\quad (\alpha \subseteq \beta)$ for $(A\zeta)\,Cp\zeta\alpha p\zeta\beta$, $\quad (\alpha \subset \beta)$ for $K(\alpha \subseteq \beta)(\alpha \neq \beta)$.

Read $(\alpha \subseteq \beta)$ as '$\alpha$ is contained in $\beta$'. The things which stand in the relation $p$ to $\alpha$ are less extensive than the things which stand in the relation $p$ to $\beta$. $\subset$ is called the *inclusion symbol.*

D 28 $\qquad \mathscr{E}m\alpha$ for $(A\xi)\,Np\xi\alpha$.

'$\alpha$ is without $p$-predecessors', or $\alpha$ is *empty*.

We can regard the predicate $p$ as being an ordering relation. If $p\alpha\beta$ we shall call $\alpha$ an *immediate p-predecessor* to $\beta$ and $\beta$ an *immediate p-successor* to $\alpha$. We can then identify $\beta$ with the class of its immediate $p$-predecessors, and so identify the relation $p$ with the membership relation. This amounts to reading $p\alpha\beta$ as '$\alpha$ stands in the relation $p$ to $\beta$' or as '$\alpha$ is a member of the class of things which stand in the relation $p$ to $\beta$'. Now $\beta$ is of type $\iota$ and $(\lambda\xi p\xi\beta)$ is of type $(o\iota)$. Let ^ be a symbol of type $\iota(o\iota)$ then $\wedge(\lambda\xi p\xi\beta)$ is of type $\iota$ and this term is uniquely fixed by $\beta$. Thus if we identity $\beta$ with $\hat{}(\boldsymbol{\lambda}\xi p\xi\beta)$ then we have formalized the above informal exposition. To make this identification we require $(\beta = \hat{}(\boldsymbol{\lambda}\xi p\xi\beta))$ more fully: $Bp\xi\beta p\xi\hat{}(\boldsymbol{\lambda}\xi p\xi\beta)$ and $Bp\beta\xi p\hat{}(\boldsymbol{\lambda}\xi p\xi\beta)\,\xi$.

In conformity with previous uses of $(\boldsymbol{\lambda}\xi\phi\{\xi\})$ we define

D 29 $\quad (\hat{\xi}\phi\{\xi\})$ for $\hat{}(\boldsymbol{\lambda}\xi\phi\{\xi\})$.

If we have a set $\mathscr{S}$ of things $\alpha, \beta, \ldots$, and are given a truth-value for each of $p\alpha\alpha, p\alpha\beta, p\beta\alpha, p\beta\beta, \ldots$, then for fixed $\beta$ we can form the class of things $\alpha$ for which $p\alpha\beta$ is $\boldsymbol{t}$. This gives us another set, say $\mathscr{S}_1^*$, whose members are classes of the members of $\mathscr{S}$, and there is a (1–1)-correspondence between $\mathscr{S}$ and $\mathscr{S}^*$. Every member of $\mathscr{S}^*$ is a class of members of $\mathscr{S}$ and there is a (1–1)-correspondence between $\mathscr{S}$ and $\mathscr{S}^*$ whereby $\alpha$ of $\mathscr{S}$ corresponds to the member $\alpha^*$ of $\mathscr{S}^*$, where $\alpha^*$ is the class of members of $\mathscr{S}$ which stand in the relation $p$ to $\alpha$, call this class $(\hat{x}px)$. The relation $p\alpha\beta$ translates into $p^*\alpha^*\beta^*$ or $(\alpha^*\epsilon\beta^*)$, read '$\alpha^*$ is a member of the class $\beta^*$'. With this interpretation $\mathscr{S}^*$ is a set of classes whose members are also classes whose members in turn are again classes, and so on, stopping, if at all, only at the null or empty class which is without members. The things $\alpha^*, \beta^*$, as defined, are subclasses of $\mathscr{S}^*$, and between these subclasses we define the relation $p^*$ which we interpret as the *membership relation*.

The members of $\mathscr{S}^*$ are classes of members of $\mathscr{S}^*$. Classes can be combined by certain operations to yield new classes. Thus two classes can combine to form their union or their intersection. From a single class we form its complement, and so on. Thus we can extend $\mathscr{S}^*$ if necessary so as to contain these and other combinations. We define

D 30 $\quad (\alpha \cap \beta)$ for $\hat{\xi}Kp\xi\alpha p\xi\beta$ *intersection*,

D 31 $\quad (\alpha \cup \beta)$ for $\hat{\xi}Dp\xi\alpha p\xi\beta$ *union*,

D 32 $\quad \bar{\alpha}$ for $\hat{\xi}Np\xi\alpha$ *complement*.

We shall require as axioms or theorems

$$Bp\xi(\alpha \cap \beta)\,Kp\xi\alpha p\xi\beta, \quad Bp\xi(\alpha \cup \beta)\,Dp\xi\alpha p\xi\beta, \quad Bp\xi\bar{\alpha}Np\xi\alpha.$$

This suggests that we have in general

$$(*) \quad Bp\eta\hat{\xi}\phi\{\xi\}\,\phi\{\eta\},$$

but this leads to an absurdity. Take $Np\eta\eta$ for $\phi\{\eta\}$, then the suggested axiom becomes: $Bp\eta\hat{\xi}Np\xi\xi Np\eta\eta$, now substitute the term $\hat{\xi}Np\xi\xi$ of type $\iota$ for $\eta$ and we get: $Bp(\hat{\xi}Np\xi\xi)(\hat{\xi}Np\xi\xi)Np(\hat{\xi}Np\xi\xi)(\hat{\xi}Np\xi\xi)$ which is absurd. If instead we take: $C(E\zeta)p\eta\zeta Bp\eta\hat{\xi}\phi\{\xi\}\phi\{\eta\}$, and proceed as before we merely arrive at $(A\zeta)Np(\hat{\xi}Np\xi\xi)\zeta$, i.e. the term $\hat{\xi}Np\xi\xi$ fails to stand in the relation $p$ to anything. It will be a maximal element in the ordering given by the relation $p$.

D 33 $\mathfrak{S}\xi$ for $(E\eta)p\xi\eta$, $\xi$ is a *set*,

D 34 $\mathfrak{P}\xi$ for $N\mathfrak{S}\xi$, $\xi$ is a *proper class*.

Thus if we adjoin the symbol ^ we obtain constant terms of type $\iota$ of two kinds, sets and proper classes. Let us then change the notation for variables of type $\iota$ and take them to be: $X$, $X'$, $X''$, .... Then we define another sort of variable, called *set variables*, by *relativization*, thus:

D 35 $(Ex)\phi\{x\}$ for $(EX)K\mathfrak{S}X\phi\{X\}$,

D 36 $\hat{x}\phi\{x\}$ for $\hat{X}K\mathfrak{S}X\phi\{X\}$.

These give $(Ax)\phi\{x\}$ for $(AX)C\mathfrak{S}X\phi\{X\}$.

We shall want some information as to whether a given class is a set or is a proper class. For convenience we use the letters $x$, $y$, $z$, $u$, $v$, $w$ with subscripts or superscripts as variables for sets and $X$, $Y$, $Z$, $U$, $V$, $W$ with subscripts or superscripts as variables for classes. We define:

D 37 $\{X, Y\}$ for $\hat{u}D(u = X)(u = Y)$, *pair class*,

D 38 $\Sigma X$ for $\hat{u}(Ev)KpuvpvX$, *union class*,

D 39 $PX$ for $\hat{u}(u \subseteq X)$, *power class*,

D 40 $\{X\}$ for $\{X, X\}$, *unit class*,

D 41 $\langle X, Y\rangle$ for $\{\{X\}, \{X, Y\}\}$, *ordered pair*,

D 42 $\langle X, Y, Z\rangle$ for $\langle X, \langle Y, Z\rangle\rangle$, *ordered triplet*, etc., in $\nu$-tuplets, $\langle x', \ldots, x^{(\nu)}\rangle$, pointed brackets are put back by association to the right,

D 43 $V$ for $\hat{u}(u = u)$, *universal class*,

D 44 $\Lambda$ for $\hat{u}(u \neq u)$, *null class*,

| | | |
|---|---|---|
| D 45 | $(UXV)$ for | $p\langle UV\rangle X$, *U stands in the relation X to V.* |
| D 46 | $(X \times Y)$ | $\hat{w}(Eu,v)\,K^2(w = \langle u,v\rangle)\,puXpvY$, *direct product.* |
| D 47 | $X^2$ | $(X \times X)$, |
| | $X^3$ | $(X^2 \times X)$, etc., |
| D 48 | $\mathscr{D}X$ | $\hat{u}(Ev)\,p\langle v,u\rangle X$, *domain,* |
| D 49 | $\mathscr{R}X$ | $\hat{v}(Eu)\,p\langle v,u\rangle X$, *range,* |
| D 50 | $\breve{X}$ or $Cnv_1X$ | $\hat{w}(Eu,v)\,K(w = \langle u,v\rangle)\,p\langle vu\rangle X$, *converse,* |
| D 51 | $RelX$ | $(X \subseteq V^2)$, *relation,* |
| D 52 | $X``Y$ | $\hat{u}(Ev)\,KpvYp\langle u,v\rangle X$, *transform of Y by X,* |
| D 53 | $UnX$ | $(Au,v,w)\,CKp\langle u,v\rangle Xp\langle w,v\rangle X(u = w)$, *one-valued relation,* |
| D 54 | $\mathscr{E}$ | $\hat{w}(Eu,v)\,K(w = \langle u,v\rangle)\,puv$, *membership relation,* |
| D 55 | $I$ | $\hat{w}(Eu,v)\,K(w = \langle u,v\rangle)\,(u = v)$, *identity relation,* |
| D 56 | $(X/Y)$ | $\hat{z}(Eu,v,w)\,K^2(z = \langle u,v\rangle)\,p\langle u,w\rangle Xp\langle w,v\rangle Y$, *product of relations,* |
| D 57 | $Cnv_2X$ | $\hat{z}(Eu,v,w)\,Kp\langle v,w,u\rangle X(z = \langle u,v,w\rangle)$, } *converses* |
| D 58 | $Cnv_3X$ | $\hat{z}(Eu,v,w)\,Kp\langle u,w,v\rangle X(z = \langle u,v,w\rangle)$, } |
| D 59 | $\Pi X$ | $\hat{x}(Ay)\,CpyXpxy$, *the intersection of the sets of X.* Clearly $\Pi V = \Lambda$ and $\Pi\Lambda = V$, |
| D 60 | $\mathscr{F}nX$ | $KRelXUnX$, *function,* |
| D 61 | $X`Y$ | $\hat{w}(Eu)\,Kpwu(Av)\,BvXY(v = u)$, *the value of the function X* for argument $Y$, everything is a set, so the value of a function is a set, $vXY$ see D 37, |
| D 62 | $(X\upharpoonright Y)$ | $X \cap (V \times Y)$, *function with domain restricted to Y,* |
| D 63 | $(Y\urcorner X)$ | $X \cap (Y \times V)$, *function with range restricted to Y,* |
| D 64 | $(X\mathscr{F}nY)$ | $K\mathscr{F}nX(\mathscr{D}X = Y)$, *function over Y,* |
| D 65 | $Un_2X$ | $KUnXUn\breve{X}$, $(1-1)$ *relation.* |

## 3.26 *Set theory*

The system $\mathscr{B}\mathscr{F}_C$ consists of the binary $\mathscr{F}_C$ with exactly one predicate $p$ and the class-forming symbol ^ of type $\iota(o\iota)$, so far we have only given definitions, if we wish to use the system as a set theory we shall have to add some axioms and rules. For instance we want some information as to which classes are sets, and conditions when a member of a class satisfies the defining statement of that class, i.e. for which $\mathscr{B}\mathscr{F}_C$-statements $\phi$ do we have: $Bp\eta\hat{\xi}\phi\{\xi\}\phi\{\eta\}$ as a $\mathscr{B}\mathscr{F}_C$-theorem? We have already seen that we are debarred from having this without qualification. Some classes must fail to be members of the universal class, the class of all sets. If all classes were sets we should have $pVV$ and the ordering relation $p$ would fail to be irreflexive. So now we add some axioms and rules and obtain a theory $\mathscr{S}$ based on $\mathscr{F}_C$ which we call set theory. In future we write $(\alpha\epsilon\beta)$ instead of $p\alpha\beta$. We take as our definition of equality:

D 66 $(X = Y)$ for $(Au)B(u\epsilon X)(u\epsilon Y)$, and $(\alpha\bar{\epsilon}\beta)$ for $N(\alpha\epsilon\beta)$.

This is the extensional approach, common in mathematics. Two classes are called equal if they have exactly the same members. We add the rule:

R 1. $$\frac{D(AU)K(X\epsilon U)(Y\epsilon U)\omega}{D(X = Y)\omega}.$$

As axioms to tell us which classes are sets we take:

Ax. 1. $\mathfrak{S}\Sigma x$,

Ax. 2. $\mathfrak{S}\{x, y\}$,

Ax. 3. $\mathfrak{S}Px$,

Ax. 4. $CUnX\mathfrak{S}X``y$,

these say that the union class of a set is a set, the pair class of two sets is a set, the power class of a set is a set and lastly the transform of a set by a one-valued relation (which may be a class) is a set. The free set variables in these axioms mean that we may only quantify with set quantifiers.

We add the following rules to tell us when a class satisfies the defining statement of that class:

R 2′. $$\frac{D(x\epsilon y)\omega}{D(\langle x, y\rangle \epsilon \mathscr{E})\omega}.$$ R 2″. $$\frac{DN(x\epsilon y)\omega}{DN(\langle x, y\rangle \epsilon \mathscr{E})\omega}.$$

R 3′. $$\frac{DD(u\epsilon X)(u\epsilon Y)\omega}{D(u\epsilon X \cup Y)\omega}.$$ R 3″. $$\frac{DND(u\epsilon X)(u\epsilon Y)\omega}{DN(u\epsilon X \cup Y)\omega}.$$

R 4′. $\dfrac{DN(u\epsilon X)\,\omega}{D(u\epsilon\overline{X})\,\omega}$.  R 4″. $\dfrac{D(u\epsilon X)\,\omega}{DN(u\epsilon\overline{X})\,\omega}$.

R 5′. $\dfrac{D(Ev)\,(\langle v,u\rangle\epsilon X)\,\omega}{D(u\epsilon\mathscr{D}X)\,\omega}$.  R 5″. $\dfrac{DN(Ev)\,(\langle v,u\rangle\epsilon X)\,\omega}{DN(u\epsilon\mathscr{D}X)\,\omega}$.

R 6′. $\dfrac{D(u\epsilon X)\,\omega}{D(\langle v,u\rangle\epsilon(V\times X))\,\omega}$.  R 6″. $\dfrac{DN(u\epsilon X)\,\omega}{DN(\langle v,u\rangle\epsilon(V\times X))\,\omega}$.

R 7′. $\dfrac{D(\langle u,v\rangle\epsilon X)\,\omega}{D(\langle v,u\rangle\epsilon\overset{\cup}{X})\,\omega}$.  R 7″. $\dfrac{DN(\langle u,v\rangle\epsilon X)\,\omega}{DN(\langle v,u\rangle\epsilon\overset{\cup}{X})\,\omega}$.

R 8′. $\dfrac{D(\langle v,w,u\rangle\epsilon X)\,\omega}{D(\langle u,v,w\rangle\epsilon Cnv_2X)\,\omega}$.  R 8″. $\dfrac{DN(\langle v,w,u\rangle\epsilon X)\,\omega}{DN(\langle u,v,w\rangle\epsilon Cnv_2X)\,\omega}$.

R 9′. $\dfrac{D(\langle u,w,v\rangle\epsilon X)\,\omega}{D(\langle u,v,w\rangle\epsilon Cnv_3X)\,\omega}$.  R 9″. $\dfrac{DN(\langle u,w,v\rangle\epsilon X)\,\omega}{DN(\langle u,v,w\rangle\epsilon Cnv_3X)\,\omega}$.

PROP. 30. *The rules R2′, ..., R 9″ are reversible.*

Take for instance rule R 4′. We have an $\mathscr{S}$-proof of $D(u\epsilon\overline{X})\,\omega$, we wish to find an $\mathscr{S}$-proof of $DN(u\epsilon X)\,\omega$. In the $\mathscr{S}$-proof of $D(u\epsilon\overline{X})\,\omega$ note the places where related occurrences of $(u\epsilon\overline{X})$ enter the $\mathscr{S}$-proof. These will be at R 4′, II$a$ or T.N.D., viz. $DN(u\epsilon\overline{X})\,(u\epsilon\overline{X})$, in the first two cases replace $(u\epsilon\overline{X})$ by $N(u\epsilon X)$ and similarly for all related occurrences, in the third case add above T.N.D. $DN(u\epsilon\overline{X})\,N(u\epsilon X)$. This is an $\mathscr{S}$-theorem, from R 4″ with $N(u\epsilon X)$ for $\omega$. Now replace related occurrences of $(u\epsilon\overline{X})$ by $N(u\epsilon X)$ and we have an $\mathscr{S}$-proof of $DN(u\epsilon X)$. Similarly for R 4″ and the other rules.

D 67. $\hat{\mathfrak{x}}\phi\{\mathfrak{x}\}$ for $\hat{y}(Ex',\ldots,x^{(\nu)})\,K(y=\langle\mathfrak{x}\rangle)\,\phi\{x',\ldots,x^{(\nu)}\}$,

where $\mathfrak{x}$ stands for $x',\ldots,x^{(\nu)}$ and $\langle\mathfrak{x}\rangle$ for $\langle x',\ldots,x^{(\nu)}\rangle$.

PROP. 31. *If all the bound variables in $\phi\,\{\mathfrak{y}\}$ are set variables and if $\mathfrak{y}$ contains the complete list of free variables in $\phi\{\mathfrak{y}\}$ then:*

$$\frac{D\phi\{\mathfrak{y}\}\,\omega}{D(\langle\mathfrak{y}\rangle\epsilon\hat{\mathfrak{x}}\phi\{\mathfrak{x}\})\,\omega}\quad\textit{and}\quad\frac{DN\phi\{\mathfrak{y}\}\,\omega}{DN(\langle\mathfrak{y}\rangle\epsilon\hat{\mathfrak{x}}\phi\{\mathfrak{x}\})\,\omega}$$

*and conversely.*

Prop. 31 says that we can replace the suggested general rule (∗) applied to $\mathscr{S}$-statements without bound class variables by 8 particular cases

R 2′, …, R 9″. If all the bound variables in $\phi\{\mathfrak{y}\}$ are set variables then $\phi\{\mathfrak{y}\}$ is called a *normal $\mathscr{S}$-statement* and $\hat{\mathfrak{x}}\phi\{\mathfrak{x}\}$ is called a *normal class*. We demonstrate the proposition by formula induction on the normal $\mathscr{S}$-statement $\phi\{\mathfrak{y}\}$. We first make a simplification; we replace $(\alpha\epsilon x)$, where $\alpha$ is different from a variable by $(Ey)\,K(y\epsilon x)\,(\alpha = y)$ and $(\alpha\epsilon\beta)$ by

$$(Ey)\,K(\beta = y)\,(\alpha\epsilon y) \quad \text{and} \quad (x\epsilon x) \quad \text{by} \quad (Ey)\,(x = y)\,(x\epsilon y).$$

Suppose $\mathfrak{y}$ is $\langle x', \ldots, x^{(\nu)}\rangle$ and $\phi\{\mathfrak{y}\}$ is $x^{(\lambda)}\epsilon x^{(\mu)}$. We have:

R 5′, 5″ (*a*) $\langle zxy\rangle \epsilon X \sim \langle xy\rangle \epsilon \mathscr{D}X$ with $z$ for $v$ and $\langle xy\rangle$ for $u$,

R 9′, 9″ $\langle xzy\rangle \epsilon X \sim \langle xyz\rangle \epsilon Cnv_3 X$,

R 8′, 8″ $\sim \langle zxy\rangle \epsilon Cnv_2 Cnv_3 X$.

R 5′, 5″ (*b*) $\sim \langle xy\rangle \epsilon \mathscr{D}\, Cnv_2 Cnv_3 X$,

R 8′, 8″ $\langle xyz\rangle \epsilon X \sim \langle zxy\rangle \epsilon Cnv_2 X$.

R 5′, 5″ (*c*) $\sim \langle xy\rangle \epsilon \mathscr{D} Cnv_2 X$.

From these we get:

R 5′, 5″ (*d*) $\langle yx' \ldots x^{(\nu)}\rangle \epsilon X \sim \langle x' \ldots x^{(\nu)}\rangle \epsilon \mathscr{D}X$,

from (*a*) with $y$ for $z$, $x'$ for $x$ and $\langle x' \ldots x^{(\nu)}\rangle$ for $y$, hence by repetition:

(*e*) $\langle y' \ldots y^{(\mu)} x' \ldots x^{(\nu)}\rangle \epsilon X \sim \langle x' \ldots x^{(\nu)}\rangle \epsilon \underbrace{\mathscr{D}(\mathscr{D}( \ldots \mathscr{D}}_{\mu\text{-times}}X)\ldots)$,

(*f*) $\langle x'y'x'' \ldots x^{(\nu)}\rangle \epsilon X \sim \langle y'x'x'' \ldots x^{(\nu)}\rangle \epsilon \mathscr{C}_2\mathscr{C}_3 X$, from (*b*) with $y'$ for $x$, $x'$ for $z$ and $\langle x'' \ldots x^{(\nu)}\rangle$ for $y$.

(*f*′) is equivalent to $\langle x' \ldots x^{(\nu)}\rangle \epsilon \mathscr{D}\mathscr{C}_2\mathscr{C}_3 X$, (writing $\mathscr{C}_2$ for $Cnv_2$ and $\mathscr{C}_3$ for $Cnv_3$), hence by repetition:

(*g*) $\langle x'y' \ldots y^{(\mu)} x'' \ldots x^{(\nu)}\rangle \epsilon X \sim \langle x' \ldots x^{(\nu)}\rangle \epsilon \mathscr{D}\mathscr{C}_2\mathscr{C}_3 (\underbrace{\mathscr{D}\mathscr{C}_2\mathscr{C}_3( \ldots \mathscr{D}\mathscr{C}_2\mathscr{C}_3}_{\mu\text{-times}} X).)$

(*h*) $\langle x'x''y' \ldots y^{(\mu)}\rangle \epsilon X \sim \langle x'x''\rangle \epsilon \mathscr{D}\mathscr{C}_2 X$, from (*c*) with $x'$ for $x$, $x''$ for $y$ and $\langle y' \ldots y^{(\mu)}\rangle$ for $z$.

(*i*) $\langle xy' \ldots y^{(\mu)}\rangle \epsilon X \sim \langle\langle y' \ldots y^{(\mu)}\rangle x\rangle \epsilon \breve{X}$

$\sim x \epsilon \mathscr{D}\breve{X}$.

Lastly (*j*) $(Ex')\,(\langle x'x'' \ldots x^{(\nu)}\rangle \epsilon X) \sim \langle x'' \ldots x^{(\nu)}\rangle \epsilon \mathscr{D}X$.

(i) $\phi$ is atomic. Thus $\phi$ is either $x^{(\lambda)} \epsilon x^{(\mu)}$ or $x^{(\lambda)} \epsilon \hat{x}\phi\{x\}$. In the first case we want to show that there is a class $\alpha$ such that:

$$\langle x' \dots x^{(\nu)}\rangle \epsilon \alpha \sim x^{(\lambda)} \epsilon x^{(\mu)}.$$

If $\lambda \leqslant \mu$ we have $x^{(\lambda)} \epsilon x^{(\mu)} \sim \langle x^{(\lambda)} x^{(\mu)}\rangle \epsilon \mathscr{E}$.

If $\mu < \lambda$ we have $x^{(\lambda)} \epsilon x^{(\mu)} \sim \langle x^{(\mu)} x^{(\lambda)}\rangle \epsilon \overset{\cup}{\mathscr{E}}$.

By ($h$) $\langle x^{(\lambda)} x^{(\mu)} x^{(\mu+1)} \dots x^{(\nu)}\rangle \epsilon X \sim \langle x^{(\lambda)} x^{(\mu)}\rangle \epsilon \mathscr{D}\mathscr{C}_2 X$.

By ($g$) $\langle x^{(\lambda)} \dots x^{(\nu)}\rangle \epsilon X \sim \langle x^{(\lambda)} x^{(\mu)} x^{(\mu+1)} \dots x^{(\nu)}\rangle \epsilon \underset{\mu-\lambda-1\text{-times}}{\mathscr{D}\mathscr{C}_2\mathscr{C}_3(\mathscr{D}\mathscr{C}_2\mathscr{C}_3}$ $(\dots \mathscr{D}\mathscr{C}_2\mathscr{C}_3 X)\dots))$.

By ($e$) $\langle x' \dots x^{(\nu)}\rangle \epsilon X \sim \langle x^{(\lambda)} \dots x^{(\nu)}\rangle (\epsilon \underset{\lambda-1\text{-times}}{\mathscr{D}(\mathscr{D}(\dots \mathscr{D}} X)\dots))$.

Hence altogether:

$$\langle x' \dots x^{(\nu)}\rangle \epsilon X \text{ is equivalent to}$$

$$\langle x^{(\lambda)} x^{(\mu)}\rangle \epsilon \underset{\lambda-1\text{-times}}{\mathscr{D}(\mathscr{D}(\dots(\mathscr{D}}(\underset{\mu-\lambda-1\text{-times}}{\mathscr{D}\mathscr{C}_2\mathscr{C}_3(\mathscr{D}\mathscr{C}_2\mathscr{C}_3 \dots (\mathscr{D}\mathscr{C}_2\mathscr{C}_3}\mathscr{D}\mathscr{C}_2 X)\dots)))\dots)).$$

Now write $\qquad \mathscr{C}^3_2 X = X \quad \text{and} \quad \mathscr{C}^2_3 X = X.$

$\mathscr{C}^2_2(V \times (\underset{\mu-\lambda-1\text{-times}}{\mathscr{C}_3\mathscr{C}^2_2 \dots (\mathscr{C}_3\mathscr{C}^2_2} V \times (\underset{\lambda-1\text{-times}}{V \times \dots (V} \times \mathscr{E})\dots))\dots))$ for $X$ and we obtain:

$$x^{(\lambda)} \epsilon x^{(\mu)} \sim \langle x' \dots x^{(\nu)}\rangle \epsilon \mathscr{C}^2_2(V \times (\underset{\mu-\lambda-1\text{-times}}{\mathscr{C}_3\mathscr{C}^2_2 \dots (\mathscr{C}_3\mathscr{C}^2_2} V \times (\underset{\lambda-1\text{-times}}{V \times \dots (V} \times \mathscr{E})\dots))\dots)).$$

Since $\mathscr{E}$ is $\hat{x}^{(\mu)}\hat{x}^{(\lambda)}(x^{(\mu)} \epsilon x^{(\lambda)})$ it follows from the definitions of $V$, $\times$, $\mathscr{C}_2$, $\mathscr{C}_3$ that $x^{(\lambda)} \epsilon x^{(\mu)} \sim \langle x', \dots, x^{(\nu)}\rangle \epsilon \langle x', \overset{\wedge}{\dots}, x^{(\nu)}\rangle (x^{(\lambda)} \epsilon x^{(\mu)})$.

Thus any two $\mathscr{S}$-statements may be put into equivalent forms with exactly the same free variables.

(ii) $\phi\{\mathfrak{y}\}$ is $D\phi'\{\mathfrak{y}\}\phi''\{\mathfrak{y}\}$. By induction hypothesis we have

$$\frac{D\phi'\{\mathfrak{x}\}\omega}{D(\langle\mathfrak{x}\rangle \epsilon \mathfrak{x}\phi'\{\mathfrak{x}\})\omega}, \qquad \frac{D\phi''\{\mathfrak{x}\}\omega}{D(\langle\mathfrak{x}\rangle \epsilon \hat{\mathfrak{x}}\phi''\{\mathfrak{x}\})\omega}.$$

Hence:

$$\cfrac{\cfrac{\cfrac{D\phi'\{\mathfrak{x}\}\phi''\{\mathfrak{x}\}}{D(\langle\mathfrak{x}\rangle \epsilon \hat{\mathfrak{x}}\phi''\{\mathfrak{x}\})(\langle\mathfrak{x}\rangle \epsilon \hat{\mathfrak{x}}\phi''\{\mathfrak{x}\})}\ \text{Hyp.}}{(\langle\mathfrak{x}\rangle \epsilon (\hat{\mathfrak{x}}\phi'\{\mathfrak{x}\} \cup \hat{\mathfrak{x}}\phi'\{\mathfrak{x}\})}\ \text{R}\,3'}{(\langle\mathfrak{x}\rangle \epsilon \hat{\mathfrak{x}} D\phi'\{\mathfrak{x}\}\phi''\{\mathfrak{x}\})}\ \text{D}\,31$$

as desired.

(iii) $\phi\{\mathfrak{x}\}$ is $N\phi'\{\mathfrak{x}\}$. By induction hypothesis we have

$$\frac{\phi'\{\mathfrak{x}\}}{(\langle\mathfrak{x}\rangle \epsilon \hat{\mathfrak{x}}\phi'\{\mathfrak{x}\})} \quad \text{and} \quad \frac{N\phi'\{\mathfrak{x}\}}{N(\langle\mathfrak{x}\rangle \epsilon \hat{\mathfrak{x}}\phi'\{\mathfrak{x}\})}.$$

Hence
$$\cfrac{\cfrac{\cfrac{N\phi'\{\mathfrak{x}\}}{N(\langle\mathfrak{x}\rangle\epsilon\hat{\mathfrak{x}}\phi'\{\mathfrak{x}\})}\ \text{Hyp.}}{(\langle\mathfrak{x}\rangle\epsilon\overline{\hat{\mathfrak{x}}\phi'\{\mathfrak{x}\}})}\ \text{R4}'}{(\langle\mathfrak{x}\rangle\epsilon\hat{\mathfrak{x}}N\phi'\{\mathfrak{x}\})}\ \text{D32}$$
as desired.

(iv) $\phi\{\mathfrak{x}\}$ is $(Ey)\,\phi'\{y,\mathfrak{x}\}$. By induction hypothesis we have:

$$\frac{\phi'\{y,\mathfrak{x}\}}{(\langle y,\mathfrak{x}\rangle\epsilon\hat{y}\hat{\mathfrak{x}}\phi'\{y,\mathfrak{x}\})}.$$

We have:
$$\cfrac{\cfrac{\cfrac{\cfrac{(Ey)\,\phi'\{y,\mathfrak{x}\}}{(Ey)\,(\langle y,\mathfrak{x}\rangle\epsilon\hat{y}\hat{\mathfrak{x}}\phi'\{y,\mathfrak{x}\})}\ *}{(\langle\mathfrak{x}\rangle\epsilon\mathscr{D}\hat{y}\hat{\mathfrak{x}}\phi'\{y,\mathfrak{x}\})}\ \text{R}5'}{(\langle\mathfrak{x}\rangle\epsilon\hat{\mathfrak{x}}(Ey)(\langle y,\mathfrak{x}\rangle\epsilon\hat{y}\hat{\mathfrak{x}}\phi'\{y,\mathfrak{x}\}))}\ \text{D}47}{(\langle\mathfrak{x}\rangle\epsilon\hat{\mathfrak{x}}(Ey)\,\phi'\{y,\mathfrak{x}\})}\ \text{Hyp.}$$
This completes the demonstration of the proposition.

## 3.27 *Ordinals*

The system $\mathscr{S}$ is very useful for model making. For instance we can define the *natural numbers* thus:

$$0 \text{ for } \Lambda,\quad 1 \text{ for } \{0\},\quad 2 \text{ for } \{0,\{0\}\},\quad 3 \text{ for } \{0,\{0\},\{0\{0\}\}\},\ldots$$

so that $S\nu$ is defined as the set of lesser natural numbers. The first transfinite number $\omega$ is then defined as the class of all natural numbers. To carry this through we have to $\mathscr{S}$-prove that the natural numbers defined as above are sets, otherwise they are debarred from membership of other sets and the process breaks down. $\mathfrak{S}\Lambda$ is easily $\mathscr{S}$-proved so are:

$$\mathfrak{S}(x\cap y), \mathfrak{S}(x\cup y), \mathfrak{S}(x\times y), \mathfrak{S}\mathscr{D}x, \mathfrak{S}\mathscr{R}x, \mathfrak{S}\overset{\cup}{x}, \text{ etc.}$$

and these suffice to show that the natural numbers as defined above are sets. But to show that $\omega$ is a set so that the process can continue into the transfinite we require another axiom. We could take $S\omega$ itself as an extra axiom, but it is usual to take:

Ax. 5. $$(Ex)\,KN\mathscr{E}mx(Ay)\,C(y\epsilon x)\,(Ez)\,K(z\epsilon x)\,(y\subset z).$$

This ensures the existence of a set containing an unending strictly increasing sequence of sets. It is called the *axiom of infinity*.

An ordinal is defined as the class of lesser ordinals and is *well-ordered* by the membership relation $\mathscr{E}$. The successor of an ordinal $x$ is then $x \cup \{x\}$, and the limit of a class of ordinals $X$ is $\Sigma X$.

D 68 $X \mathscr{W}e Y$ for $K(X^2 \subseteq Y \cup \breve{Y} \cup I)(AU)CKN\mathscr{E}mU(U \subseteq X)$
$$(Ev)(v \epsilon U)(\mathscr{E}mU \cap Y``\{v\}),$$

i.e. $X$ is well-ordered by $Y$; for any two members $x, x'$ of $X$ we have $DD(xYx')(x'Yx)(x = x')$, and every non-empty subclass of $X$ has a least member in the ordering $Y$.

The official definition of an ordinal is

D 69 $$Ord\,X \quad \text{for} \quad KX\mathscr{W}e\mathscr{E}(X \subseteq PX),$$

i.e. $X$ is well-ordered by the membership relation and members of members of $X$ are members of $X$. It is easy to show $Ord\ 0$, i.e. $Ord\ \Lambda$.

$\mathscr{O}n$ for $\hat{x}\,Ord\,x$.

$\mathscr{O}n$ is the class of all set ordinals.

D 70 $$X < Y \quad \text{for} \quad X \epsilon Y,$$
$$X \leqslant Y \quad \text{for} \quad D(X < Y)(X = Y).$$

We shall use $a, b, c, d, a', \ldots$, as variables for set ordinals.

We have
$$a \bar{\epsilon} a,\ NK(a \epsilon b)(b \epsilon a), \text{ etc.}$$
$$B(a \subset b)(a \epsilon b),$$
$$C(u \epsilon a)\,Ord\,u.$$

Any member of an ordinal is an ordinal. In fact an ordinal is the class of all lesser ordinals. The tricotomy holds
$$DD(a < b)(a = b)(b < a).$$

$a \subseteq \mathscr{O}n$ any ordinal is a subset of and a member of any larger ordinal. $\mathscr{O}n$ itself is an ordinal but is a proper class $\mathscr{P}r\,\mathscr{O}n$. Thus we are unable to form the successor of $\mathscr{O}n$ and so the antinomy of the greatest ordinal is avoided. An ordinal is either a member of $\mathscr{O}n$ or is $\mathscr{O}n$ itself.

D 71 $$Lim\,X \quad \text{or} \quad Max\,X \quad \text{for} \quad \Sigma X.$$

D 72 $$X+1 \quad \text{for} \quad X \cup \{X\}.$$

D 73 $$1 \quad \text{for} \quad 0+1,$$
$$2 \quad \text{for} \quad 1+1, \quad \text{etc.}$$

We have $C(X \subseteq \mathcal{O}n)\, Ord\, \Sigma X$. The limit of a class of ordinals is an ordinal.

We have $C(X \subseteq \mathcal{O}n)\, KOrd\, \Sigma X(Aa)\, C(a \epsilon X)\, (a \leqslant \Sigma X)$,

$$C(X \subseteq \mathcal{O}n)\, K\, Ord\, \Sigma X(Aa)\, C(X \subseteq a)\, (\Sigma X \leqslant a),$$

the properties of the limit ordinal of a class of ordinals.

We have

$$(Ax)\, B(x+1 \epsilon \mathcal{O}n)\, (x \epsilon \mathcal{O}n),$$

$$N(a < b < a+1).$$

D 74 $K_{\mathrm{I}}$ for $\hat{x}(Ea)\, D(x = a+1)\, (x = 0)$, ordinals of the first kind,

D 75 $K_{\mathrm{II}}$ for $\mathcal{O}n - K_{\mathrm{I}}$, ordinals of the second kind.

We have

$$C(a \epsilon K_{\mathrm{II}})\, K(a = \Sigma a)\, (a \neq 0),$$

$$C(a \epsilon K_{\mathrm{I}})\, D(a = \Sigma a + 1)\, (a = 0).$$

D 76 $\omega$ for $\hat{a}(a+1 \subseteq K_I)$,

the members of $\omega$ and the members of the members of $\omega$ are all of the first kind. We have $Ord\, \omega$, $S\omega$ and $\omega \in K_{\mathrm{II}}$. $\omega$ is a set ordinal of the second kind. Members of $\omega$ are called *intergers*. We use $i, j, i', \ldots$ as variables for integers. The *principle of Mathematical Induction* can be obtained in the form

$$\frac{\phi\{0\} \quad (Ai)\, C\phi\{i\}\, \phi\{i+1\}}{(Ai)\, \phi\{i\}}$$

provided that $\phi$ is without bound class variables.

D 77 $(X \doteq Y)$ for $(Ez)\, KKKUn_2 z Rel\, z(\mathcal{D}z = X)\, (\mathcal{R}z = Y)$.

$X$ is *similar* to $Y$ and both are sets.

D 78 $\mathcal{E}q$ for $\hat{x}\hat{y}(x \doteq y)$.

The *similarity relation* for sets. Then

$$(Ai,j)\, B(i = j)\, (i \doteq j)$$

D 79 $Fin$ for $\hat{x}(Ei)\, (i \doteq x)$

the class of *finite* sets.

D 80 $InFin$ for $\hat{x}(Ai)\, N(i \doteq x)$

the class of *infinite* sets.

Having defined the integers we can then define rational numbers as triplets of integers, then real numbers as Dedekind sections of rational numbers and lastly complex numbers as ordered pairs of real numbers. This is further discussed in Ch. 7, § 28. We are then ready to develop analysis and as explained in § 32 of this chapter we can introduce all topological concepts. An ordinal is either of the first kind or of the second kind or the ordinal is $\mathcal{O}n$.

If $X$ is a non-void class of ordinals then $\Pi X$ is the least member of $X$. Thus for $X \subseteq \mathcal{O}n$ we have $\Pi X \epsilon X$ and $\mathcal{E}m(X \cap \Pi X)$.

## 3.28 *Transfinite induction*

The principle of *transfinite induction* is

$$\frac{\phi\{0\} \quad (Aa)\,C\phi\{a\}\,\phi\{a+1\} \quad (Ab)\,C(Aa)\,C(a<b)\,\phi\{a\}\,\phi\{b\}}{(Aa)\,\phi\{a\}},$$

provided that $\phi$ is without bound class variables. This comes from $C(X \subseteq \mathcal{O}n)\,XW e\mathcal{E}$. This allows us to prove properties of ordinals by transfinite induction, since the class of ordinals without the property, if non-void, will have a first member. By a proof by transfinite induction we mean the reductio ad absurdam of the existence of a least ordinal without the property in question.

$$\frac{N\mathcal{E}mX \quad X \subseteq \mathcal{O}n}{(Ev)\,K(v\epsilon X)\,\mathcal{E}m(X \cap v)},$$

$$\frac{X \subseteq \mathcal{O}n \quad N(Ev)\,K(v\epsilon\mathcal{O}n - X)\,\mathcal{E}m((\mathcal{O}n - X)\cap v)}{X = \mathcal{O}n},$$

$$\frac{X \subseteq \mathcal{O}n \quad (Aa)\,C(a \subseteq X)\,(a\epsilon X)}{X = \mathcal{O}n}.$$

We also want to define functions by transfinite induction. It makes for easier reading if we use $F, G, H, F', \ldots,$ as variables for functions, corresponding small letters if they are sets, and $R, S, T, R', \ldots$ as variables for relations, corresponding small letters if they are sets. We want then to define $F‘a$ by means of the behaviour of $F$ for arguments less than $a$. Now $F \restriction a$ is the function $F$ with arguments restricted to $a$. Hence the induction should have the form

$$F‘a = G‘(F \restriction a),$$

where $G$ is a previously defined function. Thus we shall have

$$(AG)\,(E!F)\,(KF\mathscr{F}n\,\mathcal{O}n(Aa)\,(F\text{‘}a = G\text{‘}(F \upharpoonright a)).$$

The method of demonstrating this is to take the union of all partial solutions. Thus

$$H \quad \text{for} \quad \hat{f}(Eb)\,(Kf\mathscr{F}nb(Aa)\,C(a < b)\,(f\text{‘}a = G\text{‘}(f \upharpoonright a)),$$

then show that $\Sigma H$ has the required property. Something like this is done in detail in Ch. 11.

The addition, multiplication and exponentiation of ordinals are defined by transfinite induction thus:

D 81 $\quad +_b \quad$ for $\quad \hat{u}\hat{v}(DDK(u = 0)\,(v = b)\,(Ea)\,(u = a+1)\,(v = +_b\text{‘}a+1)$
$\qquad (Ea)\,K(v = a = \Sigma a)\,(v = \Sigma +_b\text{“}a),$

D 82 $\quad \times_b \quad$ for $\quad \hat{u}\hat{v}(DDK(u = 0)\,(v = 0)\,(Ea)\,K(u = a+1)\,(v = +_b \times_b\text{‘}a)$
$\qquad (Ea)\,K(u = a = \Sigma a)\,(v = \Sigma \times_b\text{“}a),$

D 83 $\quad exp_b \quad$ for $\quad \hat{u}\hat{v}(DDK(u = 0)\,(v = 1)\,(Ea)\,K(u = a+1)$
$\qquad (v = \times_b\, exp_b\text{‘}a)\,(Ea)\,K(u = a = \Sigma a)\,(v = \Sigma\, exp_b\text{“}a).$

The first of these defines the function $+_b$, the second the function $\times_b$ and the last the function $exp_b$. They are all of the form $F\text{‘}a = G\text{‘}(F \upharpoonright a)$. We usually write

$(b+a)$ for $+_b\text{‘}a$

$(b\times a)$ for $\times_b\text{‘}a$

$b^a$ for $exp_b\text{‘}a$.

$(a+b)$, $(a\times b)$ and $a^b$ are all set ordinals. They satisfy some of the usual rules of addition, multiplication and exponentiation. But some rules fail, e.g. $1+\omega = \omega$.

It can be verified that the ordinal $(a+b)$ is isomorphic as regards order to the order type obtained when we stick the order type $b$ at the end of the order type $a$, and that the ordinal $(a\times b)$ is isomorphic as regards order to the order type $\langle c,d\rangle$, $c\epsilon a$, $d\epsilon b$ ordered by last differences, i.e. $\langle c,d\rangle < \langle c',d'\rangle$ if $d < d'$ or $d = d'$ and $c < c'$. The ordinal $a^b$ is isomorphic to as regards order to the ordering by last differences of functions over $b$ with values in $a$, but with only a bounded number of non-zero values, i.e. if $f$ and $g$ are two such functions then $f < g$ if $f\text{‘}d < g\text{‘}d$, where $d$ is the greatest ordinal for which

$f'd \dotplus g'd$. In fact we could have taken these properties as definitions of addition, multiplication and exponentiation of ordinals provided we have shown

$$\frac{xWeT}{(Ea, R)\, R\, Isom\begin{pmatrix} x, a \\ T, \mathscr{E} \end{pmatrix}}$$

where

D 84 $R\, Isom\begin{pmatrix} X, Y \\ S, T \end{pmatrix}$ for

$$KKKKUn_2 R\, Rel\, R\mathscr{D}R = X\mathscr{R}R = Y(Au, v)\, C(u, v\epsilon X)\, B(uSv)\, (R'uTR'v),$$

i.e. $R$ is a (1–1)-correspondence between $X$ and $Y$ such that two members of $X$ stand in the relation $S$ if and only if their images in $Y$ by $R$ stand in the relation $T$.

Any class of ordinals is well-ordered by $\mathscr{E}$, hence a decreasing sequence of ordinals terminates, otherwise the sequence would be without first member and so would violate the condition of being well-ordered.

## 3.29 *Cardinals*

A *cardinal number* is frequently defined as the class of classes similar to a given class. We shall define a cardinal as an ordinal which is dissimilar to any lesser ordinal. This is less general than the usual definition because there may be classes dissimilar to any ordinal.

D 85 $\bar{\bar{x}}$ for $\hat{b}(Aa)\,(C(a \simeq x)\,(b\epsilon a)$,

then $\bar{\bar{x}}$ is the least ordinal which is similar to the set $x$. $\bar{\bar{x}}$ is called an *initial ordinal* or the *cardinal integer* of the set $x$ in case $x$ is finite. The class of ordinals similar to a given ordinal is non-void, because $a$ is similar to itself. Hence the class of ordinals similar to a given ordinal exists and being a class of ordinals has a least member $\bar{\bar{a}}$. But if $x$ is any set then set theory as we are developing it may turn out to be so poor in modes of expression that the (1–1)-correspondence required to show that $x$ is similar to an ordinal may be missing. Thus the concept of cardinal is relative, that is relative to the set theory used.

We have $\bar{\bar{\bar{\bar{a}}}} = \bar{\bar{a}}$.

We divide ordinals into *classes*, the members of one class being similar to each other, except that the first class is to consist of all the integers. The ordinals of the second class are similar to $\omega$, they are the denumerable ordinals. The least member of class III is denoted by $\Omega$, it is the least non-denumerable ordinal.

D 86 $$\Omega \text{ for } \Sigma\hat{b}(b \dot{\simeq} \omega).$$

Note that this use of the word 'class' is distinct from 'class' as opposed to 'set'.

D 87 $$\mathcal{N} \text{ for } \hat{a}(Eb)(a = \bar{\bar{b}}),$$

$\mathcal{N}$ is the class of integers and initial ordinals.

D 88 $$\mathcal{N}' \text{ for } \mathcal{N} - \omega,$$

$\mathcal{N}'$ is the class of initial ordinals. $\mathcal{N}'$ being a class of ordinals is well-ordered, hence there is an isomorphism between the initial ordinals and the ordinals. Let $\aleph$ be this isomorphism.

D 89 $$\omega_a \text{ or } \aleph_a \text{ for } \aleph`a.$$

Then $$\omega_0 = \omega = \aleph_0, \quad \Omega = \aleph_1, \text{etc.}$$

These cardinals are called *alephs*.

We can define addition, multiplication and exponentiation for cardinals. If $\alpha_j$, $j \in J$ is a class of cardinals then $\sum_{j \epsilon J} \alpha_j$ is the cardinal of the union of classes of cardinals $\alpha_j$ for $j \epsilon J$. In forming the union we require that the representative classes be distinct. This is achieved by using ordered pairs $\langle a, j\rangle$, $a \epsilon A_j \epsilon \alpha_j$. Then the ordered pairs are distinct for different $j \epsilon J$. If the cardinals are initial ordinals $\bar{\bar{a}}$ then $\bar{\bar{a}}$ itself is a representative class of that cardinal. The cardinal of the product of the class of cardinals $\alpha_j, j \epsilon J$ is the cardinal of the class of functions $f$ over $J$ such that $f`j \epsilon \alpha_j$. This amounts to picking out one member from each $\alpha_j$ and doing this in all possible ways. This raises the question as to whether there is any such function at all. The statement of the existence of such a function is known as the *Multiplicative Axiom*, *Axiom of Choice* (A.C.) or *Zermelo's Axiom*. If it failed then the product of an unbounded set of cardinals would be zero. The axiom is:

Ax. 6. $$(EF)(Ax) KF\mathcal{F}n V D\mathcal{E}mx(F`x \epsilon x).$$

This is a very strong form of the axiom of choice because it allows for the simultaneous choice from each set of an element of that set. The axiom of choice occurs frequently in mathematics, sometimes it is possible to avoid it by a more elaborate proof. At the end of Ch. 12 we sketch a demonstration of the independence of the axiom of choice from the other axioms of set theory.

Using the axiom of choice we can show that every set can be well-ordered, conversely if every set can be well-ordered then A.C.

$$(Ax)(Ea,f)KK(f\mathscr{F}na)\,Un_2 f(x = f``a),$$

so that $f`b$ for $b < a$ well-orders $x$. By transfinite induction we define a function $G$ so that $G\mathscr{F}n\,\mathscr{O}n$ and $(Aa)(G`a = F`(x - \mathscr{R}(G \restriction a)))$, where $F$ is the function postulated in Ax. 6. Then $G`0 = F`x$, the member chosen from $x$ by $F$, $G`1 = F`(x - \{G`0\})$, the member chosen from $x - \{G`0\}$ by $F$, etc. Then $G`b$ for $b < a$ well-orders $x$.

If we use Ax. 6 then every set can be well-ordered, hence every set is similar to an ordinal and so all cardinals are alephs, and hence the tricotamy will hold for cardinals. But without the axiom of choice there may be cardinals without any order relationship with any aleph.

The exponentiation of cardinals is defined by: $\alpha^\beta$ is the cardinal of the class of functions over $\beta$ with values in $\alpha$. Thus $2^{\aleph_0}$ is the cardinal of the real numbers. The equation $2^{\aleph_0} = \aleph_1$ is known as the *Continuum Hypothesis* (C.H.). It is now known to be independent of the other axioms of set theory and a brief sketch of this is given at the end of Ch. 12. It can be shown that the sum, product and exponent of alephs is an aleph.

The equation $2^{\aleph_\alpha} = \aleph_{S\alpha}$ is known as the *Generalized Continuum Hypothesis* (G.C.H.). Again it is now known to be independent of the other axioms of set theory and a brief sketch of this is given at the end of Ch. 12. Many statements about cardinals are now known to be independent of the axioms of set theory. But there are some important theorems about cardinals.

D 90 $\qquad \alpha < \beta \quad \text{for} \quad K(\alpha \doteqdot \beta' \subset \beta)N(\alpha \doteqdot \beta),$

the definition of order of cardinals.

PROP. 32. $\aleph_\alpha < 2^{\aleph_\alpha}$.

We shall show $\bar{\bar{x}} < \overline{\overline{Px}}$. Each member $y$ of $x$ gives rise to a subset $\{y\}$ of $x$, hence $x$ can be put into (1–1)-correspondence with a proper subset of $Px$, thus $\bar{\bar{x}} \leqslant \overline{\overline{Px}}$. Note that $x$ and $Px$ are sets.

Suppose that $\bar{\bar{x}} = \overline{\overline{Px}}$, then there is a (1–1)-correspondence between $x$ and $Px$. Let $\sigma\{y\}$ be the correlate of $y$ by this correspondence. Let $\alpha$ be the class of members of $y$ such that $y\bar{\varepsilon}\sigma\{y\}$. Let $\alpha$ be the correlate of $z$ so that $\alpha = \sigma\{z\}$. If $z\varepsilon\alpha$ then $z\bar{\varepsilon}\sigma\{z\}$ by definition of $\alpha$ and Prop. 31, but $\sigma\{z\} = \alpha$

and so $z\bar{\epsilon}\alpha$. Again if $z\bar{\epsilon}\alpha$ then by definition of $\alpha\ z\epsilon\sigma\{z\}$ and Prop. 31, i.e. $z\epsilon\alpha$. We have an absurdity in either case, hence $\bar{\bar{x}} \neq \overline{\overline{Px}}$.

Prop. 33. *If $\alpha \dot{\simeq} \beta'$ and $\beta' \subseteq \beta$ and $\beta \dot{\simeq} \alpha'$ and $\alpha' \subseteq \alpha$, then $\alpha \dot{\simeq} \beta$.*

Let $f$ map $\alpha$ (1–1) onto $\beta' \subset \beta$ and $g$ map $\beta$ (1–1) onto $\alpha' \subset \alpha$. We can clearly assume that $\alpha \cap \beta = \Lambda$. Now $(\alpha \cup \beta)$ is the disjoint union of sequences

$$\sigma: (a, f`a, g`f`a, \ldots) \quad (a\epsilon\alpha),$$

$$\sigma': (b, g`b, f`g`b, \ldots) \quad (b\epsilon\beta),$$

$$\sigma'': (\ldots, a, f`a, g`f`a, \ldots), \text{ where the sequence is without beginning.}$$

We can now map form a (1–1) map $h$, from $\alpha$ onto $\beta$. In cases $\sigma, \sigma''$ let $h`a = f`a$ for any $a\epsilon\alpha$ which occurs in a $\sigma$ or in a $\sigma''$. But if $a$ occurs in $\sigma'$ then let $h`a = g^{-1}`a$. Thus $\alpha$ is similar to $\beta$.

Prop. 34. *G.C.H. implies A.C.*

We have to show that if G.C.H. is taken as an extra axiom then A.C. is a theorem. G.C.H. says that if $\alpha \leqslant \beta \leqslant P\alpha$ then there is either a (1–1)-correspondence between $\alpha$ and $\beta$ or there is a (1–1)-correspondence between $\beta$ and $P\alpha$. This means that there are so many (1–1)-correspondences about that everything gets well-ordered.

We shall write $P_0X$ for $X$, $P_{S_\nu}X$ for $PP_\nu X$.

Lemma (i). *For any set $\alpha$ there is a well-ordered set $w$, such that*

$$w \subseteq P_4\alpha \quad and \quad \bar{\bar{w}} \nleqslant \bar{\bar{\alpha}}.$$

Consider the class $w$ of well-orderings of $\alpha$ and subsets of $\alpha$. A well-ordering of $\alpha$ is a class of ordered pairs hence is in $P_3\alpha$. Thus $w \subseteq P_4\alpha$. $w$ is isomorphic to a class of ordinals and so is well-ordered and is isomorphic to an ordinal. If $\bar{\bar{w}} \leqslant \bar{\bar{\alpha}}$ then $w$ would be order isomorphic to a well-ordered subset of $\alpha$ and so $w$ would be order isomorphic to a proper subset of itself, which is impossible. Hence lemma (i). For the moment we assume that $\overline{\overline{2P_i\alpha}} = \overline{\overline{P_i\alpha}}$.

Lemma (ii). *If $\gamma$ and $\delta$ are disjoint sets such that $\overline{\overline{\gamma \cup \delta}} = \overline{\overline{P(2\gamma)}}$ then $\bar{\bar{\delta}} \geqslant \overline{\overline{P\gamma}}$.*

$2\gamma$ denotes the union of two disjoint copies of $\gamma$, say $\gamma_1$ and $\gamma_2$ If $f$ maps $\gamma \cup \delta$ onto $P(\gamma_1 \cup \gamma_2) \dot{\simeq} P\gamma_1 \times P\gamma_2$, then the image of $\gamma$ projected into

$P\gamma_1$ is only a proper part of $P\gamma_1$, since $\overline{\overline{\gamma_1}} < \overline{\overline{P\gamma_1}}$, and hence if $\xi$ is outside the projection, $f$ must map some subset of $\delta$ onto $\xi \times P\gamma_2$, which means that $\overline{\overline{\delta}} \geqslant \overline{\overline{P\gamma}}$. Whence lemma (ii).

Now we have $\overline{\overline{P_3\alpha}} \leqslant \overline{\overline{w}} + \overline{\overline{P_3\alpha}} \leqslant \overline{\overline{P_4\alpha}} + \overline{\overline{P_3\alpha}} \leqslant \overline{\overline{P_4\alpha}}$ by the assumption. Thus by G.C.H. either:

$$\overline{\overline{w}} + \overline{\overline{P_3\alpha}} = \overline{\overline{P_4\alpha}} \quad \text{or} \quad \overline{\overline{w}} + \overline{\overline{P_3\alpha}} = \overline{\overline{P_3\alpha}}.$$

Consider the first case. $\overline{\overline{w}} + \overline{\overline{P_3\alpha}} = \overline{\overline{P_4\alpha}} = \overline{\overline{P(2P_3\alpha)}}$, by the assumption. We have, by lemma (ii), $\overline{\overline{w}} \geqslant \overline{\overline{P_4\alpha}}$, but $\overline{\overline{w}} \leqslant \overline{\overline{P_4\alpha}}$, hence $\overline{\overline{w}} = \overline{\overline{P_4\alpha}}$. Thus there is a (1–1)-correspondence between $P_4\alpha$ and $w$, thus $P_4\alpha$ is well-ordered, but $\alpha$ can be embedded in $P_4\alpha$, hence $\alpha$ can be well-ordered, and we are done. In the other case we have $\overline{\overline{w}} + \overline{\overline{P_3\alpha}} = \overline{\overline{P_3\alpha}}$ then we have $w \leqslant \overline{\overline{P_3\alpha}}$, hence $\overline{\overline{w}} = \overline{\overline{P_3\alpha}}$, and we are done, as before, or $\overline{\overline{w}} < \overline{\overline{P_3\alpha}}$, and whence by G.C.H. $w \leqslant \overline{\overline{P_2\alpha}}$, but then $\overline{\overline{w}} = \overline{\overline{P_2\alpha}}$, and we are done as before, or $\overline{\overline{w}} \leqslant \overline{\overline{P_1\alpha}}$, now by lemma (i) $\overline{\overline{w}} \nleqslant \overline{\overline{\alpha}}$, whence by G.C.H. $\overline{\overline{w}} = \overline{\overline{P_1\alpha}}$, and we are done as before.

It remains to show $\overline{\overline{2P_i\alpha}} = \overline{\overline{P_i\alpha}}$, for $0 < i < 4$. If we put $\beta = P(\alpha \cup \omega)$, where $\alpha \cap \omega = \Lambda$, then easily $\overline{\overline{2P_i\beta}} = \overline{\overline{P_i\beta}}$, for $0 < i < 4$. Let $\gamma$ be new.
$\overline{\overline{2\beta}} = \overline{\overline{P(\alpha \cup \omega \cup \{\gamma\})}} = \overline{\overline{P(\alpha \cup \omega)}} = \overline{\overline{\beta}}$. Also $\overline{\overline{\beta}} \leqslant \overline{\overline{\beta \cup \gamma}} \leqslant \overline{\overline{2\beta}}$, so $\overline{\overline{\beta \cup \gamma}} = \overline{\overline{\beta}}$. Now $2.2^{\beta} = 2^{\beta \cup \{\gamma\}} = 2^{\beta}$, and similarly $\overline{\overline{2P_i\beta}} = \overline{\overline{P_i\beta}}$, for all $i$. Hence our argument can be applied to $\beta$, and so $\beta$ can be well-ordered. But $\alpha$ can be embedded in $\beta$ in a natural manner, hence $\alpha$ can be well-ordered. Thus Prop. 33 is demonstrated.

## 3·30 *Elimination of the* $\epsilon$*-symbol*

The choice effected by A.C. can also be effected by the $\epsilon$-symbol, thus $(Ax)\,CN\mathscr{E}mx(\epsilon_y(y\epsilon x)\,\epsilon x)$, then $\hat{u}\hat{v}CN\mathscr{E}mv(u = \epsilon_y(y\epsilon v))$, is the required function that picks out a member from each set. We could also have the rule.

$$\text{C} \qquad \frac{N\mathscr{E}m\hat{x}\phi\{x\}}{\phi\{\epsilon_x\phi\{x\}\}},$$

provided that $\phi$ is a normal statement.

The system $\mathscr{S}$ with the $\boldsymbol{\epsilon}$-symbol and rule C is called the system $C\mathscr{S}$. We could also, as in D 24, define the existential quantifier in terms of the $\boldsymbol{\epsilon}$-symbol. The system $\mathscr{S}$ with rule II$d$ replaced by rule E and rule II$e'$ replaced by rule E$'$ is called the system $E\mathscr{S}$. We could then dispense with rule E$'$ in favour of a rule of substitution. The system $CE\mathscr{S}$ is the system $E\mathscr{S}$ plus rule C.

PROP. 35. *If an $\mathscr{S}$-statement is a $C\mathscr{S}$-theroem then it is an $\mathscr{S}$-theorem.*

We shall show that if the $\boldsymbol{\epsilon}$-symbol is used in a $C\mathscr{S}$-proof of an $\mathscr{S}$-statement $\phi$ then it can be eliminated leaving an $\boldsymbol{\epsilon}$-free proof of $\mathscr{S}$, which is thus an $\mathscr{S}$-proof. Thus $\phi$ is an $\mathscr{S}$-theorem. This proposition says that if $C\mathscr{S}$ is inconsistent with respect to negation then so is $\mathscr{S}$. For if we can $C\mathscr{S}$-prove the $\mathscr{S}$-statements $\phi$ and $N\phi$ then these $C\mathscr{S}$-proofs can be transformed into $\mathscr{S}$-proofs and so $\mathscr{S}$ would be inconsistent as well. This means that the axiom of choice is consistent with the other axioms of set theory.

First we replace the rule C by the rule E, this converts a $C\mathscr{S}$-proof into an $E\mathscr{S}$-proof, where $E\mathscr{S}$ is the system $\mathscr{S}$ with the rule E added. To do this, consider rule C, say, $\dfrac{D(E\xi)\,\phi\{\xi\}\,\omega}{D\phi\{\boldsymbol{\epsilon}_\xi\,\phi\{\xi\}\}\,\omega}$, and consider the places where related occurrences of $(E\xi)\,\phi\{\xi\}$ entered the C$\mathscr{S}$-proof, these will be of the form $\dfrac{D\phi'\{\alpha\}\,\omega'}{D(E\xi)\,\phi'\{\xi\}\,\omega'}$, where $\phi'\{\xi\}$ is a variant of $\phi\{\xi\}$ by $I(ii)$, replace the lower formula by $D\phi'\{\boldsymbol{\epsilon}_\xi\,\phi'\{\xi\}\}\,\omega'$, and similarly for all descendents. Entrance by II$a$ can be replaced by entrance by II$d$, the special $\mathscr{S}$-rules fail to introduce $D(E\xi)\,\phi\{\xi\}\,\omega$. The original $C\mathscr{S}$-proof-tree becomes an $E\mathscr{S}$-proof-tree of the original statement, because these related occurrences of $(E\,\xi)\,\phi\{\xi\}$ all occur in the subsidiary formulae, except $I(ii)$, from their introduction to the application of the C-rule under discussion. Thus we require to eliminate the $\boldsymbol{\epsilon}$-symbol from an E$\mathscr{S}$-proof of an $\mathscr{S}$-statement, free from the $\boldsymbol{\epsilon}$-symbol.

We next replace the special $\mathscr{S}$-rules by axioms, thus: if $\dfrac{D\phi\omega}{D\psi\omega}$ is an $\mathscr{S}$-rule replace it by $CD\phi\omega D\psi\omega$, we recover the rule by Modus Ponens, which can be eliminated from a theory in free disjunctive form. To make use of this result we replace rule R 1 by its free variable form viz. $\dfrac{DK(X\epsilon U)\,(Y\epsilon U)\,\omega}{DB(u\epsilon X)\,(u\epsilon Y)\,\omega}$ because by the reversibility of II$e'$ if we can $\mathscr{S}$-prove the upper formula of R 1 then we can $\mathscr{S}$-prove the upper formula of its free variable form. We similarly replace R 5′ and R 5″ by

$$\frac{D(\langle\alpha, u\rangle \epsilon X)\,\omega}{D(u\epsilon\mathscr{D}X)\,\omega} \quad\text{and}\quad \frac{D(\langle v, u\rangle \bar{\epsilon} X)\,\omega}{D(u\bar{\epsilon}\mathscr{D}X)\,\omega} \quad\text{respectively.}$$

This assures the reversibility of II$e'$ because $A$ can then only be introduced at II$e'$ and so the demonstration of reversibility of II$e'$ goes through as before, The axioms can be put into resolved form, viz.:

$$\text{Ax. 1. } (\Sigma x\epsilon\{\Sigma x\}), \quad \text{Ax. 2. } (\{x,y\}\epsilon\{\{x,y\}\}), \quad \text{Ax. 3. } (Px\epsilon\{Px\})$$

$$\text{Ax. 4. } CCK(\langle u,v\rangle\epsilon X)(\langle w,v\rangle\epsilon X)B(y\epsilon u)(y\epsilon w)(X``y\epsilon\{X``y\}).$$

From these the original axioms may be recovered. The system $\mathscr{S}$ can now be put into free disjunctive form, so that Modus Ponens can be eliminated. If we retain the axiom of infinity then we replace it by: $(\Omega\epsilon\{\Omega\})$, $(\Lambda\epsilon\Omega)$, $C(u\epsilon\Omega)(\{u\}\epsilon\Omega)$, then $\Omega$ contains the unending set: $\{\Lambda\}$, $\{\{\Lambda\}\}$, $\{\{\{\Lambda\}\}\}$, .... Call the resulting system $\mathscr{S}'$, then Modus Ponens can be eliminated from $\mathscr{S}'$. Let $\phi$ be an $E\mathscr{S}$-theorem then $\phi$ is an $E\mathscr{S}'$-theorem, let $\chi$ be the closure of the $\mathscr{S}'$-axioms used in the $E\mathscr{S}'$-proof of $\phi$, then by the Deduction Theorem $C\psi\phi$ is an $E\mathscr{F}_C$-theorem. From its $E\mathscr{F}_C$-proof we wish to eliminate the $\epsilon$-symbol. By hypothesis the $\epsilon$-symbol is absent from $\phi$ also the $\mathscr{S}'$-axioms are without the $\epsilon$-symbol, hence $C\chi\phi$ is without the $\epsilon$-symbol.

The method we shall adopt is to replace $\epsilon$-terms by other terms which lack the $\epsilon$-symbol in such a manner that the $E\mathscr{S}'$-proof remains correct.

Consider first the simple case when all the E-rules belong to the same $\epsilon$-term. Suppose $\dfrac{D\psi\{\alpha\}\,\omega}{D\psi\{\epsilon_\xi\psi\{\xi\}\}\,\omega}$ is an E-rule used in the $E\mathscr{S}'$-proof of $\phi$, where $\psi$ is $\epsilon$-free and closed; in this replace the $\epsilon$-term $\epsilon_\xi\psi\{\xi\}$ by $\alpha$, we are left with a tree with $\phi$ at its base because this is $\epsilon$-free. The above E-rule becomes a repetition, all applications of other rules are preserved except other applications of E-rules, which by supposition belong to the same $\epsilon$-term. For instance $\dfrac{D\psi\{\beta\}\,\omega'}{D\psi\{\epsilon_\xi\psi\{\xi\}\}\,\omega'}$ E becomes $\dfrac{D\psi\{\beta\}\,\omega'}{D\psi\{\alpha\}\,\omega'}$ which fails to be a case of any rule. But if we add $\psi\{\alpha\}$ as an extra axiom we can obtain $D\psi\{\alpha\}\,\omega'$ by II$a$, each case of rule E in the $E\mathscr{S}'$-proof of $\phi$ can be replaced by a case of II$a$; hence we obtain an $\mathscr{F}_C$-deduction of $\phi$ from the hypothesis $\psi\{\alpha\}$. Similarly we get an $\mathscr{F}_C$-deduction of $\phi$ from

$\psi\{\alpha^{(\theta)}\}$ for all $\theta$ for which $\psi\{\alpha^{(\theta)}\}$ occurs as the main upper formula of an E-rule. Since we are supposing that all E-rules belong to the same $\boldsymbol{\epsilon}$-term then $\psi$ is the same in each case. Thus we get an $\mathscr{F}_C$-deduction of $\phi$ from the hypotheses $\psi\{\alpha'\}$, $\psi\{\alpha''\}, \ldots, \psi\{\alpha^{(\nu)}\}$, where this is the complete list of main formulae in the upper formulae of E-rules used in the $E\mathscr{F}_C$-proof of $\phi$. Hence we obtain an $\mathscr{F}_C$-deduction of $\phi$ from $\sum_{\pi=1}^{\nu} \psi\{\alpha^{(\pi)}\}$.

On the other hand we have

$$\cfrac{\cfrac{N\psi\{\alpha^{(\theta)}\} \quad D\psi\{\alpha^{(\theta)}\}\,\omega^{(\theta)}}{\omega^{(\theta)}}\ \text{M.P.}}{D\psi\{\alpha\}\,\omega^{(\theta)}}\ \text{II}a$$

But Modus Ponens can be eliminated, and so we get an $\mathscr{F}_C$-deduction of $\phi$ from the hypotheses $N\psi\{\alpha'\}, N\psi\{\alpha''\}, \ldots, N\psi\{\alpha^{(\nu)}\}$, where we have the same set of $\alpha', \ldots, \alpha^{(\nu)}$ as before. Thus by the Deduction Theorem we obtain the $\mathscr{F}_C$-theorems $C\psi\{\alpha'\}\,\phi, \ldots, C\psi\{\alpha^{(\nu)}\}\,\phi$, and $CN\sum_{\theta=1}^{\nu}\{\alpha^{(\theta)}\}\,\phi$. Whence $C\sum_{\theta=1}^{\nu}\psi\{\alpha^{(\theta)}\}\,\phi$ and $CN\Sigma\psi\{\alpha^{(\theta)}\}\,\phi$ are $\mathscr{F}_C$-theorems and so by Modus Ponens is $\phi$. Thus this simple case we have eliminated the $\boldsymbol{\epsilon}$-terms, and obtained an $\mathscr{F}_C$-proof of $\phi$.

Next let us consider the case when $\phi$ is without universal quantifiers. We may suppose that $\phi$ is in prenex normal form, say $\phi$ is $(E\mathfrak{x})\,\phi\{\mathfrak{x}\}$, where $\mathfrak{x}$ stands for $\xi', \ldots, \xi^{(\kappa)}$. We may also suppose that the $E\mathscr{F}_C$-proof is without free variables of any type, because we can replace any that there may be by constants of the same type, and for this purpose we need only use one constant of each of the types required. This will fail to affect $\phi$ because it is closed, also the $E\mathscr{F}_C$-proof will remain an $E\mathscr{F}_C$-proof. Now replace each occurrence of rule II$d$ by a corresponding occurrence of the E-rule, this will replace the end formula by $\phi\{\mathfrak{a}\}$ where $\mathfrak{a}$ stands for $\alpha', \alpha'', \ldots, \alpha^{(\kappa)}$,

Before going any further we examine the structure of $\boldsymbol{\epsilon}$-terms. *The order of an $\boldsymbol{\epsilon}$-term* is the greatest numeral $\lambda$ such that we can find a sequence $\boldsymbol{\epsilon}', \boldsymbol{\epsilon}'', \ldots, \boldsymbol{\epsilon}^{(\lambda)}$ of $\boldsymbol{\epsilon}$-terms such that $\boldsymbol{\epsilon}^{(S\theta)}$ fails to occur bound in $\boldsymbol{\epsilon}^{(\theta)}$ but occurs in $\boldsymbol{\epsilon}^{(\theta)}$ i.e. occurs free in $\boldsymbol{\epsilon}^{(\theta)}$. The *rank of an $\boldsymbol{\epsilon}$-term* is defined as the greatest numeral $\mu$ such that we can find a sequence of $\boldsymbol{\epsilon}$-terms $\boldsymbol{\epsilon}', \boldsymbol{\epsilon}'', \ldots, \boldsymbol{\epsilon}^{(\mu)}$ such that $\boldsymbol{\epsilon}^{(S\theta)}$ occurs bound in $\boldsymbol{\epsilon}^{(\theta)}$, i.e. $\boldsymbol{\epsilon}^{(S\theta)}$ contains a free variable in the scope of the binding variable of $\boldsymbol{\epsilon}^{(\theta)}$. Clearly both sequences of $\boldsymbol{\epsilon}$-terms are terminating. Also if $\alpha$ is an $\boldsymbol{\epsilon}$-term occurring free in the $\epsilon$-term $\beta$ then the replacement of $\alpha$ by another $\epsilon$-term lacking the binding variable of $\beta$ has no effect on the rank of $\beta$. And if $\alpha$ is an $\epsilon$-term occurring in the $\epsilon$-term $\beta$ and containing the binding variable of $\beta$ then $\alpha$ is of lower rank than $\beta$. Also the rank of an $\epsilon$-term is unaltered when we change a free variable to a new variable.

We propose to eliminate the $\epsilon$-terms by replacing $\epsilon$-terms by other terms as we did in the simple case when all E-rules belonged to the same $\epsilon$-term, thus in the E-rule $\dfrac{D\psi\{\alpha\}\,\omega}{D\psi\{\epsilon_\xi\psi\{\xi\}\}\,\omega}$ we replace the $\epsilon$-term $\epsilon_\xi\psi\{\xi\}$ wherever it occurs by $\alpha$, then the above application of the E-rule becomes a repetition; the application of the E-rule $\dfrac{D\psi\{\beta\}\,\omega'}{D\psi\{\,\epsilon_\xi\psi\{\xi\}\}\,\omega'}$ becomes $\dfrac{D\psi\{\beta\}\,\omega'}{D\psi\{\alpha\}\,\omega'}$ which fails to be an application of any rule if $(\alpha \neq \beta)$, but we can obtain the lower formula from the hypothesis $\psi\{\alpha\}$. The subsitution of $\alpha$ for $\epsilon_\xi\psi\{\xi\}$ may alter the end formula $\phi$, say it becomes $\phi\{\mathfrak{a}_1\}$. If the only $\epsilon$-term in the $E\mathscr{F}_C$-proof of $\phi$ is $\epsilon_\xi\psi\{\xi\}$ then we should have $E\mathscr{F}_C$-deductions of $\phi\{\mathfrak{a}_1'\}$ from $\psi\{\alpha'\}, \ldots, \phi\{\mathfrak{a}_1^{(\kappa)}\}$ from $\psi\{\alpha^{(\kappa)}\}$, where the E-rule occurs only as $\dfrac{D\psi\{\alpha^{(\theta)}\}\,\omega^{(\theta)}}{D\psi\{\epsilon_\xi\psi\{\xi\}\}\,\omega^{(\theta)}}$, $1 \leqslant \theta \leqslant \kappa$. By the deduction theorem and $\mathscr{P}_C$ this would give us: $C\sum_{\theta=1}^{\kappa}\psi\{\alpha^{(\theta)}\}\sum_{\theta=1}^{\kappa}\phi\{\mathfrak{a}_1^{(\theta)}\}$, without using the E-rule belonging to the $\epsilon$-term $\epsilon_\xi\psi\{\xi\}$. On the other hand we have by Modus Ponens

$$\cfrac{\cfrac{N\psi\{\alpha^{(\theta)}\}\quad D\psi\{\alpha^{(\theta)}\}\,\omega^{(\theta)}}{\omega^{(\theta)}}\;\text{M.P.}}{D\psi\{\alpha\}\,\omega^{(\theta)}}\;\text{II}a$$

but Modus Ponens can be eliminated.

Thus if we introduce $N\psi\{\alpha'\}, N\psi\{\alpha''\}, \ldots, N\psi\{\alpha^{(\kappa)}\}$ as hypotheses we obtain an $E\mathscr{F}_C$-deduction of $\phi\{\mathfrak{a}\}$ without using the E-rule belonging to the $\epsilon$-term $\epsilon_\xi\psi\{\xi\}$. This gives us an $E\mathscr{F}_C$-proof of $C\prod_{\theta=1}^{\kappa}N\psi\{\alpha^{(\theta)}\}\phi\{\mathfrak{a}\}$.

From
$$C\sum_{\theta=1}^{\kappa}\psi\{\alpha^{(\theta)}\}\sum_{\theta=1}^{\kappa}\phi\{\mathfrak{a}_1^{(\theta)}\} \quad\text{and}\quad CN\Sigma\psi\{\alpha^{(\theta)}\}\phi\{\mathfrak{a}\}$$

we obtain by Modus Ponens $\sum_{\theta=0}^{\kappa}\phi\{\mathfrak{a}_1^{(\theta)}\}$, where $\mathfrak{a}_1^{(0)}$ is $\mathfrak{a}$. Modus Ponens can be eliminated so we obtain an $E\mathscr{F}_C$-proof of $\sum_{\theta=0}^{\kappa}\phi\{\mathfrak{a}_1^{(\theta)}\}$. The reason for replacing rule II$d$ by rule E is to avoid the existential quantifier binding an $\boldsymbol{\epsilon}$-term, thus:

$$\cfrac{\cfrac{D\phi\{\alpha,\beta\}\omega}{D\phi\{\boldsymbol{\epsilon}_\xi\phi\{\xi,\beta\},\beta\}\omega}\,\mathrm{E}}{D(E\eta)\,\phi\{\boldsymbol{\epsilon}_\xi\phi\{\xi,\eta\},\eta\}\,\omega}\,\mathrm{II}d$$

will fail to occur instead, we shall have in the lower formula

$$D\phi\{\boldsymbol{\epsilon}_\xi\phi\{\xi,\boldsymbol{\epsilon}_\eta\phi\{\boldsymbol{\epsilon}_\xi\phi\{\xi,\eta\},\eta\}\},\boldsymbol{\epsilon}_\eta\phi\{\boldsymbol{\epsilon}_\xi\phi\{\xi,\eta\},\eta\}\}\,\omega,$$

which has an $\boldsymbol{\epsilon}$-term of higher rank. But if there are other applications of the E-rule in the $E\mathscr{F}_C$-proof of $\phi\{\mathfrak{a}\}$ then the substitution of $\alpha^{(\theta)}$ for $\boldsymbol{\epsilon}_\xi\psi\{\xi\}$ may destroy them.

The following cases can arise for the E-rule

$$\frac{D\chi\{\beta\}\omega}{D\chi\{\boldsymbol{\epsilon}_\xi\chi\{\xi\}\}\omega}\ \mathrm{E}.$$

(i) $\boldsymbol{\epsilon}_\xi\psi\{\xi\}$ occurs at most only in $\beta$, the E-rule is then $\dfrac{D\chi\{\beta\{\boldsymbol{\epsilon}_\xi\psi\{\xi\}\}\}\omega}{D\chi\{\boldsymbol{\epsilon}_\eta\chi\{\eta\}\}\omega}$ this becomes $\dfrac{D\chi\{\beta\{\alpha\}\}\omega}{D\chi\{\boldsymbol{\epsilon}_\eta\chi\{\eta\}\}\omega}$ which is the E-rule.

(ii) $\boldsymbol{\epsilon}_\xi\psi\{\xi\}$ occurs in $\chi\{\eta\}$ and possibly in $\beta$ as well the E-rule is then $\dfrac{D\chi\{\beta\{\gamma\},\gamma\}\omega}{D\chi\{\boldsymbol{\epsilon}_\eta\chi\{\eta,\gamma\},\gamma\}\omega}$ this becomes $\dfrac{D\chi\{\beta\{\alpha\},\alpha\}\omega}{D\chi\{\boldsymbol{\epsilon}_\eta\chi\{\eta,\alpha\},\alpha\}\omega}$ which is the E-rule, here $\gamma$ stands for $\boldsymbol{\epsilon}_\xi\psi\{\xi\}$.

(iii) One or both of $\beta$, $\delta$ where $\delta=\boldsymbol{\epsilon}_\xi\chi\{\xi\}$ is contained in $\gamma=\boldsymbol{\epsilon}_\xi\psi\{\xi\}$, (so that in the first case $\gamma$ is of the form $\gamma'\{\beta\}$), and $\chi\{\eta\}$ is of the form $\chi'\{\gamma'\{\eta\}\}$; the E-rule is then $\dfrac{D\chi'\{\gamma'\{\beta\}\}\omega}{D\chi'\{\gamma'\{\boldsymbol{\epsilon}_\zeta\chi'\{\gamma'\{\zeta\}\}\}\}\omega}$ this becomes $\dfrac{D\chi'\{\alpha\}\omega}{D\chi'\{\gamma'\{\boldsymbol{\epsilon}_\zeta\chi'\{\gamma'\{\zeta\}\}\}\}\omega}$ which fails to be an E-rule, $\zeta$ is new to $\chi'\{\gamma'\{\Gamma_\iota\}\}$. Similarly if $\gamma$ is of the form $\gamma'\{\delta\}$, and $\chi(\eta)$ is of the form $\chi'\{\gamma'\{\eta\}\}$, the E-rule is $\dfrac{D\chi'\{\gamma'\{\beta\}\}\omega}{D\chi'\{\gamma'\{\boldsymbol{\epsilon}_\zeta\chi'\{\gamma'\{\zeta\}\}\}\}\omega}$ this is $\dfrac{D\chi'\{\gamma'\{\beta\}\}\omega}{D\chi'\{\gamma'\{\delta\}\}\omega}$ this becomes $\dfrac{D\chi'\{\gamma'\{\beta\}\}\omega}{D\chi'\{\alpha\}\omega}$ which fails to be an E-rule. Similarly if $\gamma$ is of the form $\gamma'\{\beta,\delta\}$ and $\chi$ is of one of the forms $\chi'\{\gamma'\{\eta,\delta\}\}$, $\chi'\{\gamma'\{\beta,\eta\}\}$, $\chi'\{\gamma'\{\eta,\eta\}\}$, etc. In all these cases the $\boldsymbol{\epsilon}$-term $\boldsymbol{\epsilon}_\zeta\chi\{\gamma'\{\zeta\}\}$ is of higher rank than the term $\gamma'\{\beta\}$ for which a substitution is being made.

(iv) The only remaining case is when one or both of $\beta$, $\delta$ are contained in $\gamma$ so that $\gamma = \gamma'\{\beta\}$, etc., but the variable $\eta$ in $\chi\{\eta\}$ fails to occur in a part $\gamma'\{\eta\}$ of $\chi\{\eta\}$. In this case the E-rule remains correct as can be seen from case (iii) by omitting $\gamma'$ in $\chi\{\gamma'\{\eta\}\}$ in the lower formulae. In case (ii) the $\epsilon$-term belonging to the E-rule is of higher order than $\gamma$ but is of the same rank as $\gamma$, but in case (iii) it is of higher rank than $\gamma$.

Thus if we make the substitution for an $\epsilon$-term of highest rank and among those of highest rank we choose one of highest order then cases (ii), (iii) will fail to arise; we can proceed as in the first case where all the E-rules belonged to the same $\epsilon$-term. The result of the elimination of one $\epsilon$-term is an $E\mathscr{F}_C$-proof of a disjunction $\sum_{\theta=0}^{\kappa} \phi\{\mathfrak{a}^{(\theta)}\}$. A second application will produce a similar disjunction of these disjunctions which is merely a longer disjunction of the same kind. Finally we can eliminate all the E-rules and are left with an $\mathscr{F}_C$-proof of a disjunction $\sum_{\theta=1}^{\nu} \phi\{\mathfrak{a}^{(\theta)}\}$. If there are any $\epsilon$-terms left in this disjunction then replace them all with the same new free variable, the result is an $\mathscr{F}_C$-proof of a free variable disjunction which is $\epsilon$-free. This is possible because all substitutions had been pushed back into the axioms before we started so that an $\epsilon$-term in the $\mathscr{F}_C$-proof of the disjunction $\sum_{\theta=1}^{\nu} \phi\{\mathfrak{a}^{(\theta)}\}$ could only be introduced into the $\mathscr{F}_C$-proof at an axiom and if it is replaced by a new variable axioms remain axioms and rules are preserved. From the free variable disjunction we easily obtain $(E\mathfrak{x})\,\phi\{\mathfrak{x}\}$. This completes the case when $\phi$ is without universal quantifiers.

We may suppose that $\phi$ is in Skolem $V$-normal form, i.e. is of the form: $(E\mathfrak{x})\,(A\mathfrak{y})\,\psi\{\mathfrak{x},\mathfrak{y}\}$ where the matrix $\psi\{\mathfrak{x},\mathfrak{y}\}$ is quantifier-free. Since $\phi$ is an $\mathscr{S}$-statement then it can be regarded as a binary $\mathscr{F}_C$-statement, containing various terms. From lemma (iii) Prop. 24 we see that we can $\mathscr{F}_C$-prove $\phi$ from hypotheses $\mathfrak{X}$ if and only if we can do the same for $\phi'$, where $\phi'$ is the Skolem $V$-normal form of $\phi$. Now an $\mathscr{S}$-proof is the same as an $\mathscr{F}_C$-deduction from hypotheses, say $\mathfrak{X}$. Thus if we can get an $\mathscr{F}_C$-deduction of $\phi'$ from hypotheses $\mathfrak{X}$, then we can do the same for $\phi$. Also if we have an $E\mathscr{F}_C$-deduction of $\phi$ from hypotheses $\mathfrak{X}$, then we also have an $E\mathscr{F}_C$-deduction of $\phi'$ from hypotheses $\mathfrak{X}$. Lastly, if we can convert this $E\mathscr{F}_C$-deduction of $\phi'$ from hypotheses $\mathfrak{X}$ into an $\mathscr{F}_C$-deduction of $\phi'$ from hypotheses $\mathfrak{X}$ then we can convert an $E\mathscr{F}_C$-deduction of $\phi$ from hypotheses $\mathfrak{X}$ into an $\mathscr{F}_C$-deduction of $\phi$ from hypotheses $\mathfrak{X}$. Thus a $C\mathscr{S}$-proof of $\phi$ can be converted into an $\mathscr{S}$-proof of $\phi$.

Now suppose that there are universal quantifiers in $\phi$. Introduce sufficient new functors so that we can replace each general variable by a function of its superior restricted variables. Then omit all the universal quantifiers, we are left with : $(E\mathfrak{x})\,\phi\{\mathfrak{x}, \mathfrak{f}\mathfrak{x}\}$. From the $E\mathscr{F}_C$-proof of $\phi$ we can obtain one of $(E\mathfrak{x})\,\phi\{\mathfrak{x}, \mathfrak{f}\mathfrak{x}\}$ as follows: We have the $\mathscr{F}_C$-theorem

$$C(A\mathfrak{y})\,\phi\{\mathfrak{x},\mathfrak{y}\}\,\phi\{\mathfrak{x},\mathfrak{f}\mathfrak{x}\} \quad \text{whence} \quad C(E\mathfrak{x})\,(A\mathfrak{y})\,\phi\{\mathfrak{x},\mathfrak{y}\}\,(E\mathfrak{x})\,\phi\{\mathfrak{x},\mathfrak{f}\mathfrak{x}\}.$$

Thus from the $E\mathscr{F}_C$-proof of $(E\mathfrak{x})\,(A\mathfrak{y})\,\phi\{\mathfrak{x},\mathfrak{y}\}$ we can obtain one of $(E\mathfrak{x})\,\phi\{\mathfrak{x},\mathfrak{f}\mathfrak{x}\}$. Thus from the $E\mathscr{F}_C$-proof of $\phi$ we obtain one of $(E\mathfrak{x})$ $\phi\{\mathfrak{x},\mathfrak{f}\mathfrak{x}\}$, where $\mathfrak{f}$ denotes the sequence : $f', f'', \ldots, f^{(\pi)}$ of functors and $f^{(\theta)}\mathfrak{x}$ signifies that the argument places of $f^{(\theta)}$ are filled with the superior restricted variables only and so may fail to embrace all the variables in $\mathfrak{x}$. From the case when the end formula was without universal quantifiers obtain an $\mathscr{P}_C$-proof of a disjunction

$$\sum_{\theta=0}^{\kappa} \phi\{\mathfrak{a}^{(\theta)}, \mathfrak{f}\mathfrak{a}^{(\theta)}\} \quad \text{where} \quad \mathfrak{a}', \mathfrak{a}'', \ldots, \mathfrak{a}^{(\kappa)}$$

are sequences of terms of type $\iota$, $\epsilon$-free but which may contain $f', f'', \ldots, f^{(\pi)}$. Moreover the $\mathscr{P}_C$-proof of this disjunction is a free variable $\mathscr{P}_C$-proof free from the substitution rule, because all substitutions had been pushed back into the axioms before we began. If in $\sum_{\theta=0}^{\kappa} \phi\{\mathfrak{a}^{(\theta)}, \mathfrak{f}\mathfrak{a}^{(\theta)}\}$ and its $\mathscr{P}_C$-proof we replace each atomic formula by a propositional variable, distinct formulae by distinct propositional variables, identical formulae by the same propositional variable, we are left with a $\mathscr{P}_C$-proof. Thus $\sum_{\theta=1}^{\kappa} \phi\{\mathfrak{a}^{(\theta)}, \mathfrak{f}\mathfrak{a}^{(\theta)}\}$ arises from a $\mathscr{P}_C$-theorem by substitution of atomic formulae of a theory for propositional variables. The same will hold if we replace the terms $f^{(\theta)}\mathfrak{a}^{(\mu)}$ by new free individual variables using distinct variables for distinct terms, the same variable for different occurrences of the same term. Consider the term $f^{(\theta)}\mathfrak{a}^{(\mu)}$ and the number of distinct occurrences of $f', f'', \ldots, f^{(\pi)}$ which are contained in it. Call this number the *complexity* of $f^{(\theta)}\mathfrak{a}^{(\mu)}$. Then the complexity of $f'\mathfrak{a}^{(\mu)}, \ldots, f^{(\pi)}\mathfrak{a}^{(\mu)}$ are all the same. We associate this complexity number with $\phi\{\mathfrak{a}^{(\mu)}, \mathfrak{f}\mathfrak{a}^{(\mu)}\}$. Now arrange the disjunctive terms of $\sum_{\theta=0}^{\kappa} \phi\{\mathfrak{a}^{(\theta)}, \mathfrak{f}\mathfrak{a}^{(\theta)}\}$ in order of increasing complexity from left to right. We suppose that duplicates have been omitted from the disjunction. Then $f^{(\theta)}\mathfrak{a}^{(\mu)}$ is distinct from $f^{(\theta')}\mathfrak{a}^{(\mu')}$ if and only if $\theta \neq \theta'$ or $\mu \neq \mu'$. Also $f^{(\theta)}\mathfrak{a}^{(\mu)}$ can only occur in $\mathfrak{a}^{(\mu')}$ if $\mu < \mu'$. Now replace $f'\mathfrak{a}', \ldots, f^{(\pi)}\mathfrak{a}', f\mathfrak{a}'', \ldots, f^{(\pi)}\mathfrak{a}^{(\kappa)}$ by new individual variables $\zeta', \ldots, \zeta^{(\pi\kappa)}$, so that $f^{(\theta)}\mathfrak{a}^{(\mu)}$ is replaced by $\zeta^{((\mu-1)\pi+\theta)}$, then $\sum_{\theta=0}^{\kappa} \phi\{\mathfrak{a}^{(\theta)}, \mathfrak{f}\mathfrak{a}^{(\theta)}\}$ becomes: $\sum_{\theta=0}^{\kappa} \phi\{\mathfrak{b}^{\theta)}, \zeta^{(\theta\cdot\pi+1)}, \ldots, \zeta^{(\theta\cdot\pi+\pi)}\}$. In this disjunction the variables $\zeta^{((\nu-1))\pi+1)}, \ldots, \zeta^{(\nu\cdot\pi)}$ are absent from the first $(\nu-1)$ disjunctions because $f^{(\theta)}\mathfrak{a}^{(\mu)}$ can only be a member of the sequence $\mathfrak{a}^{(\lambda)}$ or be contained in a member of the sequence $\mathfrak{a}^{(\lambda)}$ when $\mu < \lambda$, also $f^{(\theta)}\mathfrak{a}^{(\mu)}$ fails to occur in or be any member of the sequence $\mathfrak{a}^{(\mu)}$. Now we can generalize the variables $\zeta^{(\kappa-1)\pi+1}, \ldots, \zeta^{\kappa\cdot\pi}$ then apply II$d$ repeatedly to the terms in $\mathfrak{b}^{(\kappa)}$ this converts the last disjunctand to $(E\mathfrak{x})(A\mathfrak{y})\phi\{\mathfrak{x}, \mathfrak{y}\}$. We can proceed similarly with each disjunctand obtaining a disjunction of $\kappa$ disjunctions all being $(E\mathfrak{x})(A\mathfrak{y})\phi\{\mathfrak{x}, \mathfrak{y}\}$, from this $(E\mathfrak{x})(A\mathfrak{y})\phi\{\mathfrak{x}, \mathfrak{y}\}$ can be obtained by I$b$ repeatedly. This completes the demonstration of the proposition.

## 3·31 *Complete Boolean Algebras*

In some Boolean Algebras any subset has an *l.u.b.* that is, if $\mathscr{X}$ is a subset of a Boolean Algebra $\mathscr{B}$ then there is an element $\alpha$ of $\mathscr{B}$ such that $\beta \leqslant \alpha$ for each element $\beta$ of $\mathscr{X}$, and if $\beta \leqslant \gamma$ for each element $\beta$ of $\mathscr{X}$ then $\alpha \leqslant \gamma$. We denote $\alpha$ by *l.u.b.* $\mathscr{X}$. Similarly a *g.l.b.* can exist; in any case if the *l.u.b.* exists then we call the Boolean Algebra *complete*.

PROP. 36. *If $\mathscr{B}$ is a complete Boolean Algebra then*

$$\overline{l.u.b.\,\overline{\mathscr{X}}} = g.l.b.\,\mathscr{X},$$

*where $\overline{\mathscr{X}}$ is the set of complements of members of $\mathscr{X}$.*

We sometimes write $\bigcup_{i \in I} \alpha_i$ for *l.u.b.* $\{\alpha_i\}_{i \in I}$, and similarly $\bigcap_{i \in I} \alpha_i$ for *g.l.b.* $\{\alpha_i\}_{i \in I}$.

PROP. 35. $\underset{i \in I}{l.u.b.}\ \underset{j \in J}{l.u.b.}\ \alpha_{ij} = \underset{i \in I,\, j \in J}{l.u.b.}\ \alpha_{ij}$,

$$\alpha \cap \bigcup_{i \in I} \beta_i = \bigcup_{i \in I} (\alpha \cap \beta_i),$$

$$\alpha \cup \bigcap_{i \in I} \beta_i = \bigcap_{i \in I} (\alpha \cup \beta_i),$$

$$\bigcup_{i \in I} \alpha_i = 0 \quad \text{iff} \quad \alpha_i = 0 \quad \text{for} \quad i \epsilon I,$$

$$\bigcap_{i \in I} \alpha_i = 1 \quad \text{iff} \quad \alpha_i = 1 \quad \text{for} \quad i \epsilon I.$$

The '$\epsilon$' is the membership symbol.

A Boolean Algebra is said to satisfy the *countable chain condition* if every disjoint set of non-zero elements is countable. Two elements of a Boolean Algebra are said to be *disjoint* if their intersection is zero.

The *l.u.b.* acts like the union of an unbounded set and the *g.l.b.* acts like their intersection. These correspond to the Existential and Universal Quantifiers respectively.

*Distributive Laws*

In some complete Boolean Algebras there are extensions to the distributive laws corresponding to unbounded sets. Thus:

$$\bigcap_{i \in I} \bigcup_{j \in J} \alpha_{ij} = \bigcup_{\tau \in J^I} \bigcap_{i \in I} \alpha_{i\tau(i)} \quad \text{and} \quad \bigcup_{j \in J} \bigcap_{i \in I} \alpha_{ij} = \bigcap_{\tau \in I^J} \bigcup_{j \in J} \alpha_{\tau(j)j}.$$

Here $J^I$ denotes the set of functions with domain $I$ and range in $J$. But these laws can fail in some complete Boolean Algebras.

## 3.32 *Truth-definitions for set theory*

A truth-definition for a formal system can be given by formula induction. First a truth-definition is given for closed atomic statements, then the truth-definition for closed compound statements are obtained in the usual manner by truth tables, if we are seeking a standard two-valued truth-definition. If closed atomic formulae are lacking as in $\mathscr{F}_C$ then we usually give a definition of validity. In set theory the closed atomic statements are of the form

$$\hat{x}\phi\{x\} \epsilon \hat{x}\psi\{x\},$$

a closed statement of this form will be true if and only if

$$\psi\{\hat{x}\phi\{x\}\} \,\&\, (Ey)\,(\hat{x}\phi\{x\} \epsilon y),$$

this in turn will be true if and only if

$$\psi\{\hat{x}\phi\{x\}\} \,\&\, \chi\{\hat{x}\phi\{x\}\} \,\&\, (Ey)\,(\hat{x}\phi\{x\} \epsilon y),$$

for some $\chi$, but this will be true if and only if

$$\psi\{\hat{x}\phi\{x\}\} \,\&\, (Ey)\,(\hat{x}\phi\{x\} \epsilon y),$$

which is what we had before, so we must abandon this method.

Another way of giving a truth-definition is to construct a model. By this we mean a class of elements $V$ and truth-values for all atomic statements of the form $(a\epsilon b)$, where $a$ and $b$ are two elements of $V$. But we can easily be more general because we can take the truth-values to be members of a Boolean Algebra $\mathscr{B}$. Let then $\|a\epsilon b\|$ and $\|a=b\|$ be the members of the Boolean Algebra associated with the statements $(a\epsilon b)$ and $(a=b)$ respectively. From the Boolean values of these statements we can find the Boolean values of compound statements thus:

$$\|N\phi\| = \overline{\|\phi\|},$$
$$\|K\phi\psi\| = \|\phi\| \cap \|\psi\|,$$
$$\|(A\xi)\,\phi\{\xi\}\| = \bigcap_{a\in V} \|\phi\{a\}\|,$$

from these we obtain

$$\|D\phi\psi\| = \|\phi\| \cup \|\psi\|,$$
$$\|C\phi\psi\| = \|\phi\| \rightarrow \|\psi\|,$$
$$\|B\phi\psi\| = \|\phi\| \leftrightarrow \|\psi\|,$$
$$\|(E\xi)\,\phi\{\xi\}\| = \bigcup_{a\in V} \|\phi\{a\}\|.$$

We write $\models \phi$ for $\|\psi\| = 1$, and $\vdash \phi$ for $\phi$ is an $\mathscr{S}$-theorem. We want to show that, if $\vdash \phi$ then $\models \phi$, i.e. all $\mathscr{S}$-theorems take the Boolean value 1. First of all we easily show that $\models DN\phi\phi$, and, if $\models \psi$ then $\models \phi$, where $\psi/\phi$ is an $\mathscr{F}_C$-rule. Set theory is based on $\mathscr{F}_C$- so we must include the $\mathscr{F}_C$-axioms and rules among those of set theory. These are easily dealt with. For the axioms of equality and extensionality we proceed as follows: write $a\{x\}$ for $\|x\epsilon a\|$, so $a\{x\}$ is like the characteristic function of a set $a$ except that now the values are in the Boolean Algebra $\mathscr{B}$ instead of being $\boldsymbol{f}$ or $\boldsymbol{t}$. (The simplest Boolean Algebra.) $(a=b)$ is defined as $(Ax)\,B(x\epsilon a)\,(x\epsilon b)$, thus: $\|a=b\| = \bigcup_{x\in V} (a\{x\} \leftrightarrow b\{x\})$.

It is impracticable to introduce all the members of $V$ at once. We proceed by a transfinite process. We start with $V_0 = \Lambda$. The elements to be added to $V_\alpha$ to produce $V_{\alpha+1}$ will be functions over $V_\alpha$ with values in $\mathscr{B}$. These will correspond to new 'sets'. Thus $f\{x\}$ defined over $V_\alpha$ with values in $\mathscr{B}$ will be the 'characteristic function' of a new set. Clearly $V_1 = \{\Lambda\}$. If $\alpha$ is a limit ordinal $V_\alpha = \bigcup_{\beta<\alpha} V_\beta$.

The step from $V_\alpha$ to $V_{\alpha+1}$ is defined as follows: we assume that the members of $V_\alpha$ have been defined and that $V_\beta \subseteq V_\gamma$ for $\beta < \gamma \leqslant \alpha$ and that for $a, b \in V_\alpha$ we have

$$\|a\epsilon b\| = \bigcup_{x\epsilon\mathscr{D}b} \|K(x\epsilon b)(a = x)\|, \tag{1}$$

$$\|a = b\| = \bigcap_{x\epsilon\mathscr{D}a} \|C(x\epsilon a)(x\epsilon b)\| \cap \bigcap_{x\epsilon\mathscr{D}b} \|C(x\epsilon b)(x\epsilon a)\|, \tag{2}$$

and that for $a$, $b$, $c$, $\epsilon$ $V_\alpha$ we have

$$a = a, \tag{3}$$

$$C(a = b)(b = a), \tag{4}$$

$$CK(a = b)(b = c)(a = c), \tag{5}$$

$$CK(a = b)(b\epsilon c)(a\epsilon c), \tag{6}$$

$$CK(a\epsilon b)(b = c)(a\epsilon c), \tag{7}$$

and we also assume that every member of $V_\alpha$ is a function whose values are in the Boolean Algebra $\mathscr{B}$.

If

$$a\epsilon V_{\beta+1}, a\bar{\epsilon} V_\beta, \beta < \alpha \quad \text{then} \quad \mathscr{D}a = V_\beta, \tag{8}$$

if

$$a\epsilon V_\beta, x\epsilon\mathscr{D}a \quad \text{then} \quad \|x\epsilon a\| = a\{x\}, \tag{9}$$

if

$$x, y\epsilon\mathscr{D}a \quad \text{then} \quad a\{x\} \cap \|x = y\| \leqslant a\{y\}, \tag{10}$$

a function which satisfies (10) is called extensional.

We first put into $V_{\alpha+1}$ every member of $V_\alpha$. Next we generate each function $f$ from $V_\alpha$ to $\mathscr{B}$. As $\mathscr{D}f = V_\alpha$ for each new $f$, the value of $\|x = y\|$ has been determined for each $x, y \in \mathscr{D}f$. Hence for each $x, y \in \mathscr{D}f$ the value of $f\{x\} \cap \|x = y\|$ is determined, we discard all functions $f$ for which this value is $\nleqslant f\{y\}$. Thus we restrict our choice to extensional functions.

For $x\epsilon V_\alpha$ we define $\|x\epsilon a\|$ to be $a\{x\}$ for each new $a$. We now define $\|a = b\|$ by (2). If $a, b\epsilon V_\alpha$ this duplicates a known result. If $a\epsilon V_\alpha$, $b\epsilon V_{\alpha+1}$, $b\bar{\epsilon} V_\alpha$, this is an acceptable definition, since $\mathscr{D}a \subseteq V_\alpha$ $\mathscr{D}b = V_\alpha$, hence $\|x\epsilon a\|$ is determined by (1) and $\|x\epsilon b\|$ by $b\{x\}$, so $\|C(x\epsilon a)(x\epsilon b)\|$ and $\|C(x\epsilon b)(x\epsilon a)\|$ are both determined. Similarly in the other cases. So (2) holds for $V_{\alpha+1}$. Now if $a\bar{\epsilon}\mathscr{D}b$, $b\epsilon V_{\alpha+1}$, $b\bar{\epsilon} V_\alpha$ we define: $\|a\epsilon b\|$ for

$$\bigcup_{x\epsilon\mathscr{D}b} \|K(x\epsilon b)(a = x)\|$$

this is an acceptable definition since for each $x\epsilon\mathscr{D}b$ $\|K(x\epsilon b)(a = x)\|$ is determined by (9) and (2).

We have to show that (1) holds in $V_{\alpha+1}$. The only case to consider is $b \epsilon V_{\alpha+1}, b \bar{\epsilon} V_\alpha, a \epsilon \mathscr{D}b$. In this case $\mathscr{D}b = V_\alpha$, so that $a \epsilon V_\alpha$, then we have

$$\|a\epsilon b\| = \|K(a\epsilon b)\,(a = a)\| \leqslant \bigcup_{x\epsilon\mathscr{D}b} \|K(x\epsilon b)\,(a = x)\|. \tag{11}$$

Now take $x\epsilon\mathscr{D}b = V_\alpha$. Then by (9)

$$\|K(x\epsilon b)\,(a = x)\| \leqslant b\{a\} = \|a\epsilon b\|,$$

thus

$$\bigcup_{x\epsilon\mathscr{D}b} \|K(x\epsilon b)\,(a = x)\| \leqslant \|a\epsilon b\|,$$

then by (11) we infer (1) for $V_{\alpha+1}$.

It remains to check that (3)–(7) inclusive hold in $V_{\alpha+1}$. By (2) we conclude (3), since $\|C(x\epsilon a)\,(x\epsilon a)\| = 1$. Also (2) gives (4).

LEMMA (i). *If $x\epsilon\mathscr{D}b$ then $\|x\epsilon b\| \cap \|b = c\| \leqslant \|x\epsilon c\|$.*
We have

$$\|x\epsilon b\| \cap \|C(x\epsilon b)\,(x\epsilon c)\| = \|x\epsilon b\| \cap \|x\epsilon c\|,$$

thus

$$\|x\epsilon b\| \cap \|C(x\epsilon b)\,(x\epsilon c)\| \leqslant \|x\epsilon c\|,$$

whence

$$\|x\epsilon b\| \cap \bigcap_{x\epsilon\mathscr{D}b} \|C(x\epsilon b)\,(x\epsilon c)\| \leqslant \|x\epsilon c\| \quad \text{if} \quad x\epsilon\mathscr{D}b.$$

By (2) the lemma follows.

LEMMA (ii). *If $a, b\epsilon V_\alpha$ and $c\epsilon V_{\alpha+1}$ then* (7) *holds.*
If $c\epsilon V_\alpha$ then by assumption (7) holds. Thus suppose that $c\bar{\epsilon}V_\alpha$. Then $\mathscr{D}c = V_\alpha$. By (8) and $V_\alpha \subseteq V_{\alpha+1}$ we have $\mathscr{D}b \subseteq \mathscr{D}c$. Then by lemma (i)

$$\|K(x\epsilon b)\,(a = x)\| \cap \|b = c\| \leqslant \|K(x\epsilon c)\,(a = x)\|,$$

whence using (1)

$$\|a\epsilon b\| \cap \|b = c\| \leqslant \bigcup_{x\epsilon\mathscr{D}c} K\|(x\epsilon c)\,(a = x)\|$$

$$= \|a\epsilon c\| \quad \text{by (1), which is (7).}$$

We now verify (5) for $\alpha + 1$. Let $x\epsilon\mathscr{D}a$. Then, by lemma (i) we have

$$\|x\epsilon a\| \cap \|a = b\| \leqslant \|x\epsilon b\|.$$

So

$$\|K(a = b)\,(b = c)\| \cap \|x\epsilon a\| \leqslant \|x\epsilon b\| \cap \|b = c\|. \tag{12}$$

Case (i) $b\epsilon V_\alpha$, by lemma (i)

$$\|x\epsilon b\| \cap \|b = c\| \leqslant \|x\epsilon c\|. \tag{13}$$

Case (ii) $b\tilde{\epsilon}V_\alpha$. Then $\mathscr{D}b = V_\alpha$, so that $x\epsilon\mathscr{D}b$. Then by lemma (i) we again have (13). Therefore (13) holds in either case. By (12), (13) we obtain

$$\|K(a = b)(b = c)\| \cap \|x\epsilon a\| \leqslant \|x\epsilon c\|.$$

So $$\|K(a = b)(b = c)\| \cup \overline{\|x\epsilon a\|} \leqslant \|x\epsilon a\| \to \|x\epsilon c\|$$

and then $$\|K(a = b)(b = c)\| \leqslant \|C(x\epsilon a)(x\epsilon c)\|$$

whence $$\|K(a = b)(b = c)\| \leqslant \bigcap_{x\epsilon\mathscr{D}a} \|C(x\epsilon a)(x\epsilon c)\|. \tag{14}$$

One can start with $x\epsilon\mathscr{D}c$ and go through a similar argument to obtain

$$\|K(c = b)(b = a)\| \leqslant \bigcap_{x\epsilon\mathscr{D}c} \|C(x\epsilon c)(x\epsilon a)\|. \tag{15}$$

By (4), (2), (14) and (15)

$$\|K(a = b)(b = c)\| \leqslant \|a = c\|, \quad \text{which gives (5).}$$

We next verify (6) for $\alpha + 1$. Let $x\epsilon\mathscr{D}c$. By (5) we have

$$\|K(a = b)(b = x)\| \leqslant \|a = x\|.$$

Therefore $$\|a = b\| \cap \|K(x\epsilon c)(b = x)\| \leqslant \|K(x\epsilon c)(a = x)\|,$$

summing both sides over $x\epsilon\mathscr{D}c$ and using (1) we get (6).

Finally we verify (7) for $\alpha + 1$. Let $x\epsilon\mathscr{D}b$. By lemma (i)

$$\|x\epsilon b\| \cap \|b = c\| \leqslant \|x\epsilon c\|.$$

Therefore $$\|K(x\epsilon b)(a = x)\| \cap \|b = c\| \leqslant \|K(a = x)(x\epsilon c)\|.$$

Then by (6) $$\|K(x\epsilon b)(a = x)\| \cap \|b = c\| \leqslant \|a\epsilon c\|.$$

Sum on the left over $x\epsilon\mathscr{D}b$ and we obtain (7) by (1).

We have now obtained a universe $V$ and a Boolean value for each atomic statement $(a\epsilon b)$, where $a, b\epsilon V$. So we have given a generalized truth-definition for set theory.

Now write

$$\vDash \phi \quad \text{for} \quad \|\phi\| = 1;$$

$$\vdash \phi \quad \text{for} \quad \phi \text{ is an } \mathscr{S}\text{-theorem}.$$

It is possible to $\mathscr{S}$-prove that if $\vdash\phi$ then $\vDash\phi$, i.e. all $\mathscr{S}$-theorems take the value 1. Thus $C$ becomes a model for $\mathscr{S}$. The proof must take place in some system which can deal with ordinals because they are essential to the construction of $V$. The system $\mathscr{S}$ is such a system.

An interesting application of this method for defining truth for set theory is that by suitable choice of the Boolean Algebra $\mathscr{B}$, we can show that A.C., G.C.H., etc. take values different from 1 and hence are non-theorems of $\mathscr{S}$. Thus it is impossible to $\mathscr{S}$-prove them. On the other hand we can find another type of model, namely an '*inner model*', constructed as follows: we start with the null set and by a process of transfinite induction we define all the sets which can be obtained from it by repeatedly performing the operations allowed for the construction of new sets from old ones, e.g. union, complementation, etc., In this way it seems clear that all sets are given an ordinal and so $V$ is well-ordered, so that A.C. holds. It can also be shown that G.C.H. holds in this model. All this can be done in the system $\mathscr{S}$. Thus A.C. and G.C.H. are consistent with set theory if this is itself consistent. Altogether A.C. and G.C.H. are independent of the other axioms of set theory. The full details are lengthy.

### 3·33 *Predicative and impredicative properties*

In a second order predicate calculus we can have bound predicate variables and hence we can form properties $\lambda x\phi\{x, (AX)\,\psi\{X, x\}\}$. This property is defined in terms of all properties including itself. This is a circular definition and as such seems to be objectionable, so we wish to avoid its occurrence. We do so by a *ramified* second order predicate calculus. In this we have properties of various *orders* and property variables of each of these orders:

$X_1, X_1' \ldots$ *property variables of the first order*,

$X_2, X_2', \ldots$ *property variables of the second order*, etc.

The *order of a property* is the greatest of (i) the order of a free property variable in it, (ii) the successor of the order of a bound property variable in it. Then $\lambda x\phi\{x, (AX)\,\psi\{X, x\}\}$ is of order at least one greater than that of $X$ and hence the circularity is eliminated. Let $\mathscr{F}^{(2)}_{RC}$ be the ramified predicate calculus of the second order. In an $\mathscr{F}^{(2)}_{RC}$-proof a free property variable may be replaced by a property of the same order, possibly containing new free variables, to guard against *collision of variables*, i.e. to preserve the holding of conditions on variables in II*e*, II*e'*. Thus we can show, by methods already used, that we have $\dfrac{D(AX)\,\phi\{X\}\,\omega}{D\phi\{\Delta\}\,\omega}*$, the reversibility of

II$e'$, here $\Delta$ is to be a property of order the same or less than that of the variable $X$. We obtain this result from II$e'$ by everywhere replacing related occurrences of $X$ in the part of the tree above the upper formula of II$e'$ by the property $\Delta$ which must be of the same or less order as $X$.

A property is called *predicative* if it fails to be defined in terms of itself, otherwise it is called *impredicative*. Thus in $\mathscr{F}_{RC}^{(2)}$ we only have predicative properties. If we try to give a definition of validity to $\mathscr{F}_C^{(2)}$, which contains impredicative properties, then we get into trouble because we should require $(AX)\,\phi\{X\}$ to be valid if and only if $\phi\{\Delta\}$ is valid for all properties $\Delta$, but one of these is $\hat{X}\phi\{X\}$ and another is $\hat{X}\phi\{\phi\{X\}\}$, and so on, we should get referred to more and more complicated cases. Whereas if we wish to define validity in $\mathscr{F}_{RC}^{(2)}$ we get referred to simpler cases, and the process will ultimately terminate, though leaving us, as in $\mathscr{F}_C$, with an unbounded set of cases to decide; but this is the sort of thing we expect an *oracle* to do for us. Oracles will appear in Ch. 7.

## 3·34 *Topology*

The system of set theory that we have constructed is very useful for expressing topological concepts. To avoid the use of a multitude of primes we shall use $x, y, z, u, v, w$ and the same with primes to denote variables of type $\iota$. We make the following definitions:

D 91 $\mathit{Top}x$ for $(Au,v)\,(u,v\in x.\rightarrow.u\cap v\in x)\,\&\,(Ay)\,(y\subset x\rightarrow\Sigma y\in x)$; read '$x$ is a topology'.

D 92 $x\mathit{Top}y$ for $\mathit{Top}x\,\&\,y=\Sigma x$; read '$x$ is a topology for $y$'.

D 93 $x\mathit{Op}y$ for $\mathit{Top}x\,\&\,y\in x$; read '$y$ is an open set in the topology $x$'.

D 94 $x\mathit{Cl}y$ for $\mathit{Top}x\,\&\,\mathit{Op}(x-y)$; read '$y$ is a closed set in the topology $x$'.

D 95 $\mathit{TopSp}x$ for $(Ey)\,(\mathit{Top}y\,\&\,x=\Sigma y)$; read '$x$ is a topological space'.

D 96 $\mathit{IndisTop}x$ for $(Ey)\,(x=\{y,\emptyset\})$; read '$x$ is an indiscrete topology'.

D 97 $\mathit{DisTop}x$ for $(Ey)\,(x=Py)$; read '$x$ is a discrete topology'.

D 98 $uTopxNeighy$ for $(Ev)(v \subseteq u \,\&\, y \in v \,\&\, xOpv) \,\&\, u \subseteq x$;
read '$u$ is a neighbourhood of $y$ in the topology $x$'.

D 99 $yTopxLimz$ for

$$z \subseteq x \,\&\, Topx \,\&\, (Au)(uTopxNeighy \rightarrow (Ev)(v \in u \cap z \,\&\, v \neq y));$$

read '$y$ is a limit point of the subset $z$ of the topological space $x$'.

D 100 $z^{*Topx}$ for $\hat{y}(yTopxLimz) \cup z$;
read 'the closure of the subset $z$ of the topological space $x$'.

D 101 $IntzTopx$ for $\hat{y}(zTopxNeighy)$;
read 'the interior of the subset $z$ of the topological space $x$'.

D 102 $BdyzTopx$ for $z^{*Topx} \cap (x-z)^{*Topx}$;
read 'the boundary of the subset $z$ of the topological space $x$'.

D 103 $zBaseTopx$ for $z \subseteq x \,\&\, (Ay, u)(y \in x \,\&\, uTopxNeighy$
$\rightarrow (Ev)(y \in v \,\&\, v \in z \,\&\, y \subseteq u))$;

read '$z$ is a base for the topology $x$'.

D 104 $Sepx$ for $Topx \,\&\, (Ez, w)(zBaseTopx \,\&\, w``z = \omega)$;
read '$x$ is separable', $x$ is separable if $x$ has a countable base.

D 105 $zDenseTopx$ for $z^{*Topx} = x$;
read '$z$ is dense in the topological space $x$'.

D 106 $uCovw$ for $w \subseteq \Sigma u$;
read '$u$ covers $w$'.

D 107 $Sep[y,z]Topx$ for $y^{*Topx} \cap z = \Lambda \,\&\, z^{*Topx} \cap y = \Lambda$;
read '$y$ and $z$ are separated in the topology $x$'.

D 108 $ConnTopx$ for $A(y,z)(x = y \cup z \,\&\, Sep[y,z]Topx.$
$\rightarrow . y = \Lambda \vee z = \Lambda)$;

read '$x$ is a connected topological space'.

D 109 $HausSpx$ for $Topx \,\&\, (Au, v)(u, v \in \Sigma x \,\&\, u \neq v.$
$\rightarrow (Ew, w')(w \cap w' = \Lambda \,\&\, wTopxNeighu \,\&\, w'TopxNeighv))$;
read '$x$ is a Hausdorff space'.

D 110 $Compx$ for $(Ay)(Ez)(yCovx \,\&\, zCovx \,\&\, z \subseteq y \,\&\, (Ew)$
$(w``z \subset \omega)) \,\&\, Topx$;

read '$x$ is a compact topological space'.

D 111 $NWDenseyTopx$ for $y \subseteq x \,\&\, Int\, y\, Topx = \Lambda$;
read '$y$ is nowhere dense in the topology $x$',

and so on. If we are dealing with a single topology for a topological space then we can omit *Topx* in the above definitions. The whole of general topology can now be formalized without undue difficulty.

## Historical remarks to Chapter 3

Predicate calculi differ from propositional calculi by the adjunction of quantifiers, whose intended meaning always has something to do with the cardinal number of things which satisfy a certain statement. Quantifiers were first introduced by Frege (1879). Somewhat later and independently quantifiers were used by Pierce who introduced the term 'quantifier'. Thereafter their use becomes general, though the notation for them varies. The various orders of predicate calculi is due to Russells' theory of types and perhaps to Frege's Stufen and Schröder's Mannigfaltigkeiten. Löwenheim and Skolem in effect gave a treatment of the first order predicate calculus with equality. But the first explicit formulation of the classical predicate calculus of the first order as a formal system in its own right is in the first edition of the book by Hilbert and Ackermann (1928). Thereafter it was much studied as a formal system.

Many-sorted predicate calculi were discussed by Schmidt (1938) and Wang (1952). Models by Kemeny (1949) among others.

Predicative and impredicative predicate calculi arises from the unqualified use of the concept of 'all'. Russell's (1906) *P.M.* vol. 1, Ch. II, vicious-circle principle, designed to avoid paradoxes, was 'no totality can contain members defined in terms of that totality'. The term 'impredicative is due to Poincaré (1905) who condemned impredicative definitions, as did Weyl (1918).

In developing the classical predicate calculus of the first order we again use the direct formulation due to Gentzen (1934) and further studied by Schütte, (1950–1, 1960). Prop. 4, the elimination of M.P., is Gentzen's Hauptsatz, the demonstration we give is due to Lorenzen (1951).

The prenex normal form is due to Skolem who also found other normal forms. Props. 9, 10 and 11 are due to Herbrand (1930) as is the discussion on *H*-disjunctions, hence their name.

Much has been contributed to the concepts of validity and satisfaction by Tarski (1933). Prop. 12, the completeness theorem of the classical predicate calculus of the first order is due to Herbrand (1930), Gödel (1930), Löwenheim (1915) and Skolem (1920). Cor (ii) is due to Löwenheim (1915). Prop. 13, the independence of the axioms and rules of the classical predicate calculus of the first order was first considered by Gödel

(1930) and the consistency by Hilbert–Ackermann (1928). The discussion of theories is due to Tarski (1935–6).

The classical predicate calculus with equality is implicit in the work of Pierce and Schröder, but its first treatment as a system in its own right is in Hilbert and Ackermann (1928) and again in Hilbert and Bernays (1934–6).

Prop. 19, the elimination of axiom schemes, is due to Skolem (1959) who used a formulation of set theory due to Gödel (1940).

Early attempts at finding a decision procedure for the classical predicate calculus of the first order were unsuccessful (because as we shall see in a later chapter there is none), so research turned to finding decision procedures for special classes of statements. One of the earliest of these was by Löwenheim (1915). He gave a decision procedure for the monadic predicate calculus of the first order. This was followed by work by Skolem (1919, 1920) and Behmann (1922) on the monadic predicate calculus of the second order and of the first order with equality. Prop. 22 is due to Bernays and Schönfinkel (1928) and Prop. 23 is due to Gödel (1933), Kalmár (1933) and Schütte (1934) A detailed account of all known decision procedures for special classes of statements has been given by Ackermann (1954), see also Church (1956).

A related type of problem is the reduction problem. Here we try to find special classes of statements such that any statement of the predicate calculus of the first order is equivalent as regards validity to one in the special class, there is a corresponding problem for satisfiability. A simple case is the class of statements in prenex normal form. These special classes are called reduction classes. Prop. 24 is due to Löwenheim (1915) and Gödel (1933). The Skolem $V$- and $S$-normal forms are of course due to Skolem (1925) and lemma (iv), the restriction to exactly three universal quantifiers is due to Gödel (1933). Prop 25, where we have only one predicate and that one binary is due to Surányi (1943) and Kalmár (1947) whom we follow closely. Many reduction types with a variety of prefixes have been found by Kalmár and Surányi, and account of them is given by Surányi (1959), see also Church (1956). The method of semantic tableau is due to Beth (1955, 1959) and Hintikka (1953, 1955) and Prop. 26 is due to Craig (1953) and Kleene (1952, 1967).

The idea of a resolved predicate calculus is due to Hilbert [$H$–$B$, II], but the '$\eta$' symbol (written $\iota$) and its use is due to Peano (1897), Frege (1893, 1962) and (1905, 1956). The '$\epsilon$' symbol is due to Hilbert [$H$–$B$].

He demonstrated two theorems about the elimination of the $\epsilon$-symbol, the first is virtually Prop. 35.

An historical account of the predicate calculus has been given by Hermes & Scholz (1932). Many examples are to be found in Church (1956).

The system $\mathscr{B}\mathscr{F}_C$ is suggested by Kalmár & Surányi's reduction type whose only predicate is a single binary one. Definitions D 26 and 27 go back to Leibniz He took two classes to be the same if they contained exactly the same members. This is the extensional approach, and is used in classical mathematics. But one might consider two classes to be distinct, even if they contained exactly the same members on the ground that the rules for membership were different. This is called the intensional approach, we shall have to speak about it again in later chapters. The concept of set and class as defined in D 33 and 34 is due to v. Neumann (1925) who used it to avoid the syntactic paradoxes which had crept into set theory since Frege and Cantor. The definitions D 37–57 are largely from P.M. But the definition of an ordered pair has received simplification at the hands of Wiener (1912) and Kuratowski (1921). The rules R 2′–9″ are modifications of the axioms used by Gödel (1940) in his account of set theory. Prop. 31 on normal classes is due to Gödel (1940).

When P.M. first appeared it was considered to have a few blemishes. One was the complicated type theory. Several writers have tried various ways of simplifying this, notably Leon Chwistek (1921, 1927) and F. P. Ramsey (1926), with his simple theory of types. Another thing thought by some to be a blemish is the axiom of infinity. Perhaps the reason for this may be compared to the reasons for considering Euclid's axiom of parallels to be a blemish on his work in geometry (first formalized by Hilbert (1922)), namely, one might think that it should follow from the other axioms. But apparently this is not the case. In fact in Gödel's (1940) formulation of set theory the axioms previous to the axiom of infinity are consistent, this is seen by taking $\alpha\epsilon\beta$ to be always false. But with the axiom of infinity this is no longer the case, because this axiom postulates the existence of a set. There are many ways of formulating an axiom of infinity, all one requires is the existence of some set with an infinity of members. The last thing thought by some to be a blemish in P.M. is the axiom of reduction. Here this is avoided by the distinction between classes and sets.

Set theory is very useful for making models of various mathematical conceptions. Thus we easily get a model for ordinal numbers. So we give a short account of them. Fuller accounts are given in Cantor (1895, 7), Gödel (1940), Bernays–Fraenkel (1958), Sierspinski (1928, 1958), Bachmann (1955). We have indicated in Ex. 41–4 inclusive how to develop the algebra of ordinals. Ordinals and cardinals were first invented by Cantor, but finite ordinals and cardinals were known to the Greeks. Ex. 43 (xi) is known as Cantors normal form for ordinals. When we introduce ordinals we immediately require a new axiom, namely the axiom of infinity, its object is to ensure that the ordinals we define are sets, so the process of ordinal construction can proceed. We largely follow Gödel's (1940) account of ordinals and cardinals. The alephs are defined as certain ordinals, but there may be other cardinals that are incompatible with the alephs, if we define cardinals as classes of similar classes. It is at this point that we come across A.C., first introduced by Zermelo (1904). Prop. 32 is due to Cantor and Prop. 33 to Cantor and Bernstein (1905). Cantor's theorem immediately gives rise to C.H. and G.C.H. For many years the logical position of these and A.C. was unknown. Then Gödel (1940), by constructing the constructible universe, was able to find an 'inner model' in which A.C., G.H. and G.C.H. were all satisfied provided that set theory itself is consistent. Thus A.C., C.H. and G.C.H. are consistent with set theory provided set theory is consistent. The proof of this takes place in set theory. Gödel's method of inner models, as shown by Shepherdson (1951, 2, 3), is incapable of showing that the negations of A.C., C.H. or G.C.H. are consistent with set theory. Many years were to pass before Cohen (1963, 4, 5) developed an entirely new method for dealing with independence proofs, this was based on the denumerable model found by Skolem. This was a remarkable breakthrough, the original paper was couched in such strange terms that only the most resolute of professional logicians could understand it. Now however, monographs are appearing and the method is explained so that it is available to the general mathematician.

There is a monograph by Cohen (1965), Another method of doing the same things was discovered by Solovay and Scott (1970). This proceeds by forming Boolean valued models and gives a generalization of two-valued truth. This method arose because it was noticed that the main feature of Cohen's 'forcing' method was the semi-order it gave rise to. Accounts of this method are given by Rosser (1969), Scott (1966*a*, *b*), Jensen (1967). By suitable choice of Boolean Algebra models can be found in which A.C. or C.H. or G.C.H. fail. The complete Boolean Algebras required came in after Boole, they are discussed by Halmos (1963) and Sikorski (1960). We just show how to set up the model. The full proof that it is a model for set theory together with the independence proofs is given by Rosser (1969). Thus with Gödel's result that A.C., C.H. and G.C.H. are consistent with set theory the final result is that A.C., C.H. and G.C.H. are independent of the other axioms of set theory. The proof takes place in set theory so that the result just stated only holds if set theory is itself consistent. This is parallelled in mathematical history by Caley's proof in Euclidean geometry that the axiom of parallels is independent of the other axioms of Euclidean geometry provided these themselves are consistent.

The elimination of the $\epsilon$-symbol is due to Hilbert and two theorems about it are given in *H–B* (1934–6), we give one of these because it is tantamount to the consistency of A.C. with the other axioms of set theory. That is we give an effective method of converting a contradiction in set theory plus A.C. into a contradiction in set theory itself.

The chapter closes with some topological definitions, many of which occur in the latter parts of *P.M.*, they show again how useful the system $\mathscr{S}$ is for talking about all sorts of mathematical concepts.

Other works on set theory are: Suppes (1960), Halmos (1960), Skolem (1962), Sierpinski (1951), Fraenkel and Bar-Hillel (1958), Fraenkel (1953) with a complete bibliography to 1953 and Fraenkel (1946).

## Examples 3

1. Complete the demonstration of Prop. 6.
2. Complete the demonstration of Prop. 8.
3. Put into prenex normal form

$$(Ax)\,CpxC(Ax')\,p'x'x(Ax'')\,p''x'',$$

$$B(Ex)\,(Ex')\,pxx'(Ex')\,(Ex)\,pxx'.$$

4. Put the $\mathscr{F}_C$-proof of the prenex normal forms of

$$C(Ax)\,Cpxp'xC(Ex)\,px(Ex)\,p'x,$$

$$C(Ex)\,NNpxNN(Ex)\,px,$$

$$BK(Ax)\,Cpxp'x(Ax)\,Cp''xp'x(Ax)\,CDpxp''xp'x,$$

into normal form.

5. Obtain $H$-disjunctions for

$$(Ex)(Ax')(Ex'')\,DDpxx'xpxx'x''Npx''xx',$$
$$(Ax)(Ex')(Ex'')(Ax'')\,DDpxx'x''Npxxx''px''x'x,$$
$$(Ex)(Ax')(Ex'')(Ex''')\,DDpx''x'xNpx'''x'x''px'''x'x.$$

6. Obtain semantic tableau for the statements of Ex. 5.

7. Find the equivalent of

$$(Ax, x', x'')(Ey)\,KDpxx'p'x'yCKp''x''xp'x'yDpyx''p''xx''$$

according to Prop. 25.

8. Find which of the following are $\mathscr{F}_C$-theorems:

$$(Ax, x', x'')(Ex^{(3)}, x^{(4)}, x^{(5)})\,DCpxx^{(4)}x^{(3)}px'x''x^{(5)}Cpxx^{(3)}x^{(4)}$$
$$px''x'x^{(3)},$$
$$(Ax, x')(Ex'', x''')\,CCpxx''px'x'''Cpx''xpx'''x'.$$

9. Find which of the following are $\mathscr{F}_C$-theorems:

$$(Ex)(Ax', x'')(Ex''')\,CKpxx'px'x''Cpx'''xpx''x',$$
$$(Ex)(Ax', x'')(Ex''')\,CCpxx'px'x''Cpx'''x''px'x.$$

10. Reduce

$$(Ex)(Ax', x'')(Ex''')(Ax^{\text{iv}})\,CCpxx'xCpx''x'''x'Cpx''x^{\text{iv}}x'''px^{\text{iv}}x'x$$

to the form given in Prop. 24.

11. Put in resolved form, using the $\epsilon$-terms

$$(Ax)(Ex')(Ax'')\,\phi\{x, x', x''\},$$
$$(Ex)(Ax')(Ex'')\,\phi\{x, x', x''\}.$$

12. Define

$$R_\epsilon \quad \text{for} \quad \hat{\mathfrak{x}}\hat{\mathfrak{x}}'\hat{\mathfrak{x}}''(x = x'), \qquad \mathscr{D}_\epsilon \quad \text{for} \quad \hat{\mathfrak{x}}\hat{\mathfrak{x}}'\hat{\mathfrak{x}}''(x = x'').$$

Show that:

$$\mathscr{R}_\epsilon``X = \mathscr{R}X,$$
$$\breve{\mathscr{D}}_\epsilon``X = V \times X,$$
$$\Sigma\{x, x'\} = x \cup x',$$
$$\Sigma\{x\} = x,$$
$$\mathscr{E}``\{x\} = x,$$
$$\Sigma x = \mathscr{E}``X,$$
$$C(X \subseteq X')(\mathscr{D}X \subseteq \mathscr{D}X'),$$
$$CK(X \subseteq X')(X'' \subseteq X''')\,CX``X'' \subseteq X'``XX''',$$

13. Show that $(X \times X') \cap (X'' \times X^{\text{iv}}) = (X \cap X'') \times (X' \cap X^{\text{iv}})$,

$$\Sigma(X \cup X') = \Sigma X \cup \Sigma X',$$

$$\Sigma(X \cap X') \subseteq \Sigma X \cap \Sigma X',$$

$$X = \Sigma P X,$$

$$CN\mathscr{E}mX(X' \subseteq \Sigma\Sigma(X \times X')),$$

$$X \subseteq P\Sigma X,$$

$$X \times X' \subseteq PP(X \cup X'),$$

$$\Sigma\Sigma I = V,$$

$$V = \mathscr{D}\breve{\mathscr{E}}.$$

14. Construct an applied predicate calculus of the second order which has exactly one constant predicate $R$ which is binary. State axioms for order and express 'every non-empty class has a least member in the ordering $R$'.

15. Set up axioms for a group using one ternary predicate.

16. Find the Skolem $S$-normal forms and the Skolem $V$-normal forms for the following:

$$C(Ax)px(Ex)px,$$

$$C(Ax)Cpxp'xC(Ex)px(Ex)p'x,$$

$$C(Ex)(Ex')pxx'(Ex')(Ex)pxx'.$$

17. Give the demonstration of Prop. 24, lemma (iii).

18. Show that the conditions (A), (B), (C) and (D) below are necessary in order that $(A\xi, \xi', \xi'')(E\eta)\phi\{\xi, \xi', \xi'', \eta\}$ be satisfiable. There is a non-empty set $\Sigma$ of $4 \times 4$ tables which satisfy $\phi$ such that:

(A) If $\mathrm{T}_0$ is a table of $\Sigma, \nu, \nu', \nu'' = 1, 2, 3, 4$ then there is a table T of $\Sigma$ such that:

$$[\mathrm{T}/1] = [\mathrm{T}_0/1] \quad \text{and} \quad [\mathrm{T}/123] = [\mathrm{T}_0/\nu', \nu', \nu''].$$

(B) If $\mathrm{T}_0$ is a table of $\Sigma$ then there is a table T of $\Sigma$ such that:

$$[\mathrm{T}/1] = [\mathrm{T}_0 1] \quad \text{and} \quad \mathrm{T} = [\mathrm{T}/1114].$$

(C) If $\mathrm{T}_1, \mathrm{T}_2$ are tables of $\Sigma$ then there are tables $\mathrm{T}, \mathrm{T}', \mathrm{T}''$ of $\Sigma$ such that:

$$[\mathrm{T}/1] = [\mathrm{T}_1/1], \quad [\mathrm{T}/2] = [\mathrm{T}_2/1], \quad \mathrm{T} = [\mathrm{T}/1224],$$

$$[\mathrm{T}'/2] = [\mathrm{T}_1/1], \quad [\mathrm{T}'/1] = [\mathrm{T}_2/1], \quad \mathrm{T}' = [\mathrm{T}/1214],$$

$$[\mathrm{T}''/2] = [\mathrm{T}_1/1], \quad [\mathrm{T}'/1] = [\mathrm{T}_2/1], \quad \mathrm{T}'' = [\mathrm{T}''/1134].$$

(D) If $T_1, T_2, T_3$ are tables of $\Sigma$ then there is a table T of such that:

$$[T/1] = [T_1/1], \quad [T/2] = [T_2/1], \quad [T/3] = [T_3/1].$$

19. Apply Prop. 20 to decide which of the following are $\mathscr{F}_C$-theorems.

(i) $(Ex)(Ax')CBp'xpBp'x'p$,

(ii) $B(Ex)Cpxp'x(Ex,x')Cpxp'x$,

(iii) $(Ax')C(Ax)CpxCpx'p'xCpC(Ax)pxp'x'$.

20. Apply Prop. 22 to decide

$$(Ax,x',x'')(Ey,y')Cpxx'Cpyx''DKpyx'py'x'Kpxx''py'x''.$$

21. Apply Prop. 22 to decide

$$(Ax,x')(Ex'')KCpx''xKCpxx'pxx''Cpxx''CNpxx'Kpx''xpx'x''.$$

22. Find the binary $\mathscr{F}_C$ statement equivalent to

$$(Ex)(Ax')(Ex'')Cpxx'x''px'x''x$$

as regards satisfiability according to Prop. 24.

23. Continue Ex. 22 to find a binary $\mathscr{F}_C$-statement with exactly one binary predicate which is equivalent to (22) as regards satisfiability according to Prop. 25.

24. Ditto for

$$(Ex)(Ax')(Ex'')(Ax''')Cpxx'x''px'x''x''',$$

$$(Ex)(Ax')(Ex'',x''')Cpxx'x''x'''x''px''xx'x''x'''.$$

25. Demonstrate Prop. 12 for the system $I\mathscr{F}_C$.

26. Demonstrate Prop. 14 for the system $I\mathscr{F}_C$.

27. Show that the singularly $I\mathscr{F}_C$ is decidable. The only predicates in the singulary $I\mathscr{F}_C$ other than $I$ are singulary.

28. Show that

$$B(X = Y)(Au)B(u\epsilon X)(u\epsilon Y)$$

and

$$B(X = Y)(AU)(X\epsilon U)(Y\epsilon U)$$

are independent. [See Robinsohn, *J.S.L.* 4, 69.]

29. Show by formula induction

$$\frac{X = Y}{B\phi\{X\}\phi\{Y\}},$$

where $\Gamma$ is free for $X$, $Y$ in $\phi\{\Gamma\}$.

30. Show
$$B\phi\{Y\}(AX)C(X = Y)\phi\{X\}$$
and
$$B\phi\{Y\}(EX)K(X = Y)\phi\{X\}.$$

31. Show $$B(X\epsilon Y)(Eu)(u = X)K(u = X)(u\epsilon Y).$$

32. Show $$X = \hat{y}(y\epsilon X).$$

33. Show
$$\frac{(A\xi', \ldots, \xi^{(\nu)})\, B\phi\psi}{\gamma = \delta},$$

where $\delta$ is like $\gamma$ except for containing $\psi$ at some places where $\gamma$ contains $\phi$ and $\xi', \ldots, \xi^{(\nu)}$ exhaust the variables with respect to which those occurrences of $\phi, \psi$ are bound in $\gamma$, $\delta$.

34. Show
$$\frac{(A\xi', \ldots, \xi^{(\nu)})\,(\gamma = \delta)}{B\phi\psi},$$

where $\psi$ is like $\phi$, except for containing $\delta$ at some places where $\phi$ contains $\gamma$ and $\xi', \ldots, \xi^{(\nu)}$ exhaust the variables with respect to which there occurrences of $\gamma$, $\delta$ are bound in $\phi$, $\gamma$.

35. Show
$$\frac{Rel\,(X, Y) \quad (Au, v)\, BuXvuYv}{X = Y}.$$

36. Show
$$\frac{\{u, v\} = \{x, y\}}{DK(u = x)(v = y)K(u = y)(v = x)}.$$

37. Show
$$\frac{\langle u, v\rangle = \langle x, y\rangle}{K(u = x)(v = y)}.$$

38. Show $$(Ay)(y = I`y).$$

39. Show
$$\frac{(Av)\, B(v = y)\, vXx}{y = X`x}.$$

40. Show
$$(Ax)\, BKUnX(x\epsilon\mathscr{D}X)(Eu)(Av)B(u = v)\, vXx,$$
$$(Ax)\, BKUnX(x\epsilon\mathscr{D}X)(Ay)B(y = X`x)\, yXx,$$
$$\frac{Y\mathscr{F}nX}{(Aw)\, B(w\epsilon Y)\ (Eu)\, K(u\epsilon X)(w = \langle Y`u, u\rangle)},$$
$$\frac{Y\mathscr{F}nX \quad Z\mathscr{F}nX \quad (Au)\, C(u\epsilon X)(Y`u = Z`u)}{Y = Z}.$$

41. Defining the *sum* $a+b$ of two ordinals $a, b$ as the ordinal isomorphic as regards order to the order type obtained by sticking the order type $b$ at the end of the order type $a$, show that:

(i) If $0 < b$ then $a < a+b$.

(ii) $b \leqslant a+b$.

(iii) If $b < a$ then there is a unique ordinal $c$ such that $b+c = a$.

(iv) If $b \leqslant a$ and $d \leqslant c$, then $b+d \leqslant a+c$.

(v) $a = b_1+c_1 = b_2+c_2, c_1 < c_2$, then $b_2 < b_1$.

(vi) If $a = b+c$, $c$ is called a *remainder* of $a$ and $b$ is called a *segment* of $a$. Show that the number of remainders of an ordinal is finite.

(vii) An ordinal $a$ is called *decomposable* if $a = b+c$, $0 < b$, $c < a$. Otherwise *indecomposable*. Show that the least positive remainder of a positive ordinal is indecomposable.

(viii) If $c_1 < c_2$ are both remainders of an ordinal $a$, then $c_1$ is a remainder of $c_2$.

(ix) The least positive remainder of an ordinal $a$ is a remainder of every other remainder of $a$.

(x) The only positive remainder of an indecomposable ordinal is itself.

(xi) The only positive indecomposable remainder of an ordinal is its least positive remainder.

(xii) If $c$ is indecomposable and $b < c$ then $b+c = c$.

(xiii) If $c > 0$ and $b+c = c$ whenever $b < c$ then $c$ is indecomposable.

(xiv) Any ordinal $a$ is the sum of a finite decreasing sequence of decreasing indecomposable ordinals.

(xv) If $a = c_1+c_2+\ldots+c_n, c_1 > c_2 > \ldots > c_n$ and $c_1, c_2, \ldots, c_n$ indecomposable then $c_1$ is the greatest indecomposable ordinal $< a$.

(xvi) If $A$ is a set of indecomposable ordinals then $\Sigma A$ is indecomposable.

(xvii) Every ordinal can be uniquely represented as a finite sum of non-decreasing indecomposable ordinals.

42. Defining the *product* $ab$ of two ordinals $a, b$ as the ordinal order isomorphic to the set of ordered pairs $\langle c, d\rangle$ $c < a, d < b$ ordered by last differences, show that

(i) If $b_1 < b_2$ and $0 < a$ then $ab_1 < ab_2$.

(ii) If $a_1 < a_2$, $b_1 \leqslant b_2$ then $a_1 b_1 \leqslant a_2 b_2$.

(iii) We can have $a_1 < a_2$, $b > 0$ and $a_1 b = a_2 b$.

(iv) If $a = bc$, $0 < b, 1 < c$, then $b < a$ and $c \leqslant a$.

(v) If $ab_1 < ab_2$ then $b_1 < b_2$.

(vi) If $a_1 b < a_2 b$ then $a_1 < a_2$.

(vii) If $ab_1 = ab_2$, $0 < a$ then $b_1 = b_2$.

(viii) If $c < ab$ then uniquely $c = ab_1 + d, b_1 < b, d < a$.

(ix) If $0 < a$ then $b = ac + d$, $d < a$, uniquely.

(x) If $c$ is indecomposable, $1 < c$ then $ac$ is indecomposable.

(xi) If $0 < a$ then the least indecomposable ordinal greater than $a$ is $a\omega$.

(xii) If $c$ is indecomposable and positive then the next greater indecomposable ordinal is $c\omega$.

(xiii) Every indecomposable ordinal is divisible on the left by every lesser positive ordinal and the quotient is indecomposable.

(xiv) An ordinal is *prime* if it is greater than unity and is different from the product of any two lesser ordinals. Show that every ordinal $> 1$ is the product of a finite number of primes.

(xv) Show that the number of right divisors of an ordinal is finite.

43. Defining $b^a$ for ordinals $a$, $b$ as the ordinal order isomorphic to the order type of functions over $a$ with a finite number of non-zero values in $b$ ordered by last differences, show that:

(i) This order type is that of an ordinal.

(ii) If $0 < a < b$, $1 < c$ then $c^a < c^b$.

(iii) If $0 < a < b$, $0 < c$ then $a^c \leqslant b^c$.

(iv) $c^{a+b} = c^a . c^b$, $0 < a, b, c$.

(v) If $b$ is a limit ordinal, $1 < a$ then $a^b$ is the limit of all ordinals $a^c$ for $c < b$.

(vi) $(a^a)^c = a^{bc}$, $0 < a, b, c$.

(vii) $\omega^a$ is indecomposable for $a > 0$.

(viii) If $0 < a$, $1 < c$ then $a < c^a$.

(ix) If $0 < d$, $1 < c$ then there is exactly one ordinal $a$ such that $c^a \leqslant d < c^{a+1}$.

(x) Every indecomposable ordinal $\geqslant \omega$ is of the form $\omega^a$ for some $a > 0$.

(xi) Every ordinal $a > 0$ can be uniquely expressed in the following normal form:

$$a = \omega^{c_1} . n_1 + \omega^{c_2} . n_2 + \ldots + \omega^{c_m} . n_m$$

where $c_1 > c_2 > \ldots > c_m$ and $n_1, n_2, \ldots, n_m$ are integers.

44. An $\epsilon$-ordinal is an ordinal $\epsilon$ which satisfies $\epsilon = \omega^\epsilon$. Show that:

(i) If $\epsilon_0 = \omega + \omega^\omega + \omega^{(\omega^\omega)} + \ldots$ then $\epsilon_0 = \omega^{\epsilon_0}$.

(ii) $\epsilon_0$ is the least $\epsilon$-ordinal.

(iii) If $1 < c < \epsilon_0$ then $\epsilon_0 = c^{\epsilon_0}$.

(iv) If $0 < c, c_0 = c, c_{n+1} = \omega^{c_n}$ then $lim c_n$ is the least $\epsilon$-ordinal not less than $c$.

45. Show that there exist ordinals which satisfy $a = \omega_a$. [Consider

$$c_0 = \omega, \quad c_{n+1} = \omega_{c_n}, \quad c = lim c_n.]$$

46. Show that: $(Ax, x')\, D(x = x')\, K\phi x N\phi x'$ is consistent in a universe of one element but is inconsistent in any other universe.

47. $H\{z', \ldots, z^{(\pi)}\}$ is a conjunction of identities and differences $z_\nu = z_\mu$, $z_\nu \neq z_\mu$ which contains exactly one of these for each pair of $\nu, \mu, 1 \leqslant \nu$, $\mu \leqslant \pi$. By using the disjunctive normal form show that a statement $F\{\phi, \psi, \chi, \ldots, =, z', \ldots, z^{(\pi)}, x', \ldots, x^{(\theta)}\}$ can be expressed in the form $\sum_{\kappa=1}^{\lambda} H_\kappa\{z', \ldots, z^{(\pi)}\}\, F_\kappa\{\phi, \ldots, =, \mathfrak{z}, \mathfrak{x}\}$, where $\mathfrak{z}$ stands for $z', \ldots, z^{(\pi)}$ and $\mathfrak{x}$ for $x', \ldots, x^{(\theta)}$. Hence show that $(E\mathfrak{z})\,(A\mathfrak{x})\, F\{\phi, \ldots, =, \mathfrak{z}, \mathfrak{x}\}$ is equivalent to one of $(E\mathfrak{z})\,(A\mathfrak{x})\, H_\kappa F_\kappa$ for some $\kappa$, $1 \leqslant \kappa \leqslant \lambda$.

48. Show that

$$DN(Ax)\,(Ey)\,\phi\{x, y\}\,(Ex, x', x'', x''')\,(DKKKKKK\phi\{x, x'\}\,\phi\{x', x''\}\,\phi\{x'', x'''\}$$
$$\phi\{x, x''\}\,\phi\{x, x'''\}\,\phi\{x', x'''\}\,KKKK\phi\{x, x'\}\,\phi\{x', x''\}\,\phi\{x'', x'''\}\,N\phi\{x, x''\}$$
$$N\phi\{x', x'''\})$$

is generally valid. Ackermann *J.S.L.* **21** (1956), p. 197.

49. Show that

(i) $(Ex)\, DN\phi\{x, x\}\,(Ay)\, NK\phi\{x, y\}\, N\phi\{y, y\}$,

(ii) $N(Ax)\, K\phi\{x, x\}\,(Ey)\, K\phi\{x, y\}\, N\phi\{y, y\}$,

(iii) $N(Ax)\, DK\phi\{x, x\}\,(Ey)\, K\phi\{x, y\}\, N\phi\{y, y\}\, K\phi\{x, x\}\,(Ey)\, K\phi\{y, x\}$
$N\phi\{y, y\}$

are generally valid. Oglesby (1962).

Chapter 4

# A complete, decidable arithmetic. The system $\mathbf{A}_{00}$

## 4.1 *The system* $\mathbf{A}_{00}$

In this chapter we construct the formal system $\mathbf{A}_{00}$. It is a very simple arithmetic with familiar fundamental concepts. These are: the natural number *zero*, the *successor function*, the operation of repeatedly *applying* a function, the operation of forming functions by *abstraction*, *equality* and *inequality* between numerical expressions, and the logical connectives, conjunction and disjunction. The atomic statements are *equations* and *inequations* between *numerical terms*, *compound statements* are built up from atomic statements by conjunction and disjunction. Negation, *material implication* and *material equivalence* are definable, but existential quantification and universal quantification are unrepresentable.

We give definitions of $\mathbf{A}_{00}$-*truth* and of $\mathbf{A}_{00}$-*falsity* for closed $\mathbf{A}_{00}$-statements, and show that they are exclusive properties. We also show that a closed $\mathbf{A}_{00}$-statement is $\mathbf{A}_{00}$-true if and only if it is an $\mathbf{A}_{00}$-theorem. Thus the system $\mathbf{A}_{00}$ is consistent in the sense that $\mathbf{A}_{00}$-theorems are $\mathbf{A}_{00}$-true; and is complete in the sense that $\mathbf{A}_{00}$-true $\mathbf{A}_{00}$-statements are $\mathbf{A}_{00}$-theorems. We give a procedure which applied to a closed $\mathbf{A}_{00}$-statement will terminate and tell us whether it is $\mathbf{A}_{00}$-true or is $\mathbf{A}_{00}$-false. Thus the system $\mathbf{A}_{00}$ is decidable.

## 4.2 *The* $\mathbf{A}_{00}$-*rules of formation*

To construct the system $\mathbf{A}_{00}$ we first list the $\mathbf{A}_{00}$-signs and attach a type to each proper $\mathbf{A}_{00}$-symbol and give each $\mathbf{A}_{00}$-symbol a name which will assist the reader in understanding how the system was first conceived. Parentheses round type symbols are usually omitted by association to the left as explained in Ch. 1. (See table overleaf.)

The required unending sequence of variables is obtained by repeatedly attaching primes, thus: $x$, $x'$, $x''$, .... The table order of the proper and improper symbols is called their *lexicographic order*. In the system $\mathbf{A}_{00}$

| symbol | type | name |
|---|---|---|
| $0$ | $\iota$ | *zero* |
| $S$ | $\iota\iota$ | *successor function* |
| $=$ | $o\iota\iota$ | *equality* |
| $\neq$ | $o\iota\iota$ | *inequality* |
| & | $ooo$ | *conjunction* |
| $\vee$ | $ooo$ | *disjunction* |
| $\mathscr{I}$ | $\iota\iota(\iota\iota)$ | *iterator operator* |
| $x$ | $\iota$ | *variable* |

The improper symbols are, as usual:

| | |
|---|---|
| $\boldsymbol{\lambda}$ | *abstraction operator* |
| ( | *left parenthesis* |
| ) | *right parenthesis* |
| ′ | *generating sign* |

the natural numbers are represented by certain formulae of type $\iota$ called *numerals*. These formulae are defined by the following rules:

(i) 0 is a numeral,
(ii) if $\nu$ is a numeral, then so is $(S\nu)$,
(iii) these are the only numerals.

Thus $0$, $(S0)$, $(S(S0))$, $(S(S(S0)))$, ... are the numerals. Note that a numeral is a formula, distinct numerals are distinct formulae. An $\mathbf{A}_{oo}$-statement is an $\mathbf{A}_{oo}$-formula of type $o$, according to the universal rules as given in Ch. 1. The only abstracts allowed are of types $\iota\iota$, $\iota\iota\iota$, $\iota\iota\iota\iota$, etc., i.e. $(\boldsymbol{\lambda}x\mathscr{I})$, $(\boldsymbol{\lambda}x((=x)x)$, etc. are rejected.

It is convenient in talking about the system $\mathbf{A}_{oo}$ to supplement the English language with Greek letters. Thus we shall use Greek letters, with or without superscripted primes or subscripts, as follows:

| | |
|---|---|
| $\theta, \kappa, \lambda, \mu, \nu, \pi$ | to denote undetermined *numerals*, |
| $\alpha, \beta, \gamma, \delta$ | to denote undetermined formulae of type $\iota$, called *numerical terms*, |
| $\xi, \eta, \zeta, \upsilon$ | to denote undetermined variables of type $\iota$, |
| $\phi, \psi, \chi, \omega$ | to denote undetermined *statements*, |
| $\rho, \sigma, \tau$ | to denote undetermined *functors* of type $\iota\iota$, $\iota\iota\iota$, etc. |

In the last case the type can be introduced as a subscript if desired. From now on we usually omit the concatenation sign. Thus

$=\alpha\beta$ stands for an undetermined equation between numerical terms,
$\& \phi\psi$ stands for an undetermined conjunction,
$\vee \phi\psi$ stands for an undetermined disjunction.

We shall frequently write these in the more familiar manner:

D 112 $(\alpha = \beta)$ for $=\alpha\beta$, $(\alpha \neq \beta)$ for $\neq \alpha\beta$,

D 113 $(\phi \,\&\, \psi)$ for $\& \phi\psi$,

D 114 $(\phi \vee \psi)$ for $\vee \phi\psi$.

We shall frequently use the parenthesis convention, thus:

$\phi \vee \psi \vee \chi$ stands for $((\phi \vee \psi) \vee \chi)$ and hence for $((\vee ((\vee \phi) \psi)) \chi)$,

$\phi \vee (\psi \vee \chi)$ stands for $(\phi \vee (\psi \vee \chi))$ and hence for $((\vee \phi) ((\vee \psi) \chi))$.

Note that, by the parenthesis convention, $= \alpha\beta$ stands for $((= \alpha) \beta)$, but the defined expression $(\alpha = \beta)$ has its complete set of parentheses. As in Ch. 1 we shall use $\phi\{\Gamma\}$, $\alpha\{\Gamma\}$, etc. as statement-forms, and numerical-term-forms respectively, $\phi\{\xi\}$ stands for an undetermined statement containing an undetermined variable free, this arises from the statement-form $\phi\{\Gamma\}$ by substitution. We shall also use the notation $\phi\{\mathfrak{G}\}$, $\phi\{\mathfrak{x}\}$, as explained in Ch. 1, here $\mathfrak{x}$ denotes an ordered set of variables. Sometimes we subscript type symbols to $\Gamma, \Gamma', \ldots$ in formulae-forms to denote that substitutions will only be made for these types. Functors of types $\iota$, $\iota\iota$, and so on will be called functions of natural numbers or simply functions.

The iterator symbol $\mathscr{I}$ is such that $\mathscr{I}\rho\alpha\beta$ is a numerical term whenever $\alpha$, $\beta$ are numerical terms and $\rho$ is a function of type $\iota\iota$, $\mathscr{I}\rho$ is a function of type $\iota\iota$. Thus $\mathscr{I}$ converts a function of type $\iota\iota$ into another function of type $\iota\iota$.

## 4·3· *The $A_{00}$-rules of consequence*

The system $A_{00}$ has the following axiom schemes:

$Ax_{00}$. 1 $(\alpha = \alpha)$,

$Ax_{00}$. 2.1 $(S\alpha \neq 0)$,

Ax$_{oo}$. 2.2 $(0 \neq S\alpha)$,

Ax$_{oo}$. 3.1 $(\mathscr{I}\rho\alpha 0 = \alpha)$,

Ax$_{oo}$. 3.2 $(\mathscr{I}\rho\alpha(S\beta) = \rho\beta(\mathscr{I}\rho\alpha\beta))$,

where $\alpha$, $\beta$ are closed numerical terms and $\rho$ is a closed function of type $\iota\iota$.

$\underbrace{\iota \dots \iota}_{S\pi\text{-times}}$ is the result of striking out each parenthesis and the zero in the formula $(S\pi)$ and then replacing each occurrence of $S$ by a corresponding occurrence of $\iota$ and replacing the parentheses by association to the left.

Ax$_{oo}$. 4.1 $(\boldsymbol{\lambda}\xi.\rho\{\xi\})\,\beta\beta' \dots \beta^{(\pi)} = \rho\{\beta\}\beta' \dots \beta^{(\pi)}$,

Ax$_{oo}$. 4.2 $(\boldsymbol{\lambda}\xi.\rho\{\xi\})\,\beta\beta' \dots \beta^{(\pi)} = (\boldsymbol{\lambda}\xi'.\rho\{\xi'\})\,\beta\beta' \dots \beta^{(\pi)}$,

where $\rho\{\xi\}$ is of type $\underbrace{\iota \dots \iota}_{S\pi\text{-times}}$ and both sides are closed and $\xi, \xi'$ fail to occur free in $\rho\{\Gamma_\iota\}$ and $\Gamma_\iota$ is *free for* $\xi, \xi'$ in $\rho\{\Gamma_\iota\}$. When written in full the parentheses are put back by association to the left, viz.:

$$(\dots(((\boldsymbol{\lambda}\xi.\rho\{\xi\})\,\beta)\,\beta')\dots\beta^{(\pi)}).$$

This axiom allows us to apply an argument to a function.

We really only need functions of at most two arguments but it is sometimes useful to have functions of any number of arguments. This makes these two axioms more complicated. In them $\rho\{\xi\}$ is of type $\underbrace{\iota \dots \iota}_{S\pi\text{-times}}$ so it must be of the form:

$$(\boldsymbol{\lambda}\xi'(\boldsymbol{\lambda}\xi''(\dots(\boldsymbol{\lambda}\xi^{(\pi)}\alpha\{\xi, \xi', \xi'', \dots, \xi^{(\pi)}\})\dots))),$$

where $\alpha\{\xi, \xi', \xi'', \dots, \xi^{(\pi)}\}$ is a numerical term, or it could be;

$$(\boldsymbol{\lambda}\xi'(\boldsymbol{\lambda}\xi''(\dots(\boldsymbol{\lambda}\xi^{(\pi-1)}S)\dots)) \quad \text{or} \quad (\boldsymbol{\lambda}\xi'(\boldsymbol{\lambda}\xi''(\dots(\boldsymbol{\lambda}\xi^{(\pi-1)}\mathscr{I}\rho\alpha)\dots))$$

or

$$(\boldsymbol{\lambda}\xi'(\boldsymbol{\lambda}\xi''(\dots(\boldsymbol{\lambda}\xi^{(\pi-2)}\mathscr{I}\rho)\dots)),$$

with appropriate modifications if $\pi = 0, 1, 2$. By repeatedly applying Ax$_{oo}$. 4.1, 4.2 we can change any $\xi^{(\theta)}$ to a new variable.

The first axiom scheme states a familiar property of equality and the two parts of the second axiom scheme state a familiar property of the successor function. We require both parts because if we worked with only one part then some properties of inequality which we require would fail. The third axiom scheme shows how the iterator operator acts. This may become clear if Ax$_{oo}$. 3.2 is written in ordinary mathematical notation: suppose that $f$ is a function of two arguments, then

$$((\mathscr{I}f)\,m(n+1) = f(n, (\mathscr{I}f)\,mn)).$$

If we write $g[n, m]$ for $\mathscr{I}fmn$ we obtain:

$$g[n+1, m] = f[n, g[n, m]].$$

Thus
$$\begin{aligned} g[3, m] &= f[2, g[2, m]] \\ &= f[2, f[1, g[1, m]]] \\ &= f[2, f[1, f[0, g[0, m]]]] \\ &= f[2, f[1, f[0, m]]], \end{aligned}$$

since $g[0, m] = m$ by $\mathrm{Ax}_{00}$. 3.1. In general

$$g[n, m] = f[n, f[n-1, [f[n-2, \ldots, f[2, f[1, f[0, m]]] \ldots]]]].$$

The system $\mathrm{A}_{00}$ has the following rules of procedure:

R 1
$$\frac{\phi\{\alpha\} \vee \omega \quad (\alpha = \beta) \vee \omega}{\phi\{\beta\} \vee \omega},$$

where $\alpha$ and $\beta$ are closed numerical terms and the $\mathbf{A}_{00}$-statement form $\phi\{\Gamma\}$ and $\omega$ are closed. The order of the premisses is immaterial. The $\mathbf{A}_{00}$-statement $\omega$ is subsidiary and may be absent, in which case the disjunction sign is omitted, $\phi\{\alpha\}$, $(\alpha = \beta)$ and $\phi\{\beta\}$ are the main formulae

R 2
$$\frac{(\alpha \neq \beta) \vee \omega}{(S\alpha \neq S\beta) \vee \omega},$$

where $\alpha$ and $\beta$ are closed numerical terms and $\omega$ is a closed $\mathbf{A}_{00}$-statement and is subsidiary as in R 1, $(\alpha \neq \beta)$ and $(S\alpha \neq S\beta)$ are the main formulae. Note that since $\alpha$ and $\beta$ are closed in R 1 then a variable whether free or bound is unaffected by applications of R 1.

The remaining rules are labelled in a different manner because they are some of the rules of $\mathscr{P}_C$. In listing them, except in one case, we omit the condition that the $\mathbf{A}_{00}$-statements in them be closed because they can only be used in $\mathbf{A}_{00}$-proofs when this is so. In the exceptional case the condition must be stated otherwise free variables could be introduced into an $\mathbf{A}_{00}$-proof.

I. Remodelling rules

$$(a) \quad \frac{\omega' \vee \phi \vee \psi \vee \omega}{\omega' \vee \psi \vee \phi \vee \omega}.$$

permutation

II. Building rules

$$(a) \quad \frac{\chi}{\phi \vee \chi}; \qquad (b') \quad \frac{\phi \vee \omega \quad \psi \vee \omega}{(\phi \,\&\, \psi) \vee \omega}.$$

dilution composition

In II (*a*) $\phi$ is closed. In II (*b'*) the order of the premisses is immaterial. The $\mathbf{A}_{00}$-statements $\omega$, $\omega'$ are subsidiary and may be omitted, the other $\mathbf{A}_{00}$-statements are the main formulae. The $\mathbf{A}_{00}$-statement $\chi$ is secondary and must be present. We have omitted parentheses by association to the left and the outer pair is usually omitted. The rules are known by the names beneath them.

## 4·4 *Definition of* $\mathbf{A}_{00}$*-truth*

We say that a closed numerical term $\gamma$ *determines* a numeral $\nu$ under the following conditions:

(i) $\gamma$ is $\nu$.

(ii) $\nu$ can be obtained from $\gamma$ by replacements of the following kinds:

(*a*) replace an occurrence of $\mathscr{I}\rho\alpha 0$ by a corresponding occurrence of $\alpha$;

(*b*) replace an occurrence of $\mathscr{I}\rho\alpha(S\beta)$ by a corresponding occurrence of $\rho\beta(\mathscr{I}\rho\alpha\beta)$;

(*c*) replace an occurrence of $\boldsymbol{\lambda}\xi.\rho\{\xi\}\beta\beta'\ldots\beta^{(\pi)}$ by a corresponding occurrence of $\rho\{\beta\}\beta'\ldots\beta^{(\pi)}$;

(*d*) replace an occurrence of $\boldsymbol{\lambda}\xi.\rho\{\xi\}\beta\beta'\ldots\beta^{(\pi)}$ by a corresponding occurrence of $\boldsymbol{\lambda}\xi'.\rho\{\xi'\}\beta\beta'\ldots\beta^{(\pi)}$,

where $\xi$ and $\xi'$ fail to occur free in $\rho\{\Gamma_\iota\}$ and $\Gamma_\iota$ is free for $\xi$ and $\xi'$ in $\rho\{\Gamma_\iota\}$.

If $\gamma$ determines the numeral $\nu$ then $\gamma = \nu$ is an $\mathbf{A}_{00}$-theorem and the above replacements give a special type of $\mathbf{A}_{00}$-proof of $\gamma = \nu$.

We say that an $\mathbf{A}_{00}$-statement is $\mathbf{A}_{00}$-true if and only if it satisfies the following conditions $\mathscr{T}_{00}$.

(i) $\phi$ is an equation between closed numerical terms and both terms determine the same numeral,

(ii) $\phi$ is an inequation between closed numerical terms and these determine distinct numerals,

(iii) $\phi$ is a conjunction and both conjunctands are $\mathbf{A}_{00}$-true,

(iv) $\phi$ is a disjunction and a disjunctand is $\mathbf{A}_{00}$-true.

Thus the $\mathbf{A}_{00}$-truth of an $\mathbf{A}_{00}$-statement is referred to the $\mathbf{A}_{00}$-truth of shorter $\mathbf{A}_{00}$-statements until finally we arrive at closed equations or inequations between numerals.

## 4·5 *Definition of* $\mathbf{A}_{00}$*-falsity*

A closed $\mathbf{A}_{00}$-statement $\phi$ is said to be $\mathbf{A}_{00}$-false if and only if it satisfies the following conditions $\mathscr{F}_{00}$:

(i) $\phi$ is an equation between closed numerical terms and these terms determine distinct numerals,

(ii) $\phi$ is an inequation between closed numerical terms and these terms determine the same numeral,

(iii) $\phi$ is a conjunction and a conjunctand is $\mathbf{A}_{00}$-false,

(iv) $\phi$ is a disjunctand and both disjunctands are $\mathbf{A}_{00}$-false.

## 4·6 *Exclusiveness of* $\mathbf{A}_{00}$*-truth and* $\mathbf{A}_{00}$*-falsity*

We shall see later that $\mathbf{A}_{00}$-truth and $\mathbf{A}_{00}$-falsity are exclusive properties. First we show that a closed numerical term determines a numeral.

Let the *rank* of a closed numerical term $\alpha$ be the numeral defined as follows: erase all symbols in $\alpha$ except the iterator and abstraction symbols, then replace each iterator and abstraction symbol by a successor symbol then place zero on the right and add parentheses by association to the right. If $\alpha$ is without iterator or abstraction symbols then the rank of $\alpha$ is to be zero. Let the *order* of a closed numerical term $\alpha$ be the numeral defined as follows: erase all symbols of $\alpha$ except the successor symbol then place zero at the end and add parentheses by association to the right. If $\alpha$ is without successor symbols then the order of $\alpha$ is to be zero. If $\alpha$ is a closed numerical term of rank zero then $\alpha$ is a numeral and this numeral is the order of $\alpha$.

We now give a method for finding a numeral determined by a closed numerical term $\alpha$ by reference to terms of lesser rank (i.e. whose rank is a proper part of the rank of $\alpha$) or to terms of the same rank but lesser order. Thus continued application of the method will ultimately stop. If the closed numerical term $\alpha$ is of rank zero then it is a numeral and by condition (i) for determining a numeral we have found a numeral which $\alpha$ determines. Let $\alpha$ be of rank $S\pi$, where $\pi$ is a numeral, first suppose that $\alpha$ is of one of the forms $\mathscr{I}\rho\beta\gamma$ or $(\boldsymbol{\lambda}\xi.\delta\{\xi\})\beta\beta'\ldots\beta^{(\theta)}$, where $\gamma, \beta, \beta', \ldots, \beta^{(\theta)}$ are closed numerical terms, $\rho$ is a closed function of two arguments and $\delta\{\xi\}$ is a function of type $\underbrace{\iota\ldots\iota}_{S\theta\text{-times}}$ with at most $\xi$ as free variable, and $\xi$ fails to occur free in $\delta\{\Gamma_\iota\}$. The ranks of $\beta$ and $\gamma$ are less than that of $\alpha$, suppose that we have found that they determine numerals $\kappa$ and $\nu$ respectively. Then we require to find a numeral determined by $\mathscr{I}\rho\kappa\nu$ or $\delta\{\kappa\}\beta'\ldots\beta^{(\theta)}$ respectively. These are both of lower rank than $\alpha$ except that in the first case if both $\beta$ and $\gamma$ are numerals then $\mathscr{I}\rho\beta\gamma$ and its rank are unaltered. In this case if $\nu$ is zero then $\alpha$ determines $\kappa$, while if $\nu$ is $S\nu'$ then we require to find the numeral determined by $\rho\nu'(\mathscr{I}\rho\kappa\nu')$. This is of higher rank than $\alpha$ but $\mathscr{I}\rho\kappa\nu'$ is of the same rank but lower order than $\alpha$. Suppose that $\mathscr{I}\rho\kappa\nu'$ determines the numeral $\theta$ then we require to find a numeral determined by $\rho\nu'\theta$, but this is of lower rank than $\alpha$. Thus in each case we are ultimately referred to terms of lower rank than $\alpha$ or to terms of the same rank but of lower order, thus the process will ultimately stop at a term of rank zero and so produces a numeral which $\alpha$ determines. Secondly if $\alpha$ is of the form $S\ldots S\alpha'$, where $S\ldots S$ stands for a sequence of successor symbols (in full notation parentheses should be inserted by association to the right) and $\alpha'$ is a closed numerical term of one of the forms already considered, then we deal with $\alpha'$ as in the previous cases. Note that the reduction can be performed in several different ways so that it might happen that a closed numerical term determined several distinct numerals according to the method of reduction. We now show that a closed numerical term determines a unique numeral.

PROP. 1. $\mathbf{A}_{00}$-*truth and* $\mathbf{A}_{00}$-*falsity are exclusive properties*

We have to show that a closed $\mathbf{A}_{00}$-statement fails to be both $\mathbf{A}_{00}$-true and $\mathbf{A}_{00}$-false. We first show that a closed numerical term determines a unique numeral.

Suppose that the closed numerical term $\gamma$ determines the numerals $\kappa, \nu$. Then according to a previous remark we obtain $\mathbf{A}_{00}$-proofs of the equations $\gamma = \kappa$ and $\gamma = \nu$ and hence by R 1 of $\kappa = \nu$. It suffices then to show that if an equation $\kappa = \nu$, where $\kappa$ and $\nu$ are numerals, is $\mathbf{A}_{00}$-provable using only $\mathbf{A}_{00}$-axiom schemes 1, 3.1, 3.2, 4.1, 4, 2 and rule R 1 then $\kappa$ is the same numeral as $\nu$.

Given a closed numerical term $\alpha$ we define the *standard determination* (s.d.) and the *standard* $\mathbf{A}_{00}$-*proof* (s.p.) of the equation $\alpha = \nu$ inductively on the rank and order of $\alpha$.

(i) $\alpha$ is zero, the s.d. of $\alpha$ is 0, the s.p. of $\alpha = 0$ is $0 = 0$.

(ii) $\alpha$ is $S\beta$, the s.d. of $\alpha$ is $S\nu$, where $\nu$ is the s.d. of $\beta$.

If $\beta$ fails to be a numeral the s.p. of $\alpha = S\nu$ is $\dfrac{\alpha = S\beta \quad \overset{\vdots}{\overline{\beta = \nu}}}{\alpha = S\nu}$ R 1; where $\overset{\vdots}{\overline{\beta = \nu}}$ is the s.p. of $\beta = \nu$. If $\beta$ is a numeral $\nu$ the s.p. of $\alpha = S\nu$ is $S\nu = S\nu$. In this case $\alpha$ is $S\nu$.

(iii) $\alpha$ is $\mathscr{I}\rho\beta\gamma$; $\beta$, $\gamma$ are of lower rank than $\alpha$ let their s.d.'s be $\kappa$, $\pi$ respectively. $\mathscr{I}\rho\kappa\pi$ is of lower rank than $\alpha$ unless $\beta$, $\gamma$ are $\kappa$, $\pi$ respectively.

(*a*) In the first case let $\nu$ be the s.d. of $\mathscr{I}\rho\kappa\pi$ and let $\overset{\vdots}{\overline{\beta = \kappa}}$ $\overset{\vdots}{\overline{\gamma = \pi}}$ $\overset{\vdots}{\overline{\mathscr{I}\rho\kappa\pi = \nu}}$ be the s.p.'s of $\beta = \kappa$, $\gamma = \pi$ and $\mathscr{I}\rho\kappa\pi = \nu$ respectively, then:

$$\dfrac{\dfrac{\alpha = \mathscr{I}\rho\beta\gamma \quad \overset{\vdots}{\overline{\beta = \kappa}} \quad \overset{\vdots}{\overline{\gamma = \nu}}}{\alpha = \mathscr{I}\rho\kappa\pi}\ \text{R 1 twice} \quad \overset{\vdots}{\overline{\mathscr{I}\rho\kappa\pi = \nu}}}{\alpha = \nu}\ \text{R 1}$$

is the s.p. of $\alpha = \nu$. In this case if only one of $\beta$, $\gamma$ is a numeral then the rank of the other fails to be zero. If $\beta$ or $\gamma$ is a numeral then we omit $\overset{\vdots}{\overline{\beta = \kappa}}$ or $\overset{\vdots}{\overline{\gamma = \pi}}$ as the case may be.

(*b*) If $\beta$, $\gamma$ are $\kappa$, 0 respectively then the s.d. of $\alpha$ is $\kappa$ and the s.p. of $\alpha = \kappa$ is $\alpha = \kappa$ (3.1).

(*c*) If $\beta$, $\gamma$ are $\kappa$, $S\pi$ respectively then $\mathscr{I}\rho\kappa\pi$ is of the same rank but lesser order than $\alpha$ let its s.d. be $\theta$ and let the s.p. of $\mathscr{I}\rho\kappa\pi = \theta$ be $\overset{\vdots}{\overline{\mathscr{I}\rho\kappa\pi = \theta}}$ also $\rho\pi\theta$ is of lesser rank than $\alpha$ let its s.d. be $\nu$ and let the s.p. of $\rho\pi\theta = \nu$ be $\overset{\vdots}{\overline{\rho\pi\theta = \nu}}$. Then the s.d. of $\alpha$ is $\nu$ and the s.p. of $\alpha = \nu$ is

$$\dfrac{\dfrac{\alpha = \rho\pi(\mathscr{I}\rho\kappa\pi)\,(3.2) \quad \overset{\vdots}{\overline{\mathscr{I}\rho\kappa\pi = \theta}}}{\alpha = \rho\pi\theta}\ \text{R 1} \quad \overset{\vdots}{\overline{\rho\pi\theta = \nu}}}{\alpha = \nu}\ \text{R 1.}$$

(iv) $\alpha$ is $(\boldsymbol{\lambda}\xi.\rho\{\xi\})\beta\beta' \ldots \beta^{(n)}$, $\beta$ is of lower rank than $\alpha$ let its s.d. be $\kappa$

and let the s.p. of $\beta = \kappa$ be $\begin{array}{c}\vdots\\ \hline \beta = \kappa\end{array}$. $\rho\{\kappa\}\beta' \dots \beta^{(\pi)}$ is of lower rank than $\alpha$ let its s.d. be $\nu$ and the s.p. of $\rho\{\kappa\}\beta' \dots \beta^{(\pi)} = \nu$ be $\begin{array}{c}\vdots\\ \hline \rho\{\kappa\}\beta' \dots \beta^{(\pi)} = \theta\end{array}$. Then the s.d. of $\alpha$ is $\nu$ and the s.p. of $\alpha = \nu$ is

$$\cfrac{\cfrac{\alpha = \rho\{\beta\}\beta' \dots \beta^{(\pi)}\,(4.1) \qquad \begin{array}{c}\vdots\\ \hline \beta = \kappa\end{array}}{\alpha = \rho\{\kappa\}\beta' \dots \beta^{(\pi)}}\ \text{R 1 (omit if } \beta \text{ is } \kappa) \qquad \rho\{\kappa\}\beta' \dots \beta^{(\pi)} = \nu}{\alpha = \nu}\ \text{R 1.}$$

Clearly the s.d. $\nu$ of $\alpha$ and the s.p. of $\alpha = \nu$ are unique.

We now show that a closed numerical term determines a unique numeral. Suppose that the closed numerical term $\alpha$ determines the numerals $\nu$ and $\kappa$, then we have $\mathbf{A}_{00}$-proofs of $\alpha = \nu$ and of $\alpha = \kappa$ and hence of $\nu = \kappa$. The $\mathbf{A}_{00}$-proof of $\nu = \kappa$ uses $\mathbf{A}_{00}$-axiom schemes 1, 3.1, 3.2, 4.1, 4.2 and rule R 1 only.

LEMMA. *If the closed numerical terms $\gamma$ and $\gamma'$ have the same s.d.'s then so do the closed numerical terms $\alpha\{\gamma\}$ and $\alpha\{\gamma'\}$.*

If $\Gamma_\iota$ fails to occur in $\alpha\{\Gamma_\iota\}$ the result is trivial. The result is also trivial if $\gamma$ is the same as $\gamma'$. Thus we may suppose that one of $\gamma$, $\gamma'$ is distinct from a numeral and that $\Gamma_\iota$ has an occurrence in $\alpha\{\Gamma_\iota\}$. We use induction on the rank and order of $\alpha\{\nu\}$ where $\nu$ is the common s.d. of $\gamma$ and $\gamma'$. $\alpha\{\Gamma_\iota\}$ is $\Gamma_\iota$, the result is trivial. $\alpha\{\Gamma_\iota\}$ is $S\beta\{\Gamma_\iota\}$ then $\beta\{\nu\}$ is of the same rank and lesser order than $\alpha\{\nu\}$, hence by our hypothesis $\beta\{\gamma\}$ and $\beta\{\gamma'\}$ have the same s.d. By (ii) $\alpha\{\gamma\}$ and $\alpha\{\gamma'\}$ have the same s.d. $\alpha\{\Gamma_\iota\}$ is $\mathscr{I}\rho\{\Gamma_\iota\}\beta\{\Gamma_\iota\}\delta\{\Gamma_\iota\}$ then $\beta\{\nu\}$ and $\delta\{\nu\}$ are of lower rank than $\alpha\{\nu\}$, if $\Gamma_\iota$ occurs in $\beta\{\Gamma_\iota\}$ then by our hypothesis $\beta\{\gamma\}$ and $\beta\{\gamma'\}$ both have the same s.d., say $\nu'$, otherwise they trivially have the same s.d., similarly $\delta\{\gamma\}$ and $\delta\{\gamma'\}$ both have the same s.d., say $\nu''$. By (iii) *a*, the s.d. of $\alpha\{\gamma\}$ is the same as that of $\mathscr{I}\rho\{\gamma\}\nu'\nu''$, and the s.d. of $\alpha\{\gamma'\}$ is the same as that of $\mathscr{I}\rho\{\gamma'\}\nu'\nu''$, this is trivial if $\beta\{\Gamma_\iota\}$ is $\nu'$ and $\delta\{\Gamma_\iota\}$ is $\nu''$. (Note that by the lemma of Ch. 1 $\gamma$ fails to overlap $\rho$, $\beta$ or $\delta$.) If $\nu''$ is 0 then $\mathscr{I}\rho\{\gamma\}\nu'0$ and $\mathscr{I}\rho\{\gamma'\}\nu'0$, by (iii) *b*, both have s.d. $\nu'$. If $\nu''$ is $S\nu'''$ then by (iii) *c* the s.d. of $\mathscr{I}\rho\{\gamma\}\nu'(S\nu''')$ is the same as that of $\rho\{\gamma\}\nu'''\theta$ and the s.d. of $\mathscr{I}\rho\{\gamma'\}\nu'(S\nu''')$ is the same as that of $\rho\{\gamma'\}\nu'''\theta'$, where $\theta$ is the s.d. of $\mathscr{I}\rho\{\gamma\}\nu'\nu'''$ and $\theta'$ is the s.d. of $\mathscr{I}\rho\{\gamma'\}\nu'\nu'''$. Now $\mathscr{I}\rho\{\nu\}\nu'\nu'''$ is of lower rank or of the same rank and lesser order than $\alpha\{\nu\}$, (this occurs also if $\beta\{\Gamma_\iota\}$ and $\delta\{\Gamma_\iota\}$ are both $\Gamma_\iota$ or are both numerals) hence by our hypothesis $\mathscr{I}\rho\{\gamma\}\nu'\nu'''$ and $\mathscr{I}\rho\{\gamma'\}\nu'\nu'''$ both have the same s.d., thus $\theta = \theta'$. By (iii)$c$ the s.d.'s of $\mathscr{I}\rho\{\gamma\}\nu'(S\nu''')$ and $\mathscr{I}\rho\{\gamma'\}\nu'(S\nu''')$ are the same as those of $\rho\{\nu\}\nu'''\theta$ and $\rho\{\gamma'\}\nu'''\theta$ respectively. Now the rank of $\rho\{\nu\}\nu'''\theta$ is lower than that of $\alpha\{\nu\}$, hence by our hypothesis $\rho\{\gamma\}\nu'''\theta$ and $\rho\{\gamma'\}\nu'''\theta$ both have the same s.d., say $\nu$. By (iii) the s.d.'s of $\alpha\{\gamma\}$ and $\alpha\{\gamma'\}$ are both $\nu$. $\alpha\{\Gamma_\iota\}$ is

$$(\boldsymbol{\lambda}\xi.\delta\{\Gamma_\iota,\xi\})\beta\{\Gamma_\iota\}\beta'\{\Gamma_\iota\}\dots\beta^{(\pi)}\{\Gamma_\iota\},$$

where $\delta\{\Gamma_\iota,\xi\}$ is of type $\underbrace{\iota\dots\iota}_{SS\pi\text{-times}}$ and has at most $\xi$ as free variable and $\beta\{\Gamma_\iota\},\dots,\beta^{(\pi)}\{\Gamma_\iota\}$ are closed numerical term-forms. $\beta\{\nu\}$ is of lower rank than $\alpha$, hence by our hypothesis $\beta\{\gamma\}$ and $\beta\{\gamma'\}$ both have the same s.d., say $\kappa$. By (iv) $\alpha\{\gamma\}$ and $\alpha\{\gamma'\}$ have the same s.d.'s as $\delta\{\gamma,\kappa\}\beta'\{\gamma\}\dots\beta^{(\pi)}\{\gamma\}$ and $\delta\{\gamma',\kappa\}\beta'\{\gamma'\}\dots\beta^{(\pi)}\{\gamma'\}$ respectively, but $\delta\{\nu,\kappa\}\beta'\{\nu\}\dots\beta^{(\pi)}\{\nu\}$ has lower rank than $\alpha\{\nu\}$ hence by our hypothesis $\delta\{\gamma,\kappa\}\beta'\{\gamma\}\dots\beta^{(\pi)}\{\gamma\}$ and $\delta\{\gamma',\kappa\}\beta'\{\gamma'\}\dots\beta^{(\pi)}\{\gamma'\}$ both have the same s.d., say $\theta$. By (iv) $\theta$ is the s.d. of $\alpha\{\gamma\}$ and of $\alpha\{\gamma'\}$. This completes the demonstration of the lemma.

We now show that if the upper formulae of an application of R 1 are equations and if both sides of each equation have the same s.d.'s then the same holds for the lower formula. Let the application of $\mathrm{R}_{00}$ 1 be

$$\frac{\alpha\{\gamma\} = \beta\{\gamma\} \quad \gamma = \gamma'}{\alpha\{\gamma'\} = \beta\{\gamma'\}}.$$

By our supposition and the lemma $\alpha\{\gamma\}$ and $\alpha\{\gamma'\}$ have the same s.d. and so do $\beta\{\gamma\}$ and $\beta\{\gamma'\}$. But by hypothesis $\alpha\{\gamma\}$ and $\beta\{\gamma\}$ have the same s.d., hence $\alpha\{\gamma'\}$ and $\beta\{\gamma'\}$ have the same s.d.

To complete the demonstration that $\nu$ is the same as $\kappa$ we require to show that both sides of an $\mathbf{A}_{00}$-axiom 1, 3.1, 3.2, 4.1, 4.2 have the same s.d. For Ax. 1 this follows at once because the s.d. is unique. For Ax. 3.1 the result follows from (iii) $a, b$ and the uniqueness of the s.d. For Ax. 3.2, let the axiom be $\mathscr{I}\rho\alpha(S\beta) = \rho\beta(\mathscr{I}\rho\alpha\beta)$. If $\kappa$, $\pi$ are the s.d.'s of $\alpha$, $\beta$ respectively then by the lemma the l.h.s. has the same s.d. as $\mathscr{I}\rho\kappa(S\pi)$ and the r.h.s. has the same s.d. as $\rho\pi(\mathscr{I}\rho\kappa\pi)$. $\pi$ is the s.d. of $\beta$ and $\theta$ is the s.d. of $(\mathscr{I}\rho\alpha\beta)$. Now $\rho$ is of type $\iota\iota$ and hence is either $(\boldsymbol{\lambda}\xi.(\boldsymbol{\lambda}\xi'.\delta\{\xi,\xi'\}))$, $\boldsymbol{\lambda}\xi.S$, $\mathscr{I}\sigma$ or $(\boldsymbol{\lambda}\xi.\mathscr{I}\sigma\{\xi\}\delta\{\xi\})$, where $\sigma$ is of type $\iota\iota$. In the first case the s.d. of the r.h.s. is by (iv) the same as that of $(\boldsymbol{\lambda}\xi'.\delta\{\pi,\xi'\})(\mathscr{I}\rho\alpha\beta)$, and again by (iv) the s.d. of this is the same as that of $\delta\{\pi,\theta\}$. In a similar manner the s.d. of $\rho\pi\theta$ is the same as that of $\delta\{\pi,\theta\}$, so that in the first case both sides of Ax. 3.2 have the same s.d. In the second case the s.d. of

$(\boldsymbol{\lambda}\xi.S)\beta(\mathscr{I}\rho\alpha\beta)$ is by (iv) the same as that of $S(\mathscr{I}\rho\alpha\beta)$ which in turn is $S\theta$, and the s.d. of $(\boldsymbol{\lambda}\xi.S)\pi\theta$ is similarly $S\theta$, thus in the second case both sides of Ax. 3.2 have the same s.d. In the third case the s.d. of the l.h.s. is the same as that of $\rho\pi\theta$ and the s.d. of the r.h.s. viz. $\mathscr{I}\sigma\beta(\mathscr{I}\rho\alpha\beta)$ is the same as that of $\mathscr{I}\sigma\pi\theta$, i.e. the same as that of $\rho\pi\theta$. In the last case the s.d. of the l.h.s. is the same as that of $\mathscr{I}(\lambda\xi.\mathscr{I}\sigma\{\xi\}\delta\{\xi\}\alpha)(S\beta)$ and the s.d. of the r.h.s., viz.

$$(\boldsymbol{\lambda}\xi.\mathscr{I}\sigma\{\xi\}\delta\{\xi\})\beta(\mathscr{I}\rho\alpha\beta)$$

is by (iv) the same as that of $\mathscr{I}\sigma\{\pi\}\delta\{\pi\}(\mathscr{I}\rho\alpha\beta)$, the l.h.s. is by (iii) the same as that of $\mathscr{I}\sigma\{\pi\}\delta\{\pi\}\theta$, where $\theta$ is the s.d. of $\mathscr{I}\rho\alpha\beta$. By (iv) the s.d.'s of $\mathscr{I}\sigma\{\pi\}\delta\{\pi\}(\mathscr{I}\rho\alpha\beta)$ and of $\mathscr{I}\sigma\{\pi\}\delta\{\pi\}\theta$ are the same as that of $\mathscr{I}\sigma\{\pi\}\kappa\theta$ where $\kappa$ is the s.d. of $\delta\{\pi\}$. For Ax. 4.1. Let the axiom be

$$(\boldsymbol{\lambda}\xi.\gamma\{\xi\})\beta\beta'\ldots\beta^{(\pi)} = \gamma\{\beta\}\beta'\ldots\beta^{(\pi)}.$$

By (iv) the s.d. of the l.h.s. is the same as that of $\gamma\{\kappa\}\beta'\ldots\beta^{(\pi)}$ where $\kappa$ is the s.d. of $\beta$. By the lemma the s.d.'s of $\gamma\{\beta\}\beta'\ldots\beta^{(\pi)}$ and $\gamma\{\kappa\}\beta'\ldots\beta^{(\pi)}$ are the same. This completes the demonstration that if $\nu = \kappa$ is an $\mathbf{A}_{00}$-theorem then $\nu$ is $\kappa$ and hence a closed numerical term determines a unique numeral.

## 4·7 *Consistency of* $\mathbf{A}_{00}$ *with respect to* $\mathbf{A}_{00}$*-truth*

It now easily follows that a closed $\mathbf{A}_{00}$-equation is either $\mathbf{A}_{00}$-true or $\mathbf{A}_{00}$-false and fails to be both. Similarly for closed $\mathbf{A}_{00}$-inequations. It then easily follows that a closed $\mathbf{A}_{00}$-statement has the same property.

PROP. 2. *The system* $\mathbf{A}_{00}$ *is consistent with respect to* $\mathbf{A}_{00}$*-truth.*

We have to show that each $\mathbf{A}_{00}$-theorem is $\mathbf{A}_{00}$-true. We show that the $\mathbf{A}_{00}$-axioms are $\mathbf{A}_{00}$-true and that the $\mathbf{A}_{00}$-rules preserve $\mathbf{A}_{00}$-truth. Since a closed numerical term $\alpha$ determines a unique numeral then both sides of Ax. 1 determine the same numeral, hence by $\mathscr{T}_{00}$ (i) Ax. 1 is $\mathbf{A}_{00}$-true. Each $\mathbf{A}_{00}$-axiom of the schemes 2.1, 2.2 is $\mathbf{A}_{00}$-true because a closed numerical term determines a unique numeral and $S\alpha$ determines a numeral distinct from zero and zero determines zero, hence by $\mathscr{T}_{00}$ (ii) the $\mathbf{A}_{00}$-axioms 2.1, 2.2 are $\mathbf{A}_{00}$-true. Each $\mathbf{A}_{00}$-axiom of the schemes 3.1, 3.2 is $\mathbf{A}_{00}$-true by the definition of determining a numeral, for both sides of 3.1, 3.2 then determine the same numeral. Similarly for $\mathbf{A}_{00}$-axioms 4.1, 4.2.

We now show that the $\mathbf{A}_{00}$-rules preserve $\mathbf{A}_{00}$-truth. Rule R 1 preserves $\mathbf{A}_{00}$-truth. Suppose $\alpha = \beta \vee \omega$ is $\mathbf{A}_{00}$-true, where $\alpha$ and $\beta$ are closed numerical terms, and that $\phi\{\alpha\} \vee \omega$ is $\mathbf{A}_{00}$-true, where $\phi\{\alpha\}$ and $\omega$ are closed $\mathbf{A}_{00}$-statements. The $\mathbf{A}_{00}$-statement form $\phi\{\Gamma_\iota\}$ is then closed and so is $\phi\{\beta\}$. If $\alpha = \beta$ is $\mathbf{A}_{00}$-true then $\alpha$ and $\beta$ determine the same numeral, say $\nu$, and $\nu$ is unique. In finding the $\mathbf{A}_{00}$-truth-value of $\phi\{\alpha\}$ we have to replace each closed numerical term by the unique numeral it determines and we continue doing this until the only numerical terms left are numerals. $\phi\{\alpha\}$ is built up from equations and inequations of the forms $\delta\{\alpha\} = \delta'\{\alpha\}$, and $\delta''\{\alpha\} \neq \delta'''\{\alpha\}$ and $\phi\{\beta\}$ is built up according to the same pattern from $\delta\{\beta\} = \delta'\{\beta\}$, $\delta''\{\beta\} \neq \delta'''\{\beta\}$. By the lemma $\delta^{(\lambda)}\{\alpha\}$ and $\delta^{(\lambda)}\{\beta\}$, $0 \leqslant \lambda \leqslant 3$, determine the same numeral, hence the $\mathbf{A}_{00}$-truth-values of $\phi\{\alpha\}$ and $\phi\{\beta\}$ are the same. Thus the $\mathbf{A}_{00}$-truth-values of $\phi\{\alpha\} \vee \omega$ and $\phi\{\beta\} \vee \omega$ are the same. If $\omega$ is $\mathbf{A}_{00}$-true then $\phi\{\alpha\} \vee \omega$ and $\phi\{\beta\} \vee \omega$ are $\mathbf{A}_{00}$-true. Since $\alpha = \beta \vee \omega$ is $\mathbf{A}_{00}$-true then one of $\alpha = \beta$ or $\omega$ is $\mathbf{A}_{00}$-true, so in either case $\mathbf{A}_{00}$-truth is preserved.

Clearly R 2 preserves $\mathbf{A}_{00}$-truth. If $\alpha \neq \beta \vee \omega$ is $\mathbf{A}_{00}$-true then either $\omega$ is $\mathbf{A}_{00}$-true or $\alpha$ and $\beta$ determine distinct numerals. In either case $S\alpha \neq S\beta \vee \omega$ is $\mathbf{A}_{00}$-true.

It is clear that the remodelling rules preserve $\mathbf{A}_{00}$-truth from (iv) of the definition of $\mathbf{A}_{00}$-truth. Similarly for the building rules using (iii) and (iv) of the definition of $\mathbf{A}_{00}$-truth. Thus finally an $\mathbf{A}_{00}$-theorem is $\mathbf{A}_{00}$-true.

The importance of the result that a closed numerical term determines a unique numeral is shown by the following: suppose that $\alpha$ determines $\nu$ and $\kappa$ where $\nu \neq \kappa$ and that $\beta$ determines $\nu$ only and $\gamma$ determines $\kappa$ only. Consider

$$\frac{\alpha = \beta \quad \alpha = \gamma}{\gamma = \beta} \quad \text{R 1.}$$

The premisses are $\mathbf{A}_{00}$-true because $\alpha$, $\beta$ determine the same numeral. And so do $\alpha$, $\gamma$. But the conclusion is false.

## 4.8 *Completeness and decidability of* $\mathbf{A}_{00}$ *with respect to* $\mathbf{A}_{00}$-*truth*

PROP. 3. *The system* $\mathbf{A}_{00}$ *is complete with respect to* $\mathbf{A}_{00}$-*truth.*

We have to show that each $\mathbf{A}_{00}$-statement which is $\mathbf{A}_{00}$-true is an $\mathbf{A}_{00}$-theorem. We show this by formula induction. Let $\phi$ be an $\mathbf{A}_{00}$-true $\mathbf{A}_{00}$-statement then $\phi$ is closed. First suppose that $\phi$ is atomic, then

$\phi$ is of one of the forms: $\alpha = \beta$ or $\alpha \neq \beta$, where $\alpha$ and $\beta$ are closed numerical terms. In the first case $\alpha$ and $\beta$ determine the same numeral, say $\theta$ and $\theta$ is unique. Then $\alpha = \theta$ and $\beta = \theta$ are both $\mathbf{A}_{oo}$-provable; from R 1 we obtain $\alpha = \beta$. In the second case $\alpha$ and $\beta$ determine distinct numerals which are unique, say they are $\nu$ and $\kappa$ respectively. Since $\nu$ and $\kappa$ are distinct numerals then one will be a proper part of the other, let $\nu$ be $S \ldots S\kappa$, where $S \ldots S$ denotes a non-null succession of successor symbols so that $S \ldots S\kappa$ with appropriate parentheses stands for a numeral. We first $\mathbf{A}_{oo}$-prove: $S \ldots S\kappa \neq \kappa$. We have

$$\frac{S(S \ldots S0) \neq 0}{S^{***}S(S \ldots S0) \neq S^{***}S0} \quad \begin{array}{l} \text{Ax}_{oo} 2.1 \\[4pt] \text{R 2 repeatedly,} \end{array}$$

where $S^{***}S$ acts like $S \ldots S$. If $\kappa$ is $S^{***}S0$ then $S^{***}(S \ldots S)0$ is $\nu$. Thus $\nu \neq \kappa$ is an $\mathbf{A}_{oo}$-theorem. Similarly if $\kappa$ is $S \ldots S\nu$ we have the $\mathbf{A}_{oo}$-theorem $\nu \neq \kappa$ using Ax 2.2 instead of Ax 2.1. But we also have $\alpha = \nu$ and $\beta = \kappa$ and so by R 1 we obtain $\alpha \neq \beta$, as desired.

Now suppose that $\phi$ is $\mathbf{A}_{oo}$-true and is a compound $\mathbf{A}_{oo}$-statement and that we have demonstrated the result for $\mathbf{A}_{oo}$-statements with fewer logical connectives $\vee$ and &. $\phi$ being compound is of one of the forms: $\psi \vee \chi$ or $\psi \& \chi$. If $\phi$ is $\mathbf{A}_{oo}$-true and is a conjunction, say $\psi \& \chi$, then by $\mathcal{T}_{oo}$ (iii) $\psi$ is $\mathbf{A}_{oo}$-true and $\chi$ is $\mathbf{A}_{oo}$-true. But $\psi$ and $\chi$ contain fewer logical connectives than $\phi$ hence by our hypothesis $\psi$ and $\chi$ are $\mathbf{A}_{oo}$-theorems, by II $b'$ $\psi \& \chi$ is an $\mathbf{A}_{oo}$-theorem. If $\phi$ is a disjunction, say $\psi \vee \chi$ then by $\mathcal{T}_{oo}$ (iv) $\psi$ is $\mathbf{A}_{oo}$-true or $\chi$ is $\mathbf{A}_{oo}$-true, hence $\psi$ is $\mathbf{A}_{oo}$-provable or $\chi$ is $\mathbf{A}_{oo}$-provable, by II $a$ $\chi \vee \psi$ or $\psi \vee \chi$ is $\mathbf{A}_{oo}$-provable, in the first case using I $a$ $\psi \vee \chi$ is $\mathbf{A}_{oo}$-provable. Thus every $\mathbf{A}_{oo}$-true $\mathbf{A}_{oo}$-statement is an $\mathbf{A}_{oo}$-theorem.

PROP. 4. *If an $\mathbf{A}_{oo}$-disjunction is an $\mathbf{A}_{oo}$-theorem then one of the disjunctands is an $\mathbf{A}_{oo}$-theorem, and we can find which disjunctand is an $\mathbf{A}_{oo}$-theorem and we can find an $\mathbf{A}_{oo}$-proof for it.*

By Prop. 1 if an $\mathbf{A}_{oo}$-disjunction is $\mathbf{A}_{oo}$-provable then it is $\mathbf{A}_{oo}$-true. By $\mathcal{T}_{oo}$ (iv) an $\mathbf{A}_{oo}$-disjunction is $\mathbf{A}_{oo}$-true if and only if one of the disjunctands is $\mathbf{A}_{oo}$-true moreover we can find which. According to the demonstration of Prop. 2 we can find an $\mathbf{A}_{oo}$-proof of an $\mathbf{A}_{oo}$-true $\mathbf{A}_{oo}$-statement.

PROP. 5. *If an $\mathbf{A}_{oo}$-conjunctand is $\mathbf{A}_{oo}$-provable then so is each conjunctand.* This follows similarly.

PROP. 6. *The system $A_{00}$ is decidable.*

Let $\phi$ be a closed $A_{00}$-statement then we can decide whether $\phi$ is $A_{00}$-true or is $A_{00}$-false, hence we can decide whether $\phi$ is an $A_{00}$-theorem.

Since the system $A_{00}$ is decidable then we could dispense with the $A_{00}$-rules and say that an $A_{00}$-statement is an $A_{00}$-axiom if and only if it is $A_{00}$-true. The requirement that we can decide of any $A_{00}$-statement whether it is an $A_{00}$-axiom or is distinct from any $A_{00}$-axiom would then be satisfied.

## 4.9 *Negation in the system $A_{00}$*

The system $A_{00}$ lacks a negation sign, nevertheless negation can be $A_{00}$-defined as follows:

D 115 $N[``\phi"]$ for the result of replacing:

$$\begin{array}{ccc} = & \text{by} & \neq \\ \neq & \text{by} & = \\ \& & \text{by} & \vee \\ \vee & \text{by} & \& \end{array}$$

throughout $\phi$. Similarly material implication and material equivalence can be $A_{00}$-defined:

D 116 $[``\phi \to \psi"]$ for $(N[``\phi"] \vee \psi)$,

D 117 $[``\phi \leftrightarrow \psi"]$ for $((N[``\phi"] \vee \psi) \,\&\, (N[``\psi"] \vee \phi))$.

We use square brackets and inverted commas in this type of definition to denote that $\phi$ fails to occur in $N[``\phi"]$, but that $N[``\phi"]$ is effectively constructed from $\phi$. Note that $N[``N[``\phi"]"]$ is $\phi$. Thus our definition of negation is classical and different from the intuitionist negation.

PROP. 7. *$A_{00}$ is consistent and complete with respect to negation.*

We have to show that if $\phi$ is a closed $A_{00}$-statement then exactly one of $\phi$, $N[``\phi"]$ is an $A_{00}$-theorem. Let $\phi$ be a closed $A_{00}$-statement, test $\phi$ for $A_{00}$-truth. If $\phi$ is $A_{00}$-true then $\phi$ is an $A_{00}$-theorem by Prop. 3. If $\phi$ is $A_{00}$-false then $N[``\phi"]$ is $A_{00}$-true and so by Prop. 3 is an $A_{00}$-theorem. Thus $\phi$ or $N[``\phi"]$ is an $A_{00}$-theorem. It is impossible for both of them to be $A_{00}$-theorems by Prop. 1.

We can divide the closed $\mathbf{A}_{00}$-statements into two classes so that one class consists exactly of the negations of the other class. To do this we enumerate the $\mathbf{A}_{00}$-formulae by length and order those of equal length lexicographically. We then run through the list testing for being an $\mathbf{A}_{00}$-statement, when we come across a closed $\mathbf{A}_{00}$-statement we put it in the first list provided its negation is absent from the segment of that list so far obtained, otherwise we place it in the second list. We could obtain other definitions of $\mathbf{A}_{00}$-truth and $\mathbf{A}_{00}$-falsity which preserve the property of exclusiveness and such that $N[``\phi"]$ has the opposite truth-value to $\phi$ if we lay down that an $\mathbf{A}_{00}$-statement is $\mathbf{A}_{00}$-true if and only if it is in list I and is $\mathbf{A}_{00}$-false if and only if it is in list II. We should however require a new set of axioms and rules in order to obtain consistency with respect to the new truth definition. For instance interchanging $\mathbf{A}_{00}$-truth with $\mathbf{A}_{00}$-falsity we obtain:

## 4.10 *The system* $\mathbf{B}_{00}$ *(the anti-*$\mathbf{A}_{00}$*-system)*

The $\mathbf{B}_{00}$-symbols are exactly the $\mathbf{A}_{00}$-symbols, the $\mathbf{B}_{00}$-axioms are the negations of the $\mathbf{A}_{00}$-axioms, the $\mathbf{B}_{00}$-rules are:

*Remodelling rule*

I′*a* $$\frac{\omega' \,\&\, \phi \,\&\, \psi \,\&\, \omega}{\omega' \,\&\, \psi \,\&\, \phi \,\&\, \omega}.$$

*Destruction rules*

II′*a* $$\frac{\chi}{\phi \,\&\, \chi}.$$ *concentration*

*b*′ $$\frac{\phi \,\&\, \omega \quad \psi \,\&\, \omega}{(\phi \vee \psi) \,\&\, \omega}.$$ *dispersion*

R 1′ $$\frac{\phi\{\alpha\} \,\&\, \omega \quad \alpha \neq \beta \,\&\, \omega}{\phi\{\beta\} \,\&\, \omega}.$$

R 2′ $$\frac{\alpha = \beta \,\&\, \omega}{S\alpha = S\beta \,\&\, \omega}.$$

In these rules $\omega$ and $\omega'$ are subsidiary and can be absent. $\chi$ is secondary and must be present. The $\mathbf{B}_{00}$-axioms are $\mathbf{A}_{00}$-false and the $\mathbf{B}_{00}$-rules preserve $\mathbf{A}_{00}$-falsity. Thus the $\mathbf{B}_{00}$-theorems are $\mathbf{A}_{00}$-false and so $\mathbf{B}_{00}$ is consistent with respect to $\mathbf{A}_{00}$-falsity. On the other hand a closed $\mathbf{A}_{00}$-false $\mathbf{B}_{00}$-statement is a $\mathbf{B}_{00}$-theorem. Thus $\mathbf{B}_{00}$ is complete with respect to $\mathbf{A}_{00}$-falsity. To show that $\mathbf{B}_{00}$ is complete with respect to $\mathbf{A}_{00}$-falsity we use formula induction. $\phi$ is atomic and is $\mathbf{A}_{00}$-false, then

$\phi$ is an equation $\alpha = \beta$ or an inequation $\alpha \neq \beta$. In the first case $\alpha = \beta$ is $\mathbf{A}_{00}$-false and so $\alpha$ and $\beta$ determine distinct numerals $\nu$ and $\kappa$ respectively. If in the $\mathbf{A}_{00}$-proof of $\alpha = \nu$ we change each equality sign into an inequality sign then we obtain a $\mathbf{B}_{00}$-proof of $\alpha \neq \nu$, because $\mathbf{A}_{00}$-axioms become $\mathbf{B}_{00}$-axioms and an application of R 1 becomes an application of R 1′ (subsidiary formulae are absent) so that $\alpha \neq \nu$ is a $\mathbf{B}_{00}$-theorem for exactly one numeral $\nu$. We have $\mathbf{B}_{00}$-theorems $\alpha \neq \nu$, $\beta \neq \kappa$ also if $\kappa$ is $S$—$S0$ and $\nu$ is $S \ldots SS$—$S0$ then:

$$\frac{S \ldots S0 = 0 \quad (2.1')}{S\text{—}SS \ldots S0 = S\text{—}S0} \quad \text{R 2 repeatedly,}$$

i.e. $\nu = \kappa$. Hence

$$\frac{\nu = \kappa \quad \alpha \neq \nu \quad \beta \neq \kappa}{\alpha = \beta} \quad \text{R 1' repeatedly.}$$

Similarly if $\alpha \neq \beta$ is $\mathbf{A}_{00}$-false then $\alpha = \beta$ is $\mathbf{A}_{00}$-true and so $\alpha$ and $\beta$ determine the same numeral, say $\nu$. We then have:

$$\frac{\alpha \neq \nu \quad \beta \neq \nu}{\alpha \neq \beta} \;\text{R 1'}. \quad \text{Note} \quad \frac{\beta \neq \beta \quad \beta \neq \nu}{\nu \neq \beta} \;\text{R 1'}.$$

If $\phi$ is $\phi'$ & $\phi''$ and is $\mathbf{A}_{00}$-false then $\phi'$ or $\phi''$ is $\mathbf{A}_{00}$-false and hence by our induction hypothesis $\phi'$ or $\phi''$ is a $\mathbf{B}_{00}$-theorem. By II′*a* $\phi$ is a $\mathbf{B}_{00}$-theorem. If $\phi$ is $\phi' \vee \phi''$ and is $\mathbf{A}_{00}$-false then $\phi'$ and $\phi''$ are both $\mathbf{A}_{00}$-false and hence by our hypothesis are $\mathbf{B}_{00}$-theorems. By II′*b*′ $\phi' \vee \phi''$ is a $\mathbf{B}_{00}$-theorem.

Clearly the system $\mathbf{B}_{00}$ is decidable. If a $\mathbf{B}_{00}$-conjunction is a $\mathbf{B}_{00}$-theorem then so is one conjunctand and we can find which and supply its $\mathbf{B}_{00}$-proof. If a $\mathbf{B}_{00}$-disjunctand is a $\mathbf{B}_{00}$-theorem then so is each disjunctand. We shall see later that it is only because the system $\mathbf{A}_{00}$ is decidable that the system $\mathbf{B}_{00}$ is complete.

## Historical remarks to Chapter 4

A complete arithmetic was first given by Myhill (1950) who built on the basic logic of Fitch (1942). Subsequently Löb (1953) gave an equivalent system. The system $\mathbf{A}_{00}$ is a weaker but simpler system, and the system $\mathbf{A}_0$, which follows in Chs. 6 and 7 serves much the same purpose as the systems of Myhill and Löb. It is simpler in that it deals with the natural numbers rather than with sequences or chains. Myhill uses the ancestral (first occurring in *P.M.*) and Löb uses a limited universal quantifier in order to obtain primitive recursive functions. We use the iterator symbol for this purpose, this had previously been used by Goodstein (1957). Church's (1941, 1936) $\boldsymbol{\lambda}$-conversion and Curry's (1958) combinatory logic serve somewhat the same purpose as the three systems just mentioned, but they are very differently constructed and were conceived for very different reasons.

The successor symbol goes back to Peano (1897), the others we have mentioned before except the generating symbol, this has been frequently used in formal definitions of systems. It is easier reading to have many different letters for variables such as $x\ y\ z\ u\ v\ w$ and their superscripted and subscripted varieties, but this makes the definition of the system longer; later on in the book we use several different letters. Greek or Gothic letters in the metalanguage probably go back to Carnap (1937) at least.

The system $\mathbf{A}_{00}$ is set up after the fashion of Gentzen (1934, 1955). As we proceed we try to define a truth-definition for each of our systems, this gives it meaning whereas without this it is purely symbolic. Throughout the book we keep inequality as a primitive, until we have to adopt a negation symbol if we wish to express negation in the system. The proof of Prop. 1 is due to Rowbottom [by letter].

Prop. 4 is the sort of thing the intuitionists like. In fact we try to satisfy their wishes as long as possible, see Heyting (1956).

A formalization of recursive arithmetic was given by Curry (1941).

EXAMPLES 4

1. Find the numerals determined by:

$$\mathscr{I}(\boldsymbol{\lambda}x\,.\,S)\,\kappa\pi,$$
$$\mathscr{I}(\boldsymbol{\lambda}xx'\,.\,\mathscr{I}(\boldsymbol{\lambda}x''\,.\,S)\,\kappa x')\,0\pi,$$
$$\mathscr{I}(\boldsymbol{\lambda}xx'\,.\,\mathscr{I}(\boldsymbol{\lambda}x''x'''\,.\,\mathscr{I}(\boldsymbol{\lambda}x^{(\mathrm{iv})}S)\,\kappa x''')\,0x')\,1\pi,$$

for $\kappa, \pi = 0, 1, 2, 3, 4$.

2. Show that Modus Ponens is a derived rule in $\mathbf{A}_{00}$.
3. Similarly for the rule I$b$.
4. Show that if

$$\rho[\nu, \theta] = \mathscr{I}(\boldsymbol{\lambda}xx'\,.\,\sigma[\nu \dot{-} x, x'])\,\kappa\theta,$$

then $\qquad \rho[\nu, 0] = \kappa,$

$$\rho(\nu, S\theta) = \sigma[\nu \dot{-} \theta, \sigma[\nu \dot{-} (\theta \dot{-} 1), \ldots, \sigma[\nu, \kappa] \ldots]],$$

where $\qquad \nu \dot{-} \kappa = \mathscr{I}(\boldsymbol{\lambda} xx' . \mathscr{I}(\boldsymbol{\lambda} x''x''' . x'')\, 0x')\, \nu\kappa.$

5. Show that the Deduction Theorem holds in $\mathbf{A}_{oo}$.
6. Give $\mathbf{A}_{oo}$-proofs of

$$\frac{\omega' \vee \phi \vee \psi \vee \omega}{\phi \vee \omega' \vee \omega \vee \psi}, \quad \frac{\omega' \vee \phi \vee \psi \vee \omega}{\omega \vee \omega' \vee \psi \vee \phi}.$$

## Chapter 5

# $A_{00}$-Definable functions

### 5.1 *Calculable functions*

An $A_{00}$-function is an $A_{00}$-formula of type $\iota, \iota\iota, \ldots$, these are called one-, two-, ...place $A_{00}$-functions. If $\rho$ is a one-place $A_{00}$-function and if $\nu$ is a numeral then $(\rho\nu)$ is a numerical term and determines a unique numeral called the value of the function $\rho$ for the argument $\nu$. Similarly if $\rho$ is a two-place $A_{00}$-function and if $\nu$ and $\kappa$ are numerals then $((\rho\nu)\kappa)$ is a numerical term and so determines a unique numeral called the value of the function $\rho$ for the arguments $\nu$ and $\kappa$, in that order. Similarly for many-place functions. A *calculable function* of natural numbers is a rule or set of rules such that given the ordered argument set we can by following the rules effectively find a natural number called the value of the function for that argument set, the value must be unique and a given function always requires the same number of arguments. Thus the $A_{00}$-functions are a particular kind of calculable function in that the rules for finding the value given the argument set is of a particular kind. For example a set of instructions to find the value of a function of natural numbers might be: replace $(S(S\nu))$ by $\nu$ as long as possible, this would replace an even number by 0 and an odd number by $(S0)$. Clearly we could always choose or alter our notation so that the natural numbers are represented as we have done in the system $A_{00}$. We shall say that a calculable function $f$ of natural numbers is $A_{00}$-definable if there is an $A_{00}$-formula $\rho$ of type $\underbrace{\iota \ldots \iota}_{S\pi\text{-times}}$, where $\pi$ is the number of arguments of the function, such that $\rho\nu' \ldots \nu^{(\pi)}$ determines the numeral $\nu$ when and only when this is the value of the function $f$ for the arguments $\nu' \ldots \nu^{(\pi)}$. We want to investigate the kind of calculable functions that are $A_{00}$-definable. There are various known classes of calculable functions defined according to the kind of rules allowed in the calculation of the value of the function. Many of these arose from attempts to make a rigorous definition of 'calculable'. One of the earliest such definitions is the definition of *primitive recursive function*. It was, for many years, thought that the class of primitive recursive functions was co-extensive with the class of calculable functions, but Ackermann produced an example of a calculable function outside the class of primitive recursive functions. Another method due to Turing produces a class of calculable functions more extensive than the class of primitive recursive functions and including the example of Ackermann; this method is thought to form an exact definition of the vague intuitive concept of calculable function. But how does one demonstrate that an exactly defined concept is the same as a vague intuitive concept?

## 5.2 *Primitive recursive functions*

The primitive recursive functions are obtained from certain *initial functions* by applications of certain schemes which produce new functions from already acquired functions. The initial functions are:

(i) the *successor function* $S$;

(ii) the *constant function zero*.

D 118 $$O_{(S\pi)} \quad \text{for} \quad \boldsymbol{\lambda} x' \ldots x^{(S\pi)} 0:$$

(iii) the *identity functions*:

D 119 $$U'_{(S\pi)} \quad \text{for} \quad \boldsymbol{\lambda} x' \ldots x^{(S\pi)} . x', \ldots, U^{(S\pi)}_{(S\pi)} \quad \text{for} \quad \boldsymbol{\lambda} x' \ldots x^{(S\pi)} . x^{(S\pi)}.$$

Then $$U^{(\theta)}_{(S\pi)} x' \ldots x^{(S\pi)} = x^{(\theta)}, \quad 1 \leqslant \theta \leqslant S\pi.$$

The schemes are:

(i) the *scheme of substitution*:

D 120 $(\sigma/[\tau', \ldots, \tau^{(S\theta)}])_{(S\pi)}$, for

$$\boldsymbol{\lambda} x' \ldots x^{(S\pi)} . \sigma(\tau' x' \ldots x^{(S\pi)}) \ldots (\tau^{(S\theta)} x' \ldots x^{(S\pi)});$$

where the type of $\sigma$ is $\underbrace{\iota \ldots \iota}_{SS\theta\text{-times}}$. Thus

$$(\sigma/[\tau', \ldots, \tau^{(S\theta)}])_{(S\pi)} x' \ldots x^{(S\pi)} = \sigma(\tau' x' \ldots x^{(S\pi)}) \ldots (\tau^{(S\theta)} x' \ldots x^{(S\pi)}).$$

Note that we substitute the undetermined values of functions with the same number of arguments; this is quite general because otherwise, by use of the identity functions this requirement can always be satisfied. Thus

$$\rho x' x'' = \rho U'_{(S\pi)} x' \ldots x^{(S\pi)} U''_{(S\pi)} x' \ldots x^{(S\pi)}, \quad \text{etc.}$$

(ii) the *scheme of primitive recursion*:

D 121 $\widehat{\tau\sigma}_{(\pi)}$ for

$$\boldsymbol{\lambda} xx' \ldots x^{(\pi)} . \mathscr{I}(\boldsymbol{\lambda} x^{(S\pi)} x^{(SS\pi)} . \tau x^{(S\pi)} x^{(SS\pi)} x' \ldots x^{(\pi)}) (\sigma x' \ldots x^{(\pi)}) x,$$

where $\sigma$ is of type $\underbrace{\iota \ldots \iota}_{S\pi\text{-times}}$ and the type of $\tau$ is $\underbrace{\iota \ldots \iota}_{SSS\pi\text{-times}}$; if $\pi = 0$ then replace $(\sigma x' \ldots x^{(\pi)})$ by $\alpha$.

From this we obtain by Ax. 3.1, 3.2:

$$\widehat{\tau\sigma}_{(\pi)} 0 \nu' \ldots \nu^{(\pi)} = \mathscr{I}(\boldsymbol{\lambda} xx' . \tau xx' \nu' \ldots \nu^{(\pi)}) (\sigma \nu' \ldots \nu^{(\pi)}) 0$$
$$= \sigma \nu' \ldots \nu^{(\pi)}, \tag{3.1}$$

$$\widehat{\tau\sigma}_{(\pi)} (S\nu) \nu' \ldots \nu^{(\pi)} = \mathscr{I}(\boldsymbol{\lambda} xx' . \tau xx' \nu' \ldots \nu^{(\pi)}) (\sigma \nu' \ldots \nu^{(\pi)}) (S\nu)$$
$$= \tau\nu(\mathscr{I}(\boldsymbol{\lambda} xx' . \tau xx' \nu' \ldots \nu^{(\pi)}) (\sigma \nu' \ldots \nu^{(\pi)}) \nu)$$
$$= \tau\nu(\widehat{\tau\sigma}_{(\pi)} \nu \nu' \ldots \nu^{(\pi)}). \tag{3.2}$$

This is the scheme of primitive recursion with $\pi$ *parameters*. The scheme of primitive recursion without parameters is:

$$\widehat{\tau\sigma}_{(0)} \quad \text{for} \quad \boldsymbol{\lambda} x . \mathscr{I}(\lambda x' x'' . \tau x' x'') \alpha x,$$

so that $$\widehat{\tau\sigma}_{(0)} 0 = \alpha \quad \text{and} \quad \widehat{\tau\sigma}_{(0)} (S\nu) = \tau\nu\widehat{\tau\sigma}_{(0)}\nu.$$

Thus starting with the initial functions and applying the schemes of substitution and primitive recursion repeatedly we generate a class of functions. Each primitive recursive function has a *construction sequence* showing how it is built up from the initial functions by the schemes.

PROP. 1. *A primitive recursive function is* $\mathbf{A}_{oo}$*-definable.*

We have just given the required details and definitions, i.e. we have $\mathbf{A}_{oo}$-defined the initial functions: successor, constant and identity functions and have $\mathbf{A}_{oo}$-defined functions obtained from already acquired functions by the schemes of substitution and primitive recursion.

PROP. 2. *A closed* $\mathbf{A}_{oo}$*-function is primitive recursive.*

I.e. a closed $\mathbf{A}_{oo}$-function can be obtained from the initial functions by the schemes of substitution and of primitive recursion.

A closed $\mathbf{A}_{oo}$-function $\rho\{\mathfrak{k}\}$ is called $\boldsymbol{\lambda}$–0-*reduced* if the $\mathbf{A}_{oo}$-function-form $\rho\{\mathfrak{G}\}$ is without occurrences of zero and without parts of the form $((\boldsymbol{\lambda}\xi . \alpha\{\xi\xi\}) \beta)$. If we had allowed other occurrences of the abstraction symbol then we could apply the $\boldsymbol{\lambda}$-axiom and eliminate them. The order of a $\boldsymbol{\lambda}$–0-reduced $\mathbf{A}_{00}$-function is defined as follows: strike out every symbol in the term except the iterator symbol, then replace each iterator symbol by the successor symbol add zero on the right and parentheses by association to the right, the resulting numeral is the required order. If a $\boldsymbol{\lambda}$–0-reduced $\mathbf{A}_{00}$-function is without occurrences of the iterator symbol then its order is to be zero, and the function is either the successor function, a constant function zero or an identity function or the result of applying the successor function to either of these functions. These functions are primitive recursive being either initial functions or functions obtained from the initial functions by the scheme of substitution. Any other $\boldsymbol{\lambda}$–0-reduced $\mathbf{A}_{00}$-function is of the form:

$$S \dots S\mathscr{I}\rho\{\mathfrak{k}\}\gamma\{\mathfrak{k}\}\beta\{\mathfrak{k}\},$$

where $\boldsymbol{\lambda}\mathfrak{x}\beta\{\mathfrak{x}\}$, $\boldsymbol{\lambda}\mathfrak{x}\gamma\{\mathfrak{x}\}$ and $\rho\{\mathfrak{k}\}$ are $\boldsymbol{\lambda}$–0-reduced $\mathbf{A}_{00}$-functions. Clearly if $\rho\{\mathfrak{k}\}$ is a recursive function then $\mathscr{I}\rho\{\mathfrak{k}\}\nu'\nu''$ is primitive recursive in $\nu''$ with $\mathfrak{k}$, $\nu'$ as parameters by Ax. 3.1, 3.2. Thus if $\boldsymbol{\lambda}\mathfrak{x}\beta\{\mathfrak{x}\}$ and $\boldsymbol{\lambda}\mathfrak{x}\gamma\{\mathfrak{x}\}$ are primitive recursive functions then by the rule of substitution so is (i). Now $\boldsymbol{\lambda}\mathfrak{x}\beta\{\mathfrak{x}\}$, $\boldsymbol{\lambda}\mathfrak{x}\gamma\{\mathfrak{x}\}$ and $\rho\{\mathfrak{k}\}$ are of lower order than (i), and the result follows. An $\mathbf{A}_{00}$-function is *equivalent* to a $\boldsymbol{\lambda}$–0-reduced $\mathbf{A}_{00}$-function, i.e. takes the same value for the same argument.

COR. (i). *Given a closed* $\mathbf{A}_{00}$*-function we can find the primitive recursion scheme of an equivalent function.*

The necessary details have been given in Props. 1 and 2.

## 5·3 *Definitions of particular primitive recursive functions*

We now give definitions of some primitive recursive functions leading up to a primitive recursive ordering of all terminating sequences of natural numbers. An ordered sequence of natural numbers will be represented by its place number in this ordering. Zero will represent the null sequence of natural numbers. The place number of the ordered sequence $\nu', \dots, \nu^{(S\pi)}$ will be $\mathbf{A}_{00}$-represented by the numeral determined by the function $\langle \nu', \dots, \nu^{(S\pi)} \rangle$. We also give a primitive recursive ordering of all $(S\pi)$-tuplets of natural numbers for each natural number $\pi$. An $(S\pi)$-tuplet will be $\mathbf{A}_{00}$-represented by its place number in this ordering, the place number of the ordered $(S\pi)$-tuplet $\nu', \dots, \nu^{(S\pi)}$ in the ordering of all $(S\pi)$-tuplets will be $\mathbf{A}_{00}$-represented by $\{\nu', \dots, \nu^{(S\pi)}\}$.

In ordinary mathematical notation the connexion between these orderings is given by

$$\langle \nu', \ldots, \nu^{(S\pi)} \rangle = 2^{\pi}.(2.\{\nu', \ldots, \nu^{(S\pi)}\}+1),$$

where

$$\{\nu', \ldots, \nu^{(S\pi)}\}$$

$$= \frac{N_1.(N_1+1)\ldots(N_1+\pi)}{(S\pi)!} + \frac{N_2.(N_2+1)\ldots(N_2+\pi-1)}{\pi!} + \ldots + N_{(S\pi)},$$

and $\quad N_1 = \nu'+\ldots+\nu^{(S\pi)}, \quad N_2 = \nu''+\ldots+\nu^{(S\pi)}, \ldots, N_{(S\pi)} = \nu^{(S\pi)}.$

In particular $\langle \nu \rangle = 2.\nu+1$.

It is easily seen that each natural number $\nu$ is uniquely expressible in the form $\{\nu', \ldots, \nu^{(S\pi)}\}$ since

$$\sum_{\nu', \ldots, \nu^{(S\pi)}} x^{\{\nu', \ldots, \nu^{(S\pi)}\}} = \sum_{0 \leqslant N_{(S\pi)} \leqslant N\pi \leqslant \ldots \leqslant N_1} x^{\{\nu', \ldots, \nu^{(S\pi)}\}} = \frac{1}{1-x} \quad \text{for} \quad x \leqslant 1.$$

Thus a numeral $\mathbf{A}_{00}$ represents either a natural number or an $(S\pi)$-tuplet of natural numbers or an ordered set of natural numbers in a numbering of all terminating sequences of natural numbers. The context will make it clear which is intended.

Also given a natural number $\nu$ we define primitive recursive functions, called *coordinate functions*, *pt* and *Pt* such that

$$pt[\nu, S\pi, S\kappa] = \nu^{(S\kappa)}, \quad \text{where} \quad \nu = \{\nu', \ldots, \nu^{(S\pi)}\} \quad \text{and} \quad \kappa \leqslant \pi,$$

$$Pt[\nu, S\kappa] = \nu^{(S\kappa)}, \quad \text{where} \quad \nu = \langle \nu', \ldots, \nu^{(S\pi)} \rangle \quad \text{and} \quad \kappa \leqslant \pi,$$

in the last case $\pi$ is the greatest power of 2 which divides $\nu$.

We have

$$\nu = \{pt[\nu, S\pi, 1], \ldots, pt[\nu, S\pi, S\pi]\},$$

and

$$\nu = \langle Pt[\nu, 1], \ldots, Pt[\nu, S\pi] \rangle,$$

in this case $\pi$ is as described above.

D 122 $\quad$ 1 for $(S0)$, $\quad$ 2 for $(S(S0))$, $\quad \ldots,$ $\quad$ 9 for $(S8)$.

D 123 $\quad (\alpha+\beta)$ for $\mathscr{I}(\boldsymbol{\lambda}\xi S)\alpha\beta$,

the result of adding $\beta$ on the right to $\alpha$ is the result of applying $\beta$-times the successor function to $\alpha$. + is called the *addition function*.

Instead of D 123 we could have had + for $\widehat{(S/U^2_3)_3\, U^1_1}$. Then $+\beta\alpha = (\alpha+\beta)$.

Ax. 3.1, 3.2 give

$$(\alpha+0) = \alpha, \quad (\alpha+S\beta) = S(\alpha+\beta).$$

D 124 $\quad (\alpha \times \beta)$ or $(\alpha . \beta)$ for $\mathscr{I}(\boldsymbol{\lambda}\xi\eta . (\alpha+\eta))\, 0\beta.$

The result of multiplying $\alpha$ by $\beta$ on the right is the result of applying $\beta$-times addition of $\alpha$ on the left to zero. $\times$ is called the *multiplication function.*

Ax. 3.1, 3.2 give

$$(\alpha \times 0) = 0, \quad (\alpha \times S\beta) = \alpha + (\alpha \times \beta).$$

Instead of D 124 we could have had $\times$ for $(\widehat{+/U^3_3 U^2_3)_3 O}_1$.

D 125 $\quad \alpha^\beta$ or $(\alpha\, exp\, \beta)$ for $\mathscr{I}(\boldsymbol{\lambda}\xi\eta\alpha \times \eta)\, 1\beta,$

The result of raising $\alpha$ to the power $\beta$ is the result of applying $\beta$-times multiplication by $\alpha$ on the left to 1. *exp* is called the *exponentiation or power function.*

Ax. 3.1, 3.2 give

$$\alpha^0 = 1, \quad \alpha^{S\beta} = \alpha \times \alpha^\beta.$$

D 126 $\quad !\alpha$ for $\mathscr{I}(\boldsymbol{\lambda}\xi\eta . (S\xi) \times \eta)\, 1\alpha.$

In these definitions $\alpha$ and $\beta$ are numerical terms in which $\xi$ and $\eta$ fail to occur free. ! is called the *factorial function.*

D 127 $\quad P$ for $\boldsymbol{\lambda}\zeta . \mathscr{I}(\lambda\xi\eta . \xi)\, 0\zeta,$

then $\quad P0 = 0, \quad P(S\alpha) = \alpha.$

$P$ is called the *predecessor function.*

D 128 $\quad (\alpha \dot{-} \beta)$ for $\mathscr{I}(\boldsymbol{\lambda}\xi\eta . P\eta)\, \alpha\beta,$

then $\quad (\alpha \dot{-} 0) = \alpha, \quad (\alpha \dot{-} S\beta) = P(\alpha \dot{-} \beta),$

this is a *limited form of subtraction.* We have $(\alpha \dot{-} 1) = P(\alpha \dot{-} 0) = P\alpha.$

D 129 $\quad (\alpha < \beta)$ for $(\beta \dot{-} \alpha) \neq 0, \quad (\alpha > \beta)$ for $(\beta < \alpha).$

D 130 $\quad (\alpha \leqslant \beta)$ for $(\alpha \dot{-} \beta) = 0, \quad (\alpha \geqslant \beta)$ for $(\beta \leqslant \alpha).$

Instead of $(\beta \dot{-} \alpha) \neq 0$ in D 129 we could have had $(S\alpha \dot{-} \beta) = 0$. In this manner the concept of *order* is $\mathbf{A}_{00}$-defined. Using these definitions we can $\mathbf{A}_{00}$-prove the familiar properties of addition, multiplication and exponentiation and of order thus:

$$(\alpha+\beta) = (\beta+\alpha), \quad \frac{(\alpha < \beta) \quad (\beta < \gamma)}{(\alpha < \gamma)}*, \quad \text{etc.}$$

We omit the details because we have given more general results in Ch. 4 from which they can be obtained. Namely an $\mathbf{A}_{00}$-true $\mathbf{A}_{00}$-statement is an $\mathbf{A}_{00}$-theorem. For instance the closed numerical terms $\alpha$ and $\beta$ determine unique numerals, say $S\ldots S0$ and $S{-}{-}{-}S0$ respectively. Hence $(\alpha+\beta) = S\ldots S0 + S{-}{-}{-}S0$, by repeatedly using $(\gamma + S\delta) = S(\gamma+\delta)$ and then $(\gamma+0) = \gamma$ we get $(\alpha+\beta) = S{-}{-}{-}SS\ldots S0$, similarly $(\beta+\alpha) = S\ldots SS{-}{-}{-}S0$, hence, clearly $(\alpha+\beta) = (\beta+\alpha)$. To complement rule R 2 we have

$$\frac{S\alpha = S\beta}{\overline{\alpha = \beta}}.$$

For

$$\cfrac{\cfrac{\cfrac{PS\alpha = PS\alpha \qquad S\alpha = S\beta}{PS\alpha = PS\beta \quad PS\alpha = \alpha}\ \text{Ax. 1, R 1.}}{\alpha = PS\beta \quad PS\beta = \beta}\ \text{Ax. 3.2, R 1.}}{\alpha = \beta}\ \text{Ax. 3.2, R 1.}$$

We also have for closed numerical terms $\alpha$ and $\beta$:

$$\frac{(\alpha+\beta) = 0}{\alpha = 0 \quad \beta = 0}*, \quad \frac{\alpha = 0 \quad \beta = 0}{\overline{(\alpha+\beta) = 0}}.$$

When we define primitive recursive functions we need only show how they are defined from previously defined primitive recursive functions by substitution and the scheme of primitive recursion because we have already given complete instructions to obtain their explicit definition by an $\mathbf{A}_{00}$-function. Thus

D 131 $$A_1[0] = 0, \quad A_1[(S\nu)] = 1.$$

This is easily put in the form of the scheme of recursion without parameters, thus
$$A_1[0] = 0, \quad A_1[S\nu] = SO_2[\nu, A_1[\nu]].$$

Alternative definitions for $A_1[\nu]$ are $(1 \dot{-} (1 \dot{-} \nu))$, $(\nu \dot{-} P\nu)$, $\mathscr{I}(\boldsymbol{\lambda}\xi\eta\,.\,1)\,0\nu$.

D 132 $$B_1[\nu] \quad \text{for} \quad (1 \dot{-} A_1[\nu]).$$

Alternative definitions for $B_1[\nu]$ are $\mathscr{I}(\boldsymbol{\lambda}\xi\eta.0)\,1\nu$, $(1 \dot{-} \nu)$.

D 133 $\qquad A_2[\kappa,\nu]$ for $A_1[(\kappa \dot{-} \nu)+(\nu \dot{-} \kappa)]$.

D 134 $\quad B_2[\kappa,\nu]$ for $(1 \dot{-} A_2[\kappa,\nu])$, i.e. $B_1[(\kappa \dot{-} \nu)+(\nu \dot{-} \kappa)]$.

We have:

$$\frac{\kappa=\nu}{A_2[\kappa,\nu]=0}, \quad \frac{\kappa=\nu}{B_2[\kappa,\nu]=1}, \quad \frac{\kappa\neq\nu}{A_2[\kappa,\nu]=1}*, \quad \frac{\kappa\neq\nu}{B_2[\kappa,\nu]=0}*,$$

$$\frac{A_2[\kappa,\nu]=0}{\kappa=\nu}*, \quad \frac{B_2[\kappa,\nu]=1}{\kappa=\nu}*, \quad \frac{A_2[\kappa,\nu]=1}{\kappa\neq\nu}*, \quad \frac{B_2[\kappa,\nu]=0}{\kappa\neq\nu}*,$$

$\dfrac{A_1[\alpha]=0}{\alpha=0}*$ because $\alpha$ determines a unique numeral, suppose $\alpha = S\nu$ for some numeral $\nu$, then

$$\frac{A_1[\alpha]=0 \quad \alpha=S\nu}{A_1[S\nu]=0} \ \mathrm{R}\,1,$$

but this is absurd by Prop. 2, Ch. 3, hence $\alpha$ determines 0 and so $\alpha = 0$ is $\mathbf{A}_{00}$-provable.

Suppose the function $\tau$ of type $\iota$ has already been defined then we define:

D 135 $\qquad (\Sigma\tau)\,0 = \tau 0, \quad (\Sigma\tau)\,(S\nu) = (\Sigma\tau)\,\nu + \tau(S\nu).$

This gives $\qquad (\Sigma\tau)\,\nu = \mathscr{I}(\boldsymbol{\lambda}\xi\eta.(\tau(S\xi)+\eta))\,(\tau 0)\,\nu.$

We shall sometimes write:

$$\sum_{0\leqslant\xi\leqslant\nu}\tau\xi \ \text{ for } (\Sigma\tau)\nu, \ \text{ and } \ (\sum_{\kappa\leqslant\xi\leqslant\nu}\tau\xi) \ \text{ for } \ B_1[\kappa \dot{-} \nu]\times \sum_{0\leqslant\xi\leqslant(\nu\dot{-}\kappa)}\tau(\kappa+\xi),$$

so that
$$\sum_{\kappa\leqslant\xi\leqslant\nu}\tau = 0 \quad \text{if} \quad \nu<\kappa.$$

Note that $\Sigma$ acts like a symbol of type $\iota(\iota)$, it is called the *summation function.*

D 136 $\qquad (\Pi\tau)\,0 = \tau 0, \quad (\Pi\tau)\,(S\nu) = (\Pi\tau)\,\nu \times \tau(S\nu).$

This gives $\qquad (\Pi\tau)\,\nu = \mathscr{I}(\boldsymbol{\lambda}\xi\eta.(\tau(S\xi)\times\eta))\,(\tau 0)\,\nu.$

We shall sometimes write

$$\prod_{0\leqslant\xi\leqslant\nu}(\tau\xi) \quad \text{for} \quad (\Pi\tau)\,\nu$$

and $\qquad \displaystyle\prod_{\kappa\leqslant\xi\leqslant\nu}(\tau\xi)$ for $B_1[\kappa \dot{-} \nu]\times(\displaystyle\prod_{0\leqslant\xi\leqslant(\nu\dot{-}\kappa)}\tau)(\kappa+\xi)+A_1[\kappa \dot{-} \nu].$

$\Pi$ is called the *product function.*

D 137 $\quad Max[\nu, \kappa]$ for $(\nu \dot{-} \kappa) + \kappa$.

D 138 $\quad Min[\nu, \kappa]$ for $(\kappa \dot{-} (\kappa \dot{-} \nu))$.

D 139 $\quad (Max\,\rho)\,0 = 0,$

$$(Max\,\rho)\,S\nu = Max[(Max\,\rho)\,\nu, \rho(S\nu)],$$

where $\rho$ is a one-place function. Then

$$(Max\,\rho)\,\nu = \mathscr{I}(\lambda\xi\eta\,.\,Max[\rho(S\nu, \eta)])\,(\rho 0)\,\nu.$$

We often write $\underset{0\leqslant\xi\leqslant\nu}{Max}[\rho\xi]$ for $(Max\,\rho)\,\nu$.

D 140 $\quad (Min\,\rho)\,0 = \rho 0,$

$$(Min\,\rho)\,S\nu = Min[(Min\,\rho)\,\nu, \rho(S\nu)].$$

We often write $\underset{0\leqslant\xi\leqslant\nu}{Min}[\rho\xi]$ for $(Min\,\rho)\,\nu$.

D 141 $\quad \underset{0\leqslant\xi\leqslant\nu}{Max}(\alpha\{\xi\} = 0)$ for $\underset{0\leqslant\xi\leqslant\nu}{Max}[\xi\,.\,B_1[\alpha\{\xi\}]]$.

D 142 $\quad \underset{0\leqslant\xi\leqslant\nu}{Min}(\alpha\{\xi\} = 0)$ for $\underset{0\leqslant\xi\leqslant\nu}{Min}[\xi\,.\,B_1[\alpha\{\xi\}]]$.

D 143 $\quad \underset{0\leqslant\xi\leqslant\nu}{Max}(\alpha\{\xi\} \leqslant \kappa)$ for $\underset{0\leqslant\xi\leqslant\nu}{Max}[\xi\,.\,A_1[S\kappa \dot{-} \alpha\{\xi\}]]$.

D 144 $\quad \underset{0\leqslant\xi\leqslant\nu}{Min}(\alpha\{\xi\} \leqslant \kappa)$ for $\underset{0\leqslant\xi\leqslant\nu}{Min}[\xi\,.\,A_1[S\kappa \dot{-} \alpha\{\xi\}]]$.

We sometimes write

$$\underset{\kappa\leqslant\xi\leqslant\nu}{Max}[\rho\xi] \quad \text{for} \quad \underset{0\leqslant\xi\leqslant(\nu\dot{-}\kappa)}{Max}[\rho(\kappa+\xi)],$$

and similarly for other cases from D 141 ... 144 inclusive.

D 145 $\quad [\kappa/\nu]$ for $\underset{0\leqslant\xi\leqslant\kappa}{Max}[\xi\,.\,B_1[\xi\,.\,\nu \dot{-} \kappa]]$.

The *quotient* on dividing $\nu$ into $\kappa$.

D 146 $\quad \varpi\nu$ for $A_1[\nu]\,.\,S\underset{0\leqslant\xi\leqslant\nu}{Max}(2^{\xi}\,.\,(2\,.\,[\nu/2^{S\xi}]+1) = \nu)$.

The value of $\varpi\nu$ is the successor of the greatest power of 2 which divides $\nu$, but $\varpi 0 = 0$.

D 147 $\quad D[\nu, \kappa]$ for $\underset{0<\xi\leqslant\kappa}{Max}[B_2[\nu\times\xi, \kappa]]$.

Then $D[\nu,\kappa] = 1$ if and only if $\nu$ divides $\kappa$. We have $D[S\nu, 0] = 1$, $D[0, S\kappa] = 0$.

D 148 $$Rem\,[\nu,\kappa] \quad \text{for} \quad (\kappa \dot{-} [\kappa/\nu].\nu).$$

The *remainder* on dividing $\nu$ into $\kappa$. We have $Rem\,[\nu,0] = 0$, $Rem\,[0,\kappa] = \kappa$ and $\kappa = [\kappa/\nu].\nu + Rem\,[\nu,\kappa]$.

D 149 $$\Delta[\nu,\kappa] \quad \text{for} \quad [!((\nu+\kappa) \dot{-} 1)/!(\nu \dot{-} 1).!\kappa].A_1[\kappa].A_1[\nu].$$

We have $\Delta[0,\kappa] = \Delta[\nu,0] = 0$, $\Delta[\nu,1] = \nu$.

D 150 $$\sigma_2[\nu,\kappa] \quad \text{for} \quad \underset{0 \leqslant \xi \leqslant \nu}{Max}(\Delta[\xi,\kappa] \leqslant \nu).A_1[\kappa].$$

Then $\sigma_2[\nu,\kappa]$ is such that

$$\Delta[\sigma_2[\nu,\kappa],\kappa] \leqslant \nu < \Delta[S\sigma_2[\nu,\kappa],\kappa] \quad \text{for positive } \kappa.$$

As $\nu$ increases $\sigma_2[\nu,\kappa]$ increases by unity each time $\nu$ is the value of $\Delta[\pi,\nu]$ for some numeral $\pi$, provided that $\kappa$ is positive. We have $\sigma_2[\nu,0] = 0$, $\sigma_2[\nu,1] = \nu$.

D 151 $$\begin{cases} N[\nu,\kappa,0] = \nu, \\ N[\nu,\kappa,S\pi] = N[\nu,\kappa,\pi] \dot{-} \Delta[\sigma_2[N[\nu,\kappa,\nu],(\kappa \dot{-} \pi)],(\kappa \dot{-} \pi)]. \end{cases}$$

D 152 $$\sigma_3[\nu,\kappa,\pi] \quad \text{for} \quad \sigma_2[N[\nu,\kappa,\pi],(\kappa \dot{-} \pi)].$$

We have $\sigma_3[\nu,\kappa,\pi] = 0$ for $\pi \geqslant \kappa$, also $\sigma_3[0,\kappa,\pi] = \sigma_3[\nu,0,\pi] = 0$.

In ordinary mathematical notation, if

$$\theta' \geqslant \theta'' \geqslant \ldots \geqslant \theta^{(\kappa)}$$

and $$\nu = \theta'.(\theta'+1)\ldots(\theta'+\kappa-1)/!\kappa + \theta''.(\theta''+1)\ldots(\theta''+\kappa-2)/\,!(\kappa-1) + \ldots + \theta^{(\kappa)}/!1$$

then

$$N[\nu,\kappa,\pi] = \theta^{(S\pi)}.(\theta^{(S\pi)}+1)\ldots(\theta^{(S\pi)}+\kappa-S\pi)/!(\kappa-\pi) + \ldots + \theta^{(\kappa)}/!1, \quad \pi < \kappa,$$

also $$\sigma_3[\nu,\kappa,\pi] = \theta^{(S\pi)}, \quad \sigma_3[\nu,\kappa,0] = \sigma_2[\nu,\kappa] = \theta'.$$

Thus if

$$\nu = \frac{\pi'.(\pi'+1)\ldots(\pi'+\kappa-1)}{!\kappa} + \frac{\pi''.(\pi''+1)\ldots(\pi''+\kappa-2)}{!(\kappa-1)} + \ldots + \frac{\pi^{(\kappa)}}{!1},$$

then $\pi' = \sigma_2[\nu,\pi]$, so $\pi' = \theta'$. Similarly

$$\pi'' = \theta'' = \underset{0 \leqslant \xi \leqslant \nu}{Max}(\Delta[\xi,\kappa-1]) \leqslant \nu - \Delta[\theta',\kappa],$$

and so on. Thus $\pi' = \theta', \pi'' = \theta'', \ldots, \pi^{(\kappa)} = \theta^{(\kappa)}$. Thus if

$$\{\mu', \ldots, \mu^{(\kappa)}\} = \{\lambda', \ldots, \lambda^{(\kappa)}\} \quad \text{then} \quad \mu' = \lambda', \ldots, \mu^{(\kappa)} = \lambda^{(\kappa)}.$$

D 153 $\qquad pt[\nu, \kappa, \pi]$ for $(\sigma_3[\nu, \kappa, (\pi \dot{-} 1)] \dot{-} \sigma_3[\nu, \kappa, \pi])$.

Then $pt[\nu, \kappa, \pi] = \theta^{(\pi)} - \theta^{(S\pi)}$, for $1 \leqslant \pi \leqslant \kappa$, and $pt[\nu, \kappa, \kappa] = \theta^{(\kappa)}$, in the above notation, if $\kappa$ is positive. We have $pt[\nu, \kappa, \pi] = 0$ for $\pi > \kappa$, also $pt[0, \kappa, \pi] = pt[\nu, 0, \pi] = pt[\nu, \kappa, 0] = 0$. $pt$ is called a *coordinate function.* This formalizes the informal discussion of coordinate functions given earlier in this chapter.

D 154 $\qquad Pt[\nu, \pi]$ for $pt[[\nu/2^{\varpi\nu}], \varpi\nu, \pi]$.

If $\nu = 2^{(\kappa \dot{-} 1)}.(2.\theta + 1)$ then $Pt[\nu, \pi] = pt[\theta, \kappa, \pi]$. We have $Pt[\nu, \pi] = 0$ if $\pi > \varpi\nu$. Also $Pt[0, \pi] = Pt[\nu, 0] = 0$. $Pt$ is also called a *coordinate function.*

If the one-place function $\rho$ has already been defined then:

D 155 $\qquad \underset{1 \leqslant \xi \leqslant \kappa}{\{\rho\xi\}}$ for $\underset{1 \leqslant \xi \leqslant \kappa}{\Sigma} \Delta[\underset{\xi \leqslant \eta \leqslant \kappa}{\Sigma} \rho\eta, (\kappa \dot{-} (\xi \dot{-} 1))]$.

In ordinary mathematical notation, for $1 \leqslant \kappa$,

$$\underset{1 \leqslant \xi \leqslant k}{\{\rho\xi\}} = \frac{\left(\sum_{\xi=1}^{\kappa} \rho\xi\right) . \left(\sum_{\xi=1}^{\kappa} \rho\xi + 1\right) \ldots \left(\sum_{\xi=1}^{\kappa} \rho\xi + \kappa - 1\right)}{!\kappa}$$

$$+ \frac{\left(\sum_{\xi=2}^{\kappa} \rho\xi\right) . \left(\sum_{\xi=2}^{\kappa} \rho\xi + 1\right) \ldots \left(\sum_{\xi=2}^{\kappa} \rho\xi + \kappa - 2\right)}{!\kappa - 1} + \ldots + \rho\kappa.$$

D 156 $\qquad \underset{1 \leqslant \xi \leqslant \kappa}{\langle\rho\xi\rangle}$ for $2^{(\kappa \dot{-} 1)}.(2.\underset{1 \leqslant \xi \leqslant \kappa}{\{\rho\xi\}} + 1).A_1[\kappa]$.

Then $\underset{1 \leqslant \xi \leqslant 0}{\{\rho\xi\}} = \underset{1 \leqslant \xi \leqslant 0}{\langle\rho\xi\rangle} = 0$. We sometimes write: $\langle\rho 1, \rho 2, \ldots, \rho\kappa\rangle$ for $\underset{1 \leqslant \xi \leqslant \kappa}{\langle\rho\xi\rangle}$.

Thus $\underset{1 \leqslant x \leqslant 6}{\langle 9 \rangle} = \langle 9, 9, 9, 9, 9, 9\rangle$.

D 157 $(\nu^\frown\kappa)$ for $\underset{1 < \xi < \varpi\nu + \varpi\kappa}{\langle Pt[\nu, \xi].B_1[\xi \dot{-} \varpi\nu] + Pt[\kappa, \xi \dot{-} \varpi\nu].A_1[\xi \dot{-} \varpi\nu]\rangle}$.

($^\frown$ is distinct from the concatenation sign of Ch. 1.) This is the place number of the ordered sequence of natural numbers obtained by adding the $\kappa$th ordered sequence to the right of the $\nu$th ordered sequence. We have $\langle 0^\frown\kappa\rangle = \langle\kappa^\frown 0\rangle = \kappa$.

D 158
$$\underset{1 \leqslant \xi \leqslant 0}{\widehat{\rho\xi}} = 0,$$

$$\underset{1 < \xi < S\nu}{\widehat{\rho\xi}} = (\widehat{\rho\xi}^\frown \rho(S\nu)).$$

We write $\underset{\kappa\leqslant\xi\leqslant\nu}{\widehat{\rho\xi}}$ for $\underset{1\leqslant\eta\leqslant(\nu\dot{-}\kappa)}{\rho((\kappa\dot{-}1)+\eta)}$.

*Definition by cases*

We wish to define a function $\rho[\nu,\kappa]$ such that

$$\rho[\nu,\kappa] = \begin{cases} \sigma[\nu,\kappa] & \text{if} \quad \tau\nu = 0, \\ \sigma'[\nu,\kappa] & \text{if} \quad \tau'\nu = 0, \\ \sigma''[\nu,\kappa] & \text{if} \quad \tau''\nu = 0, \end{cases}$$

where

$$0 < \tau'\nu+\tau''\nu, \quad 0 < \tau''\nu+\tau\nu, \quad 0 < \tau\nu+\tau'\nu, \quad \tau\nu.\tau'\nu.\tau''\nu = 0,$$

so that exactly one of $\tau\nu$, $\tau'\nu$, $\tau''\nu$ is zero. The $\rho[\nu,\kappa]$ is defined by

$$\sigma[\nu,\kappa].B_1[\tau\nu]+\sigma'[\nu,\kappa].B_1[\tau'\nu]+\sigma''[\nu,\kappa].B_1[\tau''\nu].$$

Similarly when there are more cases.

## 5·4 *Characteristic functions*

PROP. 3. *Let $\phi\{\xi',\ldots,\xi^{(\pi)}\}$ be an* $\mathbf{A}_{00}$*-statement whose free variables are exactly $\xi',\ldots,\xi^{(\pi)}$, then there is a numerical term $\alpha\{\xi',\ldots,\xi^{(\pi)}\}$, with exactly the same free variables which only takes the values* 0 *and* 1, *such that for numerals $\nu',\ldots,\nu^{(\pi)}$*

$$\frac{\phi\{\nu',\ldots,\nu^{(\pi)}\}}{\alpha\{\nu',\ldots,\nu^{(\pi)}\}=0}* \quad \textit{and} \quad \frac{N[\phi\{\nu',\ldots,\nu^{(\pi)}\}]}{\alpha\{\nu',\ldots,\nu^{(\pi)}\}=1}*$$

*and conversely in both cases.*

The $\mathbf{A}_{00}$-function $\boldsymbol{\lambda}x'\ldots x^{(\pi)}.\alpha\{x',\ldots,x^{(\pi)}\}$ is called the *characteristic function* of the $\mathbf{A}_{00}$-statement $\phi\{\xi',\ldots,\xi^{(\pi)}\}$. Then

$$\alpha\{\nu',\ldots,\nu^{(\pi)}\}=0 \quad \text{iff} \quad \phi\{\nu',\ldots,\nu^{(\pi)}\}$$

is an $\mathbf{A}_{00}$-rule. An $\mathbf{A}_{00}$-statement is called primitive recursive if its characteristic function is primitive recursive. Since each $\mathbf{A}_{00}$-function is primitive recursive then each $\mathbf{A}_{00}$-statement is primitive recursive. The system $\mathbf{A}_{00}$ is then a formalization of primitive recursion.

An $\mathbf{A}_{00}$-statement is built up from equations and inequations by means of the logical connectives conjunction and disjunction. In the $\mathbf{A}_{00}$-statement $\phi\{\xi',\ldots,\xi^{(\pi)}\}$ make as many replacements as possible of the following sorts.

Replace $\beta = \gamma$ by $A_2[\beta, \gamma] = 0$,

$\beta \neq \gamma$ by $B_2[\beta, \gamma] = 0$.

If $\psi$ has been replaced by $\beta = 0$ and $\chi$ by $\gamma = 0$ then replace:

$$\psi \,\&\, \chi \text{ by } A_1[\beta+\gamma] = 0 \quad \text{and} \quad \psi \vee \chi \text{ by } \beta \times \gamma = 0.$$

The result now follows from Props. 1, 2, Ch. 4 (consistency and completeness of $\mathbf{A}_{00}$) and remarks after D 134. Note that an $\mathbf{A}_{00}$-statement can have distinct but, of course, equivalent characteristic functions. For instance $\rho\nu = \rho\nu + (\sigma\nu \dot{-} \sigma\nu)$ for each numeral $\nu$. The unique characteristic function which we have defined is called the *principal characteristic function.* It is unique because the construction of an $\mathbf{A}_{00}$-statement from equations and inequations is unique. Similarly we define the *principal equivalent equation.*

D 159 $\quad (E\xi)_\nu \phi\{\xi\}$ for $\sum_{\theta=0}^{\nu} \phi\{\theta\}$.

D 160 $\quad (A\xi)_\nu \phi\{\xi\}$ for $\prod_{\theta=0}^{\nu} \phi\{\theta\}$.

In this manner *limited existential quantifiers* and *limited universal quantifiers* can be $\mathbf{A}_{00}$-defined.

We frequently write $(E\xi, \eta)_\kappa$ instead of $(E\xi)_\kappa (E\eta)_\kappa$ and similarly in other cases. We have

$$\frac{N[\text{``}(E\xi)_\beta \phi\{\xi\}\text{''}]}{(A\xi)_\beta N[\text{``}\phi\{\xi\text{''}\}]}* \quad \text{and} \quad \frac{N[\text{``}(A\xi)_\beta \phi\{\xi\}\text{''}]}{(E\xi)\,[N_\beta\text{``}\phi\{\xi\}\text{''}]}*.$$

D 161 $\quad (\nu\, ISeg\, \kappa)$ for $(E\xi)_\kappa (\kappa = \nu^\frown \xi)$,

$\nu$ is an *initial segment* of $\kappa$ or the $\nu$th sequence of natural numbers is an initial segment of the $\kappa$th sequence of natural numbers.

D 162 $\quad (\nu\, ESeg\, \kappa)$ for $(E\xi)_\kappa (\kappa = \xi^\frown \nu)$,

$\nu$ is an *end segment* of $\kappa$ or the $\nu$th sequence of natural numbers is an end segment of the $\kappa$th sequence of natural numbers.

D 163 $\quad (\nu\, Part\, \kappa)$ for $(E\xi, \eta)_\kappa (\kappa = \xi^\frown \nu^\frown \eta)$,

$\nu$ is *part* of $\kappa$ or the $\nu$th sequence of natural numbers is a consecutive part of the $\kappa$th sequence of natural numbers.

D 164 $$\begin{cases} (\mu\xi)_0[\text{“}\phi\{\xi\}\text{”}] = 0, \\ (\mu\xi)_{S\nu}[\text{“}\phi\{\xi\}\text{”}] = \begin{cases} (\mu\xi)_\nu[\text{“}\phi\{\xi\}\text{”}] & \text{if} \quad (E\xi)_\nu \phi\{\xi\}, \\ S\nu & \text{if} \quad (A\xi)_\nu N[\text{“}\phi\{\xi\}\text{”}] \,\&\, \phi\{S\nu\}, \\ 0 & \text{if} \quad (A\xi)_\nu N[\text{“}\phi\{\xi\}\text{”}] \\ & \qquad \& \, N[\text{“}\phi\{S\nu\}\text{”}]. \end{cases} \end{cases}$$

The last clause is a case of definition by cases. Then $(\mu\xi)_\nu[\text{“}\phi\{\xi\}\text{”}]$ remains zero until $\nu$ becomes the least numeral for which $\phi\{\nu\}$ is $\mathbf{A}_{00}$-true, then and thereafter $(\mu\xi)_\nu[\text{“}\phi\{\xi\}\text{”}]$ remains at that value. Thus $(\mu\xi)_\nu[\text{“}\phi\{\xi\}\text{”}]$ is the least numeral $\kappa$ less than $S\nu$ for which $\phi\{\kappa\}$ is $\mathbf{A}_{00}$-true and is zero otherwise. If $\phi\{\nu\}$ is $\mathbf{A}_{00}$-false for each numeral $\nu$ then $(\mu\xi)_\nu[\text{“}\phi\{\xi\}\text{”}]$ is always zero.

The calculation of the least numeral $\nu$ such that $\rho\pi = 0$ can be put under the following scheme:

$$g[0,\kappa] = \kappa, \quad g[S\nu,\kappa] = g[\rho(S\kappa), S\kappa].$$

The calculation perpetuates if $\rho\pi$ is always distinct from zero. Otherwise the least numeral $\pi$ such that $\rho\pi = 0$ is the value of $g[\rho 0, 0]$. If the calculation fails to terminate then $g$ fails to be a calculable function. $\mu$ is called the *least number operator*.

Statements can also be defined by recursion schemes. Let $\phi$ and $\psi$ be already $\mathbf{A}_{00}$-defined, let $\chi \equiv \chi'$ denote that $\chi$ and $\chi'$ have the same $\mathbf{A}_{00}$-truth values when all free variables are replaced by numerals (same variable by same numeral at all occurrences). Consider:

$$P\{0\} \equiv \phi, \quad P\{S\nu\} \equiv \psi\{\nu, P\{\nu\}\}.$$

Replace the predicates by their principal equivalent equations. Replace $\phi$ by $\alpha = 0$ and $P\{\nu\}$ by $\rho\nu = 0$ and $\psi\{\nu, \rho\nu = 0\}$ by $\sigma[\nu, \rho\nu] = 0$. Then the recursion is

$$\rho 0 = \alpha, \quad \rho(S\nu) = \sigma[\nu, \rho\nu],$$

here $\alpha$ and $\sigma[\nu, \rho\nu]$ take only the values 0 and 1, hence so does $\rho[\nu]$. Note that $P\{\nu\}$ is free in $\psi\{\nu, P\{\nu\}\}$ because we lack means of binding any variable that occurs in $P\{\nu\}$ by anything that occurs in $\psi\{\nu, \Gamma_o\}$. In Ch. 8 we shall come across an important case where a variable free in $P\{\nu\}$ is bound in $\psi\{\nu, P\{\nu\}\}$.

### 5·5 *Other schemes for generating calculable functions*

In addition to the scheme for generating calculable functions of natural numbers, which we have called the scheme of primitive recursion, there are several other schemes for generating calculable functions of natural numbers from previously acquired functions. At first sight some of these schemes appear to be more general than the scheme of primitive recursion. We now give some schemes for generating calculable functions of natural numbers and show that they only produce functions equivalent to primitive recursive functions. Then we shall give an example of a calculable function outside the class of primitive recursive functions. A calculable function which fails to be primitive recursive requires for its calculation a search through the natural numbers for one having a certain property knowing that there is such a number, so that if we only continue the search sufficiently long we shall find the required number, but we are without a primitive recursive bound to the length of the search.

Consider the scheme

$$\rho[0,\mathfrak{k}] = \sigma\mathfrak{k}, \quad \rho[S\nu,\mathfrak{k}] = \tau[\nu, \sum_{0\leqslant\xi\leqslant\nu} \sigma'[\xi,\rho[\xi,\mathfrak{k}],\mathfrak{k}],\mathfrak{k}],$$

where $\sigma$, $\sigma'$ and $\tau$ are primitive recursive, and $\mathfrak{k}$ denotes $\kappa', \ldots, \kappa^{(\theta)}$ for some numeral $\theta$ (if $\theta = 0$ then $\mathfrak{k}$ is absent). Let $\rho'[\nu, \mathfrak{k}]$ be the primitive recursive function defined by

$$\rho'[0,\mathfrak{k}] = \sigma'[0,\sigma\mathfrak{k},\mathfrak{k}], \quad \rho'[S\nu,\mathfrak{k}] = \rho'[\nu,\mathfrak{k}] + \sigma'[S\nu,\tau[\nu,\rho'[\nu,\mathfrak{k}],\mathfrak{k}]].$$

Then $$\rho[\nu,\mathfrak{k}] = \sigma\mathfrak{k}\times B_1\nu + \tau[\nu \dot{-} 1,\rho'[\nu \dot{-} 1,\mathfrak{k}],\mathfrak{k}]\times A_1\nu,$$

thus $\rho[\nu,\mathfrak{k}]$ is primitive recursive.

Similarly for

$$\prod_{0\leqslant\xi\leqslant\nu} \sigma'[\xi,\rho[\xi,\mathfrak{k}],\mathfrak{k}], \langle\sigma'[\xi,\rho[\xi,\mathfrak{k}],\mathfrak{k}]\rangle_{1\leqslant\xi\leqslant\nu}, (\mu\xi)_\nu[\phi\{\xi\}], \quad \text{etc.}$$

in place of $$\sum_{0\leqslant\xi\leqslant\nu} \sigma'[\xi,\rho[\xi,\mathfrak{k}],\mathfrak{k}].$$

The above scheme is a form of *course of values recursion* when the value at $S\nu$ depends on a variable number of previous values.

## 5·6 *Course of values recursion*

We say that a function $\rho$ is defined by a *course of values recursion* from functions $\tau$, $\sigma$, $\sigma'$, ..., $\sigma^{(S\pi)}$ if

$$\rho[0,\mathfrak{k}] = \sigma[\mathfrak{k}], \quad \rho[S\nu,\mathfrak{k}] = \tau[\nu,\rho[\sigma'\nu,\mathfrak{k}],\ldots,\rho[\sigma^{(S\pi)}\nu,\mathfrak{k}],\mathfrak{k}],$$

where $\sigma'\nu \leqslant \nu, \ldots, \sigma^{(S\pi)}\nu \leqslant \nu$ and $\mathfrak{k}$ is as before. Note that the conditions on $\sigma', \ldots, \sigma^{(S\pi)}$ ensure that the calculation of $\rho[S\nu, \mathfrak{k}]$ depends on previously calculated values. Let $\rho'[\nu,\mathfrak{k}] = \langle\rho[\xi,\mathfrak{k}]\rangle_{0\leqslant\xi\leqslant\nu}$. The scheme for course of values recursion becomes

$$\rho'[0,\mathfrak{k}] = \langle\sigma\mathfrak{k}\rangle,$$

$$\begin{aligned}\rho'[S\nu,\mathfrak{k}] &= \rho'[\nu,\mathfrak{k}]^\frown\langle\tau[\nu,Pt[\rho'[\nu,\mathfrak{k}],S\sigma'\nu],\ldots,Pt[\rho'[\nu,\mathfrak{k}],S\sigma^{(S\pi)}\nu],\mathfrak{k}]\rangle.\\ &= \tau'[\nu,\rho'[\nu,\mathfrak{k}],\mathfrak{k}], \quad \text{say,}\end{aligned}$$

this defines the function $\rho'$ by primitive recursion from the functions $\tau, \sigma, \sigma', \ldots, \sigma^{(S\pi)}$. The required function is defined in terms of $\rho'$ by

$$\rho[\nu,\mathfrak{k}] = Pt[\rho'[\nu,\mathfrak{k}],S\nu].$$

Thus the function $\rho$ is obtained by the scheme of primitive recursion from the functions $\tau, \sigma, \sigma', \ldots, \sigma^{(S\pi)}$.

## 5·7 *Simultaneous recursion*

We say that the functions $\rho', \ldots, \rho^{(\theta)}$ are defined by *simultaneous recursion* from the functions $\sigma', \ldots, \sigma^{(\theta)}$, $\tau'$, ..., $\tau^{(\theta)}$ if $\theta \geqslant 2$ and

$$\begin{aligned}\rho'[0,\mathfrak{k}] &= \sigma'[\mathfrak{k}],\\ &\;\vdots\\ \rho^{(\theta)}[0,\mathfrak{k}] &= \sigma^{(\theta)}[\mathfrak{k}],\\ \rho'[S\nu,\mathfrak{k}] &= \tau'[\nu,\rho'[\nu,\mathfrak{k}],\ldots,\rho^{(\theta)}[\nu,\mathfrak{k}],\mathfrak{k}],\\ &\;\vdots\\ \rho^{(\theta)}[S\nu,\mathfrak{k}] &= \tau^{(\theta)}[\nu,\rho'[\nu,\mathfrak{k}],\ldots,\rho^{(\theta)}[\nu,\mathfrak{k}],\mathfrak{k}].\end{aligned}$$

Consider $\{\rho'[\nu,\mathfrak{k}],\ldots,\rho^{(\theta)}[\nu,\mathfrak{k}]\}$ call this $\rho[\nu,\mathfrak{k}]$. Then

$$\rho[0,\mathfrak{k}] = \{\sigma'[\mathfrak{k}],\ldots,\sigma^{(\theta)}[\mathfrak{k}]\} = \sigma[\mathfrak{k}] \quad \text{say,}$$

$$\begin{aligned}\rho[S\nu,\mathfrak{k}] &= \{\tau'[\nu,pt\,[\rho[\nu,\mathfrak{k}],\theta,1],\ldots,pt\,[\rho[\nu,\mathfrak{k}],\theta,\theta],\mathfrak{k}],\ldots,\\ &\qquad \tau^{(\theta)}[\nu,pt\,[\rho[\nu,\mathfrak{k}],\theta,1],\ldots,pt\,[\rho[\nu,\mathfrak{k}]\,\theta,\theta],\mathfrak{k}]\}\\ &= \tau[\nu,\rho[\nu,\mathfrak{k}],\mathfrak{k}] \quad \text{say.}\end{aligned}$$

This defines $\rho[\nu,\mathfrak{k}]$ by the scheme of primitive recursion from the functions $\sigma[\mathfrak{k}]$ and $\tau[\nu,\mu,\mathfrak{k}]$.

The functions $\rho',\ldots,\rho^{(\theta)}$ are defined from $\rho$ by

$$\rho'[\nu,\kappa] = pt[\rho[\nu,\kappa],\theta,1],$$
$$\vdots$$
$$\rho^{(\theta)}[\nu,\kappa] = pt[\rho[\nu,\kappa],\theta,\theta].$$

We can combine the last two results to cover the case of simultaneous course of values recursion. For example if $\rho[\nu,\mathfrak{k}]$ and $\rho'[\nu,\mathfrak{k}]$ are defined by

$$\rho[0,\mathfrak{k}] = \sigma[\mathfrak{k}],$$
$$\rho'[0,\mathfrak{k}] = \sigma'[\mathfrak{k}],$$
$$\rho[S\nu,\mathfrak{k}] = \tau[\nu,\rho[\sigma''\nu,\mathfrak{k}],\rho'[\sigma'''\nu',\mathfrak{k}],\mathfrak{k}],$$
$$\rho'[S\nu,\mathfrak{k}] = \tau'[\nu,\rho[\sigma^{(\mathrm{iv})}\nu,\mathfrak{k}],\rho'[\sigma^{(\mathrm{v})}\nu,\mathfrak{k}],\mathfrak{k}],$$

where $\sigma''\nu \leqslant \nu,\ldots,\sigma^{(\mathrm{v})}\nu \leqslant \nu$, then by using

$$\rho''[\nu,\mathfrak{k}] = \{\rho[\nu,\mathfrak{k}],\rho'[\nu,\mathfrak{k}]\}$$

and
$$\rho'''[\nu,\mathfrak{k}] = \langle\rho''[\xi,\mathfrak{k}]\rangle_{\theta\leqslant\xi\leqslant\nu}$$

we can reduce the definitions to a case of primitive recursion.

## 5.8 *Recursion with substitution in parameter*

We say that the function $\rho$ is defined by *recursion with substitution in parameter* when:

$$\rho[0,\mathfrak{k}] = \sigma[\mathfrak{k}],$$

$$\rho[S\nu,\mathfrak{k}] = \tau[\nu,\rho[\nu,\mathfrak{k}],\rho[\nu,\mathfrak{s}[\nu,\mathfrak{k}]],\ldots,\rho[\nu,\mathfrak{s}^{(S\theta)}[\nu,\mathfrak{k}]],\mathfrak{k}],$$

where

$$\mathfrak{s}^{(\lambda)}[\nu,\mathfrak{k}] \text{ for } \sigma_1^{(\lambda)}[\nu,\mathfrak{k}],\ldots,\sigma_{S\pi}^{\lambda}[\nu,\mathfrak{k}], \quad \mathfrak{k} \text{ for } \kappa',\ldots,\kappa^{(S\pi)} \quad 1\leqslant\lambda\leqslant S\theta.$$

We first use ordered tuplets and reduce to

$$\rho'[0,\kappa] = \overline{\sigma}'[\kappa],$$
$$\rho'[S\nu,\kappa] = \tau'[\nu,\rho'[\nu,\kappa],\rho'[\nu,\overline{\sigma}'[\nu,\kappa]],\ldots,\rho'[\nu,\overline{\sigma}^{(S\theta)}[\nu,\kappa]],\kappa],$$

where
$$\rho'[\nu,\kappa] = \rho[\nu,\mathfrak{k}] \quad \text{with} \quad \kappa = \{\kappa',\ldots,\kappa^{(S\pi)}\},$$
$$\overline{\sigma}'[\nu,\kappa] = \{\sigma_1'[\nu,\mathfrak{k}],\ldots,\sigma_{S\pi}'[\nu,\mathfrak{k}]\}, \quad \text{etc.,}$$

so that $\kappa' = pt[\kappa, S\pi, 1], \ldots, \kappa^{(S\pi)} = pt[\kappa, S\pi, S\pi]$ and similarly for $\sigma'_1, \ldots, \sigma'_{S\pi}$. The value of $\rho'[S\nu, \kappa]$ depends on the values of

$$\rho'[\nu,\kappa], \rho'[\nu, \bar{\sigma}'[\nu,\kappa]], \ldots, \rho'[\nu, \bar{\sigma}^{(S\theta)}[\nu,\kappa]].$$

Let
$$\sigma''[\nu,\kappa] = Max[\kappa, \underset{\substack{1 \leqslant \eta \leqslant S\theta \\ 0 \leqslant \xi \leqslant \kappa}}{Max} \bar{\sigma}^{(\eta)}[\nu,\xi]].$$

Thus the value of $\langle \rho'[S\nu,\xi]\rangle_{0\leqslant\xi\leqslant\kappa}$ depends on the value of $\langle\rho'[\nu,\xi]\rangle_{0\leqslant\xi\leqslant\sigma''[\nu,\kappa]}$. Let $\rho''[\nu,\kappa] = \langle\rho'[\nu,\xi]\rangle_{0\leqslant\xi\leqslant\kappa}$, then $\rho'[\nu,\kappa] = Pt[\rho''[\nu,\kappa], S\kappa]$ and

$$\rho''[0,\kappa] = \langle\rho'[0,\xi]\rangle_{0\leqslant\xi\leqslant\kappa} = \langle\sigma'[\xi]\rangle_{0\leqslant\xi\leqslant\kappa},$$

$$\begin{aligned}
\rho''[S\nu,\kappa] &= \langle\rho'[S\nu,\xi]\rangle_{0\leqslant\xi\leqslant\kappa} \\
&= \langle\tau'[\nu, \rho'[\nu,\xi], \rho'[\nu,\bar{\sigma}'[\nu,\xi]], \ldots, \rho'[\nu, \bar{\sigma}^{(S\theta)}[\nu,\xi]], \xi]\rangle_{0\leqslant\xi\leqslant\kappa} \\
&= \langle\tau'[\nu, Pt[\rho''[\nu,\sigma''[\nu,\kappa]], S\xi], Pt[\rho''[\nu,\sigma''[\nu,\kappa]], S\bar{\sigma}'[\nu,\xi]], \ldots \\
&\qquad Pt[\rho''[\nu,\sigma''[\nu,\kappa]], S\sigma^{S\theta}[\nu,\xi]], \xi]\rangle_{0\leqslant\xi\leqslant\kappa}.
\end{aligned}$$

Hence $\quad \rho''[0,\kappa] = \sigma'''[\kappa], \quad \rho''[S\nu,\kappa] = \tau''[\nu, \rho''[\nu,\sigma''[\nu,\kappa]],\kappa],$

for suitable $\sigma'''$ and $\tau''$. We can easily recover $\rho[\nu,\mathfrak{k}]$ from $\rho''[\nu,\mathfrak{k}]$.

Consider then the recursion with substitution in parameter

$$\rho[0,\kappa] = \sigma'[\kappa], \quad \rho[S\nu,\kappa] = \tau[\nu, \rho[\nu, \sigma[\nu,\kappa]], \kappa].$$

The first few values of $\rho[\nu,\kappa]$ are

$$\begin{aligned}
&\rho[0,\kappa] = \sigma'[\kappa], \\
&\rho[1,\kappa] = \tau[0, \sigma'[\sigma[0,\kappa]], \kappa], \\
&\rho[2,\kappa] = \tau[1, \tau[0, \sigma'[\sigma[0,\sigma[1,\kappa]]], \sigma[1,\kappa]], \kappa], \\
&\rho[3,\kappa] = \tau[2, \tau[1, \tau[0, \sigma'[\sigma[0,\sigma[1,\sigma[2,\kappa]]]], \sigma[1,\sigma[2,\kappa]]], \sigma[2,\kappa]], \kappa].
\end{aligned}$$

It is a little easier to see how this continues if we write

$$\tau'[\nu,\kappa,\theta] \quad \text{for} \quad \tau[\nu,\theta,\kappa],$$

that is if we put the parameter in the second argument place instead of in the third argument place. We then have

$$\begin{aligned}
\rho[S\nu,\kappa] = \tau'[\nu, \bar{\sigma}[\nu,0], \tau'[\nu \dot{-} 1, \bar{\sigma}[\nu,1], \\
\tau'[\nu \dot{-} 2, \bar{\sigma}[\nu,2], \ldots, \tau'[0, \bar{\sigma}[\nu,\nu], \sigma'[\bar{\sigma}[\nu,S\nu]]]\ldots]]],
\end{aligned}$$

for suitable $\bar{\sigma}$ ($\bar{\sigma}$ contains $\kappa$). This function can be $\mathbf{A}_{00}$-defined:

$$\rho[S\nu, \kappa] = \mathscr{I}(\boldsymbol{\lambda}\xi\eta . \tau'[\xi, \bar{\sigma}[\nu, \nu \dot{-} \xi], \eta])(\sigma'[\bar{\sigma}[\nu, S\nu]])(S\nu),$$

since by Ax 3.2 we have

$$\rho[S\nu, \kappa] = \tau'[\nu, \bar{\sigma}[\nu, 0], \mathscr{I}(\boldsymbol{\lambda}\xi\eta . \tau'[\xi, \bar{\sigma}[\nu, \nu \dot{-} \xi], \eta])(\sigma'[\bar{\sigma}[\nu, S\nu]])\nu],$$

$$\rho[S\nu, \kappa] = \tau'[\nu, \bar{\sigma}[\nu, 0], \tau'[\nu \dot{-} 1, \sigma[\nu, 1], \mathscr{I}(\boldsymbol{\lambda}\xi\eta . \tau'[\xi, \bar{\sigma}[\nu, \nu \dot{-} \xi]], \eta]])$$
$$(\sigma'[\bar{\sigma}[\nu, S\nu]])(\nu \dot{-} 1)],$$

$$= \tau'[\nu, \bar{\sigma}[\nu, 0], \tau'[\nu \dot{-} 1, \bar{\sigma}[\nu, 1], \ldots, \tau'[0, \bar{\sigma}[\nu, \nu]], \sigma'[\bar{\sigma}[\nu, S\nu]]] \ldots].$$

The function $\bar{\sigma}$ has its first $SS\nu$ values (all that are relevant) given by

$$\begin{aligned} \bar{\sigma}[\nu, 0] &= \kappa, \\ \bar{\sigma}[\nu, 1] &= \sigma[\nu, \kappa], \\ \bar{\sigma}[\nu, 2] &= \sigma[\nu \dot{-} 1, \sigma[\nu, \kappa]], \\ &\vdots \\ \bar{\sigma}[\nu, \nu] &= \sigma[1, \sigma[2, \ldots, \sigma[\nu, \kappa] \ldots]], \\ \bar{\sigma}[\nu, S\nu] &= \sigma[0, \sigma[1, \sigma[2, \ldots, \sigma[\nu, \kappa] \ldots]]]. \end{aligned}$$

The function $\bar{\sigma}$ can be $\mathbf{A}_{00}$-defined by

$$\bar{\sigma}[\nu, \theta] = \mathscr{I}(\boldsymbol{\lambda}\xi'\eta' . \sigma[\nu \dot{-} \xi', \eta'])\kappa\theta.$$

Then $\bar{\sigma}[\nu, 0] = \kappa,$

$$\bar{\sigma}[\nu, S\theta] = \sigma[\nu \dot{-} \theta, \mathscr{I}(\boldsymbol{\lambda}\xi'\eta' . \sigma[\nu \dot{-} \xi', \eta'])\kappa\theta],$$

$$= \sigma[\nu \dot{-} \theta, \sigma[\nu \dot{-} (\theta \dot{-} 1), \ldots, \sigma[\nu, \kappa] \ldots]], \text{ as desired.}$$

Thus altogether

$$\rho[\nu, \kappa] = \mathscr{I}(\boldsymbol{\lambda}\xi\eta . \tau[\xi, \eta, \mathscr{I}(\boldsymbol{\lambda}\xi'\eta' . \sigma[\nu \dot{-} S\xi', \eta'])\kappa(\nu \dot{-} S\xi)])$$
$$(\sigma'[\mathscr{I}(\boldsymbol{\lambda}\xi'\eta' . \sigma[\nu \dot{-} S\xi', \eta'])\kappa\nu])\nu,$$

where we have replaced $\tau'$ by its definition in terms of $\tau$. According to Prop. 2 of this chapter a numerical term is a primitive recursive function of all its arguments, thus $\rho[\nu, \kappa]$ is a primitive recursive function of $\nu, \kappa$.

## 5·9 *Double recursion*

We say that the function $\rho$ is defined by *double recursion* from the functions $\sigma$, $\sigma'$, $\sigma''$, $\tau$ when

$$\rho[0, \kappa, \mathfrak{p}] = \sigma'[\kappa, \mathfrak{p}], \quad \rho[S\nu, 0, \mathfrak{p}] = \sigma''[\nu, \mathfrak{p}],$$

$$\rho[S\nu, S\kappa, \mathfrak{p}] = \tau[\nu, \kappa, \rho[\nu, \sigma[\nu, \kappa, \mathfrak{p}], \mathfrak{p}], \rho[S\nu, \kappa, \mathfrak{p}], \mathfrak{p}].$$

Then the values of $\rho$ are given on the two coordinate axes, while the value at $\{S\nu, S\kappa\}$ is given in terms of the value of $\rho[\nu', \kappa', \mathfrak{p}]$ where $\{\nu', \kappa'\}$ lies to the left or below $\{S\nu, S\kappa\}$. Clearly the process of calculating the value of $\rho[\nu, \kappa, \mathfrak{p}]$ will terminate. We shall have to find $\rho[S\nu, \theta, \mathfrak{p}]$ for $\theta < S\kappa$ and $\rho[\nu, \sigma[\nu, \theta, \mathfrak{p}], \mathfrak{p}]$ for $\theta < S\kappa$, and so on. Write $\sigma'''[\nu, \kappa, \mathfrak{p}]$ for $\underset{0 \leqslant \xi \leqslant \kappa}{Max}\, \sigma[\nu, \xi, \mathfrak{p}]$ then the value of $\rho[S\nu, S\kappa, \mathfrak{p}]$ depends on the value of $\underset{0 \leqslant \xi \leqslant \sigma'''[\nu, \kappa, \mathfrak{p}]}{\langle \rho[\nu, \xi, \mathfrak{p}] \rangle}$. Similarly the value of $\underset{0 \leqslant \xi \leqslant S\kappa}{\langle \rho[S\nu, \xi, \mathfrak{p}] \rangle}$ depends on the value of $\underset{0 \leqslant \xi \leqslant \sigma'''[\nu, \kappa, \mathfrak{p}]}{\langle \rho[\nu, \xi, \mathfrak{p}] \rangle}$. Write $\rho'[\nu, \kappa, \mathfrak{p}]$ for $\underset{0 \leqslant \xi \leqslant \kappa}{\langle \rho[\nu, \xi, \mathfrak{p}] \rangle}$. Then the value of $\rho'[S\nu, S\kappa, \mathfrak{p}]$ depends on the value of $\rho'[\nu, \sigma'''[\nu, \kappa, \mathfrak{p}], \mathfrak{p}]$. Thus

$$\rho'[0, S\kappa, \mathfrak{p}] = \underset{0 \leqslant \xi \leqslant S\kappa}{\langle \sigma'[\xi, \mathfrak{p}] \rangle},$$

$$\begin{aligned} \rho'[S\nu, S\kappa, \mathfrak{p}] &= \underset{0 \leqslant \xi \leqslant S\kappa}{\langle \rho[S\nu, \xi, \mathfrak{p}] \rangle} \\ &= \langle \sigma''[\nu, \mathfrak{p}] \rangle \frown \underset{1 \leqslant \xi \leqslant S\kappa}{\langle \tau[\nu, \xi, \rho[\nu, \sigma[\nu, \xi, \mathfrak{p}], \mathfrak{p}],} \\ &\qquad\qquad \rho[S\nu, \xi, \mathfrak{p}], \mathfrak{p}] \rangle \\ &= \langle \sigma''[\nu, \mathfrak{p}] \rangle \frown \underset{1 \leqslant \xi \leqslant \kappa}{\langle \tau[\nu, \xi, Pt[\rho'[\nu, \sigma'''[\nu, \kappa', \mathfrak{p}], \mathfrak{p}], S\sigma[\nu, \xi, \mathfrak{p}]],} \\ &\qquad\qquad Pt[\rho'[S\nu, \kappa, \mathfrak{p}], S\xi], \mathfrak{p}] \rangle, \end{aligned}$$

where $\kappa' \geqslant \kappa$. Denote the r.h.s. of the last equation by

$$\begin{aligned} &\tau'[\nu, \kappa, \rho'[\nu, \sigma'''[\nu, \kappa', \mathfrak{p}], \mathfrak{p}], \rho'[S\nu, \kappa, \mathfrak{p}], \mathfrak{p}] \\ &\quad = \tau'[\nu, \kappa, \rho'[\nu, \sigma'''[\nu, \kappa', \mathfrak{p}], \mathfrak{p}], \tau'[\nu, \kappa \dot{-} 1, \\ &\qquad \rho'[\nu, \sigma'''[\nu, \kappa', \mathfrak{p}], \mathfrak{p}], \rho'[S\nu, \kappa \dot{-} 1, \mathfrak{p}], \mathfrak{p}], \mathfrak{p}] \quad \text{if} \quad \kappa > 0. \end{aligned}$$

The reason for introducing $\kappa'$ with $\kappa' \geqslant \kappa$ is so that the second occurrence of $\sigma'''$ can have the same arguments as the first occurrence of $\sigma'''$. Now take $\kappa' = \kappa$ and denote the resulting expression by

$$\tau''[\nu, \kappa, \rho'[\nu, \sigma'''[\nu, \kappa, \mathfrak{p}], \mathfrak{p}], \rho'[S\nu, \kappa \dot{-} 1, \mathfrak{p}], \mathfrak{p}].$$

Suppose that for a certain $\theta$, with $0 \leqslant \theta \leqslant \kappa$ we have defined a function $\tau^{(\theta)}$ so that

$$\begin{aligned} \rho'[S\nu, S\kappa, \mathfrak{p}] &= \tau^{(\theta)}[\nu, \kappa, \rho'[\nu, \sigma'''[\nu, \kappa, \mathfrak{p}], \mathfrak{p}], \\ &\qquad \rho'[S\nu, S(\kappa \dot{-} \theta), \mathfrak{p}], \mathfrak{p}] \quad \text{for} \quad 0 < \theta \leqslant \kappa, \end{aligned}$$

then $\rho'[S\nu, S\kappa, \mathfrak{p}] = \tau^{(\theta)}[\nu, \kappa, \rho'[\nu, \sigma'''[\nu, \kappa, \mathfrak{p}], \mathfrak{p}], \tau'[\nu, \kappa \dot{-} \theta,$

$$\rho'[\nu, \sigma'''[\nu, \kappa, \mathfrak{p}], \mathfrak{p}], \rho'[S\nu, \kappa \dot{-} \theta, \mathfrak{p}]\, \mathfrak{p}], \mathfrak{p}],$$

(we have again taken $\kappa' = \kappa$ in (i)). We denote the r.h.s. of this equation by

$$\tau^{(S\theta)}[\nu, \kappa, \rho'[\nu, \sigma'''[\nu, \kappa, \mathfrak{p}], \mathfrak{p}], \rho'[S\nu, \kappa \dot{-} \theta, \mathfrak{p}], \mathfrak{p}].$$

We have thus defined a sequence of functions $\tau', \tau'', \ldots, \tau^{(S\theta)}$. Now define by recursion with substitution in parameter

$$\begin{cases} \bar{\tau}[0, \nu, \kappa, \theta', \theta'', \mathfrak{p}] = \tau'[\nu, \kappa, \theta', \theta'', \mathfrak{p}], \\ \bar{\tau}[S\theta, \nu, \kappa, \theta', \theta'', \mathfrak{p}] = \bar{\tau}[\theta, \nu, \kappa, \theta', \tau'[\nu, \kappa \dot{-} \theta, \theta', \theta'', \mathfrak{p}], \mathfrak{p}], \end{cases}$$

so that $$\bar{\tau}[\theta, \nu, \kappa, \theta', \theta'', \mathfrak{p}] = \tau^{(\theta)}[\nu, \kappa, \theta', \theta'', \mathfrak{p}].$$

Then $$\begin{aligned} \rho'[S\nu, S\kappa, \mathfrak{p}] &= \bar{\tau}[\kappa, \nu, \kappa, \rho'[\nu, \sigma'''[\nu, \kappa, \mathfrak{p}], \mathfrak{p}], \tau'[\nu, 0, \\ &\qquad \rho'[\nu, \sigma'''[\nu, \kappa, \mathfrak{p}], \langle \sigma''[\nu, \mathfrak{p}] \rangle], \mathfrak{p}], \mathfrak{p}] \\ &= \sigma^{\mathrm{iv}}[\nu, \kappa, \rho'[\nu, \sigma'''[\nu, \kappa, \mathfrak{p}], \mathfrak{p}], \mathfrak{p}], \quad \text{say.} \end{aligned}$$

Thus altogether

$$\begin{cases} \rho'[0, S\kappa, \mathfrak{p}] = \langle \sigma'[\xi, \mathfrak{p}] \rangle_{0 \leqslant \xi \leqslant S\kappa} \\ \rho'[S\nu, S\kappa, \mathfrak{p}] = \sigma^{\mathrm{iv}}[\nu, \kappa, \rho'[\nu, \sigma'''[\nu, \kappa, \mathfrak{p}], \mathfrak{p}], \mathfrak{p}]. \end{cases}$$

This is a recursion with substitution in parameter. Finally

$$\begin{aligned} \rho[\nu, \kappa, \mathfrak{p}] = Pt\,[\rho'[\nu, \kappa, \mathfrak{p}], S\kappa] . A_1 \kappa \\ + \sigma''[\nu, \mathfrak{p}] . B_1 \kappa . A_1 \nu + \sigma'[\kappa, \mathfrak{p}] . B_1 \kappa . B_1 \nu. \end{aligned}$$

We can obtain an explicit expression for $\rho$ because we can do so for $\rho'$ by the method used for recursion with substitution in parameter.

## 5.10 *Simple nested recursion*

The scheme of simple nested recursion is:

$$\begin{cases} \rho[0, \mathfrak{k}] = \sigma[\mathfrak{k}], \\ \rho[S\nu, \mathfrak{k}] = \tau[\nu, \rho[\nu, \mathfrak{k}], \rho[\nu, \mathfrak{s}'[\nu, \rho[\nu, \mathfrak{k}], \mathfrak{k}]], \ldots, \rho[\nu, \mathfrak{s}^{(\theta)}[\nu, \rho[\nu, \mathfrak{k}], \mathfrak{k}]], \mathfrak{k}]. \end{cases}$$

The scheme of substitution in parameter is a special case of this scheme, and this scheme in turn is a special case of a similar scheme but with the omission of the second argument of $\tau$ and replacement of $\theta$ by $S\theta$. Define $\bar{\sigma}[\kappa]$ for $\sigma[\mathfrak{k}]$, where $\{\mathfrak{k}\} = \kappa$,

$$\begin{aligned} \bar{\sigma}^{(\mu)}[\nu, \lambda, \kappa] &\quad \text{for} \quad \{\mathfrak{s}^{(\mu)}[\nu, \lambda, \mathfrak{k}]\}, \\ \bar{\tau}[\nu, \lambda', \ldots, \lambda^{(S\theta)}, \kappa] &\quad \text{for} \quad \tau[\nu, \lambda', \ldots, \lambda^{(S\theta)}, \mathfrak{k}], \\ \bar{\rho}[\nu, \kappa] &\quad \text{for} \quad \rho[\nu, \mathfrak{k}]. \end{aligned}$$

Then $\begin{cases} \bar{\rho}[0,\kappa] = \bar{\sigma}[\kappa], \\ \bar{\rho}[S\nu,\kappa] = \bar{\tau}[\nu,\bar{\rho}[\nu,\bar{\sigma}'[\nu,\bar{\rho}[\nu,\kappa],\kappa]],\ldots,\bar{\rho}[\nu,\bar{\sigma}^{(S\theta)}[\nu,\bar{\rho}[\nu,\kappa],\kappa]],\kappa]. \end{cases}$

Define $$\bar{\bar{\rho}}(\nu,\kappa] \quad \text{for} \quad \langle \bar{\rho}[\nu,\xi]\rangle_{0\leqslant\xi\leqslant\kappa},$$

then $$\begin{cases} \bar{\bar{\rho}}[0,\kappa] = \langle\bar{\sigma}[\xi]\rangle_{0\leqslant\xi\leqslant\kappa} \\ \bar{\bar{\rho}}[S\nu,\kappa] = \langle\bar{\tau}[\nu,\bar{\rho}[\nu,\bar{\sigma}'[\nu,\bar{\rho}[\nu,\xi],\xi]],\ldots,\bar{\rho}[\nu,\bar{\sigma}^{(S\theta)}[\nu,\bar{\rho}[\nu,\xi],\xi]],\xi]\rangle_{0\leqslant\xi\leqslant\kappa} \\ \qquad = \langle\bar{\tau}[\nu,\bar{\rho}[\nu,\bar{\sigma}_1'[\nu,\bar{\bar{\rho}}[\nu,\kappa],\xi]],\ldots,\bar{\rho}[\nu,\bar{\sigma}_1^{(S\theta)}[\nu,\bar{\bar{\rho}}[\nu,\kappa],\xi]],\xi]\rangle_{0\leqslant\xi\leqslant\kappa}, \end{cases}$$

where $\bar{\sigma}_1^{(\lambda)}[\nu,\mu,\xi] = \bar{\sigma}^{(\lambda)}[\nu,Pt[\mu,S\xi],\xi],$

$$= \langle\bar{\tau}[\nu,Pt[\bar{\bar{\rho}}[\nu,\sigma[\nu,\bar{\bar{\rho}}[\nu,\kappa],\kappa]],S\bar{\sigma}_1'[\nu,\bar{\bar{\rho}}[\nu,\kappa'],\xi]],\ldots,$$
$$Pt[\bar{\bar{\rho}}[\nu,\sigma[\nu,\bar{\bar{\rho}}[\nu,\kappa],\kappa]],S\bar{\sigma}_1^{(S\theta)}[\nu,\bar{\bar{\rho}}[\nu,\kappa'],\xi]],\xi]\rangle_{0\leqslant\xi\leqslant\kappa}$$

where $$\sigma[\nu,\lambda,\kappa] \quad \text{for} \quad \mathop{Max}_{1\leqslant\pi\leqslant S\theta}[\mathop{Max}_{0\leqslant\xi\leqslant\kappa}[\bar{\sigma}_1^{(\pi)}[\nu,\lambda,\xi]]],$$

and $\kappa'$ can be any number greater than or equal to $\kappa$.

$$= \bar{\tau}_1[\nu,\bar{\bar{\rho}}[\nu,\kappa'],\bar{\bar{\rho}}[\nu,\bar{\sigma}''[\nu,\bar{\bar{\rho}}[\nu,\kappa],\kappa]],\kappa] \quad \text{say,}$$

where $\bar{\sigma}''$ is any function dominating $\sigma$. Now consider

$$\bar{\bar{\bar{\rho}}}[\nu,\kappa] \quad \text{for} \quad \{\nu,\kappa,\bar{\bar{\rho}}[\nu,\kappa]\}.$$

Then $\bar{\bar{\bar{\rho}}}$ will satisfy a similar first equation and also

$$\bar{\bar{\bar{\rho}}}[S\nu,\kappa] = \{S\nu,\kappa,\bar{\tau}_1[\nu,\bar{\bar{\rho}}[\nu,\kappa],\bar{\bar{\rho}}[\nu,\bar{\sigma}''[\nu,\bar{\bar{\rho}}[\nu,\kappa],\kappa]],\kappa]\}, \tag{i}$$

here $\bar{\sigma}''[\nu,\bar{\bar{\rho}}[\nu,\kappa],\kappa]$ may be replaced by

$$\{\bar{\sigma}''[\bar{\bar{\bar{\rho}}}[\nu,\kappa],\bar{\bar{\bar{\rho}}}[\nu,\kappa],\bar{\bar{\bar{\rho}}}[\nu,\kappa]],\bar{\bar{\bar{\rho}}}[\nu,\kappa]\} = \bar{\sigma}'''[\bar{\bar{\bar{\rho}}}[\nu,\kappa]] \quad \text{say.}$$

Clearly each of $\nu$, $S\nu$, $\kappa$, $\bar{\bar{\rho}}[\nu,\kappa]$, $\bar{\bar{\rho}}[\nu,\bar{\sigma}'''[\bar{\bar{\bar{\rho}}}[\nu,\kappa]]]$ is a primitive recursive function of $\bar{\bar{\bar{\rho}}}[\nu,\sigma'''[\bar{\bar{\bar{\rho}}}[\nu,\kappa]]]$.

Thus we can take the r.h.s. of (i) to be $\bar{\bar{\bar{\tau}}}[\bar{\bar{\bar{\rho}}}[\nu,\sigma'''[\bar{\bar{\bar{\rho}}}[\nu,\kappa]]]]$.

Thus it suffices to consider the scheme

$$\rho[0,\kappa] = \sigma'[\kappa], \quad \rho[S\nu,\kappa] = \tau[\rho[\nu,\sigma[\rho[\nu,\kappa]]]],$$

where $\sigma$ has a left inverse $\sigma^{\mathrm{L}}$, i.e. $\sigma^{\mathrm{L}}[\sigma[\nu]] = \nu$. Therefore it is sufficient to consider the scheme for $\rho^*$, where $\rho^*[\nu,\kappa] = \sigma[\rho[\nu,\kappa]]$,

$$\rho^*[0,\kappa] = \sigma[\sigma'[\kappa]], \quad \rho^*[S\nu,\kappa] = \sigma\tau[\sigma^{\mathrm{L}}[\rho^*[\nu,\rho^*[\nu,\kappa]]]].$$

Therefore we now consider the scheme

$$\rho[0,\kappa] = \sigma[\kappa], \quad \rho[S\nu,\kappa] = \tau[\rho[\nu,\rho[\nu,\kappa]]].$$

For each $\nu$, the function $\boldsymbol{\lambda} x\rho[\nu, x]$ is a composition of $\sigma$'s and $\tau$'s. Thus

$$\begin{aligned}\rho[0,\kappa] &= \sigma[\kappa],\\ \rho[1,\kappa] &= \tau\sigma^2[\kappa],\\ \rho[2,\kappa] &= \tau^2\sigma^2\tau\sigma^2[\kappa],\\ \rho[3,\kappa] &= \tau^3\sigma^2\tau\sigma^2\tau^2\sigma^2\tau\sigma^2[\kappa] \quad \text{etc.}\end{aligned}$$

Define a function $\chi$ by

$$\begin{cases}\chi[0,\kappa] = \kappa,\\ \chi[S\nu,\kappa] = \begin{cases}\sigma^2[\chi[\nu,\kappa]] & \text{if } S\nu \text{ is odd,}\\ \tau^{\varpi S\nu}[\chi[\nu,\kappa]] & \text{if } S\nu \text{ is even.}\end{cases}\end{cases}$$

This is a primitive recursive function and

$$\rho[0,\kappa] = \sigma[\kappa], \quad \rho[S\nu,\kappa] = \chi[2^{S\nu},\kappa].$$

Thus simple nested recursion has been reduced to primitive recursion.

## 5·11 *Alternative definitions of primitive recursive functions*

D 165 $$Rt0 = 0, \quad Rt[S\nu] = Rt\nu + B_1[(SRt\nu)^2 \dot{-} S\nu],$$

then $Rt\nu$ is the greatest natural number whose square is less than or equal to $\nu$, it is called the *square root function.*

PROP. 4. *All primitive recursive functions are obtainable by adding to the initial functions the four functions; addition, multiplication, limited subtraction and square root and using the schemes of substitution and pure iteration without parameter.*

The scheme of *pure iteration without parameter* is

$$\rho 0 = \alpha, \quad \rho[S\nu] = \sigma[\rho\nu],$$

so that $$\rho\nu = \mathscr{I}(\boldsymbol{\lambda}\xi\eta\,.\,\sigma[\eta])\,\alpha\nu.$$

We first reduce the set of parameters to a single parameter.

D 166 $E\nu$ for $\nu \dot{-} (Rt\nu)^2$,

the difference between $\nu$ and the greatest square less than or equal to $\nu$.

D 167 $J[\nu,\kappa]$ for $(\nu+\kappa)^2+\nu$,

D 168 $U\nu$ for $E\nu$,

D 169 $V\nu$ for $Rt\nu \dot{-} U\nu$.

Then $$UJ[\nu,\kappa]=\nu,\quad VJ[\nu,\kappa]=\kappa,\quad J[U\theta,V\theta]=\theta.$$

Thus if $J[\nu,\kappa]=J[\nu',\kappa']$ then $\nu=\nu'$ and $\kappa=\kappa'$. $J$ is a *pairing function*, it gives the number of an ordered pair in a numbering of all ordered pairs. It is sufficient to show how two parameters may be reduced to one parameter, because the repetition of the process will reduce any number of parameters to one parameter. Suppose then

$$\rho[\nu,\kappa,0]=\sigma[\nu,\kappa],\quad \rho[\nu,\kappa,S\theta]=\tau[\nu,\kappa,\theta,\rho[\nu,\kappa,\theta]],$$

so that $\rho$ is a primitive recursive function with two parameters. Let

$$\rho'[\nu',0]=\sigma[U\nu',V\nu'],\quad \rho'[\nu',S\theta]=\tau[U\nu',V\nu',\theta,\rho'[\nu',\theta]].$$

Then $$\rho'[J[\nu,\kappa],\theta]=\rho[\nu,\kappa,\theta].$$

Thus we have eliminated one parameter, hence $S\lambda$ parameters can be reduced to one parameter. Now suppose that

$$\rho[\kappa,0]=\sigma[\kappa],\quad \rho[\kappa,S\theta]=\tau[\kappa,\theta,\rho[\kappa,\theta]],$$

so that $\rho$ is a primitive recursive function with one parameter. Let $\rho'[\kappa,\theta]$ for $J[\kappa,\rho[\kappa,\theta]]$, then

$$\rho'[\kappa,0]=\sigma'[\kappa],\quad \rho'[\kappa,S\theta]=\tau'[\theta,\rho'[\kappa,\theta]],\tag{i}$$

where

$$\sigma'[\kappa]\ \text{ for }\ J[\kappa,\sigma[\kappa]]\ \text{ and }\ \tau'[\kappa,\nu]\ \text{ for }\ J[U\nu,\tau[U\nu,\kappa,V\nu]],$$

also $$\rho[\kappa,\theta]=V\rho'[\kappa,\theta].$$

Then $\rho$ is defined by substitution and iteration with one parameter as in (i), using addition, multiplication, limited subtraction and square root as additional initial functions.

Take $\rho''[\sigma'[\xi],\zeta]$ instead of $\rho'[\xi,\zeta]$ and we have

$$\rho''[\xi,0]=\xi,\quad \rho''[\xi,S\zeta]=\tau'[\zeta,\rho''[\xi,\zeta]].$$

Then $$\rho'[\xi,\zeta]=\rho''[\sigma'[\xi],\zeta].$$

In the construction of $\rho'$ from initial functions by schemes the two functions $\sigma'$ and $\tau'$ will have been previously constructed. So that if they can be constructed as in the proposition then so can $\rho'$. Thus

$$\rho[\xi,\zeta]=V\rho''[\sigma'[\xi],\zeta].$$

We can replace (i) by a scheme of pure iteration with one parameter, this we now give, though it is independent of what is to follow. Write $\rho'''[\xi,\zeta]$ for $J[\zeta,\rho''[\xi,\zeta]]$ then

$$\rho'''[\xi,0] = J[0,\xi], \quad \rho'''[\xi,S\zeta] = \tau''[\rho''[\xi,\zeta]], \tag{ii}$$

where $$\tau''[\eta] = J[SU\eta, \tau'[U\eta, V\eta]] \quad \text{and} \quad \rho''[\xi,\zeta] = V\rho'''[\xi,\zeta].$$

Thus $\rho'''$ and hence $\rho$ is defined by substitution and pure iteration with one parameter as in (ii), using addition, multiplication, limited subtraction and square root as additional initial functions.

We now wish to remove the single parameter. Write

$$\begin{aligned} &J_1[\xi,\eta] \quad \text{for} \quad ((\xi+\eta)^2+\xi)^2+\eta, \\ &U_1\zeta \quad \text{for} \quad URt\zeta, \\ &V_1\zeta \quad \text{for} \quad U\zeta. \end{aligned}$$

Then $$U_1 J_1[\xi,\eta] = \xi, \quad V_1 J_1[\xi,\eta] = \eta.$$

Also if $$V_1 S\zeta \neq 0 \quad \text{then} \quad U_1 S\zeta = U_1\zeta \quad \text{and} \quad V_1 S\zeta = SV_1\zeta.$$

For if $US\zeta \neq 0$ then $S\zeta$ fails to be a square and so $Rt(S\zeta) = Rt\zeta$, whence

$$U_1 S\zeta = URt(S\zeta) = URt\zeta = \zeta$$

and $$\begin{aligned} V_1 S\zeta &= S\zeta \dot{-} (Rt(S\zeta))^2 = (1+\zeta) \dot{-} (Rt\zeta)^2 \\ &= (\zeta \dot{-} (Rt\zeta)^2) + (1 \dot{-} ((Rt\zeta)^2 \dot{-} \zeta)) \\ &= (\zeta \dot{-} (Rt\zeta)^2) + 1. \end{aligned}$$

For this set of pairing functions the equations $U_1\zeta = \alpha$, $V_1\zeta = \beta$ have many solutions for $\zeta$, and $J_1[U_1\zeta, V_1\zeta] = \zeta$ can fail. Suppose

$$\rho[\xi,0] = \xi, \quad \rho[\xi,S\zeta] = \tau[\zeta,\rho[\xi,\zeta]],$$

as in (i). Write $\rho'[\zeta]$ for $\rho[U_1\zeta, V_1\zeta]$ then

$$\rho'[0] = 0, \quad \rho'[S\zeta] = \tau'[\zeta,\rho'[\zeta]],$$

where $$\tau'[\zeta,\eta] = \begin{cases} U_1 S\zeta & \text{if } V_1 S\zeta = 0, \\ \tau[V_1\zeta,\eta] & \text{if } V_1 S\zeta \neq 0. \end{cases}$$

Whence $$\tau'[\zeta,\eta] = U_1[S\zeta].B_1[V_1 S\zeta] + \tau[V_1\zeta,\eta].A_1[V_1 S\zeta].$$

Note that $A_1[0] = 0$, $A_1[S\zeta] = 1[A_1\zeta]$, where 1 is the constant function 1, i.e. $SO_1$. And $B_1[\zeta] = 1 \dot{-} A_1[\zeta]$. Thus recursion without parameter suffices for the definition of $A_1$ and $B_1$. Now write $\rho''\zeta$ for $J[\zeta, \rho'\zeta]$ then

$$\rho''0 = 0, \quad \rho''[S\zeta] = \tau''[\rho''\zeta],$$

where $\tau''$ for $J[SU_1\eta, \sigma'[U_1\eta, V_1\eta]]$, thus $\rho''$ satisfies a scheme of pure iteration without parameters, also $\rho'\zeta = V_1\rho''\zeta$.

Thus altogether if we adjoin addition, multiplication, limited subtraction and square root to the initial functions then the schemes of substitution and pure iteration without parameters suffice for the construction of all primitive recursive functions with any number of parameters.

Cor. *All primitive recursive functions are obtainable by adding to the initial functions the two functions addition and E and using only the schemes of substitution and pure iteration without parameter.*

We have to define limited subtraction, multiplication and square root in terms of addition and $E$ and the initial functions and the schemes of substitution and pure iteration without parameter. We first define multiplication and square root in terms of addition, limited subtraction and $Q$ and the schemes of substitution and pure iteration without parameter. Here

$$Q\nu = \begin{cases} 1 & \nu \text{ is a perfect square,} \\ 0 & \text{otherwise.} \end{cases}$$

Let $$F0 = 0, \quad FS\nu = SF\nu + 2.Q(S\nu).$$

Now $S\nu$ is a square if and only if $\nu = \kappa^2 + 2.\kappa$, if and only if $F\nu = \kappa^2 + 4\kappa$, if and only if $F\nu + 4$ is a square, thus $QS\nu = Q(F\nu + 4)$. Thus

$$\begin{aligned} FS\nu &= S(F\nu) + 2.Q(F\nu + 4) \\ &= BF\nu, \end{aligned}$$

where $$B\theta \quad \text{for} \quad S\theta + 2.Q(\theta + 4) \quad \text{and} \quad F\nu = \nu + 2[Rt\nu].$$

Thus $F$ is obtained by pure iteration without parameter from the functions $Q$, $S$ and addition. Again $(S\nu)^2 = SF(\nu^2)$ so that the function 'square of' is obtained by pure iteration without parameter from the initial functions $Q$, $S$ and addition. Let

$$T0 = 0, \quad TS\nu = B_1[T\nu] + 2.B_1[(T\nu \dot{-} 1) + (1 \dot{-} T\nu)].$$

Then $T\nu$ is the remainder when $\nu$ is divided by 3 and is obtained from addition and limited subtraction by pure iteration without parameter. Now let

$$G0 = 0, \quad GS\nu = S(G\nu) + T[G\nu],$$

so that $G$ is defined from $S$, $T$ and addition by pure iteration without parameter. Then we have

$$\left[\frac{\nu}{2}\right] = G\nu \dot{-} \nu,$$

$$(\nu \times \kappa) = \left[\frac{(((\nu+\kappa)^2 \dot{-} \nu^2) \dot{-} \kappa^2)}{2}\right],$$

$$Rt\nu = \left[\frac{F\nu \dot{-} \nu}{2}\right].$$

Thus we have defined multiplication and square root in terms of addition, limited subtraction and $Q$ using the initial functions and the schemes of substitution and pure iteration without parameter. Lastly we define limited subtraction and $Q$ in terms of addition, $E$ and the initial functions and the schemes of substitution and pure iteration without parameter. We have

$$Q\nu = B_1[E\nu], \quad (\nu \dot{-} \kappa) = E[(\nu+\kappa)^2 + 3.\nu + \kappa + 1],$$

since the preceding square is $(\nu+\kappa)^2 + 2.\nu + 2.\kappa + 1$.

This completes the demonstration of the Corollary.

## 5.12 *Existence of a calculable function which fails to be primitive recursive*

At the beginning of this chapter we defined a calculable function of natural numbers to be a rule or set of rules such that given a natural number we can, using the rules, effectively find another unique natural number called the value of that function for that natural number as argument. We then gave a method for obtaining rules of this type, the resulting functions were called primitive recursive functions. We also gave some other methods for obtaining calculable functions which at first sight appeared to be more general than the rules previously given, but we showed that they only produced primitive recursive functions. There are however other methods for obtaining calculable functions of natural numbers which lead outside the class of primitive recursive functions. We have already mentioned the calculable function constructed by Ackermann which failed to be primitive recursive. Ackermann constructed a calculable function of natural numbers which increased faster than any primitive recursive function. If we examine D 123, 124, 125, we see that they follow a pattern; we can continue the series by defining a sequence of functions for $\nu = 0, 1, 2, \ldots$ as follows

$$\Delta_0[\kappa, \pi] = \mathscr{I}(\boldsymbol{\lambda}\xi . S)\,\kappa\pi \qquad = \kappa + \pi,$$

$$\Delta_1[\kappa, \pi] = \mathscr{I}(\boldsymbol{\lambda}\xi\eta . \Delta_0[\kappa, \eta])\,0\pi = \kappa \times \pi,$$

$$\Delta_2[\kappa, \pi] = \mathscr{I}(\boldsymbol{\lambda}\xi\eta . \Delta_1[\kappa, \eta])\,1\pi = \kappa^{\pi},$$

and generally

$$\Delta_{S\nu}[\kappa, \pi] = \mathscr{I}(\boldsymbol{\lambda}\xi\eta . \Delta_\nu[\kappa, \eta])\,\kappa\pi \quad \text{for} \quad \nu > 1.$$

Then

$$\left\{\begin{aligned} &\Delta_0[\kappa, \pi] = \kappa + \pi, \\ &\Delta_{S\nu}[\kappa, 0] = \begin{cases} 0 & \text{for} \quad \nu = 0, \\ 1 & \text{for} \quad \nu = 1, \\ \kappa & \text{for} \quad \nu > 1, \end{cases} \\ &\Delta_{S\nu}[\kappa, S\pi] = \Delta_\nu[\kappa, \Delta_{S\nu}[\kappa, \pi]], \end{aligned}\right\} \tag{i}$$

continues the series. This is like double recursion with $\kappa$ as parameter except that the unknown function $\Delta$ is nested in itself instead of having a known function nested in the unknown function. It is easily seen that this function is calculable because the values are given outright on the co-ordinate axes and the value at $\{S\nu, S\pi\}$ is given in terms of values with lesser $\nu$ or same $\nu$ and lesser $\pi$.

In simple nested recursion the unknown function is nested in itself but then the recursion is on one variable only while here it is a double recursion. A scheme such as (i) is called *double nested recursion*. This scheme can lead outside the class of primitive recursive functions, this is why we have put $\nu$ as a subscript rather than as an argument. If the scheme (i) gave a primitive recursive function of $\nu$, $\kappa$, $\pi$ then by the scheme of substitution the function $\boldsymbol{\lambda}\xi . \Delta_\xi[\xi, \xi]$ would also be primitive recursive. But this is the function which Ackermann showed to increase faster than any primitive recursive function. The reader might like to evaluate

$$\Delta_1[1, 1], \quad \Delta_2[2, 2], \quad \Delta_3[3, 3] \quad \text{and} \quad \Delta_4[4, 4],$$

on a wet afternoon, this should convince him of the reasonableness of Ackermann's assertion!

We will, however, obtain another example of a calculable function which is outside the class of primitive recursive functions. The gist of the argument is this: we first give an effective enumeration of

$\mathbf{A}_{00}$-functions of type $\iota\iota$. This amounts to an effective enumeration of all primitive recursive functions. Let the enumeration be:

$$\rho_{|}, \rho_{||}, \rho_{|||}, \ldots, \rho_{|\kappa|}, \ldots,$$

where $|\kappa|$ denotes a sequence of tallies obtained by replacing each $S$ in $\kappa$ by a tally and omitting 0 and parentheses. We now form the sequence: $S(\rho_{|}\, 1), S(\rho_{||}\, 2), S(\rho_{|||}\, 3), \ldots$, as far as we please. Clearly this gives us the successive values of a calculable function of natural numbers. Suppose that this function is equivalent to a primitive recursive function, then by Prop. 1 of this chapter it is $\mathbf{A}_{00}$-represented by a closed $\mathbf{A}_{00}$-function of type $\iota\iota$. This $\mathbf{A}_{00}$-function will then appear in the above enumeration, say it is $\rho_{|\kappa|}$, then $\rho_{|\kappa|}\nu = S(\rho_{|\nu|}\nu)$ for each numeral $\nu$, in particular if $\nu$ is $\kappa$ we obtain: $\rho_{|\kappa|}\kappa = S(\rho_{|\kappa|}\kappa)$ as an $\mathbf{A}_{00}$-theorem, hence $\mathbf{A}_{00}$ will be inconsistent, which is absurd. Hence this set of rules leads outside the class of primitive recursive functions.

In the above we wrote $\rho_{|\kappa|}\nu$, where $|\kappa|$ stands for a sequence of tallies, rather than $\rho[\kappa, \nu]$, because the latter means that there is an $\mathbf{A}_{00}$-function-form $\rho[\Gamma_\iota, \nu]$ from which $\rho[\kappa, \nu]$ arises on replacing $\Gamma_\iota$ everywhere by $\kappa$. But the numeral $\kappa$ may fail to occur in $\rho_{|\kappa|}\nu$ so that $\rho_{|\Gamma_\iota|}\nu$ is nonsense. This actually occurs, we have defined a sequence of one-place functions rather than a two-place function, it is in this way that the absurdity is avoided. $\mathbf{A}_{00}$ is so poor in modes of expression that the enumeration fails to be $\mathbf{A}_{00}$-definable, that is to say that there fails to be an $\mathbf{A}_{00}$-function $\sigma$ of type $\iota\iota\iota$ such that $\sigma\kappa\nu = \rho_{|\kappa|}\nu$ for all numerals $\kappa$, $\nu$.

To complete the instructions to define the calculable function whose value for the numeral $\nu$ is $S(\rho_{|\nu|}\nu)$ we require to give instructions to enumerate the closed $\mathbf{A}_{00}$-functions of type $\iota\iota$, since we have already given sufficient instructions how to proceed afterwards.

## 5.13 *Enumeration of primitive recursive functions*

We replace the $\mathbf{A}_{00}$-symbols by numerals as follows

| | | | | | | | | | | |
|---|---|---|---|---|---|---|---|---|---|---|
| 0 | $S$ | $=$ | $\neq$ | & | $\vee$ | $\mathscr{I}$ | $\boldsymbol{\lambda}$ | ( | ) | $x^{(\theta)}$ |
| 0 | 1 | 2 | 3 | 4 | 5 | 6 | 7 | 8 | 9 | $(10+\theta)$ |

These numerals will act like *names* of the symbols above them, they will be called the *Gödel numerals* (*g.n.* for short) of the $\mathbf{A}_{00}$-symbols. Each numeral denotes a unique $\mathbf{A}_{00}$-symbol and each $\mathbf{A}_{00}$-symbol is denoted by exactly one numeral. We now replace the $A_{00}$-formulae by numerals as follows: The null formula is replaced by zero. The $A_{00}$-formula consisting of the single $A_{00}$-symbol $\Sigma$ is replaced by the numeral determined by $\langle\nu\rangle$ where $\nu$ is the *g.n.* of $\Sigma$. If the $A_{00}$-formula $\Phi$ has been replaced by the numeral $\nu$ and the $A_{00}$-formula $\Psi$ by the numeral $\kappa$ then the $A_{00}$-formula $\Phi^\frown\Psi$ is replaced by the numeral determined by $\nu^\frown\kappa$. These numerals are called the *g.n.*'s of the corresponding $A_{00}$-formulae. Similarly the *g.n.* of a sequence of $A_{00}$-formulae whose *g.n.*'s are respectively $\nu', \ldots, \nu^{(\pi)}$ is the numeral determined by $\langle\nu', \ldots, \nu^{(\pi)}\rangle$.

These numerals act as the names of the $A_{00}$-symbols, formulae or sequences of $A_{00}$-formulae as the case may be. The context will make it clear which is intended. Thus a numeral may just represent a natural number or act as the name of an $A_{00}$-symbol or as the name of an $A_{00}$-formula or as the name of a sequence of $A_{00}$-formulae. Given a numeral $\nu$ we can find the $A_{00}$-symbol, the $A_{00}$-formula and the sequence of $A_{00}$-formulae of which it is the name. The *g.n.*'s have been so chosen that each numeral has been employed in each of the three cases.

A numerical term $\alpha$ is called a *normal numerical term of kind* $\kappa$ if it is either (i) 0 or (ii) one of $x, x'.x'', \ldots, .x^{(2.\kappa)}$ or (iii) $S\beta$ where $\beta$ is a normal numerical term of kind $\kappa$ or (iv) $(((\mathscr{I}(\lambda x^{(2.\kappa+1)}(\lambda x^{(2.\kappa+2)}\beta)))\gamma)\delta)$, where $\delta$ and $\gamma$ are normal numerical terms of kind $\kappa$ and $\beta$ is a normal numerical term of kind $S\kappa$. To test if a numerical term $\alpha$ is a normal numerical term of kind $\kappa$ we have to test if proper parts of $\alpha$ are normal numerical terms of kinds $\kappa$ or $S\kappa$ and so on until we have to test whether a single symbol is a normal numerical term of kind $\kappa'$ for some calculable $\kappa'$. For instance if $\alpha$ satisfies case (iv) we have to test the proper part $\beta$ of $\alpha$ to see if it is a normal numerical term of kind $S\kappa$, if $\beta$ is $(((\mathscr{I}(\lambda x^{(2.\kappa+3)}(\lambda x^{(2.\kappa+4)}\beta')))\gamma')\delta')$, then we have to test the proper part $\beta'$ of $\beta$ to see if it is a normal numerical term of kind $SS\kappa$, and we have to test the proper parts $\gamma'$, $\delta'$ of $\beta$ to see if they are normal numerical terms of type $S\kappa$. The process is easily seen to terminate because we are referred to shorter and shorter formulae. The kind required is easily found at each stage of the testing. Hence to be a normal numerical of kind $\kappa$ is a calculable property of a numerical term, and hence of a numeral considered as the *g.n.* of a numerical term. A normal numerical term of kind $\kappa$ has at most $x, x'$, $x'', \ldots, x^{(2.\kappa)}$ as free variables. This is correct if the term consists of a single symbol, if it holds for terms with $\nu$ symbols and any kind then it follows at once for terms with $S\nu$ symbols and any kind.

An $\mathbf{A}_{00}$-function of type $u$ is called a *normal function of type* $u$ if it is $(\lambda x.\alpha)$ where $\alpha$ is a normal numerical term of kind 0.

Note that if $\nu$ is the *g.n.* of $\Psi$ and $f\{\nu\}$ is equal to the *g.n.* of $\Phi\{\Psi\}$, then $f\{\nu\}$ is of the form $\lambda'^\frown\nu^\frown\lambda''^\frown\ldots{}^\frown\lambda^{(\theta)\frown}\nu^\frown\lambda^{(S\theta)}$ where

$$\Phi\{\Gamma\} \quad \text{is} \quad \Phi'^\frown\Gamma^\frown\Phi''^\frown\ldots{}^\frown\Phi^{(\theta)\frown}\Gamma^\frown\Phi^{(S\theta)},$$

and $\Gamma$ fails to occur in $\Phi', \ldots, \Phi^{(S\theta)}$, and $\lambda'$ is the *g.n.* of $\Phi', \ldots, \lambda^{(S\theta)}$ is the *g.n.* of $\Phi^{(S\theta)}$.

The function $\rho_{|\kappa|}[\nu]$ qua function of $\kappa$ and $\nu$ is called an *enumerating function* for the class of primitive recursive functions. The one-place $\mathbf{A}_{00}$-functions $\rho$ and $\rho'$ are called *equivalent* if $\rho\nu = \rho'\nu$ is an $\mathbf{A}_{00}$-theorem for each numeral $\nu$. We wish to enumerate the one-place $\mathbf{A}_{00}$-functions; to do so we need only enumerate the one-place normal functions, because an $\mathbf{A}_{00}$-function is equivalent to a normal function by change of bound variables. In enumerating only the normal functions we merely eliminate some duplicates. The normal functions differ from the other functions merely by choice of bound variable.

D 170 $$t_1[\nu,\kappa] = \begin{cases} 0 & \text{if } \nu \text{ is the } g.n. \text{ of a normal numerical term of kind } \kappa, \\ 1 & \text{otherwise.} \end{cases}$$

Then $t_1[0,\kappa] = 1$, (0 is the *g.n.* of the null formula), and $t_1[S\nu,\kappa] = 0$ is equal to the principal characteristic function of

$$\begin{aligned} S\nu = 1 \vee \left(D[2, S\nu] = 0 \,\&\, 10 \leqslant \left[\frac{S\nu}{2}\right] \leqslant 10+2.\kappa\right) \\ \vee\, (E\xi)_\nu\, [S\nu = \langle 8,1\rangle^\frown\xi^\frown\langle 9\rangle \,\&\, t_1[\xi,\kappa] = 0] \\ \vee\, (E\xi,\xi',\xi'')_\nu\, [S\nu = \langle 8,8,8,6,8,7,11+2.\kappa,8,7,12+2.\kappa\rangle \\ {}^\frown\xi^\frown\langle 9,9,9\rangle^\frown\xi'^\frown\langle 9\rangle^\frown\xi''^\frown\langle 9\rangle \,\&\, t_1[\xi,S\kappa] = t_1[\xi',\kappa] = t_1[\xi'',\kappa] = 0]. \end{aligned}$$

This is a course of values recursion with substitution in parameter. We have given sufficient instructions to find an explicit $\mathbf{A}_{00}$-definition for $t_1[\nu,\kappa]$. The four clauses in the last statement are: $S\nu = 1$; i.e. $S\nu$ is the *g.n.* of an $\mathbf{A}_{00}$-formula consisting of the single symbol 0, i.e. $S\nu = \langle 0\rangle = 1$. The second clause says that $S\nu$ is the *g.n.* of a formula consisting of a single variable from among $x, x', x'', \ldots, x^{(2.\kappa)}$, i.e. $S\nu = \langle 10+\theta\rangle$, $0 \leqslant \theta \leqslant 2.\kappa$. The third clause says that $S\nu$ is the *g.n.* of an $\mathbf{A}_{00}$-formula $S\beta$ where $\beta$ has *g.n.* $\theta$ and $t_1[\theta,\kappa] = 0$. The fourth clause says that $S\nu$ is the *g.n.* of an

$\mathbf{A}_{00}$-formula $(((\mathscr{I}(\boldsymbol{\lambda}x^{(2.\kappa+1)}(\boldsymbol{\lambda}x^{(2.\kappa+2)}\beta)))\gamma)\delta)$, where $\gamma$ and $\delta$ have *g.n.*'s $\theta$ and $\theta'$ respectively and $t_1[\theta,\kappa] = t_1[\theta',\kappa] = 0$ and $\beta$ has *g.n.* $\theta''$ and $t_1[\theta'', S\kappa] = 0$. Notice that when we take the *g.n.* of an $\mathbf{A}_{00}$-formula we must first put the formula into full primitive notation without any definitional abbreviations. Also here we are dealing with $\mathbf{A}_{00}$-formulae even if they consist of a single symbol.

D 171 $t_{11}[\nu]$ is the characteristic function of

$$(E\xi)_\nu(\nu = \langle 8,7,10\rangle^\frown\xi^\frown\langle 9\rangle \;\&\; t_1[\xi,0] = 0).$$

$$\text{Then } t_{11}[\nu] = \begin{cases} 0 & \text{if } \nu \text{ is the } g.n. \text{ of a normal function,} \\ 1 & \text{otherwise.} \end{cases}$$

D 172 $$S_1[\nu] \quad \text{for} \quad \sum_{0\leqslant\xi\leqslant\nu} B_1[t_{11}[\xi]],$$

then $S_1[\nu]$ increases by unity each time $t_{11}[\nu] = 0$, i.e. each time $\nu$ is the *g.n.* of a normal function. Thus $S_1[\nu]$ is the number of normal functions with *g.n.*'s less than or equal to $\nu$. We can find an unending sequence of normal functions, namely: $(\boldsymbol{\lambda}x.x), (\boldsymbol{\lambda}x.(Sx)), (\boldsymbol{\lambda}x.(S(Sx))), \ldots$. Thus for each numeral $\kappa$ there is a least numeral $\nu$ such that $S_1[\nu] = \kappa$, and $\nu$ will be less than or equal to the *g.n.* $k[\kappa]$ of $(\boldsymbol{\lambda}x.(\underbrace{S(S\ldots(Sx)\ldots)}_{\kappa\text{-times}}))$.

D 173 $$\begin{cases} k[0] = \langle 8,7,10,10,9\rangle, \\ k[S\nu] = \langle \underset{1\leqslant\xi\leqslant 2\nu+3}{Pt}[k[\nu],\xi]\rangle^\frown\langle 8,1,10\rangle^\frown\underset{1\leqslant x\leqslant\nu+2}{\langle 9\rangle}. \end{cases}$$

Hence the least numeral $\nu$ such that $S_1[\nu] = \kappa$ is a primitive recursive function, $k[\nu]$ is the primitive recursive bound for the least number operator.

D 174 $$f[\kappa] \quad \text{for} \quad (\mu\xi)_{[k\kappa]}(S_1[\xi] = \kappa).$$

Then $f$ is a primitive recursive function which enumerates the *g.n.*'s of normal functions. Another method of enumeration is given in Ch. 12.

D 175 $$Num\,[0] = 1, \quad Num\,[S\nu] = \langle 8,1\rangle^\frown Num\,[\nu]^\frown\langle 9\rangle.$$

Then $Num\,[\nu]$ is equal to the *g.n.* of the numeral $\nu$. Note $Num\,[0] = \langle 0\rangle = 1$. The numeral $\nu$ is considered as a formula even if it consists of a single symbol.

From these definitions we see that

$$\langle 8\rangle^\frown f[\kappa]^\frown Num\,[\nu]^\frown\langle 9\rangle$$

is equal to the *g.n.* of $(\rho_{|\kappa|}\nu)$, where $\rho_{|\kappa|}$ is the $\kappa$th normal one-place function, it is a primitive recursive function of $\kappa$ and $\nu$. Again

$$\langle 8,1,8\rangle^\frown f[\kappa]^\frown Num\,[\nu]^\frown\langle 9,9\rangle$$

is equal to the *g.n.* of $(S(\rho_{|\kappa|}\nu))$, it is a primitive recursive function of $\kappa$ and $\nu$. Note that if the suffix $\kappa$ in $\rho_{|\kappa|}$ were an argument of an $\mathbf{A}_{oo}$-function rather than the place number of an $\mathbf{A}_{oo}$-function in a list of $\mathbf{A}_{oo}$-functions then it would have appeared in the *g.n.* of $(S(\rho_{|\kappa|}\nu))$ as $Num\,[\kappa]$ rather than just as $\kappa$ in the argument of $f$.

Let *Val* be that calculable function of natural numbers whose value for a numeral $\theta$ is the numeral determined by the closed numerical term whose *g.n.* is $\theta$ provided that $\theta$ is the *g.n.* of a closed numeral term and whose value is zero otherwise. Given $\theta$ we can decide whether it is the *g.n.* of a closed numerical term or otherwise, and if it is the *g.n.* of a closed numerical term then we can find the numeral which it determines. Hence given a numeral $\theta$ we can find a numeral $\nu$ such that $Val\,[\theta]$ has the value $\nu$. If *Val* were a primitive recursive function then by the scheme of substitution: $Val\,[\langle 8,1,8\rangle^\frown f[\kappa]^\frown Num\,[\kappa]^\frown\langle 9,9\rangle]$ would be a primitive recursive function of $\kappa$. But this is equivalent to the function whose value for the argument $\kappa$ we previously called $(S(\rho_{|\kappa|}\kappa))$. Hence *Val* is a calculable function of natural numbers which fails to be equivalent to any primitive recursive function. Thus we have an example of a calculable function outside the class of primitive recursive functions.

PROP. 5. *We can find a calculable function of natural numbers which fails to be primitive recursive.*

PROP. 6. $\lambda x\,.\,Val\,[\langle 8\rangle^\frown f[\kappa]^\frown Num\,[x]^\frown\langle 9\rangle]$ for $\kappa = 0,1,2,\ldots$ *is a sequence of functions, each member of the sequence is equivalent to a primitive recursive function and each primitive recursive function is equivalent to a member of the sequence.*

Since the function *Val* fails to be primitive recursive then it fails to be $\mathbf{A}_{oo}$-definable. If we want to have *Val* in a formal system then that system would have to be richer in modes of expression than the system $\mathbf{A}_{oo}$. The *g.n.*'s of the normal functions are enumerated by the function $f$ defined by D 174. But the functions themselves fail to be enumerated by a primitive recursive function in the sense that there fails to be a primitive recursive function $g$ of type $\iota\iota$ such that $g[\kappa,\nu] = (\rho_{|\kappa|}\nu)$. Thus the primitive recursive functions fail to be primitively recursively enumerable, but they are calculably enumerable.

## 5·14 *Definition of the proof-predicate for* $\mathbf{A}_{00}$

We now want to investigate how we can enrich the system $\mathbf{A}_{00}$ so that in the enriched system we can define the function *Val*. To do this we require some more definitions which will enable us to talk about the system $\mathbf{A}_{00}$ in the system $\mathbf{A}_{00}$, anyway to a limited extent.

D 176 $$Var[\nu] \quad \text{for} \quad D[2,\nu] = 0 \mathbin{\&} \left[\frac{\nu}{2}\right] \geqslant 10.$$

*Var* $[\nu]$ says that $\nu$ is the *g.n.* of an $\mathbf{A}_{00}$-formula which consists of a single variable.

D 177 $$Eq[\kappa,\nu] \quad \text{for} \quad \langle 8,8,2\rangle^\frown\kappa^\frown\langle 9\rangle^\frown\nu^\frown\langle 9\rangle.$$

If $\kappa$, $\nu$ are the *g.n.*'s of the $\mathbf{A}_{00}$-formulae $\alpha$, $\beta$ respectively then $Eq[\kappa,\nu]$ is equal to the *g.n.* of the $\mathbf{A}_{00}$-formula $((=\alpha)\beta)$.

D 178 $$Ineq[\kappa,\nu] \quad \text{for} \quad \langle 8,8,3\rangle^\frown\kappa^\frown\langle 9\rangle^\frown\nu^\frown\langle 9\rangle.$$

If $\kappa$, $\nu$ are the *g.n.*'s of the $\mathbf{A}_{00}$-formulae $\alpha$ and $\beta$ respectively then *Ineq* $[\kappa,\nu]$ is equal to the *g.n.* of the $\mathbf{A}_{00}$-formula $((\neq\alpha)\beta)$.

D 179 *Disj* $[\kappa,\nu]$ for
$$\langle 8,8,5\rangle^\frown\kappa^\frown\langle 9\rangle^\frown\nu^\frown\langle 9\rangle \times A_1[\kappa] \times A_1[\nu] + \kappa \,.\, B_1[\nu] + \nu \,.\, B_1[\kappa].$$

If $\kappa$, $\nu$ are the *g.n.*'s of the $\mathbf{A}_{00}$-formulae $\phi$ and $\psi$ respectively then *Disj* $[\kappa,\nu]$ is equal to the *g.n.* of the $\mathbf{A}_{00}$-formula $((\vee\phi)\psi)$, further *Disj* $[0,\nu] = \nu$ and *Disj* $[\kappa,0] = \kappa$, and *Disj* $[0,0] = 0$. We require D 179 in this form so as to be able to deal with subsidiary formulae in the $\mathbf{A}_{00}$-rules, these might be absent.

D 180 $$Conj[\kappa,\nu] \quad \text{for} \quad \langle 8,8,4\rangle^\frown\kappa^\frown\langle 9\rangle^\frown\nu^\frown\langle 9\rangle.$$

If $\kappa$, $\nu$ are the *g.n.*'s of the $\mathbf{A}_{00}$-formulae $\phi$ and $\psi$ respectively then *Conj* $[\kappa,\nu]$ is equal to the *g.n.* of the $\mathbf{A}_{00}$-formula $((\&\phi)\psi)$.

D 181 $$It[\nu,\kappa,\pi] \quad \text{for} \quad \langle 8,8,8,6\rangle^\frown\nu^\frown\langle 9\rangle^\frown\kappa^\frown\langle 9\rangle^\frown\pi^\frown\langle 9\rangle.$$

If $\nu$, $\kappa$, $\nu$ are respectively the *g.n.*'s of the $\mathbf{A}_{oo}$-formulae $\rho, \alpha, \beta$ then $It\,[\nu, \kappa, \pi]$ is equal to the *g.n.* of the $\mathbf{A}_{oo}$-formula $(((\mathscr{I}\rho)\,\alpha)\,\beta)$.

D 182 $$L[\kappa, \nu] \quad \text{for} \quad \langle 8, 7\rangle^\frown\kappa^\frown\nu^\frown\langle 9\rangle.$$

If $\kappa$, $\nu$ are respectively the *g.n.*'s of the $\mathbf{A}_{oo}$-formulae $\xi$ and $\alpha$ then $L[\kappa, \nu]$ is equal to the *g.n.* of the $\mathbf{A}_{oo}$-formula $(\lambda\xi\alpha)$.

D 183 $$S'[\kappa] \quad \text{for} \quad \langle 8, 1\rangle^\frown\kappa^\frown\langle 9\rangle.$$

If $\kappa$ is the *g.n.* of the $\mathbf{A}_{oo}$-formula $\alpha$ then $S'[\kappa]$ is equal to the *g.n.* of the $\mathbf{A}_{oo}$-formula $(S\alpha)$.

We now define '$\nu$ is the *g.n.* of an $\mathbf{A}_{oo}$-statement', '$\nu$ is the *g.n.* of an $\mathbf{A}_{oo}$-term' and '$\nu$ is the *g.n.* of an $\mathbf{A}_{oo}$-one-place function' by simultaneous recursion.

D 184 $Stat_{oo}[S\nu]$ for the principal characteristic function of

$$(E, \xi, \eta)_\nu\,[((S\nu = Conj\,[\xi, \eta] \vee S\nu = Disj\,[\xi, \eta])\;\&\; Stat_{oo}[\xi]\;\&\; Stat_{oo}[\eta]) \vee ((S\nu = Eq\,[\xi, \eta] \vee S\nu = Ineq\,[\xi, \eta])\;\&\; tm\,[\xi, \eta])],$$

$Stat_{oo}[0] = 1.$
If $Stat_{oo}[\nu] = 0$ then $\nu$ is the *g.n.* of an $\mathbf{A}_{oo}$-statement.

D 185 $tm\,[S\nu]$ for the principal characteristic function of

$$S\nu = 1 \vee Var\,[S\nu] \vee (E\xi, \eta)_\nu\,[S\nu = \langle 8\rangle^\frown\xi^\frown\eta^\frown\langle 9\rangle\;\&\; fn[\xi]\;\&\; tm[\eta]],$$

$tm\,[0] = 1.$
If $tm\,[\nu] = 0$ then $\nu$ is the *g.n.* of a numerical term.

D 186 $fn\,[S\nu]$ for the principal characteristic function of

$$S\nu = 3 \vee (E\xi, \eta)_\nu\,[S\nu = L[\xi, \eta]\;\&\; Var\,[\xi]\;\&\; tm[\eta]] \vee (E\xi, \eta)_\nu\,[\langle 8\rangle^\frown S\nu^\frown 1^\frown\langle 9\rangle = It[\xi, \eta, 1]\;\&\; tm[\eta]\;\&\; (E\zeta, \upsilon)_\nu\,[\xi = \langle 8, 7\rangle^\frown\zeta^\frown\upsilon^\frown\langle 9\rangle\;\&\; Var\,[\zeta]\;\&\; fn[\upsilon]]],$$

$fn\,[0] = 1.$

If $fn\,[\nu] = 0$ then $\nu$ is the *g.n.* of an $\mathbf{A}_{oo}$-one-place function. These last three definitions define the three functions $Stat_{oo}$, $tm$ and $fn$ by simultaneous course of values recursion. We have previously given sufficient indications so that we may obtain explicit definitions of these three functions.

D 187 $(\kappa \, Bd \, \nu)$ for $Var[\kappa] \, \& \, (\kappa \, Part \, \nu) \, \& \, (E\eta, \eta')_\nu [\nu = \eta^\frown\kappa^\frown\eta'$

$$\& \, (E\eta'', \zeta, \zeta', \eta''')_\nu [\eta = \eta''^\frown\langle 8, 7, [\tfrac{1}{2}\kappa]\rangle^\frown\zeta \, \& \, \eta'$$

$$= \zeta'^\frown\eta''' \, \& \, tm[\zeta^\frown\kappa^\frown\zeta'] \vee \eta = \eta''^\frown\langle 8, 7\rangle \, \& \, \eta' = \zeta'^\frown\eta'' \, \& \, tm[\zeta']]].$$

$\kappa$ is *bound* in $\nu$ if $\kappa$ is the *g.n.* of an $\mathbf{A}_{00}$-formula consisting of a variable standing alone (in which case the variable is $x^{[\frac{1}{2}\kappa]}$), the variable $x^{[\frac{1}{2}\kappa]}$ occurs in $\nu$ (hence $\nu$ is different from zero), each occurrence of the variable $x^{[\frac{1}{2}\kappa]}$ in the $\mathbf{A}_{00}$-formula whose *g.n.* is $\nu$ occurs in an occurrence of $(\boldsymbol{\lambda} x^{[\frac{1}{2}\kappa]} . \alpha)$ in $\nu$ where $\alpha$ is a term. We shall sometimes write '$\kappa$' for the $\mathbf{A}_{00}$-formula whose *g.n.* is $\kappa$. If we do this then the above explanation becomes: '$\kappa$' is a variable which occurs in '$\nu$' and each occurrence of '$\kappa$' in '$\nu$' occurs in an occurrence of $(\lambda`\kappa'\alpha)$ in '$\nu$' where $\alpha$ is a term. In the above definition we have allowed '$\nu$' to be any $\mathbf{A}_{00}$-formula.

D 188 $Cl[\nu]$ for $(A\xi)_\nu [Var[\xi] \, \& \, (\xi \, Part \, \nu) . \rightarrow (\xi \, Bd \, \nu)]$.

If all variables whose *g.n.*'s are less than $\nu$ (this includes all variables which occur in '$\nu$') and which occur in '$\nu$' are bound in '$\nu$', then we say '$\nu$' is a *closed* $\mathbf{A}_{00}$-*formula*.

We shall sometimes write $Cl \, Stat_{00}[\nu, \kappa]$ instead of

$$Cl[\nu] \, \& \, Cl[\kappa] \, \& \, Stat_{00}[\nu] \, \& \, Stat_{00}[\kappa]$$

and similarly in other cases; e.g. $Cl \, tm[\lambda, \mu, \nu]$, $Cl \, fn[\nu, \kappa]$, etc.

We now want to find the *g.n.* of an $\mathbf{A}_{00}$-formula $\Omega$ which results from the $\mathbf{A}_{00}$-formula $\Phi$ when each occurrence of the $\mathbf{A}_{00}$-formula $\Psi$ in the $\mathbf{A}_{00}$-formula $\Phi$ has been replaced by a corresponding occurrence of the $\mathbf{A}_{00}$-formula $\Xi$.

D 189
$$sub[\lambda, \theta; \nu] = \begin{cases} \theta \quad \text{if} \quad \nu = \lambda, \\ (\boldsymbol{\mu} v)_\nu [v^\frown\lambda \, Iseg \, \nu]^\frown\theta^\frown(\boldsymbol{\mu} v')_\nu [\nu = v_1{}^\frown\lambda^\frown v'], \\ \text{where } v_1 = (\boldsymbol{\mu} v)_\nu [v^\frown\lambda \, Iseg \, \nu], \\ \nu \quad \text{otherwise,} \end{cases}$$

replace the first occurrence of '$\lambda$' in '$\nu$' by a corresponding occurrence of '$\theta$'.

Write $sub\begin{bmatrix} \theta' & \theta'' \\ \lambda', & \lambda''; & \nu \end{bmatrix}$ for $sub[\theta', \lambda'; sub[\theta'', \lambda''; \nu]]$, etc.

We sometimes use the same definitions for the characteristic functions e.g. $Cl\,\nu$ could either stand for the definiens of D 188 or for its characteristic function, the context makes it clear which is intended.

We shall only use $sub\begin{bmatrix}\theta' & \theta'' \\ \lambda', & \lambda''; & \nu\end{bmatrix}$ when $\lambda'$ fails to occur in $\theta''$.

D 190 $$subst\,[\lambda,\theta,\nu,0] = \nu,$$
$$subst\,[\lambda,\theta,\nu,S\pi] = sub\,[\lambda,\theta;subst\,[\lambda,\theta,\nu,\pi]].$$

'$subst\,[\lambda,\theta,\nu,S\pi]$' is the result of replacing the first occurrence of '$\lambda$' in $subst\,[\lambda,\theta,\nu,\pi]$ by a corresponding occurrence of '$\theta$'. If we persist long enough we shall replace each occurrence of '$\lambda$' in '$\nu$' by a corresponding occurrence of '$\theta$', provided '$\lambda$' fails to occur in '$\theta$'. If '$\nu$' $= 0$ and '$\lambda$' $= 0$ and '$\theta$' is $S0$ then the substitution will go on indefinitely and we generate the numerals. We wish to avoid this so we shall take '$\lambda$' to be the ill-formed $\mathbf{A}_{00}$-formula ( ) whose *g.n.* is $\langle 8,9\rangle$, and '$\theta$' to be well-formed, then '$\lambda$' will fail to occur in '$\theta$'.

D 191 $$subst\,[\lambda,\theta,\nu] \quad \text{for} \quad subst\,[\lambda,\theta,\nu,\varpi\nu].$$

Then if '$\lambda$' fails to occur in '$\theta$' '$subst\,[\lambda,\theta,\nu]$' will be the $\mathbf{A}_{00}$-formula obtained from '$\nu$' by replacing every occurrence of '$\lambda$' by a corresponding occurrence of '$\theta$'. '$\nu$' consists of $\varpi\nu$ symbols so there will be at most $\varpi\nu$ occurrences of '$\lambda$' in '$\nu$'.

D 192 $$Subst\,[\theta,\lambda;\nu] \quad \text{for} \quad \sum_{\xi=1}^{M} subst\,[\theta,\langle 8,9\rangle,\xi]\,.\,B_2[\nu,subst\,[\lambda,\langle 8,9\rangle,\xi]]$$
$$\times \prod_{\eta,\eta'=1}^{M} A_2[\xi,\eta^\frown\lambda^\frown\eta'],$$

where $M = \langle Max_{1\leqslant\eta\leqslant\varpi\nu}\,[Pt\,[\nu,\eta],\langle 8,9\rangle]\rangle$; only one summand is different from zero. *Subst* is the result of replacing all occurrences of '$\lambda$' in '$\nu$' by '$\theta$'. The complicated bound $M$ is required to ensure that for one summand '$\nu$' is the result of replacing each occurrence of ( ) in some formula by corresponding occurrences of '$\lambda$', and that '$\lambda$' fails to occur in this formula. (Note D 143 and D 156.)

D 193 $$Ax[\nu] \quad \text{for} \quad Cl\,Stat_{00}\,\nu\,\&\,(E\xi,\xi',\xi'')_\nu\,[\nu = Eq\,[\xi,\xi]\vee$$
$$\nu = Ineq\,[1,S'\xi]\vee\nu = Ineq\,[S'\xi,1]\vee\nu = Eq\,[It[\xi'',\xi,1],\xi]\vee$$
$$\nu = Eq\,[It\,[\xi'',\xi,S'\xi'],\langle 8,8\rangle\,\xi''\xi'\langle 9\rangle\,It\,[\xi'',\xi,\xi']\langle 9\rangle]].$$

'$\nu$' is an $\mathbf{A}$-axiom 1, 2.1, 2.2, 3.1, 3.2.

If $Cl\,Stat\,\nu$ and $\nu = Eq[\pi, \pi]$ then $Cl\,tm\,\pi$, similarly in other cases, so we can omit $Cl\,tm[\xi, \xi']$, etc. as redundant.

D 194 $\quad \lambda\text{-}Ax\,(\text{i})\,[\nu]$ for $Cl\,Stat_{oo}\,\nu\,\&\,(E\xi,\xi')_\nu[\nu = Eq[\xi,\xi']\,\&$

$$(E\zeta,\zeta',\eta,\eta',\eta'')_\nu[\xi = \zeta^\frown\langle 8,8,7\rangle^\frown\eta^\frown\eta'^\frown\langle 9\rangle^\frown\eta''\langle 9\rangle^\frown\zeta'\,\&$$

$$\xi' = \zeta^\frown Subst[\eta'',\eta;\eta']^\frown\zeta']\,\&\,\zeta \underset{1\leqslant\nu\leqslant\varpi\zeta}{=} \langle 8\rangle].$$

'$\nu$' is a case of $A_{oo}$-axiom 4.1

D 195 $\quad \lambda\text{-}Ax\,(\text{ii})\,[\nu]$ for $Cl\,Stat_{oo}\,\nu\,\&\,(E\xi,\xi')_\nu[\nu = Eq[\xi,\xi']\,\&$

$$(E\zeta,\zeta',\eta,\eta',\upsilon,\upsilon')_\nu[\xi = \zeta^\frown\langle 8,7,\eta\rangle^\frown\eta'^\frown\langle 9\rangle^\frown\zeta'\,\&$$

$$\xi' = \zeta^\frown\langle 8,7,\upsilon\rangle^\frown\upsilon'^\frown\langle 9\rangle^\frown\zeta'\,\&\,\upsilon' = Subst[\upsilon,\eta;\eta']\,\&$$

$$N[\text{“}\upsilon\,Bd\,\eta'\text{”}]\,\&\,N[\text{“}\upsilon\,Bd\,\upsilon'\text{”}]\,\&\,\zeta \underset{1\leqslant\eta\leqslant\varpi\zeta}{=} \langle 8\rangle].$$

'$\nu$' is a case of $A_{oo}$-axiom 4.2.

D 196 $Rl'[\nu,\kappa]$ for $Cl\,Stat_{oo}[\nu,\kappa]\,\&\,(E\xi,\xi',\eta,\eta')_\nu$

$$[(\nu = Disj[Ineq[\xi,\xi'],\eta]\,\&\,\kappa = Disj[Ineq[S'\xi,S'\xi'],\eta]$$

$$(\vee\ \kappa = Disj[\xi,\nu]\,\&\,\nu \neq 0) \vee (\nu = Disj[Disj[Disj[\eta,\xi],\xi'],\eta']\,\&$$

$$\kappa = Disj[Disj[Disj[\eta,\xi'],\xi],\eta'])].$$

If $Rl'[\nu,\kappa]$ then '$\nu$' and '$\kappa$' are closed $A_{oo}$-statements and '$\kappa$' results from '$\nu$' by a one-premiss rule.

D 197 $\quad Rl''[\nu,\pi,\kappa]$ for $Cl\,Stat_{oo}[\nu,\pi,\kappa]\,\&\,(E\xi,\xi',\eta,\zeta,\upsilon,\upsilon')_\nu$

$$[\nu = Disj[Eq[\xi,\xi'],\eta]\,\&\,\pi = Disj[\zeta,\eta]\,\&\,\zeta = \upsilon'^\frown\xi^\frown\upsilon'\,\&$$

$$\kappa = Disj[\upsilon^\frown\xi'^\frown\upsilon',\eta]] \vee (E\xi,\xi',\eta)_\nu[\nu = Disj[\xi,\eta]\,\&$$

$$\pi = Disj[\xi',\eta]\,\&\,\kappa = Disj[Conj[\xi,\xi'],\eta]\,\&\,\xi.\xi' \neq 0].$$

If $Rl''[\nu,\pi,\kappa]$ then '$\kappa$' arises from '$\nu$' and '$\pi$' by a two-premiss $A_{oo}$-rule.

Note that the clause $Cl\,Stat_{oo}[\nu]$ has allowed us to omit a lot of other clauses in these definitions because they follow from $Cl\,Stat_{oo}[\nu]$. Thus in D 193 it is unnecessary to add the clauses $Cl\,tm[\xi,\xi']$ and $Cl\,fn[\xi'']$ because if $Cl\,Stat_{oo}[\nu]$ and $\nu = Eq[\xi,\xi]$ then we must have $Cl\,tm[\xi]$, similarly if $\nu = Eq[It[\xi'',\xi,S\xi'],\eta]$ then we must have $Cl\,fn[\xi'']$ and $Cl\,tm[\xi,\xi',\eta]$. And similarly at many places in the following definitions the requirements that the variables must satisfy are omitted because they must follow from the clause $Cl\,Stat_{oo}[\nu]$ and the hypothesis of the subclause in which they occur. These omissions decrease the complexity of the definitions which are already sufficiently complicated.

D 198 $\quad Prf_{00}[\nu, \kappa] \quad \text{for} \quad (A\xi)_{\varpi\nu \dot{-} 1}[Ax[Pt\,\nu, S\xi] \vee (E\eta)_{\xi}$
$[Rl'[Pt[\nu,\eta], Pt[\nu, S\xi]]] \vee (E\eta,\eta')[Rl''[Pt[\nu,\eta],$
$Pt[\nu,\eta'], Pt[\nu, S\xi]]]] \,\&\, Pt[\nu, \varpi\nu] = \kappa.$

'$\nu$' is a sequence of $\mathbf{A}_{00}$-statements which is an $\mathbf{A}_{00}$-proof of '$\kappa$'.

## 5.15 *The function Val*

We are now in a position to investigate what extensions could be made to the system $\mathbf{A}_{00}$ in order that we can represent the function *Val* in the extended system. If an unbounded existential quantifier were available then we could represent the equation $Val\,\kappa = \nu$ by

$$(Ex)\,Prf_{00}[x, Eq[\kappa, Num\,\nu]]. \tag{iii}$$

If we could find a primitive recursive bound for the existential quantifier in (iii) then we should have a primitive recursive predicate. Again if we had an unlimited least number operator then we could give an explicit definition of the function *Val*:

$$Val \quad \text{for} \quad \boldsymbol{\lambda} x.pt[(\mu x')[Prf_{00}[pt[x', 2, 1], Eq[x, Num[pt[x', 2, 2]]]]], 2, 2], \tag{iv}$$

here again if we could find a primitive recursive bound we would have a primitive recursive function.

According to our definitions a function of natural numbers is primitive recursive if and only if it is built up from the initial functions by the schemes of substitution and primitive recursion. If we follow the build up of a primitive recursive function then we can express it in terms of $O, U, S, /\frown$ (with appropriate subscripts and superscripts) and parentheses, and we can also $\mathbf{A}_{00}$-define an equivalent function. According to this definition the function $\boldsymbol{\lambda} x.(Val\,x \dot{-} Val\,x)$ fails to be primitive recursive even though its value is easily seen to be always zero. It fails to be primitive recursive because it is built up from the function *Val* which fails to be primitive recursive. If we wish to find the value of $Val\,\nu \dot{-} Val\,\nu$ then we must either find the value of $Val\,\nu$ or demonstrate that $\kappa \dot{-} \kappa$ has the value zero for each numeral $\kappa$. The first alternative fails to be a primitive recursive calculation and the second alternative is unavailable in the sytem $\mathbf{A}_{00}$, because this system is without general statements. Thus in either case the finding of the value of $Val\,\nu \dot{-} Val\,\nu$ is impossible in the system $\mathbf{A}_{00}$.

A closed $\mathbf{A}_{00}$-statement $\phi$ is equivalent to an $\mathbf{A}_{00}$-equation $\alpha = 0$ in the sense that if one is an $\mathbf{A}_{00}$-theorem then so is the other. Thus a closed $\mathbf{A}_{00}$-statement $\phi$ is an $\mathbf{A}_{00}$-theorem if and only if $Val\,\kappa = 0$, where $\kappa$ is the *g.n.* of $\alpha$. Hence we might be tempted to conclude that the property of a numeral of being the *g.n.* of an $\mathbf{A}_{00}$-theorem was calculable but failed to be primitive recursive. But we can only conclude that this particular way of expressing the property of being an $\mathbf{A}_{00}$-theorem fails to be primitive recursive. There might be another way which was primitive recursive. Consider a primitive recursive one-place function $\rho$, it is equivalent to the function $\boldsymbol{\lambda}x.((\rho x) \times S(Val\,x \dot{-} Val\,x))$ which fails to be primitive recursive. Whether a function which is equivalent to a primitive recursive function is itself primitive recursive depends on how the function is defined.

In classical mathematics two functions are considered to be the same if they take the same values for the same arguments. Here we have to consider two functions to be different even though they always take the same value for the same argument, on the ground that the definitions of the functions are different in some respect. A calculable function of natural numbers is given by a set of rules. The functions can, as we have seen, be equivalent yet the rules are in one case primitive recursive and fail to be so in the other case. The classical standpoint is called *extensional* the one we have to adopt is called *intensional.*

D 199 $\qquad s_{00}[\nu,\kappa]$ for $Subst\,[Num\,\nu, 21;\kappa]$.

'$s_{00}[\nu,\kappa]$' is the result of replacing each occurrence of the variable $x$ in '$\kappa$' by a corresponding occurrence of the numeral $\nu$. $s_{00}$ is called *Gödel's substitution function.* When using it we stipulate that the variable $x$ fails to occur bound in '$\kappa$'. This can always be achieved by a preliminary change of bound variable, by Ax. 4.2.

PROP. 7. *The property of being an* $\mathbf{A}_{00}$*-theorem is calculable but fails to be primitive recursive.*

We have already shown that the system $\mathbf{A}_{00}$ is decidable (Prop. 6, Ch. 4). This means that the function *th* of natural numbers whose value for the natural number $\kappa$ is zero if '$\kappa$' is an $\mathbf{A}_{00}$-theorem, and whose value is one otherwise, is a calculable function of natural numbers. Suppose that the function *th* is equivalent to a primitive recursive function, then an equivalent function is $\mathbf{A}_{oo}$-definable by an $\mathbf{A}_{oo}$-one-place function, say *th*, suppose that the variable $x$ fails to occur in *th*.

Consider $th[s_{oo}[x, x]] = 1$ and let its *g.n.* be $\kappa$, let $\nu$ be the *g.n.* of $th[s_{oo}[\kappa, \kappa]] = 1$, then $\nu$ is the numeral determined by $s_{oo}[\kappa, \kappa]$ (because '$\nu$' is the result of replacing each occurrence of the variable $x$ in '$\kappa$' ($x$ fails to occur in *th*) by the numeral $\kappa$) so that $s_{oo}[\kappa, \kappa] = \nu$ is an $\mathbf{A}_{oo}$-theorem. Suppose that the numeral determined by $th\nu$ is 0, then $th\nu = 0$ is an $\mathbf{A}_{oo}$-theorem. But then the statement '$\nu$' is an $\mathbf{A}_{oo}$-theorem, hence $th[s_{oo}[\kappa, \kappa]] = 1$ is an $\mathbf{A}_{oo}$-theorem, but then by R 1 so is $th\nu = 1$, hence $th\nu$ determines 1. But this is absurd because a closed numerical term determines a unique numeral.

Now we are supposing that the function *th* is primitive recursive and we have just shown that $th\nu$ fails to determine 0 hence it must determine 1 (it certainly determines a unique numeral and it can only determine 0 or 1). Hence $th\nu = 1$ is an $\mathbf{A}_{oo}$-theorem and hence '$\nu$' either fails to be a closed $\mathbf{A}_{oo}$-statement or is a closed $\mathbf{A}_{oo}$-statement which fails to be an $\mathbf{A}_{oo}$-theorem. But '$\nu$' is the closed $\mathbf{A}_{oo}$-statement $th[s_{oo}[\kappa, \kappa]] = 1$. Thus $th[s_{oo}[\kappa, \kappa]] = 1$ fails to be an $\mathbf{A}_{oo}$-theorem, thus $th\nu = 1$ fails to be an $\mathbf{A}_{oo}$-theorem, but $th\nu$ determines a numeral distinct from 1, hence it determines 0, but we had already found that it determined 1, this is absurd because a closed numerical term determines a unique numeral. Altogether the calculable function *th* fails to be $\mathbf{A}_{oo}$-definable and hence by Prop. 1 of this chapter the function *th* fails to be primitive recursive. Thus we have found an example of a calculable function which fails to be primitive recursive.

PROP. 8. $\mathbf{A}_{oo}$-*truth fails to be* $\mathbf{A}_{oo}$-*definable.*

The property of being an $\mathbf{A}_{oo}$-true $\mathbf{A}_{oo}$-statement coincides with the property of being an $\mathbf{A}_{oo}$-theorem (Prop. 3, Ch. 4). Thus if $\mathbf{A}_{oo}$-truth were $\mathbf{A}_{oo}$-definable then so would the property of being an $\mathbf{A}_{oo}$-theorem and we have just seen that this is absurd.

COR. (i). $\mathbf{A}_{oo}$-*falsity fails to be* $\mathbf{A}_{oo}$-*definable.*

A test of $N[\text{``}\phi\text{''}]$ for $\mathbf{A}_{oo}$-falsity, where $\phi$ is a closed $\mathbf{A}_{oo}$-statement is a test of $\phi$ for $\mathbf{A}_{oo}$-truth.

Cor. (ii). *The unlimited existential quantifier fails to be* $A_{00}$*-definable.*

If the unlimited existential quantifier were $A_{00}$-definable so that given an $A_{00}$-statement $\phi\{\xi\}$ with the variable $\xi$ as sole free variable we could construct a closed $A_{00}$-statement $(E\xi)\,\phi\{\xi\}$ which was $A_{00}$-true if and only if $\phi\{\alpha\}$ is $A_{00}$-true for some closed numerical term $\alpha$, then we could $A_{00}$-define the function *th* by means of the equation $th\nu = 0$ for $(Ex)\,[Prf_{00}[x, \nu]]$. We could then repeat the argument of Prop. 7 and again obtain an absurdity.

Cor. (iii). *The unlimited universal quantifier fails to be* $A_{00}$*-representable.* This follows the view of the definition

$$(E\xi)\,\phi\{\xi\} \quad \text{for} \quad N[\text{``}(A\xi)\,[N[\text{``}\phi\{\xi\}\text{''}]]\text{''}].$$

Cor. (iv). *The unlimited least number operator fails to be* $A_{00}$*-representable.* For if it were then we could $A_{00}$-represent *Val.*

## Historical remarks to Chapter 5

The concept of a calculable function seems to go back to Turing (1936), Church (1932) and Skolem (1923), though the concept of effectiveness, constructiveness and finiteness goes back further to Kronecker (1887), Weyl (1918), Brouwer (1908) and Hilbert (H–B). Recursive functions, or functions defined by induction go back to Dedekind (1888). The related concept of algorithm goes back to the Arabs, whence came the word 'algorithm'. Particular algorithms were known to Euclid, e.g. the algorithm for the greatest common divisor, also the sieve of Eratosthenes.

Throughout the ages mathematicians have been seeking for algorithms for this and for that. Thus, like the philosoper's stone, they sought for a universal algorithm. Algorithms were first invented for dealing with algebraic problems, so that when Descartes invented analytical geometry algorithms could be applied to geometry as well. Leibniz, as already mentioned, wanted to find a universal algorithm for all mathematics.

Certain recurrent series, like that of Fubonacci, can be considered as the true fore-runners of primitive recursive functions. Peano in his axioms did not include the assertion that functions defined by induction can be defined by recursive definitions, but took it as self-evident that this assertion is included in the axiom of induction. Dedekind (1888) noted that this fact requires proof and gave one. Since then Kalmár (1920), Landau (1930) and Lorenzen have given simpler proofs. But Skolem (1923) seems to have been the first to study primitive recursive functions as a discipline in its own right followed by Hilbert–Bernays (1934) and Péter (1951, 1967). Almost at once applications were at hand. The metamathematics of Hilbert (1922) required effective, constructive or finite proofs, so here recursive functions came into use for they are effective. In a formal system definitions, constructions and everything used must be constructive, effective and finite. Gödel (1930) immediately put recursive functions to good use in his arithmetization of syntax.

Meanwhile various writers produced accounts of calculable functions, they started on very different lines and for very different reasons. The first definition of primitive recursive function appears in Hilbert–Bernays. To them is due the functions $A_1$, $B_1$, $A_2$, $B_2$.

There are many ways of forming (1–1)-correspondences between ordered $n$-tuplets of natural numbers and the natural numbers themselves, and similarly between ordered bounded sequences of natural numbers and the natural numbers, it is a matter of choice which one takes.

Primitive recursive functions were studied by Péter (1951, 1967) who did a lot of work in showing that other schemes, apparently more general, gave equivalent functions. On the other hand schemes were found that gave calculable functions outside the class of primitive recursive functions, these will be the subject of Chs. 6 and 7. Much of Péter's work is reproduced here, but the case of simple nested recursion is due to Rowbottom and Lachlan. Prop. 4 is due to R. M. Robinson (1947) who has done much work on this line.

The diagonal argument leading to Prop. 5 originates with Cantor (1895, 1897). Combining this method with the arithmetization of syntax due to Gödel (1930) we get Prop. 5. We shall use the diagonal method many times in the course of this book, ultimately we shall find that it is not strong enough and we shall require the *priority method* of Friedberg (1957) and Mucknik (1956), later on again even this is not strong enough and at present research is being directed to find even stronger methods.

Definitions of the type of D 176–99 were first given by Gödel (1930). It is at this point that we first encounter a function which by its definition fails to be primitive recursive yet it is equivalent to a primitive recursive function. Thus we are naturally led to the consideration of the intensional and the extensional points of view.

The idea of using the ill-formed formula ( ) as a marker is due to Quine. The first calculable but non-primitive recursive function was invented by Ackermann (1928).

## EXAMPLES 5

1. $C$ is a class of functions containing the successor function and closed with respect to composition. Show that an enumerating function for the class $C$ fails to belong to $C$.

2. Do the following schemes define computable functions? If so, are the functions primitive recursive?

$$\begin{cases} f[0,\kappa] = \alpha[\kappa], \\ g[0,\kappa] = \beta[\kappa], \\ f[S\nu,\kappa] = h[\kappa,\nu,f[\nu,g[\nu,\kappa]]], \\ g[S\nu,\kappa] = k[\kappa,\nu,g[\nu,f[\nu,\kappa]]]. \end{cases}$$ $\alpha$, $\beta$, $h$, $k$ are primitive recursive.

$$\begin{cases} f[0,\kappa] = g[\kappa], \\ f[S\nu,\kappa] = g'[\nu,\kappa,f[h\nu,k[\nu,\kappa]],f[h'\nu,k'[\nu,\kappa]]], \end{cases}$$ where $h\nu$, $h'\nu \leqslant \nu$, $g$, $g'$, $h$, $h'$, $k$, $k'$ are primitive recursive.

3. Express the following functions in terms of $S, U, /, O$ and $\frown$ with numerical suffixes and parentheses:

$$exp, !, P, \dot{-}, A_1, B_1, A_2, B_2, \Sigma \text{ and } \Pi.$$

4. Find primitive recursive definitions for
$[\sqrt[3]{\nu}]$, $[\sqrt[\kappa]{\nu}]$, i.e. the integral part of the root; $p_\nu$ the $\nu$th prime; the greatest prime less than or equal to $\nu$; H.C.F.$[\nu,\kappa]$; L.C.M.$[\nu,\kappa]$ and $\nu^\nu$.

5. Define $+$, $\times$, in terms of $exp$. $\mu$, and composition.

6. The $\nu$th decimal place of the real number $\alpha$ is $f\nu$ where $f$ is a primitive recursive function, the similar function for the real number $\beta$ is $g$ which is also primitive recursive. The similar functions for $\alpha+\beta$ and $\alpha\times\beta$ are $h$ and $k$ respectively, are $h$ and $k$ computable? primitive recursive?

7. The rational real numbers can be defined as ordered triplets of natural numbers, the triplet $[\lambda,\mu,\nu]$ stands for $\dfrac{\lambda-\mu}{\nu}$, $\nu \neq 0$, define $=_r$, $<_r$, $+_r$, $\times_r$, $-_r$ and $\div_r$ for these triplets. A sequence of rational

numbers can now be defined by a function of natural numbers whose values are natural numbers, the latter being interpreted as rationals. Such a sequence is said to define a real number if it is a Cauchy sequence. Also a primitive recursive function $f$ is said to be a primitive recursive real number if

$$|f\kappa -_r f\pi| <_r \frac{1}{\nu}^r \quad \text{for} \quad \kappa, \pi > g\nu,$$

where $g$ is a primitive recursive function.

Show that if $f$, $f'$ are primitive recursive real numbers then so are $(f +_r f')$, $(f \times_r f')$, $(f -_r f')$, and $\left(\frac{f}{f'}^r\right)$ (for $f\nu' \neq 0$ for any $\nu$).

8. Is the following scheme computable? primitive recursive?

$$\begin{cases} f[0, \kappa] = \alpha[\kappa], \\ f[S\nu, \kappa] = g[\nu, \kappa, f[h\nu, k[\nu, \kappa, f[h'\nu, \kappa]]]] \quad (h\nu, h'\nu \leqslant \nu), \end{cases}$$

where $g$, $\alpha$, $h$, $h'$, $k$ are primitive recursive.

9. Show that two identity functions $U_1$, $U_2$ will suffice to define all identity functions, where $U_1$ is $\boldsymbol{\lambda}xx'.x$ and $U_2$ is $\boldsymbol{\lambda}xx'.x'$.

10. Find the characteristic functions for $Ax.\nu$, $Rl'[\nu, \kappa]$.

11. Reduce to primitive recursive schemes:

$$\begin{cases} f[0, \kappa] = g[\kappa], \\ f[S\nu, \kappa] = h[\nu, \kappa, \prod_{0 \leqslant \xi \leqslant \nu} k[\xi, f[\xi, \kappa], \kappa]]. \end{cases}$$

$$\begin{cases} f[0, \kappa] = g[\kappa], \\ f[S\nu, \kappa] = h[\nu, \kappa, \langle k[\xi, f[\xi, \kappa], \kappa]\rangle_{1 \leqslant \xi \leqslant \nu}]. \end{cases}$$

$$\begin{cases} f[0, \kappa] = g[\kappa], \\ f[S\nu, \kappa] = h[\nu, \kappa, (\mu\xi)_\nu [k[\xi, f[\xi, \kappa], \kappa] = 0]], \end{cases}$$

where $g$, $h$, $k$ are primitive recursive.

12. Is the following scheme primitive recursive?

$$\begin{cases} f[0, 0, \kappa] = g'[\kappa], \\ f[0, S\nu, 0] = g''[\nu], \\ f[0, S\nu, S\kappa] = h'[\nu, \kappa, f[\nu, k[\nu, \kappa], k'[\nu, \kappa]]], \\ f[S\pi, \nu, \kappa] = h''[\nu, \kappa, \pi, f[\nu, k''[\nu, \pi], k'''[\nu, \pi]]], \end{cases}$$

where $g'$, $g''$, $h'$, $h''$, $k$, $k'$, $k''$, $k'''$ are primitive recursive.

13. Express the primitive recursive functions: *exp*, !, $P$, $A_1$, $A_2$, $B_1$, $B_2$, $\Sigma$, *Max* in terms of $+$, $\times$, $\dot{-}$, $\surd$ using substitution and pure iteration without parameter.

14. Write down the *g.n.* of *Max* $[\nu, \kappa]$, $\varpi\nu$, $A_1[\nu]$.

15. Let $\nu$ be the *g.n.* of $\phi$ and $\kappa$ be the *g.n.* of $\psi$. Find *Ded* such that *Ded* $[\nu, \kappa]$ if and only if $\dfrac{\phi}{\psi}$. Try similarly for $\dfrac{\phi}{\psi}*$.

16. Reduce the following scheme to a primitive recursive scheme, if possible:

$$f[0, \nu] = \sigma[\nu],$$

$$f[S\pi, \nu] = \tau[\pi, \nu, f[g\pi^\frown h\pi, k\nu]],$$

where $g$, $h$, $k$ are primitive recursive functions and $g\pi$ is the ordered set $\pi$ less its first component, $h\pi$ is an ordered set whose components are less than the first component of $\pi$, if the first component of $\pi$ is zero then $h\pi$ is the null set, similarly if $\pi$ is the null set then $h\pi$ is also the null set.

17. Discuss whether the following functions are primitive recursive?

$$f[\nu] = \begin{cases} 1 & \text{if a consecutive run of exactly } \nu \text{ 7's occurs in the decimal expansion of } \pi, \\ 0 & \text{otherwise.} \end{cases}$$

$$g[\nu] = \begin{cases} 1 & \text{if a consecutive run of at least } \nu \text{ 7's occurs in the decimal expansion of } \pi, \\ 0 & \text{otherwise.} \end{cases}$$

$$h[\nu] = \begin{cases} 1 & \text{if Goldbach's conjecture is true,} \\ 0 & \text{otherwise.} \end{cases}$$

# Chapter 6

# A complete undecidable arithmetic. The system $A_0$

## 6.1 *The system* $A_0$

The system $A_0$, now to be constructed, is a primary extension of the system $A_{00}$. It is constructed from the system $A_{00}$ by adding an extra symbol $E$ of type $o(o\iota)$, called the *existential quantifier* and adding one more building rule. We define:

D 200 $$(E\xi)\,\phi\{\xi\} \quad \text{for} \quad (E(\boldsymbol{\lambda}\xi\phi\{\xi\})),$$

and we read this as 'there is a numerical term $\alpha$, for which $\phi\{\alpha\}$' or 'some numerical term has the property $\boldsymbol{\lambda}\xi\phi\{\xi\}$'. We call $\boldsymbol{\lambda}\xi\phi\{\xi\}$ a *property* where $\phi\{\xi\}$ is a type $o$. The variable $\xi$ has been abstracted from $\phi\{\xi\}$ by the $\boldsymbol{\lambda}$-symbol. An $A_0$-statement is an $A_0$-formula of type $o$ no part of which is a property except parts governed by the existential quantifier or $\mathscr{I}$. An $A_0$-statement of the form $(E\xi)\,\phi\{\xi\}$ is called an *existential* $A_0$*-statement*.

The system $A_0$ has the $A_{00}$-symbols and the new symbol $E$. It has the $A_{00}$-axioms and the $A_{00}$-rules, now used with $A_0$-statements and the additional building scheme:

II *d*. $$\frac{\phi\{\alpha\} \vee \omega}{(E\xi)\,\phi\{\xi\} \vee \omega} \quad \textit{existential dilution.}$$

The subsidiary $A_0$-statement $\omega$ together with the disjunction sign may be absent. $\phi\{\alpha\}$ and $\omega$ are closed $A_0$-statements and $\alpha$ is free in $\phi\{\alpha\}$, thus $\alpha$ is a closed numerical term. It is unnecessary to have another $\boldsymbol{\lambda}$-rule to allow us to change the bound variable $\xi$ in $(E\xi)\,\phi\{\xi\}$ to another which fails to be free in $\phi\{\Gamma_\iota\}$ because the variable in the rule can be any variable so if we wanted to change later in an $A_0$-proof then we could have introduced the required variable at the application of II *d*.

$\alpha$ must be free in $\phi\{\alpha\}$, consider:

$$(Ex)\,(x = x \,\&\, x = 0) \,\&\, (Ex)\,(x = x \,\&\, x = 1)$$

and $$(Ex')\,((Ex)\,(x = x \,\&\, x' = 0) \,\&\, (Ex)\,(x = x \,\&\, x' = 1)),$$

which arises by applying the existential quantifier to a bound variable.

The $\mathbf{A}_0$-axioms and the $\mathbf{A}_0$-rules will be denoted by Ax. 1, ..., Ax. 4.2, R 1, ..., II *d*. The $\mathbf{A}_0$-axioms are closed hence the $\mathbf{A}_0$-theorems are also closed. The definition of negation for $\mathbf{A}_0$-statements only applies to $\mathbf{A}_{00}$-statements, i.e. $\mathbf{A}_0$-statements without the $E$ symbol. The $\mathbf{A}_{00}$-definitions D 112, 113, 114, 118 ... 158, 161 ... 199 carry over into the system $\mathbf{A}_0$. D 115, 116, 117 involve negation which fails to apply to the system $\mathbf{A}_0$, and D 159, 160 involve the characteristic function which again fails to carry over into the system $\mathbf{A}_0$. If $\alpha\{\xi\} = 0$ is the principal characteristic equation for the $\mathbf{A}_{00}$-statement $\phi\{\xi\}$ then we should require an infinite product to get a characteristic function for $(E\xi)\,\phi\{\xi\}$. Formulae like $E(=0)$, though of type $o$, are unrequired.

## 6.2 $\mathbf{A}_0$-*truth*

The definition of $\mathbf{A}_0$-truth is obtained from that for $\mathbf{A}_{00}$-truth by adding:

(v) $\phi$ is of the form $(E\xi)\,\psi\{\xi\}$ and $\psi\{\nu\}$ is $\mathbf{A}_0$-true for some numeral $\nu$.

The resulting five conditions we call $\mathscr{T}_0$. An $\mathbf{A}_0$ statement which is known to satisfy $\mathscr{T}_0$ is called $\mathbf{A}_0$-true. The definition of $\mathbf{A}_0$-falsity is obtained from $\mathscr{F}_{00}$ by adding the condition:

(v) $\phi$ is of the form $(E\xi)\,\psi\{\xi\}$ and $\psi\{\nu\}$ is $\mathbf{A}_0$-false for each numeral $\nu$.

The resulting five conditions are called $\mathscr{F}_0$. An $\mathbf{A}_0$-statement which is known to satisfy $\mathscr{F}_0$ is called $\mathbf{A}_0$-false. As in the case of $\mathbf{A}_0$-truth the required knowledge might be unknown to the reader.

PROP. 1. *It is impossible to show that an* $\mathbf{A}_0$-*statement is both* $\mathbf{A}_0$-*true and* $\mathbf{A}_0$-*false.*

Suppose that $\phi$ is a closed $\mathbf{A}_0$-statement and that $\phi$ has been shown to be both $\mathbf{A}_0$-true and $\mathbf{A}_0$-false, and suppose that we have shown the required result for $\mathbf{A}_0$-statements with fewer logical symbols. By Prop. 1 of Ch. 4 the result holds for $\mathbf{A}_0$-statements without logical symbols.

(i) Suppose $\phi$ is $\phi' \vee \phi''$ and has been shown to be $\mathbf{A}_0$-false then both $\phi'$ and $\phi''$ have been shown to be $\mathbf{A}_0$-false, by induction hypothesis it is impossible to show that $\phi'$ is $\mathbf{A}_0$-true or that $\phi''$ is $\mathbf{A}_0$-true, hence by (iv) of $\mathscr{T}_0$ it is impossible to show that $\phi$ is $\mathbf{A}_0$-true.

(ii) Suppose $\phi$ is $\phi' \,\&\, \phi''$ and has been shown to be $\mathbf{A}_0$-true, then by $\mathscr{T}_0$(iii) $\phi'$ has been shown to be $\mathbf{A}_0$-true and $\phi''$ has been shown to be $\mathbf{A}_0$-true, by our induction hypothesis it is impossible to show that $\phi'$ is $\mathbf{A}_0$-false and impossible to show that $\phi''$ is $\mathbf{A}_0$-false, by $\mathscr{F}_0$(iii) it is impossible to show that $\phi$ is $\mathbf{A}_0$-false.

(iii) Suppose that $\phi$ is $(E\xi)\,\psi\{\xi\}$ and has been shown to be $\mathbf{A}_0$-true, then by $\mathscr{T}_0$(v) $\psi\{\nu\}$ has been shown to be $\mathbf{A}_0$-true for some numeral $\nu$, by our induction hypothesis it is impossible to show that $\psi\{\nu\}$ is $\mathbf{A}_0$-false, thus by $\mathscr{F}_0$(v) it is impossible to show that $\phi$ is $\mathbf{A}_0$-false.

PROP. 2. *The system* $\mathbf{A}_0$ *is consistent with respect to* $\mathbf{A}_0$*-truth.*

We have to show that each $\mathbf{A}_0$-theorem is $\mathbf{A}_0$-true. A numerical term lacks the $E$ symbol hence as in Ch. 4, Prop. 1 a closed numerical term determines a unique numeral. The rest of the demonstration of the present proposition is the same as that of Prop. 2, Ch. 4 except that we must show that rule II $d$ preserves $\mathbf{A}_0$-truth.

Suppose that $\phi\{\alpha\} \vee \omega$ is $\mathbf{A}_0$-true then $\phi\{\alpha\}$ is $\mathbf{A}_0$-true or $\omega$ is $\mathbf{A}_0$-true. In the first case $\alpha$ determines a unique numeral $\nu$ and $\phi\{\nu\}$ is $\mathbf{A}_0$-true and so $(E\xi)\,\phi\{\xi\}$ is $\mathbf{A}_0$-true and hence $(E\xi)\,\phi\{\xi\} \vee \omega$ is also $\mathbf{A}_0$-true, in the second case $(E\xi)\,\phi\{\xi\} \vee \omega$ is also $\mathbf{A}_0$-true by $\mathscr{T}_0$(iv). Thus in either case II $d$ preserves $\mathbf{A}_0$-truth.

PROP. 3. *The system* $\mathbf{A}_0$ *is complete with respect to* $\mathbf{A}_0$*-truth.*

We have to show that if a closed $\mathbf{A}_0$-statement is $\mathbf{A}_0$-true then it is an $\mathbf{A}_0$-theorem. The demonstration is the same as that of Prop. 3, Ch. 4 except that we must add the following:

If the closed $\mathbf{A}_0$-statement $\phi$ is $\mathbf{A}_0$-true and is of the form $(E\xi)\,\psi\{\xi\}$ then by $\mathscr{T}_0$(v) $\psi\{\nu\}$ is $\mathbf{A}_0$-true for some numeral $\nu$. Now $\psi\{\nu\}$ contains fewer logical connectives than $\phi$. Hence by our induction hypothesis $\psi\{\nu\}$ is an $\mathbf{A}_0$-theorem. By II $d$ $\phi$ is an $\mathbf{A}_0$-theorem.

PROP. 4. *If an* $\mathbf{A}_0$*-disjunction is an* $\mathbf{A}_0$*-theorem then one of the disjunctands is an* $\mathbf{A}_0$*-theorem and we can find which and an* $\mathbf{A}_0$*-proof for it.*

We use theorem induction. The result holds for the $\mathbf{A}_0$-axioms, for they are one-termed disjunctions. If the result holds for the upper formula or upper formulae of a rule then it holds for the lower formula. Take for instance rule II $d$, if the upper formula has the property then either $\phi\{\alpha\}$ or $\omega$ is an $\mathbf{A}_0$-theorem we can find which and supply its $\mathbf{A}_0$-proof, if $\phi\{\alpha\}$ is the one so found then from its $\mathbf{A}_0$-proof (which we can also find) we can, using II $d$, find an $\mathbf{A}_0$-proof of $(E\xi)\,\phi\{\xi\}$, if $\omega$ is the one found then we also have found a disjunctand of $(E\xi)\,\phi\{\xi\} \vee \omega$ which is an $\mathbf{A}_0$-theorem and an $\mathbf{A}_0$-proof for it.

COR. (i). *Rule* I$b$ *is a derived rule in* $\mathbf{A}_0$.

If we have an $\mathbf{A}_0$-proof of $\phi \vee \psi \vee \psi$ then by Prop. 4 we can decide whether $\phi \vee \psi$ or $\psi$ is an $\mathbf{A}_0$-theorem and supply the $\mathbf{A}_0$-proof for the one which is found to be an $\mathbf{A}_0$-theorem. Repeat, if we find that $\phi \vee \psi$ is an $\mathbf{A}_0$-theorem, so we can find which of $\phi$ and $\psi$ is an $\mathbf{A}_0$-theorem and an $\mathbf{A}_0$-proof for the one that is an $\mathbf{A}_0$-theorem. In either case we can find an $\mathbf{A}_0$-proof for $\phi \vee \psi$.

PROP. 5. *An* $\mathbf{A}_0$*-statement* $\phi\{\mathfrak{x}\}$ *is equivalent to an* $\mathbf{A}_0$*-statement of one of the forms:* $\alpha\{\mathfrak{x}\} = 0$ *or* $(E\eta)(\alpha\{\eta, \mathfrak{x}\} = 0)$, *where* $\alpha\{\mathfrak{G}_\iota\}$ *is a closed* $\mathbf{A}_0$*-numerical term-form and* $\alpha\{\eta, \mathfrak{G}_\iota\}$ *has* $\eta$ *as sole free variable. Moreover* $\alpha\{\mathfrak{n}\}$ *and* $\alpha\{\kappa, \mathfrak{n}\}$ *determine* 0 *or* 1 *only.*

Clearly $\psi\{(E\xi)(\phi\{\xi\} \,\&\, \omega)\}$ and $\psi\{(E\xi)\,\phi\{\xi\} \,\&\, \omega\}$ have the same $\mathbf{A}_0$-truth values and so do $\psi\{(E\xi)(\phi\{\xi\} \vee \omega)\}$ and $\psi\{(E\xi)\,\phi\{\xi\} \vee \omega\}$, and similarly with $\omega$ and $\phi\{\xi\}$ interchanged. By repeated application we see that an $\mathbf{A}_0$-statement is equivalent to another $\mathbf{A}_0$-statement of the form $(E\mathfrak{x})\,\psi\{\mathfrak{x}\}$, where $\psi\{\mathfrak{x}\}$ is quantifier-free so that all the quantifiers have been brought to the left until they are all initially placed. Suppose that $\mathfrak{x}$ stands for $\xi', \xi'', \ldots, \xi^{(S\pi)}$ then

$$(E\mathfrak{x})\,\phi\{\mathfrak{x}\} \quad \text{and} \quad (E\xi)\,\phi\{pt[\xi, S\pi, 1], \ldots, pt[\xi, S\pi, S\pi]\}$$

have the same $\mathbf{A}_0$-truth values, hence if the one is an $\mathbf{A}_0$-theorem then so is the other. Again $\psi\{\mathfrak{x}\}$ is equivalent to $\alpha\{\mathfrak{x}\} = 0$ for some effectively findable numerical term $\alpha\{\xi\}$. Thus altogether

$$(E\mathfrak{x})\,\phi\{\mathfrak{x}\} \quad \text{and} \quad (E\xi)(\alpha\{\xi\} = 0)$$

for some effectively findable $\alpha\{\xi\}$ have the same $\mathbf{A}_0$-truth value hence they are equivalent as desired.

We now give *g.n.*'s to $\mathbf{A}_0$-symbols, formulae and sequences of formulae as before for $\mathbf{A}_{00}$ except that the *g.n.* of $E$ shall be 10 and the *g.n.* of $x^{(\kappa)}$ shall be $11+\kappa$ instead of $10+\kappa$. We can make definitions similar to D 168 ... 199 inclusive, we omit listing them because they are constructed on exactly the same lines, we only have to note the different role of 10 and that some of the definitions will want an extra clause to account for the new symbol $E$ and the new rule II $d$. One or two of these definitions we list but omit the full details that we gave before.

D 201 $\quad Prf_0[\nu, \kappa]$ for "'$\nu$' is an $\mathbf{A}_0$-proof of '$\kappa$'".

PROP. 6. *The system* $\mathbf{A}_0$ *contains the truth definition* $\mathscr{T}_0$.

We have to show that there is an $\mathbf{A}_0$-statement $Tx$ such that $T\nu$ is an $\mathbf{A}_0$-theorem if and only if '$\nu$' is an $\mathbf{A}_0$-theorem, for only then is '$\nu$' an $\mathbf{A}_0$-true $\mathbf{A}_0$-statement. The required statement is $(Ex)\,Prf_0[x,\nu]$, i.e. $\vdash_{\mathbf{A}_0}(Ex)\,Prf_0[x,\nu]$ if and only if $\vdash_{\mathbf{A}_0}$'$\nu$', if and only if '$\nu$' is $\mathbf{A}_0$-true.

PROP. 7. *An existential* $\mathbf{A}_0$*-statement is an* $\mathbf{A}_0$*-theorem if and only if a particular instance can be found and is* $\mathbf{A}_0$*-provable.*

We have to show that if $(E\xi)\,\phi\{\xi\}$ is an $\mathbf{A}_0$-theorem then we can find a numerical term $\alpha$ such that $\phi\{\alpha\}$ is an $\mathbf{A}_0$-theorem. This is the converse of II *d*. Let $\nu$ be the *g.n.* of $(E\xi)\,\phi\{\xi\}$ then '$\nu$' is an $\mathbf{A}_0$-theorem, hence $Prf_0[\pi,\nu]$ will be $\mathbf{A}_{00}$-true for some numeral $\pi$, we only have to test $Prf_0[0,\nu], Prf_0[1,\nu], Prf_0[2,\nu], \ldots$ until we come to a numeral $\pi$ for which $Prf_0[\pi,\nu]$ is $\mathbf{A}_{00}$-true. From the numeral $\pi$ we can find the sequence of $\mathbf{A}_0$-statements of which it is the *g.n.*, we can then find the $\mathbf{A}_0$-proof-tree of $(E\xi)\,\phi\{\xi\}$, this proof-tree can only end thus:

$$\cfrac{\cfrac{\phi'\{\alpha\}}{(E\xi)\,\phi'\{\xi\}}\;\mathrm{II}\,d}{(E\xi)\,\phi\{\xi\}}\;R\,1$$

there could be several applications of R 1 or they might be absent. Note that I *b* is absent from the system $\mathbf{A}_0$. From this we obtain:

$$\frac{\phi'\{\alpha\}}{\phi\{\alpha\}}$$

using the same applications of R 1 as before. Thus we have found a numerical term $\alpha$ such that $\phi\{\alpha\}$ is an $\mathbf{A}_0$-theorem.

Note that we have used

$$\frac{(Ex)\,Prf_0[x,\nu]}{\text{'}\nu\text{'}}*.$$

We also have

$$\frac{\text{'}\nu\text{'}}{(Ex)\,Prf_0[x,\nu]}*,$$

i.e. from an $\mathbf{A}_0$-proof of '$\nu$' we can find one of $(Ex)\,Prf_{00}[x,\nu]$.

## 6.3 *Undefinability of* $A_0$*-falsity in* $A_0$

PROP. 8. $A_0$*-falsity fails to be* $A_0$*-definable.*

Suppose that we could construct an $A_0$-statement $Fals\,x$ such that $Fals\,\nu$ is an $A_0$-theorem just in case '$\nu$' is an $A_0$-false $A_0$-statement.

D 202. $s_0[x, x']$ for Gödel's substitution function for $A_0$(see D 199), then $s_0$ is a primitive recursive function and hence is $A_{00}$-definable. Consider $Fals\,[s_0[x, x]]$, let its *g.n.* be $\kappa$. Let $\nu$ be the *g.n.* of $Fals\,[s_0[\kappa, \kappa]]$, then $s_0[\kappa, \kappa] = \nu$ is an $A_{00}$-theorem. If $Fals\,\nu$ is an $A_0$-theorem then so is $(Ex)\,Prf_0[x, \lambda]$, where $\lambda$ is the *g.n.* of $Fals\,\nu$, hence $Prf_0[\pi, \lambda]$ is an $A_{00}$-theorem for some numeral $\pi$, hence we can find an $A_0$-proof of $Fals\,\nu$ so we can show that $Fals\,\nu$ is $A_0$-true. But if $Fals\,\nu$ is an $A_0$-theorem then so is $Fals\,[s_0[\kappa, \kappa]]$ by R 1 and is shown to be $A_0$-true. But if $Fals\,\nu$ is $A_0$-true then '$\nu$' has been shown to be $A_0$-false, by Prop. 1 this is absurd. Thus $Fals\,\nu$ fails to be demonstrably $A_0$-true. If $Fals\,\nu$ has been shown to be $A_0$-false then the same applies to $Fals\,[s_0[\kappa, \kappa]]$. But $Fals\,\nu$ is $A_0$-false means that ''$\nu$' is $A_0$-false' is $A_0$-false, i.e. either '$\nu$' fails to be an $A_0$-statement or '$\nu$' is $A_0$-true. But '$\nu$' is the closed $A_0$-statement $Fals\,[s_0[\kappa, \kappa]]$ hence $Fals\,[s_0[\kappa, \kappa]]$ has been shown both $A_0$-true and $A_0$-false, by Prop. 1 this is absurd. Hence *Fals* with the required properties fails to be $A_0$-definable. Thus we have a definition of $A_0$-truth but lack one for $A_0$-falsity.

COR. (i). *Negation fails to be* $A_0$*-definable.*

Suppose that to each $A_0$-statement $\phi$ we could uniquely associate another $A_0$-statement $N[\phi]$ such that $N[\phi]$ is closed if and only if $\phi$ is closed, and $N[\phi]$ is $A_0$-true just in case $\phi$ is $A_0$-false and vice versa, and that $N[\phi\{\nu\}]$ is the result of substituting $\nu$ for all free occurrences of $\xi$ in $N[\phi\{\xi\}]$.

D 203 $Cl\,Stat_0\,\nu$ the primitive recursive $A_{00}$-statement which is $A_{00}$-true just in case '$\nu$' is a closed $A_0$-statement.

Define:

$$Fals\,x \quad \text{for} \quad Cl\,Stat_0\,x\,\&\,N[(Ex')\,Prf_0[x', x]].$$

Then $Fals\,\nu$ is $A_0$-true just in case '$\nu$' is $A_0$-false, but this is impossible in the system $A_0$. Hence negation fails to be $A_0$-definable.

COR. (ii). *Material implication fails to be* $\mathbf{A}_0$*-definable.*

Suppose that we could $\mathbf{A}_0$-define an $\mathbf{A}_0$-statement $C[\phi, \psi]$ such that $C[\phi, \psi]$ is $\mathbf{A}_0$-false just in case $\phi$ is $\mathbf{A}_0$-true and $\psi$ is $\mathbf{A}_0$-false. Then we could define: $N[\phi]$ for $C[\phi, 0 = 1]$. But this is impossible.

COR. (iii). *Material equivalence fails to be* $\mathbf{A}_0$*-definable.*

Suppose that we could define a closed $\mathbf{A}_0$-statement $B[\nu, \kappa]$ such that $B[\nu, \kappa]$ is $\mathbf{A}_0$-true just in case '$\nu$' and '$\kappa$' are both $\mathbf{A}_0$-true or are both $\mathbf{A}_0$-false. Then we could define: $N[\phi]$ for $B[\phi, 0 = 1]$. But this is impossible.

COR. (iv). *The unlimited universal quantifier fails to be* $\mathbf{A}_0$*-definable.*

Suppose that given an $\mathbf{A}_0$-statement $\phi\{\xi\}$ containing exactly one free variable, namely $\xi$, we could $\mathbf{A}_0$-define a closed $\mathbf{A}_0$-statement denoted by $(A\xi)[\phi\{\xi\}]$ such that $(A\xi)[\phi\{\xi\}]$ is $\mathbf{A}_0$-true just in case each of $\phi\{0\}, \phi\{1\}, \phi\{2\}, \ldots$ is $\mathbf{A}_0$-true, then we could define:

$$Fals\, x \quad \text{for} \quad Cl\, Stat_0\, x \,\&\, (Ax')\,[N[\text{``}Prf_0[x', x]\text{''}]],$$

but this is impossible. The $N$ used in this definition is the negation defined for $\mathbf{A}_{00}$, this is allowable since $Prf_0[x', x]$ is in $\mathbf{A}_{00}$.

COR. (v). *The unlimited least number operator fails to be* $\mathbf{A}_0$*-definable.*

Suppose that we could $\mathbf{A}_0$-define an $\mathbf{A}_0$-term $\alpha\{\nu\}$ such that it determines the least numeral $\kappa$ such that $\phi\{\kappa\}$ and 0 otherwise, where '$\nu$' is $\phi\{\xi\}$. Then we could $\mathbf{A}_{00}$-define $\alpha\{\nu\}$ because $E$ fails to be used in numerical terms. But according to Prop. 8, Cor. (iv), Ch. 5 this is impossible.

## 6.4 *Enumeration of* $\mathbf{A}_0$*-theorems*

A set of numerals is said to be *primitive recursively enumerable* if and only if it is the set of values of a primitive recursive function.

PROP. 9 *The* $\mathbf{A}_0$*-theorems are primitive recursively enumerable.*

We have to produce a primitive recursive function $p$ such that for each numeral $\nu$ $p\nu$ is equal to the $g.n.$ of an $\mathbf{A}_0$-theorem and for each numeral $\kappa$ which is the $g.n.$ of an $\mathbf{A}_0$-theorem we can find a numeral $\nu$ such that

$p\nu = \kappa$. Consider the successive values of $Prf_0[pt[\nu, 2, 1], pt[\nu, 2, 2]]$. Each time this is $\mathbf{A}_{00}$-true write down the numeral determined by $pt[\nu, 2, 2]$ in a list. This list contains the *g.n.*'s of all the $\mathbf{A}_0$-theorems with repetitions but without omissions. We could do the same for $\mathbf{A}_{00}$ but there we had an effective test for $\mathbf{A}_{00}$-theoremhood so there was little use in enumerating the $\mathbf{A}_{00}$-theorems. We shall see in the next chapter that such a test is lacking for the system $\mathbf{A}_0$, so that the system $\mathbf{A}_0$ is undecidable.

The enumeration of the $\mathbf{A}_0$-theorems is given by the function $p_2$ as follows:

D 204 $p_0$ for the characteristic function of $Prf_0[pt[x, 2, 1], pt\,[x, 2, 2]]$.

D 205 $p_1 0 = 0, \quad p_1(S\nu) = (\mu x)_{\{q\nu . q\nu\}}\,[p_0 x = 0 \,\&\, x > p_1 \nu]$,

where $q\nu$ is the *g.n.* of $\nu = \nu$. ($\nu = \nu$ is an $\mathbf{A}_0$-proof of $\nu = \nu$.) Clearly $q$ is a primitive recursive function, so it gives a primitive recursive bound for the least number operator.

D 206 $$p_2 \nu \quad \text{for} \quad pt[p_1(S\nu), 2, 2].$$

Then $p_2\nu$ is the *g.n.* of the $\nu$th $\mathbf{A}_0$-theorem in a primitive recursive enumeration of $\mathbf{A}_0$-theorems, the $\mathbf{A}_0$-proof of the $\nu$th $\mathbf{A}_0$-theorem is given by: $pt[p_1(S\nu), 2, 1]$, i.e. the numeral determined by this is an $\mathbf{A}_0$-proof of the $\nu$th $\mathbf{A}_0$-theorem in the given list.

If an $\mathbf{A}_0$-statement is known to be an $\mathbf{A}_0$-theorem then we can recover its $\mathbf{A}_0$-proof by running through the list of $\mathbf{A}_0$-theorems until we come to it. If on the other hand we are ignorant whether an $\mathbf{A}_0$-statement $\phi$ is an $\mathbf{A}_0$-theorem then if we run through the list of $\mathbf{A}_0$-theorems we shall have to stop somewhere and if we have failed to come across $\phi$ then we are as ignorant as we were when we started. The reader is advised to refrain from doing this in the case of

$$(Ex, x', x'', x''')\,((x > 2 \,\&\, (Sx')^x + (Sx'')^x = (Sx''')^x),$$

which expresses the negation of Fermat's Last Theorem.

Cor. (i). *The theorems of a formal system are recursively enumerable.*

We can form $Prf_{\mathscr{L}}[\kappa, \nu]$ 'the proof predicate for the system $\mathscr{L}$' in a similar manner to $Prf_{00}[\kappa, \nu]$. The result then follows as in Prop. 9.

PROP. 10. *The cut is redundant in the system* $\mathbf{A}_0$.

This follows at once from the completeness of $\mathbf{A}_0$, and from the fact that the cut preserves $\mathbf{A}_0$-truth. The cut formula must, of course, be an $\mathbf{A}_{00}$-formula.

Since negation is absent from the system $\mathbf{A}_0$ then so is T.N.D.

## HISTORICAL REMARKS TO CHAPTER 6

The system $\mathbf{A}_0$ is similar to the systems of Myhill (1950) and of Löb (1953) but is weaker. It is simpler in that it deals with natural numbers rather than with sequences or chains.

The existential quantifier goes back to Schröder (1890) who used the symbol $\Sigma$ for it. Whitehead and Russell used $\exists$. The manner in which we introduce it in D 200 is due to Church (1932). Again we keep the development on the lines of Gentzen (1934) with the rule II *d*. The terms 'free' and 'bound' appeared soon after quantifiers were introduced, but it was some time before the restrictions on variables were correctly stated. The correct statement was made possible by Church's (1932) correct definitions of the $\lambda$-rules in his theory of $\lambda$-conversion.

Props. 4 and 7 are the sort of things the intuitionists like. Prop. 6 at first sight seems to contradict results of Tarski (1933), viz. a formal system which is consistent cannot contain its own truth-definition. But Tarski was dealing with systems which contained a negation and the system $\mathbf{A}_0$ is negationless. This is also mentioned by Myhill (1950).

Recursively enumerable sets came in with Kleene, Post, Rosser, etc.

## EXAMPLES 6

1. Develop the system $\mathbf{B}_0$ the anti-$\mathbf{A}_0$-system, related to $\mathbf{A}_0$ as $\mathbf{B}_{00}$ is to $\mathbf{A}_{00}$.
2. Find the proof-predicate for $\mathscr{F}_C$.
3. Instead of obtaining $\mathbf{A}_0$ from $\mathbf{A}_{00}$ by adjoining the symbol $E$ and the rule II *d* we could have adjoined the least number operator. Develop the resulting system, call it $\mathbf{A}_{0\mu}$.
4. Show that the symbols, axioms and rules of $\mathbf{A}_0$ are independent.
5. Discuss the systems $\mathbf{A}_{0f}$ and $\mathbf{A}_{00f}$ which are obtained from $\mathbf{A}_0$ and $\mathbf{A}_{00}$ respectively by allowing free variables in axioms and in rule II *a*.

6. An $\mathbf{A}_{0f}$-statement is equivalent to one of the form $\alpha\{\mathfrak{x}\} = 0$ or $(E\eta)\,(\alpha\{\eta, \mathfrak{x}\} = 0)$. These statements can be taken to define a class of lattice points. Thus we could express $\mathbf{A}_{0f}$ in terms of classes. Develop the corresponding system. (Cf. D 37 – 57.)

7. Show that the Deduction Theorem fails to apply to $\mathbf{A}_0$.

8. Show that Modus Ponens fails to be definable in $\mathbf{A}_0$.

9. Show that the Deduction Theorem holds in $\mathbf{A}_0$ provided the hypotheses are $\mathbf{A}_{00}$-statements.

10. Show that the cut can be eliminated in $\mathbf{A}_0$ provided the cut formula is an $\mathbf{A}_{00}$-formula.

# Chapter 7

# $A_0$-Definable functions. Recursive function theory

## 7.1 *Turing machines and Church's Thesis*

Mathematics is the art of making vague intuitive ideas precise (including the vague intuitive idea of preciseness itself), and of studying the result. We have in Ch. 1 already given a precise definition of language. In this chapter we want to give a precise definition of the vague intuitive idea of calculable function of natural numbers. In Chs. 4 and 6 we gave a precise account of natural number and of primitive recursive function of natural numbers. This will be continued in this and later chapters to give precise accounts of other types of functions of natural numbers.

The vague idea of natural number is made precise by representing them by progressions of tally marks thus:

| || ||| |||| |||||

The essential thing is that we have a starting point, here a tally mark standing alone, and a method of proceeding by discrete steps such that we continually generate new formulae, here the method of procedure is to adjoin a tally mark at the right. The arabic notation for the numerals was the first to satisfy these conditions, that is why it is one of the great achievements of the human mind. To make this procedure of adjoining precise we laid down in Ch. 1 that our signs be written on a tape divided into squares. The adjoined tally is placed on the vacant square on the right of the existing sequence (which consists of tally marks on consecutive squares of the tape with a vacant square on the left). Again we have to appeal to the goodwill of the reader to decide if a mark is intended to be a tally mark or otherwise, any two supposed tally marks differ, we need only use a magnifying glass to see how different they are. Thus the sequence of natural numbers that we are accustomed to denote by: 0, 1, 2, 3, 4 is represented precisely by:

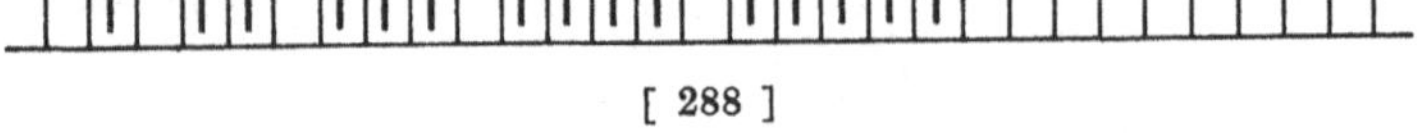

The vague intuitive idea of a calculable function of natural numbers is: a rule or set of rules such that given a terminating sequence of tally marks on consecutive squares of a tape we can after a terminating sequence of applications of the rules produce another terminating sequence of tally marks on consecutive squares of a tape which fails to foul the original sequence, this sequence we call the value of the function for the original sequence as argument, or rather the natural number represented by the function with the original sequence as argument. We can do a certain amount with this vague intuitive idea but it would be rather difficult to show that there failed to be a calculable function of natural numbers which would answer a given type of question, for instance membership of a given set of natural numbers. As an historical fact the first demonstration of such a case was only accomplished after the vague intuitive idea of a calculable function of natural numbers had been made precise. Still more difficult would be to define two functions of natural numbers whose values could be generated one after the other, and each failed to be calculable even when in the calculation we could be given, when required, the correct answer to any question 'is $\nu$ a value of the other function?'

Making a vague intuitive idea precise often leads to new mathematical disciplines, we shall see something of this sort. One way of making a vague intuitive idea precise is to set up a formal system and lay down axioms and rules which together play the part of implicit definitions of the vague idea. This has been adopted in Chs. 2, 3, 4 and 6 where we set up various formal systems to make various vague intuitive ideas precise. There are other ways of making vague intuitive ideas precise, the one we shall adopt is to make a precise description of 'set of rules' in 'a calculable function, of natural numbers is a set of rules...' We have already given a precise representation of natural numbers. The precise definition of set of rules that we shall give was first given about 30 years ago by Alan Turing of King's, and the form we give it in is due to S. C Kleene. Since then altogether at least seven different precise definitions of 'calculable function' have been given. They all had very different origins, but they have all been shown to be equivalent, that is, a function calculable by one definition is also calculable by another definition, and an effective method given to find the rules for one definition given those for the other definition. Thus the equivalence of these seven definitions of 'calculable function' and the lack of success which has attended attempts to construct a 'calculable function' which fails to come under any of the seven definitions makes it appear that we have reached an absolute concept, that is, it seems that any function that can be calculated by man can be calculated by any one of these seven methods. How can we in any wise demonstrate this? Ultimately only in some formal system where the vague intuitive concept 'calculable function' would have to be made precise before it could be handled at all, so that in any case the vague intuitive idea would be eliminated before we began, so we certainly would fail to demonstrate anything about it at all. Any form of demonstration thus being denied us we lay down the following thesis, known as *Church's Thesis*:

*A calculable function of natural numbers with natural numbers as values is a function which is calculable by a Turing machine.*

A function which is calculable by a Turing machine is calculable in the intuitive sense. Can we find a function which is calculable in the intuitive sense but which fails to be calculable by a Turing machine? As explained above we think that this is impossible. Any form of demonstration is out of the question but the evidence is overwhelming, so the only answer we can give is; produce such a function!

Turing set out to analyse what one does when one calculates the value of a function and reduced the operations to their simplest terms. In this chapter we shall explain this method of calculation in detail. In Ch. 5 we gave a vague definition of calculable function of natural numbers as a set of instructions $\mathscr{I}$ such that given a natural number $n$ on applying the instructions $\mathscr{I}$ we obtain another natural number $k$. We denote this by : $\mathscr{I}n = k$. We also found that for certain precise sets of instructions, namely those which yield primitive recursive functions, we could find an $\mathbf{A}_{00}$-function $\rho$ such that $\rho\nu = \kappa$ is an $\mathbf{A}_{00}$-theorem whenever $\mathscr{I}n = k$, where $\nu$ and $\kappa$ are the numerals which represent the natural numbers $n$ and $k$ respectively in the system $\mathbf{A}_{00}$. We further found that every $\mathbf{A}_{00}$-function was primitive recursive, so that the system $\mathbf{A}_{00}$ becomes a formalization of primitive recursive function theory. But we found a calculable function namely *Val* which failed to be primitive recursive, so that the concept of primitive recursion would fail to be any use as a definition of 'calculable function'. However we found that in the system $\mathbf{A}_0$ we could represent the function *Val* by an $\mathbf{A}_0$-statement:

D 207 $\quad VAL[\kappa, \nu] \quad \text{for} \quad (Ex)\, Prf_0\,[x, Eq[\kappa, Num\,\nu]],$ (1)

such that $Val\,\kappa = \nu$ if and only if $VAL\,[\kappa, \nu]$ is an $\mathbf{A}_0$-theorem. Thus in the system $\mathbf{A}_0$ we have representations for some functions for which we lack explicit definitions. We lack an explicit $\mathbf{A}_0$-definition for *Val* because any such explicit definition would be an $\mathbf{A}_{00}$-definition and hence the defined function would be primitive recursive and *Val* fails to be primitive recursive.

Similarly we give a vague intuitive definition of *partially calculable function of natural numbers* as a set of instructions which applied to a natural number $n$ sometimes yields a natural number $k$ called the value of that function for the argument $n$, or the calculation may fail to terminate or may terminate and yield something other than a natural number. In this case we say that the function is undefined for that argument. In this chapter we give a precise definition of partially calculable function of natural numbers. We shall also show that there is a primitive recursive function, hence $\mathbf{A}_{00}$-definable, $Un$ of type $\iota\iota\iota\iota$ such that given a function $f$ partially calculable according to our definition then we can effectively find a numeral $\theta$ such that the partially calculable function $f$ is $\mathbf{A}_{00}$-represented by the $\mathbf{A}_0$-statement:

$$(E\xi)\,(Un[\theta, \nu, \kappa, \xi] = 0), \tag{2}$$

so that (2) is an $\mathbf{A}_0$-theorem if and only if $fn = k$ where $\nu$ and $\kappa$ are the numerals which $\mathbf{A}_0$-represent the natural numbers $n$ and $k$ respectively. Thus $Un$ is a sort of *universal function.* Each function partially calculable according to our definition is $\mathbf{A}_0$-represented by (2) for some numeral $\theta$, and for each numeral $\theta$ (2) gives a partially calculable function of natural numbers, namely the function is defined for the argument $\nu$ if and only if (2) is an $\mathbf{A}_0$-theorem for some numeral $\kappa$ and the value of the function is then $\kappa$. Thus for $\theta = 0, 1, 2, \ldots$ (2) gives an enumeration, possibly with repetitions, of *partially recursive functions.*

We can obtain an explicit definition of a partially calculable function of natural numbers if we have an unlimited least number symbol $\boldsymbol{\mu}$, so that the value of $\boldsymbol{\mu}\,(\boldsymbol{\lambda}\xi\phi\{\xi\})$ is the least numeral $\kappa$ such that $\phi\{\kappa\}$ is a theorem in the system containing $\boldsymbol{\mu}$, and is undefined otherwise. The definition is:

$$pt[\boldsymbol{\mu}(\boldsymbol{\lambda}\eta(Un[\theta, \nu, \eta_1, \eta_2] = 0)), 2, 2],$$

where $\eta_1$ stands for $pt[\eta, 2, 1]$ and $\eta_2$ stands for $pt[\eta, 2, 2]$.

We similarly give a vague definition of partially calculable function of $\pi$ arguments as a set of instructions which applied to an ordered set of $\pi$

natural numbers $\{n', \ldots, n^{(\pi)}\}$ sometimes yields a natural number $k$ called the value of the function for the argument set $\{n', \ldots, n^{(\pi)}\}$, or else fails to terminate, or else yields something other than a natural number.

We wish to make these ideas precise; to do this we analyse the operations we do when we calculate the value of a function. First we will represent the natural numbers from zero onwards by sequences of tally marks thus:

| || ||| |||| |||||

these are written in squares on a tape as explained before. A sequence of natural numbers will be represented by a sequence of sequences of tally marks with a cipher between the representations of the numbers of the sequence. The null sequence is represented by a single cipher. Thus:

| | | | | | | | | | | | | | |
|---|---|---|---|---|---|---|---|---|---|---|---|---|---|
|   | 0 | \| | \| | 0 | \| | 0 | \| | \| | \| | \| | 0 |   |   |

(3)

represents the sequence $\{1, 0, 3\}$ in ordinary notation. Let us denote $\mathscr{I}[n', \ldots, n^{(\pi)}] = k$ by the sequence $\{n', \ldots, n^{(\pi)}, k\}$. Thus if $\mathscr{I}[1, 0, 3] = 2$ this is represented by

| | | | | | | | | | | | | | | | | | | |
|---|---|---|---|---|---|---|---|---|---|---|---|---|---|---|---|---|---|---|
|   | 0 | \| | \| | 0 | \| | 0 | \| | \| | \| | \| | 0 | \| | \| | \| | 0 |   |   |   |

(4)

With these conventions the calculation of $\mathscr{I}[1, 0, 3]$ then consists in replacing: (3) by (4).

We also consider the case of calculating the value of a function without arguments, this amounts to a piece of calculation beginning with a tape containing ciphers on three consecutive squares. We always put a cipher at the left of the representation of a sequence and another at the right and the null sequence is represented by a cipher, hence the three ciphers.

In performing the calculation we shall certainly have to:

(i) observe a part or the whole of the sequence of tallies and ciphers on the tape which has been produced up to that point in the calculation. For instance we certainly must observe the representation of the argument for otherwise the calculation would be independent of the argument;

(ii) write a tally on a square, write a cipher on a square;

(iii) refer to the instructions.

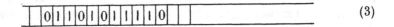

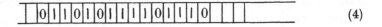

We might in addition like to do some rough work to be erased later on. This erasing could be avoided if we did all the rough work to the left of the representation of the argument, say leaving a two cipher gap between them, and just leave it there. We will suppose that we allow ourselves to do rough work, clearly this is without restriction on our calculation. For instance if we were calculating $f(gn)$ we would like to calculate $gn$ first, in fact we should have to do so, and then having found $gn$ set up the calculation to find $f(gn)$. The observation of part or whole of what is on the tape can be broken down into atomic acts of observing what is on a single square, in fact in a large calculation this would have to be the case because we would be unable to take it all in at a glance; e.g. are

and

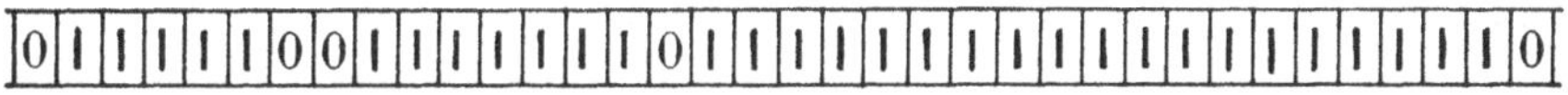

the same, or rather do they represent the same sequences? To answer this we should have to compare them square by square, certainly for very much longer sequences. Thus we can break down the observation into the atomic acts: (i) observe a single square and decide whether it contains a tally or a cipher (ii) move one square to the left or to the right. Thus altogether we have the following four atomic acts:

(i) observe a single square;

(ii) move to the square immediately to the right, if we are observing the rightmost symbol then write a cipher in the next square to the right and observe it;

(iii) similarly for the left;

(iv) interchange cipher and tally on the observed square.

We always keep a cipher at both ends, and if we go over an end and land up on an empty square we immediately write a cipher there. Alternatively we could have stipulated that each square of the tape originally was occupied by a cipher. Our calculation will then consist of instructions which will tell us which of the four atomic acts to do at each point of the calculation, that is to say, if we have just completed one atomic instruction and finish by observing a tally then we want to know which atomic instruction to refer to next.

We shall commence the calculation by observing the last tally of the representation of the argument or a cipher if the function is without argument. This we call the *standard position*. Had we decided to start in any other manner then we could easily add some preliminary instructions to reach the standard position before beginning, so this convention is without restriction. We will also finish in standard position that is observing the last tally in the representation of the value. Again this is without restriction. If a calculation terminates other than in standard position then we shall agree that the calculation has failed to produce a natural number. A calculation which terminates other than in standard position and which produces a value could easily be augmented with more instructions so that it did terminate in standard position. Thus altogether without loss of generality we can begin and end in standard position and do the calculation by a series of atomic acts. The instructions are then to be so detailed that they tell us in what order to do these atomic acts. We denote the four atomic acts by $N$, $R$, $L$, $I$ respectively. Each atomic act is initiated by one of the instructions (called an atomic instruction) which tells us which atomic act to do and which atomic in-instruction to refer to next. We denote the atomic instructions by $0, 1, 2, \ldots, \nu$. Then being under instruction $\kappa$, $0 \leqslant \kappa \leqslant \nu$ we can find out which atomic act to do and which atomic instruction to consult for the next act. To ensure that these conditions are always satisfied we lay down that the set of atomic instructions shall tell us what to do when under instruction $\kappa$, $0 \leqslant \kappa \leqslant \nu$ and observing a tally or observing a cipher. An atomic instruction can then be specified by one of:

$$\left.\begin{array}{ll} \lambda 0 N\mu & \lambda \mid N\mu \\ \lambda 0 R\mu & \lambda \mid R\mu \\ \lambda 0 L\mu & \lambda \mid L\mu \\ \lambda 0 I\mu & \lambda \mid I\mu \end{array}\right\} \quad (0 \leqslant \lambda, \mu \leqslant \nu),$$

namely, for the first one; if under instruction $\lambda$ and observing a cipher we continue observing that cipher (in fact do nothing) and then refer to atomic instruction $\mu$, and so on for the other cases. A set of atomic instructions will be called *complete* if for each $\kappa$, $0 \leqslant \kappa \leqslant \nu$, exactly one entry from each column above is listed in the set. Then we shall know what to do whatever instruction we are under and whether we are observing a tally or observing a cipher. An incomplete set of instructions might suffice to calculate a function of natural numbers because certain of the atomic instructions might be redundant, but an incomplete set can easily be made complete by adding redundant instructions and in the case just described they would fail to influence the calculation whatever they were. For instance a certain set of instructions might be such that we failed to be under instruction 3 observing a cipher so that the atomic instructions $30N\mu$, $30R\mu$, $30L\mu$, $30I\mu$ would be redundant for $0 \leqslant \mu \leqslant \nu$ and if they were all absent from a set of instructions, so that the set was incomplete, then any of them could be added to make the set complete without any effect on the calculation.

Thus without loss of generality we may take our set of atomic instructions to be complete. Our list of atomic instructions must terminate, because it is impossible to have a set that fails to terminate, the world would fail to contain them, but it would be possible in some cases to have an unlimited set of instructions such that we could generate them one after the other and so reach any one we might want. But this generation would be a subsidiary piece of calculation and so should be included in the list of atomic instructions, because we said that these were to be so detailed that they tell us which atomic act to do at each point in the calculation whereas in the case we are considering we have to do a lot of subsidiary calculation in order to find out which atomic act to do. Thus the list of atomic instructions must terminate. Next the atomic acts must be such that we can find out what to do without any calculation, because they are to be so detailed that they describe all the acts to be done, if we had to do some calculation to find out what an atomic act meant then our list of atomic instructions would fail to meet our requirements. To satisfy this requirement the atomic instructions must be displayed, then without any calculation we can see what to do.

Thus a complete set of instructions is described by a *complete table* thus:

$$\begin{matrix} 00X\pi & 0|Y\theta \\ \vdots & \vdots \\ \nu 0 X^{(\nu)}\pi^{(\nu)} & \nu | Y^{(\nu)}\theta^{(\nu)}, \end{matrix}$$

where $\pi, \pi', \ldots, \pi^{(\nu)}$ and $\theta, \theta', \ldots, \theta^{(\nu)}$ are numerals between 0 and $\nu$ inclusive, and $X, X', \ldots, X^{(\nu)}$ and $Y, Y', \ldots, Y^{(\nu)}$ are from $N$, $R$, $L$ and $I$. The calculation now proceeds as follows: we start observing the representation of the argument in standard position and refer to instruction 1.

We shall always start with instruction 1, it will be called the *initial instruction*. If we started elsewhere then we need only renumber the instructions in order to be able to start with instruction 1. Since we are observing the last tally of the representation of the argument then we refer to instruction $1 \mid Y'\theta'$, this tells us which of $N, R, L$ or $I$ to do and then refers us to instruction $\theta'$; according as we are observing a tally or a cipher after having done instruction 1, so we refer to instruction $\theta' \mid Y^{(\theta')}\theta^{(\theta')}$ or to $\theta' 0 X^{(\theta')}\theta^{(\theta')}$ to see what to do next. And so we go on until we come to an instruction of the type: $\pi 0 N \pi$ or $\pi \mid N\pi$ which are called *terminating instructions*, when we stop. If we were calculating the value of a function without argument then we would start with instruction $10X'\pi'$ because we would start observing a cipher. The instruction $\pi 0 N\pi$ $\pi 1 N\pi$ will be called the *passive instruction* and will always be given the number zero and omitted from the list of instructions, it being tacitly assumed that it is there. Any instruction of the type $\pi 0 N\pi$ or $\pi \mid N\pi$ will be replaced by the passive instruction and omitted and a single terminating instruction $\pi 0 N\pi$ or $\pi \mid N\pi$ where the other one in that line fails to be terminating will be replaced by $\pi 0 N 0$ or $\pi \mid N 0$ respectively then the only terminating instruction will be the passive instruction. Note that if a table calculates a natural number then that natural number is unique because at each stage in the calculation there is exactly one thing to do.

A table is called *$\pi$-normal* under the following circumstances: started observing a $\pi$-tuplet in standard position with two ciphers immediately to the left of the representation of the $\pi$-tuplet the leftmost of these two ciphers is unobserved during the whole calculation and the calculation terminates observing an $S\pi$-tuplet in standard position with the same two ciphers immediately to its left, the first $\pi$ members of the $S\pi$-tuplet being the representation of the original argument.

A complete table gives instructions for calculating when started in any position with anything on the tape. It is only when we are calculating the value of a function that we start it off in standard position. Again a table designed to calculate the values of a function of two arguments will do something when started off with a single argument. It might even calculate the values of a function of one argument.

We shall find it convenient to break up a table into several parts so that the output of one table is the input of the next table. In this way we shall obtain a simplified method for describing complicated tables. The number of rows in a complete table $A$ (excluding the passive instruction) is called the *order of the table* $A$. The table of order zero is the passive table consisting of the passive instruction alone. Let $A$ and $B$ be tables of orders $\nu$ and $\kappa$ respectively then $(AB)$ shall denote that table of order $(\nu+\kappa)$ whose first $\nu$ rows are the same and in the same order as those of table $A$ and whose last $\kappa$ rows are the same as those of table $B$ except that $S\nu$ is substituted for 0 in table $A$ and the instruction numbers of table $B$ are all increased by $\nu$ except 0. This makes the passive instruction of table $A$ the same as the initial instruction of table $B$. The table $(AB)$ will first do the work of table $A$ and then the work of table $B$, starting in the position where table $A$ finished. If table $A$ fails to terminate then table $(AB)$ also fails to terminate. Similarly if $A$, $B$, $C$ are tables then $(ABC)$ shall denote $((AB)\,C)$, clearly this is the same as $(A(BC))$ so we may omit parentheses. Similarly for more tables. $A^2$ stands for $(AA)$, etc.

It is sometimes convenient to write a row of a table $\nu 0 X\kappa, \nu \mathsf{I}\, Y\theta$ in the form $\nu \begin{cases} X\kappa \\ Y\theta \end{cases}$ and the row $\nu 0 X\kappa, \nu \mathsf{I}\, X\kappa$ as $\nu X\kappa$ without braces. If the rows $\kappa$ and $\theta$ are $\kappa 0 X'\kappa'$, $\kappa \mathsf{I}\; Y'\theta'$ and $\theta 0 X''\kappa''$, $\theta \mathsf{I}\; Y''\theta''$ respectively then we can depict the three rows as:

$$\nu \begin{cases} X\kappa \begin{cases} X'\kappa' \\ Y'\theta' \end{cases} \\ Y\theta \begin{cases} X''\kappa'' \\ Y''\theta'' \end{cases} \end{cases}.$$

So we can continue. Suppose that $\kappa'$ is $\nu$, then as we continue we should have to write the whole diagram over again and so on indefinitely;

$$\nu \begin{cases} X\kappa \begin{cases} X'\nu & \begin{cases} X\kappa \begin{cases} X'\nu \\ Y'\theta' \end{cases} \\ Y\theta \begin{cases} X''\kappa'' \\ Y''\theta''. \end{cases} \end{cases} \\ Y'\theta' \end{cases} \\ Y\theta \begin{cases} X''\kappa'' \\ Y''\theta'' \end{cases} \end{cases}$$

Instead of this we place a dot over $\nu$ in the first diagram, as in recurring decimals, thus:

$$\dot{\nu} \begin{cases} X\kappa \begin{cases} X'\dot{\nu} \\ Y'\theta' \end{cases} \\ Y\theta \begin{cases} X''\kappa'' \\ Y''\theta'' \end{cases} \end{cases}. \tag{3}$$

In this way the whole table can be put as a limited diagram called the *diagram of the table.* It will begin:

$$1\begin{cases} X\kappa \\ Y\theta \end{cases} \text{ or } \dot{1}\begin{cases} X\kappa \\ Y\theta \end{cases} \text{ or } 1X\kappa \text{ or } \dot{1}X\kappa.$$

A *line of a diagram* is a linear succession obtained by starting with 1 or $\dot{1}$, as the case may be, proceeding to the right and at each brace writing down either the top or the bottom row which are joined at that brace. Thus $\dot{\nu}X\kappa Y'\theta'$ in diagram (3) is part of a line. If every line of a diagram ends with a numeral carrying a dot then the calculation described by that table will always perpetuate however we start. A line can only end without a dotted numeral when that numeral is zero. Consider the complete table:

$$\left.\begin{array}{ll} 10R2 & 1\,|\,I3 \\ 20I4 & 2\,|\,N3 \\ 30L4 & 3\,|\,L4 \\ 40L5 & 4\,|\,R1 \\ 50I0 & 5\,|\,N1 \end{array}\right\} \tag{4}$$

this has diagrams:

$$\dot{1}\begin{cases} R2\begin{cases} I\dot{4} \\ N\dot{3} \end{cases} \\ I\dot{3}L\dot{4}\begin{cases} L5\begin{cases} I0 \\ N\dot{1} \end{cases} \\ R\dot{1} \end{cases} \end{cases} \quad \text{or} \quad \dot{1}\begin{cases} R2\begin{cases} I\dot{4}\begin{cases} L5\begin{cases} I0 \\ N\dot{1} \end{cases} \\ R\dot{1} \end{cases} \\ N\dot{3}L\dot{4} \end{cases} \\ L\dot{3} \end{cases}$$

We shall use the first version so that the leftmost bottom occurrence of an instruction number is followed by the description of the atomic act. Let $A$, $B$, and $C$ be tables of orders $\nu$, $\kappa$ and $\pi$ respectively then $A\begin{cases} B \\ C \end{cases}$ shall denote that table of order $S(\nu+\kappa+\pi)$ whose first $\nu$ rows are the same and in the same order as those of table $A$ except $S\nu$ is substituted for zero throughout, so that the passive instruction of table $A$ becomes instruction number $S\nu$, and we also add the new entry for instruction $S\nu$, namely:

$$(S\nu)\,0N(SS\nu), \quad (S\nu)\,|\,N(SS(\nu+\kappa)).$$

The next $\kappa$ rows are the instructions of table $B$ except that the instruction numbers are increased by $S\nu$ except zero which is unaltered, and the last $\kappa$ rows are those of table $C$ but with the instruction numbers increased by $S(\nu+\kappa)$ again, except zero which is unaltered. The table $A\begin{cases}B\\C\end{cases}$ acts as follows: first we carry out the instructions given by table $A$ and if we terminate observing a cipher then we start on table $B$, but if table $A$ terminates observing a tally then we start on table $C$. The notation $\dot{A}\begin{cases}\dot{B}\\C\end{cases}$ shall denote the same as $A\begin{cases}B\\C\end{cases}$ except that the passive instruction in table $B$ is labelled, 1, so that if table $B$ terminates then when it does we are sent back to table $A$. Similarly with a dot over $C$ instead of over $B$, if there is a dot over both $B$ and $C$ then the resulting table fails to terminate for as soon as table $B$ or $C$ has finished (if they do) we are referred back to table $A$ and so start all over again. We can use this notation when the tables are in the form of diagrams.

The diagrams can be further simplified by omitting the instruction numbers and placing different groups of dots over the repeating parts thus:

$$\dot{N}\begin{cases}R\begin{cases}\dddot{I}\\ \ddot{N}\end{cases}\\ L\ddot{L}\dddot{N}\begin{cases}L\begin{cases}I\\ \dot{N}\end{cases}\\ \dot{R}\end{cases}\end{cases}\quad\text{instead of}\quad \dot{1}\begin{cases}R2\begin{cases}I\dot{4}\\ N\dot{3}\end{cases}\\ L\dot{3}L\dot{4}\begin{cases}L5\begin{cases}I0\\ \dot{N}1\end{cases}\\ R\dot{1}\end{cases}\end{cases}$$

We have added a couple of $N$'s to carry dots. When we come to the end of a line without a dot then we stop, but if there is a group of dots on the last letter of a line then we seek a similar group of dots over some other letter in the diagram and then continue from that place; there will only be one such letter other than an end letter of a line. If the last letter of every line carries one or more dots then the calculation will fail to terminate however we start. If we use diagrams we only have to keep our place in the diagram to see what to do next. Note that the table obtained from a diagram is complete.

Now suppose that we extend table (4) by adding the entries:

$$\begin{array}{ll} 60R7 & 6\,|\,L0 \\ 70R7 & 7\,|\,I6 \end{array}$$

then the diagram would consist of two pieces, namely that already given and

$$\dot{6}\begin{cases}R\dot{7}\begin{cases}R\dot{7}\\ I\dot{6}\end{cases}\\ L0\end{cases}\quad\text{or}\quad \dot{N}\begin{cases}R\ddot{N}\begin{cases}\ddot{R}\\ \dot{I}\end{cases}\\ L\end{cases}$$

This piece is redundant if we start with instruction 1. Now suppose that we extended the complete table (4) by adding the entries

$$60R7 \qquad 6\,|\,L0$$
$$70R2 \qquad 7\,|\,I6$$

then the diagram would still consist of two pieces; namely that already given for table (4) and

$$\dot{6}\begin{cases} R7\begin{cases} R\dot{2}\begin{cases} I\dot{4}\begin{cases} L5\begin{cases} L0 \\ N1 \end{cases} \\ R1\begin{cases} R\dot{2} \\ N\dot{3} \end{cases} \end{cases} \\ N\dot{3}L\dot{4} \end{cases} \\ I\dot{6} \end{cases} \\ L0 \end{cases}$$

this piece would again be redundant, if we start with instruction 1. We shall say that a complete table is *connected* when its diagram is of one piece. It is without loss of generality to restrict ourselves to connected tables. We can test whether a table is complete and connected by forming its diagram. Two tables are called *equivalent* if they calculate the same values for the same argument set, or we should say *equivalent for $\pi$ arguments*, because they might calculate the same function of three arguments but different functions of two arguments.

## 7.2 *Some simple tables*

We now give some simple tables in the form of diagrams, each diagram is followed by a description of what the table does under the circumstances in which we shall use it. By combining these tables in the manner we have explained we shall get all the tables we shall need. Note that all these tables are complete and connected.

$A{:}\,\dot{N}\begin{cases} I \\ \dot{R} \end{cases}$ seeks the first cipher at or to the right of the observed symbol changes this to a tally and stops observing this tally.

$B_1{:}\,\dot{N}\begin{cases} L \\ \dot{R} \end{cases}$ started observing a tally belonging to a sequence of tallies finds the rightmost tally of the sequence and stops observing this tally.

$B_2{:}\,\dot{N}\begin{cases} R \\ \dot{L} \end{cases}$ started observing a tally belonging to a sequence of tallies finds the leftmost tally of the sequence and stops observing that tally.

| | | |
|---|---|---|
| $B_3 : \dot{N}$ | $\begin{cases} \dot{R} \\ N \end{cases}$ | seeks the first tally at or to the right of the observed symbol and stops observing that tally. |
| $B_4 : \dot{N}$ | $\begin{cases} \dot{L} \\ N \end{cases}$ | seeks the first tally at or to the left of the observed symbol and stops observing that tally. |
| $C : I\dot{L}$ | $\begin{cases} I \\ \dot{N} \end{cases}$ | started observing the representation of a number in standard position moves the representation of that number one place to the left. |
| $D : \dot{C}L$ | $\begin{cases} R\dot{B}_1 \\ RIRA \end{cases}$ | started observing a number $\nu$ in standard position moves this number to the left so as to leave a one cipher gap between the number $\nu$ in its new position and the next number on the left, it stops observing the number $\nu$ in its new position in standard position. |

To show how table $D$ works suppose that we start observing the representation of 2 in standard position when there is a representation of 1 on the left thus: II 000 I I $\bar{\text{I}}$, we have put a bar over the observed symbol. Table $D$ tells us to use table $C$, the results of doing this is: II 00$\bar{\text{I}}$ II. The whole calculation can be put down as a list of sequences of ciphers and tallies, the observed symbol being marked by a bar over it, at the right of a sequence stands the next operation to be done, in the next line the result of doing that operation.

| | | | | | | | | |
|---|---|---|---|---|---|---|---|---|
| I | I | 0 | 0 | 0 | I | I | $\bar{\text{I}}$ | $C$ |
| I | I | 0 | 0 | $\bar{\text{I}}$ | I | I | 0 | $L$ |
| I | I | 0 | $\bar{0}$ | I | I | I | 0 | $R$ |
| I | I | 0 | 0 | $\bar{\text{I}}$ | I | I | 0 | $B_1$ |
| I | I | 0 | 0 | I | I | $\bar{\text{I}}$ | 0 | $C$ |
| I | I | 0 | $\bar{\text{I}}$ | I | I | 0 | 0 | $L$ |
| I | I | $\bar{0}$ | I | I | I | 0 | 0 | $R$ |
| I | I | 0 | $\bar{\text{I}}$ | I | I | 0 | 0 | $B_1$ |
| I | I | 0 | I | I | $\bar{\text{I}}$ | 0 | 0 | $C$ |
| I | I | $\bar{\text{I}}$ | I | I | 0 | 0 | 0 | $L$ |
| I | $\bar{\text{I}}$ | I | I | I | 0 | 0 | 0 | $R$ |
| I | I | $\bar{\text{I}}$ | I | I | 0 | 0 | 0 | $I$ |
| I | I | $\bar{0}$ | I | I | 0 | 0 | 0 | $R$ |
| I | I | 0 | $\bar{\text{I}}$ | I | 0 | 0 | 0 | $A$ |
| I | I | 0 | I | I | $\bar{\text{I}}$ | 0 | 0 | stop. |

$E: B_2LB_4$ started observing the representation of a number in standard position moves to the left and stops observing the representation of the next number on the left in standard position.

$F: RB_3B_1$ started observing the representation of a number in standard position moves to the right and stops observing the representation of the next number on the right in standard position.

$G: \dot{R}\begin{cases}\dot{I}\\ LIL\end{cases}$ started observing the representation of a number in standard position (not the rightmost number on the tape) increases this number so as to leave just one cipher between the increased number and the next number to the right, stops observing the increased number in standard position.

$H: I\dot{L}\begin{cases}L\\ \dot{I}\end{cases}$ started observing a number in standard position will erase it and stop observing the symbol two places to the left of this number.

$J: B_2LL\begin{cases}B_3B_1\\ \dot{H}\begin{cases}B_3B_1\\ \dot{N}\end{cases}\end{cases}$ started observing a number in standard position will erase all numbers (if any) to the left of this number up to the first group of two ciphers, stops observing the original number in standard position.

$M_\nu: RBIE^{\dot{\nu}}L\begin{cases}RGF^\nu\\ RILF^\nu\dot{A}\end{cases}$ started observing the $\nu$-pile $\pi', \ldots, \pi^{(\nu)}$ in standard position with $SS\pi'$ ciphers to the right will copy $\pi'$ on the right leaving a one cipher gap and stop observing the copy in standard position. $E^{\dot{\nu}}$ means do $E$ $\nu$ times, i.e. go back to the first of the $\nu$ $E$'s, we could have written $\dot{N}E^\nu$ instead of $E^{\dot{\nu}}$.

$P_\nu: AM_\nu^\nu ILE^\nu ILF^\nu$ started observing a $\nu$-ple $\pi', \ldots, \pi^{(\nu)}$ in standard position with $\pi' + \ldots + \pi^\nu + 2.\nu + 1$ ciphers to the right will copy the $\nu$-ple to the right leaving two ciphers between the two $\nu$-ples, stops observing the copy in standard position.

We omit descriptions of what these tables do if started in positions other than those stipulated or if the conditions stated in some cases fail to be satisfied. Thus table $G$ if started as stated but without tallies to the right will go on indefinitely printing tallies on consecutive squares.

In Ch. 5 we described a certain class of calculable function of natural numbers called primitive recursive functions, which could be obtained from certain initial functions, namely: constant functions zero, successor function, and identity functions by means of two schemes, namely: the scheme of substitution and the scheme of primitive recursion. We now add to these the operation of searching through the natural numbers one after the other for one satisfying a certain condition in the form of an equation, expressible in terms of previously acquired functions. By use of the function $A_2$ (D 133) we may take the right-hand side of this equation to be zero. Let $A_{00_\mu}$ be that primary extension of the system $\mathbf{A}_{00}$ which is obtained by adding the symbol $\mathscr{H}$ of type $\iota(\iota\iota)$ and the axiom scheme:

$$(\mathscr{H}\tau)\,\alpha = \alpha\,.\,B_1[\tau\alpha] + (\mathscr{H}\tau)\,(S\alpha)\,.\,A_1[\tau\alpha],$$

where $\alpha$ is a numerical term and $\tau$ is a one-place $\mathbf{A}_{00}$-function.

We define $\quad (\mu\xi)\,(\tau\xi = 0) \quad$ or $\quad (\mu\tau) \quad$ for $\quad (\mathscr{H}\tau)\,0,$

then $\mu\tau = \kappa$ is an $\mathbf{A}_{00_\mu}$-theorem if and only if $\kappa$ is the least numeral such that $\tau\kappa = 0$ is an $\mathbf{A}_{00_\mu}$-theorem. If $\tau\kappa$ is different from zero for each numeral $\kappa$ then $\mu\tau = \kappa$ fails to be an $\mathbf{A}_{00_\mu}$-theorem for each numeral $\kappa$. In this case any attempt to find the value of $\mu\tau$ will fail to terminate, so that the system $\mathbf{A}_{00_\mu}$ would contain numerical terms which failed to determine any numeral. The operation of finding the least number which satisfies a certain condition differs from the two previous schemes in that those schemes when applied always terminate. If $\tau\kappa = 0$ for some numeral $\kappa$ then the process of finding the least such number will terminate, but if we are ignorant whether $\tau\kappa$ is ever zero then we are ignorant whether the operation will terminate or perpetuate.

A function of natural numbers whose value is calculated from the initial functions: successor, constant and identity functions by the schemes of substitution, primitive recursion and least number operator applied to previously acquired functions in the form of equations $\alpha = 0$ will be called *partial recursive*. We use the term 'partial' because the value of the function is undefined in case the least number operation perpetuates. It might happen that a partial recursive function is defined for all arguments, in that case the partial recursive function is called *general recursive*. Again it might happen that a partial recursive function is undefined for all arguments. Clearly a partial recursive function is $\mathbf{A}_{00_\mu}$-definable and an $\mathbf{A}_{00_\mu}$-definable function is partial recursive. A function which is defined everywhere is called a *total function*.

## 7·3 *Equivalence of partially calculable and partial recursive function*

PROP. 1. *A partial recursive* $(S\pi)$*-place-function is partially calculable by an* $(S\pi)$*-normal table.*

We give the diagrams of complete $(S\pi)$-normal tables for the initial functions and for functions produced by the schemes:

The successor function
$$M_1 RI.$$

Here we start observing the representation of a single number in standard position, copy it on the right leaving a one cipher gap then add a single tally on the right and stop observing that tally.

The constant function zero
$$RRI.$$

Here we start observing the representation of any $\nu$-ple in standard position move two places to the right print a tally and stop.

The identity function $U^{S\theta}_{S\pi}$
$$M_{((S\pi)\dot{-}\theta)}.$$

Here we start observing an $(S\pi)$-tuplet in standard position copy the $(S\theta)$th member to the right leaving a one cipher gap and stop observing the copy in standard position.

Clearly all these tables are $(S\pi)$-normal for any numeral $\pi$.

The scheme of substitution $\sigma_{S\kappa}/\tau'_{S\pi} \ldots \tau^{(S\kappa)}_{S\pi}$:

$$P_{S\pi} T' M^{S\pi}_{SS\pi} T'' \ldots M^{S\pi}_{SS\pi} T^{(S\kappa)} M_{(\kappa(SS\pi)+1)} M_{((\kappa\dot{-}1).(SS\pi)+2)} \ldots M_{S\kappa} \Sigma J D,$$

where $T', \ldots, T^{S\kappa}$ are the tables for the functions $\tau'_{S\pi}, \ldots, \tau^{S\kappa}_{S\pi}$ and $\Sigma$ is the table for the function $\sigma_{S\kappa}$. If these tables are complete $(S\pi)$-normal tables then so is the one given. The functions $\tau'_{S\pi}, \ldots, \tau^{S\pi}_{S\pi}$ and $\sigma_{S\kappa}$ have been previously acquired.

We start observing an $(S\pi)$-ple say $\mathfrak{n}$ in standard position, and then copy it on the right leaving a two cipher gap between the original and the copy, then we set the table $T'$ working on the copy; since it is an $(S\pi)$-normal table then it fails to foul the original $(S\pi)$-ple. When it has finished then we make a second copy of $\mathfrak{n}$ leaving a one cipher gap, then apply table $T''$ to the copy. Since it is an $(S\pi)$-normal table then it fails to foul the work of table $T'$, and so on, till we have done $T^{S\kappa}$ on a copy of $\mathfrak{n}$. Then we copy down one after the other the representations of $\tau'_{S\pi}\mathfrak{n}, \ldots, \tau^{S\kappa}_{S\pi}\mathfrak{n}$ with one cipher gap between each, then we apply the table $\Sigma$, we finish by erasing everything between the work of $\Sigma$ and the original representation of $\mathfrak{n}$, finally we close the gap by moving the work of $\Sigma$ to the left so as to leave a single cipher gap between the work of $\Sigma$ and the original representation of $\mathfrak{n}$. Clearly the table given for the substitution scheme is complete and $(S\pi)$-normal if the tables $T', \ldots, T^{S\pi}$, and $\Sigma$ are complete and $(S\pi)$-normal

The scheme of primitive recursion $\widehat{\tau_{SS\pi}\sigma_\pi}$:

$$P_{S\pi}\Sigma M_{SS\pi}IL\begin{cases} LJD \\ RRI\dot{M}_3 M^{\pi}_{(\pi+4)} T M_{(\pi+4)} IL \end{cases}\begin{cases} LJD \\ M_{(\pi+4)}\dot{A} \end{cases},$$

where $\Sigma$ is the table for $\sigma_\pi$ and $T$ is the table for $T_{SS\pi}$. We start observing the representation of $\nu$, $\mathfrak{k}$ where $\mathfrak{k}$ stands for $\kappa', \ldots, \kappa^{(\pi)}$ and is the set of parameters, if $\pi$ is zero then parameters are absent and we have a case of primitive recursion without parameter. We first copy the argument leaving a two cipher gap between the copy and the original argument, we then apply table $\Sigma$, since this is $\pi$-normal then it fails to foul the original argument. If $\pi$ is zero then table $\Sigma$ will act like a function without argument and just calculate a single number, the same for each $\nu$. We then copy $\nu$ and erase its last tally, we then test whether the result (which represents $\nu \dot{-} 1$) is zero by moving one place to the left and seeing whether there is a cipher or a tally there, if there is a cipher there then $\nu \dot{-} 1$ is zero. If we find a cipher then we have finished because the value of $\rho[0, \mathfrak{k}]$ is $\sigma_\pi \mathfrak{k}$, we end up by erasing tallies between the result and the first set of two ciphers and closing up so that we end with a one cipher gap between the representation of the argument and the representation of $\rho[0, \mathfrak{k}]$. But if we find a tally on moving one place to the left then we move two places to the right and write a tally there, we then copy $\sigma_\pi \mathfrak{k}$ and the parameters $\mathfrak{k}$. We then apply table $T$ for the $SS\pi$-place function $\tau_{SS\pi}$ thus obtaining $\rho[1, \mathfrak{k}]$, we then copy $\nu \dot{-} 1$ and take another tally from it and test if the result, $\nu \dot{-} 2$, is zero by moving one place to the left and seeing if there is a cipher there. If there is then we finish up as before with the value $\rho[1, \mathfrak{k}]$, but if there is a tally then we copy the single tally $(\pi + 4)$ places to the left and add a tally to it then copy $\rho[1, \mathfrak{k}]$ and the parameters $\mathfrak{k}$ then apply table $T$ obtaining $\rho[2, \mathfrak{k}]$. So we continue, the result is (commas between formulae denote ciphers, the other formulae denote the representation of the numerals they determine):

$$\nu,\ \mathfrak{k},,\ \nu,\ \mathfrak{k},\ \rho[0, \mathfrak{k}],\ \nu \dot{-} 1,\ 0,\ \rho[0, \mathfrak{k}],\ \mathfrak{k},\ \rho[1, \mathfrak{k}],\ \nu \dot{-} 2,\ 1,\ \rho[1, \mathfrak{k}],\ \mathfrak{k},$$
$$\rho[2, \mathfrak{k}],\ \nu \dot{-} 3,\ 2,\ \rho[2, \mathfrak{k}],\ \mathfrak{k},\ \rho[3, \mathfrak{k}],\ \nu \dot{-} 4,\ 3,$$

and so on until $\quad \nu \dot{-} \nu,\ \nu \dot{-} \nu,\ \rho[\nu \dot{-} 1, \mathfrak{k}],\ \mathfrak{k},\ \rho[\nu, \mathfrak{k}],$

we then erase all tallies between the result and the first set of two ciphers to the left and close up the result. We finish up with $\nu$, $\mathfrak{k}$, $\rho[\nu, \mathfrak{k}]$. Clearly the table we have given is complete and $S\pi$-normal.

The least number $\kappa$ such that $\tau[\nu', \ldots, \nu^{(\pi)}, \kappa] = 0$.

$$P_\pi RRI\dot{T}L \begin{cases} RM_2 JD \\ RM^{S\pi}_{SS\pi}\dot{A}, \end{cases}$$

where $T$ is a complete, $S\pi$-normal table for $\tau$. The result of applying this table is:

$\mathfrak{n}, , \mathfrak{n}, 0, \tau[\mathfrak{n}, 0], \mathfrak{n}, 1, \tau[\mathfrak{n}, 1], \ \mathfrak{n}, 2, \tau[\mathfrak{n}, 2], \ldots$ and so on till $\mathfrak{n}, \kappa, \tau[\mathfrak{n}, \kappa]$, where $\tau[\mathfrak{n}, \kappa] = 0$, it then erases from the two ciphers up to $\tau[\mathfrak{n}, \kappa]$ inclusive and closes up ending up with $\mathfrak{n}$, $\kappa$. Clearly the table we have given is $S\pi$-normal and complete if that for $\tau$ is.

Thus the initial functions have complete $S\pi$-normal tables and the result of applying the schemes to complete $S\pi$-normal tables is again a complete $S\pi$-normal table, thus a partial recursive function can be calculated by a complete $S\pi$-normal table. The object in having $S\pi$-normal tables is that they fail to destroy work already done which we shall need to refer to again.

PROP. 2. *An $S\pi$-place partially calculable function given by an $S\pi$-normal table is partial recursive.*

Suppose that we have a complete table $T$. We associate with $T$ a set of numerals each of which is called a *g.n.* of $T$. They are defined as follows:

D 208 $\quad (\nu \equiv \kappa)\, Mod\, \pi \quad$ for $\quad Rem[\pi, \nu] = Rem[\pi, \kappa]$.

A *g.n.* of table $T$ is the numeral $\kappa$ which is determined by $\langle \kappa', \ldots, \kappa^{(\varpi\kappa)} \rangle$, where $\varpi\kappa$ is the order of the table $T$, and where $\kappa^{(\theta)} = \{\kappa_1^{(\theta)}, \kappa_2^{(\theta)}\}$, $1 \leqslant \theta \leqslant \varpi\kappa$, and

$$\kappa_1^{(\theta)} = 4 \,.\, \pi_1^{(\theta)} + \begin{matrix} 0 \\ 1 \\ 2 \\ 3 \end{matrix}\, Mod\,(4 \,.\, S\varpi\kappa) \quad \text{if} \quad \theta 0 \begin{matrix} N \\ R \\ L \\ I \end{matrix}\, \pi_1^{(\theta)} \text{ is an entry in } T,$$

$$\kappa_2^{(\theta)} = 4 \,.\, \pi_2^{(\theta)} + \begin{matrix} 0 \\ 1 \\ 2 \\ 3 \end{matrix}\, Mod\,(4 \,.\, S\varpi\kappa) \quad \text{if} \quad \theta 1 \begin{matrix} N \\ R \\ L \\ I \end{matrix}\, \pi_2^{(\theta)} \text{ is an entry in } T.$$

We said before that we would avoid terminating instructions other than the passive instruction 0 and we showed that this could always be achieved by a trivial alteration. Thus to any table which has a terminating instruction other than the passive instruction there is another table which calculates the same function which has the passive instruction as sole terminating instruction. Each numeral is the *g.n.* of some complete table some of these will fail to reach a terminating instruction, e.g. the complete table: $10R1\ 1|R1$ which continually moves to the right. We shall consider a machine which reaches a terminating instruction other than the passive instruction as failing to terminate (failing to switch off, while if it gets to the passive instruction then it switches off) but continually goes on doing the same thing, e.g. $2\,0\,N\,2$ which continually observes a cipher. Having zero as sole terminating instruction simplifies the work to follow, while every numeral being the *g.n.* of some table makes the work more tidy. In this book we usually arrange things so that every numeral is a *g.n.* of one of the set of things we are considering.

Zero is the g.*n.* of the empty table, i.e. the table without entries and so consisting of the passive instruction alone. The same table is represented many times, this arises through the congruences and by renumbering the instruction numbers other than 1 and 0 and by adding redundant instructions, apart from two quite different methods of calculating a given function.

We commence our calculation by observing the representation of the argument in standard position and referring to instruction 1. Suppose that at some point in the calculation we have on the tape:

$$\epsilon'\epsilon'' \ldots \epsilon^{(\pi)}\bar{\epsilon}^{(S\pi)}\epsilon^{(SS\pi)} \ldots \epsilon^{(\pi')}\epsilon^{(S\pi')} \quad (\pi \leqslant \pi'),$$

except for ciphers at the left or right, so that $\epsilon'$ and $\epsilon^{(S\pi')}$ are tallies, but if the observed symbol is a cipher and is at the end then it must be in the list. Here each $\epsilon$ is either a cipher or a tally and $\epsilon^{(S\pi)}$ is the observed symbol, this is denoted by putting a bar over it. Suppose that we have to act according to the $\theta$th instruction. To this *situation* we give for *g.n.* the numeral determined by the ordered quartet:

$$\{\langle\epsilon', \ldots, \epsilon^{(\pi)}\rangle, \epsilon^{(S\pi)}, \theta, \langle^{(SS\pi)}, \ldots, \epsilon^{(S\pi')}\rangle\},$$

where we interpret a cipher as an even number and a tally as an odd number, and $\theta' \equiv \theta\, Mod\, (S\varpi\kappa)$, then any numeral is the *g.n.* of a situation for a given table. If $\pi$ is zero the first component of the quartet is zero, similarly if $\pi$ is $\pi'$ the last component is zero. (Zero is the numeral attached to the null set in the ordering of all ordered sets of natural numbers.) Let $\kappa$ be the *g.n.* of the complete $(S\pi)$-normal table $T$ of order $\varpi\kappa$. We refer to $Pt[\kappa, \theta]$, where $\theta = Rem\,[\theta', (S\varpi\kappa)]$ and find $\kappa_1^{(\theta)}$, $\kappa_2^{(\theta)}$ such that $Pt[\kappa, \theta] = \{\kappa_1^{(\theta)}, \kappa_2^{(\theta)}\}$. We then find $\pi_1^{(\theta)}$, $\pi_2^{(\theta)}$ where

$$4.\pi_1^{(\theta)} + \begin{matrix}0\\1\\2\\3\end{matrix} = \kappa_1^{(\theta)}, \quad 0 \leqslant \pi_1^{(\theta)} \leqslant \varpi\kappa,$$

$$4.\pi_2^{(\theta)} + \begin{matrix}0\\1\\2\\3\end{matrix} = \kappa_2^{(\theta)}, \quad 0 \leqslant \pi_2^{(\theta)} \leqslant \varpi\kappa.$$

If $\epsilon^{(S\pi)}$ is even we refer to $\kappa_1^{(\theta)}$. According as

$$Rem\,[4, \kappa_1^{(\theta)}] = \begin{matrix}0\\1\\2\\3\end{matrix}$$

so the *g.n.* of the situation changes:

$$\left.\begin{array}{ll}
\text{from} & \{\langle\epsilon', \ldots, \epsilon^{(\pi)}\rangle, 0, \theta, \langle\epsilon^{(SS\pi)}, \ldots, \epsilon^{(S\pi')}\rangle\}, \\
\text{to} & \{\langle\epsilon', \ldots, \epsilon^{(\pi)}\rangle, 0, \pi_1^{(\theta)}, \langle\epsilon^{(SS\pi)}, \ldots, \epsilon^{(S\pi')}\rangle\}, \\
\text{or} & \{\langle\epsilon', \ldots, \epsilon^{(\pi)}, 0\rangle, \epsilon^{(SS\pi)}, \pi_1^{(\theta)}, \langle\epsilon^{(SSS\pi)}, \ldots, \epsilon^{(S\pi')}\rangle\}, \\
\text{or} & \{\langle\epsilon', \ldots, \epsilon^{(\pi \dot{-} 1)}\rangle, \epsilon^{(\pi)}, \pi_1^{(\theta)}, \langle 0, \epsilon^{(SS\pi)}, \ldots, \epsilon^{(S\pi')}\rangle\}, \\
\text{or} & \{\langle\epsilon', \ldots, \epsilon^{(\pi)}\rangle, 1, \pi_1^{(\theta)}, \langle\epsilon^{(SS\pi)}, \ldots, \epsilon^{(S\pi')}\rangle\}.
\end{array}\right\} \qquad (a)$$

If $\pi$ is zero then $\epsilon^{(\pi)}$ is zero in the third alternative and the first components of all these quartets are to be zero. If $\pi = \pi'$ the last components of all these quartets are to be zero, and $\epsilon^{(SS\pi)}$ in the second alternative is to be zero. If $\pi = 1$ the first component of the third alternative is zero, and if $S\pi = \pi'$ the last component of the second alternative is zero. If $\epsilon^{(S\pi)}$ is odd we refer to $\kappa_2^{(\theta)}$, according as

$$Rem\,[4, \kappa_2^{(\theta)}] = \begin{matrix}0\\1\\2\\3\end{matrix}$$

so the $g.n.$ of the situation changes:

$$\left.\begin{array}{ll} \text{from} & \{\langle \epsilon', \ldots, \epsilon^{(\pi)}\rangle, 1, \theta, \langle \epsilon^{(SS\pi)}, \ldots, \epsilon^{(S\pi')}\rangle\}, \\ \text{to} & \{\langle \epsilon', \ldots, \epsilon^{(\pi)}\rangle, 1, \pi_2^{(\theta)}, \langle \epsilon^{(SS\pi)}. \ldots, \epsilon^{(S\pi')}\rangle\}, \\ \text{or} & \{\langle \epsilon', \ldots, \epsilon^{(\pi)}, 1\rangle, \epsilon^{(SS\pi)}, \pi_2^{(\theta)}, \langle \epsilon^{(SSS\pi)}, \ldots, \epsilon^{(S\pi')}\rangle\}, \\ \text{or} & \{\langle \epsilon', \ldots, \epsilon^{(\pi \dot- 1)}\rangle, \epsilon^{\pi}, \pi_2^{(\theta)}, \langle 1, \epsilon^{(SS\pi)}, \ldots, \epsilon^{(S\pi')}\rangle\}, \\ \text{or} & \{\langle \epsilon', \ldots, \epsilon^{(\pi)}\rangle, 0, \pi_2^{(\theta)}, \langle \epsilon^{(SS\pi)}, \ldots, \epsilon^{(S\pi')}\rangle\}. \end{array}\right\} \quad (b)$$

If $\pi$ is zero then $\epsilon^{(\pi)}$ is zero in the third alternative, and the first components of all these quartets are to be 0, $\langle 1 \rangle$, 0, 0 respectively. If $\pi$ is $\pi'$ then $\epsilon^{(SS\pi)}$ is zero in the second alternative and the last component of all these quartets are to be 0, 0, $\langle 1 \rangle$, 0 respectively. In both cases if $\pi$ is 1 the first component of the third alternative is to be zero while if $S\pi$ is $\pi'$ the last component of the second alternative is to be zero.

These changes define a function of natural numbers by cases. Let $\nu = \{\nu_1, \nu_2, \nu_3, \nu_4\}$ be the $g.n.$ of the situation, then the $g.n.$ of the situation after having made the prescribed atomic act is:

D 209

$J[\kappa, \nu] = \{\nu_1, \nu_2, \nu_3{*}{*}, \nu_4\}$ if $\nu_2$ is even and $Rem\,[4, \nu_3{*}] = 0$,

$\{\nu_1 {}^\frown \langle \nu_2 \rangle . A_1[\nu_1], Pt[\nu_4, 1], \nu_3{*}{*}, \langle Pt[\nu_4, \xi] \rangle_{2 \leqslant \xi \leqslant \varpi\nu_4}\}$ if $\nu_2$ is even and $Rem\,[4, \nu_3{*}] = 1$,

$\{\langle Pt[\nu_1, \xi] \rangle_{1 \leqslant \xi < \varpi\nu_1}, Pt[\nu_1, \varpi\nu_1], \nu_3{*}{*}, A_1[\nu_4] . \langle \nu_2 \rangle {}^\frown \nu_4\}$ if $\nu_2$ is even and $Rem\,[4, \nu_3{*}] = 2$,

$\{\nu_1, \nu_2, \nu_3{*}{*}, \nu_4\}$ if $\nu_2$ is even and $Rem\,[4, \nu_3{*}] = 3$.

There are four more cases for $\nu_2$ odd obtained from $(b)$ in a similar manner. Here $\nu_3{*}$ stands for $pt\,[Pt[\kappa, \nu_3], 2, 1]$ in the first four cases which arise from $(a)$ and $\nu_3{*}$ stands for $pt[Pt[\kappa, \nu_3], 2, 2]$ in the last four cases which arise from $(b)$. Also $\nu_3{*}{*}$ stands for

$$\left[\frac{Rem[4 . S\varpi\kappa, \nu_3{*}]}{4}\right].$$

The functions which enter into the definition of $J$ are primitive recursive hence $J$ is primitive recursive. Now define by primitive recursion

D 210 $\quad \Theta[\kappa, \nu, 0] = \nu, \quad \Theta[\kappa, \nu, S\pi] = J[\kappa, \Theta[\kappa, \nu, \pi]].$

If $\nu$ is the $g.n.$ of a situation then $\Theta[\kappa, \nu, \pi]$ is the $g.n.$ of the situation after $\pi$ atomic applications of the instructions from the table whose $g.n.$ is $\kappa$.

If the calculation terminates before $\pi$ atomic applications of the instructions then $\Theta[\kappa, \nu, \pi]$ is the *g.n.* of the terminal situation. When we have to apply the passive instruction the *g.n.* of the situation is

$$\{\langle \epsilon', \ldots, \epsilon^{(\nu)} \rangle, \epsilon^{(S\nu)}, 0, \langle \epsilon^{(SS\nu)}, \ldots, \epsilon^{(S\pi)} \rangle\}.$$

Then

$$\nu_3 = 0, Pt[\kappa, \nu_3] = 0, \quad \nu_3* = 0, \quad \nu_3** = 0, \quad Rem\,[4, \nu_3*\} = 0.$$

Thus $J[\kappa, \nu] = \nu$ and $\Theta\,[\kappa, \nu, \pi]$ remains constant at the *g.n.* of the terminal situation.

Now suppose that the instructions whose *g.n.* is $\kappa$ are for the purpose of calculating the values of a $\theta$-place function of natural numbers. We start by observing the representation of a $\theta$-tuplet in standard position and are under instruction 1. If $\theta$ is zero then we start observing a cipher. Note that in calculating the *g.n.*'s of situations we neglect ciphers at the beginning or at the end unless such a cipher is being observed. This is dealt with in D 209 by $A_1[\nu_1]$ and $A_1[\nu_4]$ in the second and third clauses respectively.

D 211 $\quad \bar{\alpha}$ for $\underset{1 \leqslant \xi \leqslant S\alpha}{\langle 1 \rangle}$, where $\alpha$ is a numerical term.

A *g.n.* of the initial situation is the numeral determined by

$$\{\overline{\nu'}{}^\frown\langle 0\rangle^\frown\ldots^\frown\langle 0\rangle^\frown\overline{\nu^{(\theta \dot{-} 1)}}{}^\frown\langle 0\rangle, 1, 1, 0\} \quad \text{if} \quad \nu^{(\theta)} = 0,$$

if $\theta$ is zero then the first component is to be zero

$$\{\overline{\nu'}{}^\frown\langle 0\rangle^\frown\ldots^\frown\langle 0\rangle^\frown\overline{\nu^{(\theta)} \dot{-} 1}, 1, 1, 0\} \quad \text{if} \quad \nu^{(\theta)} \neq 0.$$

We are observing the $\theta$-tuplet $\langle \nu', \ldots, \nu^{(\theta)} \rangle$ in standard position, under instruction 1 without symbols to the right of the observed symbol. Suppose that we do $\lambda$ atomic acts according to the table whose *g.n.* is $\kappa$ and that we have calculated the numeral $\nu$. The *g.n.* of the terminal situation is the numeral determined by :

$$\{\overline{\nu'}{}^\frown\langle 0\rangle^\frown\ldots^\frown\langle 0\rangle^\frown\overline{\nu^{(\theta)}}{}^\frown\langle 0\rangle, 1, 0, 0\} \quad \text{if} \quad \nu = 0,$$

$$\{\overline{\nu'}{}^\frown\langle 0\rangle^\frown\ldots^\frown\langle 0\rangle^\frown\overline{\nu^{(\theta)}}{}^\frown\langle 0\rangle^\frown\overline{\nu \dot{-} 1}, 1, 0, 0\} \quad \text{if} \quad \nu \neq 0.$$

We are observing the $(S\theta)$-tuplet $\langle \nu', \ldots, \nu^{(\theta)}, \nu \rangle$ in standard position, under instruction zero, without tallies to the right of the observed symbol.

D 212 $\quad \Omega^{(\theta)}[\nu', \ldots, \nu^{(\theta)}, \lambda]$

for

$$\{\overline{\nu'}^\frown\langle 0\rangle^\frown\ldots{}^\frown\langle 0\rangle^\frown\overline{\nu^{(\theta)} \dot{-} 1}, 1, \lambda, 0\}.A_1[\nu^{(\theta)}] + \{\overline{\nu'}^\frown\langle 0\rangle^\frown\ldots{}^\frown\langle 0\rangle^\frown\overline{\nu^{(\theta \dot{-} 1)}}^\frown\langle 0\rangle,$$
$$1, \lambda, 0\}\, B[\nu^{(\theta)}].$$

To describe the passage from the initial to the terminal situation we have: for sufficiently large $\lambda$ ($\lambda$ is the number of atomic acts)

$$\Theta[\kappa, \Omega^{(\theta)}[\nu', \ldots, \nu^{(\theta)}, 1], \lambda] = \Omega^{(S\theta)}[\nu', \ldots, \nu^{(\theta)}, \nu, 0].$$

The functions $\Theta$, $\Omega^{(\theta)}$ are primitive recursive hence

D 213 $\qquad Un^{(\theta)}[\kappa, \nu', \ldots, \nu^{(\theta)}, \lambda, \nu]$

for $\qquad A_2[\Theta[\kappa, \Omega^{(\theta)}[\nu', \ldots, \nu^{(\theta)}, 1], \lambda], \Omega^{(S\theta)}[\nu', \ldots, \nu^{(\theta)}, \nu, 0]].$

The partially calculable function whose table has *g.n.* $\kappa$ when applied to the $\theta$-tuplet $\langle \nu', \ldots, \nu^{(\theta)}\rangle$, if defined, has its value given by the $A_{00_\mu}$-formula

$$pt[(\mu\xi)[Un^{(\theta)}[\kappa, \nu', \ldots, \nu^{(\theta)}, \xi_1, \xi_2] = 0], 2, 2], \qquad (5)$$

where $\xi_1$ stands for $pt[\xi, 2, 1,]$ and $\xi_2$ stands for $pt[\xi, 2, 2]$, thus it is a partial recursive function. This completes the demonstration of Prop. 2. $\kappa$ is called the *index* of the set or function.

PROP. 3. *A partial recursive function can be* $A_0$*-represented.*

Suppose that $f$ is a partial recursive function of $\theta$ arguments, by Prop. 1 we can find a complete $\theta$-normal table for it, let $\kappa$ be a *g.n.* of a complete $\theta$-normal table for $f$. Suppose $f[n', \ldots, n^{(\theta)}] = n$ then if $\nu', \ldots, \nu^{(\theta)}, \nu$ are the numerals which represent the natural numbers $n', \ldots, n^{(\theta)}, n$ respectively, we have

$$(E\xi)(Un^{(\theta)}[\kappa, \nu', \ldots, \nu^{(\theta)}, \xi, \nu] = 0) \qquad (6)$$

is an $A_0$-theorem, because for some numeral $\lambda$, giving the number of atomic acts to be performed in order to reach the passive instruction,

$$Un^{(\theta)}[\kappa, \nu', \ldots, \nu^{(\theta)}, \lambda, \nu] = 0 \qquad (7)$$

is an $A_{00}$-theorem.

A calculable function is primitive recursive if there is a primitive recursive bound for the number of steps in the calculation of the value of the function. Thus if $fn$ can be shown to determine $\kappa$ in less than $gn$ steps where $g$ is primitive recursive than $f$ is primitive recursive. The result follows at once since

$$(E\xi)_{g[\nu' \ldots, \nu^{(\theta)}]}(Un^{(\theta)}[\kappa, \nu', \ldots, \nu^{(\theta)}, \xi, \nu] = 0)$$

is primitive recursive, and so is

$$pt[(\mu\xi)_{g[\nu',\ldots,\nu^{(\theta)}]}[Un^{(\theta)}[\kappa,\nu',\ldots,\nu^{(\theta)},\xi_1,\xi_2]=0],2,2],$$

where $\xi_1$ for $pt[\xi, 2, 1]$ and $\xi_2$ for $pt[\xi, 2, 2]$.

If (5) is defined or if (6) is an $\mathbf{A}_0$-theorem for some $\nu$ then we shall say that *table $\kappa$ produces $\nu$ for argument $\nu', \ldots, \nu^{(\theta)}$*, even though the table may fail to be normal.

If $f[n', \ldots, n^{(\theta)}]$ is undefined then the calculation fails to terminate, in this case (6) fails to be an $\mathbf{A}_0$-theorem for each numeral $\nu$. Note that $Un$ is primitive recursive. If $\theta$ is zero then $\nu', \ldots, \nu^{(\theta)}$ are absent.

Cor. (i). *A function which can be $\mathbf{A}_0$-represented is partial recursive.*

According to Prop. 5, Ch. 6 the $\mathbf{A}_0$-statement which says that the value of the function for the argument $\kappa$ is $\nu$ can be put in the equivalent form $(E\xi)(\beta\{\xi,\kappa,\nu\}=0)$, for some numerical term $\beta\{\xi,\kappa,\nu\}$, this is primitive recursive. Thus the value of the function for argument $\kappa$ is

$$pt[(\mu\xi)[\beta\{\xi_1,\kappa,\xi_2\}=0],2,2],$$

and this is partial recursive.

Cor. (ii). *Partial recursive functions can be enumerated.*

The $\mathbf{A}_{00_\mu}$-formula (5) for $\kappa = 0, 1, 2, \ldots$ enumerates the partial recursive $\theta$-place functions. The same office is performed in $\mathbf{A}_0$ by the $\mathbf{A}_0$-statement (6). In both cases there will be repetitions. Two $\mathbf{A}_{00_\mu}$-functions are *equivalent* if they are both undefined for the same arguments and when both are defined for an argument they both take the same value for that argument.

Cor. (iii). *There is a partial recursive function which fails to be general recursive.*

The completely undefined function given by the table:

$$10N0 \quad 11N0$$

is partial recursive and fails to be general recursive. We could get the same result by a diagonal argument thus: Consider

$$\lambda x . Spt[(\mu x')[Un'[x,x,x_1',x_2']=0],2,2],$$

where $x_1'$ stands for $pt[x', 2, 1]$ and $x_2'$ stands for $pt[x', 2, 2]$, it is partial recursive hence it is equivalent to the $\mathbf{A}_{00_\mu}$ function

$$\boldsymbol{\lambda} x . pt[(\boldsymbol{\mu} x')\,[Un'[\kappa, x, x_1', x_2'] = 0], 2, 2]$$

for some numeral $\kappa$, now apply the argument $\kappa$ and instead of getting an absurdity we see that: $(\boldsymbol{\mu} x')\,[Un'[\kappa, \kappa, x_1', x_2']] = 0$ must be undefined.

A function of $(S\theta)$ arguments is called a *universal function for $\theta$ arguments* if its value for the arguments $\kappa, \nu', \ldots, \nu^{(\theta)}$ is the value of the function of $\theta$ arguments which has the table with *g.n.* $\kappa$. From Prop. 2 we see that (5) is such a function. We have given sufficient instructions to enable us to write down a table for the function (5). A function of two arguments is called a *universal function for all arguments* if its value for the arguments $\kappa$, $\nu$ is the value of the function of $\varpi\nu$ arguments $\nu' \ldots, \nu^{(\varpi\nu)}$ which has a table of *g.n.* $\kappa$, where $\nu' = Pt[\nu, 1], \ldots, \nu^{(\varpi\nu)} = Pt[\nu, \varpi\nu]$. We see from Prop. 2 that (5) with $\theta$ replaced by 1 is such a function, because if $f$ is a one-place partial recursive function then

$$\boldsymbol{\lambda} x' \ldots x^{(S\theta)} . f[\{x', \ldots, x^{(S\theta)}\}]$$

is a partial recursive function of $S\theta$ arguments and if $f[x', \ldots, x^{(S\theta)}]$ is a partial recursive function of $(S\theta)$ arguments then

$$f[pt[x, s\theta, 1], \ldots, pt[x, s\theta, s\theta]]$$

is a partial recursive function of one argument.

## 7·4 *The $S$-$\theta$-$\theta'$ proposition*

PROP. 4. *There is a primitive recursive function $S_\theta^{(S\theta')}$ of $(SS\theta')$ arguments, such that*

$$(E\xi)\,(Un^{(\theta+S\theta')}[\kappa, \nu', \ldots, \nu^{(\theta)}, \nu^{(S\theta)}, \ldots, \nu^{(\theta+S\theta')}, \xi, \nu] = 0)$$

*is an $\mathbf{A}_0$-theorem if and only if*

$$(E\xi)\,(Un^{(\theta)}[S_\theta^{(S\theta')}[\kappa, \nu^{(S\theta)}, \ldots, \nu^{(\theta+S\theta')}], \nu', \ldots, \nu^{(\theta)}, \xi, \nu] = 0)$$

*is an $\mathbf{A}_0$-theorem.*

Suppose we apply table $\kappa$ to the ordered set $\nu', \ldots, \nu^{(\theta+S\theta')}$ and obtain the value $\nu$. We want to obtain the same value by applying some table to the ordered set $\nu', \ldots, \nu^{(\theta)}$. This amounts to having a function of $(\theta + S\theta')$ arguments fixing the last $S\theta'$ arguments and obtaining a function of the remaining $\theta$ arguments. Let $T$ be a complete table whose $g.n.$ is $\kappa$ and consider the table $T'$

$$P_\nu R(RI)^{(S\nu^{(S\theta)})} R(RI)^{(S\nu^{(\theta+2)})} \dots R(RI)^{(S\nu^{(\theta+S\theta')})} TJD. \tag{8}$$

Started observing the $\theta$-tuplet $\nu', \dots, \nu^{(\theta)}$ in standard position we print the $(\theta+S\theta')$-tuplet $\nu', \dots, \nu^{(\theta)}, \nu^{(S\theta)}, \dots, \nu^{(\theta+S\theta')}$ on the right leaving a two cipher gap between this and the original $\theta$-tuplet. We then apply table $T$ to this $(\theta+S\theta')$-tuplet obtaining the value $\nu$. We then erase everything between the representation of $\nu$ and the two cipher gap, we then move the representation of $\nu$ to the left until we are left with the representation of the $(S\theta)$-tuplet $\nu', \dots, \nu^{(\theta)}, \nu$. Clearly the $g.n.$ of table (8) is a primitive recursive function of $\kappa$, the $g.n.$ of table $T$, and of $\theta$ and $\theta'$ and $\nu^{(S\theta)}, \dots, \nu^{(\theta+S\theta')}$.

D 214 $S_\theta^{(S\theta')}[\kappa, \nu^{(S\theta)}, \dots, \nu^{(\theta+S\theta')}]$ for the primitive recursive function of Prop. 4.

We have given sufficient details to find its explicit $\mathbf{A}_{00}$-definition.

## 7·5 *The undecidability of the classical predicate calculus $\mathscr{F}_C$*

A formal system $\mathscr{L}$ is said to be *decidable* when we have a uniform method applicable to any $\mathscr{L}$-statement $\phi$ which will terminate in the answer '$\phi$ is an $\mathscr{L}$-theorem' or in the answer '$\phi$ fails to be an $\mathscr{L}$-theorem'. If $\mathscr{L}$ is decidable and if $\kappa$ is the $g.n.$ of $\phi$ then by Church's Thesis there is a general recursive function $th$ such that $th\kappa = 0$ if $\phi$ is an $\mathscr{L}$-theorem otherwise $th\kappa = 1$. Sometimes without appealing to Church's Thesis we say *recursively decidable*, leaving open the question whether there might be non-recursive methods of decision.

A theory $\mathscr{T}$ is called *essentially undecidable* if every consistent primary extension of $\mathscr{S}$ is undecidable. We might obtain a decidable extension of an undecidable theory $\mathscr{T}$ by adding extra axioms and rules. We could do this by taking a decidable theory whose theorems would form a recursive set $\mathscr{S}$ and taking a recursively enumerable subset $\mathscr{S}'$ of $\mathscr{S}$ which fails to be recursive. According to Prop. 11, Ch. 8 the set $\mathscr{S}'$ is axiomatizable and so gives a theory $\mathscr{T}'$ of which $\mathscr{T}$ is a decidable primary extension, merely add the axioms of $\mathscr{T}$ to those of $\mathscr{T}'$. This forward reference is harmless because we will fail to use this example.

**PROP. 5.** $\mathbf{A}_0$ *is essentially undecidable.*

Suppose that $\mathscr{T}$ is a consistent primary extension of $\mathbf{A}_0$ and that $\mathscr{T}$ is decidable, then $\mathscr{T}$ is a formal system and hence its statements are recursively enumerable In particular $\mathscr{T}$-statements with exactly one free variable are recursively enumerable, say $\phi'\{x\}, \phi''\{x\}, \ldots$. Now consider the property of a numeral given by $\phi^{(\nu)}\{\nu\}$ fails to be a $\mathscr{T}$-theorem, if $\mathscr{T}$ is decidable then this property is recursive and hence is $\mathbf{A}_0$-definable and so is $\mathscr{T}$-definable, hence it must be $\phi^{(\kappa)}\{x\}$ for some numeral $\kappa$, then $\phi^{(\kappa)}\{\nu\}$ is a $\mathscr{T}$-theorem if and only if $\phi^{(\nu)}\{\nu\}$ fails to be a $\mathscr{T}$-theorem, put $\kappa$ for $\nu$ and we get an absurdity since $\mathscr{T}$ is consistent.

**PROP. 6.** $\mathscr{F}_C$ and $I\mathscr{F}_C$ *are undecidable.*

**LEMMA (i).** *A theory which is complete with respect to negation is decidable.*

Let $\mathscr{T}$ be a theory, which is complete with respect to negation. Let $N[\phi]$ be the negation of $\phi$. $\mathscr{T}$ is a formal system so we can recursively enumerate its theorems say, $\phi', \phi'', \ldots$ by Ch. 6, Prop. 9, Cor (i). Now let $\psi$ be a $\mathscr{T}$-statement, run down the list of $\mathscr{T}$-theorems; since $\mathscr{T}$ is complete with respect to negation, sooner or later we shall come to $\psi$ or to $N[\psi]$, hence, we have a decision procedure for the theory $\mathscr{T}$, hence $\mathscr{T}$ is decidable.

**LEMMA (ii).** *An undecidable theory containing negation is incomplete with respect to negation.*

This follows at once from lemma (i).

Form the system $\mathbf{A}$ by adding to $\mathbf{A}_0$ the symbol $A$ of type $o(o\iota)$ and the rule $\dfrac{\phi\{\xi\} \vee \omega}{(A\xi)\,\phi\{\xi\} \vee \omega}$ *generalization*, where $(A\xi)\,\phi\{\xi\}$ for $A(\boldsymbol{\lambda}\xi\,.\,\phi\{\xi\})$; by Prop. 5, $\mathbf{A}$, being a primary extension of $\mathbf{A}_0$, is undecidable. Negation is representable in $\mathbf{A}$ (see Ch. 3) which by lemma (ii) is incomplete. Now form the system $\bar{\mathbf{A}}$ from $\mathbf{A}$ by allowing free function variables and a rule of substitution and taking as axioms instead of those of $\mathbf{A}$ the following:

$$x = x, \quad Sx \neq 0, \quad 0 \neq Sx, \quad \mathscr{I}fx(Sx') = fx'(\mathscr{I}fxx'), \quad \mathscr{I}fx0 = x$$

and T.N.D. and take the $\boldsymbol{\lambda}$-axioms as rules. $\bar{\mathbf{A}}$ is undecidable because it is a primary extension of $\mathbf{A}_0$.

LEMMA (iii). *If we suppress a bounded set of axioms from an undecidable system then we are left with another undecidable system, provided that Modus Ponens and the Deduction Theorem hold in the system.*

We show that if we add a bounded set of axioms to a decidable system then we obtain another decidable system. Let $\Phi$ be the set of axioms of a decidable system $\mathscr{T}$ and let $\psi', \psi'', \ldots, \psi^{(\theta)}$ be the bounded set of additional axioms of the system $\mathscr{T}'$. Suppose that $\chi$ is a $\mathscr{T}'$-theorem then we have

$$\Phi, \psi', \psi'', \ldots, \psi^{(\theta)} \vdash_{\mathscr{T}'} \chi,$$

then by Cor. ii, Prop. 5, Ch. 3 we obtain

$$\Phi \vdash_{\mathscr{T}} C \prod_{\lambda=1}^{\theta} \overline{\psi}^{(\lambda)} \chi,$$

where $\overline{\psi}$ is the closure of $\psi$ Hence a decision procedure for $\mathscr{T}$ would give one for $C \prod_{\lambda=1}^{\theta} \overline{\psi}^{(\lambda)} \chi$, and so by Modus Ponens one for $\mathscr{T}'$.

COR. *If a theory $\mathscr{T}$ enriched by a bounded set of new axioms is a primary extension of an essentially undecidable theory then $\mathscr{T}$ is undecidable.*

Lastly the system $I\mathscr{F}_C$ is obtained from the system $\mathbf{A}$ by discarding a bounded set of axioms, by lemma (iii) $I\mathscr{F}_C$ is undecidable.

Applications of rule R 1 in $I\mathscr{F}_C$-proofs can be pushed back into the axioms so that rule R 1 can be dispensed with. The resulting system, call it $I\mathscr{F}_C*$, is undecidable and is a primary extension of the system $\mathscr{F}_C$, having one additional axiom namely $x = x$. By lemma (iii) $\mathscr{F}_C$ is undecidable.

## 7.6 *Various undecidability results*

COR. (i). *It is impossible to find a uniform method for deciding of a primitive recursive function whether it is ever zero.*

A closed $\mathbf{A}_0$-statement is equivalent to an $\mathbf{A}_0$-statement of one of the forms: $\alpha = 0$ or $(E\xi)(\alpha\{\xi\} = 0)$, where $\alpha$ and $\alpha\{\xi\}$ are numerical terms, and these forms can be found effectively. If we had a method for deciding whether a primitive recursive function was ever zero then we would have a decision procedure for $\mathbf{A}_0$ and this is impossible.

Cor. (ii). *It is impossible to find a uniform method for deciding of a primitive recursive function whether it is ever different from zero.*

For $\alpha\{\xi\} = 0$ is equivalent to $B_1[\alpha\{\xi\}] = 1$, hence a closed $\mathbf{A}_0$-statement is equivalent to one of the forms: $\alpha = 0$ or $(E\xi)(\beta\{\xi\} = 1)$, where $\beta$ is a characteristic function, so that if we had a method for deciding of a primitive recursive function whether it was ever different from zero then we would have a method for deciding $\mathbf{A}_0$, and this is impossible.

Cor. (iii). *It is impossible to find a uniform method for deciding of a general recursive function whether it is ever zero, similarly whether it is ever different from zero.*

For any such method would provide a method for deciding the same problems for primitive recursive functions, and this is impossible.

Cor. (iv). *It is impossible to find a uniform method for deciding of a partial recursive function whether it is ever zero, similarly whether it is ever different from zero.*

As for Cor. (iii).

Cor. (v). *It is impossible to find a uniform method for deciding of two primitive recursive functions whether they are equivalent, similarly for two general recursive functions, similarly for two partial recursive functions.*

If we had such a method then, in particular we could decide whether a primitive recursive function was always zero and so whether it ever took values different from zero, but by Cor. (ii) this is impossible. Similarly for the other cases.

Cor. (vi). *It is impossible to find a uniform method for deciding of two primitive recursive functions whether one majorizes the other, similarly for general and partial recursive functions.*

If the constant function zero majorizes the primitive recursive function $\rho$ then $\rho\nu$ is always zero. Thus if we had a method for deciding if one primitive recursive function majorized another one, then we could decide whether a primitive recursive function was always zero and this by Cor. (v) is impossible.

PROP. 7. *It is impossible to find a uniform method for deciding of a complete table whether it calculates a general recursive function of $\theta$ arguments.*

Suppose that we can decide of a numeral $\kappa$ whether it is the *g.n.* of a table which calculates a general recursive function of $\theta$ arguments. Then there will be a general recursive function $\rho$ such that $\rho\kappa = 1$ if $\kappa$ is the *g.n.* of a table which calculates a general recursive function of $\theta$ arguments and $\rho\kappa = 0$ otherwise. Write $\sigma\nu$ for $(\mu\xi)\,(\sum_{0\leqslant\xi'\leqslant\xi} \rho\xi' = \nu)$, then $\sigma\nu$ is equal to the *g.n.* of the $\nu$th table, in order of increasing *g.n.*, which calculates a general recursive function. $\sigma$ is a general recursive function if $\rho$ is such, because there is an unlimited supply of distinct tables for general recursive functions, so we can find a recursive bound for the least number operator, thus $\sigma\nu$ can always be found. See D 174 for a method of calculating a bound. Now consider the $\mathbf{A}_{00_\mu}$-formula:

$$Spt[(\mu\xi)\,(Un^{(\theta)}[\sigma\nu', \nu', \nu'', \ldots, \nu^{(\theta)}, \xi_1, \xi_2] = 0), 2, 2], \tag{9}$$

where $\xi_1$ stands for $pt[\xi, 2, 1]$ and $\xi_2$ stands for $pt[\xi, 2, 2]$. It is general recursive function of $\theta$ arguments $\nu', \ldots, \nu^{(\theta)}$, because $\sigma\nu'$ is equal to the *g.n.* of a table which calculates the values of a general recursive function and $\sigma$ is itself general recursive. Hence (9) is defined for each $\theta$-tuplet $\nu', \ldots, \nu^{(\theta)}$, hence it must be equivalent to

$$pt[(\mu\xi)\,[Un^{(\theta)}[\sigma\lambda, \nu', \ldots, \nu^{(\theta)}, \xi_1, \xi_2] = 0], 2, 2]$$

for some numeral $\lambda$. Now put $\nu'$ equal to $\lambda$ and we obtain an absurdity. The absurdity can be translated into one in $\mathbf{A}_0$ using the representation in $\mathbf{A}_0$ of general recursive functions. Thus the Proposition is demonstrated.

COR. (i). *It is impossible to recursively enumerate all general recursive functions.*

The demonstration is similar to that of Prop. 7.

## 7·7 *Lattice points*

An ordered $S\pi$-tuplet of natural numbers is sometimes called a *lattice point* in space of $S\pi$-dimensions $\mathscr{R}_{S\pi}$. A set of lattice points is called primitive (general) recursive if it has a primitive (general) recursive characteristic function. If $\mathscr{S}$ is a recursive set of lattice points then we can decide of a lattice point whether it belongs to $\mathscr{S}$ or otherwise. Thus a recursive set of lattice points is solvable. A set $\mathscr{S}$ of *lattice points* is called *recursively enumerable* if there is a partial recursive function $\rho$ such that the set $\mathscr{S}$ consists exactly of the $S\pi$-tuplets $\nu', \ldots, \nu^{(\theta)}$ for which

$$\rho\lambda = \{\nu', \ldots, \nu^{(\theta)}\}$$

for some numeral $\lambda$. The function $\rho$ generates the set $\mathscr{S}$ point by point. The completely undefined function enumerates the null set. For $\mathscr{R}_1$ the set $\mathscr{S}$ is the set of values of a recursive function. A recursively enumerable set of lattice points in $\mathscr{R}_{S\pi}$ consists of the lattice points $\nu', \ldots, \nu^{(\theta)}$ for which $(E\xi)(\rho\xi = \{\nu', \ldots, \nu^{(\theta)}\})$, where $\rho$ is a partial recursive function. If the enumerating function is primitive (general) recursive then the set is called primitive (general) recursively enumerable.

PROP. 8. *An unbounded recursive set of lattice points in $\mathscr{R}_{S\pi}$ is recursively enumerable without repetitions. A recursive set of natural numbers can be enumerated in order of magnitude without repetition.*

*A non-null recursively enumerable set of natural numbers can be enumerated with repetitions by a primitive recursive function. Similarly for recursively enumerable sets of $S\pi$-tuplets.*

*A recursively enumerable set of lattice points in $\mathscr{R}_{S\pi}$ can be recursively enumerated without repetitions.*

Generate the numerals, when the numeral $\nu$ has been generated express it in the form $\{\nu', \ldots, \nu^{(S\pi)}\}$ and test whether the lattice point $\{\nu', \ldots, \nu^{(S\pi)}\}$ belongs to the recursive set $\mathscr{S}$. If it does then write down $\{\nu', \ldots, \nu^{(S\pi)}\}$ in a list. In this manner we enumerate the lattice points of $\mathscr{S}$ without repetition, and we can find a partial recursive function $\rho$ such that

$$\rho\kappa = \{\nu', \ldots, \nu^{(S\pi)}\}$$

gives the $\kappa$th lattice point in $\mathscr{S}$. If $\pi = 0$ the above process enumerates a recursive set $\mathscr{S}$ of natural numbers in order of magnitude without repetition.

Let a non-null recursively enumerable set $\mathscr{S}$ of natural numbers be recursively enumerated by the partial recursive function $\rho$ whose table has *g.n.* $\kappa$. Clearly the set $\mathscr{S}$ consists of the successive values of the primitive recursive function

$$\nu_3 \times B_1[Un'[\kappa, \nu_1, \nu_2, \nu_3]] + \nu_0 \times A_1[Un'[\kappa, \nu_1, \nu_2, \nu_3]], \tag{10}$$

where $\nu = \{\nu_1, \nu_2, \nu_3\}$ and $\nu_0$ is a fixed member of $\mathscr{S}$.

Generate the recursively enumerable set $\mathscr{S}$ and when the $\kappa$th $S\pi$-tuplet $\{\nu', \ldots, \nu^{(S\pi)}\}$ has been generated test to see if it has already been generated, if it is new then write it down in a list otherwise proceed to $S\kappa$. This generates the set $\mathscr{S}$ without repetitions.

COR. (i). *It is impossible to find a uniform method for deciding of two recursive sets of lattice points in* $\mathscr{R}_{S\pi}$ *whether they are the same.*
For this is equivalent to the problem of deciding if two characteristic functions are the same. Demonstration as for Prop. 6, Cor. (v).

COR. (ii). *It is impossible to find a uniform method for deciding of a recursive set of lattice points in* $\mathscr{R}_{S\pi}$ *whether it is contained in another recursive set of lattice points in* $\mathscr{R}_{S\pi}$.
For this is equivalent to the problem for deciding whether one characteristic function majorizes another recursive characteristic function. The demonstration that this is impossible is similar to that of Prop. 6, Cor. (vi).

COR. (iii). *It is impossible to find a uniform method for deciding of a partial recursive function whether it is bounded or unbounded.*
Consider $Prf_0[\kappa, \nu]$, given $\nu$ this is a primitive recursive function of $\kappa$. Generate the natural numbers and when the natural number $\kappa$ has been generated test whether $Prf_0[\kappa, \nu]$ is zero, if it is zero then write down $\kappa$ in a list, this gives us the successive values of a partial recursive function, this function is unbounded if and only if '$\nu$' is an $\mathbf{A}_0$-theorem, thus if we had a test for boundedness then we would have a decision procedure for $\mathbf{A}_0$ but this is impossible.

COR. (iv). *If an unbounded set* $\mathscr{S}$ *of lattice points in* $\mathscr{R}_{S\pi}$ *is recursively enumerable in order of magnitude of* $\{\nu', \ldots, \nu^{(S\pi)}\}$ *without repetitions then it is recursive.*
For to decide whether the lattice point $\{\nu', \ldots, \nu^{(S\pi)}\}$ belongs to the set $\mathscr{S}$ enumerated in order of magnitude of $\{\nu', \ldots, \nu^{(S\pi)}\}$ by the recursive function $\rho$ we need only evaluate $\rho 0, \rho 1, \rho 2, \ldots, \rho\{\nu', \ldots, \nu^{(S\pi)}\}$, because

$$\kappa \leqslant \rho\kappa < \rho(S\kappa)$$

in this case, whence

$$\{\nu', \ldots, \nu^{(S\kappa)}\} < \rho\nu \quad \text{for} \quad \{\nu', \ldots, \nu^{(S\pi)}\} < \nu.$$

COR. (v). *If a set $\mathscr{S}$ of lattice points in $\mathscr{R}_{S\pi}$ is recursively enumerable and if its complement is also recursively enumerable then $\mathscr{S}$ is a recursive set of lattice points, and conversely.*

Let $\rho$ recursively enumerate $\mathscr{S}$ and $\sigma$ recursively enumerate $\bar{\mathscr{S}}$, the complement of $\mathscr{S}$. Generate the sequence:

$$\rho 0, \sigma 0, \rho 1, \sigma 1, \rho 2, \sigma 2, \ldots$$

sooner or later we come to one which is equal to a given numeral $\nu$. Hence we have a method for deciding whether $\nu$ is in $\mathscr{S}$ or is in $\bar{\mathscr{S}}$. Thus $\mathscr{S}$ is recursive. If $\mathscr{S}$ is recursive then so is $\bar{\mathscr{S}}$, and both are recursively enumerable.

PROP. 9. (i) *Recursive sets of lattice points in $\mathscr{R}_{S\pi}$ form a field, or a denumerable Boolean Algebra with zero and unit.*

(ii) *Recursively enumerable sets of lattice points in $\mathscr{R}_{S\pi}$ form a ring, or a denumerable distributive lattice with zero and unit.*

(i) We have to show that the union, intersection and complement of recursive sets are again recursive sets. Let $\mathscr{S}$, $\mathscr{S}'$ be recursive sets of lattice points in $\mathscr{R}_{S\pi}$ and let $\rho, \rho'$ be their recursive characteristic functions, so that:

$$\rho\{\nu', \ldots, \nu^{(S\pi)}\} = \begin{cases} 0 & \text{if } \{\nu', \ldots, \nu^{(S\pi)}\} \text{ is in } \mathscr{S}, \\ 1 & \text{otherwise,} \end{cases}$$

and similarly for $\rho'$. Then $\rho, \rho'$ are general recursive functions of $S\pi$ arguments. So then are $\rho\mathfrak{n} \times \rho'\mathfrak{n}$, $A_1[\rho\mathfrak{n} + \rho'\mathfrak{n}]$ and $1 \dot{-} \rho\mathfrak{n}$ where $\mathfrak{n}$ stands for $\{\nu', \ldots, \nu^{(S\pi)}\}$, but these are the characteristic functions of the union, intersection and complement of $\mathscr{S}$ and $\mathscr{S}'$ respectively. Thus the union, intersection and complement of recursive sets are again recursive sets.

(ii) We have to show that the union and the intersection of recursively enumerable sets are again recursively enumerable sets. Let two recursively sets $\mathscr{S}$ and $\mathscr{S}'$ be given by:

$$(E\xi)(\rho\xi = \{\nu', \ldots, \nu^{(S\pi)}\}) \quad \text{and} \quad (E\xi)(\rho'\xi = \{\nu', \ldots, \nu^{(S\pi)}\}),$$

where $\rho, \rho'$ are partial recursive functions. The union of the sets $\mathscr{S}$ and $\mathscr{S}'$ is given by $(E\xi)(\sigma\xi = \{\nu', \ldots, \nu^{(S\pi)}\})$, where

$$\sigma(2.\xi) = \rho\xi \quad \text{and} \quad \sigma(2.\xi + 1) = \rho'\xi,$$

then $\sigma$ is a partial recursive function. The intersection of $\mathscr{S}$ and $\mathscr{S}'$ is given by the set of lattice points which satisfy

$$(E\xi, \xi')(\rho\xi = \rho'\xi' \,\&\, \rho\xi = \{\nu', \dots, \nu^{(S\pi)}\}). \tag{11}$$

This set is the same as the set of lattice points $\{\nu', \dots, \nu^{(S\pi)}\}$ which satisfy

$$(E\xi)(B_2[\rho\xi_1, \rho\xi_2] \times 2^{\pi-1} \times (2.\rho\xi_1 + 1) = \langle \nu', \dots, \nu^{(S\pi)} \rangle)$$

where $\xi = \{\xi_1, \xi_2\}$. The object of having $\langle \nu', \dots, \nu^{(S\pi)} \rangle$ rather than $\{\nu', \dots, \nu^{(S\pi)}\}$ is that when $B_2[\rho\kappa . \rho'\kappa] \times 2^{\pi \dot{-} 1} \times (2.\rho\kappa + 1) = 0$ then we have the null set rather than the origin. Thus the intersection of $\mathscr{S}$ and $\mathscr{S}'$ is recursively enumerable, and the enumeration is given by (11). If the intersection is finite then it is given by a partial recursive function defined on an initial segment of the natural numbers. The enumeration given by (11) is obtained thus: generate the natural numbers and when the numeral $\nu$ has been generated express it in the form $\nu = \{\nu_1, \nu_2\}$ and test whether $\rho\nu_1 = \rho'\nu_2$, if this is so then write down $\rho\nu_1$ in a list. In each case zero is the null set, and the unit is the set of all natural numbers.

We have given sufficient instructions so that given the tables for $\mathscr{S}$ and $\mathscr{S}'$ we can find the index of the table for the union and intersection of $\mathscr{S}$ and $\mathscr{S}'$ and for the complement in the case of recursive sets. We then say that the operations union, intersection and complementation are *uniformly effective* for recursive sets. Union and intersection are uniformly effective for recursively enumerable sets.

PROP. 10. (i) *We can find a recursively enumerable set of natural numbers which fails to be a recursive set.*

(ii) *We can find a set of natural numbers which fails to be recursively enumerable.*

(i) We have shown that $A_0$-theorems are recursively enumerable, Prop. 9, Ch. 6. The set of values of an enumerating function fails to be a recursive set because $A_0$ is undecidable. Thus there is a recursively enumerable but non-recursive set.

(ii) The set of *g.n.*'s of $A_0$-formulae which fail to be $A_0$-theorems fails to be recursively enumerable, for if it were then by Prop. 8, Cor. (v) the set of *g.n.*'s of $A_0$-theorems would be a recursive set and so $A_0$ would be decidable which is absurd. Thus we have found a set of natural numbers which fails to be recursively enumerable.

## 7.8 *Complete sets*

Let $\nu = \{\nu_1, \nu_2, \nu_3, \nu_4\}$ and consider the primitive recursive function $C$ defined as follows.

D 215 $\qquad C$ for $\boldsymbol{\lambda} x.\{x_1, x_4\} \times B_1[Un[x_1, x_2, x_3, x_4]]$.

Then $C\nu = \{\nu_1, \nu_4\}$ if the table whose *g.n.* is $\nu_1$ produces the value $\nu_4$ for the argument $\nu_2$ after $\nu_3$ moves, otherwise its value for the argument $\nu$ is zero. Thus the values of the function $C$ consist of zero and $\{\kappa, \theta\}$ if table $\kappa$ ever produces $\theta$. The set of positive values of the function $C$ is called the *complete set*, $\mathscr{C}$. The complete set is recursively enumerable but fails to be recursive. If the complete set were recursive then we could decide if table $\kappa$ ever produced $\theta$, i.e. we could decide any table which is absurd, in particular we could decide $\mathbf{A}_0$ whose theorems are recursively enumerable by some table. The complete set is recursively enumerated with repetitions by the primitive recursive function defined by

D 216 $\qquad C'$ for $\boldsymbol{\lambda} x.(Cx + \{\kappa_0, \theta_0\} \times B_1[Cx])$,

where table $\kappa_0$ produces $\theta_0$. $\kappa_0$ and $\theta_0$ are fixed.

## 7.9 *Simple sets*

A set of natural numbers is called *immune* if it is unbounded and fails to contain any unbounded recursively enumerable set of natural numbers. Hence an immune set fails to be recursively enumerable, because it is distinct from each recursively enumerable set. (See end of historical remarks on Ch. 7 for the motivation for introducing simple sets.)

A set of natural numbers is *simple* if it is unbounded, recursively enumerable and with immune complement.

PROP. 11. *We can* $\mathbf{A}_0$*-define a simple set.*

Form a set $\mathscr{S}$ of natural numbers as follows: we place $\theta$ in $\mathscr{S}$ if $\{\kappa, \theta\}$ is in the complete set and if $\theta > 2.\kappa$ and if among the preceding values of $C'$, say, $\{\kappa', \theta'\}, \ldots, \{\kappa^{(\nu)}, \theta^{(\nu)}\}$ whenever $\kappa^{(\pi)} = \kappa$ we always have $\theta^{(\pi)} \leqslant 2.\kappa^{(\pi)}$. Each table then contributes at most one natural number to the set $\mathscr{S}$, namely the first one produced by that table in the order given by $C'$ which is greater than twice the *g.n.* of that table. If the function calculated by the table is unbounded then it certainly contributes one member to $S$. Thus the complement of $\mathscr{S}$ fails to contain any unbounded recursively enumerable set of natural numbers.

The set $\mathscr{S}$ is recursively enumerable for it is the set of values of the primitive recursive function $S'$ (apart possibly from the value zero), defined by:

$$S' \quad \text{for} \quad \boldsymbol{\lambda} x.(pt[C'x,2,2] \times A_1[pt[C'x,2,2] \dot{-} 2.pt[C'x,2,1]]$$
$$\times \prod_{0 \leqslant x' \leqslant x} (B_2[pt[C'x',2,1], pt[C'x,2,1]] \times B_1[pt[C'x',2,2]$$
$$\dot{-} 2.pt[C'x,2,1] + A_2[pt[C'x',2,1], pt[C'x,2,1]])).$$

If $C'\nu = \{\kappa, \theta\}$, then the second factor is 1 if $\theta > 2.\kappa$, otherwise it is zero, the third factor is 1 if $C'x' = \{x_1, x_2\}$ and whenever $x_1 = \kappa$ we always have $x_2 \leqslant 2.\kappa$ for $x' < \nu$, otherwise it is zero.

The following primitive recursive function $S''$ recursively enumerates the positive values of the function $S'\xi$

$$S'' \quad \text{for} \quad \boldsymbol{\lambda} x.(S'x + \theta_0 \times B_1[S'x]),$$

where $\theta_0$ is a fixed non-zero value of $S'$.

The complement of the set $\mathscr{S}$ is unbounded, for each table contributes at most one natural number to the set $\mathscr{S}$ so that the tables with *g.n.*'s $0, 1, 2, \ldots, \kappa$ contribute at most $S\kappa$ natural numbers to the set $\mathscr{S}$ while the natural numbers contributed by the tables whose *g.n.*'s are greater than $\kappa$ are greater than $2.(S\kappa)$. Thus among the first $2.\kappa$ natural numbers at most $S\kappa$ belong to the set $\mathscr{S}$. Thus the complement of $\mathscr{S}$ is unbounded. This completes the demonstration of the proposition.

A simple set fails to be recursive, for if it were then its complement would be recursive and so recursively enumerable which is absurd.

## 7.10 *Hypersimple sets*

An *array* is a recursively enumerated sequence of bounded sets of natural numbers. An array is *discrete* if the bounded sets of natural numbers of which it is composed are mutually exclusive. A set of natural numbers is *hypersimple* if it is recursively enumerable and if it contains at least one bounded set from each discrete array and if its complement is unbounded.

Prop. 12. *We can* $A_0$*-define a hypersimple set.*

We define a set $\mathscr{H}$ by steps. At step 0 do nothing. At step $S\nu$ generate the first $\nu$ members of the complete set, let the $\nu$th member be $\{\kappa, \theta\}$ then put all the components of $\theta$ where $\theta = \langle\theta', \ldots, \theta^{(\varpi\theta)}\rangle$ in the order in which they are here listed into a set $\mathscr{H}$ provided (i) that table $\kappa$ has previously failed to contribute to $\mathscr{H}$ and (ii) that each component of $\theta$ is greater than the successor of the greatest number yet contributed to $\mathscr{H}$ up to and including the latest preceding step which has a contributing table with *g.n.* less than $\kappa$, otherwise do nothing. This recursively enumerates a set $\mathscr{H}$ of natural numbers, the function which performs the enumeration can easily be $A_0$-defined. Let $k\nu$ be the greatest numeral contributed to $\mathscr{H}$ up to and including the $\nu$th step. Let $j[\kappa, \nu]$ be the latest step before the $(S\nu)$th step which has a contributing table with *g.n.* less than $\kappa$, otherwise let $j[\kappa, \nu]$ be zero. Then at step $(S\nu)$ all the components of $\theta$ are put in order into $\mathscr{H}$ if the table $\kappa$ has been unused before and if all the components of $\theta$ are greater than $k[j[\kappa, \nu]]+1$, otherwise do nothing. A table which contributes to $\mathscr{H}$ will be called a *contributing table*, a table which produces an unending discrete array will be called a *relevant table*. We have to show that each relevant table is a contributing table and that $\bar{\mathscr{H}}$, the complement of $\mathscr{H}$ is unbounded.

Let $\kappa$ be the *g.n.* of a relevant table $T$. As we generate the complete set we shall be referred to table $\kappa$ again and again without end, because $\{\kappa, \theta\}$ will turn up for each $\theta$ produced by $T$. There is only a limited set of tables (hence of contributing tables) with *g.n.* less than $\kappa$ and each contributing table contributes only once thus only a bounded set of numbers is contributed to $\mathscr{H}$ by tables which occur before some table with *g.n.* less than $\kappa$. Hence a step $(S\nu)$ will come when $\{\kappa, \theta\}$ turns up and each component of $\theta$ is greater than $k[j[\kappa, \nu]]+1$, and so all the components of $\theta$ get put in order into $\mathscr{H}$. Thus each relevant table is a contributing table. Clearly there is an unbounded set of contributing tables and so $\mathscr{H}$ is an unbounded recursively enumerable set of natural numbers. Note that $j[\kappa, \nu]$ is constant for sufficiently large $\nu$.

Let $n0$ be the least *g.n.* of any contributing table at all, let it be used at step $m0$. Let $n(S\nu)$ be the least *g.n.* of a contributing table which occurs after step $m\nu$ and let $m(S\nu)$ be the step at which it is used. We have $n\nu < n(S\nu)$ because a table is used only once. $k[m\nu]$ is the greatest number contributed to $\mathscr{H}$ up to and including step $m\nu$. Now $n\nu$ is less than the *g.n.* of any succeeding contributing table hence these succeeding tables contribute to $\mathscr{H}$ numbers greater than $k[m\nu]+1$. Thus $k[m\nu]+1$ for $\nu = 0, 1, \ldots$ are distinct and in $\overline{\mathscr{H}}$, thus $\overline{\mathscr{H}}$ is unbounded. Note that the function $k[m\nu]+1$ fails to be recursive, $k$ is recursive but we are unable to calculate $m$.

A hypersimple set is simple because it contains at least one member from each recursively enumerable set of distinct numbers – a discrete array of singletons – and hence at least one member from each recursively enumerable set of natural numbers.

Cor. (i). *There is a simple set which fails to be hypersimple.*

The simple set $\mathscr{S}$ which we constructed in Prop. 11 had one member of its complement from among the $S\nu$ numbers

$$\nu+2, \ldots, 2.\nu+2 \quad \text{for} \quad \nu = 0, 1, 2, \ldots,$$

because at most $\nu$ of the first $2.\nu+2$ numbers are in $\mathscr{S}$. By setting $\nu = 2^\kappa \dot{-} 1, \kappa = 1, 2, \ldots$ we obtain the exclusive recursively enumerable set of bounded sets of natural numbers:

$$\langle 3, 4\rangle, \langle 5, 6, 7, 8\rangle, \ldots, \langle 2^\kappa+1, \ldots, 2^{\kappa+1}\rangle, \ldots,$$

each of which has at least one member in the complement $\overline{\mathscr{S}}$ of $\mathscr{S}$. Hence $\mathscr{S}$ fails to be hypersimple. Thus the class of hypersimple sets is a proper subclass of the class of simple sets.

Prop. 13. *If $\mathscr{D}$ is a recursively enumerable subset of the complement $\overline{\mathscr{C}}$ of the complete set $\mathscr{C}$ then we can find a member in $\overline{\mathscr{D}} \cap \overline{\mathscr{C}}$, the intersection of $\overline{\mathscr{D}}$ and $\overline{\mathscr{C}}$.*

Note that $\overline{\mathscr{D}} \cap \overline{\mathscr{C}}$ fails to be null, for if it were then since $\mathscr{D} \subseteq \overline{\mathscr{C}}$ we would have $\mathscr{D} = \overline{\mathscr{C}}$ and this would mean, by Prop. 8, Cor. v, that $\mathscr{C}$ was recursive, which is absurd.

Generate the members of the recursively enumerable set $\mathscr{D}$ and express them as ordered pairs $\{\kappa, \pi\}$, whenever $\kappa = \pi$, write down $\pi$ in a list, this gives us a recursively enumerable set $\mathscr{E}$, a partial recursive function $e$ which enumerates $\mathscr{E}$ is easily found. Let $\theta$ be the *g.n.* of a table for the function $e$. Consider $\{\theta, \theta\}$, if this is in $\mathscr{D}$ then the table $\theta$ fails to produce $\theta$ because $\mathscr{D}$ is in the complement of the complete set $\mathscr{C}$. Thus $\mathscr{E}$ whose *g.n.* is $\theta$ fails to contain $\theta$. But if $\{\theta, \theta\}$ is in $\mathscr{D}$ then $\theta$ must be in $\mathscr{E}$ by the construction of $\mathscr{E}$. This is absurd, hence $\{\theta, \theta\}$ fails to be in $\mathscr{D}$ and so is in $\overline{\mathscr{D}}$.

Now suppose that $\{\theta, \theta\}$ is in $\mathscr{C}$, then the table $\theta$ produces $\theta$, whence $\theta$ is in $\mathscr{E}$ and so by the construction of $\mathscr{E}$ $\{\theta, \theta\}$ is in $\mathscr{D}$, whence $\{\theta, \theta\}$ is in $\overline{\mathscr{C}}$. This again is absurd hence altogether $\{\theta, \theta\}$ is in $\overline{\mathscr{D}} \cap \overline{\mathscr{C}}$.

A table for $e$ is given by the partial recursive function

$$pt[d[(\mu\eta)[\sum_{0 \leqslant \xi \leqslant \eta} B_2[pt[d\xi, 2, 1], pt[d\xi, 2, 2]] = \nu]], 2, 1],$$

where the function $d$ recursively enumerates the set $\mathscr{D}$. Clearly the $g.n.$ of this table is a recursive function of the $g.n.$ for the table for $\mathscr{D}$. Let $\kappa$ be the $g.n.$ of a table for the function $d$, then $e'\kappa$ for some recursive function $e'$, which we can find as just shown, is the $g.n.$ for $e$ and $\{e'\kappa, e'\kappa\}$ is a recursive function of the $g.n.$ of $d$ and it is a number in $\overline{\mathscr{D}} \cap \overline{\mathscr{C}}$.

If $\mathscr{C}$ were recursive then so would be $\overline{\mathscr{C}}$, whence $\overline{\mathscr{C}}$ would be recursively enumerable and we could take $\mathscr{D} = \overline{\mathscr{C}}$, then $\overline{\mathscr{D}} \cap \overline{\mathscr{C}} = \emptyset$, where $\emptyset$ is the null set. Thus $e'\kappa$ which is in $\overline{\mathscr{D}} \cap \overline{\mathscr{C}}$ is a witness that $\overline{\mathscr{C}}$ fails to be a recursively enumerable set, and hence that $\mathscr{C}$ fails to be a recursive set.

## 7.11 *Creative sets*

A recursively enumerable set $\mathscr{C}r$ is said to be *creative* if there is a partial recursive function $f$ called the *production function of* $\mathscr{C}r$ such that given a recursively enumerable subset $\mathscr{D}$ of $\overline{\mathscr{C}r}$ if $\theta$ is the $g.n.$ of a table which enumerates $\mathscr{D}$ then $f\theta$ is defined and is in $\overline{\mathscr{D}} \cap \overline{\mathscr{C}r}$. Thus $f\theta$ is a witness that $\overline{\mathscr{C}r}$ fails to be recursively enumerable. Hence $\mathscr{C}r$ fails to be a recursive set.

COR. (i). *We can find a creative set.*

By Prop. 13 the complete set is creative.

COR. (ii). *We can find an unbounded recursively enumerable subset of* $\overline{\mathscr{C}r}$.

The null set is recursively enumerable by the completely undefined function and is a subset of $\overline{\mathscr{C}r}$, hence we can find a member of $\overline{\mathscr{C}r}$, say $\nu_0$. The set whose sole member is $\nu_0$ is recursively enumerable hence we can find a member $\nu_1$ of $\overline{\mathscr{C}r}$ with $\nu_0 \neq \nu_1$ and so on. Thus we can generate a sequence $\nu_0, \nu_1, \nu_2, \ldots$ of distinct members of $\overline{\mathscr{C}r}$. This gives a recursively enumerable subset $\mathscr{D}$ of $\overline{\mathscr{C}r}$, and $\mathscr{D}$ is a proper subset of $\overline{\mathscr{C}r}$. We can continue for we can find a member $\nu_\omega$ of $\overline{\mathscr{D}} \cap \overline{\mathscr{C}r}$. Add $\nu_\omega$ to $\mathscr{D}$, say as first member then we can find a number $\nu_{\omega+1} \neq \nu_\omega$ of $\overline{\mathscr{D}} \cap \overline{\mathscr{C}r}$, and so on through the *constructive ordinals*. We can continue as long as we have a recursively enumerable set. A full investigation of this would require an account of the theory of recursive ordinals.

PROP. 14. *If $\mathscr{C}r$ is a creative set and $\mathscr{D}$ a recursively enumerable set and $f$ a recursive function such that $\kappa$ is a member of $\mathscr{C}r$ if and only if $f\kappa$ is a member of $\mathscr{D}$, then $\mathscr{D}$ is a creative set.*

Let $\mathscr{X}_\nu$ be the recursively enumerable set given by the table with *g.n.* $\nu$. Now $\mathscr{C}r$ is a creative set, let $g$ be a productive function for it, so that if $\mathscr{X}_\nu \subseteq \overline{\mathscr{C}r}$ then $g\nu$ is defined and is in $\overline{\mathscr{X}}_\nu \cap \overline{\mathscr{C}r}$. (12)
Also $\kappa$ is in $\mathscr{C}r$ iff $f\kappa$ is in $\mathscr{D}$, thus

$$\kappa \text{ is in } \overline{\mathscr{C}r} \quad \text{iff} \quad f\kappa \text{ is in } \overline{\mathscr{D}}. \tag{13}$$

Now let $\mathscr{X}_\pi$ be a recursively enumerable subset of $\overline{\mathscr{D}}$ and consider the set of natural numbers $\nu$ such that $f\nu$ is in $\mathscr{X}_\pi$. Call this set $f^{-1}(\mathscr{X}_\pi)$. Then $f^{-1}(\mathscr{X}_\pi)$ is a subset of $\overline{\mathscr{C}r}$ by (13), also it is a recursively enumerable set, for generate the ordered pairs $\{\nu_1, \nu_2\}$ of numerals and when the ordered pair $\{\nu_1, \nu_2\}$ has been generated form $\Phi_\pi \nu_1$ and $f\nu_2$, where $\mathscr{X}_\pi$ is enumerated by the function $\Phi_\pi$. Whenever $\Phi_\pi \nu_1 = f\nu_2$ write $\nu_2$ down in a list, this gives an enumeration of $f^{-1}(\mathscr{X}_\pi)$. Let $f^{-1}(\mathscr{X}_\pi)$ be $\mathscr{X}_\theta$, then $\theta$ is a recursive function of $\pi$, say $\theta = k\pi$. We have just given sufficient details to find $\theta$ and the function $k$. We have also shown that $\mathscr{X}_\theta$ is a subset of $\overline{\mathscr{C}r}$.

By (12) $g\theta$ is defined and $g\theta$ is in

$$\overline{\mathscr{X}}_\theta \cap \overline{\mathscr{C}r}. \tag{14}$$

Hence $f[g\theta]$ is in $\overline{\mathscr{X}}_\pi$, because $\kappa$ is in $\mathscr{X}_\theta$ if and only if $f\kappa$ is in $\mathscr{X}_\pi$. Now $g\theta$ is in $\overline{\mathscr{C}r}$ by (14), hence by (13) $f[g\theta]$ is in $\overline{\mathscr{D}}$, thus $f[g[k\pi]]$ is in $\overline{\mathscr{X}}_\pi \cap \overline{\mathscr{D}}$. Altogether if $\mathscr{X}_\pi$ is a recursively enumerable subset of $\overline{\mathscr{D}}$ then $f[g[k\pi]]$ is in $\overline{\mathscr{X}}_\pi \cap \overline{\mathscr{D}}$, thus $\mathscr{D}$ is creative.

COR. (i). $\mathbf{A}_0$-*theorems form a creative set.*

Let $\mathscr{T}$ be the set of *g.n.*'s of $\mathbf{A}_0$-theorems and let $Cr$ be a creative set. Let $\phi\{\nu\}$ be an $\mathbf{A}_0$-statement which is an $\mathbf{A}_0$-theorem just in case $\nu$ is in $\mathscr{C}r$, so that $\phi$ expresses the set $\mathscr{C}r$. There is an $\mathbf{A}_0$-representation for each recursively enumerable set, hence given the creative set, say the complete set, then we can find $\phi$. Let $f\,Num\,\nu$ be equal to the *g.n.* of $\phi\{\nu\}$, then $f$ is a recursive function and by the completness of $\mathbf{A}_0$ $\nu$ is in $\mathscr{C}r$ if and only if $f\,Num\,\nu$ is in $\mathscr{T}$. By Prop. 14 $\mathscr{T}$ is a creative set.

COR. (ii). *If a formal system $\mathscr{L}$ can express a creative set, that is, there are $\mathscr{L}$-statements $\phi_\nu$ such that $\phi_\nu$ is an $\mathscr{L}$-theorem if and only if $\nu$ is in the creative set, then the $\mathscr{L}$-theorems form a creative set and $\mathscr{L}$ is undecidable.*

We carry through the demonstration of the previous corollary and note that a creative set fails to be recursive.

COR. (iii). *If a formal system $\mathscr{L}$ can express every recursively enumerable set then $\mathscr{L}$-theorems form a creative set and $\mathscr{L}$ is undecidable.*

For $\mathscr{L}$ can then express a creative set.

COR. (iv). *If a formal system $\mathscr{L}$ can express a creative set and if it contains a negation with respect to which it is consistent and if $\mathscr{L}$ contains the rule $\dfrac{\mathscr{N}\mathscr{N}\phi}{\phi}*$ where $\mathscr{N}\phi$ is the negation of $\phi$ then $\mathscr{L}$ contains an irresolvable statement.*

We can give *g.n.*'s to $\mathscr{L}$-symbols so that each numeral is the *g.n.* of an $\mathscr{L}$-formula. By Cor. (ii) the *g.n.*'s of $\mathscr{L}$-theorems form a creative set $\mathscr{T}$ and the *g.n.*'s of $\mathscr{L}$-statements whose negations are $\mathscr{L}$-theorems together with the *g.n.*'s of $\mathscr{L}$-formulae which fail to be $\mathscr{L}$-statements form a recursively enumerable subset $\mathscr{F}$ of $\overline{\mathscr{T}}$. Hence we can effectively find a member of $\overline{\mathscr{F}} \cap \overline{\mathscr{T}}$. This number yields an $\mathscr{L}$-statement $\phi$. Let $\mathscr{N}\phi$ be the negation of $\phi$, then $\mathscr{N}\phi$ will also be in $\overline{\mathscr{F}} \cap \overline{\mathscr{T}}$, for if it were in $\mathscr{F}$ then $\phi$ would be in $\mathscr{T}$ by the rule of double negation, and if it were in $\mathscr{T}$ then $\phi$ would be in $\mathscr{F}$ by the definition of $\mathscr{F}$. Thus $\phi$ is an irresolvable $\mathscr{L}$-statement.

The complete set $\mathscr{C}$ consists of the numerals determined by the ordered pairs of numerals $\{\kappa, \nu\}$, where table $\kappa$ produces $\nu$; thus $\{\kappa, \nu\}$ belongs to $\mathscr{C}$ if and only if $(E\xi)(Un'[\kappa, \xi_1, \xi_2, \nu] = 0)$ where $\xi = \{\xi_1, \xi_2\}$. Thus the decision whether the recursively enumerable set $\mathscr{X}_\kappa$, ever produces $\nu$ is settled by the decision whether $\{\kappa, \nu\}$ is in the complete set $\mathscr{C}$ or otherwise. We express this by saying that the complete set $\mathscr{C}$ has the *highest degree of unsolvability* for recursively enumerable sets. Recursive sets are solvable sets, hence of the lowest degree of unsolvability.

Two sets are *isomorphic*, $\mathscr{A} \equiv \mathscr{B}$, if there is a *recursive permutation function* $f$ (i.e. $f\nu' \neq f\nu''$ if and only if $\nu' \neq \nu''$ and for any $\theta$ we have $f\pi = \theta$ for some $\pi$) such that $\nu$ is a member of $\mathscr{A}$ if and only if $f\nu$ is a member of $\mathscr{B}$.

We denote the partial recursive function of $\theta$ arguments whose table has *g.n.* $\kappa$ by: $\Phi[\kappa; x', \ldots, x^{(\theta)}]$. Thus

$$\Phi[\kappa; \nu', \ldots, \nu^{(\theta)}] = \nu \quad \text{iff} \quad (E\xi)\,(Un^{(\theta)}[\kappa; \nu', \ldots, \nu^{(\theta)}, \xi, \nu] = 0).$$

We let $\mathscr{X}_\kappa$ denote the *range* of $\Phi[\kappa; \nu]$, we write this as

$$\mathscr{X}_\kappa = \hat{x}(Ex')\,(\Phi[\kappa; x'] = x).$$

## 7.12 *Productive sets*

A set $\mathscr{S}$ is *productive* if there is a partial recursive function $\rho$ such that whenever $\mathscr{X}_\kappa \subset \mathscr{S}$ then $\rho\kappa$ is defined and $\rho\kappa \in \mathscr{S} - \mathscr{X}$. $\rho$ is called a *production function* for $\mathscr{S}$. Thus the set $\mathscr{C}r$ is creative if it is recursively enumerable and its complement is productive.

PROP. 15. *If $\mathscr{C}r$ is a creative set then we can find a recursive* (1–1) *production function for it.*

Let $f$ be a (partial) recursive productive function for the creative set $\mathscr{C}r$. We define a function $f'$ as follows: $\mathscr{X}_0$ is the null set, hence $f0$ is defined, $(\mathscr{X}_0 \subset \overline{\mathscr{C}r})$, we set $f'0 = f0$. Suppose that $f'0 < f'1 < f'2 < \ldots < f'\kappa$ have been defined, we give the following procedure for calculating $f'(S\kappa)$: Let table $A$ calculate the recursively enumerable set $\mathscr{X}_{S\kappa} \cap \mathscr{C}r$ and let table $B$ do the following calculations:

Start calculating $f(S\kappa)$, if this is found, find a *g.n.* for the recursively enumerable set $\{f(S\kappa)\} \cup \mathscr{X}_{S\kappa}$ say $h_1(S\kappa)$ so that

$$\{f(S\kappa)\} \cup \mathscr{X}_{S\kappa} = \mathscr{X}_{h_1(S\kappa)}.$$

Then start calculating $f[h_1(S\kappa)]$, if this is found, find a *g.n.* for the recursively enumerable set $\{f[h_1(S\kappa)]\} \cup \mathscr{X}_{h_1(S\kappa)}$, say $h_2(S\kappa)$, then start calculating $f[h_2(S\kappa)]$, and so on until we have found

$$f(S\kappa), f[h_1(S\kappa)], \ldots, f[h_{S(f'\kappa)}(S\kappa)]. \tag{15}$$

Table $A$ is set to stop when it produces a numeral, e.g. let $\pi$ be a *g.n.* for a table which recursively enumerates $\mathscr{X}_{S\kappa} \cap \mathscr{C}r$, then generate the natural numbers and when the natural number $\nu$ has been generated find $\nu_1, \nu_2$ so that $\nu = \{\nu_1, \nu_2\}$, then set going table $\pi$ on argument $\nu_1$ for $\nu_2$ moves, if it produces a natural number $\mu$ then table $A$ stops, if not proceed to $S\nu$. Table $B$ is set to stop when it has calculated the series (15). Table $A$ will eventually stop unless $\mathscr{X}_{S\kappa} \subset \overline{\mathscr{C}r}$ in which case $\mathscr{X}_{S\kappa} \cap \mathscr{C}r$ is the null set. In this case since $f$ is a productive function for $\mathscr{C}r$

$$f(S\kappa) \quad \text{exists and} \quad f(S\kappa) \in \overline{\mathscr{X}}_{S\kappa} \cap \overline{\mathscr{C}r}, \tag{16}$$

now
$$\mathscr{X}_{h_1(S\kappa)} = \{f(S\kappa)\} \cup \mathscr{X}_{S\kappa} \tag{17}$$

thus
$$\mathscr{X}_{h_1(S\kappa)} \subset \overline{\mathscr{C}r}.$$

Again $f[h_1(S\kappa)]$ exists and

$$f[h_1(S\kappa)] \in \overline{\mathscr{X}}_{h_1(S\kappa)} \cap \overline{\mathscr{C}r}. \tag{18}$$

Now
$$\mathscr{X}_{h_2(S\kappa)} = \{f[h_1(S\kappa)]\} \cup \mathscr{X}_{h_1(S\kappa)},$$

so
$$\mathscr{X}_{h_2(S\kappa)} \subset \overline{\mathscr{C}r},$$

and so on. Thus the calculation performed by table $B$ will eventually stop. Hence either table $A$ or table $B$ will eventually stop. We set

$$f'(S\kappa) = \begin{cases} S(f'\kappa), & \text{if table } A \text{ stops first,} \\ Max\,[f(S\kappa), f[h_1(S\kappa)], \ldots, f[h_{S(f'\kappa)}]] & \text{if table } B \text{ stops first,} \end{cases} \tag{19}$$

thus $f'$ is a recursive function.

Further if $\mathscr{X}_{S\kappa} \subset \overline{\mathscr{C}r}$ we see from (16), (17), (18) that

$$f(S\kappa), f[h_1(S\kappa)], \ldots, f[h_{S(f'\kappa)}(S\kappa)]$$

are all distinct and are in $\overline{\mathscr{X}}_{S\kappa} \cap \overline{\mathscr{C}r}$, thus the maximum of these $SS(f'\kappa)$ numbers is greater than or equal to $S(f\kappa)$. Thus $f'$ is strictly increasing. Also the maximum is in $\overline{\mathscr{X}}_{S\kappa} \cap \overline{\mathscr{C}r}$. Hence $f'$ is a recursive production function for $\mathscr{C}r$, it is (1–1) because it is increasing strictly. The choice of value of $f'(S\kappa)$ given in (19) is settled for us if we let the two tables do one move alternately beginning with table $A$. This completes the demonstration of the proposition.

We say '$\mathscr{A}$ *is* (1–1)-*reducible to* $\mathscr{B}$' for 'there is a (1–1) recursive function $f$ such that $\nu \in \mathscr{A}$ if and only if $f\nu \in \mathscr{B}$', $f$ may omit some values of $\mathscr{B}$. If $\mathscr{A}$ is (1–1)-reducible to $\mathscr{B}$ we write $\mathscr{A} \leqslant_1 \mathscr{B}$.

LEMMA. *For any numeral $\kappa$ we can find a numeral $\theta$ such that*

$$\Phi[\kappa; \nu, \theta] = \Phi[\theta; \nu].$$

$\Phi[\kappa; \nu, S_1'[\pi, \pi]]$ is a partial recursive function of $\nu, \pi$ effectively given by $\kappa$, hence we can find a $g.n.$ for a table for it, say $g\kappa$, where $g$ is a recursive function. Then

$$\Phi[\kappa; \nu, S_1'[\pi, \pi]] = \Phi[g\kappa; \nu, \pi],$$

whence $$\Phi[\kappa;\nu,S_1'[g\kappa,g\kappa]] = \Phi[g\kappa;\nu,g\kappa],$$

but from Prop. 4 $$\Phi[g\kappa;\nu,g\kappa] = \Phi[S_1'[g\kappa,g\kappa];\nu],$$

put $\theta = S_1'[g\kappa,g\kappa]$ and the result follows.

## 7·13 *Isomorphism of creative sets*

PROP. 16. *Any two creative sets are isomorphic.*

The demonstration falls into two parts.

(i) any recursively enumerable set is (1–1)-reducible to any creative set.

(ii) If $\mathscr{A} \leqslant_1 \mathscr{B}$ and $\mathscr{B} \leqslant_1 \mathscr{A}$ then $\mathscr{A}$ and $\mathscr{B}$ are isomorphic.

Ad (i). Consider the creative set $\mathscr{C}r$ with (1–1) recursive productive function $f$, we show that $\mathscr{X}_\theta$ is (1–1)-reducible to $\mathscr{C}r$.

We define a set
$$\mathscr{Y}_\nu^\mu = \begin{cases} \{f\mu\} & \text{if } \nu \text{ is in } \mathscr{X}_\theta, \\ \emptyset & \text{otherwise.} \end{cases}$$

Then given $\mu$ and $\nu$ the set $\mathscr{Y}_\nu^\mu$ can be effectively enumerated. (Define a function $y$ as follows: generate the numerals and when the numeral $\pi$ has been generated find $\pi_1$ and $\pi_2$ so that $\pi = \{\pi_1,\pi_2\}$, then set going table $\theta$ for argument $\pi_1$ for $\pi_2$ moves if this has produced $\nu$ then set $y\pi = f\mu$, otherwise $y\pi$ undefined.) Let $g_1[\mu,\nu]$ be a $g.n.$ for a table for $\mathscr{Y}_\nu^\mu$.

Now $\Phi[g_1[x',\nu];x]$ is an undetermined value of a function of $x, x'$ which can be effectively found given $\nu$, thus

$$\Phi[g_1[\mu,\nu];\lambda] = \Phi[g_2\nu;\lambda,\mu]$$

for some recursive function $g_2$. By the lemma

$$\Phi[g_2\nu;\lambda,\pi] = \Phi[\pi;\lambda],$$

where $\pi = S_1'[g[g_2\nu],g[g_2\nu]]$, $g$ being as in the lemma. Hence $\mathscr{Y}_\nu^\mu$ is $\mathscr{X}_\pi$, with this value of $\pi$. If $\nu \in \mathscr{X}_\theta$ then $f\pi \in \mathscr{Y}_\nu^\mu$, by definition of $\mathscr{Y}_\nu^\mu$, hence $f\pi \in \mathscr{X}_\pi$ and so $\mathscr{X}_\pi \not\subset \overline{\mathscr{C}r}$, since $\mathscr{X}_\pi = \{f\pi\}$ then $f\pi \,\bar{\in}\, \overline{\mathscr{C}r}$ and so $f\pi \in \mathscr{C}r$. (If $X_\pi \subset \overline{\mathscr{C}r}$ then since $f$ is a productive function for $\overline{\mathscr{C}r}$ we should have $f\pi \in \overline{\mathscr{C}r} \cap \overline{\mathscr{X}}_\pi$ whence $f\pi \in \overline{\mathscr{X}}_\pi$.) If $\nu \in \overline{\mathscr{X}}_\theta$ then $\mathscr{X}_\pi$ and $\mathscr{Y}_\nu^\pi$ are both null, by the definition of $\mathscr{Y}_\nu^\pi$, whence $f\pi \in \overline{\mathscr{C}r}$ by property of $f$, $(\mathscr{X}_\pi \subset \overline{\mathscr{C}r})$. Thus

$$\nu \in \mathscr{X}_\theta \quad \text{iff} \quad f\pi \in \mathscr{C}r,$$

i.e. $$\nu \in \mathscr{X}_\theta \quad \text{iff} \quad h\nu \in \mathscr{C}r,$$

where $$h\nu = fS_1'[g[g_2\nu],g[g_2\nu]] = f\pi.$$

It remains to show that the recursive function $h$ can be made (1–1). Given a $g.n.$ of a table we can effectively enumerate an unending sequence of $g.n.$'s of equivalent tables, for instance we can easily enumerate the $g.n.$'s for the sequence of tables for the functions: $f, PSf, PSPSf, \ldots$. Now the values of the functions $g, g_2$ are defined as $g.n.$'s of certain functions so using the just mentioned fact we can define new functions $*S_1'$, $*g$, $*g_2$ which are strictly increasing and thus (1–1), and they will play the same roles as $S_1'$, $g, g_2$. Thus the composition of these functions with the function $f$ (which by the lemma is (1–1)) as in the construction of the function $h$ will be a (1–1) function $*h$ satisfying

$$\nu \epsilon \mathscr{X}_\theta \quad \text{iff} \quad *h\nu \epsilon \mathscr{C}r.$$

This completes the demonstration of (i).

Ad (ii). Let $\mathscr{A} \leqslant_1 \mathscr{B}$ and $\mathscr{B} \leqslant_1 \mathscr{A}$, then there are (1–1) recursive functions $f, g$ such that

$$\nu \epsilon \mathscr{A} \quad \text{iff} \quad f\nu \epsilon \mathscr{B},$$

$$\nu \epsilon \mathscr{B} \quad \text{iff} \quad g\nu \epsilon \mathscr{A}.$$

We wish to construct a recursive permutation function $h$ such that

$$\nu \epsilon \mathscr{A} \quad \text{iff} \quad h\nu \epsilon \mathscr{B}.$$

It will suffice to show that we can effectively enumerate a set $\Gamma$, of pairs $\{\lambda, \mu\}$ with the following properties:

(i) if $\{\lambda, \mu\}$ is in $\Gamma$ then $\lambda \epsilon \mathscr{A}$ if and only if $\mu \epsilon \mathscr{B}$,
(ii) $\{\lambda, \mu'\}, \{\lambda, \mu''\} \epsilon \Gamma$ then $\mu' = \mu''$,
(iii) $\{\lambda', \mu\}, \{\lambda'', \mu\} \epsilon \Gamma$ then $\lambda' = \lambda''$,
(iv) every numeral occurs as first member of a pair of $\Gamma$,
(v) every numeral occurs as second member of a pair of $\Gamma$.

Suppose that the first $2.\pi$ pairs have been chosen and that (i), (ii), (iii) are satisfied by them. We choose the $(2.\pi+1)$st and the $(2.\pi+2)$nd pairs as follows:

$(2.\pi+1)$. We let the first member be the least numeral $\lambda'$ which has failed to appear as yet as first member of a pair (this ensures (iii) and (iv)). Then if $f\lambda'$ has also failed to appear as second member we take the second member to be $f\lambda'$. If, however, $f\lambda'$ has been a second member let the corresponding first member be $\lambda''$; in this way we form a sequence

$$\lambda', f\lambda', \lambda'', f\lambda'', \ldots, f\lambda^{(\theta)}, \lambda^{(S\theta)}, f\lambda^{(S\theta)},$$

until we come to an $f\lambda^{(S\theta)}$ which is either the same as $f\lambda^{(S\kappa)}$ for $\kappa < \theta$ or has as yet failed to appear as second member. (Since only $2.\pi$ second members have been chosen at this stage then one of these possibilities materializes for $\theta < 2\pi$.) Only the second case can occur, because by (iii) if $f\lambda^{(S\theta)} = f\lambda^{(S\kappa)}$ for $\kappa < \theta$ then $\lambda^{(S\theta)} = \lambda^{(S\kappa)}$ since $f$ is (1–1). By (ii) this makes $f\lambda^{(\theta)} = f\lambda^{(\kappa)}$, this is impossible by our assumption that $(S\theta)$ was the first numeral such that $f\lambda^{(S\theta)} = f\lambda^{(S\kappa)}$ for some $\kappa < \theta$. Thus only the second case can occur, we take $f\lambda^{(S\theta)}$ as the second member of the $(2.\pi+1)$st pair. We see by this construction that property (ii) has been preserved because the first member of the $(2.\pi+1)$st pair has previously been absent. Also $\lambda' \in \mathscr{A}$ if and only if $f\lambda' \in \mathscr{B}$ (by the property of $f$) and this is so if and only if $\lambda'' \in \mathscr{A}$ (by the property of $\Gamma$) and this is so if and only if $f\lambda'' \in \mathscr{B}$ (by the property of $f$) and so on until if and only if $f\lambda^{(S\theta)} \in \mathscr{B}$. Thus $\lambda' \in \mathscr{A}$ if and only if $f\lambda^{(S\theta)} \in \mathscr{B}$ and the property (i) is preserved.

$(2.\pi+2)$. The construction is the same with the following obvious changes:

| second member | for | first member | |
|---|---|---|---|
| $g$ | ,, | $f$ | and vice versa. |
| $\mathscr{A}$ | ,, | $\mathscr{B}$ | |
| (ii) | ,, | (iii) | |
| (iv) | ,, | (v) | |

The construction of the recursive permutation $h$ is now obvious. This completes the demonstration of (ii).

Now let $\mathscr{C}r'$ and $\mathscr{C}r''$ be creative sets. Since $\mathscr{C}r'$ is recursively enumerable and $\mathscr{C}r''$ is creative then by (i) $\mathscr{C}r' \leqslant_1 \mathscr{C}r''$, similarly $\mathscr{C}r'' \leqslant_1 \mathscr{C}r'$. By (ii) $\mathscr{C}r'$ is isomorphic to $\mathscr{C}r''$. This completes the demonstration of the proposition.

## 7.14 *Fixed point proposition*

PROP. 17. *If $f$ is a recursive function then a numeral $\theta$ can be found such that $\Phi[f\theta; \nu] = \Phi[\theta; \nu]$ for all $\nu$.*

(I.e. both sides are undefined, or both sides are defined and have the same value.)

For fixed $\pi$, given $\nu$, find $\Phi[\pi; \pi]$ if this is defined find $\Phi[\Phi[\pi; \pi]; \nu]$. These instructions give a partial recursive function of $\nu$ whose table *g.n.* is a recursive function of $\pi$. viz. $PRRRIC_\pi JT_\pi JDUJD$, say $g\pi$, where $C_\pi$

is the table for the constant function $\pi$, $T_\pi$ is the table whose *g.n.* is $\pi$, $U$ is the universal table. Thus

$$\Phi[g\pi; \nu] = \Phi[\Phi[\pi\pi]; \nu].$$

The function $f \odot g$, $(\boldsymbol{\lambda} x.f[gx])$, is a recursive function, let $\lambda$ be a *g.n.* of a table for it. We have

$$\Phi[g\lambda; \nu] = \Phi[\Phi[\lambda; \lambda]; \nu]$$

but

$$\Phi[\lambda; \lambda] = f[g\lambda],$$

so that

$$\Phi[g\lambda; \nu] = \Phi[f[g\lambda]; \nu].$$

Now take $\theta = g\lambda$ and we have finished.

This proposition is known as the *fixed point proposition.*

Cor. (i). *There exists a recursive function h such that for any numeral $\mu$ if $\mu$ is the g.n. of a table for a recursive function f, then*

$$\Phi[f[h\mu]; \nu] = \Phi[h\mu; \nu]$$

*for all $\nu$.*

(I.e. both sides undefined or both sides defined and have the same value.)

From the proof of the fixed point Prop. we have:

$$\Phi[f[g\lambda]; \nu] = \Phi[g\lambda; \nu],$$

where $\lambda$ is a *g.n.* of a table for the recursive function $f \odot g$. Clearly this *g.n.* is a recursive function $k$ of the *g.n.* of a table for the recursive function $f$. Now take $g \odot k$ for $h$.

## 7.15 *Completely productive sets*

A set $\mathscr{A}$ is said to be *completely productive* if there exists a recursive function $f$ such that $f\pi \in (\mathscr{A} \cap \overline{\mathscr{X}}_\pi) \cup (\overline{\mathscr{A}} \cap \mathscr{X}_\pi)$ for each numeral $\pi$. A completely productive set fails to be recursively enumerable. If $\mathscr{A} = \mathscr{X}_\pi$ for some $\pi$ then

$$(\mathscr{X}_\pi \cap \overline{\mathscr{X}}_\pi) \cup (\overline{\mathscr{X}}_\pi \cap \mathscr{X}_\pi) = \emptyset,$$

and so $f\pi$ fails to be a member. A completely productive set fails to be recursively enumerable in a strongly constructive sense in that the counter example, $f\pi$, can be produced to witness that $\mathscr{A}$ is distinct from $\mathscr{X}_\pi$. A completely productive set is productive.

If $\mathscr{X}_\pi \subset \mathscr{A}$ then $\mathscr{X}_\pi \cap \overline{\mathscr{A}} = \emptyset$, whence $f\pi \in \overline{\mathscr{X}}_\pi \cap \mathscr{A}$, and so $\mathscr{A}$ is productive.

Prop. 18. *If $\mathscr{A}$ is a productive set, then $\mathscr{A}$ is a completely productive set.*

Let $f$ be a productive function for the productive set $\mathscr{A}$. Let $\pi$ be fixed. Take any numeral $\nu$. Enumerate $\mathscr{X}_\pi$ and stop when we find $f\nu$ (e.g.

generate the numerals and when the numeral $\theta$ has been generated then evaluate $Un'[\pi; \theta_1, \theta_2, \theta_3]$, where $\theta = \{\theta_1, \theta_2, \theta_3\}$, if it is zero then test if $\theta_3 = f\nu$ if this is so then stop, otherwise proceed to $S\theta$). The *g.n.* of this table is a recursive function of $\nu$, say $g\nu$. By the fixed point Prop. the *g.n.* of a table for the same calculation will be $\nu$ itself for some numeral $\nu$. From the corollary to the fixed point Prop. we see that $\nu$ is a recursive function of the *g.n.* of $g$, then we have $\nu = h\pi$ for some recursive function $h$. Thus

$$\mathscr{X}_{h\pi} = \begin{cases} \{f[h\pi]\} & \text{if } f[h\pi] \in \mathscr{X}_{h\pi}, \\ \emptyset & \text{otherwise.} \end{cases}$$

We now show that $\mathscr{A}$ is completely productive with productive function $f \odot h$. If

$$f[h\pi] \epsilon \overline{\mathscr{X}}_\pi \quad \text{then} \quad \mathscr{X}_{h\pi} = \emptyset \subset \mathscr{A},$$

hence $f[h\pi] \epsilon \mathscr{A} \cap \overline{\mathscr{X}}_{h\pi}$ since $f$ is a productive function for $\mathscr{A}$. Thus

$$f[h\pi] \epsilon \mathscr{A}, \quad \text{altogether} \quad f[h\pi] \epsilon \mathscr{A} \cap \overline{\mathscr{X}}_\pi.$$

If $$f[h\pi] \epsilon \mathscr{X}_\pi \quad \text{then} \quad \mathscr{X}_{h\pi} = \{f[h\pi]\},$$

if $$f[h\pi] \epsilon \mathscr{A} \quad \text{then} \quad \mathscr{X}_{h\pi} \subset \mathscr{A},$$

but then $f[h\pi] \epsilon \mathscr{A} \cap \overline{\mathscr{X}}_{h\pi}$, since $f$ is a productive function for $\mathscr{A}$, which is absurd, hence

$$f[h\pi] \epsilon \overline{\mathscr{A}} \quad \text{and so} \quad f[h\pi] \epsilon \overline{\mathscr{A}} \cap \mathscr{X}_\pi.$$

In either case $$f[h\pi] \epsilon (\mathscr{A} \cap \overline{\mathscr{X}}_\pi) \cup (\overline{\mathscr{A}} \cap \mathscr{X}_\pi).$$

## 7.16 *Oracles*

At the beginning of this chapter we defined a calculable function of natural numbers as a set of instructions which applied to a natural number produced another natural number. Similarly for application to sets of natural numbers. We analysed this vague intuitive idea and gave a precise definition of calculable function, namely a function calculable by a Turing machine from a complete set of atomic instructions. We also considered the case when the calculation failed to terminate or terminated in something other than a natural number, such functions we called partially calculable. We now extend the Turing machine definition of calculable function of natural numbers to include the case where at some point in the calculation we require to know the value of a certain function $f$ for a certain argument $\nu$.

If $f$ is a calculable function then we can merely add the extra piece of calculation to find $f\nu$ and we have the case we have been considering. But if $f$ fails to be a calculable function then we have a new situation. We appeal to an *Oracle* to tell us correctly the value of $f\nu$. We could suppose that we are supplied with a second tape on which are written the successive values: $f0, f1, f2, \ldots$, so that we need only copy down the value required when wanted. A function $g$ which can be calculated given the correct values of a function $f$ as required will be called an *$f$-calculable function*, or a *partially $f$-calculable function* if the calculation fails to terminate or terminates in something other than a natural number. Similarly for *$f', \ldots, f^{(\pi)}$-calculable functions* if we require to know the values of several functions.

We have defined primitive, general and partial recursive functions and we showed that partial recursive functions were partially calculable and conversely. Now add to the initial functions the functions $f', \ldots, f^{(\pi)}$, the resulting functions will be called *primitive, general or partial $f', \ldots, f^{(\pi)}$-recursive functions* respectively. If the functions $f', \ldots, f^{(\pi)}$ are general recursive and if $g$ is general $f', \ldots, f^{(\pi)}$-recursive then $g$ is general recursive and similarly in the other cases.

We have shown that a partial recursive function is $\mathbf{A}_0$-representable. Suppose that we add to $\mathbf{A}_0$ new axioms giving the values of the functions $f', \ldots, f^{(\pi)}$ for all arguments, this list will be unbounded and will fail to be given by an axiom scheme unless $f', \ldots, f^{(\pi)}$ are recursive, but this case is without interest. This system will then fail to be formal, we call it $\mathbf{A}_0[f', \ldots, f^{(\pi)}]$.

In Prop. 3 of this chapter we showed that given a partial recursive function of $\kappa$ arguments we could find a numeral $\theta$ such that the function $g$ was $\mathbf{A}_0$-represented by the $\mathbf{A}_0$-statement:

$$(E\xi)\,(Un^{(\kappa)}[\theta, \nu', \ldots, \nu^{(\kappa)}, \xi, \nu] = 0),$$

where the value, of $g[\nu', \ldots, \nu^{(\kappa)}]$ is $\nu$ and where $Un^{(\kappa)}$ is a primitive recursive function. We now obtain a similar result for $f', \ldots, f^{(\pi)}$-recursive functions.

PROP. 19. *If the function $g$ with $\theta$ arguments is partially $f', \ldots, f^{(\pi)}$ recursive and if the functions have $\theta', \ldots, \theta^{(\pi)}$ arguments respectively then $g$ can be $\mathbf{A}_0[f', \ldots, f^{(\pi)}]$-represented by*

$$(E\xi)\,(Un^{(\theta)}_{\theta', \ldots, \theta^{(\pi)}}[\kappa, \nu', \ldots, \nu^{(\theta)}, \mathfrak{f}'\xi, \ldots, \mathfrak{f}^{(\pi)}\xi, \xi, \nu] = 0) \qquad (20)$$

*for some numeral* $\kappa$, *conversely for each numeral* $\kappa$ (20) *represents a partial* $f', \dots, f^{(\pi)}$*-recursive function of* $\theta$ *arguments whose value for the arguments* $\nu', \dots, \nu^{(\kappa)}$ *is* $\nu$.

Here

D 217 $$\mathfrak{f}\xi \text{ for } \langle f[\xi', \dots, \xi^{(\theta')}]\rangle_{0 \leqslant \{\xi', \dots, \xi^{(\theta')}\} \leqslant \xi},$$

this is a *register* of the initial values of the function $f$, where $f$ is a $\theta$-place function and $Un^{\theta}_{\theta', \dots, \theta^{(\pi)}}$ is a primitive recursive $(\theta + \pi + 3)$-place function. $\kappa$ is called the *index* of the function $g$.

We proceed as in Prop. 2 of this chapter except that the complete table now has extra lines which correspond to compound acts. We have extra instructions $q', \dots, q^{(\pi)}$ such that when under instruction $q^{(\lambda)}$, $1 \leqslant \lambda \leqslant \pi$, and observing a cipher we do nothing and pass to the passive instruction. But if observing a tally we test whether this is the last tally of the representation of a $\theta^{(\lambda)}$-tuplet $\{\mu', \dots, \mu^{(\theta^{(\lambda)})}\}$ and if the test is affirmative we write down two ciphers and then $S\{\mu', \dots, \mu^{(\theta^{(\lambda)})}\}$ tallies and then, after a one cipher gap, $Sf[\mu', \dots, \mu^{(\theta^{(\lambda)})}]$ tallies and stop observing the last of these tallies and pass to instruction $S\lambda$. But if writing down these tallies we encounter a square already containing a tally then we pass to the passive instruction. If under instruction $q^{(\lambda)}$ and the test just mentioned fails then we pass to the passive instruction. In other words if under instruction $q^{(\lambda)}$ and observing a $\theta^{(\lambda)}$-tuple $\{\mu', \dots, \mu^{(\theta^{(\lambda)})}\}$ in standard position with sufficient ciphers to the right so that we fail to foul a tally then we write down the representation of $\{\mu', \dots, \mu^{(\theta^{(\lambda)})}\}, f[\mu', \dots, \mu^{(\theta^{(\lambda)})}]$ leaving a two cipher gap otherwise the machine stops. The machine just described is called a Turing $f', \dots, f^{(\pi)}$-machine and is said to $f', \dots, f^{(\pi)}$-calculate.

The extract from the 'table' for the function $g$ which involves $q^{(\lambda)}$ (the function $g$ fails to have a table in the sense used previously) is

$$q^{(\lambda)} \begin{cases} N0 \\ K^{(\lambda)} \begin{cases} N0 \\ F^{(\lambda)}\ ((S\lambda) \quad \text{or} \quad 0), \quad \text{according as we foul a tally,} \end{cases} \end{cases}$$

where $K^{(\lambda)}$ is the table for the test described above and $F^{(\lambda)}$ directs us to write: $,,\{\mu', \dots, \mu^{(\theta^{(\lambda)})}\},f^{(\lambda)}[\mu', \dots, \mu^{(\theta^{(\lambda)})}]$ where commas denote ciphers and the test affirmed that we were observing the tuplet $\{\mu', \dots, \mu^{(\theta^{(\lambda)})}\}$ in standard position. The instruction we are referred to after doing $F^{(\lambda)}$ must be unique (we finish $F^{(\lambda)}$ observing a tally) we call it instruction $(S\lambda)$, $F^{(\lambda)}$ fails to be a table unless $f^{(\lambda)}$ is partial recursive, it is the oracle, telling us something we are unable to find out for ourselves. A complete table for the calculation of a partially $f',\dots,f^{(\pi)}$-recursive function $g$ is a table of entries:

$$\begin{matrix} & N & & N & \\ \lambda 0 & R\pi_1^{(\lambda)} & \lambda 1 & R\pi_2^{(\lambda)} & (1 \leqslant \lambda \leqslant \pi'), \\ & L & & L & \\ & I & & I & \end{matrix}$$

where $$0 \leqslant \pi_1^{(\lambda)}, \pi_2^{(\lambda)} \leqslant \pi' + \pi.$$

The passive instruction is omitted as usual and so are entries corresponding to the instructions $q',\dots,q^{(\pi)}$ here numbered $S\pi',\dots,\pi'+\pi$. The order of the table is $\pi'$. The *g.n.* of a complete table for the calculation of a partially $f',\dots,f^{(\pi)}$-recursive function is the numeral $\kappa$ determined by $\langle \kappa',\dots,\kappa^{(\pi')}\rangle$, where

$$\kappa^{(\mu)} = \{\kappa_1^{(\mu)}, \kappa_2^{(\mu)}\} \quad (1 \leqslant \mu \leqslant \pi'),$$

$$\left.\begin{aligned} \kappa_1^{(\mu)} &= 4.\pi_1^{(\mu)} + \begin{matrix}0\\1\\2\\3\end{matrix} Mod[4.S(\pi+\pi')], \\ \kappa_2^{(\mu)} &= 4.\pi_2^{(\mu)} + \begin{matrix}0\\1\\2\\3\end{matrix} Mod[4.S(\pi+\pi')]. \end{aligned}\right\} \text{according as there is } \begin{matrix}N\\R\\L\\I\end{matrix} \text{ in the corresponding entry.}$$

Every numeral is the *g.n.* of some such table. Note that the *g.n.* is independent of the functions $f',\dots,f^{(\pi)}$, and is primitive recursive in $\pi$.

We define the function $J$ exactly as before in Prop 2 except that we require additional cases corresponding to the compound intructions $q',\dots,q^{(\pi)}$, here numbered $S\pi',\dots,\pi'+\pi$. If the *g.n.* of the situation is

$$\{\mu^\frown 1^\frown \overline{\mu'}^\frown 1^\frown \dots {}^\frown 1^\frown \overline{\mu^{(\theta')} \dot{-} 1}, 1, S\pi', \lambda\}$$

(if $\mu$ is zero then the first component begins $\mu'$).

Then this changes to (modification as above if $\mu$ is zero)

$$\{\mu^\frown 1^\frown \overline{\mu'}^\frown 1^\frown \dots {}^\frown 1^\frown \overline{\mu^{(\theta')}}^\frown \langle 0,0\rangle^\frown \overline{\{\mu',\dots,\mu^{(\theta')}\}}^\frown 1^\frown \overline{f'[\mu',\dots,\mu^{(\theta')}] \dot{-} 1}, 1, 2, \lambda'\},$$

provided $\lambda = 0$ or $\lambda = \underset{1 \leqslant \xi \leqslant m}{\langle 0 \rangle}^\frown \lambda'$, where $m = 4 + \{\mu',\dots,\mu^{(\theta')}\} + f'[\mu',\dots,\mu^{(\theta')}]$,

otherwise $\{\mu, \mu', S\pi', \mu''\}$ changes to $\{\mu, \mu', 0, \mu''\}$. There are similar cases when the third component of the quartet is $\pi'+2,\dots,\pi'+\pi$. The previous eight cases receive the qualification that the third component is to be less than or equal to $\pi'$. The *g.n.* $\kappa$ of the table will be $\langle \kappa', \ldots, \kappa^{(\pi')} \rangle$ where $\kappa', \ldots, \kappa^{(\pi')}$ are as just described. Then $J^{(\pi)}[\kappa, \nu]$ is the *g.n.* of the situation which arises from the situation whose *g.n.* is $\nu$ after one act (atomic or one of the new compound acts).

In the new clauses write

$$Pt[\mathfrak{f}'\alpha, \{\mu', \ldots, \mu^{(\theta')}\}] \quad \text{instead of} \quad f'[\mu', \ldots, \mu^{(\theta)'}],$$

where $\alpha$ is sufficiently large (greater than any $\{\mu', \ldots, \mu^{(\theta')}\}$ that occur). This leaves $J^{(\pi)}[\kappa, \nu]$ unaltered in value, it is a primitive recursive function of $\kappa$, $\nu$, and $\mathfrak{f}'\alpha, \ldots, \mathfrak{f}^{(\pi)}\alpha$.

D 218
$$J_{\theta', \ldots, \theta^{(\pi)}}[\kappa, \nu, \mathfrak{f}'\alpha, \ldots, \mathfrak{f}^{(\pi)}\alpha]$$

for the function $J^{(\pi)}[\kappa, \nu]$ just described. Note that $J^{(\pi)}[\kappa, \nu]$ depends on $\theta', \ldots, \theta^{(\pi)}$. The term $\alpha$ has to be sufficiently large so that the required components of the ordered sets are in the register. Thus as the calculation proceeds $\alpha$ may have to change, it may have to depend on the number of moves already done or on the *g.n.*'s of past situations. Note that the functions $f', \ldots, f^{(\pi)}$ enter into the expression D 107 in such a way that they fail to occur in $J_{\theta', \ldots, \theta^{(\pi)}}[\eta, \xi, \zeta', \ldots, \zeta^{(\pi)}]$, this is a primitive recursive function of the $(\pi+2)$ variables $\eta, \xi, \zeta', \ldots, \xi^{(\pi)}$.

We now define

D 219
$$\begin{cases} \Theta[\kappa, \nu, \nu', \ldots, \nu^{(\pi)}, 0] = \nu \\ \Theta[\kappa, \nu, \nu', \ldots, \nu^{(\pi)}, S\lambda] = J_{\theta', \ldots, \theta^{(\pi)}}[\kappa, \Theta[\kappa, \nu, \nu', \ldots, \nu^{(\pi)}, \lambda], \nu', \ldots, \nu^{(\pi)}], \end{cases}$$

$\Theta$ will depend on $\theta', \ldots, \theta^{(\pi)}$, they should have been appended as suffixes. In this replace $\nu'$ by $\mathfrak{f}'\alpha, \ldots, \nu^{(\pi)}$ by $\mathfrak{f}^{(\pi)}\alpha$, where $\alpha$ is sufficiently large, then $\Theta[\kappa, \nu, \mathfrak{f}'\alpha, \ldots, \mathfrak{f}^{(\pi)}\alpha, \lambda]$ is the *g.n.* of the situation after $\lambda$ applications of the table whose *g.n.* is $\kappa$ to the initial situation whose *g.n.* is $\nu$ Suppose the calculation terminates after $\lambda$ moves having calculated $\nu$ from the argument set $\nu', \ldots, \nu^{(\theta)}$. We then have

$$\Theta[\kappa, \Omega^{(\theta)}[\nu', \ldots, \nu^{(\theta)}, 1], \mathfrak{f}'\alpha, \ldots, \mathfrak{f}^{(\pi)}\alpha, \lambda] = \Omega^{(S\theta)}[\nu', \ldots, \nu^{(\theta)}, \nu, 0],$$

provided $\alpha$ is sufficiently large.

It remains to choose $\alpha$. If in the course of the calculation we require the value of $f'[\mu', \ldots, \ldots, \mu^{(\theta')}]$ then we have to write down

$$,,\overline{\{\mu', \ldots, \mu^{(\theta')}\}}, \quad \overline{f'[\mu', \ldots, \mu^{(\theta')}]}$$

at some place on the tape, and finish this compound act by observing the last tally written. Now the calculation terminates in the calculation of $\nu$

so that the second of the two ciphers written in the compound act must ultimately be reached because the final result is: $\nu', \ldots, \nu^{(\pi)}, \nu$ and this is without consecutive ciphers. Thus either we shall have to erase the representation of $\{\mu', \ldots, \mu^{(\theta')}\}$ and $f'[\mu', \ldots, \mu^{(\theta')}]$ or we shall have to replace the aforementioned cipher by a tally. If we erase $\{\mu', \ldots, \mu^{(\theta')}\}$ we shall have to reach the aforementioned second cipher in order to ascertain that we have reached the end of the erasion. In order to reach this position after having done the compound act we must do a further $\{\mu', \ldots, \mu^{(\theta')}\}+3$ atomic acts at least, namely atomic acts of moving one place to the left, whatever other acts the calculation calls for. Thus if we take $\alpha$ equal to the total number of moves then $\alpha$ will certainly be sufficiently large.

Thus if the calculation terminates in the calculation of $\nu$ then

$$\Theta[\kappa, \Omega^{(\theta)}[\nu', \ldots, \nu^{(\theta)}, 1], \mathfrak{f}'\lambda, \ldots, \mathfrak{f}^{(\pi)}\lambda, \lambda] = \Omega^{(S\theta)}[\nu', \ldots, \nu^{(\theta)}, \nu, 0]$$

for sufficiently large $\lambda$.

D 220 $$Un^{(\theta)}_{\theta', \ldots, \theta^{(\pi)}}[\kappa, \eta', \ldots, \eta^{(\theta)}, \upsilon', \ldots, \upsilon^{(\pi)}, \xi, \zeta]$$

for $$A_2[\Theta[\kappa, \Omega^{(\theta)}[\eta', \ldots, \eta^{(\theta)}, 1], \upsilon', \ldots, \upsilon^{(\pi)}, \xi], \Omega^{(S\theta)}[\eta', \ldots, \eta^{(\theta)}, \zeta, 0]].$$

Then $Un^{(\theta)}_{\theta', \ldots, \theta^{(\pi)}}[\eta, \eta', \ldots, \eta^{(\theta)}, \upsilon', \ldots, \upsilon^{(\theta)}, \xi, \zeta]$ is a primitive recursive function of the $(\theta+\pi+3)$ variables $\eta, \eta', \ldots, \eta^{(\theta)}$, $\upsilon', \ldots, \upsilon^{(\pi)}$, $\xi, \zeta$. The value of the function $g$ can be expressed as

$$pt[(\mu\xi)\,[Un^{(\theta)}_{\theta', \ldots, \theta^{(\pi)}}[\kappa, \nu', \ldots, \nu^{(\theta)}, \mathfrak{f}'\xi, \ldots, \mathfrak{f}^{(\pi)}\xi, \xi_1, \xi_2] = 0], 2, 2],$$

if we use the unlimited least number symbol, but this expression is outside the system $\mathbf{A}_0[f', \ldots, f^{(\pi)}]$. We can represent the function $g$ in the system $\mathbf{A}_0[f', \ldots, f^{(\pi)}]$ by

$$(E\xi)\,(Un^{(\theta)}_{\theta', \ldots, \theta^{(\pi)}}[\kappa, \nu', \ldots, \nu^{(\pi)}, \mathfrak{f}'\xi, \ldots, \mathfrak{f}^{(\pi)}\xi, \xi, \nu] = 0).$$

The machines described in the demonstration of Prop. 19 are called $\mathfrak{f}$-oracle machines, or $\mathfrak{f}$-oracle Turing machines.

Cor. (i). *An enumeration with repetitions of partial $f', \ldots, f^{(\pi)}$-recursive functions is given by D220 for $\kappa = 0, 1, 2, \ldots$.*

Cor. (ii). *If the function $g$ of $(\lambda+\lambda')$ arguments is partial $f', \ldots, f^{(\pi)}$-recursive where the functions have $\theta', \ldots, \theta^{(\pi)}$ arguments respectively and if the* *last $\lambda'$ arguments of $g$ are held constant so that $g$ gives rise to a function $g'$ of $\lambda$ arguments then $g'$ can be represented in the system $A_0[f', \ldots, f^{(\pi)}]$ by*

$$(E\xi)\,(Un^{(\lambda)}_{\theta', \ldots, \theta^{(\pi)}}[S^{(\lambda')}_{\lambda, \theta', \ldots, \theta^{(\pi)}}[\kappa, \nu^{(\lambda+1)}, \ldots, \nu^{(\lambda+\lambda')}], \nu', \ldots, \nu^{(\lambda)}, \mathfrak{f}'\xi, \ldots, \mathfrak{f}^{(\pi)}\xi, \xi, \nu] = 0),$$

*where*
$$g[\nu', \ldots, \nu^{(\lambda)}, \nu^{(S\lambda)}, \ldots, \nu^{(\lambda+\lambda')}] = \nu$$

*and where $S^{(\lambda')}_{\lambda, \theta', \ldots, \theta_{g^{(\pi)}}}[\kappa, \nu^{(S\lambda)}, \ldots, \nu^{(\lambda+\lambda')}]$ is a primitive recursive function of $\kappa, \nu^{(S\lambda)}, \ldots, \nu^{(\lambda+\lambda')}$ and the function $g$ is represented in the system $A_0[f', \ldots, f^{(\pi)}]$ by D220 with $(\lambda+\lambda')$ instead of $\theta$.*

The demonstration is similar to that of Prop. 4 of this chapter, and so is the definition of the function $S$.

D 221
$$S^{(\lambda')}_{\lambda, \theta', \ldots, \theta^{(\pi)}}[\eta, \eta', \ldots, \eta^{(\lambda')}]$$

for the primitive recursive function of Cor (ii).

The numeral $\kappa$ is the *g.n.* of a table which calculates the function $g$ of $(\lambda+\lambda')$ arguments. The numeral determined by

$$S^{(\lambda')}_{\lambda, \theta', \ldots, \theta^{(\pi)}}[\kappa, \nu^{(S\lambda)}, \ldots, \nu^{(\lambda+\lambda')}]$$

is the *g.n.* of a table which calculates the function $g'$ of $\lambda$ arguments where

$$g'[\nu', \ldots, \nu^{(\lambda)}] = g[\nu', \ldots, \nu^{(\lambda)}, \nu^{(S\lambda)}, \ldots, \nu^{(\lambda+\lambda')}],$$

whatever the numerals $\nu', \ldots, \nu^{(\lambda)}$. Thus the *g.n.* of some table for the function $g'$ is a primitive recursive function of the *g.n.* of a table for $g$ and of the values of the fixed arguments.

We have defined primitive, general, and partial $\mathfrak{f}$-recursive functions, where $\mathfrak{f}$ stands for $f', \ldots, f^{(\pi)}$, if $\pi$ is zero then omit $\mathfrak{f}$. Similarly an *$\mathfrak{f}$-recursive set of natural numbers* is a set of natural numbers with an $\mathfrak{f}$-recursive characteristic function, similarly for an $\mathfrak{f}$-recursive statement. Similarly an *$\mathfrak{f}$-recursive enumerable set of natural numbers* is the range of values of an $\mathfrak{f}$-recursive function. Similarly for $\mathfrak{f}$-recursively enumerable set of lattice points in $\mathscr{R}_{S\pi}$.

We can now repeat a lot of the work we have done on recursive and recursively enumerable sets of natural numbers, using instead $\mathfrak{f}$-recursive and $\mathfrak{f}$-recursively enumerable sets of natural numbers. Thus the *$\mathfrak{f}$-complete* set of natural numbers is the set $\{\kappa, \theta\}$ where the $\mathfrak{f}$-machine with *g.n.* $\kappa$ produces $\theta$. It is of the highest degree of unsolvability for $\mathfrak{f}$-recursively enumerable sets, that is to say, if some oracle would tell us if $\{\kappa, \theta\}$ was in the $\mathfrak{f}$-complete set then we could decide of any natural number whether it is in a given $\mathfrak{f}$-recursively enumerable set of natural numbers. The $\mathfrak{f}$-complete set is $\mathfrak{f}$-recursively enumerable but fails to be $\mathfrak{f}$-recursive, the demonstration is as for the complete set.

In Prop. 19 we have reproduced a lot of Props. 1–4 incl. for $\mathfrak{f}$-recursive sets. In a similar manner we can produce Props. 8–18 for $\mathfrak{f}$-recursive sets and $\mathfrak{f}$-recursively enumerable sets, here we use $\mathfrak{f}$-*decidable* instead of decidable. For instance $\mathscr{C}$-recursive sets contain the field of sets generated by recursively enumerable sets, e.g. to decide if $\kappa$ is in $\mathscr{X}_\theta \cup \bar{\mathscr{X}}_{\theta'}$ we only have to decide whether $\{\kappa, \theta\}$ is in $\mathscr{C}$ or whether $\{\kappa, \theta\}$ is in $\bar{\mathscr{C}}$, if $\mathscr{C}$ is decided for us by an oracle then so is $\bar{\mathscr{C}}$ and so we can $\mathscr{C}$-decide $\mathscr{X}_\theta \cup \bar{\mathscr{X}}_{\theta'}$. We omit the details of the extensions for $\mathfrak{f}$-recursive functions of Props. 8–18 incl. If we need them we shall refer to them as Prop. X for $\mathfrak{f}$-recursive functions. Thus we have $\mathfrak{f}$-*simple*, $\mathfrak{f}$-*hypersimple*, $\mathfrak{f}$-*creative* sets, etc. Note that by Prop. 16 $\mathscr{C}$-recursive and $\mathscr{C}r$-recursive sets are the same, where $\mathscr{C}r$ is a creative set.

Church's $\mathfrak{f}$-thesis is: an $\mathfrak{f}$-calculable function is calculable by an $\mathfrak{f}$-oracle machine.

$\mathscr{X}^f_\kappa$, $\mathscr{X}^{\mathscr{A}}_\kappa$ denote the set of natural numbers calculated by $f$-oracle or $\mathscr{A}$-oracle machines.

$\Phi^f_\kappa$, $\Phi^{\mathscr{A}}_\kappa$ denote the functions calculated by $f$-oracle or $\mathscr{A}$-oracle machines. $\Phi^f_\kappa\theta$, $\Phi^{\mathscr{A}}_\kappa\theta$ denote the value of these functions for argument $\theta$. $\Phi^{[f]}_\nu$, $\Phi^{[f]}_\mu\theta$ denotes that in the calculation only arguments for which $f$ is defined are used.

## 7.17 *Degrees of unsolvability*

'$f$ is recursive in $g$ and $g$ is recursive in $f$' written $f \equiv_T g$, is a reflexive, transitive and symmetric relation. Hence this relation may be used to divide functions into disjoint equivalence classes. We say that $f$ and $g$ have the same *degree of unsolvability* if they both belong to the same equivalence class. Hence we identify degrees of unsolvability with these equivalence classes.

We write $\mathbf{a} \leqslant \mathbf{b}$ for the degrees $\mathbf{a}, \mathbf{b}$ of $f$, $g$ respectively, if $f$ is recursive in $g$. Then each $f_1$ of the same degree as $f$ is recursive in each $g_1$ of the same degrees as $g$. Under this relation degrees form a partially ordered system.

We have for degrees $\mathbf{a}$, $\mathbf{b}$, $\mathbf{c}$:

$$\mathbf{a} \leqslant \mathbf{a} \quad \text{if} \quad \mathbf{a} \leqslant \mathbf{b} \quad \text{and} \quad \mathbf{b} \leqslant \mathbf{c} \quad \text{then} \quad \mathbf{a} \leqslant \mathbf{c},$$
$$\mathbf{a} \leqslant \mathbf{b} \quad \text{and} \quad \mathbf{b} \leqslant \mathbf{a} \quad \text{iff} \quad \mathbf{a} = \mathbf{b}.$$

We define $\mathbf{a} < \mathbf{b}$ for $\mathbf{a} \leqslant \mathbf{b}$ and $\mathbf{a} \neq \mathbf{b}$.

$$\mathbf{a} \geqslant \mathbf{b} \text{ for } \mathbf{b} \leqslant \mathbf{a}, \quad \mathbf{a} > \mathbf{b} \text{ for } \mathbf{b} < \mathbf{a},$$

we write $\mathbf{a}/\mathbf{b}$ when both $\mathbf{a} \leqslant \mathbf{b}$ and $\mathbf{b} \leqslant \mathbf{a}$ fail, we then say that *degree* $\mathbf{a}$ *is incomparable with degree* $\mathbf{b}$, $\mathbf{a} \nleqslant \mathbf{b}$ for $\mathbf{a}$ fails to be recursive in $\mathbf{b}$, i.e. any function of degree $\mathbf{a}$ fails to be recursive in any function of degree $\mathbf{b}$.

Then for any degrees $\mathbf{a}$ and $\mathbf{b}$ we have the quadricotomy

$$\mathbf{a} < \mathbf{b} \quad \text{or} \quad \mathbf{a} = \mathbf{b} \quad \text{or} \quad \mathbf{a} > \mathbf{b} \quad \text{or} \quad \mathbf{a}/\mathbf{b}.$$

Recursive functions form degree $\mathbf{O}$. For any degree $\mathbf{a}$, $\mathbf{O} \leqslant \mathbf{a}$. Thus $\mathbf{O}$ is the lowest degree of unsolvability, it is the *degree of solvability*.

Now suppose that (i) $f$ and $g$ are both $h$-recursive and (ii) that $h$ is $f$, $g$-recursive. For example: $h[2.\nu] = f\nu, h[2.\nu+1] = g\nu$. Further if $f_1, g_1, h_1$ are respectively of the same degrees as $f, g, h$ then (i), (ii) hold for $f_1, g_1, h_1$. Thus the functions which satisfy (i), (ii) for given $f, g$ belong to the same degree. In fact they constitute that degree, since any function $h_1$ of the same degree as $h$ also satisfies (i), (ii) for the same $f$, $g$. Further this degree is determined by the degree of $f$, $g$, for we still have (i), (ii) if we replace $f$, $g$ by $f_1, g_1$ where $f_1$ is of the same degree as $f$ and $g_1$ is of the same degree as $g$. We denote the degree of $h$ by $(\mathbf{a} \cup \mathbf{b})$ where $\mathbf{a}, \mathbf{b}$ are the degrees of $f, g$ respectively. We call it the *l.u.b* (*least upper bound*) *of* $\mathbf{a}$ *and* $\mathbf{b}$.

We have $$\mathbf{a} \leqslant \mathbf{a} \cup \mathbf{b}, \quad \mathbf{b} \leqslant \mathbf{a} \cup \mathbf{b},$$

if $$\mathbf{a} \leqslant \mathbf{c} \text{ and } \mathbf{b} \leqslant \mathbf{c} \text{ then } \mathbf{a} \cup \mathbf{b} \leqslant \mathbf{c}.$$

Thus degrees form an *upper semi-lattice*.

Further we have
$$\mathbf{a} \leqslant \mathbf{b} \text{ iff } \mathbf{a} \cup \mathbf{b} = \mathbf{b},$$
$$\mathbf{a} \cup \mathbf{a} = \mathbf{a}, \quad \mathbf{O} \cup \mathbf{a} = \mathbf{a},$$
$$\mathbf{a} \cup \mathbf{b} = \mathbf{b} \cup \mathbf{a},$$
$$\mathbf{a} \cup (\mathbf{b} \cup \mathbf{c}) = (\mathbf{a} \cup \mathbf{b}) \cup \mathbf{c}.$$

Thus parentheses may be omitted when writing l.u.b.'s, in fact

$$\mathbf{a}_1 \cup \mathbf{a}_2 \cup \ldots \cup \mathbf{a}_\nu$$

is exactly the degree of a function $h$ such that if $g_i$ is of degree $\mathbf{a}_i$ then $g_i$ is $h$-recursive and $h$ is $g_1, \ldots, g_\nu$-recursive.

If $\nu$ is zero we agree that $\mathbf{a}_1 \cup \ldots \cup \mathbf{a}_\nu$ is $\mathbf{O}$ the solvable set. If $f$ is of degree $\mathbf{b}$ and $g_i$ is of degree $\mathbf{a}_i$ $1 \leqslant i \leqslant \nu$, then $f$ is $g_1, \ldots, g_\nu$-recursive if and only if $\mathbf{b} \leqslant \mathbf{a}_1 \cup \ldots \cup \mathbf{a}_\nu$.

We say that $g_1, \ldots, g_\nu$ are *recursively independent* if none of them is recursive in the others; in degrees, if $\mathbf{a}_\kappa \leqslant \mathbf{a}_1 \cup \ldots *_\kappa \ldots \cup \mathbf{a}_\nu$ fails for $1 \leqslant k \leqslant \nu$, where $*_\kappa$ denotes the omission of $\mathbf{a}_\kappa$. If $\mathbf{a}_1, \ldots, \mathbf{a}_\nu$ are recursively independent then so is any subset of them. For $\nu = 1$ independence is non-recursiveness, for $\nu = 2$ independence is incomparability, for $\nu > 2$ independence implies pairwise incomparability but the converse fails.

What we have said about degrees of functions carries over to degrees of statements or of sets, for these are determined by their characteristic functions.

If $g$ is of the same degree as $f$ then $g$ is $f$-recursive and a $g$-recursively enumerable set (the range of a $g$-recursive function) is an $f$-recursively enumerable set, hence, the $g$-complete set, which is $g$-recursively enumerable is also $f$-recursively enumerable and so is recursive in the $f$-complete set. Similarly the $f$-complete set is recursive in the $g$-complete set. Thus the degree of the $f$-complete set depends only on the degree of $f$. If $\mathbf{a}$ is the degree of $f$ the degree of the $f$-complete set is denoted by $\mathbf{a}'$ and is called a *complete degree*. $\mathbf{a}'$ is called the jump of $\mathbf{a}$.

*If* $\mathbf{a} < \mathbf{b}$ *then* $\mathbf{a}' \leqslant \mathbf{b}'$.

If $g$ is of degree $\mathbf{a}$ and $f$ is of degree $\mathbf{b}$ and if $g$ is $f$-recursive then the $g$-complete set being $g$-recursively enumerable is also $f$-recursively enumerable and hence is recursive in the $f$-complete set, thus $\mathbf{a}' \leqslant \mathbf{b}'$.

$\mathbf{a} < \mathbf{a}'$

Because the $f$-complete set fails to be $f$-recursive, this follows from the analogue of the complete set §7.8.

$\mathbf{a}' \cup \mathbf{b}' \leqslant (\mathbf{a} \cup \mathbf{b})'$

We have $\mathbf{a} \leqslant \mathbf{a} \cup \mathbf{b}$ hence $\mathbf{a}' \leqslant (\mathbf{a} \cup \mathbf{b})'$, similarly $\mathbf{b}' \leqslant (\mathbf{a} \cup \mathbf{b})'$ and so $\mathbf{a}' \cup \mathbf{b}' \leqslant (\mathbf{a} \cup \mathbf{b})'$. Thus altogether degrees form an upper semi-lattice with a jump operation superimposed on it.

A degree is called *recursively enumerable* if it is the degree of a recursively enumerable set. If $R^f[x, x', \ldots, x^{(S\theta)}]$ is an $f$-recursive statement then $(Ex)\, R^f[x, x', \ldots, x^{(S\theta)}]$ is $f$-recursively enumerable. For generate the

$(SS\theta)$-tuplets $\{\nu, \nu', \ldots, \nu^{(S\theta)}\}$ and when the above tuplet has been generated $f$-decide whether $R^f[\nu, \nu', \ldots, \nu^{(S\theta)}]$, if this holds then write

$$\{\nu', \ldots, \nu^{(S\theta)}\}$$

down in a list, this enumerates the $(S\theta)$-tuplets $\{\nu', \ldots, \nu^{(S\theta)}\}$ for which $(Ex)\, R^f[x, \nu,' \ldots, \nu^{(S\theta)}]$. Hence this set is recursive in the $f$-complete set. If the degree of $f$ is $\mathbf{a}$ then the degree of $(Ex)\, R^f[x, \nu', \ldots, \nu^{(S\nu)}]$ is $\leqslant \mathbf{a}'$. Note the $f$-analogue of Cor. (v), Prop. 8.

*If a set $\mathscr{S}$ and its complement are both f-recursively enumerable then $\mathscr{S}$ is f-recursive.*

Thus $\mathbf{O}'$ besides containing the field generated by recursively enumerable sets also contains all sets $\mathscr{S}$ such that $\mathscr{S}$ and $\bar{\mathscr{S}}$ are both $\mathbf{O}'$-recursively enumerable and the field generated by them and recursively enumerable sets.

Again a complete degree $\mathbf{a}'$ contains all $\mathbf{a}$-recursively enumerable sets and all sets $\mathscr{S}$ such that $\mathscr{S}$ and $\bar{\mathscr{S}}$ are both $\mathbf{a}'$-recursively enumerable and the field generated by these sets.

The *projection of a set $\mathscr{S}$ of lattice points in* $\mathscr{R}_{SS\pi}$ is the set of lattice points $\{\nu', \ldots, \nu^{(S\pi)}\}$ in $\mathscr{R}_{S\pi}$, where $\{\nu, \nu', \ldots, \nu^{(S\pi)}\}$ is in $\mathscr{S}$.

A set of lattice points is called *arithmetic* if it is obtained from a recursive set of lattice points by repeatedly taking complements and projections. A degree is called *arithmetic* if it is the degree of an arithmetic set. These degrees will contain functions representable in the system $\mathbf{A}$, to be introduced in the next chapter, this system amounts to the system $\mathbf{A}_0$ plus negation, which failed to be $\mathbf{A}_0$-representable.

Recursively enumerable degrees form a sub-semi-lattice of the semi-lattice of all degrees, because if $f$ and $g$ are recursively enumerable then so is $h$ defined by $h(2.\nu) = f\nu, h(2.\nu+1) = g\nu$. Again arithmetic degrees form a sub-semi-lattice of the semi-lattice of all degrees, for a similar reason. The above function $h$ is of degree $(\mathbf{a} \cup \mathbf{b})$ where $\mathbf{a}$ and $\mathbf{b}$ are the degrees of $f$ and $g$ respectively, so that if $\mathbf{a}$ and $\mathbf{b}$ are recursively enumerable degrees then so is $(\mathbf{a} \cup \mathbf{b})$.

## 7.18 *Structure of the upper semi-lattice of degrees of unsolvability*

The next proposition contains a construction which can be used in a wide range of cases to construct degrees standing in such and such relation to each other. We shall give several such propositions and also some extensions of the method of construction.

PROP. 20. *Given a non-recursive set* $\mathscr{A}$ *of natural numbers, sets* $\mathscr{B}$, $\mathscr{C}$ *of natural numbers can be found such that if* **a**, **b**, **c** *are the degrees of* $\mathscr{A}$, $\mathscr{B}$, $\mathscr{C}$ *respectively, where* $\mathbf{a} \neq \mathbf{O}$, *then* $\mathbf{b} \cup \mathbf{c} \leqslant \mathbf{a}'$ *and*

$$(0)\ \mathbf{b} \nleqslant \mathbf{a},\quad \mathbf{b} \nleqslant \mathbf{c};$$

$$(1)\ \mathbf{c} \nleqslant \mathbf{a},\quad \mathbf{c} \nleqslant \mathbf{b};$$

$$(2)\ \mathbf{a} \nleqslant \mathbf{b},\quad \mathbf{a} \nleqslant \mathbf{c}.$$

Each of these conditions gives rise to an unending progression of conditions $[0, \kappa]$, $[1, \kappa]$, $[2, \kappa]$, where $[\nu, \kappa]$ says that the recursivity in question fails for the machine with *g.n.* $\kappa$.

Let $\alpha, \beta, \gamma$ be the characteristic functions of the sets $\mathscr{A}$, $\mathscr{B}$, $\mathscr{C}$ respectively. We shall define $\beta$ and $\gamma$ by stages so that at stage $\theta$ exactly the first $g\theta$ values of $\beta$, $\gamma$ will have been chosen where

$$g\theta < g(S\theta). \tag{21}$$

Instead of saying that the first $g\theta$ values of $\beta$ have been chosen we can say equivalently that the number

$$\mathfrak{b}\theta = \langle \beta 0, \beta 1, \beta 2, \ldots, \beta(g\theta \dot{-} 1)\rangle$$

has been chosen. Then $\beta\nu = Pt[\mathfrak{b}\theta, S\nu]$ for $\nu < g\theta$, whence

$$\beta\nu = Pt[\mathfrak{b}(S\nu), S\nu], \quad \text{since} \quad \nu < g(S\nu).$$

At stage 0 $g0 = 0$ and $\mathfrak{b}0 = 0$ (the null set).
In the step from stage $\theta$ to stage $S\theta$ the $g(S\theta) \dot{-} g\theta$ additional values of $\beta, \gamma$ will be chosen so that however the definitions of $\beta, \gamma$ are subsequently completed the $(S\theta)$th of the conditions

$$[0, 0], [1, 0], [2, 0], [0, 1], [1, 1], [2, 1], [0, 2], \ldots, [0, \kappa], [1, \kappa], [2, \kappa], \ldots$$

will be satisfied. Thus when the definitions of $\beta$, $\gamma$ are completed all of these conditions will be satisfied, and so (0), (1), (2), will be satisfied.

We shall pay careful attention to the form of the operations employed so that when we are done we shall be able to say that $\beta$, $\gamma$ and $g$ are of degree $\leqslant \mathbf{a}'$ and so $\mathbf{b} \cup \mathbf{c} \leqslant \mathbf{a}'$.

We now describe, by cases, how the $g(S\theta) \dot{-} g\theta$ additional values of $\beta, \gamma$ are to be chosen. In proceeding from stage $\theta$ to stage $S\theta$, we can think of the process as '*extending*' $\mathfrak{b}\theta$ (which incorporates the first $g\theta$ values of $\beta$) to $\mathfrak{b}(S\theta)$ (which incorporates the first $g(S\theta)$ values of $\beta$), and similarly for $\mathfrak{c}$.

At any stage $\theta$ (when only $g\theta$ values to be given finally to $\beta, \gamma$ have been chosen), it will be convenient to call any function $\beta_1$ which is a characteristic function of a set and which possesses the first $g\theta$ values of $\beta$ an '*extension*' of $\beta$, similarly for $\gamma$.

*Case* 0. *Rem* $[3, \theta] = 0$, then $\theta = 3.[\theta/3]$, let $[\theta/3] = \kappa$.
We must extend $\mathfrak{b}\theta$ to $\mathfrak{b}(S\theta)$ and $\mathfrak{c}\theta$ to $\mathfrak{c}(S\theta)$ so that condition $[0, \kappa]$ will hold for all extensions $\beta_1$ of $\beta$ and $\gamma_1$ of $\gamma$. That is we must render it impossible, for every such pair of extensions, that

$$\beta_1 \nu = [(\mu\xi)[Un^1_{1,1}[\kappa, \nu, \mathfrak{a}\xi, \mathfrak{c}\xi, \xi_1, \xi_2] = 0]]_2, \tag{22}$$

where $\xi = \{\xi_1, \xi_2\}$, hold for each numeral $\nu$, i.e. the r.h.s. of (22) is defined, for which the condition is

$$(E\xi)(Un^{1,1}_1[\kappa, \nu, \mathfrak{a}\xi, \mathfrak{c}\xi, \xi_1, \xi_2] = 0),$$

and equal in value to the l.h.s. of (22). In fact we shall render this impossible for $\nu = g\theta$.

*Subcase* 0.1. For some extensions $\gamma_2$ of $\gamma$ and some $\lambda$

$$Un^1_{1,1}[\kappa, g\theta, \mathfrak{a}\lambda, \mathfrak{c}_2\lambda, \lambda_1, \lambda_2] = 0.$$

This is equivalent to saying that there is a pair $\lambda, \pi$ of numerals such that $\pi \neq 0$ and $\mathfrak{c}\theta^\frown\pi$ is an extension of $\mathfrak{c}\theta$, in which case $(A\xi)_\lambda[Pt[\pi, \xi] < 2]$, so that $\pi = \langle \epsilon_1, \epsilon_2, \ldots, \epsilon_{S\mu} \rangle$, for some numeral $\mu$, where $\epsilon_1, \ldots, \epsilon_{S\mu}$ are 0 or 1, so that the extension represents a characteristic function of a set. Thus

$$Un^1_{1,1}[\kappa, g\theta, \mathfrak{a}\lambda, \mathfrak{c}\theta^\frown\pi, \lambda_1, \lambda_2] = 0.$$

Thus the subcase hypothesis becomes

$$(E\xi)(\xi_2 \neq 0 \,\&\, \varpi\xi_2 = \xi_1 \dot{-} g\theta \,\&\, \xi_{1,2} < 2 \,\&\, (A\eta)_{\xi_1}[Pt[\xi_2, \eta] < 2]$$
$$\& \, Un^1_{1,1}[\kappa, g\theta, \mathfrak{a}\xi, \mathfrak{c}\theta^\frown\xi_2, \xi_{1,1}, \xi_{1,2}] = 0).$$

This is of the form $\qquad (E\xi)\, R^\alpha_0\{\theta, g\theta, \mathfrak{c}\theta, \xi\}$,

where $R_0\{\lambda, \mu, \nu, \kappa\}$ is a primitive recursive $\mathbf{A}_{00}$-statement. Let us write

$$X_0 \quad \text{for} \quad (\mu\xi)[R^\alpha_0\{\theta, g\theta, \mathfrak{c}\theta, \xi\}], \tag{23}$$

then $X_{0,1}, X_{0,2}$ are the $\lambda$, $\pi$ with the above properties.

Now we first extend (if necessary, i.e. if $X_{0,1} > g\theta$) the choice of values of $\gamma$ so that

$$\mathfrak{c}X_{0,1} = \mathfrak{c}[g\theta]^\frown X_{0,2},$$

and secondly we choose for $\beta[g\theta]$ a value $\leqslant 1$ different from that given by the r.h.s. of (22) when $\nu = g\theta$. Then as desired (22) will fail when $\nu = g\theta$ for any $\beta, \gamma$ having the values already chosen (including those chosen at stage $\theta$). What this gives as value for $\beta[g\theta]$ is

$$\beta[g\theta] = 1 \dot{-} X_{0,1,2}.$$

After taking
$$g(S\theta) = Max\,[S(g\theta), X_{0,1}], \tag{24}$$

which ensures (21), thirdly we choose all values (if any) of $\beta\nu$, $\gamma\nu$ for $\nu < g(S\theta)$ not already chosen (either at stage $\theta$ or in the further choices just described) to be 0. Then

$$\begin{aligned}\mathfrak{b}(S\theta) = (\mu\xi)\,[\xi \neq 0 \,\&\, (A\eta)_{g\theta}[Pt[\xi,\eta] = Pt[\mathfrak{b}\theta,\eta]]\\ \&\, Pt[\xi, g\theta] = 1 \dot{-} X_{0,1,2} \,\&\, \tilde{\omega}\xi = g(S\theta)],\\ \mathfrak{c}(S\theta) = (\mu\xi)\,[\xi \neq 0 \,\&\, (A\eta)_{g\theta}[Pt[\xi,\eta] = Pt[\mathfrak{c}\theta,\eta]]\\ \&\, (A\zeta)_{\varpi X_{0,2}}[Pt[\xi, g\theta+\zeta] = Pt[X_{0,2},\zeta]]\\ \&\, \varpi\zeta = g(S\theta)].\end{aligned}$$

These give with (23) us equations of the form

$$\left.\begin{aligned}\mathfrak{b}(S\theta) = \chi^{\alpha}_{0,1}\{\theta, g\theta, \mathfrak{b}\theta, \mathfrak{c}\theta\},\\ \mathfrak{c}(S\theta) = \chi^{\alpha}_{0,2}\{\theta, g\theta, \mathfrak{b}\theta, \mathfrak{c}\theta\},\end{aligned}\right\} \tag{25}$$

where $\chi^{\alpha}_{0,1}\{\lambda,\mu,\nu,\kappa\}$ and $\chi^{\alpha}_{0,2}\{\lambda,\mu,\nu,\kappa\}$ are partial $\alpha$-recursive being defined exactly when: $(E\xi)\,R^{\alpha}_{0}\{\lambda,\mu,\kappa,\xi\}$.

*Subcase* 0.2. Otherwise. Then the values already chosen at stage $\theta$ render (22) impossible for $\nu = g\theta$ no matter how $\beta, \gamma$ are completed as extensions of $\mathfrak{b}\theta$, $\mathfrak{c}\theta$. However to ensure (21) we take

$$g(S\theta) = S(g\theta), \tag{26}$$

and extend $\beta, \gamma$ by choosing

$$\beta[g\theta] = \gamma[g\theta] = 0,$$

so that
$$\mathfrak{b}(S\theta) = (\mathfrak{b}\theta)^{\frown}1, \quad \mathfrak{c}(S\theta) = (\mathfrak{c}\theta)^{\frown}1.$$

Combining the two subcases using (24), (25) with our formulation of the subcase 0.1 hypothesis

$$(\mathfrak{b}S\theta) = \left\{\begin{matrix}\chi^{\alpha}_{0,1}\{\theta, g\theta, \mathfrak{b}\theta, \mathfrak{c}\theta\} & \text{if} & (E\xi)\,R^{\alpha}_{0}\{\theta, g\theta, \mathfrak{c}\theta, \xi\},\\ (\mathfrak{b}\theta)^{\frown}1 & \text{otherwise,} & \end{matrix}\right\} \tag{27}$$

and similarly for $\mathfrak{c}(S\theta)$. Thus

$$\left.\begin{aligned}\mathfrak{b}(S\theta) &= \chi^{\alpha'}_{0,1}\{\theta, g\theta, \mathfrak{b}\theta, \mathfrak{c}\theta\},\\ \mathfrak{c}(S\theta) &= \chi^{\alpha'}_{0,2}\{\theta, g\theta, \mathfrak{b}\theta, \mathfrak{c}\theta\},\end{aligned}\right\} \tag{28}$$

where $\chi^{\alpha'}_{0,1}\{\lambda, \mu, \nu, \kappa\}$ and $\chi^{\alpha'}_{0,2}\{\lambda, \mu, \nu, \kappa\}$ are partial $\alpha'$-recursive, but since it is completely defined then it is $\alpha'$-recursive. Similarly, from (23), (24), (26):

$$g(S\theta) = \chi^{\alpha'}_{0,0}\{\theta, g\theta, \mathfrak{c}\theta\}, \tag{29}$$

where $\chi^{\alpha'}_{0,0}\{\lambda, \mu, \nu\}$ is $\alpha'$-recursive.

*Case* 1. $Rem\,[3, \theta] = 1$. Similar to case 0.

*Case* 2. $Rem\,[3, \theta] = 2$. Then $\theta = 3.[\theta/3]+2$, let $\kappa = [\theta/3]$.
We must extend $\mathfrak{b}\theta$ to $\mathfrak{b}(S\theta)$ and $\mathfrak{c}\theta$ to $\mathfrak{c}(S\theta)$ so that $\{2, \kappa\}$ will hold for all extensions $\beta_1$ of $\beta$ and $\gamma_1$ of $\gamma$, i.e. we must render it impossible, for every such pair of extensions, that:

$$\alpha\nu = [(\mu\xi)\,[Un^1_{1,1}[\kappa, \nu, \mathfrak{b}\xi, \mathfrak{c}\xi, \xi_1\,\xi_2] = 0]]_2 \tag{30}$$

hold for all $\nu$.

*Subcase* 2.1. For some $\nu$ and some extensions $\beta_1, \gamma_1$, of $\beta, \gamma$ respectively the r.h.s. of (30) is defined with value opposite to $\alpha\nu$. This is equivalent to there being numerals $\lambda, \pi', \pi''$, $\nu$ such that

$$\pi', \pi'' \neq 0, \quad (A\xi)_\lambda\,[Pt[\pi', \xi] < 2], \quad (A\xi)_\lambda[Pt[\pi'', \xi] < 2],$$

and $\quad Un^1_{1,1}[\kappa, \nu, (\mathfrak{b}\theta)^\frown\pi', (\mathfrak{c}\theta)^\frown\pi'', \lambda_1, \lambda_2] = 0 \quad$ and $\quad \lambda_2 \neq \alpha\nu$.

This amounts to

$$(E\xi)\,(\xi_2, \xi_3 \neq 0\,\&\,(A\eta)_{\xi_1}[Pt[\xi_2, \eta] < 2]\,\&\,(A\eta)_{\xi_1}[Pt[\xi_3, \eta] < 2]$$
$$\&\,Un^1_{1,1}[\kappa, \xi_4, (\mathfrak{b}\theta)^\frown\xi_2, (\mathfrak{c}\theta)^\frown\,\xi_3, \xi_{1,1}, \xi_{1,2}] = 0\,\&\,\xi_{1,2} \neq \alpha\xi_4),$$

where

$$\xi_\mu = pt[\xi, 4, \mu] \quad 1 \leqslant \mu \leqslant 4, \quad \text{so that} \quad \xi = \{\xi_1, \xi_2, \xi_3, \xi_4\}.$$

This is of the form $\qquad (E\xi)\,R^\alpha_2\{\theta, g\theta, \mathfrak{b}\theta, \mathfrak{c}\theta, \xi\}$,

where $R^\alpha_2$ is $\alpha$-recursive. Put: (31)

$$X_2 \quad \text{for} \quad (\mu\xi)\,R^\alpha_2\{\theta, g\theta, \mathfrak{b}\theta, \mathfrak{c}\theta, \xi\}.$$

We first extend (if necessary) the choice of values of $\beta, \gamma$ so that

$\mathfrak{b}X_{2,1} = (\mathfrak{b}\theta)^{\frown} X_{2,2}$, $\mathfrak{c}X_{2,1} = (\mathfrak{c}\theta)^{\frown} X_{2,3}$. Then (30) will fail for $\nu = X_{2,4}$ for all $\beta, \gamma$ having the values thus far chosen. After taking

$$g(S\theta) = max\,[S(g\theta), X_{2,1}], \tag{32}$$

secondly we choose the further values (if any) of $\beta\nu, \gamma\nu$ for $\nu < g(S\theta)$ to be 0. Then:

$$\mathfrak{b}(S\theta) = (\mu\xi)\,[\xi \neq 0\,\&\,(A\eta)_{g\theta}[Pt[\xi,\eta] = Pt[\mathfrak{b}\theta,\eta]]\,\&\,(A\eta)_{X_{2,1}}[Pt[\xi,\eta] = Pt[X_{2,1},\eta]]], \tag{33}$$

whence
$$\mathfrak{b}(S\theta) = \chi^{\alpha}_{2.1}\{\theta, g\theta, \mathfrak{b}\theta, \mathfrak{c}\theta\}, \tag{34}$$

and similarly for $\mathfrak{c}$, where $\chi^{\alpha}_{2.1}\{\lambda, \mu, \nu, \kappa\}$ is partial $\alpha$-recursive being defined exactly when $(E\xi)\,R^{\alpha}_{2}\{\theta, \lambda, \mu, \nu, \xi\}$.

*Subcase* 2.2. Otherwise. We shall show that in this subcase, for some $\nu$, the r.h.s. of (30) is undefined for every pair of extensions $\beta_1, \gamma_1$ of $\beta, \gamma$ respectively, after showing which the subcase can be treated similarly to subcase 0.2. Accordingly, suppose that (for reductio ad absurdam), for every $\nu$, the r.h.s. of (30) is defined for some extensions $\beta_1, \gamma_1$ of $\beta, \gamma$ respectively. That is, for each $\nu$, there is $\lambda, \pi', \pi''$, such that

$$\pi', \pi'' \neq 0, (A\xi)_{\lambda}\,[Pt[\pi', \xi] < 2],$$
$$(A\xi)_{\lambda}\,[Pt[\pi'', \xi] < 2], \quad \varpi\pi' = \varpi\pi'', \varpi\pi' + \varpi\mathfrak{c}\theta = \lambda$$

and
$$Un^{1}_{1,1}[\kappa, \nu, \mathfrak{b}\theta^{\frown}\pi', \mathfrak{c}\theta^{\frown}\pi'', \lambda_1, \lambda_2] = 0.$$

This can be expressed in the form

$$(E\xi)\,R\{\theta, g\theta, \mathfrak{b}\theta, \mathfrak{c}\theta, \nu, \xi\},$$

where $R$ is primitive recursive. Put

$$X \quad \text{for} \quad (\mu\xi)\,R\{\theta, g\theta, \mathfrak{b}\theta, \mathfrak{c}\theta, \nu, \xi\}.$$

Then $X_{1,2}$ is the value of the r.h.s. of (30) for some extensions $\beta_1$, $\gamma_1$ of $\beta, \gamma$ respectively for which that r.h.s. is defined, But since subcase 2.1 is excluded, the r.h.s. of (30) when defined has the value $\alpha\nu$. Thus writing out the $X$ in $X_{1,2}$ in full, for each $\nu$

$$\alpha\nu = [(\mu\xi)\,R\{\theta, g\theta, \mathfrak{b}\theta, \mathfrak{c}\theta, \nu, \xi\}]_{1,2}$$

for the fixed $\theta$ under consideration, which makes $\alpha$ recursive, contrary to hypothesis.

Combining the two subcases

$$\mathfrak{g}(S\theta) = \chi_{0,2}^{\alpha'}\{\theta, \mathfrak{g}\theta, \mathfrak{b}\theta, \mathfrak{c}\theta\},$$

$$\mathfrak{b}(S\theta) = \chi_{1,2}^{\alpha'}\{\theta, \mathfrak{g}\theta, \mathfrak{b}\theta, \mathfrak{c}\theta\},$$

$$\mathfrak{c}(S\theta) = \chi_{2,2}^{\alpha'}\{\theta, \mathfrak{g}\theta, \mathfrak{b}\theta, \mathfrak{c}\theta\},$$

where $\chi_{\pi,2}^{\alpha'}\{\lambda, \mu, \nu, \kappa\}, \pi = 0, 1, 2$ are $\alpha'$-recursive.

Combining the three cases

$$\mathfrak{g}(S\theta) = \chi_0^{\alpha'}\{\theta, \mathfrak{g}\theta, \mathfrak{b}\theta, \mathfrak{c}\theta\},$$

$$\mathfrak{b}(S\theta) = \chi_1^{\alpha'}\{\theta, \mathfrak{g}\theta, \mathfrak{b}\theta, \mathfrak{c}\theta\},$$

$$\mathfrak{c}(S\theta) = \gamma_2^{\alpha'}\{\theta, \mathfrak{g}\theta, \mathfrak{b}\theta, \mathfrak{c}\theta\}.$$

These equations and $g0 = \beta 0 = \gamma 0 = 0$ define $g, \beta, \gamma$ simultaneously by recursion on $\theta$ from functions $\chi_0^{\alpha'}, \chi_1^{\alpha'}, \chi_2^{\alpha'}$. By first setting up a recursion for the triplet $\{\mathfrak{g}\theta, \mathfrak{b}\theta, \mathfrak{c}\theta\}$ and then taking components it follows that $\mathfrak{g}, \mathfrak{b}, \mathfrak{c}$ are recursive in the functions $\chi_0^{\alpha'}, \chi_1^{\alpha'}, \chi_2^{\alpha'}$. Since the latter are recursive in $\alpha'$ so are $\mathfrak{g}, \mathfrak{b}, \mathfrak{c}$ and so are $\beta, \gamma$. This completes the demonstration of the proposition.

The above type of demonstration can be used when the conditions required to be fulfilled can be put into a progression of conditions each stating that one of the conditions fails for table $\theta$, $\theta = 0, 1, 2, \ldots$. But the demonstration of the next proposition is of a different type. Here the required functions are obtained as the 'limit' of a decreasing sequence of functions.

In the system $\mathscr{F}_C$ we shall write

$$\phi \to \psi \quad \text{for} \quad C\phi\psi,$$

$$\phi \leftrightarrow \psi \quad \text{for} \quad B\phi\psi.$$

### 7.19 *Example of the priority method. Solution of Post's problem*

PROP. 21. *Two recursively enumerable sets of natural numbers neither of which is recursive in the other can be* $A_0$*-defined.*

It is clear that both such sets are neither recursive nor complete. Thus they will be of incomparable degrees lying between the degrees **O** and **O′**, also they are recursively enumerable degrees.

We wish to define two recursively enumerable sets of natural numbers $\mathscr{A}$ and $\mathscr{B}$ such that $\mathscr{A} \not\leqslant_T \mathscr{B}$ and $\mathscr{B} \not\leqslant_T \mathscr{A}$, i.e.

$$\mathscr{A} \neq \mathscr{X}_\theta^{\mathscr{B}} \quad \text{and} \quad \mathscr{B} \neq \mathscr{X}_\theta^{\mathscr{A}} \quad \text{for} \quad \theta = 0, 1, 2, \ldots.$$

These are respectively equivalent to

$$(Ex)(x\epsilon\mathscr{A} \leftrightarrow x\bar{\epsilon}\mathscr{X}_\theta^{\mathscr{B}}) \quad \text{and} \quad (Ex)(x\epsilon\mathscr{B} \leftrightarrow x\bar{\epsilon}\mathscr{X}_\theta^{\mathscr{A}}) \quad \text{for} \quad \theta = 0, 1, 2, \ldots.$$

It suffices then to show that there are two functions $f$ and $g$ such that

$$f\theta\epsilon\mathscr{A} \leftrightarrow f\theta\bar{\epsilon}\mathscr{X}_\theta^{\mathscr{B}} \quad \text{and} \quad g\theta\epsilon\mathscr{B} \leftrightarrow g\theta\bar{\epsilon}\mathscr{X}_\theta^{\mathscr{A}} \quad \text{for} \quad \theta = 0, 1, 2, \ldots. \qquad \text{(i)}$$

The method we use is known as the priority method because in the course of the construction certain things are given priority over other things in order that the latter will be unable to upset what has been achieved by the former. It is more complicated than the diagonal method which we have used many times.

The construction proceeds by stages. We start with two lists of numerals the $\mathscr{A}$-list and the $\mathscr{B}$-list. From time to time we shall place a $(+)$ or a $(-)$ against selected members of either list. Once a $(+)$ is placed against a member of either list then it will remain unchanged thereafter, but a $(-)$ may later on be changed to a $(+)$. The members of the $\mathscr{A}$-list which receive a $(+)$ shall constitute the set $\mathscr{A}$ and the members of the $\mathscr{B}$-list which recive a $(+)$ shall constitute the set $\mathscr{B}$. It will be apparent from the construction and the manner in which we place the $(+)$ signs that the sets $\mathscr{A}$ and $\mathscr{B}$ are recursively enumerable.

We shall also have two lists of *markers* the $\mathscr{A}$-markers and the $\mathscr{B}$-markers. These are respectively denoted by

$$0', 1', 2', \ldots, \quad \text{and} \quad 0'', 1'', 2'', \ldots$$

These are, as their name implies, used to keep the place. In the course of the construction we shall put markers against members on the like list of numerals and at other times we shall move some markers down the list in which they are. An essential feature of the construction is that each marker ultimately comes to rest. The numeral $f\theta$ against which the marker $\theta'$ comes to rest will be the $f\theta$ in (i), and the numeral $g(\theta)$ against which the marker $\theta''$ comes to rest will be the $g\theta$ in (i).

The numerals which receive a $(-)$ in the course of the construction are candidates for $\bar{\mathscr{A}}$ or $\bar{\mathscr{B}}$, according to which list they are in. A list like

$$0', 0'', 1', 1'', 2', \ldots$$

is called a *priority list* because in the course of the construction if the position occupied by any marker gets a $(+)$ against it then all other markers of the other kind which have so far been used and are further down the priority list are moved still further down the list of numerals they are associated with in order that what is achieved by the ( − ) signs used at that stage will never thereafter be fouled.

The construction takes place by stages. We define

$\mathscr{A}_0 = \emptyset$, $\mathscr{B}_0 = \emptyset$,

$\mathscr{A}_\nu$ = the set of natural numbers in the $\mathscr{A}$-list which have received ( + ) at the end of stage $\nu$,

$\mathscr{B}_\nu$ is similarly defined.

At any stage in the construction we shall call a member of either list of numerals *free* if neither it nor any numeral further down its list has any mark or marker against it. At any stage in the construction we shall say that a member of either list of numerals is *vacant* if it is without a ( + ).

We now give the construction.

*Stage* 1. Place marker $0'$ against 0 in the $\mathscr{A}$-list.

*Stage* 2. Place marker $0''$ against 0 in the $\mathscr{B}$-list.

..................................................................

*Stage* $2\nu+1$. Place marker $\nu'$ against the first free numeral in the $\mathscr{A}$-list. Let $a_0^{(\nu)}, a_1^{(\nu)}, a_2^{(\nu)}, \ldots, a_\nu^{(\nu)}$ be the current positions of the markers in the $\mathscr{A}$-list. Let condition $C_1\{\theta, \nu\}$ be:

(i) $a_\theta^{(\nu)}$ is vacant,

(ii) $Un_2^1[\theta, a_\theta^{(\nu)}, \mathscr{B}_{2\nu}, \nu, 1] = 0$.

Place ( + ) against the least $a_\mu^{(\nu)}$, $0 \leqslant \mu \leqslant \nu$, say $a_{\mu_0}^{(\nu)}$, which satisfies condition $C_1\{\mu, \nu\}$, place ( − ) against each vacant numeral in the $\mathscr{B}$-list whose membership of $\overline{\mathscr{B}}_{2\nu}$ is used in the evaluation of $Un_2^1[\theta, a_\theta^{(\nu)}, \mathscr{B}_{2\nu}, \nu, 1]$, then move all markers $\mu''$ in the $\mathscr{B}$-list for which $\mu_0 \leqslant \mu'' < \nu$ (this is all the markers $\mu''$ as yet in the $\mathscr{B}$-list for which $\mu_0 \leqslant \mu$) in order down to the first available free places in the $\mathscr{B}$-list. If each $a_\mu^{(\nu)}$, $0 \leqslant \mu \leqslant \nu$, fails to satisfy condition $C\{\mu, \nu\}$ then pass to stage $2\nu+2$.

*Stage* $2\nu+2$. Place marker $\nu''$ against the first free numeral in the $\mathscr{B}$-list. Let $b_0^{(\nu)}, b_1^{(\nu)}, b_2^{(\nu)}, \ldots, b_\nu^{(\nu)}$ be the current positions of all the markers in the $\mathscr{B}$-list. Let condition $C_2\{\theta, \nu\}$ be:

(i) $b_\theta^{(\nu)}$ is vacant,

(ii) $Un_2^1[\theta, b_\theta^{(\nu)}, \mathscr{A}_{2\nu+1}, \nu, 1] = 0$.

Place ( + ) against the least $b_\mu^{(\nu)}$ $0 \leqslant \mu \leqslant \nu$, say $b_{\mu_0}^{(\nu)}$, which satisfies condition $C_2\{\mu, \nu\}$, place ( − ) against each vacant numeral in the $\mathscr{A}$-list whose membership of $\overline{\mathscr{A}_{2\nu+1}}$ is used in the evaluation of $Un_2^1[\mu_0, b_{\mu_0}^{(\nu)}, \overline{\mathscr{A}_{2\nu+1}}, \nu, 1]$, then move all markers $\mu'$ in the $\mathscr{A}$-list for which $\mu_0 < \mu \leqslant \nu$ (this is all the markers $\mu'$ as yet in the $\mathscr{A}$ list for which $\mu_0 < \mu$) in order down to the first available free places in the $\mathscr{A}$-list. If each $b_\mu^{(\nu)}$ fails to satisfy condition $C_2\{\mu, \nu\}$ then pass to stage $2\nu+3$.

The success of the construction depends on the following:

LEMMA. *Each marker ultimately comes to rest.*

We proceed by induction. Marker $0'$ never moves. Marker $0''$ can move at most once. Marker $\nu'$ can move at most once for each position of markers $0'', 1'', \ldots, (\nu \dot{-} 1)''$ whence if these only move a bounded number of times then marker $\nu'$ can move only a bounded number of times. Similarly marker $\nu''$ can move at most once for each position of markers $0', 1', 2', \ldots, (\nu \dot{-} 1)'$; whence if these markers can only move a bounded number of times then marker $\nu''$ can only move a bounded number of times.

Let $f\theta$ be the final position of marker $\theta'$, and let $g\theta$ be the final position of marker $\theta''$. We show that

$$f\theta \epsilon \mathscr{A} \leftrightarrow f\theta \bar{\epsilon} \mathscr{X}_\theta^{\mathscr{B}} \quad \text{and} \quad g\theta \epsilon \mathscr{B} \leftrightarrow g\theta \bar{\epsilon} \mathscr{X}_\theta^{\mathscr{A}}.$$

Suppose that $f\theta \epsilon \mathscr{A}$ then $f\theta$ gets $(+)$ at stage $2\nu+1$ for some $\nu$. The signs introduced at this stage remain unchanged thereafter. For if $\theta < \mu$ marker $\mu''$ is moved below all these $(-)$ signs and thereafter can only move further down and new markers are introduced after all markers present in the appropriate list, thus markers $\mu''$ for which $\theta < \mu$ never foul these $(-)$ signs. Markers for which $\mu < \theta$ are already at rest, for if such a marker moved then by our construction marker $\theta'$ would have to move contrary to hypothesis that it had already reached its final position $f\theta$. Thus the evaluation of $Un_2^1[\theta, f\theta, \mathscr{B}_{2\nu}, \nu, 1]$ is the same as that of $Un_2^1[\theta, f\theta, \mathscr{B}, \nu, 1]$ because the answers to 'is $\lambda$ in $\mathscr{B}_{2\nu}$?' will be the same as the answers to 'is $\lambda$ in $\mathscr{B}$?'. Thus $f\theta \bar{\epsilon} \mathscr{X}_\theta^{\mathscr{B}}$, in order to give $f\theta$ a $(+)$ condition $C_1\{\theta, \nu\}$ must be satisfied, and this means that $f\theta \bar{\epsilon} \mathscr{X}_\theta^{\mathscr{B}_{2\nu}}$.

Now suppose that $f\theta \bar{\epsilon} \mathscr{X}_\theta^{\mathscr{B}_{2\nu}}$, then there is a $\mu$ such that for all

$$\nu \geqslant \mu \quad Un_2^1[\theta, f\theta, \mathscr{B}, \nu, 1] = Un_2^1[\theta, f\theta, \mathscr{B}_{2\nu}, \nu, 1],$$

because if $\nu$ is sufficiently large then the resister of $\mathscr{B}_{2\nu}$ will contain all the information that is required of $\mathscr{B}$ for the calculation of

$$Un_2^1[\theta, f\theta, \mathscr{B}, \nu, 1].$$

Thus if $f\theta \bar{\epsilon} \mathscr{X}_\theta^{\mathscr{B}}$ then we shall have $Un_2^1[\theta, f\theta, \mathscr{B}_{2\nu}, \nu, 1] = 0$ for all $\nu \geqslant \mu$ whence condition $C_1\{\theta, \nu\}$ part (ii) will be satisfied, hence at a certain stage $f\theta$ will be the least such marker position and so will satisfy the whole of condition $C_1\{\theta, \nu\}$, thus $f\theta$ will get a $(+)$ and so will belong to $\mathscr{A}$. Thus (i) is satisfied and so $\mathscr{A} \nleqslant_T \mathscr{B}$. Similarly $\mathscr{B} \nleqslant_T \mathscr{A}$. This completes the demonstration of the proposition. It is known as Posts' problem.

Note that $f$ and $g$ fail to be recursive functions. This is seen as follows: Let $\theta$ be in $\mathscr{A}$ if and only if $a\theta = 0$. Then if $\mathscr{C}$ is a recursively enumerable set there is a numeral $\theta$ such that $\nu$ is in $\mathscr{C}$ if and only if

$$(E\xi)\, Un_1^1[\theta, \nu, \mathfrak{b}\xi, \xi, 1] = 0,$$

where $b$ is the characteristic function of the set $\mathscr{B}$, because $\mathscr{C}$ being recursively enumerable there is a partial recursive function, $t$, (and a fortiori one partial in $b$) such that

$$t\nu = \begin{cases} 1 & \text{if } \nu \text{ is in } \mathscr{C}, \\ \text{undefined otherwise.} \end{cases}$$

The function $t$ is given by the rule: for argument $\nu$ generate $\mathscr{C}$ and if $\nu$ appears in the generation let $t\nu = 1$, otherwise $t$ is undefined, we can find a table which does this, let its *g.n.* be $\theta$. But $f\theta$ is in $\mathscr{A}$ if and only if

$$(E\xi)\,(Un_1^1[\theta, f\theta, b\xi, \xi, 1] = 0),$$

hence $f\theta$ is in $\mathscr{A}$ if and only if $f\theta$ is in $\mathscr{C}$, by definition of $\mathscr{C}$. But if $\mathscr{C} \subset \bar{\mathscr{A}}$ then $f\theta$ is in $\bar{\mathscr{C}} \cap \bar{\mathscr{A}}$. If $f$ is recursive then $\mathscr{A}$ is creative, by definition of creative, this is impossible because then $\mathscr{A}$ would be of degree $\mathbf{O}'$ and hence $\mathscr{B}$ would be $A$-recursive.

## 7.20 *Complete degrees*

PROP. 22. *A degree* $\mathbf{a}$ *is complete if and only if* $\mathbf{a} \geqslant \mathbf{O}'$.

We first show:

LEMMA. *To any degree* $\mathbf{a}$ *we can find a degree* $\mathbf{b}$ *such that*

$$\mathbf{b}' = \mathbf{b} \cup \mathbf{O}' = \mathbf{a} \cup \mathbf{O}'.$$

Let $h$ be a function of degree **a**. Define a function $k$ of degree $\leqslant \mathbf{O}'$ as follows:

$$k[\nu,\nu'] = \begin{cases} (\mu x)\,[Un_1^1[\nu,\nu,x,(\varpi x)_1,(\varpi x)_2] = 0 \,\&\, Comp\,[x,\nu']], \\ 0 \quad \text{if undefined.} \end{cases}$$

where

D 222 $\quad Comp\,[\nu,\nu']$ for $(Ax)_{Min\,[\varpi\nu,\,\varpi\nu']}\,[Pt[\nu,x] = Pt[\nu',x]]$.

The function $k$ is of degree $\leqslant \mathbf{O}'$, because to find the value of $k[\nu,\nu']$ we first have to decide whether

$$(Ex)\,(Un_1^1[\nu,\nu',x,(\varpi x)_1,(\varpi x)_2] = 0 \,\&\, Comp\,[x,\nu']) \tag{35}$$

and this is of degree $\leqslant \mathbf{O}'$, if the decision is favourable then it is a recursive process to find the least numeral which satisfies, if the decision is unfavourable then the value of $k[\nu,\nu']$ is zero.

Define a function $g$ as follows:

$$\begin{cases} g0 = 0, \\ g(S\nu) = Max\,[g\nu^\frown\langle h\nu\rangle, k[\nu, g\nu^\frown\langle h\nu\rangle]]. \end{cases}$$

Then $\quad g0 = 0,\; g1 = \langle h0\rangle^\frown\gamma 0,\; g2 = \langle h0\rangle^\frown\gamma 0^\frown\langle h1\rangle^\frown\gamma 1, \ldots,$

$$g(S\nu) = \langle h0\rangle^\frown\gamma 0^\frown\ldots^\frown\langle h\nu\rangle^\frown\gamma\nu \quad \text{for some} \quad \gamma 0,\gamma 1,\gamma 2,\ldots,\gamma\nu.$$

Now let $fx = Pt[g(Sx),x]$, let **b** be the degree of $f$. Then for each $\nu$

$$Un_1^1[\nu,\nu,\underset{1\leqslant x'\leqslant\pi}{\langle fx'\rangle},\pi_1,\pi_2] = 0 \rightarrow k[\nu,g\nu^\frown\langle h\nu\rangle] = \underset{1\leqslant x'\leqslant\pi'}{\langle fx'\rangle} \neq 0$$

for some $\pi' \leqslant \pi$ (we have $Comp\,[\underset{1\leqslant x'\leqslant\pi}{\langle fx'\rangle}, g\nu^\frown\langle h\nu\rangle]$ and

$$k[\nu,g\nu^\frown\langle h\nu\rangle] \neq 0 \rightarrow Un_1^1[\nu,\nu,\underset{1\leqslant x'\leqslant\pi}{\langle fx'\rangle},\pi_1,\pi_2] = 0,$$

where $\pi = \varpi k[\nu, g\nu^\frown h\nu]$).

(Note that if (35) holds then $k[\nu,\nu'] \neq 0$, because if

$$(\mu x)\,[Un_1^1[\nu,\nu,x,(\varpi x)_1,(\varpi x)_2] = 0] \quad \text{and} \quad x = (\varpi x)_1 = (\varpi x)_2 = 0$$

and table $\nu$ with input $\nu$ after zero moves fails to produce zero (it produces the null set)) the *g.n.*'s of the initial and final situation are different if input is $\nu$ and output is zero. Therefore

$$k[\nu,g\nu^\frown\langle h\nu\rangle] \neq 0 \leftrightarrow (Ex)\,Un_1^1[\nu,\nu,\underset{1\leqslant x'\leqslant x}{\langle fx'\rangle},x_1,x_2] = 0. \tag{36}$$

Now $g$ is $h, k$-recursive, therefore the product on the r.h.s. of (36) is $h, k$-recursive or

$$\mathbf{b}' \leqslant \mathbf{a} \cup \mathbf{O}', \tag{37}$$

On the other hand, since $h\nu = f[S(\varpi[g\nu])]$ for each $\nu$, we can substitute $f[(S\varpi[g\nu])]$ for $h$ in the definition of $g$ to obtain a formula according to which $g$ is $f, k$-recursive. Then using $h\nu = f[(S\varpi[g\nu])]$ again, we see that $h$ is also $f, k$-recursive or

$$\mathbf{a} \leqslant \mathbf{b} \cup \mathbf{O}'. \tag{38}$$

From (38) it immediately follows that

$$\mathbf{a} \cup \mathbf{O}' \leqslant \mathbf{b} \cup \mathbf{O}', \tag{39}$$

and it is elementary that

$$\mathbf{b} \cup \mathbf{O}' \leqslant \mathbf{b}' \tag{40}$$

($\mathbf{O} \leqslant \mathbf{b}$ hence $\mathbf{O}' \leqslant \mathbf{b}'$ hence $\mathbf{b} \cup \mathbf{O}' \leqslant \mathbf{b}'$).
From (37), (39), (40) we obtain:

$$\mathbf{b}' = \mathbf{b} \cup \mathbf{O}' = \mathbf{a} \cup \mathbf{O}' \tag{41}$$

and the lemma is demonstrated. We now turn to the demonstration of the proposition. Since $\mathbf{O} \leqslant \mathbf{a}$ then $\mathbf{O}' \leqslant \mathbf{a}'$ so that a complete degree is $\geqslant \mathbf{O}'$, if $\mathbf{O}' \leqslant \mathbf{a}$ then $\mathbf{a} \cup \mathbf{O}' = \mathbf{a}$ and (41) becomes

$$\mathbf{b}' = \mathbf{b} \cup \mathbf{O}' = \mathbf{a}$$

and so $\mathbf{a}$ is complete.

PROP. 23. *Given a degree* $\mathbf{d}$ *there are degrees* $\mathbf{a}$ *and* $\mathbf{b}$ *such that*

$$\mathbf{a}' \cup \mathbf{b}' = \mathbf{a} \cup \mathbf{b} = \mathbf{a}' = \mathbf{b}' = \mathbf{d}'$$

*and* $\mathbf{a}/\mathbf{b}$ *and* $\mathbf{d} < \mathbf{a} < \mathbf{d}', \mathbf{d} < \mathbf{b} < \mathbf{d}'$.

We give a demonstration for the case $\mathbf{d} = \mathbf{O}$ and then show how it can be modified for $\mathbf{d} > \mathbf{O}$.

We construct functions $\alpha$ and $\beta$ of degree $\mathbf{a}$ and $\mathbf{b}$ respectively such that

$$\mathbf{a}' \cup \mathbf{b}' \leqslant \mathbf{a} \cup \mathbf{b} \leqslant \mathbf{O}' \tag{42}$$

and then show that the other properties follow.

Let $\gamma\nu = \{\alpha\nu, \beta\nu\}$, $\alpha, \beta$ are characteristic functions so $\gamma$ will only take the values 0, 1, 2, 4. We proceed by stages as in Prop. 21 at stage $\theta$ the first $g\theta$ values of $\alpha, \beta$ have been chosen.

$$g0 = 0 \quad \text{so} \quad \mathfrak{g}0 = 0.$$

$\mathfrak{a}, \mathfrak{b}, \mathfrak{c}, \mathfrak{g}$ are the register functions for $\alpha, \beta, \gamma, g$ respectively.

For all $\theta$, $\kappa$ we shall have

$$\alpha\nu = \beta\nu \quad \text{iff} \quad \nu \neq g\theta \dot{-} 1, \tag{43}$$

$$(Ex)\, Un_1^1[\kappa, \kappa, \mathfrak{a}x, x_1, x_2] = 0 \quad \text{iff} \quad \alpha[g(2 \,.\, \kappa+1) \dot{-} 1] = 0, \tag{44}$$

$$(Ex)\, Un_1^1[\kappa, \kappa, \mathfrak{b}x, x_1, x_2] = 0 \quad \text{iff} \quad \beta[g(2 \,.\, \kappa+2) \dot{-} 1] = 0, \tag{45}$$

$$g0 = 0 \quad g(S\theta) = S(\mu x)\,[\alpha x \neq \beta x \,\&\, x > g\theta]. \tag{46}$$

From (44) $\quad \mathbf{a}' \leqslant \mathbf{a} \cup \mathbf{g}$, where $\mathbf{g}$ is the degree of $g$.

From (45) $\quad \mathbf{b}' \leqslant \mathbf{b} \cup \mathbf{g}$,

hence $\quad \mathbf{a}' \cup \mathbf{b}' \leqslant \mathbf{a} \cup \mathbf{b} \cup \mathbf{g}$,

but from (46) $\mathbf{g} \leqslant \mathbf{a} \cup \mathbf{b}$ and so $\mathbf{a}' \cup \mathbf{b}' \leqslant \mathbf{a} \cup \mathbf{b}$ whence $\mathbf{a} \cup \mathbf{b} = \mathbf{a}' \cup \mathbf{b}'$.

Let $C_{2 \,.\, \kappa+1}$ be the condition that $\gamma$ satisfies (44) and (43) for

$$g\theta \leqslant \nu < g(S\theta).$$

Suppose that $g(2 \,.\, \kappa)$ and $\gamma[g(2 \,.\, \kappa)]$ have been defined.

*Case* 1. $g(2 \,.\, \kappa+1)$ and $\mathfrak{a}[g(2 \,.\, \kappa+1)]$ can be defined so that

$$Un_1^1[\kappa, \kappa, \mathfrak{a}\pi, \pi_1, \pi_2] = 0 \quad \text{for some } \pi = \{\pi_1 \pi_2\} < g(2 \,.\, \kappa+1).$$

This is equivalent to

$$(Ex)\,(Un_1^1[\kappa, \kappa, \mathfrak{a}[g(2 \,.\, \kappa)]^\frown x_1, x_{2,1}, x_{2,2}] = 0 \,\&\, (Ax')_{\varpi x_1}[Pt[x_1, x'] < 2] \\ \&\, x_{2,1} = g(2 \,.\, \kappa) + \varpi x_1).$$

This is of the form $\quad (Ex)\,\chi\{\kappa, \mathfrak{a}[g(2 \,.\, \kappa)], g(2 \,.\, \kappa), x\}$,

where $X\,\{x, x', x'', x'''\}$ is recursive. Write

$$X \quad \text{for} \quad (\mu x)\,\chi\{\kappa, \mathfrak{a}[g(2 \,.\, \kappa)], g(2 \,.\, \kappa), x\}.$$

Set $g(2 \,.\, \kappa+1) = SY$, where $Y$ for $Max\,[g(2 \,.\, \kappa), X_2]$ and

$$\mathfrak{c}[g(2 \,.\, \kappa+1)] = \mathfrak{c}[g(2 \,.\, \kappa)]^\frown \langle \{Pt[X_1, x'], Pt[X_1, x']\}\rangle_{1 \leqslant x' \leqslant \varpi X_1}{}^\frown \langle 0, 1\rangle$$

i.e. $\quad \alpha\nu = \beta\nu = Pt[X_1, S\gamma\,[g(2 \,.\, \kappa)]] \quad$ for $\quad g(2 \,.\, \kappa) \leqslant \nu \leqslant X_2$,

(unnecessary if $g(2 \,.\, \kappa) \geqslant X_2$) and $\alpha Y = 0$, $\beta Y = 1$. This ensures condition $C_{2 \,.\, \kappa+1}$.

Thus $\quad h[2 \,.\, \kappa+1] = \gamma[g(2 \,.\, \kappa+1)] = \chi_1\{\kappa, h(2 \,.\, \kappa), g(2 \,.\, \kappa)\}$,

being defined exactly when $(Ex)\,\chi\{\kappa, h(2 \,.\, \kappa), g(2 \,.\, \kappa), x\}$.

Also $g(2.\kappa+1) = \chi_2\{\kappa, h(2.\kappa), g(2.\kappa)\}$ being defined under the same conditions.

*Case* 2. Otherwise. Then $\gamma[g(2.\kappa+1)] = \gamma[g(2.\kappa)]^\frown\langle 0,1\rangle$,

$$g(2.\kappa+1) = Sg(2.\kappa), \quad \text{so} \quad h(2.\kappa+1) = h(2.\kappa)^\frown\langle 0,1\rangle,$$

then condition $C_{2.\kappa+1}$.

*Cases* 3, 4. $h(2.\kappa+2)$ and $g(2.\kappa+2)$ are defined similarly but reversing the roles of $\alpha$ and $\beta$, then $C_{2.\kappa+2}\,\gamma$ satisfies (45) and (43) etc. Altogether

$$h(S\kappa) = \chi_3^{\mathbf{O}'}\{\kappa, h\kappa, g\kappa\}, \quad g(S\kappa) = \chi_4^{\mathbf{O}'}\{\kappa, h\kappa, g\kappa\},$$

whence $h$ and $g$ are of degree $\leqslant \mathbf{O}'$, hence so are $\alpha$ and $\beta$. Now

$$\mathbf{O}' \leqslant \mathbf{a}' \leqslant \mathbf{a}' \cup \mathbf{b}', \quad \mathbf{O}' \leqslant \mathbf{b}' \leqslant \mathbf{a}' \cup \mathbf{b}'$$

which with (42) yields

$$\mathbf{a}' \cup \mathbf{b}' = \mathbf{a} \cup \mathbf{b} = \mathbf{a}' = \mathbf{b}' = \mathbf{O}'. \tag{47}$$

If any of $\mathbf{a}/\mathbf{b}$, $\mathbf{O} < \mathbf{a} < \mathbf{O}'$, $\mathbf{O} < \mathbf{b} < \mathbf{O}'$ fail then we shall have $\mathbf{a} = \mathbf{a}'$ or $\mathbf{b} = \mathbf{b}'$ which is absurd. If $\mathbf{a}/\mathbf{b}$ fails then say $\mathbf{a} \leqslant \mathbf{b}$, whence $\mathbf{a}' \leqslant \mathbf{b}'$ and so $\mathbf{b} = \mathbf{b}'$, which is absurd. If $\mathbf{O}' \leqslant \mathbf{a}$ then $\mathbf{O}' < \mathbf{O}'' \leqslant \mathbf{a}'$ which is absurd. If $\mathbf{a}/\mathbf{O}'$ then $\mathbf{a} \cup \mathbf{O}' > \mathbf{O}'$ whence $\mathbf{O}' = \mathbf{a}' \cup \mathbf{O}' > \mathbf{a} \cup \mathbf{O}' > \mathbf{O}'$, which is absurd. If $\mathbf{a} = \mathbf{O}$ then $\mathbf{b} = \mathbf{b}'$ which is absurd.

For arbitrary degree $\mathbf{d}$. Let $\lambda x.\alpha\{2.x\}$, $\lambda x\beta\{2.x\}$ be characteristic functions of degree $\mathbf{d}$. The values of $\alpha, \beta$ for odd arguments are determined as in the case when $\mathbf{d} = \mathbf{O}$, except for obvious changes to take account of the fact that $\alpha(2.\nu)$ and $\beta(2.\nu)$ are determined prior to the construction. Letting $\mathbf{a}, \mathbf{b}$ be the degrees of $\alpha$, $\beta$ respectively we have

$$\mathbf{d} \leqslant \mathbf{a} \leqslant \mathbf{d}', \quad \mathbf{d} \leqslant \mathbf{b} \leqslant \mathbf{d}'$$

and the demonstration of the proposition is complete.

Cor. (i). *If* $\mathbf{a}' = \mathbf{b}'$ *we can have any of* $\mathbf{a} < \mathbf{b}$, $\mathbf{a} = \mathbf{b}$, $\mathbf{b} < \mathbf{a}$, $\mathbf{a}/\mathbf{b}$.

Prop. 23 gives a case of $\mathbf{a}/\mathbf{b}$ and $\mathbf{a}' = \mathbf{b}'$, also of $\mathbf{d} < \mathbf{a}$ and $\mathbf{d}' = \mathbf{a}'$.

Cor. (ii). *If* $\mathbf{a}/\mathbf{b}$ *we can have* $\mathbf{a}' \cup \mathbf{b}' < (\mathbf{a} \cup \mathbf{b})'$.

Prop. 23 gives a case of $\mathbf{a}' \cup \mathbf{b}' = \mathbf{a} \cup \mathbf{b} < (\mathbf{a} \cup \mathbf{b})'$ and $\mathbf{a}/\mathbf{b}$.

Cor. (iii). *Each complete degree is the l.u.b. of the set of lesser degrees.*

By Prop. 23 we have degrees $\mathbf{a}, \mathbf{b}$ such that $\mathbf{d} < \mathbf{a} < \mathbf{d}'$, $\mathbf{d} < \mathbf{b} < \mathbf{d}'$ and $\mathbf{a} \cup \mathbf{b} = \mathbf{d}'$.

Cor. (iv). *If* $\mathbf{a} < \mathbf{b}$ *we can have either* $\mathbf{a}' = \mathbf{b}'$ *or* $\mathbf{a}' < \mathbf{b}'$ *only. If* $\mathbf{a}/\mathbf{b}$ *we can have* $\mathbf{a}' = \mathbf{b}'$.

Prop. 23 gives a case of $\mathbf{a}/\mathbf{b}$ and $\mathbf{a}' = \mathbf{b}'$ and of $\mathbf{d} < \mathbf{a}$ and $\mathbf{d}' = \mathbf{a}'$.

Cor. (v). *If* $\mathbf{a}$ *is a degree and* $\mathbf{a} > \mathbf{O}'$ *then there fails to be a degree* $\mathbf{c}, \mathbf{c} > \mathbf{a}$, *such that if* $\mathbf{b} < \mathbf{c}$ *then* $\mathbf{b} \leqslant \mathbf{a}$.

By Prop. 22 $\mathbf{c}$ is a complete degree. By Cor. (iii) above $\mathbf{c}$ is the *l.u.b.* of degrees less than $\mathbf{c}$. Hence there fails to be a degree $\mathbf{c}$ with the properties described.

## 7.21 *Sequences of degrees*

Prop. 24. *A sequence of recursively independent, simultaneously recursively enumerable sets is* $\mathrm{A}_0$*-definable.*

We define functions $f[\lambda, \mu, \kappa]$ and $h[\lambda, \mu, \theta]$ such that $f[\lambda, \mu, \kappa]$ is always either 0 or 1 and decreases as $\lambda$ increases. Then the functions $g_\mu$ for $\boldsymbol{\lambda} x . \underset{\lambda\to\infty}{\mathrm{L}} f[\lambda, \mu, x]$ and $g^\mu$ for $\boldsymbol{\lambda} xx' . \underset{\lambda\to\infty}{\mathrm{L}} f[\lambda, x + A_1[Sx \dot{-} \mu], x']$ are recursively enumerable. We require $g_\mu$ to fail to be recursive in $g^\mu$ for each $\mu$. The witness that this is so will be $\underset{\lambda\to\infty}{\mathrm{L}} h[\lambda, \mu, \theta]$, where $h[\lambda, \mu, \theta]$ increases with $\lambda$ but is bounded so that it is ultimately constant. We write $k[\mu, \theta]$ for $\underset{\lambda\to\infty}{\mathrm{L}} h[\lambda, \mu, \theta]$, the final value of $h[\lambda, \mu, \theta]$. Then we shall show that

$$g_\mu[k[\mu, \theta]] = 0 \quad \text{iff} \quad Un_2^1[\theta, k[\mu, \theta], \mathfrak{g}^\mu[\lambda + A_1[S\lambda \dot{-} \mu], \lambda], \lambda, 1] = 0,$$

i.e. $\Phi_\theta^{g^\mu}[k[\mu, \theta]] = 1$, so that $k[\mu, \theta]$ is a witness to the fact that $g_\mu$ fails to be recursive in $g^\mu$ by table $\theta$. $\mathfrak{g}$ is the register for $g^\mu$.

We define these functions by stages $\lambda = 0, 1, 2, \ldots$. The values of $\lambda$ for which $Pt[\lambda, 1] = \theta$ and $Pt[\lambda, 2] = \mu$ will be devoted to showing that $g_\mu$ fails to be recursive in $g^\mu$ by table $\theta$. And we shall have

$$k[\mu, \theta] = \langle \mu, \theta \rangle ^\frown \pi$$

for some $\pi$. We start with

$$f[0, \mu, \kappa] = 1, \quad h[0, \mu, \kappa] = \langle \mu, \theta \rangle,$$

$$Pt[S\lambda, 2] = \mu; f[S\lambda, \mu, \kappa] = 0$$

if $$f[\lambda, \mu, \kappa] = 1, \kappa = h[\lambda, \mu, \theta], \theta = Pt[\lambda, 1]$$

and $$Un_2^1[Pt[\lambda, 1], \kappa, \mathfrak{f}[\lambda, \lambda + A_1[S\lambda \dot{-} \mu], \lambda], \lambda, 1] = 0$$

call this condition $C_\mu \lambda$;

$$f[S\lambda, \mu, \kappa] = f[\lambda, \mu, \kappa] \quad \text{otherwise},$$

$$f[S\lambda, \mu', \kappa] = f[\lambda, \mu', \kappa], \mu' \neq \mu,$$

$$h[S\lambda, \mu, \theta] = h[\lambda, \mu, \theta],$$

$$h[S\lambda, \mu', \theta] = h[\lambda, \mu, \theta]^\frown\langle\lambda\rangle \quad \text{if} \quad \mu' \neq \mu \quad \text{and} \quad C_\mu \lambda$$

and

$$= h[\lambda, \mu', \theta] \text{ otherwise}$$

$$\theta > Pt[\lambda, 1], \mu' > Pt[S\lambda, 2] = \mu.$$

LEMMA. *For fixed $\mu, \theta, h(\lambda, \mu, \theta]$ is bounded as $\lambda$ increases.*

For fixed $\mu, \theta, h[\lambda, \mu, \theta]$ will be unbounded as $\lambda$ increases if and only if $C_{\mu'} \lambda'$ occurs unendingly often with $\mu > \mu' = Pt[S\lambda', 2]$ and $\theta > Pt[\lambda', 1]$, say for pairs $\{\mu_1, \theta_1\}, \{\mu_2, \theta_2\}, \ldots$. Then $h[S\lambda', \mu, \theta] = h[\lambda', \mu', \theta]^\frown\langle\lambda'\rangle$. If this is so then $C_{\mu'} \lambda'$ occurs unendingly often with $Pt[\lambda', 1] = \theta' < \theta$ and $Pt[S\lambda', 2] = \mu' < \mu$ for some fixed $\mu', \theta'$. There will then be an unending sequence of changes in $f[\lambda', \mu', \kappa]$ with $\kappa = h[\lambda', \mu', \theta']$. But for fixed $\mu, \kappa, f[\lambda', \mu, \kappa]$ can only change once, hence there must be an unending sequence of changes in $h[\lambda', \mu', \theta']$ as $\lambda'$ increases and $\theta' < \theta, \mu' < \mu$. But this is absurd if we had taken $\{\mu, \theta\}$ to be the pair earliest in the ordering of all ordered pairs for which this occurs.

Now we want to show

$$g_\mu[k[\mu, \theta]] = 0 \quad \text{iff} \quad Un_2^1[\theta, k[\mu, \theta], \mathfrak{g}^\mu[\lambda + A_1[S\lambda \dot{-} \mu], \lambda], \lambda, 1] = 0.$$

Suppose r.h.s. then

$$g[\lambda + A_1[S\lambda \dot{-} \mu], \lambda] = f[\lambda', \lambda + A_1[S\lambda \dot{-} \mu], \lambda]$$

for $\lambda'$ sufficiently large and we can take

$$Pt[\lambda', 1] = \theta, Pt[S\lambda', 2] = \mu,$$

then we shall have

$$Un_2^1[Pt[\lambda', 1], k[\mu, \theta], \mathfrak{f}[\lambda', \lambda' + A_1[S\lambda' \dot{-} \mu], \lambda'], \lambda', 1] = 0.$$

But this is $C_\mu \lambda'$ so that if $f[\lambda', \mu, \kappa] = 1$ then $f[S\lambda', \mu, \kappa] = 0$ and so $g_\mu[k[\lambda, \mu]] = 0, \kappa = k[\mu, \theta]$, as desired. If $f[\lambda', \mu, \kappa] = 0$ then again $g_\mu \kappa = 0$ as desired.

Suppose l.h.s. then condition $C_\mu \lambda$ must have arisen for some $\lambda$ for which

$$h[\lambda, \mu, \theta] = k[\mu, \theta] \quad \text{(its final value)},$$

$$Pt[\lambda, 1] = \theta, Pt[S\lambda, 2] = \mu, \kappa = k[\mu, \theta]$$

and $$Un_2^1[\theta, k[\mu,\theta], \mathfrak{f}[\lambda, \lambda + A_1[S\lambda \dot{-} \mu], \lambda], \lambda, 1] = 0.$$

Now for $\lambda' > \lambda$ this last condition still holds, for it can only fail as $\lambda$ increases by a change in the value of $f[\lambda',\mu',\nu]$ for $\mu' \neq \mu, \mu', \nu \leqslant \lambda$ because these are the only values of $f[\lambda,\mu,\nu]$ which are used in the calculation of $\Phi_\theta^{f[\lambda,\cdot,\cdot]}[k[\mu,\theta]]$, but as $\lambda'$ increases such a change can only occur if $C_{\mu'}\lambda'', \mu' \neq \mu$ arises for $\lambda < \lambda'' < \lambda'$, and if this occurs and

$$Pt[\lambda'', 1] < \theta \quad \text{and} \quad \mu < Pt[S\lambda'', 2] = \mu'$$

then $h[\lambda'', \mu, \theta]$ will change contrary to the hypothesis that it had already reached its final value $k[\mu,\theta]$. If $C_{\mu'}\lambda'', \mu' \neq \mu$ arises for $\lambda < \lambda'' < \lambda'$ and $Pt[\lambda'', 1] \neq \theta = Pt[\lambda, 1]$ or $\mu \leqslant Pt[S\lambda'', 2] = \mu'$, then this will fail to affect the values of $f[\lambda', \mu', \kappa]$ for $\mu', \kappa \leqslant \lambda$, $\mu' \neq \mu$ because $f[\lambda',\mu',\kappa]$ can only change when $\kappa = h[\lambda', \mu', \theta]$ and $\theta = Pt[\lambda', 1]$. Now $C_\mu \lambda$ has occurred hence $h[\lambda'', \mu', Pt[\lambda'', 1]] \geqslant h[S\lambda, \mu, \theta] \geqslant \lambda$, so the change in $f[\lambda', \mu', \kappa]$ if any, can only occur outside the range concerned. Hence

$$Un_2^1[\theta, k[\mu,\theta], \mathfrak{g}^\mu[\lambda + A_1(S\lambda \dot{-} \mu), \lambda], \lambda, 1] = 0,$$

as desired. This completes the demonstration of the proposition.

Let $\mathcal{N}$ be the set of natural numbers and let $<_{\mathrm{R}}$ be a recursive partial ordering of $\mathcal{N}$, i.e. $\nu <_{\mathrm{R}} \kappa$ is a recursive predicate which gives a partial ordering of $\mathcal{N}$.

Cor. *$\mathcal{N}$ with the recursive partial ordering $<_{\mathrm{R}}$ is embedable in the upper semi-lattice of recursively enumerable degrees.*

Let $\mathscr{A}_0, \mathscr{A}_1, \mathscr{A}_2 \ldots$ be a sequence of recursively enumerable independent simultaneously recursively enumerable sets. That is $A[\mu,\nu] = 0, 1$ is such that the set whose characteristic function is $\lambda x . A[\mu, x]$ is recursively enumerable and fails to be recursive in $\lambda x x' . A[x + A_1[Sx \dot{-} \mu], x']$. From these we can form a sequence of recursively independent, disjoint, simultaneously enumerable sets, for instance let $\mathscr{B}_\mu$ be the set of ordered pairs $\{\nu, \mu\}$ such that $\nu \epsilon \mathscr{A}_\mu$ having characteristic function $\lambda x B[\mu, x]$. Let $\mathscr{C}_\mu$ be the union of the sets $\mathscr{B}_\kappa$ with $\kappa \leqslant_{\mathrm{R}} \mu$, then $\mathscr{C}_\mu$ is recursively enumerable because the $\mathscr{B}_\mu$'s are simultaneously enumerable and $\leqslant_{\mathrm{R}}$ is recursive (in fact the $\mathscr{C}$'s are simultaneously recursively enumerable

$$\nu \epsilon \mathscr{C}_\mu \leftrightarrow (E\xi, \eta)(\nu = \{\eta, \xi\} \; \& \; A[\xi, \eta] = 0 \; \& \; \xi \leqslant_{\mathrm{R}} \mu)).$$

We will now show that $\mathscr{C}_\nu$ is recursive in $\mathscr{C}_\mu$ if and only if $\nu \leqslant_{\mathrm{R}} \mu$, it will then be clear that $\mathcal{N}$ with the partial ordering $\leqslant_R$ is embedable in the upper semi-lattice of recursively enumerable degrees.

Suppose $\nu \nleqslant_R \mu$, then $\mathscr{B}_\nu \cap \mathscr{C}_\mu = \emptyset$, and $\mathscr{C}_\mu$ is recursive in

$$\boldsymbol{\lambda} xx'.B[x+A_1[Sx \dot{-} \nu], x']$$

since $$\xi \epsilon C_\mu \leftrightarrow (E\zeta, \eta)(\xi = \{\eta, \zeta\} \,\&\, \zeta \leqslant_R \mu \,\&\, \xi \epsilon \mathscr{B}_\zeta).$$

It follows from the recursive independence of $\mathscr{B}_0, \mathscr{B}_1, \mathscr{B}_2, \ldots$ that $\mathscr{B}_\nu$ fails to be recursive in $\mathscr{C}_\mu$. But then $\mathscr{C}_\nu$ fails to be recursive in $\mathscr{C}_\mu$, since $\mathscr{B}_\nu$ is recursive in $\mathscr{C}_\nu$.

Suppose $\nu \leqslant_R \mu$. Then $\mathscr{C}_\nu \subseteq \mathscr{C}_\mu$ and $\mathscr{C}_\nu$ is recursive in $\mathscr{C}_\mu$ since

$$\xi \epsilon C_\nu \leftrightarrow (E\zeta, \eta)(\xi = \{\eta, \zeta\} \,\&\, \zeta \leqslant_R \nu \,\&\, \xi \epsilon \mathscr{B}_\zeta)$$
$$\leftrightarrow (E\zeta, \eta)(\xi = \{\eta, \zeta\} \,\&\, \zeta \leqslant_R \nu \,\&\, \xi \epsilon \mathscr{C}_\mu),$$

the last line follows since the $B$'s are disjoint.

## 7.22 *Non-recursively separable recursively enumerable sets*

PROP. 25. *We can find two recursively enumerable sets of natural numbers which fail to be separable by a recursive set.*

We shall show: recursively enumerable sets $\mathscr{A}$ and $\mathscr{B}$ can be $\mathbf{A}_0$-defined such that if $\mathscr{A} \subseteq \mathscr{C}$ and $\mathscr{B} \subseteq \mathscr{D}$ where $\mathscr{C}$ and $\mathscr{D}$ are recursively enumerable then we can find a numeral $\theta$ such that $\theta$ fails to be in the union of $\mathscr{C}$ and $\mathscr{D}$. Hence $\mathscr{C} \cup \mathscr{D}$ fails to be the universal set of natural numbers and so $\mathscr{C}$ and $\mathscr{D}$ fail to be recursive.

Let $\mathscr{A}$ be $\hat{x}(Ex')\,U[x,x']$ and $\mathscr{B}$ be $\hat{x}(Ex')\,V[x,x']$, where $U[x,x']$ for

$$Un'[Pt[x,2], x_1', x_2', x] = 0 \,\&\, (Ax'')(x'' < x' \rightarrow Un'[Pt[x,1], x_1'', x_2'', x] \neq 0)$$

and $x' = \{x_1', x_2'\}$ and $V[x,x']$ for

$$Un'[Pt[x,1], x_1', x_2', x] = 0 \,\&\, (Ax'')(x'' < x' \rightarrow Un'[Pt[x,2], x_1'', x_2'', x] \neq 0).$$

$U$ and $V$ are primitive recursive, hence $\mathscr{A}$ and $\mathscr{B}$ are recursively enumerable sets. From $U[x,x'] \,\&\, V[x,x'']$ we get $x' < x'' \,\&\, x'' < x'$, hence $((Ex')\,U[x,x'] \,\&\, (Ex')\,V[x,x'])$ fails, thus $\mathscr{A}$ and $\mathscr{B}$ are disjoint.

Now consider two disjoint recursively enumerable sets $\mathscr{C}$, $\mathscr{D}$ with $\mathscr{A} \subseteq \mathscr{C} \,\&\, \mathscr{B} \subseteq \mathscr{D}$. Let them be $\hat{x}(Ex')\,R\{x,x'\}$, $\hat{x}(Ex')\,S\{x,x'\}$ respectively, where $R$ and $S$ are primitive recursive. We now find a numeral which is in $\mathscr{C} \cap \overline{\mathscr{D}}$. We have

$$(Ex')\, R\{x, x'\} \leftrightarrow (Ex')\,(Un'[\lambda, x_1', x_2', x] = 0),$$

$$(Ex')\, S\{x, x'\} \leftrightarrow (Ex')\,(Un'[\mu, x_1', x_2', x] = 0),$$

for some $\lambda$ and $\mu$. Let $\theta = \{\lambda, \mu\}$, assume that $\theta$ is in $\mathscr{C}$, i.e. $(Ex')\, R\{\theta, x'\}$, since $\mathscr{C}$ and $\mathscr{D}$ are disjoint then $\theta$ is in $\overline{\mathscr{D}}$, thus $(Ex)\,(Un'[\lambda, x_1', x_2', \theta] = 0)$, but $(Ex')\, S\{\theta, x'\}$ fails, whence $(Ax')\,[Un[\mu, x_1', x_2', \theta] \neq 0]$, hence

$$(Ex')\,(Un[\lambda, x_1'', x_2'', \theta] = 0)\;\&\;(Ax'')_{x'}\,[(Un[\mu, x_1'', x_2'', \theta] \neq 0)],$$

i.e. $(Ex')\, V[\theta, x']$, i.e. $\theta$ is in $\mathscr{B}$ whence $\theta$ is in $\mathscr{D}$ which is absurd, hence $\theta$ is in $\overline{\mathscr{C}}$. Similarly $\theta$ is in $\overline{\mathscr{D}}$.

## 7.23 *Cohesive sets*

A set of natural numbers is *cohesive* if it is unbounded and if either its intersection with a recursively enumerable set is bounded or its intersection with the complement of that recursively enumerable set is bounded. Thus a set $\mathscr{A}$ is cohesive if it is unbounded and if for each recursively enumerable set $\mathscr{B}$ either $\mathscr{A} \cap \mathscr{B}$ or $\mathscr{A} \cap \overline{\mathscr{B}}$ is bounded.

A set of natural numbers is *maximal* if it is unbounded and recursively enumerable and its complement is cohesive.

Prop. 26. *Every unbounded set of natural numbers possesses a cohesive subset.* •

Let $\mathscr{A}$ be an unbounded set of natural numbers. Define a sequence of subsets of $\mathscr{A}$ as follows

$$\mathscr{A}_0 = \mathscr{A},$$

$$\mathscr{A}_{\nu+1} = \begin{cases} \mathscr{A}_\nu \cap \mathscr{X}_\nu & \text{if this is unbounded,} \\ \mathscr{A}_\nu \cap \overline{\mathscr{X}}_\nu & \text{otherwise.} \end{cases}$$

Then $\mathscr{A}_0 \supseteq \mathscr{A}_1 \supseteq \mathscr{A}_2 \supseteq \ldots$.

Assume that $\mathscr{A}$ is without a cohesive subset. Then for each $\nu$ there is a $\mu \geqslant \nu$ such that $\mathscr{X}_\mu$ divides $\mathscr{A}_\nu$ into two unbounded subsets, otherwise $\mathscr{A}_\nu$ would be a cohesive subset of $\mathscr{A}$. Hence the sequence $\mathscr{A}_0, \mathscr{A}_1, \ldots$ contains a strictly decreasing subsequence $\mathscr{B}_0, \mathscr{B}_1, \ldots$ otherwise $\mathscr{A}$ would be constant from $\nu$ onwards contrary to being divided into two unbounded subsets by $\mathscr{X}_\mu$. Thus we can define a sequence of numerals as follows

$$\nu_\theta = \mu\pi[\pi\epsilon\mathscr{B}_{\bar{\theta}} - \mathscr{B}_{\theta+1}].$$

Denote this sequence by $\mathscr{C}$. By our construction, $\mathscr{A}_{\nu+1} \subseteq \mathscr{X}_\nu$ or $\mathscr{A}_{\nu+1} \subseteq \bar{\mathscr{X}}_\nu$. Hence $\mathscr{B}_{\nu+1} \subseteq \mathscr{X}_\nu$ or $\mathscr{B}_{\nu+1} \subseteq \bar{\mathscr{X}}_\nu$. Since all but a bounded number of members of $\mathscr{C}$ must lie in $\mathscr{B}_{\nu+1}$, either $\mathscr{C} \cap \mathscr{X}_\nu$ or $\mathscr{C} \cap \bar{\mathscr{X}}_\nu$ is bounded. This holds for all $\nu$, and thus $\mathscr{C}$ is a cohesive subset of $\mathscr{A}$ which is absurd.

## 7.24 *Maximal sets*

PROP. 27. *A maximal set can be* $\mathbf{A}_0$*-defined.*

We describe a procedure for enumerating a set of natural numbers $\mathscr{A}$ by stages, and then we show that $\mathscr{A}$ is maximal.

$\mathscr{X}_\nu^\mu$ is the set of natural numbers produced by table $\nu$ after $\mu$ steps of the following:

Do one step in the calculation by table $\nu$ with input 0, then do one step in the calculation by table $\nu$ with input 1 then do the second step in the calculation by table $\nu$ with input 0, etc., every time an output appears write it down in a list.

For any $\pi$, any stage $S\nu$, and any numeral $\theta$, we define the $\pi$-state of $\theta$ at stage $S\nu$ to be $\epsilon_0\epsilon_1\epsilon_2 \ldots \epsilon_\pi$, where

$$\epsilon_\lambda = \begin{cases} 1 & \text{if} \quad \theta\epsilon\mathscr{X}_\lambda^{S\nu} \quad \text{and} \quad \lambda \leqslant \nu, \\ 0 & \text{if} \quad \theta\bar{\epsilon}\mathscr{X}_\lambda^{S\nu} \quad \text{or} \quad \nu < \lambda, \end{cases}$$

for $0 \leqslant \lambda \leqslant \pi$.

Clearly there are only $2^{S\pi}$ $\pi$-states. We order them lexicographically $(0 < 1)$.

Note the following two properties of this ordering:

(i) for fixed $\pi$ and $\theta$, and for $\nu < \mu$, the $\pi$-state of $\theta$ at stage $\mu$ must be at least as high as the $\pi$-state of $\theta$ at stage $\nu$. $\epsilon_\lambda^{(\nu)} \leqslant \epsilon_\lambda^{(S\nu)}$;

(ii) for fixed stage $\nu$, for given $\theta$ and $\theta'$, and for $\pi < \pi'$ if the $\pi$-stage of $\theta'$ is higher than the $\pi$-stage of $\theta$ then the $\pi'$-state of $\theta'$ is higher than the $\pi'$-state of $\theta$.

To enumerate the set $\mathscr{A}$ we proceed as follows: We begin with a list of numerals each associated with a marker, $\nu$ is associated with the marker $\nu^*$. We then move these markers down the list of numerals according to certain rules. If at any time a numeral loses a marker then thereafter it remains markerless. The set $\mathscr{A}$ will consist of the numerals which are ultimately markerless.

*Stage* $S\nu$. For each $\mu$, let $\mu^{(\nu)}$ be the position of marker $\mu^*$ at the end of stage $\nu$. Compute $\mathscr{X}_\lambda^{S\nu}$ for all $\lambda \leqslant \nu$. Find the least numeral $\mu_0$, if any, such that for some $\mu > \mu_0$, $\mu^{(\nu)}$ is in a higher $\mu_0$- state than $\mu_0^{(\nu)}$ at stage $S\nu$. If this search is unsuccessful then pass on to stage $(\nu+2)$. The search terminates because if $\theta > \nu$ then $\epsilon_\lambda = 0$ since if $\theta$ occurs as output it will require at least $\theta$ steps. If the search is successful, let $\mu_1$ be the least $\mu$ such that $\mu^{(\nu)}$ is in a higher $\mu_0$-state than $\mu_0^{(\nu)}$ at stage $S\nu$. Move marker $\mu_0^*$ down to $\mu_1^{(\nu)}$, and for each $\mu > \mu_0$, move marker $\mu^*$ down to $(\mu+\mu_1-\mu_0)^{(\nu)}$. Place in $\mathscr{A}$ all the numerals $\mu$ such that $\mu_0^{(\nu)} \leqslant \mu < \mu_1^{(\nu)}$ that are not yet in $\mathscr{A}$. (These are the only numerals to lose markers in stage $S\nu$.) Then go on to stage $(\nu+2)$.

We now show that $\mathscr{A}$ is maximal. Clearly $\mathscr{A}$ is recursively enumerable.

LEMMA (i). *$\bar{A}$ is unbounded.*

It suffices to show that each marker ultimately comes to rest, then there will be an unbounded set of numerals which ultimately retain markers and so belong to $\bar{\mathscr{A}}$. Assume otherwise. Let $\mu_0^*$ be the least marker which moves unendingly often. By construction and fundamental property (i) of $\pi$-states, after $(\mu_0-1)^*$ reaches its final position $\mu_0^*$ must move to positions of higher and higher $\mu_0$-states. But there is only a bounded number of different $\mu_0$-states.

LEMMA (ii). *For every $\pi$, either $\mathscr{X}_\pi \cap \bar{\mathscr{A}}$ is bounded or $\bar{\mathscr{X}}_\pi \cap \bar{\mathscr{A}}$ is bounded.*

Fix $\pi$. For each numeral $\theta$ in $\bar{\mathscr{A}}$, $\theta$ must reach a final $\pi$-state $\beta$ as stage $\nu$ increases. We say that $\theta$ terminates in $\beta$. Since $\bar{\mathscr{A}}$ is unbounded, at least one $\pi$-state $\beta$ must be associated with an unbounded number of members of $\bar{\mathscr{A}}$ which terminate in $\beta$. We show that exactly one $\pi$-state is associated with an unbounded number of members of $\bar{\mathscr{A}}$. Assume otherwise, let $\beta$ be the lowest $\pi$-state associated with an unbounded number of members of $\bar{\mathscr{A}}$, and let $\beta'$ be another $\pi$-state associated with an unbounded number of members of $\bar{\mathscr{A}}$. It follows that there must be numerals $\mu$, $\nu$, $\theta$, $\theta'$ such that $\pi < \mu < \nu$, $\theta$ is the final position of marker $\mu^*$ and $\theta'$ is the final position of marker $\nu^*$, $\theta$ terminates in the $\pi$-state $\beta$ and $\theta'$ terminates in the $\pi$-state $\beta'$. By fundamental property (ii) of $\pi$-states $\theta'$ reaches and terminates in a higher $\mu$-state than $\theta$. But this means, by construction, that marker $\mu^*$ must be moved, which is absurd.

Thus all but a bounded set of members of $\bar{\mathscr{A}}$ (final positions of markers) have $\pi$-states different from a certain $\pi$-state $\epsilon_0 \epsilon_1 \epsilon_2 \ldots \epsilon_\pi$. If $\theta \epsilon \bar{\mathscr{A}}$ and $\epsilon_\pi = 1$ then $\theta \epsilon \mathscr{X}_\pi$, if $\theta \epsilon \bar{\mathscr{A}}$ and $\epsilon_\pi = 0$ then $\theta \epsilon \bar{\mathscr{X}}_\pi$. Thus all but a bounded number of members of $\bar{\mathscr{A}}$ are in $\mathscr{X}_\pi$, or all but a bounded number of members of $\bar{\mathscr{A}}$ are in $\bar{\mathscr{X}}_\pi$. This completes the demonstration of the proposition.

In this construction marker $\mu^*$ moves to positions of higher and higher $\mu$-states. When it comes to rest all markers further down the list have states at most as high as that of marker $\mu^*$. This gives the crucial property of there being only one $\pi$-state which is associated with an unbounded number of members of $\bar{\mathscr{A}}$. At each stage we move the least marker $\mu$ for which there is a later marker of higher $\mu$-state. Ultimately each marker comes to rest, so, as the construction proceeds we treat all markers in turn, some possibly several times.

## 7.25 *Minimal degrees*

PROP. 28. *There is a non-recursive degree such that any degree of lower unsolvability is the solvable degree.*

It suffices to show that there is a non-recursive set $\mathscr{A}$ such that if a set $\mathscr{B}$ is recursive in $\mathscr{A}$ then either $\mathscr{B}$ is recursive or $\mathscr{A}$ is recursive in $\mathscr{B}$.

We construct the characteristic function $g$ of a set $\mathscr{A}$ by stages, each stage will ensure that one of the requirements is met by one table. In toto each table will satisfy all the requirements.

For each natural number $\nu$, we shall have, at the end of stage $\nu$, three functions $h^\nu, f_1^\nu, f_2^\nu$, such that the following conditions are satisfied:

For each natural number $\nu$, $h^\nu, f_1^\nu, f_2^\nu$ are recursive, and $f_1^\nu, f_2^\nu$ are characteristic functions.

For each natural number $\nu$, $h^\nu$ is a strictly monotonic increasing function. $f_1^\nu \mu = f_2^\nu \mu$ for $\mu < h^\nu 0$, and for every $\lambda$, $f_1^\nu \mu \neq f_2^\nu \mu$ for some $\mu$, $h^\nu \lambda \leqslant \mu < h^\nu(S\lambda)$. We call these intervals, the intervals determined by $h^\nu$. Thus for all natural numbers $\nu$, $f_1^\nu$ and $f_2^\nu$ are identical on the first interval determined by $h^\nu$, and differ on every other interval determined by $h^\nu$.

For all $\mu < \nu$ the range of $h^\mu$ contains the range of $h^\nu$, so that the intervals determined by $h^\nu$ are unions of bounded sets of consecutive intervals determined by $h^\mu$.

In each interval determined by $h^\mu$, $f_1^\nu$ is identical with either $f_1^\mu$ or with $f_2^\mu$, and similarly for $f_2^\nu$. (In particular all four functions $f_1^\mu, f_2^\mu, f_1^\nu, f_2^\nu$ are identical on $[0, h^\mu 0)$, i.e. the interval including 0 but excluding $h^\mu 0$. Finally for all $\mu < \nu$, $h^\mu 0 < h^\nu 0$. These are all the conditions.

Suppose that we have defined $h^\nu, f_1^\nu, f_2^\nu$ for all $\nu$. We define $g_\nu$ to be the initial segment of length $h^\nu 0$ such that $g_\nu \mu = f_1^\nu \mu = f_2^\nu \mu$, for $0 \leqslant \mu < h^\nu 0$.

From the conditions we see that $g_0, g_1, \ldots$ is a chain of initial segments whose union is everywhere defined. Let $g = \bigcup_\nu g_\nu$.

Note that for any given $\nu$ and any interval determined by $h^\nu$, either $g$ is identical with $f_1^\nu$ or $g$ is identical with $f_2^\nu$ on that interval. $g$ is the characteristic function of the desired set $\mathscr{A}$.

It now remains to define $h^\nu, f_1^\nu, f_2^\nu$ and to show that $g$ has the required properties. Note that the functions $h^\nu, f_1^\nu, f_2^\nu$ are recursive for each $\nu$ (but non-recursive in $\nu$).

*Stage* 0. Let $h^0 = \lambda x . x, f_1^0 = \lambda x . 0, f_2^0 = \lambda x . 1$.

Note that the conditions are trivially satisfied. The intervals determined by $h^0$ are all unit intervals.

*Stage* $2\nu+1$. Let $\mu = 2\nu$. By induction the conditions are satisfied for $h^\mu$, $f_1^\mu, f_2^\mu$. Then $f_1^\mu$ and $f_2^\mu$ differ on the interval $[h^\mu 0, h^\mu 1)$. See whether, on this interval $\Phi_\nu \neq f_1^\mu$. If so define $h^{S\mu} = \lambda x . h^\mu(Sx)$, $f_1^{S\mu} = f_2^{S\mu} = f_1^\mu$ on $[0, h^{S\mu}0)$, $(h^{S\mu}0 = h^\mu 1)$, and let $f_1^{S\mu}, f_2^{S\mu}$ be otherwise unchanged.

Otherwise, then $\Phi_\nu \neq f_2^\mu$ in the interval $[h^\mu 0, h^\mu 1)$ since $f_1^\mu, f_2^\mu$ differ in that interval, we define $h^{S\mu}, f_1^\mu, f_2^\mu$ as before except that $f_1^{S\mu} = f_2^{S\mu} = f_2^\mu$ on $[0, h^{S\mu}0)$.

It follows that $h^{S\mu}$, $f_1^\mu$, $f_2^\mu$ satisfy the conditions. This stage ensures that $g \neq \Phi_\nu$, hence $g$ will be non-recursive.

*Stage* $2\nu+2$. Let $\mu = 2\nu+1$. By induction the conditions are satisfied for $h^\mu, f_1^\mu, f_2^\mu$. Call $\tilde{g}$ an available segment at $\mu$ if $\tilde{g}$ is defined on the segment $[0, h^\mu\theta)$ for some $\theta$, and if, on each interval determined by $h^\mu$ in this segment $\tilde{g}$ is identical either with $f_1^\mu$ or with $f_2^\mu$ in that interval. If $\tilde{g}$ is an available segment at $\mu$, we call $g^*$ an available extension of $\tilde{g}$ if $g^*$ is an available segment at $\mu$ and is an extension of $\tilde{g}$. In the interval $[0, h^\mu 0)$ $\tilde{g}$ can have only one value because $f_1^\mu$ and $f_2^\mu$ are identical on that interval.

*Substage a.* See whether there exists a natural number $\kappa$ and an available extension at $\mu$, $\tilde{g}$ such that for all available extensions at $\mu$, $g^*$ of $\tilde{g}$, $\Phi_\nu^{[g^*]}\kappa$

is undefined. Recall that $\Phi_\nu^{[f]}\kappa$ is defined if and only if the oracle is only required to consult arguments for which $f$ is defined and gives an output. So that if $f^*$ is an extension of $f$ then $\Phi_\nu^{[f]}\kappa = \Phi_\nu^{[f^*]}\kappa$, whenever the left-hand member is defined. If so, define $h^{S\mu} = \lambda x.h\,(x+\theta)$, where $[0, h^\mu\theta)$ is the segment of $g$, define $f_1^{S\mu} = f_2^{S\mu} = g$ on $[0, h^\mu\theta)$, $f_1^{S\mu}, f_2^{S\mu}$ unchanged otherwise, then go on to stage $2\nu+3$. This ensures that $\Phi_\nu^g$ is undefined somewhere, namely at $\kappa$. Otherwise go to

*Substage b*. See whether there is some available segment at $\mu$, $\tilde{g}$ such that for all $\kappa$, $\Phi_\nu^{[g^*]}\kappa$ takes at most one value as $g^*$ varies over all available extensions at $\mu$ of $\tilde{g}$. Now $\tilde{g}$ is an available extension of $\tilde{g}$, so this means that $\Phi_\nu^{[\tilde{g}]}\kappa$ is defined for all $\kappa$. If so, define $h^{S\mu}, f_1^{S\mu}, f_2^{S\mu}$ from $\tilde{g}$ as in the last part of substage $a$, and go on to stage $2\nu+3$. The conditions hold, and we have ensured that $\Phi_\nu^g$ must be recursive if it is total. In fact to calculate $\Phi_\nu^g$ we need only calculate $\Phi_\nu^{[\tilde{g}]}$. Otherwise go to

*Substage c*. If this substage is reached, we know that for every available segment $\tilde{g}$ at $\mu$ (i) for every $\kappa$, there exists an available extension $g^*$ such that $\Phi_\nu^{[g^*]}\kappa$ is defined (substage $a$) and (ii) there exist $\kappa$ and available extensions $g_1^*$, $g_2^*$ such that $\Phi_\nu^{[g^*]}\kappa$ and $\Phi_\nu^{[g^*_2]}\kappa$ are defined and unequal (substage $b$). We now define
$$h^{S\mu}0 = h^\mu 1,$$
$$f_1^{S\mu} = f_2^{S\mu} = f_1^\mu \quad \text{on} \quad [0, h^{S\mu}0), \quad h^{S\mu}0 = h^\mu 1.$$

In this case $\Phi_\nu^{[\tilde{g}]}\kappa$ is undefined, otherwise, $\Phi_\nu^{[g^*_1]}\kappa$ and $\Phi_\nu^{[g^*_2]}\kappa$ would be the same. This part of the definition ensures that $h^{S\mu}0 > h^\mu 0$. (We could have taken $f_2^\mu$ instead of $f_1^\mu$.)

Assume that $h^{S\mu}\lambda$ has been defined for all $\lambda \leqslant \theta$, and that $f_1^{S\mu}$, $f_2^{S\mu}$ have been defined on $[0, h^{S\mu}0)$. $h^{S\mu}\theta$ determines $S\theta$ intervals on $[0, h^{S\mu}\theta)$, hence there will be $2^\theta$ different segments of length $h^{S\mu}\theta$ that will be available at $S\mu$. (Note that on $[0, h^{S\mu}0)$ the functions $f_1^{S\mu}, f_2^{S\mu}$ are identical.) Denote these segments by $\tilde{g}_1, \tilde{g}_2, \ldots, \tilde{g}_{2^\theta}$ in some order, say lexicographic. Note that however $g$ is eventually defined one of these segments will be an initial segment of $g$. We proceed through $2^\theta$ subsets in order to reach our definitions of $h^{S\mu}(S\theta)$, $f_1^{S\mu}$ and $f_2^{S\mu}$ on $[h^{S\mu}\theta, h^{S\mu}(S\theta))$. Then repeat with $S\theta$ for $\theta$ and so on.

*Substep* 1. By remark (ii) above there exist available at $\mu$ extensions $s_1$ and $t_1$ of segment $\tilde{g}_1$ and there exists a $\kappa_1$ such that $\Phi_{\nu}^{[s_1]}\kappa$ and $\Phi_{\nu_1}^{[t_1]}\kappa$

are defined and unequal; furthermore $s_1$ and $t_1$ can be taken to be of equal length. $s_1, t_1$ and $\kappa_1$ can be found effectively. We can recursively enumerate the triplets $\langle s_1, t_1, \kappa_1\rangle$; so run through these triplets until we come to one that satisfies, we know that there is such a one if substage $c$ is reached. Define

$$\tilde{g}_1^\frown u_1 = s_1, \quad \tilde{g}_1^\frown v_1 = t_1.$$

*Substep $S\lambda$.* By remark (ii), there exist available at $\mu$ extensions $s'$ and $s''$ of $\tilde{g}_{S\lambda} \cup u_\lambda$ and there exists a $\kappa_{S\lambda}$ such that $\Phi_\nu^{[s']}\kappa_{S\lambda}$ and $\Phi_\nu^{[s'']}\kappa_{S\lambda}$ are defined and unequal. By remark (i), there exists an available extension $t$ of $\tilde{g}_{S\lambda} \cup v_\lambda$ such that $\Phi_\nu^{[t]}\kappa_{S\lambda}$ is defined. Hence we can get $s_{S\lambda}$ and $t_{S\lambda}$ so that $s_{S\lambda}$ is an available extension of $\tilde{g}_{S\lambda} \cup u_\lambda$ and $t_{S\lambda}$ is an available extension of $\tilde{g}_{S\lambda} \cup v_\lambda$ and $\Phi_\nu^{[s_{S\lambda}]}\kappa_{S\lambda}$ and $\Phi_\nu^{[t_{S\lambda}]}\kappa_{S\lambda}$ are defined and unequal; furthermore $s_{S\lambda}$ and $t_{S\lambda}$ can be taken to be of equal length. As at substep 1, $\kappa_{S\lambda}, s_{S\lambda}, t_{S\lambda}$ can be found effectively. Define $\tilde{g}_{S\lambda}^\frown u_{S\lambda} = s_{S\lambda}$, $\tilde{g}_{S\lambda}^\frown v_{S\lambda} = t_{S\lambda}$. Observe that $u_{S\lambda}$ is an extension of $u_\lambda$ and similarly for $v_{S\lambda}$.

Let $\kappa$ be the least natural number greater than all members of the domain of $u_{2^\theta}$, (it is easily seen that the domain of $u_{2^\theta}$ is non-null). We define

$$h^{S\mu}(S\theta) = \kappa,$$

$$f_1^{S\mu} = u_{2^\theta} \quad \text{on} \quad [h^{S\mu}\theta, h^{S\mu}(S\theta)),$$

$$f_2^{S\mu} = v_{2^\theta} \quad \text{on} \quad [h^{S\mu}\theta, h^{S\mu}(S\theta)).$$

This completes the definition of $h^{S\mu}, f_1^{S\mu}, f_2^{S\mu}$.

It is easily seen that the conditions hold. Furthermore if substage $c$ is reached, then $g$ is recursive in $\Phi_\nu^g$. For we can decide $g$ if we can decide, for each interval $[h^{S\mu}\theta, h^{S\mu}(S\theta))$, $\theta = 0, 1, \ldots$ whether $g$ agrees with $f_1^{S\mu}$ or with $f_2^{S\mu}$ on that interval (see note $a$). Assume that $g$ has been computed on $[0, h^{S\mu}\theta)$ then we can find $\lambda \leqslant \theta$ such that $\tilde{g}_\lambda \subset g$ and we can compute $\Phi_\nu^{[s_\lambda]}\kappa_\lambda$ and $\Phi_\nu^{[t_\lambda]}\kappa_\lambda$. One and only one of these must agree with $\Phi_\nu^g\kappa_\lambda$, because by construction of $f_1^{S\mu}$ and $f_2^{S\mu}$ either $s_\lambda \subset g$ or $t_\lambda \subset g$ and in the computation of $\Phi_\nu^{[s_\lambda]}\kappa_\lambda$ only arguments for which $s_\lambda$ is defined are used. If the former agrees then $g$ must agree with $f_1^{S\mu}$ on the interval; otherwise $g$ must agree with $f_2^{S\mu}$ on the interval in question. This completes stage $2\nu+2$.

Let $g$ be the characteristic function of $\mathscr{A}$ It remains to show that $\mathscr{A}$ has the desired properties. First $\mathscr{A}$ is non-recursive; otherwise $g = \Phi_{\scriptscriptstyle\prime}$

for some $\nu$, contrary to the construction at stage $2\nu+1$. Assume that $\mathscr{B}$ is recursive in $\mathscr{A}$; that is $c_{\mathscr{B}} = \Phi_\nu^g$ for some $\nu$. If stage $2\nu+2$ terminates in substage $b$, then $\Phi_\nu^g$ is recursive, hence $\mathscr{B}$ is recursive. If substage $c$ is used then $g$ is recursive in $\Phi_\nu^g$ and so $\mathscr{A}$ is recursive in $\mathscr{B}$. This completes the demonstration of the proposition.

## 7.26 *Degrees of theories*

PROP. 29. *For every set $\mathscr{A}$ there exists a theory $\mathscr{T}_{\mathscr{A}}$ of the same degree as $\mathscr{A}$; furthermore, if $\mathscr{A}$ is recursively enumerable, then $\mathscr{T}_{\mathscr{A}}$ is axiomatizable.*

By the degree of a theory we mean the degree of the set of the *g.n.*'s of the theorems in the theory. We shall find the required theories in the pure calculus of identity $\mathscr{I}d$; this is an applied predicate calculus with identity as sole predicate. Clearly the truth or falsity of a statement in this calculus depends solely on the cardinal number of members in a model. We consider only non-empty models. We denote by $\mathscr{S}p\nu$ the class of cardinal numbers for which the statement $\phi$ whose *g.n.* is $\nu$ is true in a model of that cardinality. We call this class the spectrum of $\phi$. For instance the spectrum of $(Ax)(Ey)(x \neq y)$ is $\hat{x}\{2 \leqslant x\}$. We list some properties of spectra.

(i) Every spectrum is either bounded or its complement is bounded.

(ii) There is a uniform effective procedure for going from a statement to its spectrum, i.e. recursive functions $f$ and $g$ can be found such that for a statement whose *g.n.* is $\nu$, if $\mathscr{S}_p\nu$ is bounded then $g[\nu] = 0$ and $f[\nu]$ gives the spectrum in the ordering of all tuplets, and if the complement of the spectrum is bounded then $g[\nu] = 1$ and $f[\nu]$ gives the spectrum of the complement.

(iii) For every statement $\phi$, $\phi$ is true for some unbounded cardinal if and only if $\phi$ is true for every unbounded cardinal if and only if the complement of the spectrum of $\phi$ is bounded.

(iv) Every bounded set of natural numbers can occur as a spectrum, and so can any set of natural numbers whose complement is bounded.

(v) There is a uniform effective procedure for going from a given set of natural numbers which is bounded or whose complement is bounded to a statement in $\mathscr{I}d$ having that set as spectrum.

Ad (i) and (ii). To test if $\nu$ is in the spectrum of $\phi$ we replace universal quantifiers by conjunctions over $1, \ldots, \nu$ and existential quantifiers by disjunctions, thus

$$(A\xi)\,\phi\{\xi\} \quad \text{is replaced by} \quad \prod_{\theta=1}^{\nu} \phi\{\theta\},$$

$$(E\xi)\,\phi\{\xi\} \quad \text{is replaced by} \quad \sum_{\theta=1}^{\nu} \phi\{\theta\}.$$

Now use truth tables. This gives a uniform effective method of testing whether $\nu$ is in the spectrum. If there are $\nu$ quantifiers and if $\nu < \mu$ then the table for $\mu$ will give the same result as the table for $\nu$. We demonstrate this by induction on the construction of $\phi$. If the matrix of $\phi$ is atomic then it is $x = x$, for the case of one quantifier or is $x = y$ for the case of two quantifiers, the result follows at once. To deal with negation it is simplest to take $\neq$ as atomic as we did in the system $\mathbf{A}_{00}$, etc. The result follows at once. Let $\phi$ be $\phi' \vee \phi''$ and suppose the result holds for $\phi'$ and for $\phi''$. Then if $\phi' \vee \phi''$ has $\nu$ in its spectrum, $\nu$ must also be in the spectrum of $\phi'$ or of $\phi''$, whence $\mu$ is in the spectrum of $\phi'$ or $\phi''$. If $\phi$ has $\nu$ quantifiers then $\phi'$ and $\phi''$ have fewer than $\nu$ quantifiers, but by adding dummies they may be taken to have $\nu$ quantifiers, and so the induction hypothesis holds for them. Similarly for the case when $\phi$ is a conjunction. If $\phi$ is an existential statement $(E\xi)\,\psi\{\xi\}$ then our induction hypothesis is that the result holds for each of $\psi\{1\}, \psi\{2\}, \ldots$, if $\phi$ has $\nu$ quantifiers and $\nu$ is in the spectrum of $\phi$ then $\nu$ will be in the spectrum of one of

$$\psi\{1\}, \psi\{2\}, \ldots, \psi\{\nu\},$$

but these have one fewer quantifier than $\phi$, but by adding a dummy may be considered to have $\nu$ quantifiers. Thus one of $\psi\{1\}$, $\psi\{2\}, \ldots, \psi\{\nu\}$, by our induction hypothesis, has $\mu$ in its spectrum. Thus $\phi$ has $\mu$ in its spectrum. Similarly for the case when $\phi$ is a universal statement.

Now to find the spectrum of $\phi$ with $\nu$ quantifiers we need only test whether $1, 2, \ldots, \nu$ are in the spectrum of $\phi$, if $\nu$ is in the spectrum of $\phi$ then so is $\mu$ where $\nu < \mu$. If $\nu$ fails to be in the spectrum of $\phi$ then it is in the spectrum of $\sim \phi$ which has the same number of quantifiers as $\phi$. Thus $\mu > \nu$ is in the spectrum of $\sim \phi$ and so fails to be in the spectrum of $\phi$. Thus (i) and (ii).

Ad (iii). See above discussion on number of quantifiers.

Ad (iv). Let $F_\nu$ stand for

$$(Ex', x'', \ldots, x^{(\nu)}) \prod_{1 \leqslant \theta' < \theta < \nu} (x^{(\theta)} \neq x^{(\theta')})$$

and let $G_\nu$ stand for $F_\nu$ and $\sim F_{S\nu}$. Then

$$G_{\nu'} \vee G_{\nu''} \vee \ldots \vee G_{\nu^{(\kappa)}}$$

has spectrum the bounded set $\langle \nu', \nu'', \ldots, \nu^{(\kappa)} \rangle$, its negation will have the complement of this bounded set in its spectrum.

By $\mathscr{F}_C$ we see that if we take a set $\mathscr{B}$ of statements of $\mathscr{I}d$ as axioms and if $\psi$ is deducible from this set of axioms then spectrum $\psi \supset \bigcap_{\phi\epsilon\mathscr{B}}$ spectrum $\phi$. The completeness of $\mathscr{F}_C$ implies that if $\psi$ is true for all models for which the axioms are true then $\psi$ can be deduced from the axioms.

Now let $\mathscr{A}$ be any set of natural numbers. We can assume without loss of generality, that $0\epsilon\mathscr{A}$. Let $G_0$ be $(Ax)(x \neq x)$. Then $\sim G_\nu$ has $\mathscr{N}$-$\{0, \nu\}$ as its spectrum. Let $\mathscr{B}$ be the set of the negations of $G_\nu$ for $\nu\epsilon\mathscr{A}$. Let $\mathscr{T}_\mathscr{A}$ be the set of statements deducible from $\mathscr{B}$. Then '$\nu$'$\epsilon\mathscr{T}_\mathscr{A}$ if and only if '$\nu$' is deducible from $\mathscr{B}$ if and only if spectrum '$\nu$' $\supset \bigcap_{\psi\epsilon\mathscr{B}}$ spectrum $\psi$. But $\bigcap_{\psi\epsilon\mathscr{B}}$ spectrum $\psi$ is the complement of $\mathscr{A}$. Therefore '$\nu$'$\epsilon\mathscr{T}_\mathscr{A}$ if and only if spectrum '$\nu$' $\supset \bar{\mathscr{A}}$.

Using the recursive functions $f$ and $g$ from (ii), we have

$$\text{'}\nu\text{'}\epsilon\mathscr{T}_\mathscr{A} \quad \text{iff} \quad g[\nu] = 0 \quad \text{and} \quad f[\nu] \supset \bar{\mathscr{A}},$$

or

$$g[\nu] = 1 \quad \text{and} \quad f[\nu] \subset \mathscr{A}.$$

By $f[\nu] \subset \mathscr{A}$ we mean that the tuplet whose number is $f[\nu]$ is a subset of $\mathscr{A}$. From this we see that for all natural numbers $\nu$, $\nu\epsilon\mathscr{A}$ if and only if $\sim G_\nu \in \mathscr{T}_\mathscr{A}$; hence $\mathscr{A}$ is reducible to $\mathscr{T}_\mathscr{A}$. Again $\nu\epsilon\mathscr{T}_\mathscr{A}$ if and only if a certain sequence effectively determined by $\nu$ is a subset of $\mathscr{A}$ or contains $\bar{\mathscr{A}}$. Thus $\mathscr{T}_\mathscr{A}$ is reducible to $\mathscr{A}$.

Finally $\mathscr{T}_\mathscr{A}$ is recursively enumerable if $\mathscr{B}$ is. Hence $\mathscr{T}_\mathscr{A}$ is recursively enumerable if $\mathscr{A}$ is. Thus by Prop. 15, Ch. 8 $\mathscr{T}_\mathscr{A}$ is axiomatizable if $\mathscr{A}$ is recursively enumerable. (Strictly this remark should be listed as a corollary to Prop. 15, Ch. 8.)

## 7.27 *Chains of degrees*

PROP. 30. *There exist non-denumerable chains of degrees.*

By Zorn's lemma there exists a maximal chain of degrees. This chain is without a greatest element, otherwise the jump of that element could be used to form a longer chain. Assume that this chain is countable, then by Mult. Ax. it contains a sequence of degrees $\mathbf{a}_0, \mathbf{a}_1, \ldots$. Again by Mult.

Ax. choose a set from each of these degrees, say $\mathscr{A}_0, \mathscr{A}_1, \dots$ where $\mathscr{A}_\nu$ is in $\mathbf{a}_\nu$. Now form the set $\mathscr{B}$ of ordered pairs $\langle \nu, \mu \rangle$ where $\nu$ is in $\mathscr{A}_\mu$. This set is above all the sets $\mathscr{A}_\nu$, hence the degree of $\mathscr{B}$, say $\mathbf{b}$, satisfies $\mathbf{a}_\nu < \mathbf{b}$, for all natural numbers $\nu$, contrary to the maximality of the chain.

The demonstration of this proposition is highly non-constructive.

Bear in mind that there are at most a denumerable set of degrees beneath any given degree, because there are only a denumerable set of tables. The ordinal numbers of the second class have the property of being non-denumerable yet there are only a denumerable number of ordinals less than a given ordinal in this class.

## 7.28 *Recursive real numbers*

We finish this chapter with a brief account of *recursive real numbers* and *recursive analysis*.

*Rational numbers* are easily introduced into the system $A_{00}$ as ordered triplets of natural numbers. Thus

$\{\lambda, \mu, \nu\}$ will play the role of $\dfrac{\lambda - \mu}{\nu + 1}$ (in ordinary mathematical notation).

From this idea definitions of $+_r \times_r -_r =_r <_r 0_r 1_r \dots \nu_r, \|_r$ etc., are easily found, e.g.

D 223 $\{\lambda . S\nu' + \lambda' . S\nu, \mu . S\nu' + \mu' . S\nu, S\nu . S\nu' \dot{-} 1\}$

for $\{\lambda, \mu, \nu\} +_r \{\lambda', \mu', \nu'\}$.

We leave the rest of the definitions to the reader, also the demonstration that they are satisfactory in that the laws of algebra are obeyed. Having got the rational numbers the recursive real numbers can be defined in any one of four different ways as follows:

A recursively enumerable sequence of rational numbers given by the recursive function $f$ is called a *recursive fundamental sequence* if and only if

$$|f\mu -_r f\lambda|_r <_r \frac{1_r}{\nu_r}{}_r \quad \text{for} \quad g\nu < \mu, \lambda,$$

where $g$ is a recursive function, called the *convergence function* of the sequence.

(i) A recursive fundamental sequence of rational numbers is called a *recursive real number*.

From this definition it is easily shown, as in analysis, that if the recursive functions $f$ and $f'$ both define recursive fundamental sequences of rational numbers then so do $f \times_r f', f +_r f', f -_r f'$, etc., and define the product, sum and difference of the recursive real numbers defined by the recursively enumerable sequences $f$ and $f'$. It is easily shown that these definitions are satisfactory, in that the laws of algebra are obeyed.

(ii) If $g$ is a recursive characteristic function then the real number whose binary expansion is $\nu . g1g2g3\ldots$ is called a *recursive real number*.

It is possible to define the product, sum etc., of two recursive real numbers on this definition, details are left to the reader.

(iii) If a recursive set of rationals is a left Dedekind section then the corresponding real number is called a *recursive real number*.

Definitions of product, sum, etc. follow as for analysis, they are denoted by $\times_s$, $+_s$, etc.

(iv) If $f$, $g$ are recursive functions and if $f\mu$, $g\mu$ are interpreted as rational numbers and if $f\mu <_r g\mu$ and $g\mu -_r f\mu <_r \frac{1_r}{\nu_r}$ for $\mu > h\nu$, where $h$ is a recursive function. Then the nested sequence of intervals is called a *recursive real number*.

Definitions of product, sum, etc., are easily supplied and shown to be satisfactory.

These definitions give us four classes of recursive real numbers and these four clases are all the same if we interpret recursive as general recursive, but they differ if we interpret recursive as primitive recursive.

It is easy to show that recursive real numbers form a field. Call this $\mathscr{E}$.

We can define a recursive fundamental sequence of recursive real numbers as for a recursive fundamental sequence of rationals. For this purpose we require two recursive functions $f$ and $g$, both two-place. Here $f[\mu, \kappa]$ enumerates the $\mu$th sequence of rationals for $\kappa = 1, 2, \ldots$ with $g[\mu, \kappa]$ as convergence function. We also require another recursive function as convergence function for the fundamental sequence of recursive real numbers $\rho_1$ given by $f[1, \kappa]$, $\rho_2$ given by $f[2, \kappa]$, .... By considering the sequence $f[\kappa, g[\kappa, \kappa]]$, which is easily shown to be a recursive fundamental sequence defining a recursive real number $\rho$, we get

PROP. 31. *Every recursive fundamental sequence of real numbers has a limit in* $\mathscr{E}$.

Having defined recursive real numbers we can define recursive complex numbers in the usual way as ordered pairs of recursive real numbers with special definitions for $\times_c$ $+_c$ $-_c$ $=_c$ $0_c$ $1_c \ldots ||_c$ and show that the definitions are satisfactory in that the laws of algebra hold. The recursive complex numbers are easily shown to form a field $\mathscr{E}(\iota)$. We get the pair by $\lambda x.f\{2x\}, \lambda x.f\{2x+1\}$.

PROP. 32. *$\mathscr{E}(\iota)$ is algebraically closed.*

We indicate the demonstration. Start with a constructive proof of the fundamental theorem of algebra, and formalize it within recursive function theory. One point which requires note is in finding the maximum of a terminating sequence of recursive real numbers. Let $\rho_1, \rho_2, \ldots, \rho_\pi$ be a terminating sequence of real numbers defined by the respective sequences $f[1,\kappa], f[2,\kappa], \ldots, f[\pi,\kappa]$ of rational numbers with convergence functions $g[1,\kappa], g[2,\kappa], \ldots, g[\pi,\kappa]$ respectively. Then the recursively enumerable sequence $Max[f[1,\kappa], \ldots, f[\pi,\kappa]]$ defines the recursive real number $Max_s[\rho_1, \ldots, \rho_\pi]$ with convergence function $Max[g[1,\kappa], \ldots, g[\pi,\kappa]]$. Analoguously for $Min$.

Another point that requires comment is the formation of a recursive function of recursive complex numbers. This is achieved, if the function is $h$, say, and $f[\lambda]$ is a recursive fundamental sequence defining the complex number $\rho$, we form $h[f[\lambda]]$ and if $h$ is continuous we see that $h[f[\lambda]]$ is a recursive fundamental sequence defining $h[\rho]$.

As a basis for analysis $\mathscr{E}$ and $\mathscr{E}(\iota)$ are a bit unsatisfactory, for one thing they are enumerable, this may be a liability, or it might have little effect. But the Bolzano–Weirstrass theorem: Every bounded set of members of $\mathscr{E}$ has a limit point in $\mathscr{E}$ fails.

Consider a recursively enumerable set $\mathscr{A}$ which fails to be recursive. Its characteristic function will correspond to a non-recursive real number $\rho$. If $g$ enumerates $\mathscr{A}$ the sequence

$$\frac{1}{2^{g1}}, \frac{1}{2^{g1}} + \frac{1}{2^{g2}}, \ldots$$

is a bounded sequence of rational numbers (hence members of $\mathscr{E}$) with $\rho$ as sole limit point. But the sequence fails to be a recursive fundamental sequence. For if it were with convergence function $h$, where $h$ is recursive, then we should have

$$\frac{1}{2^{g\mu}} + \frac{1}{2^{gS\mu}} + \ldots + \frac{1}{2^{g[\mu+\nu]}} < \frac{1}{\lambda} \quad \text{for} \quad h\lambda < \mu \text{ and all } \nu.$$

Hence
$$\frac{\log \lambda}{\log 2} < Min[g\mu, \ldots] \quad \text{for} \quad h\lambda < \mu,$$

hence to decide whether $g\kappa = \theta$ we need only consider $g1, ..., gh2^{\theta}$ whence $g$ defines a recursive set contrary to construction.

But if we are allowing only recursive real numbers, then perhaps we should only allow recursive sets of real numbers.

The theory of recursive real numbers may be extended to real numbers of degree **a**. This again would give a denumerable set of real numbers, this theory remains unexplored.

## Historical remarks to Chapter 7

Turing machines, as the name implies, were invented by Turing (1936) and were described by him in detail. Post (1936) also had the same idea, but only sketched it, Turing wanted to analyse into atomic acts what one does when one computes. The term 'partial recursive function' is due to Kleene (1938). The account of Turing machines we give is due to Kleene (1952). About the same time as Turing invented his machine several other methods for defining calculable functions were invented independently and for different reasons. The system of **λ**-conversion due to Church (1941) was one of these. Turing showed that the class of functions definable by his machines was identical with the class of **λ**-definable functions. Curry (1958) produced a Combinatory Logic in order to invent a logic without variables and so avoid the difficulties of substitution. Post (1943) used a theory of productions, this was the result of analysing the concept of substitution. General recursive functions as defined by a system of equations were due to Herbrand, Gödel (1933, 1934) and Kleene (1963). More recently Markow (1951) has given a theory of algorithms which is a significant modification of the productions of Post. More recently Smullyan (1961) has produced an equivalent theory of elementary formal systems. All these systems have led to the same class of calculable functions, hence Church's Thesis. Partial recursive functions are largely due to Kleene (1938); allowing them makes the theory of recursive functions more elegant, rather like homogeneous co-ordinates in analytic geometry as opposed to Cartesian co-ordinates.

The idea of a tape, which we use from the beginning, is due to Turing (1936). 'Passive instruction' appears to be due to Kleene (1952), as is the bracket and dot notation in describing tables of machines. The demonstrations Props. 1 and 2 are due to Kleene (1952). The idea of a universal machine goes back to Leibniz, as already noted, but Turing (1936) was the first to give full details as to how to build one. But his universal machine is not so powerful as that envisaged by Leibniz, but apparently it can do anything that can be done by man.

The undecidability of the classical predicate calculus of the first order was first demonstrated by Church (1936); Turing (1936) gave another demonstration as an application of his machines. The concept 'essentially undecidable' and lemmas (i), (ii), (iii) and the corollary are due to Tarski (1949, 1953), Prop. 7 is due to Turing (1936). Prop. 8 was first stated by Post (1944) and Cor. v is often called Post's theorem. Props. 9 and 10 are Post's (1944).

The theory of complete, creative, simple and hypersimple sets was initiated by Post (1944) and the constructions in Props. 11, 12, 13 and Cor. (i), (ii) are due to him (1944). Prop. 14 and corollaries (ii), (iii) and (iv) are due to Myhill.

The concept of degree of unsolvability is Post's (1944), who defined the various classes of sets just listed in order to try and find a set whose degree of unsolvability was strictly between the degree of solvability and the degree of a complete set. This problem which remained unsolved for several years, required a new method for its solution, it is known as Post's (1944) problem. The theory of degrees was further developed by Kleene & Post (1954) and an authoritative account was given by Sachs (1963). The term 'productive' is due to Dekker (1955). Props. 15 and 16 are due to Myhill (1955). Prop. 17 is sometimes called 'the recursion theorem' or the 'fixed point theorem' is due to Kleene (1952). Prop. 18 is again due to Myhill. The term 'oracle' is due to Turing (1936) and Prop. 19 is Kleene's (1952) formulation of the analogues of Prop. 2 for oracle machines.

Prop. 20 is due to Kleene & Post (1954), the construction here is fairly straightforward, but that in Prop. 21, which solves Post's problem is quite a different matter. The method of proof is due to Friedberg (1957) and Mucknik (1956) independently, it is now called the *priority method*, a term due to Sacks (1963). It is so called because in the course of the construction certain things are given priority over others in order that part of the construction which satisfies the required conditions may not be upset as the construction proceeds. Prop. 22 is due to Friedberg (1957), Prop. 23 and its corollaries are due to Spector (1956). Prop. 24 is due to Shoenfeld, Kleene and Post, and Prop. 25 is due to Kleene. The notion of cohesiveness is due to Rose & Ullian (1963), Prop. 26 to Dekker & Myhill (1960), the demonstration of Prop. 27 comes from Yates (1965) who gave an elegant simplification to a result of Friedberg (1958).

An account of computability has been written by Davis (1958) and another by Hermes. A very detailed account of recursive functions has been given by Hartley Rogers jun. (1967), and one on degrees of unsolvability by Sacks (1963).

The theory of degrees of unsolvability is in a very untidy state at present. Most results state that the upper semi-lattice of degrees of unsolvability lacks such and such a nice property. The methods used are very ingenious, it seems that further progress requires some new method, more powerful than the diagonal method of Cantor and the priority method of Friedberg (1957).

The motivation for the construction of simple and hypersimple sets is as follows. Post (1944) considered the problem of the reducibility of recursively enumerable sets of natural numbers. He invented several different kinds of reducibility. A set $\mathscr{A}$ is reducible to a set $\mathscr{B}$ if the problem of membership of $\mathscr{A}$ can be reduced to that of membership of $\mathscr{B}$. The kinds of reducibility considered by Post are:

(*a*) Many-one reducibility, $(m-1)$-reduciblity.

$$\nu \epsilon \mathscr{A} \leftrightarrow f\{\nu\} \epsilon \mathscr{B},$$

where $f$ is a recursive function, if $f$ is a $(1-1)$-function then this type of reducibility is called one-one reducibility.

(*b*) Truth-table reducibility. (m–1)-reducibility can be put down in the form of a table thus:

| $\nu \epsilon \mathscr{A}$ | $f\{\nu\} \epsilon \mathscr{B}$ |
|---|---|
| + | + |
| − | − |

A generalization of this is:

| $\nu \epsilon \mathscr{A}$ | $f_1\{\nu\} \epsilon \mathscr{B}$, | $f_2\{\nu\} \epsilon \mathscr{B}$, | ... | $f_{g\{\nu\}}\{\nu\} \epsilon \mathscr{B}$ |
|---|---|---|---|---|
| + | + | + | ... | + |
| ........ | ........ | ........ | ........ | ........ |
| − | − | − | ... | − |

where there are $2^{g\{\nu\}}$ rows, giving all the possible ways of placing $(+)$ or $(-)$ in the $g\{\nu\}$ columns. The functions $f, \ldots$ are recursive, then if we can decide $\mathscr{B}$ we can, using the table, decide $\mathscr{A}$. If $g$ is a bounded function then this method of reduction is called reduction by bounded truth-tables, otherwise reduction by unbounded truth-tables.

(*c*) Turing reducibility, (the only one we have considered). This is a Turing oracle machine depending on the characteristic function of a set $\mathscr{B}$. This machine with input $\nu$ will produce output 0 if $\nu \epsilon \mathscr{A}$ and output 1 otherwise. Then membership of $\mathscr{A}$ has been reduced to membership of $\mathscr{B}$ in the most general way possible.

We can develop a theory of degrees for each of these methods of reduction. Post (1944) wished to find if there was a degree of unsolvability between the degree of the complete set and the degree of solvability. He found that a N. and S.C. that a recursively enumerable set be (1–1)-complete was that it be unbounded and that its complement contain an unbounded recursively enumerable set. Hence he set out to find if there were a simple set, such a set would then be of (1–1)-degree between the degree of the complete set and the degree of solvability by (1–1)-reducibility. He was able to construct a simple set but found that it was of the same degree of unsolvability as the complete set with regard to reducibility by unbounded truth-tables. This particular simple set is obtained as follows:

Consider the unbounded sequence of mutally exclusive bounded sequences:

$$\sigma: \quad (3, 4)\ (5, 6, 7, 8) \ldots (2^{\nu}+1, 2^{\nu}+2, \ldots, 2^{\nu+1}) \ldots.$$

Given a creative set $\mathscr{C}$, generate the elements of the simple set $\mathscr{S}$ which we constructed, placing each in a set $\mathscr{S}_1$. Also generate the elements of $\mathscr{C}$, and as the element $\nu$ is generated, place all the natural numbers in the $\nu$th sequence of $\sigma$ in $\mathscr{S}_1$. The resulting set $\mathscr{S}_1$ is a generated set, and hence is recursively enumerable. Since $\mathscr{S}_1$ contains $\mathscr{S}$, $\bar{\mathscr{S}}$ contains $\bar{\mathscr{S}}_1$. As $\mathscr{S}$ is simple, $\bar{\mathscr{S}}$, and hence $\bar{\mathscr{S}}_1$, is without a recursively enumerable subset. Moreover $\bar{\mathscr{S}}_1$ is unbounded because $\bar{\mathscr{C}}$ is unbounded. For each element of $\bar{\mathscr{C}}$, the corresponding sequence in $\sigma$ has only those of its members that are already in $\mathscr{S}$ also in $\mathscr{S}_1$, and hence at least one element in $\bar{\mathscr{S}}_1$. Hence $\mathscr{S}_1$ is simple.

In like manner we see that a natural number $\nu$ is in $\mathscr{C}$ or $\bar{\mathscr{C}}$ according as all of the integers in the $\nu$th sequence of $\sigma$ are in $\mathscr{S}_1$, or at least one is in $\bar{\mathscr{S}}_1$. If then we make correspond to each positive integer $\nu$ the sequence of $2^{\nu}$ natural numbers $(2^{\nu}+1, 2^{\nu}+2, \ldots, 2^{\nu+1})$ and the truth-table of order $2^{\nu}$ in which the sign under $\nu$ is $(+)$ in that row in which the signs under the $2^{\nu}\, f\{\mu\}$'s are all $(+)$, and in every other row the sign under $\nu$ is $(-)$, we have a reduction of $\mathscr{C}$ to $\mathscr{S}_1$ by unbounded truth-tables. Thus simple sets are useless as candidates for sets of lower degree of unsolvability than the degree of the complete set. The counter example just given suggested to Post to see if there was a hypersimple set. He was able to construct one, in fact the construction we have given, and showed that it was of lower degree of unsolvability than the complete set by reducibility using unbounded truth-tables. But he was dubious that they were of lower degree by Turing reducibility. Post ended his remarkable paper with the following remark.

'Indeed, if general recursive function is the formal equivalent of effective calculability, its formulation may play a role in the history of combinatory mathematics second only to that of the formulation of the concept of natural number.'

Prop. 27 on the existence of a maximal set is due to Yates (1965).

Prop. 28 on the existence of a minimal set is due to Spector (1956), we rely on the demonstration given in Rogers (1967).

Recursive real numbers have been considered by Turing (1936), Rice (1954), Klaua (1961), Goodstein (1961) and Specker (1949).

We have omitted to say much about recursive ordinals and recursive cardinals, for the latter see Dekker and Myhill (1960).

A vast number of examples on the subject matter of this chapter can be found in Rogers (1967) and in Shoenfield (1967).

Much of the matter discussed in this chapter and in chapter 5 has been worked out by Smullyan (1961) for general systems, he uses the methods of Post (1944).

## Examples 7

1. Find diagrams for the following machines:

(i) Started observing a tally replaces it by a cipher seeks the first cipher to the right replaces it by a tally and stops.

(ii) Started observing a number in standard position will decide if this is the last member of the representation of a $\kappa$-tuplet and will then write down $\kappa$ after a one cipher gap.

(iii) Started observing a number in standard position will find the first cipher to the left will move the number two places to the right and replace the ciph er by , I, i.e. it replaces, $\bar{\nu}$, by , $\bar{0}$ , $\bar{\nu}$,.

(iv) Started observing a number in standard position will decide if it is even, if so prints cipher followed by a tally, otherwise prints cipher followed by two tallies.

2. Write down the diagrams for the tables which calculate the following functions:

$$+,\ \times,\ exp,\ P,\ \dot{-},\ A_1,\ B_1,\ A_2,\ B_2,\ \Sigma\tau,\ \Pi\tau.$$

3. Similarly for the function $Un$.

4. (i) Show that it is impossible to find a uniform method for deciding if two primitive recursive functions ever take the same value.
   (ii) Show that it is impossible to find a uniform method for deciding of two recursive functions whether they are interlocked, i.e. $f\nu < g\nu < fS\nu$ for all numerals $\nu$.
   (iii) Show that it is impossible to find a uniform method for deciding of two recursive functions whether they are asymptotically equal, i.e. $f\nu = g\nu$ for all sufficiently large $\nu$.

5. Show that the family of recursive permutations form a denumerable group under composition.

6. Show that isomorphism of sets is an equivalence relation. $\mathscr{A}$ and $\mathscr{B}$ are sets of natural numbers. Write:

$\mathscr{A} \sim \mathscr{B}$ if there is a partial function $f$ such that $f\mathscr{A} = \mathscr{B}$,

$\mathscr{A} \simeq \mathscr{B}$ if there is a partial (1–1) function $f$ such that $f\mathscr{A} = \mathscr{B}$,

$\mathscr{A} \cong \mathscr{B}$ $\mathscr{A}$ is isomorphic to $\mathscr{B}$.

Show that $\mathscr{A} \cong \mathscr{B} \rightarrow \mathscr{A} \simeq \mathscr{B}$ and $\mathscr{A} \simeq \mathscr{B} \rightarrow \mathscr{A} \sim \mathscr{B}$.

$\mathscr{A} \sim \mathscr{B}$ fails to imply $\mathscr{A} \simeq \mathscr{B}$,

$\mathscr{A} \simeq \mathscr{B}$ fails to imply $\mathscr{A} \cong \mathscr{B}$.

If $\mathscr{A}$ and $\mathscr{B}$ are bounded then $\mathscr{A} \cong \mathscr{B}$, $\mathscr{A} \simeq \mathscr{B}$ and $\mathscr{A} \sim \mathscr{B}$ are equivalent.

7. If $\mathscr{A}$, $\mathscr{B}$ are recursively enumerable then

$$\mathscr{A} \simeq \mathscr{B} \quad \& \quad \bar{\mathscr{A}} \simeq \bar{\mathscr{B}}. \leftrightarrow \mathscr{A} \cong \mathscr{B}.$$

8. Let $\mathscr{A} \simeq \mathscr{B}$ then, $\mathscr{A}$ recursively enumerable $\leftrightarrow \mathscr{B}$ recursively enumerable, $\mathscr{A}$ immune $\leftrightarrow \mathscr{B}$ immune, $\mathscr{A}$ productive $\leftrightarrow \mathscr{B}$ productive.

9. Give the analogue of Prop. 17 for $f$-recursive functions.

10. Cor. (i), (ii), (iii), (iv), (v), (vi), (vii), Prop. 5, Cor. (i), Prop. 7,

Cor. (i), (ii), (iii), Prop. 8, each give rise to an oracle function, e.g. Cor. (i), Prop. 5 gives rise to $f$, where $f\kappa = \begin{cases}0\\1\end{cases}$ according as $f_{|\kappa|}\nu = 0$ for some numeral $\nu$ or otherwise. Find the relations of order between the degrees of these functions.

11. Show that it is impossible to find a uniform method for deciding whether a recursive function is never zero for sufficiently large argument.

12. Show that the set of *g.n.*'s of tables for recursive functions fails to be recursively enumerable.

13. Show that an unbounded recursively enumerable set $\mathscr{A}$ is of the form $\mathscr{A} = \mathscr{C}r \cup \mathscr{P}$, where $\mathscr{C}r \cap \mathscr{P} = \emptyset$, $\mathscr{C}r$ is creative and $\mathscr{P}$ is productive. [Consider $\mathscr{A} \cap \mathscr{C}$, $\mathscr{A} \cap \bar{\mathscr{C}}$.]

14. Let $\mathscr{P}$ be productive and $\mathscr{A}$ be recursively enumerable, show that $\mathscr{A} \subset \mathscr{P} \rightarrow (\mathscr{P} - \mathscr{A}$ is productive), $\mathscr{P} \subset \mathscr{A} \rightarrow (\bar{\mathscr{A}} \cup \mathscr{P}$ is productive).

15. Let $\mathscr{C}_f$ be the $f$-complete set, show that $\mathscr{C}_f$ fails to be $f$-recursive.

16. Show that $\mathscr{C}_f$ is $f$-productive.

17. $\mathscr{A}$ is the set of natural numbers $\nu$ such that $\mathscr{X}_\nu$ is bounded, show that $\mathscr{A}$ and $\bar{\mathscr{A}}$ are productive.

18. Give details of an oracle machine in which the characteristic function of the 'oracle' set is written down on a second tape.

19. Give a demonstration of Prop. 21. on the lines of Prop. 24.

20. Give a demonstration of Prop. 24 on the lines of Prop. 21.

21. Find the table for the machine to calculate $\mathscr{X}_\nu^\mu$ as described in Prop. 27.

22. Show that it is impossible to decide if the intersection of two recursively enumerable sets of lattice points is void, or is bounded or is unbounded.

23. Show that it is impossible to decide of a recursive function $\rho$ whether $S\rho\nu = \rho(S\nu)$ for some numeral $\nu$.

24. Find the table for the function of Prop. 18, whose *g.n.* is $\nu$.

25. Show that the following problems are recursively unsolvable:

(*a*) To decide if $\Phi_\nu$ is a constant function.

(*b*) To decide if $\nu$ is in $\mathscr{X}_\kappa$.

(*c*) To decide if two tables give the same function.

(*d*) To decide if $\mathscr{X}_\kappa = \mathscr{X}_\nu$.

(*e*) To decide if $\mathscr{X}_\kappa$ is unbounded.

26. The direct product of two sets $\mathscr{A}$ and $\mathscr{B}$ is the set of ordered pairs $\langle \nu, \mu \rangle$, where $\nu \epsilon \mathscr{A}$ and $\mu \epsilon \mathscr{B}$. Show that the operation of direct product

is uniformly effective for recursive sets and for recursively enumerable sets.

27. Show that:

(i) $\leqslant_1$ and $\leqslant_m$ are reflexive and transitive;

(ii) if $\mathscr{A} \leqslant_1 \mathscr{B}$ then $\mathscr{A} \leqslant_m \mathscr{B}$;

(iii) if $\mathscr{A} \leqslant_1 \mathscr{B}$ then $\bar{\mathscr{A}} \leqslant_1 \bar{\mathscr{B}}$;

(iv) if $\mathscr{A} \leqslant_m \mathscr{B}$ then $\bar{\mathscr{A}} \leqslant_m \bar{\mathscr{B}}$;

(v) if $\mathscr{A} \leqslant_m \mathscr{B}$ and $\mathscr{B}$ is recursively enumerable then $\mathscr{A}$ is recursively enumerable.

28. Show that there exist two non-recursive sets which are incomparable with respect to $\leqslant_m$.

29. The $m$-reducibility ordering is an upper semi-lattice; i.e. any two degrees have a unique least upper bound. Furthermore the least upper bound of two recursively enumerable degrees is a recursively enumerable degree.

30. Show that $\mathscr{A} \leqslant_m \mathscr{B}$ may hold while $\mathscr{A} \leqslant_m \bar{\mathscr{B}}$ fails.

31. Show that the complete set is 1-complete (i.e. complete w.r.t. $\leqslant_1$).

32. Show that $\hat{x}(\Phi_x$ is total) is productive.

33. Show that $\mathscr{A}$ is 1-complete if and only if $\mathscr{A}$ is $m$-complete.

34. Show that:

(i) If $\mathscr{A}$ is simple then $\mathscr{A}$ is non-recursive;

(ii) if $\mathscr{A}$ is simple then $\mathscr{A}$ is non-creative;

(iii) if $\mathscr{A}$ is simple then $\mathscr{A}$ fails to be $m$-complete;

(iv) if $\mathscr{A}$ is simple then $\mathscr{A}$ fails to be a cylinder; (a *cylinder* is a set of the form $\mathscr{B} \times \mathscr{N}$, where $\mathscr{N}$ is the set of all natural numbers).

35. Show that the following classification is mutually exclusive and exhaustive:

(i) The class of recursively enumerable sets.

(ii) The class of immune sets.

(iii) The class of non-recursively enumerable sets which are the union of an unbounded recursively enumerable set and an immune set.

(iv) The class of sets $\mathscr{A}$ such that if $\mathscr{B}$ is a recursively enumerable subset of $\mathscr{A}$ then there is a recursively enumerable unbounded subset $\mathscr{C}$ of $\mathscr{A} \cup \overline{\mathscr{B}}$, and it is impossible to find a uniform effective method for finding an index for $\mathscr{C}$ from an index for $\mathscr{B}$.

(v) The class of productive sets.

36. Show that the join and the direct product of two simple sets is a simple set (the *join of two sets* $\mathscr{A}$ and $\mathscr{B}$ is the set

$$\hat{y}\{y = 2x \,\&\, x\epsilon\mathscr{A} \vee y = 2x+1 \,\&\, x\epsilon\mathscr{B}\}).$$

37. Show that '$\mathscr{A}$ recursive in $\mathscr{B}$' is transitive, but that '$\mathscr{A}$ recursively enumerable in $\mathscr{B}$' fails to be transitive.

38. Show that the intersection of two hypersimple sets is hypersimple.

39. Show that the four definitions of recursive real numbers are the same.

40. Supply the definitions of sum, product, etc., for recursive real numbers defined by method (ii).

41. Supply the demonstration of Prop. 31.

42. Supply the demonstration of Prop. 32.

43. Show that there is a primitive recursive function $g$ such that

$$\Phi[x; \Phi[y; z]] = \Phi[g[x, y]; z].$$

## Chapter 8

# An incomplete undecidable arithmetic. The system **A**

### 8.1 *The system* **A**

The system $\mathbf{A}_0$ is complete with respect to $\mathbf{A}_0$-truth hence it is impossible to extend it in order to get more $\mathbf{A}_0$-true closed $\mathbf{A}_0$-statements as theorems in the extended system, but it is possible to extend it so as to be able to express concepts which are unrepresentable in $\mathbf{A}_0$. The system **A** is a primary extension of the system $\mathbf{A}_0$ obtained by adjoining the universal quantifier and a rule for its use. Thus **A** is richer in modes of expression than $\mathbf{A}_0$. We accordingly introduce the symbol $A$ of type $o(o\iota)$ which we call the *universal quantifier.*

D 224 $\qquad (A\xi)\,\phi\{\xi\}$ for $A(\boldsymbol{\lambda}\xi\phi\{\xi\})$.

We also add the rule:

II $d'$ $$\frac{\phi\{\eta\} \vee \omega}{(A\xi)\,\phi\{\xi\} \vee \omega} \quad \text{generalization.}$$

$\Gamma_\iota$ is free for $\eta, \xi$ in $\phi\{\Gamma_\iota\}$, the variable $\eta$ must fail to occur free in the lower formula and $\xi$ fail to occur free in $\phi\{\Gamma_\iota\}$, $\omega$ is subsidiary and may be omitted.

We also allow free variables in **A**-axioms and rules, otherwise II $d'$ would be useless. In II $d$ $\alpha$ can be a variable.

The chief advantage of the system **A** over the system $\mathbf{A}_0$ is that some $\mathbf{A}_0$-*metatheorems* become **A**-theorems. For instance $(Ax)\,(x = x)$ is an **A**-theorem while in the system $\mathbf{A}_0$ we have instead: $\alpha = \alpha$ is an $\mathbf{A}_0$-theorem for each closed numerical term $\alpha$.

Rule II $d'$ has restrictions on variables, these are required for the following reasons.

We require $\Gamma_\iota$ to be free for $\eta$ in $\phi\{\Gamma_\iota\}$ to avoid

$$\frac{(Ax)\,(x = x)}{(Ax')\,(Ax)\,(x' = x)}.$$

We require $\xi$ to fail to occur free in $\phi\{\Gamma_\iota\}$ in order to have

$$\frac{(Ax)\,\phi\{x\} \vee \omega}{\phi\{x'\} \vee \omega}*$$

the reversibility of II $d'$. This would fail if $\phi\{\Gamma_\iota\}$ was $\Gamma_\iota = x$. For instance we object to

$$\frac{(Ax)\,(x = x) \vee \omega}{x' = x \vee \omega}*.$$

We require $\eta$ to fail to occur free in $\omega$ to avoid

$$\frac{x = 0 \vee x \neq 0}{(Ax)\ (x = 0) \vee x \neq 0}.$$

We require $\eta$ to fail to occur free in $\phi\{\Gamma_\iota\}$ to avoid

$$\frac{x = x}{(Ax')\,(x' = x)}.$$

Note that the restrictions on variables are effective, we can always easily decide whether they are satisfied. Also note that once a quantifier is introduced into an A-proof then it remains in that A-proof till the end. Thus an A-proof of a closed $A_0$-statement is an $A_0$-proof of that statement and an A-proof of a closed $A_{00}$-statement is an $A_{00}$-proof of that statement. Also note that the class of closed numerical terms is the same as in $A_{00}$.

## 8.2 *Definition of* **A**-*truth*

The *g.n.*'s of the **A** symbols will be the same as before except that the *g.n.* of **A** will be 11 and that of $x^{(\theta)}$ will be $(12+\theta)$.

We give a truth-definition $\mathscr{T}$ for closed **A**-statements in that we augment the definition $\mathscr{T}_0$ of $A_0$-truth by the addition of:

(vi) $\phi$ is of the form $(A\xi)\,\psi\{\xi\}$ and $\psi\{\nu\}$ is **A**-true for each numeral $\nu$. Here $\xi$ fails to occur free in $\psi\{\nu\}$. An **A**-statement will be called **A**-valid if:

(vii) Let $\xi', \xi'', \ldots, \xi^{(\nu)}$ be exactly all the free variables in $\psi\{\xi', \xi'', \ldots, \xi^{(\nu)}\}$ and let these variables fail to be free in the statement-form $\psi\{\Gamma'_\iota, \Gamma''_\iota, \ldots, \Gamma^{(\nu)}_\iota\}$, then $\psi\{\xi', \xi'', \ldots, \xi^{(\nu)}\}$ is **A**-valid if and only if $\psi\{\kappa', \kappa'', \ldots, \kappa^{(\nu)}\}$ is **A**-true whenever $\kappa', \kappa'', \ldots, \kappa^{(\nu)}$ are numerals. If $\nu$ is zero the **A**-statement is already closed.

We define Falsity for closed A-statements by the addition of the following to the definition $\mathscr{F}_0$ of $A_0$ $A_0$-falsity:

(vi) $\phi$ is of the form $(A\xi)\,\psi\{\xi\}$ and $\psi\{\nu\}$ is A-false for some numeral $\nu$. Here $\xi$ fails to occur free in $\psi\{\nu\}$.

We say that the A-statements $\phi'\{\xi', \xi'', \ldots, \xi^{(\mu)}\}$ and $\phi''\{\xi', \xi'', \ldots, \xi^{(\lambda)}\}$ containing exactly the free variables shown are *T-equivalent* if for any numerals $\kappa', \kappa'', \ldots, \kappa^{(\nu)}$, where $\nu = Max\,[\lambda, \mu]$, $\phi'\{\kappa', \kappa'', \ldots, \kappa^{(\mu)}\}$ is A-true if and only if $\phi''\{\kappa', \kappa'', \ldots, \kappa^{(\lambda)}\}$ is A-true.

Negation can be A-represented if we add to the definition of $N[``\phi"]$ given for the system $A_{00}$ the following:

replace $A$ by $E$ and $E$ by $A$.

D 225 $N[``\phi"]$ is the result of everywhere interchanging throughout $\phi$

$$= \text{ and } \neq,$$
$$\& \text{ and } \vee,$$
$$E \text{ and } A.$$

With this definition a closed A-statement $\phi$ is A-true if and only if its negation $N[``\phi"]$ is A-false.

PROP. 1. *It is impossible to show that a closed* A*-statement is both* A*-true and* A*-false.*

We proceed as in Prop. 1 of Ch. 6 except that we must add the following:

(v) suppose $\phi$ is a closed universal statement $(A\xi)\,\psi\{\xi\}$ and has been shown to be A-false, then, $\psi\{\nu\}$ has been shown to be A-false for some numeral $\nu$. By our induction hypothesis it is impossible to show that $\psi\{\nu\}$ is A-true, hence it is impossible to show that $\phi\{\mu\}$ is A-true for each numeral $\mu$, thus it is impossible to show that $(A\xi)\,\psi\{\xi\}$ is A-true.

PROP. 2. *The system* A *is consistent with respect to* A*-validity.*

We have to show that each A-theorem is A-valid. The A-axioms are A-valid. We replace free variables by numerals and proceed as in Prop. 2, Ch. 4. The result then follows by (vii). Secondly we have to show that the rules preserve validity. In the case of the new rule II $d'$ the conclusion is A-valid if and only if the premiss is A-valid so that II $d'$ preserves A-validity. For the other rules we replace free variables by numerals, same variable by same numeral, and proceed as before, except in the case of R 1:

$$\frac{\alpha\{\xi\} = \beta\{\xi\} \vee \omega \quad (A\xi)\,\phi\{\alpha\{\xi\},\xi\} \vee \omega}{(A\xi)\,\phi\{\beta\{\xi\},\xi\} \vee \omega}. \tag{1}$$

Replace free variables by numerals throughout and this becomes

$$\frac{\alpha'\{\theta\} = \beta'\{\theta\} \vee \omega' \quad (A\xi)\,\phi'\{\alpha''\{\xi\},\xi\} \vee \omega'}{(A\xi)\,\phi'\{\beta''\{\xi\},\xi\} \vee \omega'}. \tag{2}$$

Some variables free in $\alpha\{\xi\}$ may be bound in $\phi\{\alpha\{\xi\},\xi\}$. We show that R 1 preserves validity by induction on the construction of the main formula. If this is atomic then on replacing variables by numerals we are left with R 1 applied in $\mathbf{A}_{00}$ and the result follows. The other cases are easily dealt with except the one just mentioned when the main formula is a universal statement. By our induction hypothesis

$$\frac{\alpha'\{\theta\} = \beta'\{\theta\} \vee \omega' \quad \phi'\{\alpha''\{\theta\},\theta\} \vee \omega'}{\phi'\{\beta''\{\theta\},\theta\} \vee \omega'}$$

for all numerals $\theta$, but this says that if the upper premisses in (1) are valid then so is the lower premiss. The result now follows.

## 8.3 *Incompleteness and undecidability of the system* A

PROP. 3. *The system* **A** *is incomplete with respect to* **A**-*validity.*

We have to find an **A**-valid **A**-statement which fails to be an **A**-theorem.

D 226 $Prf_{\mathbf{A}}[\xi,\eta]$ for the primitive recursive $\mathbf{A}_{00}$-statement such that $Prf_{\mathbf{A}}[\kappa,\nu]$ is $\mathbf{A}_{00}$-true if and only if $\kappa$ is the *g.n.* of an **A**-proof of the **A**-statement whose *g.n.* is $\nu$.

This is constructed as in the previous systems. But note that the *g.n.*'s of variables have been increased.

D 227 $s[\xi,\eta]$ Gödel's substitution function for **A**.

This is constructed as before and is primitive recursive. Consider $N[\text{“}Prf_{\mathbf{A}}[x',s[x,x]]\text{”}]$ let its *g.n.* be $\kappa$. (We suppose that $Prf_{\mathbf{A}}$ fails to contain $x$.) Let $\nu$ be the *g.n.* of $N[\text{“}Prf_{\mathbf{A}}[x',s[\kappa,\kappa]]\text{”}]$. Then $\nu = s[\kappa,\kappa]$ is an $\mathbf{A}_{00}$-theorem. Suppose $N[\text{“}Prf_{\mathbf{A}}[x',\nu]\text{”}]$ is an **A**-theorem, then so is $N[\text{“}Prf_{\mathbf{A}}[x',s[\kappa,\kappa]]\text{”}]$ by R 1. There will then be a numeral $\pi$ such that $Prf_{\mathbf{A}}[\pi,\nu]$, here $\pi$ is the *g.n.* of an **A**-proof of the **A**-statement whose *g.n.*

is $\nu$. But from the **A**-proof of the **A**-theorem $N[\text{“}Prf_{\mathbf{A}}[x', \nu]\text{”}]$ we can obtain an **A**-proof of $N[\text{“}Prf_{\mathbf{A}}[\pi, \nu]\text{”}]$ by everywhere replacing free occurrences of the variable $x'$ by the numeral $\pi$. But it is absurd that $Prf_{\mathbf{A}}[\pi, \nu]$ and $N[\text{“}Prf_{\mathbf{A}}[\pi, \nu]\text{”}]$ are both **A**-theorems, since $\mathbf{A}_{00}$ is consistent, if they are **A**-theorems then they are $\mathbf{A}_{00}$-theorems. Thus $N[\text{“}Prf_{\mathbf{A}}[x', \nu]\text{”}]$ fails to be an **A**-theorem.

We now show that $N[\text{“}Prf_{\mathbf{A}}[x', \nu]\text{”}]$ is **A**-valid, we have to show that $N[\text{“}Prf_{\mathbf{A}}[\pi, \nu]\text{”}]$ is **A**-true for each numeral $\pi$. Now $N[\text{“}Prf_{\mathbf{A}}[\pi, \nu]\text{”}]$ is a closed $\mathbf{A}_{00}$-statement so if it is **A**-true then it is $\mathbf{A}_{00}$-true hence $\mathbf{A}_{00}$-provable. $Prf_{\mathbf{A}}[\xi, \eta]$ is a primitive recursive $\mathbf{A}_{00}$-statement hence exactly one of $Prf_{\mathbf{A}}[\pi, \nu]$ and $N[\text{“}Prf_{\mathbf{A}}[\pi, \nu]\text{”}]$ is $\mathbf{A}_{00}$-provable. Suppose that $Prf_{\mathbf{A}}[\pi, \nu]$ is $\mathbf{A}_{00}$-provable, then $\pi$ is the *g.n.* of an **A**-proof of the **A**-statement whose *g.n.* is $\nu$. Thus $N[\text{“}Prf_{\mathbf{A}}[x', s[\kappa, \kappa]]\text{”}]$ is **A**-provable, whence so is $N[\text{“}Prf_{\mathbf{A}}[x', \nu]\text{”}]$, and hence so is $N[\text{“}Prf_{\mathbf{A}}[\pi, \nu]\text{”}]$ for each numeral $\pi$. $N[\text{“}Prf_{\mathbf{A}}[\pi, \nu]\text{”}]$ is closed and lacks quantifiers hence its **A**-proof is an $\mathbf{A}_{00}$-proof. Since $\mathbf{A}_{00}$ is consistent we must have $N[\text{“}Prf_{\mathbf{A}}[\pi, \nu]\text{”}]$ for each numeral $\pi$. Thus $N[\text{“}Prf_{\mathbf{A}}[x', \nu]\text{”}]$ is **A**-valid but fails to be **A**-provable. Thus **A** is incomplete with respect to **A**-validity.

PROP. 4. *The system* **A** *is undecidable.*

A decision-procedure for **A** would give one for $\mathbf{A}_0$, because an **A**-proof of a closed $\mathbf{A}_0$-statement is an $\mathbf{A}_0$-proof of that statement. Since $\mathbf{A}_0$ is undecidable then so is **A**.

COR. (i). **A**-*theorems form a recursively enumerable set which fails to be a recursive set.*

**A**-theorems can be recursively enumerated similarly to the $\mathbf{A}_0$-theorems. If they formed a recursive set than **A** would be decidable.

## 8.4 *Various properties of the system* **A**

PROP. 5. *If an* **A**-*disjunctand is an* **A**-*theorem then one of the disjunctands is an* **A**-*theorem and by examining the* **A**-*proof of the disjunction we can find which disjunctand is an* **A**-*theorem, and find an* **A**-*proof for it.*

The demonstration is similar to that of Prop. 4, Ch. 6, by theorem induction.

Cor. (i). *$\phi \vee N[``\phi"]$ is an* **A**-*theorem if and only if one of $\phi$* or *$N[``\phi"]$ is an* **A**-*theorem.*

Thus T.N.D. fails in **A**.

Cor. (ii). *$x = 0 \vee x \neq 0$ fails to be an* **A**-*theorem.*

Cor. (iii). I *b is a derived* **A**-*rule.*

Prop. 6. *An existential* **A**-*statement is an* **A**-*theorem if and only if a particular instance can be found and is* **A**-*provable.*

We have to show that if $(E\xi)\,\phi\{\xi\}$ is an **A**-theorem then so is $\phi\{\alpha\}$ for some numerical term $\alpha$. The demonstration is similar to that of Prop. 7, Ch. 6. If $\phi\{\xi\}$ contains a free variable $\eta$ then so may $\alpha$. Note that I *b* is still absent from the rules.

Prop. 7. *The scheme* II *d′ is reversible.*

We have to show that if $(A\xi)\,\phi\{\xi\} \vee \omega$ is an **A**-theorem then so is $\phi\{\xi\} \vee \omega$. It suffices to show that if $(A\xi)\,\phi\{\xi\}$ is an **A**-theorem then so is $\phi\{\xi\}$. Because by Prop. 5 if $(A\xi)\,\phi\{\xi\} \vee \omega$ is an **A**-theorem then so is one of $(A\xi)\,\phi\{\xi\}$ or $\omega$ and we can find which and find its **A**-proof. If $\omega$ is found to be an **A**-theorem then the result follows at once by II *a*. If $(A\xi)\,\phi\{\xi\}$ is found to be an **A**-theorem then we proceed as in Prop. 6 and obtain an **A**-proof of $\phi\{\xi\}$. By II *a* the result follows.

Cor. (i). *If $(A\xi)\,\phi\{\xi\}$ is an* **A**-*theorem then so is $\phi\{\nu\}$ for each numeral $\nu$ and these proofs are obtained the one from the other merely by substituting the numeral $\nu$ in an* **A**-*proof-form. Here $\xi$ fails to occur free in $\phi\{\Gamma_\iota\}$.*

By Prop. 7 $\phi\{\xi\}$ is an **A**-theorem. In the **A**-proof of $\phi\{\xi\}$ replace all occurrences of $\xi$ which correspond to the free occurrences of $\xi$ in $\phi\{\xi\}$ by corresponding occurrences of the numeral $\nu$. This leaves us with an **A**-proof of $\phi\{\nu\}$ because axioms are converted into axioms and applications of rules remain applications of the same rules. Consider, for example, rule II *d*:

$$\frac{\phi\{\xi\} \vee \omega\{\xi\}}{(E\xi)\,\{\xi\} \vee \omega\{\xi\}} \quad \text{II}\,d,$$

this becomes

$$\frac{\phi\{\nu\} \vee \omega\{\nu\}}{(E\xi)\,\{\xi\} \vee \omega\{\nu\}},$$

which is still an application of II *d*.

A similar result holds for $(A\xi)\,\phi\{\xi\} \vee \omega$. Hence the corollary.

Each primitive recursive function $f$ is explicitly $\mathbf{A}_{00}$-definable by an $\mathbf{A}_{00}$-formula $\rho$ of type $\iota\iota, \iota\iota\iota, \ldots$ according to the argument set, such that $\rho\nu = \kappa$ is an $\mathbf{A}_{00}$-theorem if and only if $fn = k$ where $\nu$ and $\kappa$ are the numerals which $\mathbf{A}_{00}$-represent the natural numbers $n$ and $k$. The same holds in the system $\mathbf{A}_0$. Also each partial recursive function is $\mathbf{A}_0$-represented in the sense that if the value of $g\{n', n'', \ldots, n^{(\theta)}\}$ is $k$ then there is an $\mathbf{A}_{00}$-statement $\phi\{\eta', \eta'', \ldots, \eta^{(\theta)}, \zeta, \xi\}$ such that $(E\xi)\,\phi\{\nu', \nu'', \ldots, \nu^{(\theta)}, \kappa, \xi\}$ is an $\mathbf{A}_0$-theorem, where $\nu', \nu'', \ldots, \nu^{(\theta)}, \kappa$, are the numerals which $\mathbf{A}_{00}$-represent the natural numbers $n', n'', \ldots, n^{(\theta)}, k$, respectively. All this still holds in the system **A** because $\mathbf{A}_{00}$-theorems and $\mathbf{A}_0$-theorems are **A**-theorems. But in the system **A** we have the universal quantifier and we have:

PROP. 8. *The recursion equations of the primitive recursive function* $\widehat{\tau\sigma}$ *are* **A**-*provable.*

We have: $\widehat{\tau\sigma}[\xi, \mathfrak{y}] = \mathscr{I}(\boldsymbol{\lambda}\zeta\zeta'.\tau[\zeta, \zeta', \mathfrak{y}])\,\sigma[\mathfrak{y}]\,\xi$ by D 121, where $\mathfrak{y}$ stands for $\eta', \eta'', \ldots, \eta^{(\theta)}$, absent if $\theta = 0$.

Hence

$$\widehat{\tau\sigma}[0, \mathfrak{y}] = \sigma[\mathfrak{y}] \quad \text{by} \quad \text{Ax. } 3.1.$$

Again

$$\begin{aligned}\widehat{\tau\sigma}[S\xi, \mathfrak{y}] &= \tau[\xi, \mathscr{I}(\boldsymbol{\lambda}\zeta\zeta'.\tau[\zeta, \zeta', \mathfrak{y}])\,\sigma[\mathfrak{y}]\,\xi, \mathfrak{y}] \quad \text{by Ax. } 3.2\\ &= \tau[\xi, \widehat{\tau\sigma}[\xi, \mathfrak{y}], \mathfrak{y}] \quad \text{by R } 1.\end{aligned}$$

Thus the recursion equations are **A**-theorems. We can apply II $d'$ and generalize the free variables $\xi, \eta', \eta'', \ldots, \eta^{(\theta)}$.

In the case of the general recursive function $\mathbf{A}_0$-represented in the form $(E\xi)\,\phi\{\mathfrak{n}, \kappa, \xi\}$ we inquire whether we can **A**-prove the corresponding general **A**-statement $(A\mathfrak{y})\,(E\zeta, \xi)\,\phi\{\mathfrak{y}, \zeta, \xi\}$. But if this was an **A**-theorem then so would be $(E\zeta, \xi)\,\phi\{\mathfrak{y}, \zeta, \xi\}$ by Prop. 7 and then by Prop. 6 so would be $(E\xi)\,\phi\{\mathfrak{y}, \rho[\mathfrak{y}], \xi\}$, where $\rho[\mathfrak{y}]$ is a numerical term and hence a primitive recursive function of $\mathfrak{y}$ by Prop. 2 of Ch. 5. But by Prop 5 of Ch. 5 there are general recursive functions which fail to be primitive recursive and in this case $(E\xi)\,\phi\{\mathfrak{y}, \rho[\mathfrak{y}], \xi\}$ would fail to arise. Thus the general statement $(A\mathfrak{y})\,(E\zeta, \xi)\,\phi\{\mathfrak{y}, \zeta, \xi\}$ is **A**-true but fails to be an **A**-theorem. This again demonstrates the incompleteness of the system **A**. In particular consider

$$(A\eta)\,(E\zeta, \xi)\,Prf_{00}[\xi, Eq[\langle 8, 1, 8\rangle^\frown f[\eta]^\frown Num[\eta]^\frown\langle 9, 9\rangle, Num[\zeta]]],$$

if this were an **A**-theorem then so would be

$$(E\xi)\,Prf_{00}[\xi, Eq[\langle 8.1.8\rangle^\frown f[\eta]^\frown Num[\eta]^\frown\langle 9, 9\rangle, Num[\rho[\eta]]]],$$

for some primitive recursive functions $\rho$. This says that

$$Val\langle 8,1,8\rangle^\frown f[\kappa]^\frown Num[\kappa]^\frown\langle 9,9\rangle$$

is equivalent to a primitive recursive function $\rho$, but this we have found to be absurd. Another example of an **A**-true **A**-statement which fails to be an **A**-theorem is $(Ax)(Ex', x'')(Un^1[\kappa, x, x', x''] = 0)$, if the numeral $\kappa$ is chosen as the *g.n.* of the table of a general recursive function which fails to be primitive recursive.

PROP. 9. *If* $(A\xi)(E\eta)\,\phi\{\xi,\eta\}$ *is an* **A**-*theorem then so is* $\phi\{\xi,\rho\{\xi\}\}$ *for some primitive recursive function* $\rho$.

We stated before that if $(Ax)\,\phi\{x\}$ was an **A**-theorem then so is $\phi\{\nu\}$ and that the proofs of $\phi\{\nu\}$ are all of the same length. An example of a universal **A**-statement which fails to be an **A**-theorem for this reason is $(Ax)(x+0 = 0+x)$, because the lengths of the $\mathbf{A}_{00}$-proofs of $\nu+0 = 0+\nu$ are unbounded as $\nu$ increases.

PROP. 10. *The system* **A** *contains an irresolvable statement.*

We have to find a closed **A**-statement $\phi$ such that both $\phi$ and $N[``\phi"]$ fail to be **A**-theorems. Consider $(Ax')\,N[``Prf_{\mathbf{A}}[x',\nu]"]$ where $\nu$ is as in Prop. 3, there we found that $(Ax')\,N[``Prf_{\mathbf{A}}[x',\nu]"]$ was valid but failed to be an **A**-theorem, its negation then takes the truth value $f$ and so by Prop. 2 fails to be an **A**-theorem.

Note that $(Ax')\,N[``Prf_{\mathbf{A}}[x',\nu]"]$ says that '$\nu$' fails to be an **A**-theorem, i.e. that $N[``Prf_{\mathbf{A}}[x', s[\kappa,\kappa]]"]$ fails to be an **A**-theorem and hence that $N[``Prf_{\mathbf{A}}[x',\nu]"]$ fails to be an **A**-theorem because $\nu = s[\kappa,\kappa]$ is an **A**-theorem, but $N[``Prf_{\mathbf{A}}[x',\nu]"]$ is an **A**-theorem if and only if $(Ax')\,N[``Prf_{\mathbf{A}}[x',\nu]"]$ (is an **A**-theorem, thus $(Ax')N[``Prf_{\mathbf{A}}[x',\nu]"]$ states its own unprovability in **A**.

COR. (i). $(Ex)\,Prf_{\mathbf{A}}[x,\nu] \vee (Ax')\,N[``Prf_{\mathbf{A}}[x,\nu]"]$ *fails to be an* **A**-*theorem.*

For if it were then by Prop. 5 so would be one disjunctand, and this we have just seen is absurd. Thus T.N.D. fails in **A**.

PROP. 11. **A**-*falsity fails to be* **A**-*definable.*

We have to show that **A** lacks a statement $Fals\{x\}$ with exactly one free variable, such that:

(i) $Fals\{\nu\}$ can be shown to be either **A**-true or **A**-false,

(ii) $Fals\{\nu\}$ is **A**-true just in case $\nu$ is the *g.n.* of an **A**-false closed **A**-statement,

(iii) $Fals\{\nu\}$ is **A**-false just in case $\nu$ is the *g.n.* of an **A**-true closed **A**-statement or fails to be the *g.n.* of a closed **A**-statement.

Suppose that $Fals\{\nu\}$ is **A**-definable and has the above properties. Consider $Fals\{s[x,x]\}$, let its *g.n.* be $\kappa$, and let $\nu$ be the *g.n.* of $Fals\{s[\kappa,\kappa]\}$, then $\nu = s[\kappa,\kappa]$ is an $\mathbf{A}_{00}$-theorem. Thus $Fals\{\nu\}$ states its own falsity.

If $Fals\{\nu\}$ is **A**-true then '$\nu$' is **A**-false, i.e. $Fals\{s[\kappa,\kappa]\}$ is **A**-false whence $Fals\{\nu\}$ is **A**-false, this is absurd by Prop. 1.

If $Fals\{\nu\}$ is **A**-false then '$\nu$' is **A**-true or fails to be a closed **A**-statement, but '$\nu$' is $Fals\{s[\kappa,\kappa]\}$ which is a closed **A**-statement hence $Fals\{s[\kappa,\kappa]\}$ is **A**-true whence $Fals\{\nu\}$ is **A**-true, but this is absurd. Thus altogether *Fals* fails to be in the system **A**.

D 228 $\quad Clstat_{\mathbf{A}}\,\nu$ for '$\nu$' is a closed **A**-statement.

COR. (i). ***A**-truth fails to be **A**-definable.*

Suppose that $T\nu$ is **A**-definable so that $T\nu$ is **A**-true just in case '$\nu$' is an **A**-true closed **A**-statement. Then $Clstat_{\mathbf{A}}\,\nu\ \&\ N[\text{“}T\nu\text{”}]$ would have the properties required of *Fals*, which is absurd.

The set of **A**-theorems is recursively enumerable and so is **A**-definable, but the set of **A**-true closed **A**-statements fails to be **A**-definable and hence fails to be recursively enumerable. The theorems of any formal system are recursively enumerable, hence the **A**-true closed **A**-statements fail to be the theorems of any formal system.

The argument of Prop. 11 can be applied to any formal system which contains the recursive function $s[x,x]$, or in which this function can be represented, and which contains R 1.

Apply the argument to $\mathbf{A}_{00}$. Here $\mathbf{A}_{00}$-falsity fails to be $\mathbf{A}_{00}$- definable because $\mathbf{A}_{00}$-truth is general recursive and fails to be primitive recursive, negation is available so is $Clstat_{00}$ which is primitive recursive so *Fals* is general recursive and hence unavailable in $\mathbf{A}_{00}$.

Apply the argument to the system $\mathbf{A}_0$. Here the $\mathbf{A}_0$-true statements form a recursively enumerable set which fails to be a recursive set. If *Fals* were $\mathbf{A}_0$-definable then $\mathbf{A}_0$-false closed statements would form a recursively enumerable set, we could then argue as in Prop. 8, Cor. (v) of Ch. 7 that the set of **A**-true closed **A**-statements would be recursive, which is absurd.

## 8.5 *Modus Ponens*

PROP. 12. *Modus Ponens is a derived rule in the system* $\bar{\mathbf{A}}$, $\mathbf{A}$ *plus T.N.D. plus* I(*c*).

First we make a modification to the system $\bar{\mathbf{A}}$. We base it on $\mathscr{F}_C$ instead of on $\mathscr{F}'_C$. This means that we take $N$ as a primitive symbol and use the definitions of $+$, & and $A$ in terms of $=$, $\vee$, $N$ and $E$. Prop. 5 fails in $\bar{\mathbf{A}}$.

We proceed as in Prop. 4, Ch. 3 by induction on the cut formula. We have to show:

$$\frac{\omega \vee \phi \quad N\phi \vee \chi}{\omega \vee \chi}*,$$

where $\chi$ is present, but $\omega$ can be absent.

(*a*) $\phi$ is atomic. Then $\phi$ is an equation $\alpha = \beta$. We proceed by theorem induction on $N\phi \vee \chi$. If this is an axiom T.D.N. then $\chi$ is $\phi$ and we are finished. Use theorem induction on $\Sigma N\phi \vee \chi$, this is to account for all uses of I *b* which occur between $N\phi \vee \chi$ and the next building scheme above it. Let this rule be

$$\frac{\Sigma N\phi \vee \chi'}{\Sigma N\phi \vee \chi},$$

except for permutations. Consider the case when $\Sigma N\phi$ is in the subsidiary formula. Suppose the result holds for the upper formula; then we have

$$\cfrac{\cfrac{\omega \vee \phi \quad \Sigma N\phi \vee \chi'}{\omega \vee \chi'}*}{\omega \vee \chi}$$

using the same rules as before but with $\omega$ as the subsidiary formula instead of $\Sigma N\phi$.

Since $\phi$ is atomic it is an equation and so no part disjunction of $\Sigma N\phi$ can be in the main formula of any rule except II *a*. In this case we shall have

$$\frac{\Sigma' N\phi \vee \chi'}{\Sigma N\phi \vee \chi},$$

where $\Sigma' N\phi$ is a part disjunction of $\Sigma N\phi$ and $\chi$ is $\chi' \vee \chi''$, we have diluted with $\Sigma'' N\phi \vee \chi''$, where $\Sigma N\phi$ is $\Sigma' N\phi \vee \Sigma'' N\phi$.

If the result holds for the upper formula then we have

$$\cfrac{\cfrac{\omega \vee \phi \quad \Sigma' N\phi \vee \chi'}{\omega \vee \chi'}*}{\omega \vee (\chi' \vee \chi'')}\ \text{II}\,a,$$

as desired.

The formula above $\Sigma N\phi \vee \chi$ might be just $\chi'$, where $\chi$ is $\chi' \vee \chi''$, in this case we have

$$\frac{\chi'}{\omega \vee \chi' \vee \chi''} \quad \text{II}\,a,$$

as desired.

Lastly the formula above $\Sigma N\phi \vee \chi$ might be just $\Sigma' N\phi$. In this case $\phi$ is invalid. We now show that if $\omega \vee \phi$ is an $\bar{\mathbf{A}}$-theorem and $\phi$ is an invalid equation then $\omega$ is an $\bar{\mathbf{A}}$-theorem.

Consider the $\bar{\mathbf{A}}$-proof-tree of $\omega \vee \phi$. We define an *invalid ancestor* of $\phi$ as follows: ($\phi$ must be of the form $S\alpha = 0$.)

(i) the occurrence of $\alpha = \beta$ in $\omega \vee \alpha = \beta$ is an occurrence of an invalid ancestor of $\alpha = \beta$,

(ii) if $\dfrac{\chi' \vee \alpha' = \beta'}{\chi \vee \alpha' = \beta'}$ is a case of a one premiss rule, other than cancellation of $\alpha' = \beta'$ by I $b$, and if $\alpha' = \beta'$ is an occurrence of an invalid ancestor of $\alpha = \beta$ then so is the occurrence of $\alpha' = \beta'$ in the upper formula.

(iii) if $\dfrac{\chi' \vee \alpha' = \beta' \vee \alpha' = \beta'}{\chi' \vee \alpha' = \beta'}$ is a cancellation of $\alpha' = \beta'$ by I $b$ and if $\alpha' = \beta'$ is an occurrence of an invalid ancestor in the lower formula then there are two occurrences of an invalid ancestor of $\alpha = \beta$ in the upper formula, etc. for more cancellations,

(iv) if $\dfrac{\chi' \vee \alpha' = \beta' \quad \chi'' \vee \alpha' = \beta'}{\chi''' \vee \alpha' = \beta'}$ is a case of rule II $b$ and if $\alpha' = \beta'$ is an occurrence of an invalid ancestor of $\alpha = \beta$ in the lower formula, then $\alpha' = \beta'$ are two occurrences of an invalid ancestor of $\alpha = \beta$ in the two upper formulae,

(v) if $\dfrac{\chi' \vee \chi \vee \alpha''' = \beta''' \quad \alpha'' = \beta'' \vee \chi}{\chi'' \vee \chi \vee \alpha' = \beta'}$ is a case of R 1, and if $\alpha' = \beta'$ in the lower formula is an invalid ancestor of $\alpha = \beta$, then $\alpha''' = \beta'''$, if invalid, is an invalid ancestor of $\alpha = \beta$, but if $\alpha''' = \beta'''$ is valid then $\alpha'' = \beta''$ must be invalid and is an occurrence of an invalid ancestor of $\alpha = \beta$.

We now omit all invalid ancestors of $\alpha = \beta$. Clearly we are left with a tree. At each node in this tree we have an application of the same rule that was used at that node before in case of rules I $a$, II $a$, $b$, $c$, $d$, $e$ and I $b$ provided that an invalid ancestor is in the subsidiary formula, otherwise we get a repetition. Rule R 1 either remains a case of the same rule, or if an invalid ancestor is a main formula it can become a case of rule II *a*. A-Axioms, being valid are unaffected. T.N.D. reduces to an axiom $S\alpha \neq 0$.

Altogether the new tree is an **A**-proof-tree of $\omega$, as desired. Note that the invalid ancestors of $\alpha = \beta$ must be in the subsidiary formulae of all rules except I *b* and R 1, because otherwise $\alpha = \beta$ would get governed by *ND*, *N*, *E* or *NE* and this is impossible.

This completes the case when $\phi$ is atomic.

The cases when $\phi$ is compound are dealt with exactly as in Prop. 4, Ch. 3. This completes the demonstration of Prop. 12. (See Ex. 8, 3.)

## 8.6 *Consistency*

In Ch. 1 we defined a formal system $\mathscr{L}$ as an ordered quartet $(\mathscr{S}, \mathscr{F}, \mathscr{A}, \mathscr{C})$, where $\mathscr{S}$ is a displayed list of distinct signs, $\mathscr{F}$ is a list of rules of formation, $\mathscr{A}$ is a list of axioms and $\mathscr{C}$ is a list of rules of consequence. In each case it must be possible to decide by a fixed procedure in a limited amount of time whether an object belongs to one of these lists or is foreign to them. The list of $\mathscr{L}$-symbols can be replaced by a list of natural numbers, and $\mathscr{L}$-formulae by ordered sets of natural numbers. Then to be an $\mathscr{L}$-axiom or an $\mathscr{L}$-rule of consequence or to be a well-formed $\mathscr{L}$-formula are, by Church's Thesis, recursive properties of natural numbers and hence expressible by recursive predicates.

Thus a formal system $\mathscr{L}$ can be translated into arithmetic, the natural numbers playing the part of the names of $\mathscr{L}$-formulae or of sequences of $\mathscr{L}$-formulae. We have already carried this out for the systems $\mathbf{A}_{00}$, $\mathbf{A}_0$, **A**. We say that a *formal system* $\mathscr{L}$ *is definable in a formal system* $\mathscr{L}'$ if $\mathscr{L}'$ contains an $\mathscr{L}'$-formula $\mathscr{M}[\phi]$ for every $\mathscr{L}$-formula $\phi$, and an $\mathscr{L}'$-statement $Thm_{\mathscr{L}}\{\xi\}$ with exactly one free variable $\xi$, such that $Thm_{\mathscr{L}}[\mathscr{M}[\phi]]$ is an $\mathscr{L}'$-theorem if and only if $\phi$ is an $\mathscr{L}$-theorem. Here $\mathscr{M}[\phi]$ acts as a name for $\phi$.

If every formal system is $\mathscr{L}$-definable then $\mathscr{L}$ is called a *basic* formal system.

PROP. 13. $\mathbf{A}_0$ *is a basic formal system.*

Let $\mathscr{L}$ be a formal system. We translate $\mathscr{L}$ into arithmetic in the manner just described. Since to be an $\mathscr{L}$-proof is a recursive property then there is a recursive $\mathbf{A}_0$-statement $Prf_{\mathscr{L}}[x', x]$ with exactly two free variables such that $Prf_{\mathscr{L}}[\kappa, \nu]$ is $\mathbf{A}_0$-true if and only if '$\kappa$' is an $\mathscr{L}$-proof of '$\nu$', being $\mathbf{A}_0$-true it is then an $\mathbf{A}_0$-theorem. The result now follows.

COR. (i). *The system* $\mathbf{A}_0$ *is* $\mathbf{A}_0$*-definable.*

We have already given the required $\mathbf{A}_0$-statement $Prf_0[\kappa, \nu]$.

We have just shown that any formal system is $\mathbf{A}_0$-definable and we have shown that $\mathbf{A}_0$ contains all recursive arithmetic. Thus we might attempt to reformulate all the work we have so far done in $\mathbf{A}_0$. In particular we might attempt to formulate in $\mathbf{A}_0$ the results we have obtained about consistency. Anything we have said so far which fails to be $\mathbf{A}_0$-formulable will also fail to be constructive, that is, if, as we do, we identify constructiveness with recursiveness, as in Church's Thesis.

Consistency for a formal system $\mathscr{L}$ can be defined in several different ways.

$Con_{\mathscr{L}}$ (i) Some particular closed $\mathscr{L}$-statement fails to be an $\mathscr{L}$-theorem. If $\mathscr{L}$ contains arithmetic one usually chooses the statement $(0 = 1)$, or rather its counterpart in $\mathscr{L}$.

$Con_{\mathscr{L}}$ (ii) For each closed $\mathscr{L}$-statement $\phi$, either $\phi$ fails to be an $\mathscr{L}$-theorem or $N[``\phi"]$ fails to be an $\mathscr{L}$-theorem, or both may fail to be $\mathscr{L}$-theorems. $N[``\phi"]$ is the negation of $\phi$.

$Con_{\mathscr{L}}$ (iii) Each $\mathscr{L}$-theorem is $\mathscr{L}$-true according to some definition of $\mathscr{L}$-truth, such that $\mathscr{L}$-truth and $\mathscr{L}$-falsity are disjoint properties.

$Con_{\mathscr{L}}$ (iv) $\mathscr{L}$ has a model.

We postpone discussion of models. We now consider the first three kinds of consistency for the systems $\mathbf{A}_{00}$, $\mathbf{A}_0$ and $\mathbf{A}$.

(i) Let $\kappa$ be the *g.n.* of $(0 = 1)$ (this is the same number in all three systems even though some symbols have different *g.n.*'s in the three systems). Then $Con_{\mathscr{L}}$ (i) is $N[``Thm_{\mathscr{L}}[\kappa]"]$.

For the system $\mathbf{A}_{00}$ this fails to be $\mathbf{A}_{00}$-definable but can be $\mathbf{A}_0$-defined and $\mathbf{A}$-defined. It fails to be $\mathbf{A}_{00}$-definable because $Thm_{00}[x]$ and hence its negation $N[``Thm_{00}[x]"]$ are both general recursive and fail to be primitive recursive, they are then $\mathbf{A}_0$- and $\mathbf{A}$-definable. Now $\mathbf{A}_0$ is complete, hence $N[``Thm_{00}[\kappa]"]$ is $\mathbf{A}_0$-provable since it is $\mathbf{A}_0$-true.

(i) $Con_{\mathbf{A}_{00}}$(i) *is* $\mathbf{A}_0$- *and* **A**-*provable but fails to be* $\mathbf{A}_{00}$-*provable.*

For the system $\mathbf{A}_0$, $Con_{\mathbf{A}_0}$(i) fails to be $\mathbf{A}_0$-definable because $N[\text{“}Thm[x]\text{”}]$ fails to correspond to a recursively enumerable set. If it were a recursively enumerable set then $\mathbf{A}_0$ would be decidable. But $Con_{\mathbf{A}_0}$ (i) is **A**-definable, namely $(Ax)\, N[\text{“}Prf_0[x, \kappa]\text{”}]$. Now each of $N[\text{“}Prf_0[\nu, \kappa]\text{”}]$ is an $\mathbf{A}_0$-theorem because it is $\mathbf{A}_0$-true. If $Con_{\mathbf{A}_0}$ (i) were **A**-provable then $N[\text{“}Prf_0[\nu, \kappa]\text{”}]$ would be $\mathbf{A}_{00}$-provable in the same number of steps for each numeral $\nu$. This is absurd because to show that '$\nu$' fails to be an $\mathbf{A}_0$-proof of '$\kappa$' we shall have to find the components of $\nu$, in particular the first component and this operation is unbounded as $\nu$ increases. On the other hand $N[\text{“}Prf_0[\nu, \kappa]\text{”}]$ is an $\mathbf{A}_{00}$-theorem for each $\nu$, to see this we refer to Prop. 1, Ch. 4. There we gave a standard method for proving equations, in this we reduced the proof to the use of Ax. 1 and R 1 only so that only equations between identical terms arose, thus $(0 = 1)$ fails to arise. The system **A** is too weak in rules of proof to be able to prove $Con_{\mathbf{A}_0}$(i).

(ii) $Con_{\mathbf{A}_0}$(i) *fails to be* $\mathbf{A}_0$-*definable, it is* **A**-*definable but fails to be* **A**-*provable.*

For the system **A**, $Con_{\mathbf{A}}$(i) is $N[\text{“}Thm_{\mathbf{A}}[\kappa]\text{”}]$. This is an **A**-statement. Now $(0 = 1)$ is an **A**-theorem if and only if it is an $\mathbf{A}_{00}$-theorem and hence an $\mathbf{A}_0$-theorem, hence if we could **A**-prove $N[\text{“}Thm_{\mathbf{A}}[\kappa]\text{”}]$ then we could **A**-prove $N[\text{“}Thm_{\mathbf{A}_0}[\kappa]\text{”}]$ which we have shown to be absurd.

(iii) $Con_{\mathbf{A}}$(i) *is definable only in* **A** *and fails to be an* **A**-*theorem.*

We note the following relations between the first three definitions of consistency, where we suppose that the $\mathscr{L}$-truth definition is such that $\phi$ and $N[\text{“}\phi\text{”}]$ have opposite truth values and that $0 \neq S\nu$ is an $\mathscr{L}$-theorem:

If $Con_{\mathscr{L}}$(iii) then $Con_{\mathscr{L}}$(ii).
If $Con_{\mathscr{L}}$(ii) then $Con_{\mathscr{L}}$(i).

D 229 *Neg* $\nu$ for the *g.n.* of the result of making the interchanges listed in D 113 to '$\nu$'.

This applies whether '$\nu$' is an **A**-statement or otherwise. Clearly *Neg* $\nu$ is a primitive recursive function of $\nu$.

(ii) $Con_{\mathscr{L}}$(ii) is $(Ax)\,(N[\text{“}Thm_{\mathscr{L}}[x]\text{”}] \vee N[\text{“}Thm_{\mathscr{L}}[Neg\, x]\text{”}])$.

$Con_{\mathbf{A}}$(ii) requires a universal quantifier to express it so can only be **A**-defined. But it fails to be an **A**-theorem because if it were then so would be $N[\text{“}Thm_{\mathbf{A}}[\kappa]\text{”}] \vee N[\text{“}Thm_{\mathbf{A}}[Neg[\kappa]]\text{”}]$ and hence $N[\text{“}Thm_{\mathbf{A}}[\kappa]\text{”}]$

would be an **A**-theorem, which we have shown to be absurd. $Thm_{\mathbf{A}}[Neg[\kappa]]$ is an **A**-theorem so its negation fails to be an **A**-theorem by Prop. 2 hence by Prop. 5 $N[\text{``}Thm_{\mathbf{A}}[\kappa]\text{''}]$ is an **A**-theorem if the disjunction is an **A**-theorem. But this is absurd.

(iii) $Con_{\mathbf{A}}$(ii) *is* **A**-*definable but fails to be an* **A**-*theorem.*

(ii) $Con_{\mathbf{A}_0}$(ii) *fails to arise* because $\mathbf{A}_0$ is without negation.

(i) $Con_{\mathbf{A}_{00}}$(ii) *is* **A**-*definable but fails to be an* **A**-*theorem*, for if it were an **A**-theorem then so would be one of $N[\text{``}Thm_{00}[x]\text{''}]$, $N[\text{``}Thm_{00}[Neg[x]]\text{''}]$ by Prop. 5 and this is absurd.

(iii) $Con_{\mathscr{L}}$(iii) is $(Ax)\,(N[\text{``}Thm_{\mathscr{L}}[x]\text{''}] \vee \mathscr{T}_{\mathscr{L}}[x])$, where $\mathscr{T}_{\mathscr{L}}[\nu]$ says that '$\nu$' is $\mathscr{L}$-true, i.e. $\mathscr{T}_{\mathscr{L}}[\nu]$ is $\mathscr{L}$-true if and only if '$\nu$' is $\mathscr{L}$-true. For the systems $\mathbf{A}_{00}$ and $\mathbf{A}_0$ $\mathscr{T}_{\mathscr{L}}[\nu]$ is $Thm_{\mathscr{L}}[\nu]$ and so for these systems $Con_{\mathscr{L}}$(iii) becomes $(Ax)\,(N[\text{``}Thm_{\mathscr{L}}[x]\text{''}] \vee Thm_{\mathscr{L}}[x])$ this can be **A**-defined only but fails to be an **A**-theorem, for if it were then so would be $N[\text{``}Thm[x]\text{''}]$ or $Thm[x]$ by Prop. 5. But this is absurd. For the system **A** $\mathscr{T}_{\mathbf{A}}$ fails to be **A**-definable by Prop. 11, Cor. (i).

## 8.7 *Truth definitions*

A truth definition for a formal system $\mathscr{L}$ is a statement $T[\nu]$, in an extension $\mathscr{L}'$ of $\mathscr{L}$, such that $\vdash_{\mathscr{L}'} T[\nu] \leftrightarrow \phi$, where $\nu$ is the *g.n.* of $\phi$, i.e. 'it snows' is true if and only if it snows.

PROP. 14. *A formal system which contains a negation and recursive function theory and is consistent fails to contain its truth definition.*

Suppose that the formal system $\mathscr{L}$ contains a negation and recursive function theory and that $T[\nu]$ is an $\mathscr{L}$-statement such that

$$\vdash_{\mathscr{L}} T[\nu] \leftrightarrow \phi. \tag{i}$$

Consider $N[\text{``}T[s[x,x]\text{''}]$, where $s$ is Gödel's substitution function. Let its *g.n.* be $\kappa$, and let $\nu$ be the *g.n.* of $N[\text{``}T[s[\kappa,\kappa]]\text{''}]$, then $\nu = s[\kappa,\kappa]$ is an $\mathbf{A}_{00}$-theorem. We have $\vdash_{\mathscr{L}} T[\nu] \leftrightarrow N[\text{``}T[s[\kappa,\kappa]]\text{''}]$ whence

$$\vdash_{\mathscr{L}} T[\nu] \leftrightarrow N[\text{``}T[\nu]\text{''}]$$

and so $\mathscr{L}$ is inconsistent. Thus if $\mathscr{L}$ is consistent then $T$ fails to be in $\mathscr{L}$.

In order that $T[\nu] \leftrightarrow \phi$, be a theorem of the system $\mathscr{L}'$ it is clear that the system $\mathscr{L}'$ must be somehow related to the system $\mathscr{L}$.

We have already shown that $\mathbf{A}_0$ contains a truth definition for itself and that **A** fails to contain a truth definition for itself. Let us now investigate more closely to see if we can find how the system **A** might be extended so that in the extended system we could have a definition of **A**-truth.

Consider the following primitive recursive properties and functions, all of which are explicitly **A**-definable.

D 230 $pr\nu$: '$\nu$' is an **A**-statement and '$pr\nu$' is equal to that variant of a prenex normal form of '$\nu$' which is obtained from '$\nu$' by moving the quantifiers which occur in '$\nu$' to initial positions in the order in which they occur in '$\nu$' from left to right and relabelling the bound variables with different scopes as $x', x'', \ldots, x^{(\kappa)}$ and relabelling the free variables from left to right as $x^{(S\kappa)}, \ldots, x^{(\kappa+\pi)}$. (If $\kappa$ is zero the first set is absent, if $\pi$ is zero the second set is absent.)

$\nu$ fails to be the *g.n.* of an **A**-statement and $pr\nu$ is zero.

D 231 $st[\nu, \nu', \nu'']$: '$\nu'$' is a variable, '$\nu''$' is a numerical term, '$st[\nu, \nu', \nu'']$' is that **A**-formula $\phi$ which arises from '$\nu$' when in '$\nu$' we everywhere replace the variable '$\nu'$' by the numerical term '$\nu''$'.

$\nu'$ and $\nu''$ fail to be as stated above and $st[\nu, \nu', \nu'']$ is 0.

It follows at once that a closed **A**-statement has the same truth value as any of its prenex normal forms.

D 232 If $\nu = \langle \kappa', \ldots, \kappa^{(\theta)} \rangle$ then $m\nu = \langle \kappa^{(6)}, \ldots, \kappa^{(\theta \dot{-} 2)} \rangle$, if $\theta < 8$ then $m\nu = 0$.

$T_0 \nu$: '$\nu$' is an $\mathbf{A}_{00}$-true $\mathbf{A}_{00}$-statement. This has already been defined. Thus '$\nu$' is without free variables, $E$ or $A$.

$T_\kappa \nu$: '$\nu$' is an **A**-true closed **A**-statement whose prenex normal form has exactly $\kappa$ initially placed quantifiers.

We can define $T_1, T_2, \ldots$ successively by:

$T_{(S\kappa)}\nu$: '$pr\nu$' is $(A\xi)\phi\{\xi\}$ and each of $\phi\{0\}, \phi\{1\}, \ldots$ are closed **A**-statements and are **A**-true, or '$pr\nu$' is $(E\xi)\phi\{\xi\}$ and each of $\phi\{0\}, \phi\{1\}, \ldots$ is a closed **A**-statement and one of them is **A**-true. Also $\phi\{\nu\}$ is in prenex normal form with exactly $\kappa$ initially placed quantifiers.

We can put this down in terms of the recursive predicates and functions we have just defined thus:

$$T_{(S\kappa)}x\colon Clstatx \,\&\, ((Pt[prx, 2] = 12 \,\&\, (Ax'')\, T_\kappa[st[m[prx], 14, Numx'']]) \\ \vee (Pt[prx, 2] = 11 \,\&\, (Ex'')\, T_\kappa[st[m[prx], 14, Numx'']])). \qquad \text{(A)}$$

'11' is $E$, '12' is $A$, '14' is $x'$.

This scheme defines a sequence of **A**-statements for $\kappa = 0, 1, 2, \ldots$. The prenex normal form of $T_\kappa x$ contains $2.\kappa$ initially placed quantifiers. Thus the number of quantifiers increases with $\kappa$, and if we write down $T_\kappa x$ in full primitive notation then the numeral $\kappa$ will fail to occur.

We wish to define a statement $T^*\nu$ which says that '$\nu$' is an **A**-true **A**-statement, i.e. $T^*\nu$ is **A**-true if and only if '$\nu$' is **A**-true. To do this we require to have a statement $T[x, x']$ such that $T[x, \kappa]$ is $T_\kappa x$. Suppose we add such a statement to the system **A**, if necessary, then we can proceed as follows:

$$T^*x \quad \text{for} \quad (Ex')(x' < x \,\&\, prx \,\&\, T[prx, x']).$$

Then $T^*$ is the required truth-predicate. We already know that such a statement is absent from the system **A**. The only place in the above set of definitions where we were uncertain about being in the system **A** is the replacement of the scheme (A) by a similar scheme with a free variable instead of the numeral $\kappa$. Thus we are unable to **A**-define some statements $P$ defined by the scheme:

$$\left.\begin{aligned} P[x, 0] &\equiv Q[x], \\ P[x, Sx'] &\equiv R[x, x', P[\rho[x, x', x''], x']], \end{aligned}\right\} \qquad \text{(B)}$$

where $Q$ and $R$ are previously defined **A**-statements, and the variable $x''$ is bound in $R$ and $\rho$ is a previously **A**-defined function. The symbol $\equiv$ is to mean that if one component is an $\mathscr{L}$-theorem then so is the other, the system $\mathscr{L}$ being one in which $P$ can be defined. In Ch. 11 we shall show that predicates defined by the scheme (B) can be explicitly defined in a primary extension of **A**, in which the equivalences (B) are theorems.

The truth-definitions for the systems $\mathbf{A}_{00}$, $\mathbf{A}_0$ and **A** amount to these systems having a *standard model*. In fact the consistency of these systems with respect to their truth definitions simply says that they have a standard model. In Ch. 12 we shall say something about *non-standard models*.

A formal system is said to be of degree **a** if the *g.n.*'s of its theorems form a set of degree **a**.

## 8.8 *Axiomatizable sets of statements*

A set $\mathscr{S}$ of $\mathscr{L}$-statements is said to be *axiomatizable* if a formal system $\mathscr{L}'$ can be found such that the set $\mathscr{S}$ is contained among the $\mathscr{L}'$-theorems.

Prop. 15. *A set $\mathscr{S}$ of $\mathscr{L}$-statements is axiomatizable if and only if it is recursively enumerable.*

Let $\mathscr{C}$ be the closure of a recursively enumerable set $\mathscr{B}$ under some relation $R$. Thus

$$\mathscr{C} = \bigcap_{\mathscr{X}} (\mathscr{B} \subseteq \mathscr{X} \,\&\, R``\mathscr{X} \subseteq \mathscr{X}),$$

where $R``\mathscr{X}$ is the set of things which stand in the relation $R$ to some member of $\mathscr{X}$, and $\bigcap_{\mathscr{X}} \mathscr{Y}$ (see D 43, 74) is the intersection of the classes $\mathscr{X}$ which satisfy $\mathscr{Y}$. Suppose that there is a primitive recursive relation $Q$ such that $Q$ is a symmetric subrelation of $R$, i.e. $Q\{\nu, \mu\} \rightarrow (Q\{\mu, \nu\} \,\&\, R\{\nu, \mu\})$ for all $\nu, \mu$, and such that for each $\nu \in \mathscr{B}$, $Q\{\nu, \mu\}$ for an unbounded set of $\mu$ (then $R\{\nu, \mu\}$ for an unbounded set of $\mu$ if $\nu \in \mathscr{B}$). Then there exists a primitive recursive set $\mathscr{A}$ such that $\mathscr{C}$ is the closure of $\mathscr{A}$ under $R$. For instance

$\mathscr{A} = \hat{x}(Ex')_x[Q\{fx', x\}]$ (i.e. the set of things which satisfy the displayed condition)

will do, where $f$ enumerates $\mathscr{B}$, and $f$ is primitive recursive. For each $\nu \in \mathscr{B}$ there is a $\mu \in \mathscr{A}$ such that $Q\{\nu, \mu\}$ and hence $Q\{\mu, \nu\}$ therefore the closure of $\mathscr{A}$ under $Q$ and hence under $R$, includes $\mathscr{B}$, hence the closure of $\mathscr{A}$ under $R$ includes $\mathscr{C}$. Conversely since $Q$ is a subrelation of $R$, $\mathscr{A}$ is included in $\mathscr{C}$. Finally $\mathscr{A}$ is primitive recursive. Thus the closure of $\mathscr{A}$ under $R$ is $\mathscr{C}$.

Let $\mathscr{S}$ be a formal system and let $R$ be the relation of deducibility in $\mathscr{S}$ so that $R\{\mu, \nu\}$ if and only if '$\mu$' is a proof of '$\nu$'. Suppose that $\mathscr{S}$ contains conjunction and that $\phi$ and $\phi \,\&\, \phi \,\&\, \phi \,\&\, \ldots \,\&\, \phi$ are interdeducible in $\mathscr{S}$. Let $Q\{\mu, \nu\}$ be that primitive recursive relation which holds if and only if '$\mu$' is $\phi$ and '$\nu$' is $\phi \,\&\, \phi \,\&\, \ldots \,\&\, \phi$. Consider any recursively enumerable set $\mathscr{B}$ of formulae of $\mathscr{S}$, and let $\mathscr{C}$ be the closure of $\mathscr{B}$ under deduction in $\mathscr{S}$. Then, according to the above, there exists a primitive recursive set $\mathscr{A}$ of formulae of $\mathscr{S}$, such that, if $\mathscr{A}$ is added as axiom set to $\mathscr{S}$, then the theorems of the resulting system constitute $\mathscr{C}$. $\mathscr{C}$ is called primitively recursively axiomatizable in $\mathscr{S}$.

In particular suppose that $\mathscr{C}$ is the set of those formulae of $\mathscr{S}$ which are theorems of a system $\mathscr{T}$, where $\mathscr{T}$ contains all the axioms and rules of $\mathscr{S}$. Then, provided that the set of formulae of $\mathscr{S}$ and the set of theorems of $\mathscr{T}$ are recursively enumerable, $\mathscr{C}$ can be primitively recursively axiomatized in $\mathscr{S}$, and hence formalized without the aid of additional symbols or rules of inference. For example, if $\mathscr{T}$ is a system employing higher types which express an analytic theory of numbers, then there exists a system which expresses the corresponding elementary theory of numbers, its theorems being those theorems of $\mathscr{T}$ which are without higher types. Also, for example, if $\mathscr{R}$ is any recursive set of non-logical (individual, function or predicate) constants containing at least one constant predicate, then there exists a system whose theorems are exactly those theorems of $\mathscr{T}$ in which no constants other than those of $\mathscr{R}$ occur. To take a final example, suppose that $\mathscr{S}$ is completable and hence that there exists a complete and consistent system $\mathscr{T}$ whose set of theorems is recursively enumerable and whose axioms and rules include those of $\mathscr{S}$. Then, provided that the set of formulae of $\mathscr{S}$ is recursively enumerable, $\mathscr{S}$ can be completed without use of additional symbols or rules.

If $\mathscr{T}$ is the set of **A**-true **A**-statements and if $R$ is the deducibility relation in **A** and if $\mathscr{B}$ is a recursively enumerable subset of $\mathscr{T}$ and if $\mathscr{C}$ is its closure under $R$, then we can find a primitive recursive subset $\mathscr{A}'$ of $\mathscr{B}$ such that if we form a formal system with $\mathscr{A}'$ as axiom set and $R$ as deducibility relation then $\mathscr{C}$ is its set of theorems, hence $\mathscr{C}$ is recursively enumerable and contains $\mathscr{B}$. Of course in this way we fail to obtain $\mathscr{C} = \mathscr{T}$.

PROP. 16. *The set of* **A**-*true* **A**-*statements is productive.*

We have to give a procedure for finding an **A**-true **A**-statement outside a given recursively enumerable subset of **A**-true **A**-statements.

Let $\mathscr{B}$ be a recursively enumerable subset of $\mathscr{T}$, the set of **A**-true **A**-statements, then we can find a primitive recursive subset $\mathscr{A}$ of $\mathscr{B}$ such that we obtain a formal system $\mathscr{S}$ with $\mathscr{A}$ as axiom set, rules those of **A** and a recursively enumerable set $\mathscr{C}$ of $\mathscr{S}$-theorems. We add the **A**-axioms if necessary so that $\mathscr{S}$ contains recursive number theory. Let $Prf_{\mathscr{S}}$ be the proof predicate for $\mathscr{S}$ and let $s$ be Gödel's substitution function for $\mathscr{S}$. We now proceed exactly as in Prop. 3 and Prop. 10, Cor. (i) and find that $(Ax)\,N[\text{“}Prf_{\mathscr{S}}[x, \nu]\text{”}]$ is an **A**-true **A**-statement which fails to be in the set $\mathscr{C}$ and hence fails to be in the set $\mathscr{B}$. Thus given a recursively enumerable subset $\mathscr{B}$ of $\mathscr{T}$ we have found a member of $\mathscr{T} - \mathscr{B}$. Thus $\mathscr{T}$ is productive.

Cor. (i). *An arithmetic based on the predicate calculus of the first order and containing recursive number theory is incomplete.*

It can be seen from the systems $\mathbf{A}_{00}$, $\mathbf{A}_0$ and **A** that the blame for incompleteness must be given to the universal quantifier, recursive function theory is comparatively blameless.

In the system **A** we have an irresolvable statement $G$, $G$ is **A**-true but fails to be an **A**-theorem. Now **A** is consistent with respect to negation by Prop. 1, thus we can add the statement $G$ as an extra axiom, because $N$["$G$"] fails to be an **A**-theorem, this gives us a system $\mathbf{A}'$ which is a formal system and in which we can repeat what we did for **A** and obtain an irresolvable **A**-statement $G'$, we can then add $G'$ as an extra axiom obtaining a formal system $\mathbf{A}''$ in which we again can find an irresolvable **A**-statement $G''$, we can then add $G''$ as an extra axiom and so on. In this way we generate a sequence of formal systems: $\mathbf{A}, \mathbf{A}', \mathbf{A}'', \ldots$, from these we can form a formal system $\mathbf{A}^{(\omega)}$ obtained from **A** by adding all of $G, G', G'', \ldots$ as extra axioms. But then we can begin all over again and find an irresolvable **A**-statement in $\mathbf{A}^{(\omega)}$, say $G^{(\omega)}$, this we can add to $\mathbf{A}^{(\omega)}$ as an extra axiom obtaining a formal system $\mathbf{A}^{(\omega+1)}$, and so we can continue as long as we continue to produce formal systems. We shall have to stop somewhere in the constructive ordinals. Of course, in this way we fail to obtain a complete formal system. This argument applies to any consistent arithmetic.

The consistency of **A** with respect to negation can be expressed in **A** as: $(Ax)\,N$["$(Thm_{\mathbf{A}}\,x \,\&\, Thm_{\mathbf{A}}[Neg\,x])$"], this is $Con_{\mathbf{A}}$(ii). We have indicated that this is unprovable in **A**.

The completeness of **A** can be expressed in **A** as:

$$(Ax)\,(Stat_{\mathbf{A}}\,x \rightarrow (Thm_{\mathbf{A}}\,x \vee Thm_{\mathbf{A}}\,Neg\,x)),$$

where $Stat_{\mathbf{A}}$ is the primitive recursive predicate such that $Stat_{\mathbf{A}}\,\nu$ is **A**-true if and only if '$\nu$' is an **A**-statement.

Let $G$ be an irresolvable **A**-statement, then both $G$ and $N$["$G$"] fail to be **A**-theorems, hence we can add either of them as an extra axiom. Now $G$ is **A**-valid so it seems natural to add $G$ as an extra axiom, but we can add $N$["$G$"] and still have a consistent system according to Con (ii). Call **A** with $N$["$G$"] as an extra axiom the system $\mathbf{A}_{ng}$. This **A**-statement is **A**-false so to make it $\mathbf{A}_{ng}$-true we should have to add some new elements to the numerals. We should have to add a new element $\alpha$ such that

$Prf_{\mathbf{A}}[\alpha, \nu]$ was $\mathbf{A}_{ng}$-true then we should require as new elements $\rho\alpha$ for every primitive recursive function $\rho$, and so on. We shall see what these new elements are in a later chapter. This gives rise to what are called *non-standard arithmetics*. Here '$\nu$' is $N[``G"]$

Note that rule I *b* is only required when some axioms are disjunctions, for instance if T.N.D. is among the axioms.

## Historical remarks to Chapter 8

We introduce the universal quantifier in a manner similar to the way we introduced the existential quantifier. Instead of the universal quantifier we could have introduced negation. The rule II *d'* is again after Gentzen (1934), but it also occurs in many other formulations of various systems in which universal quantification is required. It means that we have to use free variables. With the advent of universal quantification, or negation, we immediately get incompleteness and other undesirable properties. Prop. 3, the incompleteness of the system **A**, is ultimately due to Gödel (1933, 1934), and the method of demonstration follows his use of the diagonal method of Cantor. Props. 5 and 6 are again the sort of things the intuitionists like, it is the absence of T.N.D. which allows them. Prop. 10, the existence of an unsolvable statement, is again due to Gödel (1933, 1934). It was this result which dashed the hopes of Hilbert (1922) of finding a proof for any true statement.

As the theory of recursive functions progressed the structure of the set of theorems and of the set of true statements was more clearly disclosed, the first is recursively enumerable, the second productive and this fails to be recursively enumerable. Prop. 11, the impossibility of an **A**-definition of **A**-falsehood, and its corollary are due to Tarski (1933). The corollary depends on the system having a negation. Prop. 9 is related to some results of Kreisel (1951) and Prop. 8 is peculiar to the system **A**, it arises from the absence of any form of induction. Considering a formal system as an ordered quartet is due to Carnap (1937).

In Prop. 13 the term 'basic' is due to Fitch (1942) who was the first to produce a basic system of logic. $\mathbf{A}_0$ is a much simpler system than Fitch's.

Consistency was first studied by Hilbert (1904, 1934–6) who wished to show that mathematics is consistent, but according to a theorem of Gödel (1933, 1934) the consistency of a system cannot be proved in that system itself, provided the system is consistent and contains a certain amount of recursive number theory. The discussion that follows the definitions of various kinds of consistency is motivated by Gödel's theorem (Gödel's second incompleteness theorem). The concept of one formal system containing a truth-definition for another formal system is due to Tarski (1933). The investigation to find how the system **A** might be extended in order to contain the truth-definition for the system **A** is due to Bernays (*H–B*, 1936). We shall consider it again in Ch. 11.

Prop. 14 is due to Tarski (1956). Prop. 15 is due to Craig (1953). Prop. 16 is substantially due to Dekker (1955).

Non-standard arithmetic was first invented by Skolem (1934). We consider non-standard arithmetic in Ch. 12.

## Examples 8

1. Show that the scheme II $b'$ is reversible.
2. Show that the Deduction Theorem fails for the system **A**.
3. Show that Modus Ponens is a derived **A**-rule. (Use Prop. 5.)
4. Show that $\dfrac{\phi\{\xi\} \vee \omega}{\phi\{\alpha\} \vee \omega}*$ holds in the system **A** provided that variables free in $\alpha$ are free in $\phi\{\alpha\}$ and that $\xi$ fails to occur in $\omega$.
5. Investigate the system **A** plus T.N.D. and I $b$.
6. Investigate the system **B**, the anti-**A** system.

# Chapter 9

# A-definable sets of lattice points

## 9.1 *The hierarchy of* **A**-*definable sets of lattice points*

Recursive sets of lattice points in $\mathscr{R}_\kappa$ will be called $\mathscr{P}_0^\kappa$ or $\mathscr{Q}_0^\kappa$ sets. These sets are $\mathbf{A}_0$-definable by an $\mathbf{A}_0$-statement of the form $(E\xi)\,\phi\{\xi,\nu,\ldots,\nu^{(\kappa)}\}$, where $\phi\{\xi,\nu,\ldots,\nu^{(\kappa)}\}$ is quantifier-free.

We define a *hierarchy of sets* in **A** as follows:

$$\mathscr{P}_{S\pi}^\kappa\text{-sets are the projections of } \mathscr{Q}_\pi^{S\kappa}\text{-sets,}$$

$$\mathscr{Q}_{S\pi}^\kappa\text{-sets are the complements of } \mathscr{P}_\pi^\kappa\text{-sets.}$$

The complement of a recursive set is another recursive set, thus $\mathscr{Q}_0$ is the same as $\mathscr{P}_0$, the projection of a set in $\mathscr{R}_{S\kappa}$ is given by a statement of the form $(E\xi)\,\phi\{\xi,\nu',\ldots,\nu^{(\kappa)}\}$, the complement of a set defined by $\phi\{\nu',\ldots,\nu^{(\kappa)}\}$ is defined by $N[\text{“}\phi\{\nu',\ldots,\nu^{(\kappa)}\}\text{”}]$. Thus since recursive sets are **A**-definable then so is any set of lattice points obtainable from recursive sets by repeatedly taking complements and projections. For $\pi$ positive:

$\mathscr{P}_\pi^\kappa$-sets are **A**-definable by **A**-statements of the form

$$(E\xi')\,(A\xi'')\ldots\phi\{\xi',\xi'',\ldots,\xi^{(\pi)},\nu',\ldots,\nu^{(\kappa)}\}.$$

$\mathscr{Q}_\pi^\kappa$-sets are **A**-definable by **A**-statements of the form

$$(A\xi')\,(E\xi'')\ldots\phi\{\xi',\xi'',\ldots,\xi^{(\pi)},\nu',\ldots,\nu^{(\kappa)}\}.$$

In each case there are $\pi$ initially placed alternating quantifiers and $\phi$ is quantifier-free, the initial set of quantifiers are alternately universal and existential, the matrix is primitive recursive. Any **A**-statement is equivalent to one of these forms. First, an **A**-statement is equivalent to any of its prenex normal forms. Secondly, a batch of consecutive quantifiers of the same kind can be replaced by a single quantifier of that kind leaving an equivalent **A**-statement, thus

$$(A\xi')\,(A\xi'')\ldots(A\xi^{(\lambda)})\,\phi\{\xi',\xi'',\ldots,\xi^{(\lambda)}\}$$

is **A**-equivalent to

$$(A\xi)\,\phi\{pt[\xi,\lambda,1],\ldots,pt[\xi,\lambda,\lambda]\}.$$

This process is called *contraction of quantifiers.*

PROP. 1. (i) *If $\phi$ is a $\mathscr{P}^{\kappa}_{\pi}$-statement then $N[``\phi"]$ is a $\mathscr{Q}^{\kappa}_{\pi}$-statement, if $\phi$ is a $\mathscr{Q}^{\kappa}_{\pi}$-statement then $N[``\phi"]$ is a $\mathscr{Q}^{\kappa}_{\pi}$-statement.*

(ii) *Suppose that $\phi\{\eta', \eta'', ..., \eta^{(\kappa)}\}$ is equivalent to $\psi\{\eta^{(\theta')}, ..., \eta^{(\theta^{(\kappa)})}\}$ where $\theta', \theta'', ..., \theta^{(\kappa)}$ is a permutation of $1, 2, ..., \kappa$ then $\phi$ is a $\mathscr{P}^{\kappa}_{\pi}$-statement if and only if $\psi$ is a $\mathscr{P}^{\kappa}_{\pi}$-statement, and $\phi$ is a $\mathscr{Q}^{\kappa}_{\pi}$-statement if and only if $\psi$ is a $\mathscr{Q}^{\kappa}_{\pi}$-statement.*

(iii) *If $\phi$ is a $\mathscr{P}^{\kappa}_{\pi}$-statement then $\phi$ & $(\eta^{(S\kappa)} = \eta^{(S\kappa)})$ and $\phi \vee (\eta^{(S\kappa)} \neq \eta^{(S\kappa)})$ are equivalent $\mathscr{P}^{(S\kappa)}_{\pi}$-statements, and conversely.*

(iv) *If $\phi\{\eta, \eta', ..., \eta^{(\kappa)}\}$ is a $\mathscr{P}^{(S\kappa)}_{\pi}$-statement then $(E\eta)\,\phi\{\eta, \eta', ..., \eta^{(\kappa)}\}$ is a $\mathscr{P}^{\kappa}_{\pi}$-statement.*

(i) Follows at once by definition and the fact that the negation of a recursive statement is another recursive statement.

(ii) $\psi\{\eta^{(\theta')}, ..., \eta^{(\theta^{(\kappa)})}\}$ is equivalent to

$$\phi\{U^{\lambda'}_{\kappa}[\eta^{(\theta')}, ..., \eta^{(\theta^{(\kappa)})}], ..., U^{\lambda^{(\kappa)}}_{\kappa}[\eta^{(\theta')}, ..., \eta^{(\theta^{(\kappa)})}]\},$$

where $U^{\kappa'}_{\lambda}, ..., U^{\lambda^{(\kappa)}}_{\kappa}$ are identity functions and $\lambda', ..., \lambda^{(\kappa)}$ is the inverse permutation to the given one so that $\theta^{(\lambda')} = 1, ..., \theta^{(\lambda^{(\kappa)})} = \kappa$. Thus by the scheme of substitution $\psi$ is primitive recursive if $\phi$ is.

(iii) and (iv) follow at once by equivalence of prenex normal form and contraction of quantifiers.

PROP. 2. (i) *A $\mathscr{P}^{\kappa}_{\pi}$-statement is equivalent to a $\mathscr{P}^{\kappa}_{S\pi}$-statement and to a $\mathscr{Q}^{\kappa}_{S\pi}$-statement.*

(ii) *A $\mathscr{Q}^{\kappa}_{\pi}$-statement is equivalent to a $\mathscr{P}^{\kappa}_{S\pi}$-statement and to a $\mathscr{Q}^{\kappa}_{S\pi}$-statement.*

Let $\psi\{\xi', ..., \xi^{(\pi)}, \eta', ..., \eta^{(\kappa)}\}$ be the matrix of a $\mathscr{P}^{\kappa}_{\pi}$-statement, this is equivalent to

$$(A\xi^{(S\pi)})\,(\psi\{\xi', ..., \xi^{(\pi)}, \eta', ..., \eta^{(\kappa)}\} \;\&\; (\xi^{(S\pi)} = \xi^{(S\pi)}))$$

and to $$(E\xi^{(S\pi)})\,(\psi\{\xi', ..., \xi^{(\pi)}, \eta', ..., \eta^{(\kappa)}\} \vee (\xi^{(S\pi)} \neq \xi^{(S\pi)})).$$

Thus the matrix of a $\mathscr{P}^{\kappa}_{\pi}$-statement can be changed to either of the above, we chose the first if $\pi$ is odd and the second if $\pi$ is even. This will convert the $\mathscr{P}^{\kappa}_{\pi}$-statement into an equivalent $\mathscr{P}^{\kappa}_{S\pi}$-statement. A similar replacement will convert a $\mathscr{Q}^{\kappa}_{\pi}$-statement into an equivalent $\mathscr{Q}^{\kappa}_{S\pi}$-statement. By Prop. 1 $\phi$ & $(\eta^{(S\kappa)} = \eta^{(S\kappa)})$ is a $\mathscr{P}^{S\kappa}_{\pi}$-statement if and only if $\phi$ is a $\mathscr{P}^{\kappa}_{\pi}$-statement, also by Prop. 1 $N[``\phi"] \vee (\eta^{(S\kappa)} \neq \eta^{(S\kappa)})$ is a $\mathscr{Q}^{S\kappa}_{\pi}$-statement if and only if $\phi$ is a $\mathscr{P}^{\kappa}_{\pi}$-statement. Thus $(E\eta^{(S\kappa)})\,(N[``\phi"] \vee (\eta^{(S\kappa)} \neq \eta^{(S\kappa)}))$ is a $\mathscr{P}^{\kappa}_{S\pi}$-statement, it is equivalent to $N[``\phi"]$. Thus $\phi$ is equivalent to a $\mathscr{Q}^{\kappa}_{S\pi}$-statement. A similar argument shows that a $\mathscr{Q}^{\kappa}_{\pi}$-statement is equivalent to a $\mathscr{P}^{\kappa}_{S\pi}$-statement. This completes the demonstration of the proposition.

We showed in Ch. 7 that a general recursive set of lattice points in $\mathscr{R}_{\kappa}$ is $\mathbf{A}_0$-representable by the $\mathbf{A}_0$-statement

$$(E\xi)\,(Un^{(\kappa)}[\theta, \nu', \ldots, \nu^{(\kappa)}, \xi, 0] = 0),$$

for some numeral $\theta$, where $Un^{(\kappa)}$ is a primitive recursive function. Thus, by contraction of quantifiers a $\mathscr{P}_1$-statement is equivalent to an **A**-statement of the form

$$(E\xi)\,(Un^{(S\kappa)}[\theta, pt[\xi, 2, 1], \nu', \ldots, \nu^{(\kappa)}, pt[\xi, 2, 2], 0] = 0).$$

Generally a $\mathscr{P}^{\kappa}_{\pi}$-statement is equivalent to an **A**-statement of the form

$$(E\xi')\,(A\xi'')\ldots(Un^{(\kappa+\pi+1)}[\theta, \xi^{(\pi)}_1, \xi', \ldots, \xi^{(\pi \dot- 1)}, \nu', \ldots, \nu^{(\kappa)}, \xi^{(\pi)}_2, 0] = 0),$$

similarly a $\mathscr{Q}^{\kappa}_{\pi}$-statement is equivalent to an **A**-statement of the form

$$(A\xi')\,(E\xi'')\ldots(Un^{(\kappa+\pi+1)}[\theta', \xi^{(\pi)}_1, \xi', \ldots, \xi^{(\pi \dot- 1)}, \nu', \ldots, \nu^{(\kappa)}, \xi^{(\pi)}_2, 0] \neq 0).$$

Conversely statements of the two above forms are $\mathscr{P}^{\kappa}_{\pi}$- $\mathscr{Q}^{\kappa}_{\pi}$-statements respectively. Here $\theta$ is a numeral. Thus we have an enumeration of $\mathscr{P}^{\kappa}_{\pi}$- and of $\mathscr{Q}^{\kappa}_{\pi}$-statements. These sets will be called $\mathscr{P}^{\kappa}_{\pi}[\theta]$, $\mathscr{Q}^{\kappa}_{\pi}[\theta]$ respectively. If we wish to exhibit the arguments we write $\mathscr{P}^{\kappa}_{\pi}[\theta; \nu', \ldots, \nu^{(\kappa)}]$, etc.

PROP. 3. *We can find a $\mathscr{P}^{\kappa}_{\pi}$-statement which fails to be equivalent to any $\mathscr{Q}^{\kappa}_{\pi}$-statement, we can find a $\mathscr{Q}^{\kappa}_{\pi}$-statement which fails to be equivalent to any $\mathscr{P}^{\kappa}_{\pi}$-statement.*

Consider the **A**-statement

$$N[``\mathscr{P}^{\kappa}_{\pi}[\eta; \eta, \eta'', \ldots, \eta^{(\kappa)}]"]. \tag{1}$$

This is

$$(A\xi')\,(E\xi'')\ldots(Un^{(\kappa+\pi)}[\eta, \xi^{(\pi)}_1, \xi', \ldots, \xi^{(\pi \dot- 1)}, \eta, \eta'', \ldots, \eta^{(\kappa)}, \xi^{(\pi)}_2, 0] \neq 0),$$

where there are $\pi$ initially placed quantifiers. This **A**-statement is equivalent to $\mathscr{Q}^{\kappa}_{\pi}[\theta; \eta, \eta'', \ldots, \eta^{(\kappa)}]$ for some numeral $\theta$. Suppose that (1) is equivalent to $\mathscr{P}^{\kappa}_{\pi}[\theta''; \eta, \eta'', \ldots, \eta^{(\kappa)}]$, replace $\eta$ by $\theta''$ and we obtain an absurdity. Hence (1) is a $\mathscr{Q}^{\kappa}_{\pi}$-statement which fails to be equivalent to any $\mathscr{P}^{\kappa}_{\pi}$-statement. In a similar manner we can find a $\mathscr{P}^{\kappa}_{\pi}$-statement which fails to be equivalent to a $\mathscr{Q}^{\kappa}_{\pi}$-statement.

Cor. (i). *We can find a $\mathscr{P}^{\kappa}_{S\pi}$-statement which fails to be equivalent to any $\mathscr{P}^{\kappa}_{\pi}$-statement, we can find a $\mathscr{Q}^{\kappa}_{S\pi}$-statement which fails to be equivalent to any $\mathscr{Q}^{\kappa}_{\pi}$-statement.*

The $\mathscr{Q}^{\kappa}_{\pi}$-statement $\mathscr{Q}^{\kappa}_{\pi}[\theta; \eta, \eta'', \ldots, \eta^{(\kappa)}]$ of Prop. 3 is equivalent to a $\mathscr{P}^{\kappa}_{S\pi}$-statement. Hence we have found a $\mathscr{P}^{\kappa}_{S\pi}$-statement which fails to be equivalent to any $\mathscr{P}^{\kappa}_{\pi}$-statement. In a similar manner we can find a $\mathscr{Q}^{\kappa}_{S\pi}$-statement which fails to be equivalent to any $\mathscr{Q}^{\kappa}_{\pi}$-statement. Altogether we have:

Prop. 4. *There exists a hierarchy of* **A**-*definable sets of lattice points in $\mathscr{R}_{\kappa}$ defined by* **A**-*statements of the forms*

$$\mathscr{P}^{\kappa}_{0}[\theta]\ (\mathscr{Q}^{\kappa}_{0}[\theta])\ \ \mathscr{P}^{\kappa}_{1}[\theta]\ \ \mathscr{P}^{\kappa}_{2}[\theta] \ldots \kappa = 0, 1, 2, \ldots$$

$$\mathscr{Q}^{\kappa}_{1}[\theta]\ \ \mathscr{Q}^{\kappa}_{2}[\theta] \ldots \theta = 0, 1, 2, \ldots.$$

*Each type contains a set which fails to be equivalent to any of the earlier types and to the other type in the same column (if any).*

Prop. 5. *$\mathscr{P}^{\kappa}_{\pi}$-sets form a ring, so do $\mathscr{Q}^{\kappa}_{\pi}$-sets.*

We have already shown this for $\mathscr{P}^{\kappa}_{0}$-sets for all $\kappa$, Prop. 9 (i), Ch. 7. Suppose that the result holds for $\mathscr{P}^{\kappa}_{\pi}$-sets, and $\mathscr{Q}^{\kappa}_{\pi}$-sets for all $\kappa$.

$$\mathscr{P}^{\kappa}_{S\pi}[\theta; \eta', \ldots, \eta^{(\kappa)}] \vee \mathscr{P}^{\kappa}_{S\pi}[\theta'; \eta', \ldots, \eta^{(\kappa)}]$$

is equivalent to

$$(E\xi')\ \mathscr{Q}^{S\kappa}_{\pi}[\theta''; \xi', \eta', \ldots, \eta^{(\kappa)}] \vee (E\xi'')\ \mathscr{Q}^{S\kappa}_{\pi}[\theta'''; \xi'', \eta', \ldots, \eta^{(\kappa)}],$$

and this in turn is equivalent to

$$(E\xi)\,(\mathscr{Q}^{S\kappa}_{\pi}[\theta''; \xi_1, \eta', \ldots, \eta^{(\kappa)}] \vee \mathscr{Q}^{S\kappa}_{\pi}[\theta'''; \xi_2, \eta', \ldots, \eta^{(\kappa)}]) \quad \text{where} \quad \xi = \{\xi_1, \xi_2\},$$

by induction hypothesis this is equivalent to

$$(E\xi)\ \mathscr{Q}^{S\kappa}_{\pi}[\theta^{\text{iv}}; \xi, \eta', \ldots, \eta^{(\kappa)}],$$

and this is a $\mathscr{P}^{\kappa}_{S\pi}$-statement. Again

$$\mathscr{P}^{\kappa}_{S\pi}[\theta; \eta', \ldots, \eta^{(\kappa)}]\ \&\ \mathscr{P}^{\kappa}_{S\pi}[\theta'; \eta', \ldots, \eta^{(\kappa)}]$$

is equivalent to

$$(E\xi')\ \mathscr{Q}^{S\kappa}_{\pi}[\theta''; \xi', \eta', \ldots, \eta^{(\kappa)}]\ \&\ (\xi'' = \xi'')$$
$$\&\ (E\xi'')\ \mathscr{Q}^{S\kappa}_{\pi}[\theta'''; \xi'', \eta', \ldots, \eta^{(\kappa)}]\ \&\ (\xi' = \xi')$$

which is equivalent to

$$(E\xi', \xi'')\, \mathscr{Q}_{\pi}^{SS\kappa}[\theta^{\text{iv}}; \xi', \xi'', \eta', \ldots, \eta^{(\kappa)}] \quad \text{by induction hypothesis,}$$

but this is equivalent to a $\mathscr{P}_{S\pi}^{\kappa}$-statement by contraction of quantifiers. In a similar manner we deal with $\mathscr{Q}_{S\pi}^{\kappa}$-statements. This completes the demonstration of the proposition.

## 9.2 $\Delta_{\pi}^{\kappa}$-*sets*

An **A**-statement which is equivalent to a $\mathscr{P}_{\pi}^{\kappa}$-statement and also is equivalent to a $\mathscr{Q}_{\pi}^{\kappa}$-statement will be called a $(\mathscr{P}_{\pi}^{\kappa} \cap \mathscr{Q}_{\pi}^{\kappa})$-statement, and denoted by $\Delta_{\pi}^{\kappa}$.

COR. (i). *$(\mathscr{P}_{\pi}^{\kappa} \cap \mathscr{Q}_{\pi}^{\kappa})$-statements form a field.*

PROP. 6. *A $(\mathscr{P}_{S\pi}^{\kappa} \cap \mathscr{Q}_{S\pi}^{\kappa})$-statement is $\mathscr{P}_{\pi}^{S\kappa}$-recursive also $\mathscr{Q}_{\pi}^{S\kappa}$-recursive.*
Suppose that $\phi\{\xi', \ldots, \xi^{(\kappa)}\}$ is equivalent to a $\mathscr{P}_{S\pi}^{\kappa}$-statement

$$(E\eta)\, \psi\{\eta, \xi', \ldots, \xi^{(\kappa)}\}$$

and also is equivalent to a $\mathscr{Q}_{S\pi}^{\kappa}$-statement

$$(A\eta)\, \chi\{\eta, \xi', \ldots, \xi^{(\kappa)}\}.$$

Then $N[``\phi\{\xi', \ldots, \xi^{(\kappa)}\}"]$ is equivalent to the $\mathscr{P}_{S\pi}^{\kappa}$-statement

$$(E\eta)\, N[``\chi\{\eta, \xi', \ldots, \xi^{(\kappa)}\}"].$$

Now

$$\phi\{\xi', \ldots, \xi^{(\kappa)}\} \vee N[``\phi\{\xi', \ldots, \xi^{(\kappa)}\}"]$$

is **A**-valid (it may fail to be an **A**-theorem) whence

$$(E\eta)\, (\psi\{\eta, \xi', \ldots, \xi^{(\kappa)}\} \vee N[``\chi\{\eta, \xi', \ldots, \xi^{(\kappa)}\}"])$$

is **A**-valid. Hence

$$(\mu\eta)\, [\psi\{\eta, \xi', \ldots, \xi^{(\kappa)}\} \vee N[``\chi\{\eta, \xi', \ldots, \xi^{(\kappa)}\}]"]$$

may be used if we adjoin the unlimited least number symbol to the system **A**. Call this term $f[\xi', \ldots, \xi^{(\kappa)}]$.

If $\psi\{f[\nu', \ldots, \nu^{(\kappa)}], \nu', \ldots, \nu^{(\kappa)}\}$ is **A**-true so is $(E\eta)\, \psi\{\eta, \nu', \ldots, \nu^{(\kappa)}\}$ whence so is $\phi\{\nu', \ldots, \nu^{(\kappa)}\}$. Again

$$(E\eta)\, (\psi\{\eta, \nu', \ldots, \nu^{(\kappa)}\} \vee N[``\chi\{\eta, \nu', \ldots, \nu^{(\kappa)}\}"]) \tag{2}$$

is **A**-true, whence so is

$$\psi\{f[\nu',\ldots,\nu^{(\kappa)}],\nu',\ldots,\nu^{(\kappa)}\} \vee N[\text{``}\chi\{f[\nu',\ldots,\nu^{(\kappa)}],\nu',\ldots,\nu^{(\kappa)}\}\text{''}] \tag{3}$$

whence so is

$$\psi\{f[\nu',\ldots,\nu^{(\kappa)}],\nu',\ldots,\nu^{(\kappa)}\} \vee (E\eta)\, N[\text{``}\chi\{\eta,\nu',\ldots,\nu^{(\kappa)}\}\text{''}]$$

whence so is

$$\psi\{f[\nu',\ldots,\nu^{(\kappa)}],\nu',\ldots,\nu^{(\kappa)}\} \vee N[\text{``}\phi\{\nu',\ldots,\nu^{(\kappa)}\}\text{''}]$$

whence, if $\phi\{\nu',\ldots,\nu^{(\kappa)}\}$ is **A**-true then so is $\psi\{f[\nu',\ldots,\nu^{(\kappa)}],\nu',\ldots,\nu^{(\kappa)}\}$. Thus $\phi\{\nu',\ldots,\nu^{(\kappa)}\}$ and $\psi\{f[\nu',\ldots,\nu^{(\kappa)}],\nu',\ldots,\nu^{(\kappa)}\}$ have the same **A**-truth value.

In a similar manner we show that $N[\text{``}\phi\{\nu',\ldots,\nu^{(\kappa)}\}\text{''}]$ has the same **A**-truth value as $N[\text{``}\chi\{f[\nu',\ldots,\nu^{(\kappa)}],\nu',\ldots,\nu^{(\kappa)}\}\text{''}]$, whence $\phi\{\nu',\ldots,\nu^{(\kappa)}\}$ has the same **A**-truth value as $\chi\{f[\nu',\ldots,\nu^{(\kappa)}],\nu',\ldots,\nu^{(\kappa)}\}$. Now $\psi\{\nu,\nu',\ldots,\nu^{(\kappa)}\}$ is a $\mathcal{Q}_\pi^{S\kappa}$-statement and $\chi\{\nu,\nu',\ldots,\nu^{(\kappa)}\}$ is a $\mathcal{P}_\pi^{S\kappa}$-statement. Thus a $(\mathcal{P}_{S\pi}^{\kappa} \cap \mathcal{Q}_{S\pi}^{\kappa})$-statement is obtained from a $\mathcal{P}_\pi^{S\kappa}$-statement by substituting $f[\nu',\ldots,\nu^{(\kappa)}]$ for $\nu$, it can be similarly obtained from a $\mathcal{Q}_\pi^{S\kappa}$-statement. Now the function $f$ is recursive in $\mathcal{Q}_\pi^{S\kappa}$-statements, because its value can be found if we are given the truth-values of certain $\mathcal{Q}_\pi^{S\kappa}$-statements, in our case the statement (3) with $0, 1, 2, \ldots$ successively replacing $f[\nu',\ldots,\nu^{(\kappa)}]$, the search terminates since (2) is **A**-true. Thus $\phi$ is recursive in $\mathcal{Q}_\pi^{S\kappa}$-statements, similarly it is recursive in $\mathcal{P}_\pi^{S\kappa}$-statements.

COR. (i). *A* $(\mathcal{P}_1^{\kappa} \cap \mathcal{Q}_1^{\kappa})$*-statement is recursive.*

This is Prop. 8, Cor. (v), Ch. 7.

COR. (ii). *A* $(\mathcal{P}_{S\pi}^{\kappa} \cap \mathcal{Q}_{S\pi}^{\kappa})$*-statement has degree* $\leqslant O^{(\pi)}$.

$\mathcal{P}_\pi^{\kappa}$-statements and $\mathcal{Q}_\pi^{\kappa}$-statements are of degree $\leqslant O^{(\pi)}$.

PROP. 7. *An* **A**-*statement* $\mathcal{P}_\pi^{\kappa}$*-recursive or* $\mathcal{Q}_\pi^{\kappa}$*-recursive is equivalent to a* $(\mathcal{P}_{S\pi}^{\kappa} \cap \mathcal{Q}_{S\pi}^{\kappa})$*-statement.*

Let the **A**-statement $\phi$ be $\psi$-recursive, where $\psi$ is a $\mathcal{Q}_\pi^{\kappa}$-statement. Then the **A**-truth-value of $\phi$ is a recursive function of the **A**-truth-value of $\psi$. Thus for numerals $\nu',\ldots,\nu^{(\kappa)}$ $\phi\{\nu',\ldots,\nu^{(\kappa)}\}$ is equivalent to:

$$(E\xi)\,(Un_1^{(\kappa)}[\theta,\nu',\ldots,\nu^{(\kappa)},\mathfrak{f}[\xi],0]=0),$$

for some $\theta$, where

$$\mathfrak{f}[\xi] = \langle f[\xi',\ldots,\xi^{(\kappa)}]\rangle_{0\leqslant\{\xi',\ldots,\xi^{(\kappa)}\}\leqslant\xi},$$

and $\mathfrak{f}$ is the characteristic function of $\psi\{\xi',\ldots,\xi^{(\kappa)}\}$. Thus $\phi\{\nu',\ldots,\nu^{(\kappa)}\}$ is equivalent to

$$(E\xi,\eta)\,((Un^{(\kappa)}[\theta,\nu',\ldots,\nu^{(\kappa)},\eta,\xi,0]=0)$$
$$\&\,(A\zeta)_{(S\xi)}Pt[\eta,\zeta]=0\leftrightarrow\psi\{pt[\zeta,\kappa,1],\ldots,pt[\zeta,\kappa,\kappa]\})). \qquad (4)$$

Now $\psi$ is a $\mathcal{Q}_\pi^\kappa$-statement, hence its prefix begins with a universal quantifier, say it is $A\,E\,A\,E\,A$ (omitting variables), so that $\psi$ is a $\mathcal{Q}_5^\kappa$-statement, then $N[\text{``}\psi\text{''}]$ has the prefix $E\,A\,E\,A\,E$.

Now $\chi'\leftrightarrow\chi''$ is an abbreviation for $(N[\text{``}\chi'\text{''}]\vee\chi'')\,\&\,(N[\text{``}\chi''\text{''}]\vee\chi')$. If $\chi'$ is quantifier-free and if $\chi''$ is $\psi$ then $\chi'\leftrightarrow\psi$ is of the form:

$$(N[\text{``}\chi'\text{''}]\vee A\,E_1\,A\,E_2\,A\psi')\,\&\,(\chi'\vee E_3\,A\,E_4\,A\,E_5\,N[\text{``}\psi'\text{''}]), \qquad (5)$$

where $\psi'$ is quantifier-free. $E_\theta$ is short for $(Ev^{(\theta)})$. (5) is equivalent to

$$E_3\,A\,E_1\,E_4\,A\,E_2\,E_5\,A((N[\text{``}\chi'\text{''}]\vee\psi')\,\&\,(\chi'\vee N[\text{``}\psi'\text{''}])).$$

By contraction of quantifiers this is equivalent to an **A**-statement of the form: $E\,A\,E\,A\,E\,A\omega$, where $\omega$ is quantifier-free. The second clause in (4) is of the form

$$(A\zeta)_{S\xi}(E\eta)\,A\,E\,A\,E\,A\omega\{\zeta,\eta\}.$$

This is equivalent to

$$(E\eta)\,(A\zeta)_{S\xi}\,A\,E\,A\,E\,A\omega\{\zeta,pt[\eta,\xi,\zeta]\}$$

and this is equivalent to $E\,A\,E\,A\,E\,A\omega'$, where $\omega'$ is quantifier-free, by bringing quantifiers to the left and contracting.

Thus (4), by contraction of quantifiers is equivalent to $E\,A\,E\,A\,E\,A\omega''$ where $\omega''$ is quantifier-free. Thus (4) is equivalent to a $\mathcal{P}_6^\kappa$-statement. In general if $\psi$ is a $\mathcal{Q}_\pi^\kappa$-statement then (4), and hence $\phi$, is equivalent to a $\mathcal{P}_{S\pi}^\kappa$-statement. A $\mathcal{P}_\pi^\kappa$-statement is recursive in a $\mathcal{Q}_\pi^\kappa$-statement, namely its negation. Hence we need only consider the case when $\phi$ is recursive in $\mathcal{Q}_\pi^\kappa$-statements. Again $N[\text{``}\phi\text{''}]$ is recursive in $\phi$ and hence in $\mathcal{Q}_\pi^\kappa$-statements. Thus $N[\text{``}\phi\text{''}]$ is equivalent to a $\mathcal{P}_{S\pi}^\kappa$-statement, whence $\phi$ is equivalent to a $\mathcal{Q}_{S\pi}^\kappa$-statement. Altogether $\phi$ is equivalent to a $\mathcal{P}_{S\pi}^\kappa$-statement and to a $\mathcal{Q}_{\pi S}^\kappa$- statement, and so is equivalent to a $(\mathcal{P}_{\pi S}^\kappa\cap\mathcal{Q}_{S\pi}^\kappa)$-statement.

We proceed similarly if $\phi$ is recursive in more than one $\mathcal{Q}_\pi^\kappa$-statement, the only alteration to the quantifiers is that $(E\xi,\eta)$ in (4) becomes $(E\xi,\eta',\ldots,\eta^{(\mu)})$ which can be contracted into a single quantifier.

## 9·3 *Sets undefinable in* A

PROP. 8. *There is a set of natural numbers which fails to be* **A**-*definable.*

Consider the set of numerals which satisfy $N[``\mathscr{P}^1_{\nu_1}[\nu_2, \nu]"]$, where $\nu = \{\nu_1, \nu_2\}$, if it is **A**-definable then it is equivalent to $\mathscr{P}^1_\theta[\kappa, \nu]$ for some numerals $\theta, \kappa$. We should then have for each numeral $\nu$ $N[``\mathscr{P}^1_{\nu_1}[\nu_2, \nu]"]$ is **A**-true if and only if $\mathscr{P}^1_\theta[\kappa, \nu]$ is **A**-true, if we take $\nu = \{\theta, \kappa\}$ we obtain an absurdity. Thus we have given a description of a set which fails to be **A**-definable. The condition that a numeral $\nu$ satisfies the conditions requires varying numbers of quantifiers according to $\nu_1$ and this number is unbounded. The system **A** is too poor in modes of expression to be able to define such a set.

The complete set has degree $\mathbf{O}'$, it is the set $\{\theta, \kappa\}$, where table $\theta$ produces $\kappa$. The set $\mathbf{O}''$ is the set $\{\theta, \kappa\}$ where table $\theta^{\mathbf{O}'}$, i.e. table with *g.n.* $\theta$ recursive in the complete set of degree $\mathbf{O}'$, produces $\kappa$, and so on. $\mathbf{O}^{(\omega)}$ is the degree of the function $f^{(\omega)}$ such that $f^{(\omega)}\nu = f^{(\nu_1)}\,\nu_2$, where $f^{(\nu_1)}$ gives the complete set of degree $\nu_1$.

$$f'\nu = 0 \leftrightarrow (E\eta)\,(Un^1[\nu_1, \eta_1, \eta_2, \nu_2] = 0),$$

then table $\nu_1$ produces $\nu_2$.

$$f''\nu = 0 \leftrightarrow (E\eta)\,(Un^1_1[\nu_1, \eta_1, \mathfrak{f}'\eta_2, \eta_2, \nu_2] = 0),$$

then table $\nu_1$ recursive in $f'$ produces $\nu_2$.

$$\leftrightarrow (E\eta)\,(E\xi)\,(Un^1_1[\nu_1, \eta_1, \xi, \eta_2, \nu_2] = 0$$
$$\& \, (A\zeta)_{\eta_2}(Pt[\xi, \zeta] = 0 \leftrightarrow (E\zeta')\,(Un^1[\zeta_1, \zeta_1', \zeta_2', \zeta_2] = 0)))$$
$$\leftrightarrow (E\eta)\,(A\zeta')\,R\{\eta, \zeta'\}, \quad \text{where } R \text{ is recursive.}$$

Generally
$$f^{(S\mu)}\nu = 0 \leftrightarrow (E\eta)\,(A\eta')\ldots(Q\eta^{(\mu)})\,R^{(\mu)}\{\eta, \eta', \ldots, \eta^{(\mu)}\},$$

where $R^{(\mu)}$ is recursive and $Q$ is $E$ or $A$ according as $\mu$ is odd or even. We define $f^{(\omega)}$ by: $f^{(\omega)}\nu = f^{(\nu_1)}\nu_2$, so that $f^{(\omega)}\nu = 0$ involves a varying number of quantifiers as $\nu$ varies, and this number is unbounded.

Having obtained $f^{(\omega)}$ we can continue and get $f^{(\omega+1)}, f^{(\omega+2)}, \ldots$ and so on into the constructive ordinals. $f^{(\omega)}$ is called the *transfinite jump* of $f$.

COR. (i). *The set of g.n.'s of closed* **A**-*true* **A**-*statements fails to be arithmetically definable.*

For if it were then **A**-truth would be **A**-definable.

By using Props. 6 and 8 of Ch. 3 together with contraction of quantifiers and the equivalence of statements obtained the one from the other by change of bound variable (without of course collision of bound variable) we can obtain an algorithm for finding the position of any **A**-statement in the arithmetic hierarchy. In general one is interested in finding the lowest place in the arithmetic hierarchy. In general the equivalences mentioned above fail to give a unique result, just as the prenex normal form fails to be unique. However the lowest place in the hierarchy can always be found because there are only a bounded number of ways of applying the equivalences.

As an example consider the set of natural numbers $\nu$ such that the set of values of the function given by table number $\nu$ is recursive. The set of values calculated by table $\nu$ is recursively enumerable. A set is recursive if and only if it and its complement are recursively enumerable. Thus the condition defining our set is

$$(Ey)\,(\bar{\mathscr{X}}_\nu = \mathscr{X}_y),$$

where $\mathscr{X}_\nu$ is the set of values produced by table $\nu$. More fully this becomes

$$(Ey)\,(Ax)\,(x \bar{\epsilon} \mathscr{X}_\nu \leftrightarrow x \epsilon \mathscr{X}_y).$$

This again is

$$(Ey)\,(Ax)\,(N[\text{``}(Eu, u')\, Un'[\nu, u, u', x]\text{''}] \leftrightarrow (Ev, v')\, Un'[y, v, v', x]),$$

where $Un'$ is defined by D 122. Contracting quantifiers this becomes

$$EA(EE \,\&\, AA),$$

where we have left out the scopes of the quantifiers. This in turn is equivalent to a statement of the form:

$$EAAAEE$$

and this in turn to one of the form $EAE$. Thus the set of table numbers of recursive sets is in $\mathscr{P}^1_3$.

## 9·4 *f-definable sets of lattice points*

We can also develop the theory of $f$-definable sets of natural numbers. We give some results. The procedure is much the same as for the sets we have already considered. $\mathscr{P}^{\kappa,\mathscr{A}}_\pi$ and $\mathscr{Q}^{\kappa,\mathscr{A}}_\pi$ are the classes in the hierarchy of $\mathscr{A}$-definable sets of lattice points in $\mathscr{R}^\kappa$. Similarly for functions we use $\mathscr{P}^{\kappa,f}_\pi$ and $\mathscr{Q}^{\kappa,f}_\pi$. The individual members of these classes are

$$(\mathfrak{Q}\mathfrak{x})\,(Un^{\kappa+\pi+1}_\pi[\theta, \xi^2_1, \xi', \ldots, \xi^{(\pi \dot{-} 1)}, \nu', \ldots, \nu^{(\kappa)}, \mathfrak{f}\xi, \xi^2_2, 0] = 0), \qquad \text{(i)}$$

where the sequence of alternating quantifiers begins with $E$ for a $\mathscr{P}_\pi^{\kappa,\mathfrak{f}}$-set. A $\mathscr{Q}_\pi^{\kappa,\mathfrak{f}}$-set is just the negation of (i). These we denote by $\mathscr{P}_\pi^{\kappa,\mathfrak{f}}[\theta; \nu', \ldots, \nu^{(\kappa)}]$ for the statement, and $\mathscr{P}_\pi^{\kappa,\mathfrak{f}}[\theta]$ for the set. $\theta$ is called the index of the set. We now collect some useful results. We often omit the dimension number $\kappa$. Similarly for $\mathfrak{A}$ and $\mathfrak{f}$. $\mathfrak{A}'$ is the $\mathscr{A}$-complete set.

PROP. 9. (i) *$\mathscr{B}$ is recursive in $\mathscr{A}$ if and only if $\mathscr{B}$ and $\overline{\mathscr{B}}$ are recursively enumerable in $\mathscr{A}$.*

(ii) *$\mathscr{A}'$ is recursively enumerable in $\mathscr{A}$.*
(iii) *$\mathscr{A}'$ fails to be recursive in $\mathscr{A}$.*
(iv) *$\mathscr{B}$ is recursively enumerable in $\mathscr{A}$ iff $\mathscr{B} \leqslant_1 \mathscr{A}'$.*
(v) *$\mathscr{B}$ is recursive in $\mathscr{A}$ iff $\mathscr{B}' \leqslant_1 \mathscr{A}'$.*
(vi) *$\mathscr{B} \in \mathscr{P}_{S\pi}^{\mathscr{A}}$ if and only if $\mathscr{B}$ is recursively enumerable in $\mathscr{A}^{(\pi)}$.*
(vii) *$\mathscr{B} \in \mathscr{P}_{S\pi}^{\mathscr{A}} \cap \mathscr{Q}_{S\pi}^{\mathscr{A}}$ if and only if $\mathscr{B}$ is recursive in $\mathscr{A}^{(\pi)}$.*
(viii) *If $\mathscr{B} \in \mathscr{P}_{S\pi}^{\mathscr{A}}$ then an index for $\mathscr{B}$ as a set recursively enumerable in $\mathscr{A}^{(\pi)}$ can be found from any $\mathscr{P}_{S\pi}^{\mathscr{A}}$ index for $\mathscr{B}$, and conversely.*

(i) The demonstration is as for Prop. 8, Cor. v, Ch. 7.

(ii), (iii) are dealt with as for the complete set $\mathscr{C}$.

(iv) $\mathscr{B}$ is the range of a function recursive in $\mathscr{A}$, thus $\mathscr{B} = \mathscr{X}_\theta^{\mathscr{A}}$ for some $\theta$, thus $\lambda \epsilon \mathscr{B}$ iff $\langle\theta, \lambda\rangle \epsilon \mathscr{A}'$, so $\mathscr{B} \leqslant_1 \mathscr{A}'$. Conversely if $\mathscr{B} \leqslant_1 \mathscr{A}'$, so that $\lambda \epsilon \mathscr{B}$ iff $h\lambda \epsilon \mathscr{A}'$, where $h$ is a 1-1 function. We have $\mu \epsilon \mathscr{A}'$ iff $\mu = \langle\theta, \lambda\rangle$ and table $\theta$ in $\mathscr{A}$ produces $\lambda$. Thus generate the natural numbers, and when the natural number $\nu$ has been generated, set table $\nu_1^3$ going for $\nu_2^3$ moves with input $\nu_3^3$ if this produces the natural number $\lambda$ then write $h^{-1}\lambda$ down in a list, this enumerates $\mathscr{B}$.

(v) If $\mathscr{B}$ is recursive in $\mathscr{A}$, by (ii) $\mathscr{B}'$ is recursively enumerable in $\mathscr{B}$. By (iii) $\mathscr{B}' \neq \varnothing$. Hence $\mathscr{B}'$ is the range of some function $f$ recursive in $\mathscr{B}$. But if $f$ is recursive in $\mathscr{B}$ then $f$ is recursive in $\mathscr{A}$. Hence $\mathscr{B}'$ is recursively enumerable in $\mathscr{A}$, and, by (iv) $\mathscr{B}' \leqslant_1 \mathscr{A}'$.

If $\mathscr{B}' \leqslant_1 \mathscr{A}'$. Trivially $\mathscr{B}$ is recursively enumerable in $\mathscr{B}$ and $\overline{\mathscr{B}}$ is recursively enumerable in $\mathscr{B}$. Hence $\mathscr{B} \leqslant_1 \mathscr{B}'$ and $\overline{\mathscr{B}} \leqslant_1 \mathscr{B}'$ by (iv). Hence $\mathscr{B} \leqslant_1 \mathscr{A}'$ and $\overline{\mathscr{B}} \leqslant_1 \mathscr{A}'$. Therefore by (iv), both $\mathscr{B}$ and $\overline{\mathscr{B}}$ are recursively enumerable in $\mathscr{A}$. By (i) $\mathscr{B}$ is recursive in $\mathscr{A}$.

If (vi) then (vii).

(vi) By induction on $\pi$; the result is trivial for $\pi = 0$, $\mathscr{P}_1$-sets are recursively enumerable, similarly $\mathscr{P}_1^{\mathscr{A}}$-sets are recursively enumerable in $\mathscr{A}$.

Suppose the result for $\pi$. Suppose that $\mathscr{B} = \mathscr{X}_\theta^{\mathscr{A}^{(S\pi)}}$, then we want to show that $\mathscr{B}\epsilon\mathscr{P}_{SS\pi}^{\mathscr{A}}$. We have

$$\lambda\epsilon\mathscr{X}_\theta^{\mathscr{A}^{(S\pi)}} \leftrightarrow (Ez)\,(Un_1^1[\theta, \lambda, \mathfrak{f}z, z, 0] = 0)$$
$$\leftrightarrow (Ez, w)\,(Un_1^1[\theta, \lambda, w, z, 0] = 0$$
$$\& \, (Au)_{\varpi z}(w_u^{\varpi z} = 0 \leftrightarrow u\epsilon\mathscr{A}^{(S\pi)}) \,\&\, (Au)_{\varpi z}(w_u^{\varpi z} \leqslant 1)).$$

Now $u\epsilon\mathscr{A}^{(S\pi)} \leftrightarrow u_2^2\epsilon\mathscr{X}_{u_1^2}^{\mathscr{A}^{(\pi)}}$

$$\leftrightarrow (Ew)\,(Un_1^1[u_1^2, u_2^2, \mathfrak{g}w, w, 0] = 0$$
$$\& \, (Au)_{\varpi w}(\mathfrak{g}u = 0 \leftrightarrow u\epsilon\mathscr{A}^{(\pi)}) \,\&\, (Au)_{\varpi w}(\mathfrak{g}u \leqslant 1)).$$

We next show that $u\epsilon\mathscr{A}^{(\pi)}$ is a $\mathscr{P}_\pi$-statement. Clearly $u\epsilon\mathscr{A}'$ is a $\mathscr{P}_1$-statement. If $u\epsilon\mathscr{A}^{(\pi)}$ is a $\mathscr{P}_\pi$-statement then by (ii) $u\epsilon\mathscr{A}^{(S\pi)}$ is a $\mathscr{P}_{S\pi}$-statement. Substituting this in (ii) and reducing to prenex normal form we see that $\lambda\epsilon\mathscr{X}_\theta^{\mathscr{A}^{(S\pi)}}$ is a $\mathscr{P}_{SS\pi}$-statement.

Now suppose that $\mathscr{B}\epsilon\mathscr{P}_{SS\pi}^{\mathscr{A}}$. Then $\mathscr{B} = \hat{x}(Ey)\,S\{x, y\}$, where $S\{x, y\}$ is a $\mathscr{Q}_{S\pi}^{\mathscr{A}}$-statement, then $\bar{S}$ is a $\mathscr{P}_{S\pi}^{\mathscr{A}}$-statement, hence by induction hypothesis $\bar{S}$ is recursively enumerable in $\mathscr{A}^{(\pi)}$. Hence $\bar{S}$ is recursive in $\mathscr{A}^{(S\pi)}$, by (iv), whence so is $S$, hence by projection $\mathscr{B}$ is recursive in $\mathscr{A}^{(S\pi)}$. Hence (vi).

(viii) Again by induction on $\pi$. For $\pi = 0$ the indices are the same. Assume the result for $S\pi$, the result follows from detailed consideration of the construction in (vi).

## 9·5 *Computing degrees of unsolvability*

$\varnothing$ is the null set,

$\varnothing'$ is the complete recursively enumerable set,

$\varnothing''$ is the complete set which is recursively enumerable in $\varnothing'$, generally

$\varnothing^{(S\lambda)}$ is the complete set which is recursively enumerable in $\varnothing^{(\lambda)}$.

The degress of these sets are $\mathbf{O}, \mathbf{O}', \mathbf{O}'', \ldots, \mathbf{O}^{(S\lambda)}, \ldots$, respectively.

Note that by the construction the $g.n.$ of $\nu\epsilon\varnothing^{(\lambda)}$ is a primitive recursive function of $\nu, \lambda$.

These degrees form a useful measuring rod in the upper semi-lattice of degrees. Note that the degree of any recursively enumerable set is $\leqslant \mathbf{O}'$, and so is the degree of the complement of any recursively enumerable set.

One frequently wants to compute the degree of a set. By the place of the defining statement of the set in the hierarchy of Prop. 4 one can find an upper bound to the degree of a set. To find a lower bound is more troublesome. We give one or two results and techniques.

First, note that the sequence of degrees $\mathbf{O}^{(\lambda)}$ can be extended into the constructive transfinite. We first define $\mathbf{O}^{\omega}$ as the degree of the set of natural numbers $\nu$ for which $\nu_1^2 \in \mathbf{O}^{(\nu_2^2)}$.

Prop. 10. *The set of g.n.'s of closed* **A**-*true statements is of degree* $\mathbf{O}^{\omega}$.

The sequence of degrees $\mathrm{O}^{(\lambda)}$ is formed on a recursive plan. Hence the *g.n.*'s of $\nu_1^2 \in \mathrm{O}^{(\nu_2^2)}$ are the values of a primitive recursive function of $\nu$. For instance

$$\nu \epsilon \mathrm{O}' \leftrightarrow (Ew)\,(Un^1[\nu_1^2, w_1^2, w_2^2, \nu_2^2] = 0) \leftrightarrow (Ew)\,H'[w, \nu] \quad \text{say},$$

$$\nu \epsilon \mathrm{O}'' \leftrightarrow (Ew)\,(Un_1^1[\nu_1^2, w_1^2, \mathfrak{k}'w, w_1^2, \nu_2^2] = 0),$$

where $\mathfrak{k}'$ is the characteristic function of the complete set. Thus

$$\begin{aligned} \nu \epsilon \mathrm{O}'' &\leftrightarrow (Ew)\,(Ez)\,(Un_1^1[\nu_1^2, w_1^2, z, w_2^2, \nu_2^2] = 0\ \&\ (Au)_{\varpi w}(z_u^{\varpi w} = 0 \\ &\leftrightarrow (Ev)\,(Un^1[u_1^2, v_1^2, v_2^2, u_2^2] = 0)\ \&\ (Au)_{\varpi w}(z_u^{\varpi w} \leqslant 1)). \\ &\leftrightarrow (Eu)\,(Av)\,H''[u, v, \nu], \quad \text{say}, \end{aligned}$$

where $H$ is quantifier free. Generally if

$$\nu \epsilon \mathrm{O}^{(\lambda)} \leftrightarrow (Eu')\,(Au'') \ldots \begin{matrix}(Eu^{(\lambda)})\\(Au^{(\lambda)})\end{matrix}\, H^{(\lambda)}[u', \ldots, u^{(\lambda)}, \nu],$$

then

$$\begin{aligned} \nu \epsilon \mathrm{O}^{(S\lambda)} &\leftrightarrow (Ew, z)\,(Un_1^1[\nu_1^2, w_1^2, z, w_2^2, \nu_2^2] = 0\ \&\ (Au)_{\varpi w}(z_u^{\varpi w} = 0 \\ &\leftrightarrow (Ev, y)\,(Un_1^1[u_1^2, v_1^2, y, v_2^2, u_2^2] = 0\ \&\ (Au)_{\varpi w}(z_u^{\varpi w} \leqslant 1)) \\ &\&\ (Ax)_{\varpi v}(y_x^{\varpi v} = 0 \leftrightarrow (Eu')\,(Au'') \ldots \begin{matrix}(Eu^{(\lambda)})\\(Au^{(\lambda)})\end{matrix}\, H^{(\lambda)}[u' \ldots u, x] \\ &\qquad\qquad \&\ (Ax)_{\varpi v}(u_x^{\varpi v} \leqslant 1)))), \end{aligned}$$

reducing this to prenex normal form we see that

$$\nu \epsilon \mathrm{O}^{(S\lambda)} \leftrightarrow (Eu')\,(Au'') \ldots \begin{matrix}(Eu^{(\lambda)})\,(Au^{(S\lambda)})\\(Au^{(\lambda)})\,(Eu^{(S\lambda)})\end{matrix}\, H^{(S\lambda)}[u', u'', \ldots, u^{(S\lambda}, \nu].$$

Let $g[\nu, \lambda]$ be equal to the *g.n.* of $\nu \epsilon C^{(\lambda)}$, the complete set of degree $\lambda$ then $g$ is a primitive recursive function of $\nu$, $\lambda$. We have

$$\langle \nu, \lambda \rangle \epsilon\, \varnothing^{\omega} \leftrightarrow \nu \epsilon C^{(\lambda)} \leftrightarrow g[\nu, \lambda] \epsilon \mathscr{T},$$

where $\mathscr{T}$ is the set of *g.n.*'s of closed **A**-true **A**-statements. Thus $C^\omega$ is reducible to $\mathscr{T}$.

Again, if $\nu$ is the *g.n.* of an **A**-true closed **A**-statement $\phi$, let $(\mathfrak{Q}\mathfrak{x})\,\psi(\mathfrak{x}\}$ be obtained from a prenex normal form of $\phi$ after contraction of quantifiers. Consider $(Ey)\,(\mathfrak{Q}\mathfrak{x})\,(z = z \;\&\; y = y \;\&\; \psi\{\mathfrak{x}\})$, this is a $\mathscr{P}_{S\lambda}$-statement which defines a set $\mathscr{B}$. Then $\mathscr{B}$ is $\mathscr{N}$ if $\phi$ is **A**-true and $\mathscr{B}$ is $\varnothing$ if $\phi$ is **A**-false. By (viii) Prop 9 given an index of $\mathscr{B}$ as a set in $\mathscr{P}_{S\lambda}$ we can find an index of $\mathscr{B}$ as set recursively enumerable in $C^\lambda$. Thus

$$\phi \epsilon \mathscr{T} \leftrightarrow 0 \epsilon \mathscr{B} \leftrightarrow g[0]\,\epsilon C^\lambda \leftrightarrow \langle g[0], \lambda\rangle\,\epsilon C^\omega,$$

hence $\mathscr{T}$ is reducible to $C^\omega$. Altogether $\mathscr{T}$ and $C^\omega$ are of the same degree.

## Historical remarks to Chapter 9

The discussion of the hierarchy of **A**-definable sets of lattice-points is due to Kleene (1943) and Mostowski (1946) independently. We use the notation of Mostowski. Another notation is: $\overset{0}{\underset{\pi}{\Sigma}}, \overset{0}{\underset{\pi}{\Pi}}$ which can be further extended when we allow property variables in the definition of sets of lattice points. We prefer the Mostowski notation in this chapter because the $\Sigma$–$\Pi$ notation leaves out the dimension number $\kappa$. Thus most of this chapter is due to Kleene (1943) and Mostowski (1946), but Props. 6 and 7 are ultimately due to Post (1944).

The algorithm for finding the place of an arithmetic set in the hierarchy of arithmetic sets is due to Kuratowski and Tarski.

Addison has pointed out analogies in the Borel, Lusin and Kleene hierarchies (1955*a*, 1959*b*, 1962*a*, 1962*b*).

## Examples 9

1. Let $C$ and $P$ represent the operations of complementation and projection respectively. Let a sequence of such symbols denote a corresponding sequence of operations in reverse order of application. Thus $PPC$ represents complementation followed by two projections.

Show that $CPPCP$ applied to a relation $R$ can be expressed by a prefix $AAE$ applied to $R$.

Find prefixes corresponding to $PCPCPC$ and $CCCP$.

Find sequences of operations corresponding to $EAEA$ and $EAEEAA$.

2. Show that a bounded quantifier can be moved to the right past an adjacent universal quantifier, and that a bounded existential quantifier can be moved to the right past an adjacent existential quantifier.

3. Obtain a classification in the hierarchy for the following sets of natural numbers:

(i) the numbers of the tables of simple sets,
(ii) the numbers of tables of hypersimple sets,
(iii) the numbers of tables of creative sets,
(iv) the numbers of tables of maximal sets,
(v) the numbers of tables of unbounded sets.

4. Carry out the details of defining the hierarchy of $f$-definable sets of natural numbers.

# Chapter 10

# Induction

## 10.1 *Limitations of the system* A

There are some universal **A**-statements which in classical mathematics are normally proved by the Principle of Mathematical Induction. Many of the so-called Laws of Algebra are statements of this kind. These laws can be stated in the system **A** and are **A**-true but many of them fail to be **A**-theorems.

According to Cor. (i), Prop. 7, Ch. 8 if $(A\xi)\,\phi\{\xi\}$ is an **A**-theorem then so is $\phi\{\nu\}$ for each numeral $\nu$ and the **A**-proofs have the same number of steps, in fact they are obtained by substitution in an **A**-proof-form. Thus if $\phi\{\nu\}$ for $\nu = 0, 1, 2, \ldots$ are all **A**-theorems but the lengths of their shortest **A**-proofs are unbounded then $(A\xi)\,\phi\{\xi\}$ fails to be an **A**-theorem.

This occurs in the case of the **A**-statement

$$(Ax)\,(x+0 = 0+x), \tag{1}$$

written in fuller notation this is

$$(Ax)\,(\mathscr{I}(\boldsymbol{\lambda}x'x''.\,Sx'')\,x0 = \mathscr{I}(\boldsymbol{\lambda}x'x''.\,Sx'')\,0x). \tag{2}$$

If (2) were an **A**-theorem then each of

$$\mathscr{I}(\boldsymbol{\lambda}x'x''.\,Sx'')\,\nu 0 = \mathscr{I}(\boldsymbol{\lambda}x'x''.\,Sx'')\,0\nu$$

for $\nu = 0, 1, 2, \ldots$ would be **A**-provable in the same number of steps. By axiom 3.1

$$\mathscr{I}(\boldsymbol{\lambda}x'x''.\,Sx'')\,\nu 0 = \nu.$$

Hence by one more step we obtain by R 1

$$\nu = \mathscr{I}(\boldsymbol{\lambda}x'x''.\,Sx'')\,0\nu. \tag{3}$$

Thus if (1) is an **A**-theorem then (3) can be **A**-proved in the same number of steps for each numeral $\nu$. Now (3) is an **A**-true $\mathbf{A}_{00}$-statement, hence an $\mathbf{A}_{00}$-true $\mathbf{A}_{00}$-statement and so is an $\mathbf{A}_{00}$-theorem. In fact an **A**-proof of 3 is an $\mathbf{A}_{00}$-proof of (3). The $\mathbf{A}_{00}$-proof of (3) will use axioms 1,

3.1, 3.2, 4.1, 4.2, and rule R 1 only. The other axioms and rules are absent because (3) is without the symbols $\neq$ & $\vee$ and once these are introduced into an $\mathbf{A}_{00}$-proof without redundant parts they remain in that $\mathbf{A}_{00}$-proof from that place to the end and so occur in the $\mathbf{A}_{00}$-theorem proved.

To $\mathbf{A}_{00}$-prove (3) we have to find the numeral determined by $\mathscr{I}(\boldsymbol{\lambda}x'x''.Sx'')0\nu$ and to do this we have to use axioms 3.1, 3.2 to get rid of the iterator symbol and axioms 4.1, 4.2 to get rid of the abstraction symbol, axiom 1 might be used as a starting point, we also use rule R 1.

If $\nu$ is zero then we can finish in one step by axiom 3.1. If $\nu$ is $S\nu'$ then we can use axiom 1 or axiom 3.2 only; axiom 1 gets us nowhere, axiom 3.2 gives us

$$\mathscr{I}(\boldsymbol{\lambda}x'x''.Sx'')0(S\nu') = (\boldsymbol{\lambda}x'x''.Sx'')\nu'(\mathscr{I}(\boldsymbol{\lambda}x'x''.Sx'')0\nu').$$

Write $F$ for $(\boldsymbol{\lambda}x'x''.Sx'')$, then if at any time we have arrived at

$$(F^*(F^* \ldots (F^*(\mathscr{I}F0(S\nu''))) \ldots)), \tag{3.1}$$

where $F^*$ is either $F\mu$ for some numeral $\mu$ or is $S$, then we have a choice of two moves only; we could either replace one of the initially placed $F^*$'s which is $F\mu$ by $S$ using axiom 4.1 and R 1, or we could use axiom 3.2 and replace the part $\mathscr{I}(F0(S\nu''))$ by $F^*(\mathscr{I}F0\nu'')$ using R 1. We finish up with an expression of the same form as (3.1) or possibly with $S\nu''$ replaced by 0.

Let the order of (3.1) be the number of $\boldsymbol{\lambda}$'s and $\mathscr{I}$'s it contains. We start with

$$\mathscr{I}(\boldsymbol{\lambda}x'x''.Sx'')0(S\nu')$$

of order 2. We apply one of the two possible moves, the order either is decreased by one or is increased by one, such an increase can only happen $\nu'$ times and a decrease can only follow a previous increase. Thus it is clear that we shall require at least $2.\nu'+1$ steps to obtain an expression of order zero which is a numeral. Extra steps can be added by applying axioms 4.1 and 4.2 alternately. The last step is an application of axiom 3.1 to get rid of $\mathscr{I}$. Thus the shortest proof of (3) is unbounded in $\nu$.

## 10.2 *Possible ways of extending the system* $\mathbf{A}_0$

We now wish to extend the system **A** so that more **A**-true **A**-statements become theorems in the extended system. We wish to do this in such a way that directness of proof is maintained, for then we can easily see, by arguments outside the system, that the resulting system is consistent in the sense of *Con* (i), in that $(0 = 1)$ will fail to be provable.

The **A**-statement (1) says that two $\mathbf{A}_{00}$-functions, that is two primitive recursive functions, are equivalent, that is they take the same value for the same argument. We might then consider a new rule which allows us to prove the equivalence of two equivalent primitive recursive functions.

Primitive recursive functions are built up from the initial functions by the schemes of substitution and primitive recursion. The rule R 1 enables us to prove the equivalence of two primitive recursive functions which are built up from functions, whose equivalence has been proved, by substitution of functions, whose equivalence has been proved. For instance, if we have proved $\rho x = \rho' x$ and $\sigma x = \sigma' x$ then by R 1 we obtain $\rho(\sigma x) = \rho'(\sigma' x)$. Thus we require a rule that will enable us to prove the equivalence of two primitive recursive functions, formed by the scheme of primitive recursion, from functions whose equivalence has been proved.

Suppose that the primitive recursive functions $\rho$ and $\rho'$ satisfy

$$\rho[0,\eta] = \alpha[\eta], \quad \rho[S\xi,\eta] = \tau[\xi,\rho[\xi,\eta],\eta], \tag{4}$$

$$\rho'[0,\eta] = \beta[\eta], \quad \rho'[S\xi,\eta] = \sigma[\xi,\rho'[\xi,\eta],\eta], \tag{5}$$

where
$$\alpha[\eta] = \beta[\eta] \quad \text{and} \quad \tau[\xi,\zeta,\eta] = \sigma[\xi,\zeta,\eta]. \tag{6}$$

We require a rule which enables us to prove $\rho[\xi,\eta] = \rho'[\xi,\eta]$. Given such a rule we could then prove the equivalence of two primitive recursive functions which are built up from the same construction sequence but using functions of proved equivalence at corresponding places in the construction.

From (5) and (6) and R 1 we obtain

$$\rho'[0,\eta] = \alpha[\eta], \quad \rho'[S\xi,\eta] = \tau[\xi,\rho'[\xi,\eta],\eta]. \tag{7}$$

Conversely from (6) and (7) we obtain (5). Thus our new rule should allow us to obtain

$$\rho[\xi,\eta] = \rho'[\xi,\eta] \tag{8}$$

from (4) and (7).

Note that (8) might be A-valid but it might be impossible to find $\alpha$, $\beta$, $\tau$, $\sigma$ such that (4), (5) and (6). For instance the primitive recursive function whose value is zero for the argument $\nu$ if $(2.\nu+6)$ is the sum of two odd primes and whose value is one otherwise might be such that we are unable to show that it is built up using the same construction sequence as the constant function zero, but using instead functions of proved equivalence at corresponding places in the construction.

As a first suggestion consider the rule

$$\frac{(\rho[0,\mathfrak{y}] = \sigma[\mathfrak{y}]) \vee \omega \quad (\rho[S\xi,\mathfrak{y}] = \tau[\xi,\rho[\xi,\mathfrak{y}],\mathfrak{y}]) \vee \omega}{(\rho[\alpha,\mathfrak{y}] = \widehat{\tau\sigma}[\alpha,\mathfrak{y}]) \vee \omega} \qquad \text{R } 3',$$

where $\widehat{\tau\sigma}$ is defined by D 121, $\mathfrak{y}$ comprises exactly all the free variables in $\sigma[\mathfrak{y}]$; $\xi, \zeta, \mathfrak{y}$ exactly all the free variables in $\tau[\xi,\zeta,\mathfrak{y}]$, $\omega$ is subsidiary and can be omitted; $\xi$ fails to occur free in $\sigma[\mathfrak{y}]$, $\tau[\Gamma_\iota,\zeta,\mathfrak{y}]$, $\omega$ and $\rho[\Gamma_\iota,\mathfrak{y}]$: $\alpha$ is free in $\rho[\alpha,\mathfrak{y}]$ and in $\widehat{\tau\sigma}[\alpha,\mathfrak{y}]$. Rule R 3′ is more powerful when we allow any numerical term $\alpha$ instead of just a variable $\xi$. From (4), (7), R 3′ and R 1 we can obtain (8) as desired. The system A with the new rule R 3′ will be called the system $\mathbf{A}_{I'}$. In the system $\mathbf{A}_{I'}$ we can prove the equivalence of two primitive recursive functions which satisfy the same recursion equations.

As for the system A we can show that if a disjunction is provable then so is one disjunctant, we can find which and find a proof for it, we can also obtain an irresolvable $\mathbf{A}_{I'}$-statement, thus T.N.D. fails in the system $\mathbf{A}_{I'}$. For if T.N.D. held in $\mathbf{A}_{I'}$, then G $\vee$ $N$["G"], where G is the irresolvable $\mathbf{A}_{I'}$-statement, would be an $\mathbf{A}_{I'}$-theorem hence so would be one disjunctand, which is absurd.

A particular case of rule R 3′ is

$$\frac{(\rho[0,\mathfrak{y}] = 0) \vee \omega \quad (\rho[S\xi,\mathfrak{y}] = 0) \vee \omega}{(\rho[\alpha,\mathfrak{y}] = 0) \vee \omega}. \qquad \text{R } 3''$$

This can be generalized to

$$\frac{\phi\{0\} \vee \omega \quad \phi\{S\xi\} \vee \omega}{\phi\{\alpha\} \vee \omega}. \qquad \text{R } 3'''$$

Now consider the rule

$$\frac{\phi\{0\} \vee \omega \quad \dfrac{\phi\{\xi\} \vee \omega}{\phi\{S\xi\} \vee \omega}}{\phi\{\alpha\} \vee \omega}, \qquad \text{R } 3^{(\text{iv})}$$

where $\xi$ is free in $\phi\{\xi\}$ and fails to occur free in $\phi\{\Gamma_\iota\}$, and $\alpha$ is free in $\phi\{\alpha\}$.

R $3^{(iv)}$ is properly stronger than R‴ because we can R $3^{(iv)}$-prove $(0+x = x+0)$, but we are unable to R‴-prove this.
The R $3^{(iv)}$-proof of $(0+x) = (x+0)$ is as follows

$$\cfrac{\cfrac{\cfrac{\cfrac{S(0+x) = S(0+x)\, Ax\, 1 \quad 0+x = x+0 \quad Hyp}{S(0+x) = S(x+0)} \quad x+0 = x\, Ax\, 3.1 \quad \cfrac{Sx+0 = Sx\, Ax\, 3.1 \quad Sx+0 = Sx+0\, Ax\, 1}{Sx = Sx+0}\, \text{R}\,1}{S(0+x) = Sx}\;\text{R}\,1}{S(0+x) = Sx+0 \quad 0+Sx = S(0+x)\, Ax\, 3.2}\;\text{R}\,1}{0+Sx = Sx+0}\;\text{R}\,1.$$

Thus we have $\quad 0+0 = 0+0 \quad \dfrac{0+x = x+0 \quad Hyp}{0+Sx = Sx+0},$

whence by R $3^{(iv)}$ we obtain $0+x = x+0$.

If we have the premisses of R 3‴ then we have the premisses of R $3^{(iv)}$—if we can prove $\phi\{S\xi\} \vee \omega$ then we can deduce $\phi\{S\xi\} \vee \omega$ from the hypothesis $\phi\{\xi\} \vee \omega$—hence if $\phi$ is R 3‴-provable then $\phi$ is R $3^{(iv)}$-provable. Suppose we have an **A**-proof of $0+x = x+0$ using R 3‴; in each application of R 3‴ replace $\alpha$ by $S\xi$, the result is an **A**-proof of $0+Sx = Sx+0$; whence we can obtain $\mathbf{A}_{00}$-proofs of $0+S\nu = S\nu+0$ in the same number of steps for each numeral $\nu$ which is absurd.

The weakness of R $3^{(iv)}$ is that we might want to show

$$\frac{\phi\{\xi\} \vee \omega}{\phi\{S\xi\} \vee \omega}$$

by induction; we can overcome this difficulty by introducing a new symbol $\rightarrow$ of type *ooo* called the *implication symbol* and writing a one premiss rule $\dfrac{\phi}{\psi}$ as $((\rightarrow\phi)\psi)$ which we usually abbreviate as $(\phi \rightarrow \psi)$. A two-premiss rule would then become $((\phi \,\&\, \psi) \rightarrow \chi)$, this converts rules into axiom schemes, we should then require a new rule in order to be able to proceed at all. The rule

$$\frac{\phi \quad \phi \rightarrow \psi}{\psi} \quad \textit{Modus Ponens}$$

suggests itself. With this modification rule R $3^{(iv)}$ becomes

$$\frac{\phi\{0\} \vee \omega \quad (\phi\{\xi\} \vee \omega) \rightarrow (\phi\{S\xi\} \vee \omega)}{\phi\{\alpha\} \vee \omega} \qquad \text{R}\,3^{(v)}$$

or even $\quad ((\phi\{0\} \vee \omega) \,\&\, ((\phi\{\xi\} \vee \omega) \rightarrow (\phi\{S\xi\} \vee \omega))) \rightarrow (\phi\{\alpha\} \vee \omega),$

with the same conditions on $\xi$ and $\alpha$ as before. This is an axiom scheme. The disadvantage of this method is that proofs cease to be direct, in Modus Ponens the whole build up of $\phi$ disappears at one step.

We have the possibility of defining

$$(\phi \to \psi) \quad \text{for} \quad N[\text{“}\phi\text{”}] \vee \psi,$$

and leaving the rules in their original form, and we should have a truth-value for A-statements of the form $(\phi \to \psi)$, but then in R $3^{(v)}$ we again lose directness of proof. This definition of implication is called *material implication.*

All these suggested rules have this in common: if we are able to prove the upper formulae then we can prove successively

$$\phi\{0\} \vee \omega,\ \phi\{1\} \vee \omega,\ \phi\{2\} \vee \omega, \ldots$$

moreover the proofs are on a general plan which can be described and we are given a demonstration that the plan is correct. This leads us to consider the scheme

$$\frac{\phi\{\nu\} \vee \omega \quad \text{for each numeral } \nu}{\phi\{\alpha\} \vee \omega} \qquad \text{R } 3^{(vi)}$$

where $\alpha$ is free in $\phi\{\alpha\}$ and where the proofs of $\phi\{\nu\} \vee \omega$ are on a general plan which is described and a demonstration is given that this plan is correct. We must add these conditions because it is essential that a proof be checkable; it is useless to say that the plan is such and such, if we are without a demonstration that the plan is correct; then to check the proof we shall have to supply such a demonstration ourselves. Hence we require that the demonstration be given. Now what is to be the nature of this demonstration?

We wish to remain with a formal system, it must then be a recursive process to test formulae to see if they are well-formed and if so whether they are statements, to test statements to see if they are axioms and to test applications of rules. This is all complied with if we include the conditions we have appended to rule R $3^{(vi)}$ provided the demonstration is in some system other than the one we are constructing, otherwise we may get into serious trouble.

We require $\alpha$ to be free in $\phi\{\alpha\}$ to avoid:

$$\frac{(\boldsymbol{\lambda} x\,.\,\nu)\,\kappa = \nu}{(\boldsymbol{\lambda} x\,.\,x)\,\kappa = x\text{'}}$$

which could otherwise arise. Rule R $3^{(vi)}$ has an unbounded number of premisses, but this is unobjectionable when the appended conditions are added.

It is possible to check an unending sequence of statements if the checking is on a general plan for which a checkable demonstration of correctness has been given. We have done this sort of thing many times before, for instance in the system $A_{00}$ we showed that $\alpha+\beta = \beta+\alpha$ for any closed numerical terms $\alpha$, $\beta$; we sketched a general method whereby any particular case could be dealt with and gave a demonstration that the method was correct. The resulting meta-theorem failed to be an $A_{00}$-theorem because the methods used in obtaining it were unavailable in the system $A_{00}$ itself. The rule R $3^{(vi)}$ has conditions attached to it, this is harmless, we have had conditions attached to rules many times before, chiefly connected with variables, but it was so easy to see whether they were satisfied that we failed to require that a demonstration be given that this was so. But in the case of rule R $3^{(vi)}$ it might be really difficult to see that such and such a plan was in fact correct, hence we demand that the demonstration of correctness be given. Then we are able to supply the proofs of $\phi\{0\} \vee \omega$, $\phi\{1\} \vee \omega, \dots$ as far as we like and know that though we have to stop somewhere yet the ones we have left out would be correct if written down, but without the proof of correctness of the general plan we might doubt whether some of those proofs which we omitted to write down were in fact correct.

The proofs which we have encountered anywhere as yet have given rise to proof-trees; these have axioms at the tops of the branches and at each node a bounded number of branches unite by an application of a rule. Each *thread* of the proof-tree contains a bounded number of nodes. We wish to keep this last feature of a proof-tree when we use rule R $3^{(vi)}$. We shall require that the number of applications of rule R $3^{(vi)}$ in any thread be less than a fixed number. The *order of a proof* using R $3^{(vi)}$ will be defined as the maximum number of *inductions*—applications of R $3^{(vi)}$—in any thread of the proof. We then restrict ourselves to proofs of bounded order. Without this restriction we should have great difficulty in defining a proof-predicate. An example of a proof of unbounded order is obtained by taking a proof of bounded order and adding redundant parts, e.g. by II *a*, R $3^{(vi)}$ and I *b*. (We shall add the rule I *b*.)

## 10.3 *The system* E

We will now construct a formal system in which the demonstration of correctness required by rule R $3^{(vi)}$ is to be carried out. The system **E** is an *equation calculus*, it will be used to prove the equivalence of primitive recursive functions. The symbols of the system **E** are

$$0\ S\ \mathscr{I}\ \boldsymbol{\lambda} = x\ (\ )\ '$$

where the prime is a generating symbol, the types are as in the system $\mathbf{A}_{00}$. The axioms of the system **E** are the $\mathbf{A}_{00}$-axioms

1   3.1   3.2   4.1   4.2

and the rules are the $\mathbf{A}_{00}$-rules

R 1   R 3′   without subsidiary formula.

Prop. 1. *We can find a primitive recursive function whose value is always one but it is impossible to* **E**-*prove that it is equivalent to the constant function one. Thus the system* **E** *is incomplete.*

Let $prf_{\mathbf{E}}$ be the proof-predicate for the system **E**, this is constructed as for other systems and is primitive recursive. Let $s_{\mathbf{E}}[x', x]$ be Gödel's substitution function for the system **E**. Consider $prf_{\mathbf{E}}[x', s_{\mathbf{E}}[x, x]] = 1$ and let its *g.n.* be $\kappa$, let the *g.n.* of $prf_{\mathbf{E}}[x', s_{\mathbf{E}}[\kappa, \kappa]] = 1$ be $\nu$, then $\nu = s[\kappa, \kappa]$ is an $\mathbf{A}_{00}$-theorem and hence is an **E**-theorem. Suppose that $prf_{\mathbf{E}}[x', \nu] = 1$ is an **E**-theorem, then by R 1 so is $prf_{\mathbf{E}}[x', s_{\mathbf{E}}[\kappa, \kappa]] = 1$. There will then be a numeral $\pi$ such that $prf_{\mathbf{E}}[\pi, \nu] = 0$, here $\pi$ is the *g.n.* of an **E**-proof of $prf_{\mathbf{E}}[x', s_{\mathbf{E}}[\kappa, \kappa]] = 1$ which is the **E**-statement with *g.n.* $\nu$. But if $prf_{\mathbf{E}}[x', \nu] = 1$ is an **E**-theorem then so is $prf_{\mathbf{E}}[\pi, \nu] = 1$, because **E**-theorems are **A**-valid hence $prf_{\mathbf{E}}[\pi, \nu] = 1$ is $\mathbf{A}_{00}$-true hence is an $\mathbf{A}_{00}$-theorem and hence is an **E**-theorem. But this is absurd because a closed numerical term determines a unique numeral. Hence $prf_{\mathbf{E}}[x', \nu] = 1$ fails to be an **E**-theorem.

We now show that $prf_{\mathbf{E}}[x', \nu] = 1$ is **A**-valid. Suppose that $prf_{\mathbf{E}}[\pi, \nu] = 0$ is $\mathbf{A}_{00}$-true for some numeral $\pi$, then the **E**-statement '$\nu$' is an **E**-theorem. But this **E**-statement is $prf_{\mathbf{E}}[x', s_{\mathbf{E}}[\kappa, \kappa]] = 1$ in which case $prf_{\mathbf{E}}[x', \nu] = 1$ is also an **E**-theorem, and so $prf_{\mathbf{E}}[\pi, \nu] = 1$ is $\mathbf{A}_{00}$-true. But this is absurd because a closed numerical term determines a unique numeral. Hence $prf_{\mathbf{E}}[\pi, \nu] = 1$ is $\mathbf{A}_{00}$-true hence is an **E**-theorem for each numeral $\pi$. Thus $\lambda x . prf_{\mathbf{E}}[x, \nu]$ is the required function.

Prop. 1 shows that we must be very careful in saying that $fx = 0$ is an **E**-theorem when we have somehow shown that $f\pi = 0$ for each numeral $\pi$. We should require something stronger than rule R 3′ to prove this in all cases. To **E**-prove that $fx = 0$ we must show that $f$ satisfies the same recursion equations as the constant function zero, and this we are sometimes unable to do.

COR. (i) **E**-*validity fails to be a primitive recursive property.*

In other words **E**-validity fails to be **E**-definable, because the only **E**-definable concepts are **E**-defined by an equation of the form $fx = 0$ where $f$ is a primitive recursive function. Suppose that $val_{\mathbf{E}}$ is **E**-definable so that $val_{\mathbf{E}}[\nu] = 0$ if and only if '$\nu$' is an **E**-valid **E**-statement. Consider $val_{\mathbf{E}}[s_{\mathbf{E}}[x, x]] = 1$, let its *g.n.* be $\kappa$ and let the *g.n.* of $val_{\mathbf{E}}[s_{\mathbf{E}}[\kappa, \kappa]] = 1$ be $\nu$, then $s_{\mathbf{E}}[\kappa, \kappa] = \nu$ is an $\mathbf{A}_{00}$-theorem, hence is an **E**-theorem. Suppose that $val_{\mathbf{E}}[\nu] = 0$ then '$\nu$' is **E**-valid hence $val_{\mathbf{E}}[s_{\mathbf{E}}[\kappa, \kappa]] = 1$ is **E**-valid and so is $val_{\mathbf{E}}[\nu] = 1$ but this is absurd because a closed numerical term determines a unique numeral. Suppose that $val_{\mathbf{E}}[\nu] = 1$ then '$\nu$' is **E**-invalid or fails to be an **E**-statement but '$\nu$' is the **E**-statement $val_{\mathbf{E}}[s_{\mathbf{E}}[\kappa, \kappa]] = 1$, hence this is invalid and so $val_{\mathbf{E}}[s_{\mathbf{E}}[\kappa, \kappa]] = 0$ because $val_{\mathbf{E}}$ is an $\mathbf{A}_{00}$-function whose value is either zero or one and fails to be both. But this again is absurd as before. Thus $val_{\mathbf{E}}$ with the required properties fails to be **E**-definable.

COR. (ii). **E** *is essentially undecidable.*

The demonstration is exactly the same as for $\mathbf{A}_0$. Prop. 5, Ch. 7.

By Cor. (ii), Prop. 6, Ch. 7 **E**-validity is undecidable.

COR. (iii). **E**-*valid statements fail to form a recursively enumerable set.*

If **E**-valid statements formed a recursively enumerable set then we could find a formal system in which they were the theorems by Prop. 15, Ch. 8. But then by the argument of Prop. 1 above we should be able to find an **E**-valid statement outside this enumeration.

When we use R 3′

$$\frac{\rho 0 = \alpha \quad \rho(Sx) = \tau[x, \rho x]}{\rho\beta = \widehat{\tau\alpha\beta}}$$

or the same with parameter we sometimes say '*by induction on x' with substitution*, if $\beta$ is $x$ we omit 'with substitution'. Similarly when we use R 3′ to prove $fx = gx$.

PROP. 2. *The rule:*

$$\frac{\sigma 0 = 0 \quad \dfrac{\sigma x = 0}{\overline{\overline{\sigma(Sx) = 0}}}}{\sigma\alpha = 0} \quad x \textit{ held constant} \tag{9}$$

*is a derived* **E**-*rule.*

We say that $x$ *is held constant* in a deduction if the deduction is without any induction on $x$.

Write out the deduction of $\sigma(Sx) = 0$ from the hypothesis $\sigma x = 0$ in tree-form and multiply both sides of each equation in the deduction by $B_1[\sigma x]$. The hypothesis becomes $\sigma x . B_1[\sigma x] = 0$, but this is an easily proved **E**-theorem. ($xB_1[x] = 0$ by induction on $x$, whence $\sigma x . B_1[\sigma x] = 0$ because we are allowed a substitution in the lower formula in rule R 3′.) The axioms become easily proved **E**-theorems (($x . \alpha = x . \beta$) is easily **E**-proved from $\alpha = \beta$) and the rules become derived **E**-rules. For rule R 1 ($b$ for $B_1[x]$)

$$\frac{\rho\{\alpha\} = \tau\{\alpha\} \quad \alpha = \beta}{\rho\{\beta\} = \tau\{\beta\}}$$

becomes
$$\frac{b . \rho\{\alpha\} = b . \tau\{\alpha\} \quad b . \alpha = b . \beta}{b . \rho\{\beta\} = b . \tau\{\beta\}}. \tag{10}$$

We have by induction on $x$

$$B_1[x] . \tau[x'] = B_1[x] . \tau[B_1[x] . x']. \tag{11}$$

Hence
$$\cfrac{\cfrac{\cfrac{b . \rho\{\alpha\} = b . \tau\{\alpha\}}{b . \rho\{b . \alpha\} = b . \tau\{b . \alpha\}} \quad b . \alpha = b . \beta}{b . \rho\{b . \beta\} = b . \tau\{b . \beta\}}}{b . \rho\{\beta\} = b . \tau\{\beta\}.}$$

Thus (10) is a derived **E**-rule.

An application of R 3′

$$\frac{\rho[0, \eta] = \sigma[\eta] \quad \rho[S\xi, \eta] = \tau[\xi, \rho[\xi, \eta], \eta]}{\rho[\alpha, \eta] = \widehat{\tau\sigma}[\alpha, \eta]}$$

becomes
$$\frac{b . \rho[0, \eta] = b . \sigma[\eta] \quad b . \rho[S\xi, \eta] = b . \tau[\xi, \rho[\xi, \eta], \eta]}{b . \rho[\alpha, \eta] = b . \widehat{\tau\sigma}[\alpha, \eta]}. \tag{12}$$

Using (11) this becomes

$$\frac{b . \rho[0, \eta] = b . \sigma[\eta] \quad b . \rho[S\xi, \eta] = b . \tau[\xi, b . \rho[\xi, \eta], \eta]}{b . \rho[\alpha, \eta] = b . \widehat{\tau(b . \sigma)}\,[\alpha, \eta]} \tag{13}$$

since (see D 121) $b.\widehat{\tau\sigma}[\alpha,\eta] = b.\widehat{\tau(b.\sigma)}[\alpha,\eta]$ and since (13) is an application of R 3′ then (12) is a derived **E**-rule.

Thus we obtain an **E**-proof of $B_1[\sigma\alpha].\sigma(S\alpha) = 0$ and hence an **E**-proof of $\sigma(S\alpha) = A_1[\sigma\alpha].\sigma(S\alpha)$, since $A_1[x]+B_1[x] = 1$ is easily **E**-proved by induction on $x$. Now consider the functions $f$ which satisfy the scheme of primitive recursion:

$$f0 = 0, \quad f(Sx) = A_1[fx].\sigma(Sx).$$

One such function is $Ox$ and another is $\sigma[x]$. Hence by R 3′ $\sigma[x] = 0$, as desired, thus (9)

PROP. 3. $$x+(x' \dot{-} x) = x'+(x \dot{-} x')$$
*is an* **E**-*theorem.*

This may be expressed as

$$Max\,[x,x'] = Max\,[x',x].$$

The result could be very eaily proved if we had the rule that two functions which satisfied the same scheme of double recursion are equivalent, but we require this **E**-theorem to show that the equivalence of two functions which satisfy the same scheme of double recursion is a derived **E**-rule.

We have $x' \dot{-} x = \sum_{0 \leqslant x'' \leqslant x'} A_1[x'' \dot{-} x]$ by induction on $x'$

and $x = (x \dot{-} 1)+A_1[x]$ by induction on $x$,

hence
$$x' = \sum_{0 \leqslant x'' \leqslant x'} A_1[x'']$$
$$= \sum_{0 \leqslant x'' \leqslant x+x'} A_1[x'' \dot{-} x],$$

thus
$$x = \sum_{0 \leqslant x'' \leqslant x+x'} A_1[x'' \dot{-} x']$$
$$= \sum_{0 \leqslant x'' \leqslant x'} A_1[x'' \dot{-} x'] + \sum_{Sx \leqslant x'' \leqslant x+x'} A_1[x'' \dot{-} x']$$
$$= (x \dot{-} x') + \sum_{0 \leqslant x'' \leqslant x+x'} A_1[x'' \dot{-} x'].A_1[x'' \dot{-} x],$$

similarly
$$x' = (x' \dot{-} x) + \sum_{0 \leqslant x'' \leqslant x+x'} A_1[x'' \dot{-} x].A_1[x'' \dot{-} x'],$$

the result now follows. The familiar properties of summations which we have used are easily **E**-proved.

COR. (i).

$$\frac{\rho x = \sigma x}{\rho 0 = \sigma 0 \quad \rho(Sx) = \sigma(Sx)} \quad x \text{ held constant}$$

$$\frac{\rho 0 = \sigma 0 \quad \rho(Sx) = \sigma(Sx)}{\rho\alpha = \sigma\alpha}$$

*is a derived* **E**-*rule.*

We have

$$\frac{(x \dot{-} x') + (x' \dot{-} x) = 0}{x = x'} \qquad (a)$$

and

$$\frac{x = x'}{(x \dot{-} x') + (x' \dot{-} x) = 0} \qquad (b)$$

are **E**-provable. The first follows from $(x+x') \dot{-} x' = x$ (by induction on $x$), thus if $x+x' = 0$ then $x = 0$ and $x' = 0$, this gives

$$\frac{(x \dot{-} x') + (x' \dot{-} x) = 0}{(x \dot{-} x') = 0 \quad (x' \dot{-} x) = 0}$$

whence by Prop. 3 $x = x'$ as desired. The second follows from

$$\frac{\dfrac{x = x'}{(x \dot{-} x') = 0 \quad (x' \dot{-} x) = 0}}{(x \dot{-} x') + (x' \dot{-} x) = 0.}$$

From the hypotheses we obtain

$$(\rho 0 \dot{-} \sigma 0) + (\sigma 0 \dot{-} \rho 0) = 0 \qquad \frac{(\rho x \dot{-} \sigma x) + (\sigma x \dot{-} \rho x) = 0}{(\rho(Sx) \dot{-} \sigma(Sx)) + (\sigma(Sx) \dot{-} \rho(Sx)) = 0}$$

whence by Prop. 2 $\quad (\rho\alpha \dot{-} \sigma\alpha) + (\sigma\alpha \dot{-} \rho\alpha) = 0.$

By $(a)$ the result follows.

PROP. 4. *If two functions of natural numbers satisfy the same scheme* $(\Sigma)$, *etc., or course of values recursion or if two sets of functions of natural numbers satisfy the same scheme of simultaneous recursion or if two functions of natural numbers satisfy the same scheme of recursion with substitution in parameter or of double recursion or of simple nested recursion then their equivalence can be* **E**-*proved.*

If $\rho$ is a function satisfying the scheme of course of values recursion or of one of the schemes $(\Sigma)$, etc., or if a set of functions satisfy the same scheme of simultaneous recursion then we showed in Ch. 5 that there was a primitive recursive function $f$ such that $f\rho$ satisfied a scheme of primitive recursion and hence $f\rho$ can be **E**-proved to be unique; further we showed that there was another primitive function $g$ such that $gf\rho x = \rho x$, whence $\rho$ is unique. For the schemes of recursion with substitution in parameter, double recursion or of simple nested recursion we need only consider the scheme of simple nested recursion because the scheme of recursion with substitution in parameter is a particular case of the scheme of simple nested recursion and the scheme of double recursion is settled when we have settled the scheme of substitution in parameter (see Ch. 5). As regards the scheme of simple nested recursion we need only show that two functions which satisfy the scheme

$$\rho[0,\kappa] = \sigma[\kappa],\ \rho[S\nu,\kappa] = \tau[\rho[\nu,\rho[\nu,\kappa]]] \tag{14}$$

are equivalent (see the account of simple nested recursion given in Ch. 5) because if $\rho'$ satisfies a scheme of simple nested recursion then we can find a primitive recursive function $f$ such that $f\rho'$ satisfies (14) and we can find another primitive recursive function $g$ such that $gf\rho' x = \rho' x$. Thus if the solution to (14) is unique then so is a function which satisfies a scheme of simple nested recursion.

Let $\rho$ be a function which satisfies (14) and let $\chi$ be the primitive recursive function such that

$$\begin{cases} \chi[0,\kappa] = \kappa, \\ \chi[S\nu,\kappa] = \begin{cases} \sigma^2[\chi[\nu,\kappa]] & \text{if } S\nu \text{ is odd,} \\ \tau^{\varpi(S\nu)}[\chi[\nu,\kappa]] & \text{if } S\nu \text{ is even.} \end{cases} \end{cases}$$

Then
$$\chi[x,\kappa] = \mathscr{I}(\boldsymbol{\lambda}\xi\eta\,.\,\tau_1[\xi,\eta])\,\kappa x,$$

where
$$\tau_1[\xi,\eta] = e(S\xi)\,\sigma^2[\eta] + (1 \dot{-} e(S\xi))\,\tau^{\varpi(S\xi)}[\eta],$$

so that
$$\chi[Sx,\kappa] = \tau_1[x,\chi[x,\kappa]] \quad \text{and} \quad \chi[0,\kappa] = \kappa,$$

here $e\nu = 0$ if $\nu$ is even, $e\nu = 1$ if $\nu$ is odd, further $\chi$ is unique.

LEMMA. *If* $\quad f[0,\eta] = \sigma[\eta],\quad f[Sx,\eta] = \tau[x, f[x,\eta],\eta],$

*then* $\quad f[x+x',\eta] = \mathscr{I}(\boldsymbol{\lambda}\xi\zeta\,.\,\tau[x+\xi,\zeta,\eta])f[x,\eta]\,x'.$

Let $g[x,x',\eta]$ stand for

$$\mathscr{I}(\boldsymbol{\lambda}\xi\zeta\,.\,\tau[x+\xi,\zeta,\eta])f[x,\eta]\,x',$$

then $$g[x, 0, \eta] = f[x, \eta]$$

and $$g[x, Sx', \eta] = \tau[x+x', g[x, x', \eta], \eta],$$

also $$f[x+0, \eta] = f[x, \eta]$$

and $$f[x+Sx', \eta] = \tau[x+x', f[x+x', \eta], \eta].$$

Thus by R 3′ we get $g[x, x', \eta] = f[x+x', \eta]$, and the lemma follows.

Applying the lemma to $\chi$ we get

$$\chi[2^{Sx}, \kappa] = \mathscr{I}(\boldsymbol{\lambda}\xi\zeta.\tau_1[2^x+\xi, \zeta])\,\chi[2^x, \kappa]\,2^x,$$

also $$\chi[2^x, \kappa] = \mathscr{I}(\boldsymbol{\lambda}\xi\zeta.\tau_1[\xi, \zeta])\,\kappa 2^x.$$

Now $$\tau_1[2^x+\xi, \zeta] = \tau_1[\xi, \zeta] \quad \text{for} \quad S\xi < 2^x,$$

hence $$\chi[2^{Sx}, \kappa] = \tau_1[2^{Sx} \dot{-} 1, \mathscr{I}(\boldsymbol{\lambda}\xi\zeta.\tau_1[2^x+\xi, \zeta])\,\chi[2^x, \kappa]\,(2^x \dot{-} 1)]$$

$$= \tau^{Sx}[\mathscr{I}(\boldsymbol{\lambda}\xi\zeta.\tau_1[2^x+\xi, \zeta])\,\chi[2^x, \kappa]\,(2^x \dot{-} 1)].$$

Also $$\chi[2^x, \kappa] = \tau_1[2^x \dot{-} 1, \mathscr{I}(\boldsymbol{\lambda}\xi\zeta.\tau_1[\xi, \zeta])\,\kappa(2^x \dot{-} 1)]$$

$$= \tau^x[\mathscr{I}(\boldsymbol{\lambda}\xi\zeta.\tau_1[\xi, \zeta])\,\kappa(2^x \dot{-} 1)].$$

Thus $$\chi[2^{Sx}, \kappa] = \tau[\chi[2^x, \chi[2^x, \kappa]]].$$

We now show $$\frac{\rho[Sx, \kappa] = \chi[2^{Sx}, \kappa]}{\rho[SSx, \kappa] = \chi[2^{SSx}, \kappa].} \quad x \text{ held constant.}$$

We have $$\rho[SSx, \kappa] = \tau[\rho[Sx, \rho[Sx, \kappa]]]$$

$$= \tau[\chi[2^{Sx}, \chi[2^{Sx}, \kappa]]] \quad \text{by hypothesis,}$$

$$= \chi[2^{SSx}, \kappa] \quad \text{as we have just shown.}$$

We also have $$\rho[1, \kappa] = \chi[2, \kappa] = \tau[\sigma^2\kappa].$$

Hence by R 3′ we obtain $\rho[Sx, \kappa] = \chi[2^{Sx}, \kappa]$, thus $\rho$ is unique because $\chi$ is unique. Thus two functions which satisfy the same scheme of simple nested recursion can be **E**-proved to be equivalent. This completes the demonstration of the proposition.

COR. (i). *The recursive equations of a primitive recursive function are* **E**-*provable.*

For instance if $\rho\{0\} = \alpha, \quad \rho\{S\nu\} = \gamma\{\nu, \rho\{\nu\}\}$

then $\rho\{Sx\} = \gamma\{x, \rho\{x\}\}$ is an **E**-theorem. $\rho\{x\}$ and $\gamma\{x \dot{-} 1, \gamma\{x \dot{-} 1, \rho\{x' \dot{-} 1\}\}\}$ satisfy the same scheme of primitive recursion.

PROP. 5. *If* $\tau 0 = 0$ *is an* $\mathbf{A}_{00}$-*theorem and if* $\dfrac{\tau\Gamma = 0}{\tau(S\Gamma) = 0}$ *is an* $\mathbf{A}_{00}$-*deduction-form, then* $\tau x = 0$ *is an* **E**-*theorem.*

If $\tau 0 = 0$ is an $\mathbf{A}_{00}$-theorem then it is an **E**-theorem. Note that if $\tau\{\nu\} = \sigma\{\nu\}$ is an $\mathbf{A}_{00}$-axiom for all $\nu$ then $\tau\{x\} = \sigma\{x\}$ is an **E**-axiom, and that if

$$\frac{\alpha\{\nu, \gamma\{\nu\}\} = \beta\{\nu, \gamma\{\nu\}\} \quad \gamma\{\nu\} = \delta\{\nu\}}{\alpha\{\nu, \delta\{\nu\}\} = \beta\{\nu, \delta\{\nu\}\}} \quad \text{for all } \nu$$

is a case of R 1 in $\mathbf{A}_{00}$ then the same with $x$ instead of $\nu$ is an **E**-rule.

Thus if $\dfrac{\Gamma\tau = 0}{\tau(S\Gamma) = 0}$ is an $\mathbf{A}_{00}$-deduction-form then the same with $x$ instead of $\Gamma$ is an **E**-deduction. By Prop. 2 the result follows.

COR. (i). *If* $\tau 0 = \sigma 0$ *is an* $\mathbf{A}_{00}$-*theorem and if* $\dfrac{\tau\Gamma = \sigma\Gamma}{\tau(S\Gamma) = \sigma(S\Gamma)}$ *is an* $\mathbf{A}_{00}$-*deduction-form then* $\tau x = \sigma x$ *is an* **E**-*theorem.*

The demonstration is similar to that of Prop. 5, but uses Cor. (i), Prop. 3, instead of Prop. 2.

COR. (ii). *If* $\tau[0, 0] = 0$, $\tau[S\Gamma, 0] = 0$ *and* $\tau[0, S\Gamma'] = 0$ *are* $\mathbf{A}_{00}$-*theorem-forms and if* $\dfrac{\tau[\Gamma, \Gamma'] = 0}{\tau[S\Gamma, S\Gamma'] = 0}$ *is an* $\mathbf{A}_{00}$-*deduction-form then* $\tau[x, x'] = 0$ *is an* **E**-*theorem.*

Similarly using Prop. 4.

## 10.4 *The system* $\mathbf{A}_I$

The system $\mathbf{A}_I$ is a primary extension of the system $\mathbf{A}_0$. It has additional symbols $A$ of types $o(o\iota)$, $o(o\iota\iota)$, $o(o\iota\iota\iota)$, ... and we use abbreviation D 224. The symbols $A$ are called the *universal quantifiers*, they are introduced into $\mathbf{A}_I$-proofs by a new rule R 3 called *Induction*. The system $\mathbf{A}_I$ fails to be a primary extension of the system $\mathbf{A}$ because free variables are forbidden in the system $\mathbf{A}_I$ and the rule for introducing the universal quantifier is different. The *g.n.* of $A$ is to be 11, the type of $A$ is then determined uniquely in a well-formed formula. The *g.n.* of $x^{(\theta)}$ is to be $12+\theta$.

We wish to formalize a rule of induction to the effect

$$\frac{\phi\{\mathfrak{n}\}\vee\omega}{(A\mathfrak{x})\,\phi\{\mathfrak{a}\{\mathfrak{x}\}\}\vee\omega}\text{'}\quad \begin{array}{l}\text{for all } \mathfrak{n}\text{, where } \mathfrak{n} \text{ is a } \pi\text{-tuplet,}\\ \text{where } A \text{ is of type } o(\underbrace{o\iota\ldots\iota}_{\iota\ \pi\ \text{times}}),\end{array} \qquad \text{R 3}$$

where the proofs of $\phi\{\nu\}\vee\omega$ are on a primitive recursive plan which can be **E**-proved to be correct. When this is so we say '$\phi\{\nu\}\vee\omega$ for all $\nu$, **E**-*correct*'. To strengthen the system $\mathbf{A}_I$ we have allowed a substitution corresponding to the substitution rule in the system **E**. Rule R 3 is to the effect that if we have proved $\phi\{\mathfrak{n}\}\vee\omega$ for all $\pi$-tuplets of numerals $\mathfrak{n}$ and have demonstrated in the system **E** that the plan of proof is correct then we are entitled to have $\phi\{\mathfrak{a}\{\mathfrak{n}'\}\}\vee\omega$ for all $\pi$-tuplets $\mathfrak{a}$ and for all $\pi'$-tuplets of numerals $\mathfrak{n}'$ and to have $(A\mathfrak{x})\,\phi\{\mathfrak{a}\{\mathfrak{x},\mathfrak{n}'\}\}\vee\omega$ for similar sets of functions $\alpha$ and of numerals $\mathfrak{n}'$. It may be that the rule of substitution in the systems **E** and $\mathbf{A}_I$ is redundant, anyway with them the systems are more amenable, also the rule of substitution is intuitively acceptable. We also use rule I *b*.

In Ch. 1 we defined a formal system as an ordered quarter $\{\mathscr{S},\mathscr{F},\mathscr{A},\mathscr{P}\}$, $\mathscr{S}$ is a displayed set of signs, $\mathscr{F}$ is a bounded recursive set of rules of formation, $\mathscr{A}$ is a recursive set of statements called axioms, and $\mathscr{P}$ is a recursive set of rules of procedure which we depicted as $\dfrac{\phi}{\chi}$, $\dfrac{\phi\ \psi}{\chi}$. In the system $\mathbf{A}_I$ we want something more general. The main point about a formal system is that we must be able to check a proof, so that we are independent of pronunciations of powerful and strong-willed persons, also of blind faith. The rule of induction in the system $\mathbf{A}_I$ is different from the rules we have had so far in that it has an unbounded set of premisses, thus an application of the rule can only be described, it is impossible to write it down in detail, still less is it possible to write down in detail the $\mathbf{A}_I$-proofs of the premisses, but it is to be possible to describe them, say by a proof-form, from which by substitution we can obtain the proofs in detail as far as we wish.

An $\mathbf{A}_I$-proof is an unbounded tree-like figure and we shall be given sufficient instructions so that we can write down any bounded part of it, but we shall have to stop somewhere and are unable to write down the whole thing. Anyway it will be such that if $\phi\{\mathfrak{x}\}$ is an $\mathbf{A}_I$-theorem in which each variable of the set $\mathfrak{x}$ is general in $\phi\{\mathfrak{x}\}$ and only these, then we can write down the full $\mathbf{A}_0$-proof of $\phi^*\{\mathfrak{n}\}$, for any set of numerals $\mathfrak{n}$, where $\phi^*\{\Gamma_\iota\}$ is $\phi\{\Gamma_\iota\}$ shorn of its universal quantifiers.

We also require that for each $\mathbf{A}_I$-proof there be a numeral $\kappa$ such that each thread of that $\mathbf{A}_I$-proof encounters at most $\kappa$ inductions. The numeral $\kappa$ will vary from proof to proof.

The $\mathbf{A}_I$-proofs of the upper formulae of an induction can only be described as being on such and such a plan, it is impossible to display them as in the systems we have constructed up till now. For instance a description of an induction might consist of the $\mathbf{A}_0$-proof of $\phi\{0\} \vee \omega$ together with the $\mathbf{A}_0$-deduction-form

$$\frac{\phi\{\Gamma_\iota\} \vee \omega}{\phi\{S\Gamma_\iota\} \vee \omega}, \qquad (a)$$

from this by replacing $\Gamma_\iota$ by numerals in all possible ways we can obtain the $\mathbf{A}_0$-proofs of the premisses $\phi\{\nu\} \vee \omega$ for $\nu = 0, 1, 2, \ldots$. Similarly we might have the $\mathbf{A}_0$-proof of $\phi\{0, 0\} \vee \omega$, descriptions of $\mathbf{A}_0$-proofs of

$$\phi\{S\nu, 0\} \vee \omega \quad \text{and} \quad \phi\{0, S\nu'\} \vee \omega,$$

possible like $(a)$ above, and the $\mathbf{A}_0$-deduction-form

$$\frac{\phi\{\Gamma_\iota, \Gamma'_\iota\} \vee \omega}{\phi\{S\Gamma_\iota, S\Gamma'_\iota\} \vee \omega}, \qquad (b)$$

from which we can obtain the $\mathbf{A}_0$-proofs of the premisses

$$\phi\{\nu, \nu'\} \vee \omega \quad \text{for} \quad \nu, \nu' = 0, 1, 2, \ldots.$$

The position is somewhat similar to the use of axiom-schemes; rules normally are schemes. It is impossible to write down all the $\mathbf{A}_0$-axioms, but we can describe them.

Along with (*a*), (*b*), and any other method of description that we might use, we require a demonstration that the method is correct. We have decided that this demonstration take place in the system **E**. The proofs and deductions above are in the system $\mathbf{A}_0$ because they are without inductions. Let $\tau\mathfrak{n}$ be equal to the *g.n.* of an $\mathbf{A}_0$-proof of $\phi\{\mathfrak{n}\} \vee \omega$ and let $\sigma\{\mathfrak{Num}\,\nu\}$ be equal to the *g.n.* of $\phi\{\mathfrak{n}\} \vee \omega$ (if $\mathfrak{n}$ is $\nu', \ldots, \nu^{(\pi)}$ then $\sigma\{\mathfrak{Num}\,\nu\}$ is $\sigma\{Num\,\nu', \ldots, Num\,\nu^{(\pi)}\}$) then we require an **E**-proof of $prf_0[\tau\mathfrak{x}, \sigma[\mathfrak{Num}\,\mathfrak{x}]] = 0$ to be given, this can be done by giving its *g.n.* From this *g.n.* we can recover $\sigma[\mathfrak{Num}\,\mathfrak{x}]$ and hence $\phi\{\mathfrak{n}\} \vee \omega$ and hence we can find

$$(A\mathfrak{x})\,\phi\{\mathfrak{x}\} \vee \omega, \quad \phi\{\mathfrak{a}\{\mathfrak{n}'\}\} \vee \omega \quad \text{and} \quad (A\mathfrak{x})\,\phi\{\mathfrak{a}\{\mathfrak{x}, \mathfrak{n}'\}\} \vee \omega.$$

## 10.5 *Definition of an* $\mathbf{A}_I$*-proof*

An $\mathbf{A}_I$-proof consists of a bounded number of *levels*, say $S\lambda$ levels. Level one consists of a bounded number of *g.n.*'s of **E**-proofs of equations of the form

$$prf_0[\tau\mathfrak{x}, \sigma[\mathfrak{Num}\,\mathfrak{x}]] = 0,$$

where $\tau\mathfrak{n}$ is equal to the *g.n.* of an $\mathbf{A}_0$-proof of a closed $\mathbf{A}_0$-statement whose *g.n.* is equal to $\sigma[\mathfrak{Num}\,\mathfrak{n}]$. Let these $\mathbf{A}_0$-statements be

$$\left.\begin{array}{c} \phi_0'\{\mathfrak{n}, \mathfrak{n}', \ldots, \mathfrak{n}^{(\lambda)}\} \vee \omega_0'\{\mathfrak{n}', \ldots, \mathfrak{n}^{(\lambda)}\}, \\ \vdots \\ \phi_0^{(\mu_0)}\{\mathfrak{n}, \mathfrak{n}', \ldots, \mathfrak{n}^{(\lambda)}\} \vee \omega_0^{(\mu_0)}\{\mathfrak{n}', \ldots, \mathfrak{n}^{(\lambda)}\}, \end{array}\right\} \tag{15}$$

where $\mathfrak{n}', \ldots, \mathfrak{n}^{(\lambda)}$ are parameters. They indicate the variables which will be generalized in later levels. Then level one gives us the $\mathbf{A}_0$-proofs of these $\mathbf{A}_0$-statements for each set of numerals $\mathfrak{n}', \ldots, \mathfrak{n}^{(\lambda)}$ for all $\mathfrak{n}$ **E**-correct. But we shall demand more. We want (15) to be **E**-correct for all $\mathfrak{n}, \mathfrak{n}', \ldots, \nu^{(\lambda)}$. Let

$$\left.\begin{array}{c} \tau_0'[\mathfrak{n}, \mathfrak{n}', \ldots, \mathfrak{n}^{(\lambda)}], \\ \vdots \\ \tau_0^{(\mu_0)}[\mathfrak{n}, \mathfrak{n}', \ldots, \mathfrak{n}^{(\lambda)}], \end{array}\right\} \tag{16}$$

be equal to the *g.n.*'s of the $\mathbf{A}_0$-proofs of the respective $\mathbf{A}_0$-statements (15). Let

$$\left.\begin{array}{c} \sigma_0'[\mathfrak{Num}\,\nu, \mathfrak{Num}\,\nu', \ldots, \mathfrak{Num}\,\nu^{(\lambda)}], \\ \vdots \\ \sigma_0^{(\mu_0)}[\mathfrak{Num}\,\nu, \mathfrak{Num}\,\nu', \ldots, \mathfrak{Num}\,\nu^{(\lambda)}], \end{array}\right\} \tag{17}$$

be equal to the *g.n.*'s of the respective statements (15). Then level one will consist of the *g.n.*'s of the **E**-proofs of

$$\left.\begin{aligned} &prf_0[\tau_0'[\mathfrak{x},\mathfrak{x}',\ldots,\mathfrak{x}^{(\lambda)}],\sigma_0'[\mathfrak{Num}\,\mathfrak{x},\mathfrak{Num}\,\mathfrak{x}',\ldots,\mathfrak{Num}\,\mathfrak{x}^{(\lambda)}]]=0,\\ &\qquad\vdots\\ &prf_0[\tau_0^{(\mu_0)}[\mathfrak{x},\mathfrak{x}',\ldots,\mathfrak{x}^{(\lambda)}],\sigma_0^{(\mu_0)}[\mathfrak{Num}\,\mathfrak{x},\mathfrak{Num}\,\mathfrak{x}',\ldots,\mathfrak{Num}\,\mathfrak{x}^{(\lambda)}]]=0.\end{aligned}\right\} \quad (18)$$

A particular case is when $\mathfrak{x},\ldots,\mathfrak{x}^{(\lambda)}$ are absent and we are left with a single $\mathbf{A}_0$-proof.

The *production of level one* consists of a bounded selection of **A**-statements of the forms

$$\left.\begin{aligned} &(A\mathfrak{x})\,\phi_0'\{\mathfrak{a}'\{\mathfrak{x},\mathfrak{n}^*\},\mathfrak{n}',\ldots,\mathfrak{n}^{(\lambda)}\}\vee\omega_0'\{\mathfrak{n}',\ldots,\mathfrak{n}^{(\lambda)}\},\\ &\qquad\vdots\\ &(A\mathfrak{x})\,\phi_0^{(\mu_0)}\{\mathfrak{a}_0^{(\mu_0)}\{\mathfrak{x},\mathfrak{n}^*\},\mathfrak{n}',\ldots,\mathfrak{n}^{(\lambda)}\}\vee\omega_0^{(\mu_0)}\{\mathfrak{n}',\ldots,\mathfrak{n}^{(\lambda)}\},\\ &\phi_0'\{\mathfrak{a}'\{\mathfrak{n}^*\},\mathfrak{n}',\ldots,\mathfrak{n}^{(\lambda)}\}\vee\omega_0'\{\mathfrak{n}',\ldots,\mathfrak{n}^{(\lambda)}\},\\ &\qquad\vdots\\ &\phi_0^{(\mu_0)}\{\mathfrak{a}^{(\mu_0)}\{\mathfrak{n}^*\},\mathfrak{n}',\ldots,\mathfrak{n}^{(\lambda)}\}\vee\omega_0^{(\mu_0)}\{\mathfrak{n}',\ldots,\mathfrak{n}^{(\lambda)}\}.\end{aligned}\right\} \quad (19)$$

A statement is in level $S\lambda$ if it is at or above the production of level $S\lambda$ but below the production of level $\lambda$. An $\mathbf{A}_I$-proof of level one is either an $\mathbf{A}_0$-proof or ends in a single application of R 3.

The idea is that if we have proved $\phi\{\mathfrak{n}\}$ for all $\pi$-tuplets of numerals $\mathfrak{n}$ then we have proved $\phi\{\mathfrak{a}\{\mathfrak{n}^*\}\}$ for all $\pi$-tuplets of functions $\mathfrak{a}$ and all $\pi^*$-tuplets of numerals $\mathfrak{n}^*$, and that we have proved $(A\mathfrak{x})\,\phi\{\mathfrak{a}\{\mathfrak{x},\mathfrak{n}^*\}\}$ for all similar tuplets of functions $\mathfrak{a}$ and of numerals $\mathfrak{n}^*$. Thus level one produces sufficient evidence that the $\mathbf{A}_0$-statements (19) are $\mathbf{A}_0$-theorems for all appropriate sets of numerals $\mathfrak{n},\mathfrak{n}',\ldots,\mathfrak{n}^{(\lambda)}$. The set $\mathfrak{n}^*$ of numerals may be any, it may overlap the sets $\mathfrak{n}',\ldots,\mathfrak{n}^{(\lambda)}$, or it may contain fresh numerals. The production may contain several different $\pi$-tuplets of numerical terms $\mathfrak{a}',\ldots$, several different numerical terms $\mathfrak{a}^{(\mu_0)}$. We say that the $\pi$-tuplet of numerals $\mathfrak{n}$ is *treated* in level one.

Having defined level $\theta$ we proceed to define level $S\theta$.

D 233 $ded_0[\lambda,\mu,\nu]=0$ if and only if $\mu$ is the *g.n.* of an $\mathbf{A}_0$-deduction of the closed **A**-statement whose *g.n.* is $\nu$ from hypotheses whose *g.n.* is $\lambda$. Level $S\theta$ consists of the *g.n.*'s of **E**-proofs of equations of the form

$$\left.\begin{aligned} &ded_0[\rho_\theta[\mathfrak{x}^{(S\theta)},\ldots,\mathfrak{x}^{(\lambda)}],\tau_{S\theta}'[\mathfrak{x}^{(S\theta)},\ldots,\mathfrak{x}^{(\lambda)}],\sigma_{S\theta}'[\mathfrak{Num}\,\mathfrak{x}^{(S\theta)},\ldots,\mathfrak{Num}\,\mathfrak{x}^{(\lambda)}]]=0,\\ &\qquad\vdots\\ &ded_0[\rho_\theta[\mathfrak{x}^{(S\theta)},\ldots,\mathfrak{x}^{(\lambda)}],\tau_{S\theta}^{(\mu_{S\theta})}[\mathfrak{x}^{(S\theta)},\ldots,\mathfrak{x}^{(\lambda)}],\sigma_{S\theta}^{(\mu_{S\theta})}[\mathfrak{Num}\,\mathfrak{x}^{(S\theta)},\ldots,\mathfrak{Num}\,\mathfrak{x}^{(\lambda)}]]=0.\end{aligned}\right\} \quad (20)$$

Here
$$\rho_\theta[\mathfrak{n}^{(S\theta)}, \ldots, \mathfrak{n}^{(\lambda)}] \tag{21}$$

is equal to the *g.n.* of the sequence of those **A**-statements similar to (19) which form the production of level $\theta$. These **A**-statements form the hypotheses of the deductions in level $S\theta$, we may suppose that each deduction in level $S\theta$ is from the same set of hypotheses, because a deduction from a set of hypotheses $\Phi$ is also a deduction from hypotheses $\Phi$ and $\Psi$, those in $\Psi$ being dummies.

$$\tau^{(\kappa)}_{S\theta}[\mathfrak{n}^{(S\theta)}, \ldots, \mathfrak{n}^{(\lambda)}] \tag{22}$$

for $1 \leqslant \kappa \leqslant \mu_{S\theta}$, is equal to the *g.n.* of the $\mathbf{A}_0$-deductions of the **A**-statements (24) from the hypotheses whose *g.n.* is $\rho_\theta[\mathfrak{n}^{(S\theta)}, \ldots, \mathfrak{n}^{(\lambda)}]$.

$$\sigma^{(\kappa)}_{S\theta}[\mathfrak{Num}\,\nu^{(S\theta)}, \ldots, \mathfrak{Num}\,\nu^{(\lambda)}] \tag{23}$$

for $1 \leqslant \kappa \leqslant \mu_{S\theta}$, is equal to the *g.n.* of the $\kappa$th statement (24)

$$\left.\begin{array}{c} \phi'_{S\theta}\{\mathfrak{n}^{(S\theta)}, \ldots, \mathfrak{n}^{(\lambda)}\} \vee \omega'_{S\theta}\{\mathfrak{n}^{(SS\theta)}, \ldots, \mathfrak{n}^{(\lambda)}\}, \\ \vdots \\ \phi^{(\mu_{S\theta})}_{S\theta}\{\mathfrak{n}^{(S\theta)}, \ldots, \mathfrak{n}^{(\lambda)}\} \vee \omega^{(\mu_{S\theta})}_{S\theta}\{\mathfrak{n}^{(SS\theta)}, \ldots, \mathfrak{n}^{(\lambda)}\}. \end{array}\right\} \tag{24}$$

The production of level $S\theta$ consists of a bounded number of **A**-statements of the forms:

$$\left.\begin{array}{l} (A\xi^{(S\theta)})\,\phi'_{S\theta}\{\mathfrak{a}'\{\mathfrak{x}^{(S\theta)}, \mathfrak{n}^*\}, \mathfrak{n}^{(SS\theta)}, \ldots, \mathfrak{n}^{(\lambda)}\} \vee \omega^{(\mu_{S\theta})}_{S\theta}\{\mathfrak{n}^{(SS\theta)}, \ldots, \mathfrak{n}^{(\lambda)}\}, \\ \qquad\vdots \\ (A\xi^{(S\theta)})\,\phi^{(\mu_{S\theta})}_{S\theta}\{\mathfrak{a}^{(\mu_{S\theta})}\{\mathfrak{x}^{(S\theta)}, \mathfrak{n}^*\}, \mathfrak{n}^{(SS\theta)}, \ldots, \mathfrak{n}^{(\lambda)}\} \vee \omega^{(\mu_{S\theta})}_{S\theta}\{\mathfrak{n}^{(SS\theta)}, \ldots, \mathfrak{n}^{(\lambda)}\}, \\ \phi'_{S\theta}\{\mathfrak{a}'\{\mathfrak{n}^*\}, \mathfrak{n}^{(SS\theta)}, \ldots, \mathfrak{n}^{(\lambda)}\} \vee \omega'_{S\theta}\{\mathfrak{n}^{(SS\theta)}, \ldots, \mathfrak{n}^{(\lambda)}\}, \\ \qquad\vdots \\ \phi^{(\mu_{S\theta})}_{S\theta}\{\mathfrak{a}^{(\mu_{S\theta})}\{\mathfrak{n}^*\}, \mathfrak{n}^{(SS\theta)}, \ldots, \mathfrak{n}^{(\lambda)}\} \vee \omega^{(\mu_{S\theta})}_{S\theta}\{\mathfrak{n}^{(SS\theta)}, \ldots, \mathfrak{n}^{(\lambda)}\}, \end{array}\right\} \tag{25}$$

where $\mathfrak{n}^*$ is any set of numerals.

We can arrange the notation so that $\mathfrak{n}^{(\theta)}$ is treated at the $\theta$th level. In (25) as in (19) we allow several different $\mathfrak{a}', \ldots,$ several different $\mathfrak{a}^{(\mu_{S\theta})}$. The levels continue until all the variable numerals have been generalized. If new variable numerals are introduced by substitution at each level then the process fails to terminate and so fails to give an $\mathbf{A}_I$-proof.

The last level consists of a single $\mathbf{A}_0$-deduction from the production of the penultimate level as hypotheses (case when $\mathfrak{x}$ is absent in (20) and (20) consists of only one line), or is a single application of R 3. In either case the closed **A**-statement thus produced is the $\mathbf{A}_I$-theorem proved.

$\mathbf{A}_I$-proofs are complicated, but they can be checked. Suppose that an $\mathbf{A}_I$-proof has $S\lambda$ levels, then we are given $S\lambda$ sets of numerals, the last set containing only one member. The members of the first set are *g.n.*'s of **E**-proofs of equations of the form (18), the members of the other sets are *g.n.*'s of **E**-proofs of equations of the form (20). From these *g.n.*'s we can recover the **E**-proofs and from these in turn we can recover the **A**-statements $\mathbf{A}_0$-proved or $\mathbf{A}_0$-deduced, lastly from the hypotheses of level $S\theta$ we can find the production of level $\theta$. Thus we can see the cases of generalization and substitution that have been used. Moreover if the theorem proved is $\phi\{\mathfrak{x}\}$, where each variable of the set $\mathfrak{x}$ is generalised in $\phi\{\mathfrak{x}\}$, and only these, then we can write out in full the $\mathbf{A}_0$-proof of $\phi^*\{\mathfrak{n}\}$ for any set of numerals $\mathfrak{n}$, where $\phi^*\{\mathfrak{x}\}$ is $\phi\{\mathfrak{x}\}$ minus its general quantifiers.

Note that a production of level $\theta$ can be a production of any higher level, because $\chi$ is a deduction from $\chi$ as hypothesis. In this case we say that $\chi$ is *carried forward*.

An $\mathbf{A}_I$-proof having only one level is either an $\mathbf{A}_0$-proof, case when $\mathfrak{x}$ is absent in $prf_0[\tau\mathfrak{x}, \sigma[\mathfrak{Num}\,\mathfrak{x}]]$, or finishes with a single application of R 3.

We could have written (15) as

$$\phi_0'\{\nu, \alpha', \nu', \ldots, \alpha^{(\lambda)}, \nu^{(\lambda)}, \alpha^{(S\lambda)}\},$$

indicating the places which are going to be restricted as well as those which are to be generalized. In the resolved form of a theorem the restricted variables are replaced by certain functions of the superior general variables. The only functions available to us are primitive recursive functions, so that if we demand proofs of (15) for all $\nu, \nu', \ldots, \nu^{(\lambda)}$ **E**-correct then we restrict ourselves to cases when in the resolved form the functions required were all primitive recursive. Other types of functions required in the resolution can be replaced by their values which are expressed by numerals when the superior general variables are replaced by numerals, and so available to us; but we are without means of dealing with these cases in proving (15) for all $\nu, \nu', \ldots, \nu^{(\lambda)}$ **E**-correct.

Furthermore the proof-predicate for $\mathbf{A}_I$ becomes undefinable if we only demand (15) for all $\mathfrak{n}$ **E**-correct.

We could depict an $\mathbf{A}_I$-proof as follows:

$$
\begin{array}{llll l}
 & \vdots & \vdots & \vdots & \\
\text{P.L. 1} & \dfrac{\phi_1\{\lambda,\mu,\nu\} \vee \omega_1\{\mu,\nu\}}{(A\xi)\,\phi_1\{\xi,\mu,\nu\} \vee \omega_1\{\mu,\nu\}} & \dfrac{\phi_2\{\lambda,\mu\} \vee \omega_2\{\mu\}}{(A\xi)\,\phi_2\{\xi,\mu\} \vee \omega_2\{\mu\}} & \dfrac{\phi_3\{\lambda\} \vee \omega_3}{(A\xi)\,\phi_3\{\xi\} \vee \omega_3} & \text{R 3} \\
 & & & \text{carried forward} & \\
\text{P.L. 2} & \dfrac{(A\xi)\,\phi'\{\xi,\mu,\nu\} \vee \omega_1'\{\mu,\nu\} \vee \omega_1''\{\nu\}}{(A\eta)\,((A\xi)\,\phi_1'\{\xi,\eta,\nu\} \vee \omega_1'\{\eta,\nu\}) \vee \omega_1''\{\nu\}} & \dfrac{(A\xi)\,\phi_2'\{\xi,\mu\} \vee \omega_2'\{\mu\} \vee \omega_2''}{(A\eta)\,((A\xi)\,\phi_2'\{\xi,\eta\} \vee \omega_2'\{\eta\}) \vee \omega_2''} & \dfrac{(A\xi)\,\phi_3'\{\xi\} \vee \omega_3}{(A\xi)\,\phi_3'\{\xi\} \vee \omega_3} & \text{R 3} \\
 & & \text{carried forward} & \text{carried forward} & \\
\text{P.L. 3} & \dfrac{(A\eta)\,((A\xi)\,\phi_1''\{\xi,\eta,\nu\} \vee \omega_1'\{\eta,\nu\}) \vee \omega_1'''\{\nu\} \vee \omega_1^{\text{iv}}}{(A\zeta)\,((A\eta)\,((A\xi)\,\phi_1''\{\xi,\eta,\zeta\} \vee \omega_1^{\text{vi}}\{\eta,\zeta\}) \vee \omega_1'''\{\zeta\}) \vee \omega_1^{\text{iv}}} & \dfrac{(A\xi)\,\phi_2''\{\xi,\mu\} \vee \omega_2''\{\mu\} \vee \omega_2'''}{(A\eta)\,((A\xi)\,\phi_2''\{\xi,\eta\} \vee \omega_2''\{\eta\}) \vee \omega_2'''} & \dfrac{(A\xi)\,\phi_3'\{\xi\} \vee \omega_3}{(A\xi)\,\phi_3'\{\xi\} \vee \omega_3} & \text{R 3}
\end{array}
$$

$$\psi$$

This depicts an $\mathbf{A}_I$-proof with 4 levels the last one being an $\mathbf{A}_0$-deduction of $\psi$ from the production of level 3 as hypotheses. Similarly with batches of quantifiers and with substitutions. $\phi_1''$ can only be a variant of $\phi_1'$ by R 1, etc. in other cases of $\phi_\mu^{(\nu)}$.

The second displayed line consists of the hypotheses in the $\mathbf{A}_0$-deductions of the statements in the third displayed line, etc.

In this scheme level 1 gives the proof-schemes for

$$\begin{array}{lll} \phi_1\{\lambda,\mu,\nu\} \vee \omega_1\{\mu,\nu\} & \text{for all } \lambda,\mu,\nu, & \textbf{E}\text{-correct} \\ \phi_2\{\lambda,\mu\} \vee \omega_2\{\mu\} & \text{for all } \lambda,\mu, & \textbf{E}\text{-correct} \\ \phi_3\{\lambda\} \vee \omega_3 & \text{for all } \lambda, & \textbf{E}\text{-correct} \end{array}$$

this is indicated by the column of dots above each. Written out in full it would be as in the example to follow. Level 1 ends with an application of R 3 giving the production (P.L. 1) of level 1. Similarly for the other levels. Of course we could apply II *d* to P.L. 1 obtaining an existential quantifier between $(A\eta)$ and $(A\xi)$, etc. for II *b*, *c*.

## 10.6 *Theorem induction*

We have frequently in previous chapters used formula induction and theorem induction. We can still do so in an $\mathbf{A}_I$-proof. Formula induction is exactly as before because the formulae are as before except for extra symbols, namely the universal quantifiers. But $\mathbf{A}_I$-proofs are of a different character in that we now allow rules with an unbounded number of premisses. Theorem induction is as follows:

If (i) Property P holds for axioms,

(ii) if property P holds for the upper formulae of a rule then it holds for the lower formula of that rule,

then (iii) property P holds for theorems.

This requires amplification when property P is of the form

$$\text{if } \phi \text{ is an } \mathbf{A}_I\text{-theorem then so is } \mathrm{R}[\phi],$$

where $\mathrm{R}[\phi]$ is an **A**-statement obtained from $\phi$ by a recursive process. The amplification is required because in the case of R 3 we have the requirement 'for all $\mathfrak{n}$, **E**-correct', so that we should have to show that $\mathrm{R}[\phi]$ for all $\mathfrak{n}$ **E**-correct, then we can apply R 3 and the result will hold for the lower formula provided that

$$(A\xi)\,\mathrm{R}[\phi] \quad \text{is the same as} \quad \mathrm{R}[(A\xi)\,\phi],$$

which is easily checked, the process of forming $\mathrm{R}[\phi]$ being recursive.

We deal with this situation as follows: (we confine ourselves to the introduction of a single universal quantifier). The $\mathbf{A}_I$-proof of $\phi$ consists of a bounded number of **E**-theorems of the types

$$prf_0[\tau x, \sigma[Num\, x]] = 0, \quad ded_0[\rho x, \tau x, \sigma[Num\, x]] = 0.$$

If property $P$ is of the kind under consideration then each statement $\psi$ in the $\mathbf{A}_I$-proof-tree is replaced by $\mathrm{R}[\psi]$ and possibly new ones are intercalcated. An application of R 3, say

$$\frac{\chi\{\nu\} \vee \omega}{(A\xi)\,\chi'\{\xi\} \vee \omega} \quad \text{for all } \nu, \textbf{E}\text{-correct},$$

becomes $$\underline{\mathrm{R}[\chi\{\nu\} \vee \omega]},$$

as far as the upper formulae are concerned, we require to show that

$$\mathrm{R}[\chi\{\nu\} \vee \omega] \quad \text{for all } \nu, \textbf{E}\text{-correct.}$$

This amounts to obtaining **E**-proofs for

$$prf_0[\tau^*x, \sigma^*[Num\, x]] = 0,$$

and $$ded_0[\rho^*x, \tau^*x, \sigma^*[Num\, x]] = 0,$$

where the asterisk denotes what these functions have become after application of the operation R.

Let $\mathrm{R}'\pi$ be the ordered set whose components are the *g.n.*'s of the result of operating R on the formulae whose *g.n.*'s are the components of $\pi$.

D 234 $\langle \nu \rangle_\kappa$ for the ordered $\kappa$-tuplet consisting of the first $\kappa$ components of $\nu$, or $\nu$ if $\kappa \geqslant \varpi\nu$.

By hypothesis we have $prf_0[\tau x, \sigma[Num\, x]] = 0$ in an **E**-theorem, hence we have

$$prf_0[\langle \tau x \rangle_\kappa, Pt[\tau x, \kappa]] = 0 \text{ is an } \textbf{E}\text{-theorem.}$$

This follows at once from the definition of $prf_0$, namely

$$\begin{aligned} prf_0[\pi, \nu] = A_2[\nu, Pt[\pi, \varpi\pi]] + \sum_{1 \leqslant \xi \leqslant \varpi\pi} \Big(&Ax_0[Pt[\pi, \xi]] \\ &\times \prod_{1 \leqslant \xi' \leqslant \xi} Rl^1[Pt[\pi, \xi], Pt[\pi, \xi']] \\ &\times \prod_{1 \leqslant \xi < \xi' < \xi''} Rl^2[Pt[\pi, \xi], Pt[\pi, \xi'], Pt[\pi, \xi'']]\Big) \end{aligned}$$

and the **E**-theorem $$\sum_{1 \leqslant x' \leqslant x''} \rho x' \leqslant \sum_{1 \leqslant x' \leqslant x'' + x'''} \rho x',$$

which is easily proved.

We now show

$$\frac{\begin{array}{c} prf_0[\mathrm{R}'[\langle\tau x\rangle_1], Pt[\mathrm{R}'[\langle\tau x\rangle_1], \varpi\mathrm{R}'[\langle\tau x\rangle_1]]] = 0, \\ prf_0[\mathrm{R}'[\langle\tau x\rangle_{x'}], Pt[\mathrm{R}'[\langle\tau x\rangle_{x'}], \varpi\mathrm{R}'[\langle\tau x\rangle_{x'}]]] = 0, \end{array}}{prf_0[\mathrm{R}'[\langle\tau x\rangle_{Sx'}], Pt[\mathrm{R}'[\langle\tau x\rangle_{Sx'}], \varpi\mathrm{R}'[\langle\tau x\rangle_{Sx'}]]] = 0}$$

whence by Prop. 2 we obtain

$$prf_0[\mathrm{R}'[\langle\tau x\rangle_{x'}], Pt[\mathrm{R}'[\langle\tau x\rangle_{x'}], \varpi\mathrm{R}'[\langle\tau x\rangle_{x'}]]] = 0,$$

now take $\varpi(\tau x)$ for $x'$ and we obtain

$$prf_0[\mathrm{R}'[\tau x], \mathrm{R}'\langle\tau[Num\,x]\rangle] = 0$$

as desired.

This settles the matter for $\mathbf{A}_I$-proofs of one level. For $\mathbf{A}_I$-proofs having $S\lambda$ levels we proceed by induction on the number of levels. Assume then that theorems having $\lambda$ levels have property P, we show that $\mathbf{A}_I$-theorems having $S\lambda$ levels also have property P. We proceed to treat $ded_0$ exactly as we treated $prf_0$. The only difference is that there is one more factor in the terms of the summation in the definition of $prf_0$, namely the factor

$$Hyp[Pt[\pi, \kappa]],$$

which is zero if and only if $Pt[\pi, \kappa]$ is a production of level $\lambda$. $ded_0[\rho x, \tau x, \pi[Num\,x]]$ becomes $ded_0[\mathrm{R}'[\rho x], \mathrm{R}'[\tau x], \mathrm{R}'[\langle\sigma[Num\,x]\rangle]]$. We **E**-prove this to be zero exactly as for $prf_0$ and our hypothesis that P'$\nu$' is an $\mathbf{A}_I$-theorem if '$\nu$' is in the production of level $\lambda$, for then

$$Hyp[Pt[x, x']] = Hyp[\mathrm{R}'[\langle Pt[x, x']\rangle]],$$

which, by our induction hypothesis is an $\mathbf{A}_I$-theorem.

In rule R 3 we have allowed the simultaneous introduction of a batch of universal quantifiers all at once. This is done to make $\mathbf{A}_I$ easier to use, see the worked example to follow. It is probable that the system $\mathbf{A}_I$ with rule R 3 restricted to the introduction of a single quantifier at a time is equivalent to the system proposed, but this investigation is omitted.

The intention is that once a batch of quantifiers has been introduced with scope $\Phi$ then all other batches of quantifiers that are introduced by R 3 have a scope $\Psi$ which properly contains $\Phi$ and its batch of quantifiers or the intersection of $\Phi$ and $\Psi$ is void. Thus between the introduction of a batch of quantifiers to $\Phi$ and the introduction of another batch of quantifiers having $\Phi$ in its scope there will be some building schemes to build up $\Phi$ and its quantifiers to $\Psi'$ which properly contains $\Phi$.

It would be possible to introduce other batches of quantifiers. For instance from $2[\nu/2] \leqslant \nu \leqslant 2[\nu/2]+1$ for all $\nu$, **E**-correct, we might allow

$$(Ax)(Ex')(2x \leqslant x' \leqslant 2x+1).$$

We omit further discussion on these lines.

Our method of introducing quantifiers by batches is without loss of generality. Because if we introduce a batch of quantifiers in two sub-batches, possibly with other building schemes between (these must act so that our original batch is in the subsidiary formula, except for R 1), then we could have introduced them all at once either at the first place or at the second place.

The notation we have used in (15)–(25) inclusive is to be interpreted as follows: in level 1 we generalize the $\pi$-tuplet $\mathfrak{n}$. In level $S\theta$ we generalize the $\pi^{(\theta)}$-tuplet $\mathfrak{n}^{(\theta)}$. Thus we require (15)–(25) to be theorems for all $\mathfrak{n}, \ldots, \mathfrak{n}^{(\lambda)}$. But in level 1 the tuplets $\mathfrak{n}', \ldots, \mathfrak{n}^{(\lambda)}$ may be absent or only partly present, but the tuplet $\mathfrak{n}$ is present, normally in full. If one or more of the places in the tuplet $\mathfrak{n}$ is always void then, when we generalize, we get a vacuous generalization.

If level $S\lambda$ is a single $\mathbf{A}_0$-deduction from the production of level $\lambda$ as hypothesis, then the tuplet $\mathfrak{n}^{(\lambda)}$ is always absent.

If fresh numerals are introduced by substitution, then it is intended that these be ultimately generalized (in one notation others are just omitted). If fresh numerals $\mathfrak{n}^*$ are introduced, then those of them that are already in one of the tuplets $\mathfrak{n}', \ldots, \mathfrak{n}^{(\lambda)}$ will get generalized when that tuplet gets generalized. Thus the notation must be such that all the introduced numerals are in one of the tuplets displayed in an $\mathbf{A}_I$-proof of $S\lambda$ levels.

## 10.7 *The* $\mathbf{A}_I$*-proof-predicate*

We wish to define an $\mathbf{A}_0$-statement $\chi\{\kappa, \nu\}$ such that $\chi\{\kappa, \nu\}$ is $\mathbf{A}_0$-true if and only if $\kappa$ is the *g.n.* of an $\mathbf{A}_I$-proof of the closed **A**-statement whose *g.n.* is $\nu$.

Suppose that the $\mathbf{A}_I$-proof of '$\nu$' has $S\lambda$ levels, then we shall require that $\kappa$ has $SS\lambda$ components

$$\kappa = \langle \kappa', \ldots, \kappa^{(S\lambda)}, \nu \rangle,$$

where $$\kappa^{(\theta)} = \langle \kappa^{(\theta,1)}, \ldots, \kappa^{(\theta, S\pi^{(\theta)})} \rangle, \quad 1 \leqslant \theta \leqslant S\lambda.$$

Thus $\kappa$ tells us the number of levels in the $\mathbf{A}_I$-proof, namely there are $\varpi\kappa \dot{-} 1$ levels.

$\kappa^{(\theta)}$ is to give us the deductions in level $\theta$. Thus $\kappa^{(1,\theta')}$ is to be the *g.n.* of the **E**-proof of the $\theta'$th line of (18), and $\kappa^{(S\theta,\theta')}$, for $1 \leqslant \theta < S\lambda$, is to be the *g.n.* of the **E**-proof of the $\theta'$th line of (20), $\kappa^{(S\lambda)}$ is to have only one component, namely the *g.n.* of an **E**-proof of a line of (20) with $\lambda$ instead of $\theta$, or the *g.n.* of an $\mathbf{A}_0$-deduction from the production of level $\lambda$ as hypotheses, and the last component of $\kappa$ is to be $\nu$. We can easily tell which is the case with level $S\lambda$, because we can decide of a sequence of formulae whether they form an **E**-proof or an $\mathbf{A}_0$-deduction or neither.

D 235 $$(\kappa_\theta^\pi) \quad \text{for} \quad pt[\kappa, \pi, \theta],$$
$$(\kappa_\theta^\infty) \quad \text{for} \quad Pt[\kappa, \theta],$$
$$(l.c.\,\kappa) \quad \text{for} \quad Pt[\kappa, \varpi\kappa].$$

We frequently omit the outer parentheses.

D 236 $\Theta\kappa = 0$ if $(\kappa_1^\infty)_{\theta'}^\infty$ is the *g.n.* of an **E**-proof of an equation (18), for $1 \leqslant \theta' \leqslant \varpi\kappa_1^\infty$, and $(\kappa_{S\theta}^\infty)_{\theta'}^\infty$ is the *g.n.* of an **E**-proof of an equation (20), for $1 \leqslant \theta \leqslant \varpi\kappa \dot{-} 2$ and $1 \leqslant \theta' \leqslant \varpi(\kappa_\theta^\infty)$, and $\kappa_{\varpi\kappa \dot{-} 1}^\infty$ has only one component and $\kappa_{\varpi\kappa \dot{-} 1}^\infty$ is the *g.n.* of an **E**-proof of an equation (20) or is *g.n.* of an $\mathbf{A}_0$-deduction and $(l.c.\,\kappa) = \nu$,
1 otherwise.

Clearly $\Theta$ is a recursive function because each of the clauses can effectively be tested.

In an $\mathbf{A}_I$-proof the production of level $\theta$ forms the hypotheses of level $S\theta$, and the hypotheses in the various deductions in level $S\theta$ are all the same. We have to incorporate this into $\chi\{\kappa, \nu\}$.

D 237 $$\Xi\delta = \begin{cases} \mu & \text{if} \quad \delta = g.n. \text{ of } ded_0[\pi, \lambda, \mu], \\ 0 & \text{otherwise.} \end{cases}$$

D 238 $$\Delta\delta = \begin{cases} \pi & \text{if} \quad \delta = g.n. \text{ of } ded_0[\pi, \lambda, \mu], \\ 0 & \text{otherwise.} \end{cases}$$

Clearly $\Xi$ and $\Delta$ are recursive functions. Note that $ded_0[0, \lambda, \mu]$ is the same as $prf_0[\lambda, \mu]$.

The hypotheses of the various deductions in level $\theta$, $1 < \theta \leqslant \varpi\kappa \dot{-} 1$ are the same is expressed by

$$(A\xi, \xi')_{\varpi\kappa^{(\theta)}}(\Delta(l.c.\,\kappa^{(\theta,\,\xi)}) = \Delta(l.c.\,\kappa^{(\theta,\,\xi')})), \tag{26}$$

for $1 < \theta < \varpi\kappa$.

We say that a line of (19) and a line of (15) are *related* if both are formed using the same $\phi_0$ and $\omega_0$, similarly we say that a line of (25) and a line of (24) are related, if they are both formed using the same $\phi_{S\theta}$ and $\omega_{S\theta}$. A line of (24) may be related to several lines of (25), but a line of (25) can be related to only one line of (24). We are naturally supposing that the lines of (24) are without duplicates and similarly for the other cases.

D 239 $H[\pi,\pi'] = \begin{cases} 0 & \text{if } \pi \text{ is the } g.n. \text{ of a line of (19) or (25) and } \pi' \text{ is the } g.n. \text{ of a related line of (15) or (24) respectively,} \\ 1 & \text{otherwise.} \end{cases}$

Clearly $H$ is a recursive function, because given two numerals we can effectively decide whether they are related in the manner described above, or otherwise. We now require

$$(A\xi)_{\delta}(E\eta)_{\delta'}(H[(\Delta(l.c.\,\kappa^{(S\theta,\,1)})^{\infty}_{\xi}, \Xi(l.c.\,\kappa^{(\theta,\,\eta)})] = 0), \tag{27}$$

for $1 \leqslant \theta < \tilde{\omega}\kappa \dot{-} 1$, where $\delta = \tilde{\omega}\Delta(l.c.\,\kappa^{(S\theta,\,1)})$ and $\delta' = \tilde{\omega}\kappa^{(\theta)}$.

(27) says that each hypothesis in a deduction in level $S\theta$ is related to one of the statements deduced in level $\theta$. This allows some statements deduced in level $S\theta$ to be redundant, but this is immaterial.

We still require a clause to the effect that '$\nu$' is related to the statement proved in level $S\lambda$ or is the statement deduced in level $S\lambda$ in case level $S\lambda$ is an $\mathbf{A}_0$-deduction. This is

$$H\left[\nu, \left(l.c.\left[\frac{\kappa^{\infty}_{\varpi\kappa \dot{-} 1}}{2}\right]\right)\right] \vee \nu = \left(l.c.\left[\frac{\kappa^{\infty}_{\varpi\kappa \dot{-} 1}}{2}\right]\right). \tag{28}$$

Division by 2 is required because the components of $\kappa$, except the last are considered as ordered sets and the penultimate set is a singleton. The $\mathbf{A}_I$-proof-predicate is

$$\Theta\kappa = 0 \,\&\, \sum_{1<\theta<\varpi\kappa} (26) \,\&\, \sum_{1\leqslant\theta<\varpi\kappa \dot{-} 1} (27) \,\&\, (28), \tag{29}$$

this is complicated but nevertheless it is recursive.

PROP. 6. $A_I$ *is consistent with respect to its truth definition.*

We have to show that $A_I$-theorems are A-true. It is clear that the $A_I$-axioms are A-true and that if the upper formulae of an A-rule are A-true then so is the lower formula. Thus $A_I$-theorems are A-true.

**10.8   *An example of an $A_I$-proof***

EXAMPLE.   $(Ax, x')\, (x = x' \vee x \neq x')$ *is an $A_I$-theorem.*

We give a one level proof without using I $b$. This example shows the advantage we gain by introducing batches of universal quantifiers all at once.

First we have to give a description of $A_{oo}$-proofs of

$$\nu = \nu' \vee \nu \neq \nu' \quad \text{for all numerals } \nu, \nu'$$

on a general plan and then we have to show that this plan is E-correct. That is to say we have to give the E-proof of

$$prf_{oo}[\tau[x, x'], \sigma[Num\, x, Num\, x']] = 0,$$

where $\sigma[Num\, \nu, Num\, \nu']$ is equal to the *g.n.* of $\nu = \nu' \vee \nu \neq \nu'$ and $\tau[\nu, \nu']$ is equal to the *g.n.* of an $A_{oo}$-proof of this disjunction.

Consider the following displayed scheme:

$$\{0, 0\} \qquad\qquad \mathrm{II}\,a \;\; \frac{\dfrac{0 = 0}{0 \neq 0 \vee 0 = 0}}{0 = 0 \vee 0 \neq 0} \quad \begin{array}{l} Ax\,1 \\ {} \end{array} \qquad \begin{array}{l} \mathrm{a} \\ \mathrm{b} \\ \mathrm{c} \end{array}$$

$$\{S\nu, 0\} \qquad\qquad \mathrm{II}\,a \;\; \frac{S\nu \neq 0}{S\nu = 0 \vee S\nu \neq 0} \quad Ax\,2.1 \qquad \begin{array}{l} \mathrm{d}[Num\,\nu] \\ \mathrm{e}[Num\,\nu] \end{array}$$

$$\{0, S\nu'\} \qquad\qquad \mathrm{II}\,a \;\; \frac{0 \neq S\nu'}{0 = S\nu' \vee 0 \neq S\nu'} \quad Ax\,2.2 \qquad \begin{array}{l} \mathrm{f}[Num\,\nu'] \\ \mathrm{g}[Num\,\nu'] \end{array}$$

$\{S\nu, S\nu'\}$

$$\mathrm{I}\,a \;\; \frac{\nu = \nu' \vee \nu \neq \nu'}{} \qquad Hyp. \quad \mathrm{h}$$

$$\begin{array}{ll} \mathrm{m} \\ \quad \mathrm{II}\,a \end{array} \frac{S\nu = S\nu \quad Ax\,1}{} \qquad \mathrm{R}\,2 \;\; \frac{\nu \neq \nu' \vee \nu = \nu'}{} \quad \mathrm{i}$$

$$\begin{array}{ll} \mathrm{n} \\ \quad \mathrm{I}\,a \end{array} \frac{S\nu \neq S\nu' \vee S\nu = S\nu}{} \qquad \mathrm{I}\,a \;\; \frac{S\nu \neq S\nu' \vee \nu = \nu'}{} \quad \mathrm{j}$$

$$\begin{array}{ll} \mathrm{p} \\ \quad \mathrm{R}\,1 \end{array} \frac{S\nu = S\nu \vee S\nu \neq S\nu'}{} \qquad \frac{\nu = \nu' \vee S\nu \neq S\nu'}{} \quad \mathrm{k}$$

$$\frac{}{S\nu = S\nu' \vee S\nu \neq S\nu'} \qquad\qquad \mathrm{l}$$

From these schemes $\tau[\nu, \nu']$ can be found, this is the $g.n.$ of an $\mathrm{A}_{00}$-proof of

$$\nu = \nu' \vee \nu \neq \nu'.$$

Let the $g.n.$'s of the various statements in the above schemes be as shown at the sides. h–p inclusive should be in full $\mathrm{h}[Num\,\nu, Num\,\nu']$, etc. We have

$$\tau[0, 0] = \langle \mathrm{a, b, c} \rangle,$$

$$\tau[S\nu, 0] = \langle \mathrm{d}[Num\,\nu], \mathrm{e}[Num\,\nu] \rangle = \mathrm{r}[Num\,\nu], \quad \text{say},$$

$$\tau[0, S\nu'] = \langle \mathrm{f}[Num\,\nu'], \mathrm{g}[Num\,\nu'] \rangle = \mathrm{s}[Num\,\nu'], \quad \text{say},$$

$$\tau[S\nu, S\nu'] = \tau[\nu, \nu']^\frown \langle \mathrm{i, j, k, m, n, p, l} \rangle$$

$$= \tau[\nu, \nu']^\frown \mathrm{t}[Num\,\nu, Num\,\nu'], \quad \text{say}.$$

This defines a primitive recursive function $\tau[\nu, \nu']$. Let $\sigma[Num\,\nu, Num\,\nu']$ be equal to the $g.n.$ of $\nu = \nu' \vee \nu \neq \nu'$. Then we have

$$prf_{00}[\tau[0, 0], \sigma[Num\,0, Num\,0]] = 0,$$

$$prf_{00}[\tau[S\nu, 0], \sigma[Num\,S\nu, Num\,0]] = 0,$$

$$prf_{00}[\tau[0, S\nu'], \sigma[Num\,0, Num\,S\nu']] = 0,$$

and

$$\frac{prf_{00}[\tau[\nu, \nu'], \sigma[Num\,\nu, Num\,\nu']] = 0}{prf_{00}[\tau[S\nu\,S\nu'], \sigma[Num\,S\nu, Num\,S\nu']] = 0.}$$

We require to show that

$$prf_{00}[\tau[0, 0], \sigma[Num\,0, Num\,0]] = 0, \tag{i}$$

$$prf_{00}[\tau[Sx, 0], \sigma[Num\,Sx, Num\,0]] = 0, \tag{ii}$$

$$prf_{00}[\tau[0, Sx], \sigma[Num\,0, Num\,Sx]] = 0, \tag{iii}$$

are **E**-theorems, and that:

$$\frac{prf_{00}[\tau[x, x'], \sigma[Num\,x, Num\,x']] = 0}{prf_{00}[\tau[Sx, Sx'], \sigma[Num\,Sx, Num\,Sx']] = 0,} \tag{iv}$$

is an **E**-deduction. Then by Prop. 4 we shall have

$$prf_{00}[\tau[x, x'], \sigma[Num\,x, Num\,x']] = 0 \quad \text{as desired.}$$

(i) is an $\mathbf{A}_{00}$-true equation, hence is an **E**-theorem. To deal with the others, we note that

$$prf_{00}[\kappa,\nu] = A_2[Pt[\kappa,\varpi\kappa],\nu] + \sum_{1\leqslant\xi\leqslant\varpi\kappa} (Ax_{00}\, Pt[\kappa,\xi]$$

$$\times \prod_{1\leqslant\xi'<\xi} Rl^1[Pt[\kappa,\xi'],Pt[\kappa,\xi]] \times \prod_{1\leqslant\xi''<\xi'<\xi} Rl^2[Pt[\kappa,\xi''],Pt[\kappa,\xi'],Pt[\kappa,\xi]]),$$

that is to say; the last component of $\kappa$ is $\nu$, and each component of $\kappa$ is either an axiom or arises from an earlier component by a one-premiss rule or arises from two earlier components by a two-premiss rule.

Using the easily proved **E**-theorems:

$$0+0=0 \quad\text{and}\quad 0\times x = 0 \quad\text{and if}\quad \rho x = 0 \quad\text{then}\quad \sum_{1\leqslant x'\leqslant x} \rho x' = 0,$$

it suffices to show that $A_2[Pt[\kappa,\varpi\kappa],\nu] = 0$

and that one of the factors in the product is zero for each $1 \leqslant \xi \leqslant \varpi\kappa$.

Ad (ii). We have to **E**-prove

$$prf_{00}[\mathrm{r}[Num\, x], disj[\mathrm{d}'[Num\, x], \mathrm{d}[Num\, x]]] = 0,$$

where $\mathrm{d}'[Num\,\nu]$ is equal to the *g.n.* of $S\nu = 0$, now,

$$\mathrm{r}[Num\, x] = \langle \mathrm{d}[Num\, x], disj[\mathrm{d}'[Num\, x], \mathrm{d}[Num\, x]]\rangle.$$

Thus $A_2[Pt[\mathrm{r}[Num\, x], \varpi\mathrm{r}[Num\, x]], disj[\mathrm{d}'[Num\, x], \mathrm{d}[Num\, x]]]$

$$= A_2[disj[\mathrm{d}'[Num\, x], \mathrm{d}[Num\, x]], disj\,[\mathrm{d}'[Num\, x], \mathrm{d}[Num\, x]]] = 0,$$

in virtue of the easily proved **E**-theorem $A_2[x,x] = 0$ and substitution. Also we have $Ax^{2.1}[Pt[\mathrm{r}[Num\, x], 1]] = 0$ because $Ax^{2.1}[\nu]$ is

$$\nu = \langle 8,8,3,8,1\rangle^\frown\kappa^\frown\langle 9,9,0,9\rangle \;\&\; tm\kappa$$

or $$Ax^{2.1}\nu = \prod_{1\leqslant\xi<\nu} (A_2[\nu, \langle 8,8,3,8,1\rangle^\frown\xi^\frown\langle 9,9,0,9\rangle] + tm\xi),$$

where $tm\kappa$ for '$\kappa$ is of type $\iota$'.

In our case we have

$$Pt[\mathrm{r}[Num\, x], 1] = \mathrm{d}[Num\, x]$$

and $Ax^{2.1}[\mathrm{d}[Num\, x]] = 0$ is an **E**-theorem, because

$$\mathrm{d}[Num\, x] = \langle 8,8,3,8,1\rangle^\frown Num\, x^\frown\langle 9,9,0,9\rangle \;\&\; tm\, Num\, x.$$

This follows since

$$Num\, x = \mathscr{I}(\boldsymbol{\lambda} x'x''.\langle 8,1\rangle^\frown x''^\frown\langle 9\rangle)\,0x$$

and $$tm\kappa = A_2[\kappa, 0] \times \prod_{1\leqslant\xi<\kappa} (A_2[\kappa, \langle 8, 1\rangle^\frown\xi^\frown\langle 9\rangle] + tm\xi)$$
$$\times \prod_{1\leqslant\xi,\xi',\xi''<\kappa} (A_2[\kappa, \langle 8, 8, 6\rangle^\frown\xi^\frown\xi'^\frown\langle 9\rangle^\frown\xi''^\frown\langle 9\rangle] + Var\,\xi + tm[\xi', \xi''])$$
$$\times \prod_{1\leqslant\xi,\xi',\eta,\eta',\eta'',<\kappa} (A_2[\kappa, \langle 8, 8, 7, 8, 6\rangle^\frown\xi^\frown\langle 8, 6\rangle^\frown\xi'^\frown\eta^\frown \langle 9, 9, 9\rangle^\frown\eta'^\frown\langle 9\rangle^\frown\eta''^\frown\langle 9\rangle] + Var[\xi, \xi'] + tm[\eta, \eta', \eta'']).$$

We have $$tm\,Num\,0 = tm\,0 = 0,$$

and $$\frac{tm\,Num\,x = 0}{tm\,Num\,Sx = 0}.$$

$Num\,Sx = \langle 8, 1\rangle^\frown Num\,x^\frown\langle 9\rangle$, hence if $tm\,Num\,x = 0$ then $tm\,Num\,Sx = 0$, since the second clause in the definition of $tm$ is satisfied. Thus from Prop. 4 we have $tm\,Num\,x = 0$. Thus $Ax^{2.1}[\mathrm{d}[Num\,x]] = 0$.

In our case there are only two components in $\tau[Sx, 0]$, viz. $\mathrm{d}[Num\,x]$ and $disj[\mathrm{d}'[Num\,x], \mathrm{d}[Num\,x]]$ and we have

$$Rl^{\mathrm{IIa}}[\mathrm{d}'[Num\;\;x], disj[\mathrm{d}'[Num\,x], \mathrm{d}[Num\,x]]] = 0.$$

This follows since:

$$Rl^{\mathrm{IIa}}[\kappa, \pi] = \prod_{1\leqslant\xi<\pi} A_2[\pi, disj[\xi, \kappa]].$$

In our case this is

$$\prod_{1\leqslant\xi\leqslant disj[\mathrm{d}'[Num\,x], \mathrm{d}[Num\,x]]} A_2[disj[\mathrm{d}'[Num\,x], \mathrm{d}[Num\,x]], disj[\xi, \mathrm{d}[Num\,x]]],$$

Now $$d'[Num\,x] < disj[d'[Num\,x], d[Num\,x]]$$

as follows from D 139, also we have the **E**-theorem

$$\prod_{1\leqslant x'\leqslant x} \rho x' = \prod_{1\leqslant x'<x''} \rho x' \times \rho x'' \times \prod_{x''<x''<x} \rho x' \quad \text{for} \quad x'' < x.$$

Thus $$Rl^{\mathrm{IIa}}[\mathrm{d}'[Num\,x], disj[\mathrm{d}'[Num\,x], \mathrm{d}[Num\,x]]] = 0,$$

and so $$Rl^{1}[Pt[\mathrm{r}[Num\,x], 1], Pt[\mathrm{r}[Num\,x], 2]] = 0.$$

Hence (ii).

(iii) Similarly.

(iv) This follows from

$$Rl^{\mathrm{Ia}}[\mathrm{h}', \mathrm{i}'] + Rl^{\mathrm{R2}}[\mathrm{i}', \mathrm{j}'] + Rl^{\mathrm{Ia}}[\mathrm{j}', \mathrm{k}'] + Ax^{1}\mathrm{m}' + Rl^{\mathrm{IIa}}[\mathrm{m}', \mathrm{n}'] + Rl^{\mathrm{Ia}}[\mathrm{n}', \mathrm{p}'] + Rl^{\mathrm{RI}}[\mathrm{l}', \mathrm{p}', \mathrm{k}'] + A_2[Pt[\tau[Sx, Sx], \varpi\tau[Sx, Sx]], \mathrm{l}'] = 0. \quad \text{(v)}$$

So that (iv). Here i′, ..., l′ are the same as i, ..., l but with $x$ instead of $\nu$ and $x'$ instead of $\nu'$. It remains to show that each of the summands in (v) is zero.

$Rl^{\mathrm{I}a}[\mathrm{h}',\mathrm{i}'] = (A_2[\mathrm{i}', disj[\alpha,\beta]] + A_2[\mathrm{h}', disj[\beta,\alpha]]) \times \ldots$ for some $\alpha$, $\beta$. The remainder of the products refer to cases when subsidiary formulae are present. This is easily seen to be an **E**-theorem in virtue of the **E**-theorem $A_2[x,x] = 0$ and substitution.

The remainder follow similarly, $Ax^1$ m follows as did $Ax^{2.1}[\mathrm{d}[Num\, x]]$, it involves the **E**-theorem $tm\, Num\, x = 0$. This completes the **E**-proof of $prf_{\mathrm{oo}}[\tau[x,x'], \sigma[Num\, x, Num\, x']] = 0$. Thus $(Ax,x')\,(x = x' \vee x \neq x')$ is an $\mathbf{A}_I$-theorem.

We use the terminology

$$\text{‘}\phi\{\alpha\{\mathfrak{n}\}\} \quad \text{for all } \mathfrak{n}\ \mathbf{E}\text{-correct’}$$

when $prf_{\mathrm{o}}[\tau\mathfrak{x}, \sigma[\mathfrak{Num}\,\mathfrak{x}]] = 0$ or $ded_{\mathrm{o}}[\rho\mathfrak{x}, \tau\mathfrak{x}, \sigma[\mathfrak{Num}\,\mathfrak{x}]] = 0$, are **E**-theorems, where $\tau\mathfrak{n}$ is equal to the *g.n.* of an $\mathbf{A}_0$-proof or -deduction respectively of the **A**-statement whose *g.n.* is equal to $\sigma[\mathfrak{Num}\,\mathfrak{n}]$ from hypotheses (in the second case) whose *g.n.* is equal to $\rho\mathfrak{n}$.

## 10.9 *Relations between $\mathbf{A}_0$-theorems and* **E**-*correctness*

PROP. 7. *If $\phi\{\mathfrak{G}\}$ is an $\mathbf{A}_0$-axiom-form then $\phi\{\mathfrak{n}\}$ for all $\mathfrak{n}$ **E**-correct. If $\phi\{\mathfrak{n}\}$, $\psi\{\mathfrak{n}\}$ for all $\mathfrak{n}$ **E**-correct and if*

$$\frac{\phi\{\mathfrak{G}\}}{\chi\{\mathfrak{G}\}} \qquad\qquad \frac{\phi\{\mathfrak{G}\} \quad \psi\{\mathfrak{G}\}}{\chi\{\mathfrak{G}\}}$$

*are $\mathbf{A}_0$-rule-forms then $\chi\{\mathfrak{n}\}$ for all $\mathfrak{n}$ **E**-correct.*

The demonstration is similar to the cases treated in the example. There we dealt with one axiom in detail and one rule in detail. The others are dealt with similarly.

COR. (i). *If $\phi\{\mathfrak{G}_\iota\}$ is an $\mathbf{A}_0$-theorem-form then $\phi\{\mathfrak{n}\}$ for all $\mathfrak{n}$ **E**-correct.*

We put down the $\mathbf{A}_0$-proof-tree-form for $\phi\{\mathfrak{G}_\iota\}$. By Prop. 6 the axioms hold for all $\mathfrak{n}$ **E**-correct and if the upper formulae of an $\mathbf{A}_0$-rule hold for all $\mathfrak{n}$ **E**-correct then so does the lower formula, thus by a simple induction the same holds for the theorem at the base.

Cor. (ii). *If we have an* $\mathbf{A}_0$*-deduction-form of* $\phi\{\mathfrak{G}_\iota\}$ *from hypotheses-forms* $\psi\{\mathfrak{G}_\iota\}, \ldots, \psi^{(\pi)}\{\mathfrak{G}_\iota\}$, *where* $\psi\{\mathfrak{n}\}, \ldots, \psi^{(\pi)}\{\mathfrak{n}\}$ *hold for all* $\mathfrak{n}$ **E**-*correct, then* $\phi\{\mathfrak{n}\}$ *for all* $\mathfrak{n}$ **E**-*correct.*

The demonstration is similar to that of Cor. (i).

Cor. (iii). *If* $\phi\{0\}$ *is an* $\mathbf{A}_0$*-theorem and if* $\dfrac{\phi\{\Gamma_\iota\}}{\phi\{S\Gamma_\iota\}}$ *is an* $\mathbf{A}_0$*-deduction-form, then* $\phi\{\nu\}$ *for all* $\nu$, **E**-*correct.*

This follows from Prop. 5, because the hypotheses give

$$prf_0[\tau 0, \sigma[Num\, 0]] = 0 \quad \text{and} \quad \frac{prf_0[\tau\Gamma_\iota, \sigma[Num\,\Gamma_\iota]] = 0}{prf_0[\tau S\Gamma_\iota, \sigma[Num\, S\Gamma_\iota]] = 0}.$$

Cor. (iv). *If* $\phi\{\mathfrak{n}\}$ *and* $\psi\{\mathfrak{n}\}$ *are deduced from* $\omega\{\mathfrak{n}\}$ *and if the deductions hold for all* $\mathfrak{n}$, **E**-*correct, and if* $\dfrac{\phi\{\mathfrak{n}\}}{\chi\{\mathfrak{n}\}}$ *or* $\dfrac{\phi\{\mathfrak{n}\} \quad \psi\{\mathfrak{n}\}}{\chi\{\mathfrak{n}\}}$ *are* $\mathbf{A}_0$*-rules, then* $\chi\{\mathfrak{n}\}$ *is a deduction from* $\omega\{\mathfrak{n}\}$ *for all* $\mathfrak{n}$, **E**-*correct.*

The demonstration is similar to that of Prop. 7.

This corollary shows the advantage we get from dividing an $\mathbf{A}_I$-proof up into a sequence of $\mathbf{A}_0$-deductions. For instance the deductions in level $S\theta$ are all $\mathbf{A}_0$-deductions from the productions of level $\theta$ as hypotheses. The fact that these hypotheses are themselves $\mathbf{A}_I$-theorems appears only in the requirement that they are related to the results of $\mathbf{A}_0$-deductions in level $\theta$ in the manner described in D 239. The $\mathbf{A}_I$-proof of $\phi\{\mathfrak{n}\}$ for all $\mathfrak{n}$, **E**-correct is the **E**-proof of $prf_0[\tau x, \sigma[Num\, x]] = 0$. This drops out and is replaced by the relation requirement of D 239. Otherwise we would be getting the *g.n.* of $prf_0[\tau x, \sigma[Num\, x]] = 0$ occurring in the $\mathbf{A}_I$-predicate, then the *g.n.* of this *g.n.* and so on.

From these corollaries we see that if $(A\xi)\,\phi\{\xi\}$ is proved in the system $\mathbf{A}_0$ plus the principle of Mathematical Induction then $(A\xi)\,\phi\{\xi\}$ is an $\mathbf{A}_I$-theorem.

We now show that the substitution in the induction rule is redundant.

Lemma. *If* $\psi\{\nu\}$ *is an* $\mathbf{A}_0$*-theorem for all* $\nu$, **E**-*correct, then* $\psi\{\alpha\{\nu\}\}$ *is an* $\mathbf{A}_0$*-theorem for all* $\nu$, **E**-*correct, where* $\alpha$ *is a primitive recursive function.*

Let $\tau\nu$ be equal to the *g.n.* of an $\mathbf{A}_0$-proof of $\psi\{\nu\}$ and let $\sigma[Num\,\nu]$ be equal to the *g.n.* of $\psi\{\nu\}$.

We are given $prf_0[\tau x, \sigma[Num\, x]] = 0$ is an **E**-theorem, hence so is

$$prf_0[\tau\gamma x, \sigma[Num\,\gamma x]] = 0, \tag{i}$$

where $\gamma$ is a primitive recursive function.

Now $prf_0[\tau\gamma\nu, \sigma[Num\,\gamma\nu]] = 0$ says that $\tau\gamma\nu$ is equal to the $g.n.$ of an $\mathbf{A}_0$-proof of $\psi\{S^*\gamma\nu\}$, where

$$S^*0 = 0, \quad S^*S\nu = SS^*\nu.$$

$S^*\gamma\nu$ is the $g.n.$ of the numeral determined by '$\gamma\nu$'. It is $\underbrace{S\ldots S}_{\text{'}\gamma\nu\text{'-times}}0$. We have $\gamma\nu = S^*\gamma\nu$, generally $\gamma x = S^*\gamma x$ is an **E**-theorem, Cor. (i), Prop. 4. The identity function and $S^*$ obey the same scheme of primitive recursion.

We require to show

$$prf_0[\tau' x, \sigma[\beta[Num\, x]]] = 0 \quad \text{is an } \mathbf{E}\text{-theorem, for some } \tau',$$

where $\beta[Num\,\nu]$ is equal to the $g.n.$ of $\alpha\{\nu\}$. $Num\,\alpha\{\nu\}$ is equal to the $g.n.$ of $\underbrace{S\ldots S}_{\text{'}\alpha\{\nu\}\text{'-times}}0$; from $\alpha\{\nu\} = S^*\alpha\{\nu\}$ we obtain $Eq[\beta[Num\,\nu], [Num\,\alpha\{\nu\}]] = 0$; we require

$$prf_0[\tau'' x, Eq[\beta[Num\, x], Num\,\alpha\{x\}]] = 0 \quad \text{for some } \tau'', \tag{ii}$$

to be an **E**-theorem. Take $\alpha$ for $\gamma$ in (i), then from (i) and (ii) we obtain the **E**-theorem:

$$prf_0[\tau\alpha\{x\}^\frown\tau'' x^\frown\langle\sigma[\beta[Num\, x]]\rangle, \sigma[\beta[Num\, x]]] = 0,$$

as required.

It remains to find $\tau''$ and to show that (ii) is an **E**-theorem. To do this we refer to the s.p. of $\alpha\{\nu\} = S^*\alpha\{\nu\}$ as given in Ch. 4. We have three cases:

(a) $\qquad \alpha\{\nu\} = \nu,$

(b) $\qquad \alpha\{\nu\} = S\alpha'\{\nu\},$

(c) $\qquad \alpha\{\nu\} = \mathscr{I}\rho\{\nu\}\gamma\{\nu\}\delta\{\nu\}.$

Here $\alpha'\{\nu\}$ is of lower order than $\alpha\{\nu\}$; $\gamma\{\nu\}$, $\delta\{\nu\}$, $\rho\{\nu\}$ are of lower rank than $\alpha\{\nu\}$. We suppose that the result holds for primitive recursive functions of lower order or rank than $\alpha\{\nu\}$.

*Case* (*a*). We have to show $prf_0[\tau'' x, Eq[Num\, x, Num\, x]] = 0$ is an **E**-theorem for some $\tau''$. Clearly $\tau''\nu = Eq[Num\,\nu, Num\,\nu]$ and we are finished.

*Case* (*b*). We are given, by hypothesis

$$prf_0[\tau''x, Eq[\beta'[Num\, x], Num\, \alpha'\{x\}]] = 0,$$

where $\beta'[Num\, \nu]$ is equal to the *g.n.* of $\alpha'\{\nu\}$, is an **E**-theorem for some primitive recursive function $\tau''$. We want to show:

$$prf_0[\tau'''x, Eq[\langle 8, 1\rangle^\frown \beta'[Num\, x]^\frown \langle 9\rangle, Num\, S\alpha'\{x\}]] = 0$$

is an **E**-theorem for some primitive recursive function $\tau'''$.

By hypothesis we have an $\mathbf{A}_0$-proof of $\alpha'\{\nu\} = S^*\alpha'\{\nu\}$ for all $\nu$, **E**-correct. To this we add

$$\frac{S\alpha'\{\nu\} = S\alpha'\{\nu\}}{S\alpha'\{\nu\} = SS^*\alpha'\{\nu\}}\ \text{R}\,1 \quad \text{Ax}_{00}\,1, \qquad \frac{SS^*\alpha'\{\nu\} = S^*S\alpha'\{\nu\}}{S\alpha'\{\nu\} = S^*S\alpha'\{\nu\}}\ \text{R}\,1$$

Cor. (i), Prop. 4

all these hold for all $\nu$, **E**-correct, hence using Prop. 7, Case (*b*).

*Case* (*c*). $\gamma\{\nu\}$ and $\delta\{\nu\}$ are of lower rank than $\alpha\{\nu\}$, hence by the induction hypothesis we have:

$$\left.\begin{aligned} &\gamma\{\nu\} = S^*\gamma\{\nu\}, \text{ for all } \nu, \textbf{E}\text{-correct},\\ \text{and}\quad &\delta\{\nu\} = S^*\delta\{\nu\}, \text{ for all } \nu, \textbf{E}\text{-correct}.\end{aligned}\right\} \qquad \text{(iii)}$$

$\mathscr{I}\rho S^*\gamma\{\nu\}\, S^*\delta\{\nu\}$ is of lower rank than $\alpha\{\nu\}$ unless $\gamma\{\nu\}$ and $\delta\{\nu\}$ are both numerals. We get three subcases:

*Subcase* ($c_1$). One of $\gamma\{\nu\}$ or $\delta\{\nu\}$ is different from a numeral,

*Subcase* ($c_2$). $\gamma\{\nu\}$, $\delta\{\nu\}$ are $\kappa$, 0 respectively,

*Subcase* ($c_3$). $\gamma\{\nu\}$, $\delta\{\nu\}$ are $\kappa$, $S\pi$ respectively.

*Subcase* ($c_1$). We have (iii) and by R 1,

$$\mathscr{I}\rho\{\nu\}\,\gamma\{\nu\}\,\delta\{\nu\} = \mathscr{I}\rho\{\nu\}\, S^*\gamma\{\nu\}\, S^*\delta\{\nu\} \quad \text{for all } \nu, \textbf{E}\text{-correct},$$

also by induction hypothesis,

$$\mathscr{I}\rho\{\nu\}\, S^*\gamma\{\nu\}\, S^*\delta\{\nu\} = S^*\mathscr{I}\rho\{\nu\}\, S^*\gamma\{\nu\}\, S^*\delta\{\nu\} \quad \text{for all } \nu, \textbf{E}\text{-correct},$$

hence using Prop. 7, Subcase ($c_1$).

*Subcase* ($c_2$). We have ($\kappa$ might be $S \ldots S\nu$)

$$\mathscr{I}\rho\{\nu\}\,\kappa 0 = \kappa \quad \text{Ax}_{00}\,3.1,$$

$$\kappa = S^*\kappa,$$

hence by R 1, $\mathscr{I}\rho\{\nu\}\kappa 0 = S^*\mathscr{I}\rho\{\nu\}\kappa 0,$

these hold for all $\nu$, **E**-correct, hence, using Prop. 7, Subcase $(c_2)$.

*Subcase* $(c_3)$. We have

$$\mathscr{I}\rho\{\nu\}\kappa(S\pi) = \rho\{\nu\}\pi(\mathscr{I}\rho\{\nu\}\kappa\pi) \quad \mathrm{Ax}_{00}\,3.2,$$

$\mathscr{I}\rho\{\nu\}\kappa\pi$ is of lower order than $\alpha\{\nu\}$, hence by induction hypothesis

$$\mathscr{I}\rho\{\nu\}\kappa\pi = S^*\mathscr{I}\rho\{\nu\}\kappa\pi \quad \text{for all } \nu, \mathbf{E}\text{-correct},$$

hence, by R 1 $\mathscr{I}\rho\{\nu\}\kappa(S\pi) = \rho\{\nu\}\pi S^*(\mathscr{I}\rho\{\nu\}\kappa\pi),$

this is of lesser rank than $\alpha\{\nu\}$, hence by induction hypothesis

$$\begin{aligned}\rho\{\nu\}\pi S^*(\mathscr{I}\rho\{\nu\}\kappa\pi) &= S^*\rho\{\nu\}\pi S^*(\mathscr{I}\rho\{\nu\}\kappa\pi)\\ &= S^*\mathscr{I}\rho\{\nu\}\kappa(S\pi).\end{aligned}$$

These hold for all $\nu$, **E**-correct, hence, using Prop. 7, Subcase $(c_3)$.

This completes the demonstration of the lemma.

## 10.10 *Some properties of the system* $\mathbf{A}_I$

PROP. 8. *The system* $\mathbf{A}_I$ *contains an irresolvable statement.*

The $\mathbf{A}_I$-proof-predicate is of the form: $(Ex'')\,P[\kappa, \nu, x'']$, where $P$ is primitive recursive, because '$\kappa$ is the *g.n.* of an $\mathbf{A}_I$-proof of the closed **A**-statement whose *g.n.* is $\nu$' is recursive, and any recursive statement can be expressed in this form.

Consider

$$(Ax')\,N[\text{``}(Ex'')\,P[x', s[x,x], x'']\text{''}],$$

where $s$ is Gödel's substitution function, let its *g.n.* be $\kappa$, and let $\nu$ be the *g.n.* of $(Ax')\,N[(\text{``}Ex'')\,P[x', s[\kappa,\kappa], x'']\text{''}]$, then $\nu = s[\kappa,\kappa]$ is an $\mathbf{A}_{00}$-theorem.

Now consider $(Ax')\,N[\text{``}(Ex'')\,P[x', \nu, x'']\text{''}]$, if it is an $\mathbf{A}_I$-theorem then so is $(Ax')\,N[\text{``}(Ex'')\,P[x', s[\kappa,\kappa], x'']\text{''}]$, whence by the properties of the $\mathbf{A}_I$-proof-predicate $(Ex'')\,P[\pi, \nu, x'']$, for some numeral $\pi$, is an $\mathbf{A}_0$-theorem, and so $(Ex'')\,P[\pi, s[\kappa,\kappa], x'']$ is also an $\mathbf{A}_0$-theorem, here the numeral $\pi$ is the *g.n.* of an $\mathbf{A}_I$-proof of $(Ax')\,N[\text{``}(Ex'')\,P[x', s[\kappa,\kappa], x'']\text{''}]$, but if this is an $\mathbf{A}_I$-theorem then so is $N[\text{``}(Ex'')\,P[\pi, s[\kappa,\kappa], x'']\text{''}]$ for each numeral $\pi$.

But $(Ex'')\,P[\pi, s[\kappa,\kappa], x]$ is recursive, hence so is its negation. Hence only one of them is correct, but under the supposition that

$$(Ax')\,N[\text{``}(Ex'')\,P[x',\nu,x'']\text{''}]$$

is an $\mathbf{A}_I$-theorem both of them must be correct. Thus it must be that $(Ax')\,N[\text{``}(Ex'')\,P[x',\nu,x'']\text{''}]$ fails to be an $\mathbf{A}_I$-theorem.

We now show that $(Ax')\,N[\text{``}(Ex'')\,P[x',\nu,x'']\text{''}]$ is **A**-valid, thus its negation fails to be an $\mathbf{A}_I$-theorem, this means that it is irresolvable. If $(Ax')\,N[\text{``}(Ex'')\,P[x',\nu,x'']\text{''}]$ fails to be **A**-valid then it will take the value **f** and so $(Ex'')\,P[\pi,\nu,x'']$ will take the value **t** for some numeral $\pi$. But this is an $\mathbf{A}_0$-statement and if it takes the value **t** then it is an $\mathbf{A}_0$-theorem, hence $\pi$ is the *g.n.* of an $\mathbf{A}_I$-proof of the **A**-statement whose *g.n.* is $\nu$. Thus

$$(Ax')\,N[\text{``}(Ex'')\,P[x',\nu,x'']\text{''}]$$

is an $\mathbf{A}_I$-theorem, and so is

$$N[\text{``}(Ex'')\,P[\pi,\nu,x'']\text{''}]$$

and so $(Ex'')\,P[\pi,\nu,x'']$ takes the value **f**, but it is recursive, so we have an absurdity, thus $(Ax')\,N[\text{``}(Ex'')\,P[x',\nu,x'']\text{''}]$ is **A**-valid but fails to be an $\mathbf{A}_I$-theorem.

We have thus found an example of a universal **A**-statement $(Ax)\,\phi\{x\}$ which fails to be an $\mathbf{A}_I$-theorem while each of $\phi\{\nu\}$ for $\nu = 0, 1, 2, \ldots$ are $\mathbf{A}_I$-theorems. This comes about as follows: $(Ex'')\,P[\pi,\nu,x'']$ is recursive, hence so is $N[\text{``}(Ex'')\,P[\pi,\nu,x'']\text{''}]$, this is then of the form $(Ex'')\,Q[\pi,\nu,x'']$, where $Q$ is primitive recursive because any recursive statement can be so expressed. Thus we have

$$(Ex'')\,Q[\pi,\nu,x''] \quad \text{for} \quad \pi = 0, 1, 2, \ldots \tag{30}$$

while $(Ax')\,(Ex'')\,Q[x',\nu,x'']$ fails to be an $\mathbf{A}_I$-theorem. Thus the conditions for an induction fail to be satisfied, in fact we omitted to show that the proofs of $(Ex'')\,Q[\pi,\nu,x'']$ for $\pi = 0, 1, 2, \ldots$ are on a primitive recursive plan **E**-correct, and this must fail to be the case. In obtaining (30) we used arguments of quite a different kind, the consistency of $\mathbf{A}_0$, etc.

This proposition shows that we must be very careful in saying that $(Ax)\,\phi\{x\}$ is an $\mathbf{A}_I$-theorem when we have somehow shown that $\phi\{\nu\}$ is an $\mathbf{A}_I$-theorem for all numerals $\nu$. To have a formal system we have to make the method of 'showing' definite. We have done this by laying down that the proofs of $\phi\{\nu\}$ for $\nu = 0, 1, 2, \ldots$ be on a primitive recursive plan which can be proved to be correct in the system **E**.

PROP. 9. *T.N.D. holds in* $\mathbf{A}_I$.

We have to show that $\phi \vee N[``\phi"]$ is an $\mathbf{A}_I$-theorem for any closed **A**-statement $\phi$. We proceed by induction on the construction of $\phi$. We shall show the stronger result that $\phi\{\mathfrak{n}\} \vee N[``\phi\{\mathfrak{n}\}"]$ for all $\mathfrak{n}$ **E**-correct.

(*a*) $\phi$ is atomic. We have to show

$$\alpha\{\mathfrak{n}\} = \beta\{\mathfrak{n}\} \vee \alpha\{\mathfrak{n}\} \neq \beta\{\mathfrak{n}\} \quad \text{for all } \mathfrak{n}\ \mathbf{E}\text{-correct.}$$

This comes from the example by the lemma. Thus Case (*a*).

(*b*) $\phi$ is a disjunction, say $\phi\{\mathfrak{n}\}$ is $\psi\{\mathfrak{n}\} \vee \psi'\{\mathfrak{n}\}$ and the result holds for $\psi\{\mathfrak{n}\}$ and for $\psi'\{\mathfrak{n}\}$. Then

$$\psi\{\mathfrak{n}\} \vee N[``\psi\{\mathfrak{n}\}"] \quad \text{for all } \mathfrak{n},\ \mathbf{E}\text{-correct,}$$

$$\psi'\{\mathfrak{n}\} \vee N[``\psi'\{\mathfrak{n}\}"] \quad \text{for all } \mathfrak{n},\ \mathbf{E}\text{-correct.}$$

We have

$$\frac{\psi\{\mathfrak{n}\} \vee N[``\psi\{\mathfrak{n}\}"]}{} \ \text{II}\,a, \text{I}\,a \ \frac{\psi'\{\mathfrak{n}\} \vee N[``\psi'\{\mathfrak{n}\}"]}{}$$

$$\frac{N[``\psi\{\mathfrak{n}\}"] \vee (\psi\{\mathfrak{n}\} \vee \psi'\{\mathfrak{n}\}) \quad N[``\psi'\{\mathfrak{n}\}"] \vee (\psi\{\mathfrak{n}\} \vee \psi'\{\mathfrak{n}\})}{N[``(\psi\{\mathfrak{n}\} \vee \psi'\{\mathfrak{n}\})"] \vee (\psi\{\mathfrak{n}\} \vee \psi'\{\mathfrak{n}\})} \ \text{II}\,b.$$

Thus by Cor. (ii), Prop. 7, the result holds for all $\mathfrak{n}$ **E**-correct.

(*c*) $\phi$ is a conjunction, say $\phi\{\mathfrak{n}\}$ is $\psi\{\mathfrak{n}\} \,\&\, \psi'\{\mathfrak{n}\}$ and the result holds for $\psi\{\mathfrak{n}\}$ and for $\psi'\{\mathfrak{n}\}$. Then

$$\psi\{\mathfrak{n}\} \vee N[``\psi\{\mathfrak{n}\}"] \quad \text{for all } \mathfrak{n},\ \mathbf{E}\text{-correct,}$$

$$\psi'\{\mathfrak{n}\} \vee N[``\psi'\{\mathfrak{n}\}"] \quad \text{for all } \mathfrak{n},\ \mathbf{E}\text{-correct.}$$

We have

$$\frac{\psi\{\mathfrak{n}\} \vee N[``\psi\{\mathfrak{n}\}"]}{} \ \text{II}\,a, \text{I}\,a \ \frac{\psi'\{\mathfrak{n}\} \vee N[``\psi'\{\mathfrak{n}\}"]}{}$$

$$\frac{\psi\{\mathfrak{n}\} \vee N[``(\psi\{\mathfrak{n}\} \,\&\, \psi'\{\mathfrak{n}\})"] \quad \psi'\{\mathfrak{n}\} \vee N[``(\psi\{\mathfrak{n}\} \,\&\, \psi'\{\mathfrak{n}\})"]}{(\psi\{\mathfrak{n}\} \,\&\, \psi'\{\mathfrak{n}\}) \vee N[``(\psi\{\mathfrak{n}\} \,\&\, \psi'\{\mathfrak{n}\})"]} \ \text{II}\,b'.$$

Thus by Cor. (ii), Prop. 7, the result holds for $\psi\{\mathfrak{n}\} \,\&\, \psi'\{\mathfrak{n}\}$ for all $\mathfrak{n}$ **E**-correct.

(*d*) $\phi$ is a general statement, say $(A\xi)\,\psi\{\xi\}$, and the result holds for $\psi\{\kappa\}$. Then $\psi\{\kappa, \mathfrak{n}\} \vee N[``\psi\{\kappa, \mathfrak{n}\}"]$ for all $\kappa$, $\mathfrak{n}$, **E**-correct. We have

$$\frac{\psi\{\kappa, \mathfrak{n}\} \vee N[``\psi\{\kappa, \mathfrak{n}\}"]}{\overline{\psi\{\kappa, \mathfrak{n}\} \vee (E\xi)\,N[``\psi\{\xi, \mathfrak{n}\}"]}} \quad \text{I}a, \text{II}d, \text{I}a,$$

$$\frac{\psi\{\kappa, \mathfrak{n}\} \vee (E\xi)\,N[``\psi\{\xi, \mathfrak{n}\}"]}{(A\xi)\,\psi\{\xi, \mathfrak{n}\} \vee N[``(A\xi)\,\psi\{\xi, \mathfrak{n}\}"]} \quad \begin{array}{l}\text{for all } \mathfrak{n}, \textbf{E}\text{-correct by Prop. 7, Cor. (ii)} \\ \text{by induction, for all } \mathfrak{n}\ \textbf{E}\text{-correct.}\end{array}$$

Thus by Prop. 7 the result holds for $(A\xi)\,\psi\{\xi, \mathfrak{n}\}$.

(*e*) $\phi$ is an existential statement, the demonstration is similar.

This completes the demonstration of the proposition.

PROP. 10. *If $\phi$ is a closed $I\mathscr{F}'_C$-theorem and if $\psi$ is obtained from $\phi$ by replacing each predicate $\pi$ in $\phi$ by an* $\mathbf{A}_{00}$*-equation with exactly the same number of free variables and if the same* $\mathbf{A}_{00}$*-equation or a variant thereof is used to replace other occurrences of $\pi$, so that $\pi$ and its replacement always have the same set of free variables, then $\psi$ is an* $\mathbf{A}_I$*-theorem.*

T.N.D. is an $\mathbf{A}_I$-theorem and all the $I\mathscr{F}'_C$-rules are $\mathbf{A}_I$-rules or are derived $\mathbf{A}_I$-rules, except II$e'$. In the $I\mathscr{F}'_C$-proof-tree of $\phi$ make the replacements as described in the ennunciation of the proposition, add the $\mathbf{A}_I$-proofs of the cases of T.N.D. used, and we obtain an $\mathbf{A}_I$-proof-tree except that we have II$e'$ instead of induction and there may be free variables. Replace the free variables by numerals in all possible ways, we obtain an $\mathbf{A}_I$-proof-tree when we replace applications of II$e'$, say

$$\frac{\phi\{\xi\} \vee \omega}{(A\xi)\,\phi\{\xi\} \vee \omega}$$

by

$$\frac{\phi\{\nu\} \vee \omega}{(A\xi)\,\phi\{\xi\} \vee \omega} \quad \text{for all } \nu, \textbf{E}\text{-correct.}$$

This follows since the proofs of the upper formulae of the above inductions are exactly the same except for change of numeral, hence by Cor. (i), Prop. 6, they are **E**-correct. This completes the demonstration of the proposition.

PROP. 11. *The rule* I$b$ *is independent.*

Consider the **A**-statement

$$(Ex)\,(Ax')\,(Ex'', x''')\,(Ax^{\text{iv}})\,(N[``\phi\{x''', x', x^{\text{iv}}\}"] \vee \phi\{x, x'', x'''\}), \qquad \text{(i)}$$

from results in Ch. 3 and Prop. 10 above we know that (i) is an $\mathbf{A}_I$-theorem. If we are denied the use of rule I *b* then the only possible way of proving (i) would be from T.N.D.:

$$N[``\phi\{\xi,\eta,\zeta\}"] \vee \phi\{\xi,\eta,\zeta\}$$

by use of R 3, II *d* twice, R 3, II *d*. But this would involve generalizing on one occurrence of the variable $\zeta$, and this is impossible. Thus I *b* is independent.

Note that Prop. 5 of Ch. 8 fails in $\mathbf{A}_I$, for if G is an irresoluble $\mathbf{A}_I$-statement then G $\vee$ $N$["G"] is an $\mathbf{A}_I$-theorem but neither disjunctand is an $\mathbf{A}_I$-theorem.

## 10.11 *Reversibility of rules*

PROP. 12. *Rule* II $b'$ *is reversible.*

We have to show that if $(\phi \,\&\, \psi) \vee \omega$ is an $\mathbf{A}_I$-theorem then so are $\phi \vee \omega$ and $\psi \vee \omega$. We may suppose that the main formula of II *a* is atomic. First consider an $\mathbf{A}_I$-proof of $(\phi \,\&\, \psi) \vee \omega$ of level one and which ends in a single induction. An $\mathbf{A}_0$-proof is dealt with as in Prop. 2, Ch. 2. We have then the **E**-theorem: $prf_0[\tau x, \sigma[Num\, x]] = 0$, where $\tau\nu$ is equal to the *g.n.* of an $\mathbf{A}_0$-proof of $(\phi \,\&\, \psi) \vee \omega' \vee \omega''\{\nu\}$, whose *g.n.* is equal to $\sigma[Num\,\nu]$. Here $\omega$ is $\omega' \vee (Ax)\,\omega''\{\alpha\{x\}\}$. In the $\mathbf{A}_0$-proofs of $(\phi \,\&\, \psi) \vee \omega' \vee \omega''\{\nu\}$ consider the corresponding occurrences of &, these can only enter the proofs at applications of II $b'$. Suppose that one of these is

$$\frac{\phi' \vee \omega^*\{\nu\} \quad \psi' \vee \omega^*\{\nu\}}{(\phi' \,\&\, \psi') \vee \omega^*\{\nu\}} \;\; \text{II}\, b'.$$

Let the *g.n.* of the $\mathbf{A}_0$-proof of the left upper formula be $\tau_1'\nu$ and let $\tau_1\nu$ be $\tau_1'\nu^\frown\ldots{}^\frown\tau_1^\lambda\nu$, where the $\tau_1^\lambda\nu$'s refer to all the $\mathbf{A}_0$-proofs of left upper formulae of II $b'$ that contain corresponding occurrences of &. Note that $(\phi' \,\&\, \psi')$ can only be a variant of $(\phi \,\&\, \psi)$ by R 1. Similarly define $\tau_2\nu$ using the right upper formula. Let $\tau_3\nu$ be $\tau\nu$ less the components of $\tau_1\nu$ and $\tau_2\nu$. Then $\tau^*\nu = \tau_1\nu^\frown\tau_2\nu^\frown\tau_3\nu$ is equal to the *g.n.* of an $\mathbf{A}_0$-proof of $(\phi \,\&\, \psi) \vee \omega$, it may differ from $\tau\nu$ by having some repetitions. Now let $\tau_3'\nu$ be the *g.n.* of the result of substituting $\phi'$ for $(\phi' \,\&\, \psi')$ at all occurrences of this conjunction which are related to $(\phi \,\&\, \psi)$ in $\tau_3\nu$. Let $\tau'\nu = \tau_1\nu^\frown\tau_2\nu^\frown\tau_3'\nu$. Let $\sigma'[Num\,\nu]$ be equal to the *g.n.* of $\phi \vee \omega' \vee \omega''\{\nu\}$. We want to show that: $prf_o[\tau' x, \sigma'[Num\, x]] = 0$ is an **E**-theorem, then $\phi \vee \omega$ will be an $\mathbf{A}_I$-theorem of level one.

By hypothesis $prf_o[\tau x, \sigma[Num\, x]] = 0$ is an **E**-theorem. (0)

We have
$$(\tau^* x_\xi^\infty \quad \text{for} \quad (\tau^* x)_\xi^\infty)$$

$$\begin{aligned} prf_o[\tau^* x, \sigma[Num\, x]] = {} & A_2[l.c.\, \tau^* x, \sigma[Num\, x]] + \sum_{1 \leqslant \xi \leqslant a} Ax_o[\tau^* x_\xi^\infty] \\ & \times \prod_{1 \leqslant \eta < \xi} Rl_o^1[\tau^* x_\eta^\infty, \tau x_\xi^\infty] \times \prod_{1 \leqslant \zeta < \eta < \xi} Rl_o^2[\tau^* x_\zeta^\infty, \tau^* x_\eta^\infty, \tau^* x_\xi^\infty] \\ & + \sum_{a < \xi \leqslant b} Ax_o[\tau^* x_\xi^\infty] \times \prod_{a < \eta < \xi} Rl_o^1[\tau^* x_\eta^\infty, \tau^* x_\xi^\infty] \\ & \times \prod_{a < \zeta < \eta < \xi} Rl_o^2[\tau^* x_\zeta^\infty, \tau^* x_\eta^\infty, \tau^* x_\xi^\infty] + \sum_{b < \xi \leqslant c} Ax_o[\tau^* x_\xi^\infty] \\ & \times \prod_{0 < \eta < \xi} Rl_o^1[\tau^* x_\eta^\infty, \tau^* x_\xi^\infty] \\ & \times \prod_{0 < \zeta < \eta < \xi} Rl_o^2[\tau^* x_\zeta^\infty, \tau^* x_\eta^\infty, \tau^* x_\xi^\infty], \end{aligned} \tag{i}$$

where $a = \varpi \tau_1 x$, $b = a + \varpi \tau_2 x$, $c = b + \varpi \tau_3 x$. This follows from the definitions of $\tau_1 x$, $\tau_2 x$, $\tau_3 x$.

We have four similar summands for $prf_o[\tau' x, \sigma'[Num\, x]]$ which we call (ii). We wish to show that $prf_o[\tau' x, \sigma'[Num\, x]] = 0$.

We have $l.c.\, \tau' x = \sigma'[Num\, x]$ by definition, hence

$$A_2[l.c.\, \tau' x, \sigma'[Num\, x]] = 0,$$

from the easily proved **E**-theorem $A_2[x, x] = 0$ and R 1. Also

$$\langle \tau' x \rangle_a = \langle \tau x \rangle_a,$$

by definition (D 234); hence the second summand in (ii) is zero since the second summand in (i) is zero, and the **E**-deduction

$$\frac{x + x' = 0}{x = 0},$$

and (0). Similarly for the third summand.

For the fourth summand of (ii) we have

$$\tau' x_\xi^\infty = disj[\alpha, \beta] \quad \text{and} \quad \tau x_{\xi+b}^\infty = disj[\alpha, \beta'] \quad \text{for} \quad b < \xi \leqslant b + c.$$

But we have

$$\left.\begin{aligned} Ax_o[disj[\alpha, \beta]] &= Ax_o[disj[\alpha, \beta']], \\ Rl_o^1[disj[\alpha, \beta], disj[\gamma, \beta]] &= Rl_o^1[disj[\alpha, \beta'], disj[\gamma, \beta']] \end{aligned}\right\} \tag{iii}$$

and similarly for $Rl_0^2$. The terms $\alpha$ and $\beta$ can be explicitly defined in terms of $\tau'x$, and similarly for $\alpha$ and $\beta'$. To show that (iii) are **E**-theorems it suffices to show that

$$\begin{aligned} Ax_0^1[disj[\alpha,\beta]] &= Ax_0^1[disj[\alpha,\beta']], \\ &\vdots \\ Ax_0^{4\cdot2}[disj[\alpha,\beta]] &= Ax_0^{4\cdot2}[disj[\alpha,\beta']], \end{aligned} \tag{iv}$$

$$\begin{aligned} Rl^{\mathrm{I}a}[disj[\alpha,\beta],disj[\gamma,\beta]] &= Rl^{\mathrm{I}a}[disj[\alpha,\beta'],disj[\gamma,\beta']], \\ &\vdots \\ Rl^{\mathrm{II}b'}[disj[\alpha,\beta],disj[\gamma,\beta],disj[\delta,\beta]] & \\ &= Rl^{\mathrm{II}b'}[disj[\alpha,\beta'],disj[\gamma,\beta'],disj[\delta,\beta']]. \end{aligned} \tag{v}$$

In showing that (ii) is zero we are dealing with the only application of R 3 which occurs in an $\mathbf{A}_I$-proof with only one level.

Take the various axioms in turn. The term $\beta$ is of the form $conj[\beta\dagger,\beta^*]$ and the term $\beta'$ is $\beta\dagger$, or else both are null formulae, this follows from the construction of $\beta'$. If both are null then the first axiom becomes $Ax_0^1[\alpha] = Ax_0^1[\alpha]$, which is an **E**-theorem. If one of $\beta$, $\beta'$ is present then so is the other, and since axioms are atomic, then both sides of the first equation in (iv) are equal to one. This is somewhat heuristic, what we require is

$$Ax_0^1[disj[x,conj[x',x'']]] = Ax_0^1[disj[x,x']] \quad \text{if} \quad x' \neq 0,$$

and

$$Ax_0^1[x] = Ax_0^1[x],$$

are **E**-theorems. Even this is insufficient because the two cases are run together in (iv). This is effected by proving that if $x \neq 0$,

$$Ax_0^1[disj[x,A_1[x']\times conj[x',x'']]] = Ax_0^1[disj[x,x']], \tag{vi}$$

if $x'$ is replaced by zero this becomes $Ax_0^1[x] = Ax_0^1[x]$ (see D 174), while if $x'$ is replaced by something different from zero both sides are equal to one. Actually we need only **E**-prove (vi) under the hypothesis $stat[x,x',x'']$, but it will suffice to **E**-prove the more general statement. Now

$$Ax_0^1[x] = A_1[A_2[x,Eq[\alpha,\alpha]]+tm\alpha],$$

the term $\alpha$ can be explicitly defined. To **E**-prove (vi) it suffices to **E**-prove

$$A_1[x]\times Ax_0^1[disj[x,x']] = A_1[x]\times(A_1[x']+B_1[x']\times Ax_0^1[x]), \tag{vii}$$

and

$$\begin{aligned} A_1[x]\times Ax_0^1[disj[x,A_1[x']\times conj[x',x'']]] & \\ = A_1[x]\times(A_1[x']+B_1[x']\times Ax_0^1[x]). \end{aligned}$$

Both sides of the first equation of (vii) take the same value if $x = 0$ or if $x' = 0$ also

$$A_2[disj[Sx, Sx'], Eq[\alpha, \alpha]] = 1$$

because $\quad A_2[\langle 8,8,5\rangle^\frown\beta^\frown\langle 9\rangle^\frown\gamma^\frown\langle 9\rangle, \langle 8,8,2\rangle^\frown\alpha^\frown\langle 9\rangle^\frown\alpha^\frown\langle 9\rangle] = 1$

is an **E**-theorem, from the **E**-theorem $A_2[x^\frown x', x^\frown(x'+Sx'')] = 1$.

The second equation of (vii) is dealt with similarly. Thus the first equation of (iv) is settled. The other equations in (iv) are treated similarly. The equations in (v) are **E**-theorems because alteration of the subsidiary formula fails to affect a rule; thus

$$Rl^{\mathrm{I}b}[disj[x, x'], disj[x'', x']] = Rl^{\mathrm{I}b}[x, x'']$$

is an **E**-theorem, thus

$$Rl^{\mathrm{I}b}[disj[x, x'], disj[x'', x']] = Rl^{\mathrm{I}b}[disj[x, x'''] \, disj[x'', x''']]$$

and similarly for the other rules.

The set (v) should also include the cases when $\beta$ and $\beta'$ occur in the main formula of I$a$, I$b$, R 1. In the case of a cancellation of $(\phi' \mathbin{\&} \psi')$ by I$b$ we have:

$$Rl^{\mathrm{I}b}[disj[disj[x', x'], x''], disj[x', x'']]$$
$$= Rl^{\mathrm{I}b}[disj[disj[x, x], x''], disj[x, x'']].$$

Similarly in the other cases. The full details are somewhat lengthy but enough has been displayed to show how they can be obtained.

This completes the case of $\mathbf{A}_I$-proofs of one level. Now suppose that the result holds for $\mathbf{A}_I$-proofs having $\lambda$ levels and we will show that it holds for $\mathbf{A}_I$-proofs having $S\lambda$ levels. An $\mathbf{A}_I$-proof having $S\lambda$ levels finishes with an **E**-proof of

$$ded_0[\rho x, \tau x, \sigma[Num\, x]] = 0.$$

$\rho\nu$ is the *g.n.* of a sequence of $\mathbf{A}_I$-theorems, some of which are of the forms $\omega \vee (\phi' \mathbin{\&} \psi')$, the remainder are without a related occurrence of $(\phi \mathbin{\&} \psi)$. Replace this by $\rho' x$, which by our induction hypothesis can be taken as the *g.n.* of a sequence of $\mathbf{A}_I$-theorems of the respective forms $\omega \vee \phi'$ or just $\omega$.

As before write $\tau x = \tau_1 x^\frown \tau_2 x^\frown \tau_3 x$, where $\tau_1\nu$, $\tau_2\nu$, $\tau_3\nu$ are as before except that now they are deductions from the **A**-statements in $\rho\nu$ as hypotheses. We form $\tau' x$ as before and we require to obtain an **E**-proof of the equation

$$ded_0[\rho' x, \tau' x, \sigma'[Num\, x]] = 0.$$

The details of this **E**-proof are almost the same as in the case of an $\mathbf{A}_I$-proof having only one level. The main difference is that we have another conjunctive clause to add to (ii) namely: $Hyp[\tau'^{\infty}_{\xi}]$, with an obvious notation. To deal with this we have extra equations to add to (iv), namely

$$Hyp[disj[\alpha, \beta]] = Hyp[disj[\alpha, \beta']].$$

But this is

$$\prod_{1 \leqslant x' \leqslant \varpi\rho x} A_2[\rho x^{\infty}_{x'}, disj[\alpha, \beta]] = \prod_{1 \leqslant x' \leqslant \varpi\rho' x} A_2[\rho' x^{\infty}_{x'}, disj[\alpha, \beta']].$$

It suffices to show that

$$A_2[\rho x^{\infty}_{x'}, disj[\alpha, \beta]] = A_2[\rho' x^{\infty}_{x'}, disj[\alpha, \beta']]. \qquad \text{(viii)}$$

is an **E**-theorem. But this follows at once by definition because

$$\rho x^{\infty}_{x'} = disj[\gamma, \beta] \quad \text{and} \quad \rho' x^{\infty}_{x'} = disj[\gamma, \beta'],$$

so both sides of (viii) are equal to $A_2[\alpha, \gamma]$, and the result follows.

The full details of the rest of this case are as for the case when there is only one level. This completes the demonstration of Prop. 12.

Cor. *If* $(\phi \,\&\, \psi) \vee \omega\{\nu\}$ *for all* $\nu$, **E**-*correct, then* $\phi \vee \omega\{\nu\}$ *for all* $\nu$, **E**-*correct.*

Prop. 13. *Rule R 3 is reversible.*

By this we mean that if $(A\xi)\,\phi\{\xi\} \vee \omega$ is an $\mathbf{A}_I$-theorem then so are $\phi\{\nu\} \vee \omega$ for all $\nu$, **E**-correct. Note that if $(A\xi)\,\phi\{\alpha\{\xi\}\} \vee \omega$ is an $\mathbf{A}_I$-theorem then $\phi\{\nu\} \vee \omega$ for all $\nu$, **E**-correct may be wrong. We proceed by induction on the number of levels in the $\mathbf{A}_I$-proof of $(A\xi)\,\phi\{\xi\} \vee \omega$.

First suppose that there is only one level. $(A\xi)\,\phi\{\xi\}$ can only enter the $\mathbf{A}_I$-proof at II *a* or R 3. We may suppose that II *a* only introduces atomic statements, thus $(A\xi)\,\phi\{\xi\}$ can only enter the $\mathbf{A}_I$-proof of $(A\xi)\,\phi\{\xi\} \vee \omega$ at R 3. There can only be one application of R 3, because if there were two or more then we should require to use some rule after R 3 and such an $\mathbf{A}_I$-proof would require at least two levels. Thus an $\mathbf{A}_I$-proof of $(A\xi)\,\phi\{\xi\} \vee \omega$ with one level must end with an application of R 3 thus

$$\frac{\psi\{\nu\} \vee \omega \quad \text{for all } \nu, \mathbf{E}\text{-correct}}{(A\xi)\,\phi\{\xi\} \vee \omega}, \quad \text{where } \psi\{\alpha\{\nu\}\} \text{ is } \phi\{\nu\},$$

hence the result follows at once from the lemma after Prop. 7. If $\omega$ contains $\kappa$ and the lower formula holds for all $\kappa$, **E**-correct then the upper formula holds for all $\kappa$, $\nu$, **E**-correct.

Now suppose that we have obtained the result for all $\kappa$, **E**-correct for $\mathbf{A}_I$-proofs with $\lambda$ levels and we will show that it also holds for all $\kappa$, **E**-correct for $\mathbf{A}_I$-proofs with $S\lambda$ levels. The $S\lambda$th level is either an $\mathbf{A}_0$-deduction from the production of the $\lambda$th level as hypotheses or else it is a single application of R 3. In the first case the hypotheses are of the forms:

$$(A\xi)\,\phi'\{\xi\} \vee \omega',$$

$$(A\xi)\,\chi\{\xi\} \vee \omega'' \vee (A\xi)\,\phi'\{\xi\},$$

$$(A\xi)\,\chi\{\xi\} \vee \omega'',$$

where $(A\xi)\,\phi'\{\xi\}$ is a related occurrence of $(A\xi)\,\phi\{\xi\}$, and where $\omega''$ is without related occurrences of $(A\xi)\,\phi\{\xi\}$.

We modify the $\mathbf{A}_0$-deduction which constitutes the $S\lambda$th level as follows: replace the hypotheses and their related descendants by

$$\phi'\{\nu\} \vee \omega' \quad \text{for all } \nu, \textbf{E}\text{-correct},$$

$$(A\xi)\,\chi\{\xi\} \vee \omega'' \vee \phi'\{\nu\} \quad \text{for all } \nu, \textbf{E}\text{-correct},$$

$$(A\xi)\,\chi\{\xi\} \vee \omega''.$$

The structure of the deduction is otherwise unaltered. The result is an $\mathbf{A}_0$-deduction of $\phi\{\nu\} \vee \omega$ for all $\nu$, **E**-correct. This follows from Prop. 7 and our induction hypothesis. By our induction hypothesis $\phi'\{\nu\} \vee \omega'$ holds for all $\nu$, **E**-correct, because the production of level $\lambda$ are $\mathbf{A}_I$-theorems having $\lambda$ levels. $(A\xi)\,\chi\{\xi\} \vee \omega'' \vee \phi'\{\nu\}$ holds for all $\nu$, **E**-correct because $(A\xi)\,\chi\{\xi\} \vee \omega'' \vee (A\xi)\,\phi'\{\xi\}$ is an $\mathbf{A}_I$-theorem having $\lambda$ levels so that $\chi\{\kappa\} \vee \omega'' \vee (A\xi)\,\phi'\{\xi\}$ holds for all $\kappa$, **E**-correct, whence so does $(A\xi)\,\phi'\{\xi\} \vee \chi\{\kappa\} \vee \omega''$ for all $\kappa$, **E**-correct; but these are $\lambda$ level $\mathbf{A}_I$-proofs, hence by our induction hypothesis $\phi'\{\nu\} \vee \chi\{\kappa\} \vee \omega''$ for all $\kappa$, $\nu$, **E**-correct, thus $\chi\{\kappa\} \vee \omega'' \vee \phi'\{\nu\}$ for all $\kappa$, $\nu$, **E**-correct, thus by R 3

$$(A\xi)\,\chi\{\xi\} \vee \omega'' \vee \phi'\{\nu\}$$

for all $\nu$, **E**-correct. Thus the modified hypotheses are $\mathbf{A}_I$-theorems of $\lambda$ levels, and so $\phi\{\nu\} \vee \omega$ is an $\mathbf{A}_I$-theorem having $S\lambda$ levels. (If $(A\xi)\,\chi\{\xi\}$ is absent throughout then $\phi\{\nu\} \vee \omega$ is an $\mathbf{A}_I$-theorem with $\lambda$ levels.)

If the $S\lambda$th level consists of a single application of R 3 then it must end

$$\frac{\psi\{\nu\} \vee \omega}{(A\xi)\,\phi\{\xi\} \vee \omega} \quad \text{for all } \nu, \textbf{E}\text{-correct, where } \phi\{\xi\} \text{ is } \psi\{\alpha\{\xi\}\},$$

and as in the first case we have finished.

## 10.12 *Deduction theorem*

PROP. 14. *The deduction theorem holds in* $\mathbf{A}_I$.

We have to show that if $\psi$ can be $\mathbf{A}_I$-deduced from hypotheses $\phi', \phi'', \ldots, \phi^{(S\kappa)}$ then $N[``\phi^{(S\kappa)}"] \vee \psi$ can be $\mathbf{A}_I$-deduced from hypotheses $\phi', \phi'', \ldots, \phi^{(\kappa)}$. We proceed by induction on the number of levels in the $\mathbf{A}_I$-deduction.

First suppose that there is only one level. This deduction is either an $\mathbf{A}_0$-deduction of $\psi$ from hypotheses $\phi', \phi'', \ldots, \phi^{(S\kappa)}$ or is a single application of R 3. In the first case we are finished because the result holds for $\mathbf{A}_0$-deductions. In the second case we have an $\mathbf{A}_I$-deduction of $(A\xi)\,\psi\{\xi\} \vee \omega$ and

$$ded_0[\alpha^\frown\langle\alpha'\rangle, \tau x, \sigma[Num\,x]] = 0 \tag{i}$$

is an $\mathbf{E}$-theorem, where $\alpha$ is the $g.n.$ of the sequence $\phi', \phi'', \ldots, \phi^{(\kappa)}$ and $\alpha'$ is the $g.n.$ of $\phi^{(S\kappa)}$, $\tau\nu$ is equal to the $g.n.$ of an $\mathbf{A}_I$-deduction of $\psi\{\nu\} \vee \omega$ from the hypotheses whose $g.n.$ is equal to $\alpha^\frown\langle\alpha'\rangle$, and $\sigma[Num\,\nu]$ is equal to the $g.n.$ of $\psi\{\nu\} \vee \omega$. Now $\tau x = \tau_1 x^\frown\alpha^\frown\langle\alpha'\rangle^\frown\tau_2 x$, where $\tau_1\nu$ is equal to the $g.n.$ of the axioms used in the deduction of $\psi\{\nu\} \vee \omega$.

In this deduction add the disjunctive clause $N[``\phi^{(S\kappa)}"]$ to the subsidiary formula in each statement in the deduction, then add the $\mathbf{A}_I$-proof of $\phi^{(S\kappa)} \vee N[``\phi^{(S\kappa)}"]$ at the top, and the axioms used in the original deduction and the hypotheses $\phi', \phi'', \ldots, \phi^{(\kappa)}$. The result gives us $\mathbf{A}_I$-deductions of $N[``\phi^{(S\kappa)}"] \vee \psi\{\nu\} \vee \omega$ for all $\nu$, we wish to show that this is $\mathbf{E}$-correct. In our case $\psi$ is $(A\xi)\,\psi\{\alpha\{\xi\}\} \vee \omega$. To show $\mathbf{E}$-correctness we have to show that

$$ded_0[\alpha, \tau' x, \sigma'[Num\,x]] = 0 \quad \text{is an } \mathbf{E}\text{-theorem}, \tag{ii}$$

here $\tau' x = \tau_1 x^\frown\alpha^\frown\beta^\frown\tau^* x$, where $\beta$ is the $g.n.$ of the $\mathbf{A}_I$-proof of

$$N[``\phi^{(S\kappa)}"] \vee \phi^{(S\kappa)},$$

and $\tau^* x$ is the result of replacing each component of $\tau_2 x$, say $\delta$, by $disj[\delta, \alpha']$. Referring to the definition of $ded_0$, which is similar to (i) in Prop. 12, we see that the gist of the matter we wish to $\mathbf{E}$-prove is:

$$Rl^1[x, x'] = Rl^1[disj[x, x''], disj[x', x'']], \tag{iii}$$

but we have already seen that this is an $\mathbf{E}$-theorem, it says that an application of a rule remains an application of that rule when something is added to the subsidiary formulae throughout. Thus the first case is settled.

Now suppose that we have demonstrated the result for $\mathbf{A}_I$-deductions of $\lambda$ levels and we will show that it holds for $\mathbf{A}_I$-deductions with $S\lambda$ levels. An $\mathbf{A}_I$-deduction with $S\lambda$ levels is either an $\mathbf{A}_0$-deduction from the production of level $\lambda$ as hypotheses or it is a single application of R 3 from the production of level $\lambda$ as hypotheses. In the first case if we replace the productions of level $\lambda$ by their disjunctions with $N[$"$\phi^{(S\kappa)}$"$]$ then they are, by our induction hypothesis, $\mathbf{A}_I$-deductions from $\phi', \phi'', \ldots, \phi^{(\kappa)}$ as hypothesis, thus if we alter the $\mathbf{A}_I$-deduction of $\psi$ as in the first case then we are left with an $\mathbf{A}_I$-deduction of $N[$"$\phi^{(S\kappa)}$"$] \vee \psi$ from $\phi', \phi'', \ldots, \phi^{(\kappa)}$ as hypotheses, thus this subcase is settled.

If the $\mathbf{A}_I$-deduction of level $S\lambda$ is a single application of R 3, then we proceed as in the first case. We first have to show that if $\psi\{\nu\}$ is deduced from the hypotheses $\phi', \phi'', \ldots, \phi^{(\kappa)}$ and that this holds for all $\kappa$, **E**-correct, then $N[$"$\phi^{(S\kappa)}$"$] \vee \psi\{\nu\}$ can be $\mathbf{A}_I$-deduced from $\phi', \phi'', \ldots, \phi^{(\kappa)}$ for all $\nu$, **E**-correct. This is dealt with as in the first case but writing $\mathfrak{x}$ instead of $x$. Thus we may suppose that the productions of level $\lambda$, namely $\chi\{\nu\}$, etc. hold for all $\nu$, **E**-correct and that the same holds for $N[$"$\phi^{(S\kappa)}$"$] \vee \chi\{\nu\}$. The only difference between this case and the first is that instead of (i) we now have

$$ded_0[\rho x^\frown \alpha^\frown \langle \alpha' \rangle, \tau x, \sigma[Num\, x]] = 0 \quad \text{is an } \mathbf{E}\text{-theorem,}$$

we modify this as before by adding on the disjunctive clause $N[$"$\phi^{(S\kappa)}$"$]$ to $\rho x$, $\tau x$, $\sigma[Num\, x]$. We then require

$$ded_0[\rho' x^\frown \alpha, \tau' x, \sigma'[Num\, x]] = 0 \quad \text{to be an } \mathbf{E}\text{-theorem,}$$

again the gist of the proof that this is so is (iii). Thus the second case is settled. This completes the demonstration of Prop. 14.

## 10.13 *Cuts with an* $\mathbf{A}_{00}$*-cut formulae*

We now come to the problem whether the system $\mathbf{A}_I$ is weaker than the system $\mathbf{A}_I$ plus the cut, call this system $\mathbf{A}_{IC}$. That is to say, is there an **A**-true statement which is an $\mathbf{A}_{IC}$-theorem but lacks an $\mathbf{A}_I$-proof?

An $\mathbf{A}_I$-proof of level 1 is either an $\mathbf{A}_0$-theorem or else is a single application of R 3, i.e. is of the form

$$\frac{\phi\{\mathfrak{n}\} \vee \omega}{(A\mathfrak{x})\, \phi\{\alpha\{\mathfrak{x}\}\} \vee \omega} \qquad \text{are } \mathbf{A}_0\text{-theorems for all } \mathfrak{n}, \mathbf{E}\text{-correct.}$$

Now M.P. can be eliminated from an $\mathbf{A}_0$-proof (the cut formula must be an $\mathbf{A}_{00}$-formula because its negation is required in the cut). Thus if

$\phi\{\mathfrak{n}\} \vee \omega$ can be $A_{0C}$-proved then it can be $A_0$-proved. Here $A_{0C}$ is the system $A_0$ plus the cut. Thus if we have

$$\phi\{\mathfrak{n}\} \vee \omega \quad \text{can be } A_{0C}\text{-proved for all } \mathfrak{n}, \text{ E-correct} \tag{i}$$

then we shall have

$$\phi\{\mathfrak{n}\} \vee \omega \quad \text{can be } A_0\text{-proved for all } \mathfrak{n}. \tag{ii}$$

The question whether this is **E**-correct remains open. If the cut is eliminable from $A_{IC}$ then to apply R 3 we must have (ii) **E**-correct.

If we have (i), then the $A_{0C}$-proofs of $\phi\{\mathfrak{n}\} \vee \omega$ are on a primitive recursive plan, **E**-correct. Removal of cuts in these $A_{0C}$-proofs so that they become $A_0$-proofs (which would be on a certain plan) might destroy the primitive recursiveness of the plan, though it might be general recursive, or it might upset the possibility of proving **E**-correctness, if the plan remained primitive recursive.

An $A_{0C}$-proof is easily converted into an $A_0$-proof. The cut formulae are $A_{00}$-formulae, hence they are $A_{00}$-true or $A_{00}$-false; this is a recursive decision, unfortunately it violates the required condition of being primitive recursive. In the $A_{0C}$-proof we omit all false ancestors of false cut formulae and all false ancestors of negations of true cut formulae. This will convert a cut into a case of II *a* (or rather a sequence of cases of II *a* if we require that II *a* only dilutes with atomic formulae) other rules remain applications of the same rules. If we do this and also remove all branches above $A_{00}$-true ancestors at places where branches join, then we are left with an $A_0$-proof of the formula produced by the original cut. (Removal of these branches can be dispensed with, they are merely redundant, it is technically simple to retain them.)

It is easily seen that if we do this elimination of a cut then higher cuts have their cut formulae unaffected. Thus simultaneous removal of false ancestors can take place, or they can be removed in any order. A false ancestor of one cut is distinct from any false ancestor of another cut.

Since the test for $A_{00}$-truth, though recursive, fails to be primitive recursive, we must seek some other test. Let us remove all ancestors of cut formulae and of their negations. The two upper branches above a cut will become two new trees, by the completeness of $A_0$ one of them at least will be a proof-tree of its final formulae, namely the one above the cut formulae if this is $A_{00}$-false, otherwise the other one. From the original proof-tree we obtain the new tree by a primitive recursive process. It is a primitive recursive process to decide if a number is the $g.n.$ of an $\mathbf{A}_0$-proof. Thus we can decide, on the information available to us by a primitive recursive process, which branch to take. Before, when we were held up by a decision, though recursive, failing to be primitive recursive, we were using only part of the information available to us.

Having removed the cuts we are left with an $\mathbf{A}_0$-proof-tree of the same formula as before, the $g.n.$ of this $\mathbf{A}_0$-proof is a primitive recursive function of the $g.n.$ of the original $\mathbf{A}_{0C}$-proof. Let $\tau\mathfrak{n}$ be equal to the $g.n.$ of the original $\mathbf{A}_{0C}$-proof of $\phi\{\mathfrak{n}\} \vee \omega$ and let $\sigma[\mathfrak{Num}\,\mathfrak{n}]$ be equal the $g.n.$ of $\phi\{\mathfrak{n}\} \vee \omega$. Then we have:

$$prf_{0C}[\tau\mathfrak{x}, \sigma[\mathfrak{Num}\,\mathfrak{x}]] = 0 \quad \text{is an } \mathbf{E}\text{-theorem.} \qquad \text{(i)}$$

Let $\tau$ become $\tau'$ after the removal of cuts, then we wish to show

$$prf_0[\tau'\mathfrak{x}, \sigma[\mathfrak{Num}\,\mathfrak{x}]] = 0 \quad \text{is an } \mathbf{E}\text{-theorem.} \qquad \text{(ii)}$$

This comes about because if in the original proof a certain rule was used then at the corresponding part of the new proof the same rule is used or becomes a repetition except that cases of the cut become cases of II $a$.

Thus, except for considerable technical detail, we have settled the case of level 1.

LEMMA A. *The cut can be removed from* $\mathbf{A}_{IC}$*-proofs of level* 1.

Suppose that the cut formula is $\mathbf{A}_{00}$-false, the argument is similar if its negation is $\mathbf{A}_{00}$-false. For simplicity we give the demonstration in the case when the cut formula is an $\mathbf{A}_{00}$-false equation. The argument is similar in the general case. We merely have to add the following to the definition of *false ancestor*. Let the cut formula be $\phi$.

Suppose that $\phi'$ is a false ancestor of $\phi$ and that $\phi'$ arises from $\phi''$ and $\phi'''$ by II $b$ thus

$$\frac{\phi'' \vee \omega \quad \phi''' \vee \omega}{(\phi'' \,\&\, \phi''') \vee \omega},$$

where $(\phi'' \,\&\, \phi''')$ is $\phi'$, since $\phi'$ is $\mathbf{A}_{00}$-false then one or both of $\phi''$, $\phi'''$ is $\mathbf{A}_{00}$-false, again it is a primitive recursive decision as to which is the case, this one, or $\phi''$ if both are $\mathbf{A}_{00}$-false is then to be a false ancestor of $\phi$. $\phi'$ is a disjunction and arises from $\phi''$ and $\phi'''$ by II $a$ thus

$$\frac{\phi'' \vee \omega}{(\phi'' \vee \phi''') \vee \omega},$$

where $\phi'$ is $(\phi'' \vee \phi''')$, since $\phi'$ is $\mathbf{A}_{00}$-false then both $\phi''$ and $\phi'''$ must be $\mathbf{A}_{00}$-false. Then $\phi''$ is to be the false ancestor of $\phi$.

We assume then that the cut formula $\phi$ is an equation, because if $\phi$ is an inequation and $\omega$ is present then we need only interchange the roles of $\omega$ and $\chi$, while if $\omega$ is absent then $\phi$ is true, and so $N[``\phi"]$ is a false equation.

Hence suppose that $\phi$ is an equation $\alpha = \beta$, where $\alpha$ and $\beta$ are closed numerical terms. We have to show how to obtain an $\mathbf{A}_I$-proof of $\omega\{\mathfrak{n}\} \vee \chi\{\mathfrak{k}\}$ for all $\mathfrak{n}$, $\mathfrak{k}$, **E**-correct, when we are given $\mathbf{A}_I$-proofs of $\omega\{\mathfrak{n}\} \vee \alpha = \beta$, for all $\mathfrak{n}$, **E**-correct, and $\mathbf{A}_I$-proofs of $\alpha \neq \beta \vee \chi\{\mathfrak{k}\}$ for all $\mathfrak{k}$, **E**-correct. First consider the case when $\alpha = \beta$ is $\mathbf{A}_{00}$-false. We proceed similarly with $\alpha \neq \beta$ if this is $\mathbf{A}_{00}$-false.

We define an occurrence of a false ancestor of the false equation $\alpha = \beta$ as follows: (for a false inequation we add a clause for R 2)

(i′) the occurrence of $\alpha = \beta$ in $\omega\{\mathfrak{n}\} \vee \alpha = \beta$ is an occurrence of a false ancestor of $\alpha = \beta$;

(ii′) if $\dfrac{\psi \vee \alpha' = \beta'}{\psi' \vee \alpha' = \beta'}$ is except for remodellings a one premise rule or is R 3 in the $\mathbf{A}_I$-proof of $\omega\{\mathfrak{n}\} \vee \alpha = \beta$ and if the occurrence of $\alpha' = \beta'$ in the lower formula is an occurrence of a false ancestor of $\alpha = \beta$, then the occurrence of $\alpha' = \beta'$ in the upper formula is an occurrence of a false ancestor of $\alpha = \beta$;

(iii′) if
$$\frac{\psi \vee \alpha' = \beta' \quad \psi' \vee \alpha' = \beta'}{\psi'' \vee \alpha' = \beta'}$$
is except for remodellings a case of rule II $b'$ and if the occurrence of $\alpha' = \beta'$ in the lower formula is an occurrence of a false ancestor of $\alpha = \beta$ then both occurrences of $\alpha' = \beta'$ in the upper formulae are occurrences of false ancestors of $\alpha = \beta$;

(iv′) if
$$\frac{\alpha'' = \beta'' \vee \psi' \vee \psi \quad \gamma = \delta \vee \psi}{\psi'' \vee \alpha' = \beta' \vee \psi}$$
is except for remodellings a case of rule R 1 and if the occurrence of $\alpha' = \beta'$ in the lower formula is an occurrence of a false ancestor of $\alpha = \beta$, then if $\alpha'' = \beta''$ is $\mathbf{A}_{00}$-false it is an occurrence of a false

ancestor of $\alpha = \beta$, but if $\alpha'' = \beta''$ is $\mathbf{A}_{00}$-true, then $\gamma = \delta$ must be $\mathbf{A}_{00}$-false, and is an occurrence of a false ancestor of $\alpha = \beta$. Again it is a primitive recursive decision to decide which is the case.

Note that it is impossible for related occurrences of $\alpha = \beta$ to occur in the main formulae of II $b'$, $d$, otherwise they would occur in the theorem in the scope of & or $E$. By I $b$ $\alpha = \beta$ might have two false ancestors in the upper formula.

We now omit all false ancestors of $\alpha = \beta$ in the $\mathbf{A}_0$-proofs of $\omega\{\mathfrak{n}\} \vee \alpha = \beta$. By this $\tau\mathfrak{x}$ becomes $\tau'\mathfrak{x}$ and $\sigma[\mathfrak{Num}\,\mathfrak{x}]$ becomes $\sigma'[\mathfrak{Num}\,\mathfrak{x}]$, $\sigma'[Num\,\mathfrak{n}]$ is equal to the $g.n.$ of $\omega\{\mathfrak{n}\}$. We wish to show: If (i) then (ii), i.e.

$$prf_0[\tau'\mathfrak{x}, \sigma'[\mathfrak{Num}\,\mathfrak{x}]] = 0, \quad \text{is an } \mathbf{E}\text{-theorem},$$

then we shall have $\omega\{\mathfrak{n}\}$ for all $\mathfrak{n}$, $\mathbf{E}$-correct, whence by Prop. 7 and II $a$ we obtain: $\omega\{\mathfrak{n}\} \vee \chi\{\mathfrak{k}\}$ for all $\mathfrak{n}$, $\mathfrak{k}$, $\mathbf{E}$-correct, as desired.

By the definition of $\tau x$ and $\tau' x$ we have the $\mathbf{E}$-theorems:

$$clflma\tau x^{\infty}_{x'} \quad \text{and} \quad clflma\tau' x^{\infty}_{x'} \quad \text{for} \quad 0 \leqslant x' \leqslant \varpi\tau x,$$

so we can disregard these terms in the definition of $Rl^1_0$ and $Rl^2_0$. Note that by retaining redundant proofs we have $\varpi\tau x = \varpi\tau' x$.

We have the $\mathbf{E}$-theorems:

(a) $Rl^{X}[\tau x^{\infty}_{x'}, \tau x^{\infty}_{x''}] = Rl^{X}[\tau' x^{\infty}_{x'}, \tau' x^{\infty}_{x''}] \times A_2[\tau' x^{\infty}_{x'}, \tau' x^{\infty}_{x''}]$, where $X$ is a one-premiss rule,

(b) $Rl^{\mathrm{R}1}[\tau x^{\infty}_{x'}, \tau x^{\infty}_{x''}, \tau x^{\infty}_{x'''}] = Rl^{\mathrm{R}1}[\tau' x^{\infty}_{x'}, \tau' x^{\infty}_{x''}, \tau' x^{\infty}_{x'''}]$
$\times Rl^{\mathrm{II}\,a}_0[\tau' x^{\infty}_{x''}, \tau' x^{\infty}_{x'''}] \times A_2[\tau' x^{\infty}_{x'}, \tau' x^{\infty}_{x''}]$,

(c) $Rl^{\mathrm{II}\,b'}[\tau x^{\infty}_{x'}, \tau x^{\infty}_{x''}, \tau x^{\infty}_{x'''}] = Rl^{\mathrm{II}\,b'}[\tau' x^{\infty}_{x'}, \tau' x^{\infty}_{x''}, \tau' x^{\infty}_{x'''}]$.

These follow because we show that they hold in each of 500 subcases; these come about as follows. We have

$$\tau x^{\infty}_{x'} = disj[Sp_1[x, x'], Sp_2[x, x']],$$

and

$$\tau' x^{\infty}_{x'} = disj[Sp'_1[x, x'], Sp'_2[x, x']]$$

or $\tau x^{\infty}_{x'}$ fails to be a disjunction. (See D 179.)

$$NDv = \begin{cases} 0 & \text{if } \nu \text{ fails to be } g.n. \text{ of a disjunction,} \\ 1 & \text{otherwise.} \end{cases}$$

In the first case we have the subcases:

$$ND\tau x^{\infty}_{x'} = 0,$$

$$\rho_1[x,x'] = \rho'_1[x,x'] \quad \text{and} \quad \rho_2[x,x'] = \rho'_2[x,x'],$$

$$\rho_1[x,x'] \neq \rho'_1[x,x'] \quad \text{and} \quad \rho_2[x,x'] = \rho_2[x,x'],$$

$$\rho_1[x,x'] = \rho'_1[x,x'] \quad \text{and} \quad \rho_2[x,x'] \neq \rho'_2[x,x'],$$

$$\rho_1[x,x'] \neq \rho'_1[x,x'] \quad \text{and} \quad \rho_2[x,x'] \neq \rho'_2[x,x'],$$

there are similar cases for $x''$ giving 25 cases for $Rl^{\mathrm{X}}[\tau x_{x'}, \tau x_{x''}]$, X has 5 possible values thus 125 subcases for one-premiss rules. The cases are mutually exclusive and exhaustive. We show for each X

$$Rl^{\mathrm{X}}[\tau x^{\infty}_{x'}, \tau x^{\infty}_{x''}] = Rl^{\mathrm{X}}[\tau' x^{\infty}_{x'}, \tau' x^{\infty}_{x''}] \times A_2[\tau' x^{\infty}_{x'}, \tau' x^{\infty}_{x''}],$$

in each of the 25 cases, whence we obtain for each X

$$Rl^{\mathrm{X}}[\tau x^{\infty}_{x'}, \tau x^{\infty}_{x''}] = Rl^{\mathrm{X}}[\tau' x^{\infty}_{x'}, \tau' x^{\infty}_{x''}] \times A_2[\tau' x^{\infty}_{x'}, \tau' x^{\infty}_{x''}],$$

without condition.

The 25 cases are distinguished by:

Let
$$\alpha_1[x,x'] = A_1[A_2[\rho_1[x,x'],\rho'_1[x,x']] + A_2[\rho_2[x,x'],\rho'_2[x\,x']]],$$
$$\alpha_2[x,x'] = A_1[B_2[\rho_1[x,x'],\rho'_1[x,x']] + A_2[\rho_2[x,x'],\rho'_2[x,x']]],$$
$$\alpha_3[x,x'] = A_1[A_2[\rho_1[x,x'],\rho'_1[x,x']] + B_2[\rho_2[x,x'],\rho'_2[x,x']]],$$
$$\alpha_4[x,x'] = A_1[B_2[\rho_1[x,x'],\rho'_1[x,x']] + B_2[\rho_2[x,x'],\rho'_2[x,x']]].$$

Let
$$\beta_1[x,x',x''] = B_1[\alpha_1[x,x'] + \alpha_1[x,x'']],$$
$$\beta_2[x,x',x''] = B_1[\alpha_1[x,x'] + \alpha_2[x,x'']],$$
$$\vdots$$
$$\beta_5[x,x',x''] = B_1[\alpha_1[x,x'] + ND[\tau x^{\infty}_{x''}]],$$
$$\vdots$$
$$\beta_{25}[x,x',x''] = B_1[ND[\tau x^{\infty}_{x'}] + ND[\tau x^{\infty}_{x''}]].$$

Then $\beta_\theta[x,x',x''] = 1$ when the $\theta$th condition is satisfied.

We now show that

$$\beta_\theta[x,x',x''] \times Rl^{\mathrm{X}}_{\mathrm{o}}[\tau x^{\infty}_{x'}, \tau x^{\infty}_{x''}] = \beta_\theta[x,x',x''] \times Rl^{\mathrm{X}}_{\mathrm{o}}[\tau' x^{\infty}_{x'}, \tau' x^{\infty}_{x''}] \times A_2[\tau' x^{\infty}_{x'}, \tau' x^{\infty}_{x''}], \qquad \text{(iii)}$$

for $1 \leqslant \theta \leqslant 25$, and for each rule.

And
$$\sum_{1\leqslant\theta\leqslant 25} \beta_\theta[x,x',x''] = 1, \qquad \text{(iv)}$$

since the conditions are exhaustive and mutually exclusive, from (iii) by addition for $\theta = 1$ to $\theta = 25$ and (iv) we get

$$Rl_0^{\mathrm{X}}[\tau x_{x'}^{\infty}, \tau x_{x''}^{\infty}] = Rl_0^{\mathrm{X}}[\tau' x_{x'}^{\infty}, \tau' x_{x''}^{\infty}] \times A_2[\tau' x_{x'}^{\infty}, \tau' x_{x''}^{\infty}]. \tag{v}$$

This is (*a*); (*b*) and (*c*) are obtained similarly but require 125 cases each. From (*a*), (*b*) and (*c*), and the **E**-rule,

if $\qquad \rho x = \rho' x \quad \text{then} \quad \sum_{1 \leqslant x' \leqslant x} \rho x' = \sum_{1 \leqslant x' \leqslant x} \rho' x',$

we get $\qquad prf_0[\tau x, \sigma[Num\, x]] = prf_0[\tau' x, \sigma'[Num\, x]],$

as desired. It remains to show that the equations (iii), (iv) are **E**-theorems. The situations in the 25 cases can be described as follows: we here speak of $\tau x_{x'}^{\infty}$ as the upper formula and $\tau x_{x''}^{\infty}$ as the lower formula. We speak of $S\rho_1[x, x']$ as the main formula of the upper formula and $S\rho_2[x, x']$ as the subsidiary formula of the upper formula, $S\rho_1[x, x'']$ and $S\rho_2[x, x'']$ similarly refer to the lower formula. We use this manner of speaking because $Rl_0^{\mathrm{X}}[\kappa, \nu] = 0$ if and only if $\kappa$ is the *g.n.* of the upper formula of rule X and $\nu$ is the *g.n.* of the lower formula of that rule. We use the following abbreviations:

- *f.a.* false ancestor,
- *l.f.* lower formula,
- *n.d.* a formula other than a disjunction, i.e. atomic or a conjunction or an existential statement,
- *m.f.* main formula,
- *s.f.* subsidiary formula,
- *r.s.f.* right subsidiary formula.

We now give a table of the 25 cases and against each we note the only possible rules that can be satisfied in such a situation, in the last column we add conditions which may have to apply if the rule is to hold.

The asterisk denotes that the formula concerned contains an occurrence of a false ancestor of $\alpha = \beta$.

Take, for instance, (10). Here $\tau x_{x'}^{\infty} = disj[S\rho_1[x, x'], S\rho_2[x, x']]$, the first component of the disjunction contains a false ancestor of $\alpha = \beta$, while the second component is without occurrences of a false ancestor. $\tau x_{x''}^{\infty}$ is either atomic or is a conjunction or is an existential statement. The rule then looks like $\dfrac{\phi^* \vee \psi}{\chi}$ where the lower formula fails to be a disjunction.

The only possible rule is II $d$ without subsidiary formula. But then this would be $\dfrac{\phi^* \vee \psi}{(E\xi)\,(\bar{\phi}^* \vee \bar{\psi})}$, but $\phi^*$ contains an occurrence of a false ancestor, but a false ancestor once governed by $E$ would have an occurrence governed by $E$ in the theorem and this is impossible. The bar denotes a substitution of a variable for a numerical term. Hence this situation fails to arise and so $\beta_{10}[x, x', x''] = 0$, and so (iii) holds for $\theta = 10$. Similarly in the other cases.

| no. | *u.f.*, *m.f.* | *u.f.*, *s.f.* | *l.f.*, *m.f.* | *l.f.*, *s.f.* | poss. rules | remarks |
|---|---|---|---|---|---|---|
| 1 | | | | | any | |
| 2 | | | * | | II$a$ | *l.f.*, *m.f.* is the *f.a.* |
| 3 | | | | * | none | |
| 4 | | | * | * | none | |
| 5 | | | *n.d.* | | II$d$, I$b$ | I$b$ without *s.f.* and *m.f.* is *n.d.* |
| 6 | * | | | | none | |
| 7 | * | | * | | I$a$, $b$, II$d$ | II$d$ is impossible, an *f.a.* fails to occur governed by $E$ |
| 8 | * | | | * | I$a$, II$a$ | I$a$ with right *s.f.* missing, II$a$ with *f.a.* in secondary *f.* only |
| 9 | * | | * | * | I$a$ | I$a$ with right *s.f.* missing, *f.a.* in left *s.f.* and 2nd *m.f.* |
| 10 | * | | *n.d.* | | II$d$ | II$d$ impossible as in (7) |
| 11 | | * | | | none | |
| 12 | | * | * | | I$a$ | I$a$ with *r.s.f.* missing |
| 13 | | * | | * | any | |
| 14 | | * | * | * | II$a$ | |
| 15 | | * | *n.d.* | | II$d$ | II$d$ impossible as in (7) |
| 16 | * | * | | | none | |
| 17 | * | * | * | | I$a$ | I$a$ with *r.s.f.* missing |
| 18 | * | * | | * | I$b$ | I$b$ with *s.f.* missing |
| 19 | * | * | * | * | any | |
| 20 | * | * | *n.d.* | | II$d$ | II$d$ impossible as in (7) |
| 21 | *n.d.* | | | | R2, II$a$, $d$ | R2 with *s.f.* missing |
| 22 | *n.d.* | | * | | II$a$ | |
| 23 | *n.d.* | | | * | none | |
| 24 | *n.d.* | | * | * | none | |
| 25 | *n.d.* | | *n.d.* | | II$d$, R2 | both with *s.f.* missing |

This argument is somewhat heuristic, what we want is **E**-proofs. Let us first find the **E**-proof of (iv). For this purpose it is convenient to use $y$ and $z$ as extra symbols for variables of type $\iota$ to avoid a plethora of primes. We require a generalization of Cor. (ii), Prop. 5, namely:

If $\rho[0,0,0,0,0,0] = 0$ and if $\rho[\alpha,\alpha',\beta,\beta',\gamma,\gamma'] = 0$ whenever at least one argument is zero and if

$$\frac{\rho[x,x',y,y',z,z'] = 0}{\rho[Sx,Sx',Sy,Sy',Sz,Sz'] = 0}$$

then $\rho[x,x',y,y',z,z'] = 0$.

We shall define a certain function $\gamma$ and show that

$$\gamma[x,x',y,y',z,z'] = 0$$

using the above generalization; we will then make substitutions.

Replace

$$\begin{aligned}
x &\text{ by } (\rho_1[x,x'] \dot{-} \rho_1'[x,x']) + (\rho_1'[x,x'] \dot{-} \rho_1[x,x']),\\
x' &\text{ by } (\rho_2[x,x'] \dot{-} \rho_2'[x,x']) + (\rho_2'[x,x'] \dot{-} \rho_2[x,x']),\\
y &\text{ by } ND.\tau x_{x'}^{\infty},\\
y' &\text{ by } ND.\tau x_{x''}^{\infty},\\
z &\text{ by } (\rho_1[x,x''] \dot{-} \rho_1'[x,x'']) + (\rho_1'[x,x''] \dot{-} \rho_1[x,x'']),\\
z' &\text{ by } (\rho_2[x,x''] \dot{-} \rho_2'[x,x'']) + (\rho_2'[x\ x''] \dot{-} \rho_2[x,x'']).
\end{aligned}$$

Define:

$$\begin{aligned}
\alpha_1^*[x,x',y] &= A_1[A_1y + A_1x + A_1x'],\\
\alpha_2^*[x,x',y] &= A_1[A_1y + B_1x + A_1x'],\\
\alpha_3^*[x,x',y] &= A_1[A_1y + A_1x + B_1x'],\\
\alpha_4^*[x,x',y] &= A_1[A_1y + B_1x + B_1x'],\\
\alpha_5^*[y] &= B_1y,
\end{aligned}$$

$$\begin{aligned}
\beta_\theta^*[x,x',y,y',z,z'] &= B_1[\alpha_1^*[x,x',y] + \alpha_\theta^*[z,z',y']] && (1 \leqslant \theta \leqslant 5),\\
\beta_\theta^*[x,x',y,y',z,z'] &= B_1[\alpha_2^*[x,x',y] + \alpha_{\theta \dot{-} 5}^*[z,z',y']] && (6 \leqslant \theta \leqslant 10),\\
\beta_\theta^*[x,x',y,y',z,z'] &= B_1[\alpha_3^*[x,x',y] + \alpha_{\theta \dot{-} 10}^*[z,z',y']] && (11 \leqslant \theta \leqslant 15),\\
\beta_\theta^*[x,x',y,y',z,z'] &= B_1[\alpha_4^*[x,x',y] + \alpha_{\theta \dot{-} 15}^*[z,z',y']] && (16 \leqslant \theta \leqslant 20),\\
\beta_\theta^*[x,x',y,y',z,z'] &= B_1[\alpha_5^*[x,x',y] + \alpha_{\theta \dot{-} 20}^*[z,z',y']] && (21 \leqslant \theta \leqslant 25).
\end{aligned}$$

Then $$\gamma[x,x',y,y',z,z'] = \sum_{1 \leqslant \theta \leqslant 25} \beta_\theta^*[x,x',y,y',z,z'].$$

We wish to show that

$$\gamma[x,x',y,y',z,z'] = 1.$$

By computation $$\gamma[0,0,0,0,0,0] = 1.$$

Similarly
$$\gamma[0,Sx',Sy,Sy',Sz,Sz'] = 1,$$
$$\vdots$$
$$\gamma[Sx,Sx',Sy,Sy',Sz,0] = 1,$$

similarly
$$\gamma[0,0,Sy,Sy',Sz,Sz'] = 1,$$
$$\vdots$$

and so on, in fact it is easily seen that as long as one of $x$, $x'$, $y$, $y'$, $z$, $z'$ is replaced by zero then exactly one of $\alpha^*_\theta[x,x',y]$ is zero and exactly one of $\alpha^*_\theta[z,z',y']$ is zero. Thus the first requirement of our generalization of Cor. (ii), Prop. 5, is satisfied. The second requirement follows at once because exactly one of $\alpha^*_\theta[Sx,Sx',Sy]$ and exactly one of $\alpha^*_\theta[Sz,Sz',Sy']$ is zero so that exactly one of $\beta^*_\theta[Sx,Sx',Sy,Sy',Sz,Sz']$ is one. This completes the **E**-proof of (iv).

We now return to the **E**-proof of (iii). If $\beta_\theta[x,x',x''] = 0$ then (iii) holds, if $\beta_\theta[x,x',x''] = 1$ then there are some relations between $\tau x^\infty_{x'}$, $\tau x^\infty_{x''}$, $\tau' x^\infty_{x'}$, $\tau' x^\infty_{x''}$. For instance if $\beta_1[x,x',x''] = 1$ then $\tau x^\infty_{x'} = \tau' x^\infty_{x'}$ and $\tau x^\infty_{x''} = \tau' x^\infty_{x''}$ and so (iii) holds for $\theta = 1$. But we want an **E**-proof that (iii) holds. What we have just said becomes in the system **E**:

$$\frac{\beta_1[x,x',x''] = 0}{\overline{\quad(\text{iii})_1\quad}} \qquad\qquad \frac{\dfrac{\beta_1[x,x',x''] = 1}{\overline{\tau x^\infty_{x'} = \tau' x^\infty_{x'} \quad \tau x^\infty_{x''} = \tau' x^\infty_{x''}}}}{\overline{(\text{iii})_1}},$$

where $(\text{iii})_1$ denotes (iii) with $\theta = 1$. We now get an **E**-proof of $(\text{iii})_1$ as follows: we have

$$\beta_1[x,x',x''] \times (1 \dot{-} \beta_1[x,x',x'']) = 0,$$

and $$\beta_1[x,x',x''] + (1 \dot{-} \beta_1[x,x',x'']) = 1,$$ are **E**-theorems,

thus we get from the above

$$\frac{\dfrac{\dfrac{\beta_1[x,x',x''] \times (1 \dot{-} \beta_1[x,x',x'']) = 0}{\overline{(\text{iii})_1 \times \beta_1[x,x',x''] \quad (\text{iii})_1 \times (1 \dot{-} \beta_1[x,x',x''])}}}{(\text{iii})_1 \times (\beta_1[x,x',x''] + (1 \dot{-} \beta_1[x,x',x'']))}}{(\text{iii})_1}.$$

For $\theta = 2$ the conditions are:

$\tau x^{\infty}_{x'}$ is equal to the *g.n.* of $\phi \vee \omega$ and is without *f.a.*'s,

$\tau x^{\infty}_{x''}$ is equal to the *g.n.* of $\phi'^{*} \vee \omega'$ and $\phi'^{*}$ contains an *f.a.* but $\omega'$ is without *f.a.*'s,

$\tau' x^{\infty}_{x'}$ is equal to the *g.n.* of $\phi \vee \omega$,

$\tau' x^{\infty}_{x''}$ is equal to the *g.n.* of $\phi' \vee \omega'$ where $\phi'$ is $\phi'^{*}$ after removal of *f.a.*'s.

The figure to be tested for being a rule is:

$$\frac{\phi \vee \omega}{\phi'^{*} \vee \omega'} \quad \text{before removal of } f.a.\text{'s},$$

$$\frac{\phi \vee \omega}{\phi' \vee \omega'} \quad \text{after removal of } f.a.\text{'s},$$

here $\phi$, $\omega$, $\phi'$, $\omega'$ are all present. The only possible rule is II *a*

$$\frac{(\phi \vee \omega)}{\phi'^{*} \vee (\phi \vee \omega)} \quad \text{before} \qquad \frac{(\phi \vee \omega)}{\phi' \vee (\phi \vee \omega)} \quad \text{after},$$

this requires $\omega'$ to be $(\phi \vee \omega)$. If we restrict II *a* to atomic statements then $\phi'$ is null. To **E**-prove (iii)$_2$ we take the rules in turn. $Rl_0^{\mathrm{I}a}$. We shall require the following:

Lemma (i). *If $\rho$ and $\sigma$ are characteristic functions, i.e.* $(1 \dot{-} \rho x) \times \rho x = 0$ *is an* **E**-*theorem, and if* $\dfrac{\rho x = 0}{\overline{\overline{\sigma x = 0}}}$ *is an* **E**-*deduction then so is* $\dfrac{\sigma x = 1}{\overline{\overline{\rho x = 1}}}$.

In the **E**-deduction $\dfrac{\rho x = 0}{\overline{\overline{\sigma x = 0}}}$ multiply both sides of all equations by $(1 \dot{-} \rho x)$, and above the axioms thus multiplied add that axiom, and the **E**-proof that we can pass from the axiom to the multiplied form. We are left with an **E**-proof of $(1 \dot{-} \rho x) \times \sigma x = 0$ because $(1 \dot{-} \rho x) \times \rho x = 0$ is an **E**-theorem by hypothesis. From the **E**-theorem $(1 \dot{-} \rho x) \times \sigma x = 0$ and the hypothesis $\sigma x = 1$ we obtain by R 1 $\rho x = 1$, as desired.

Now

$$Rl_0^{\mathrm{I}a}[x, x'] = clstat_0[x, x'] + A_2[x, disj[disj[disj[y, Sz], Sz'], y']] + A_2[x', disj[disj[disj[y, Sz'], Sz], y']].$$

We suppose the variables $y, y', z, z'$ replaced by explicitly defined primitive recursive functions of $x$. For instance instead of $y$ we could have

$$\underset{0<y<x}{Min} \left(\langle 8,8,4,8,8,4,8,8,4\rangle^\frown y InSegx + clstaty = 0\right).$$

Remember that when we take $g.n.$ of $\phi$, then $\phi$ must be written out in full.

LEMMA (ii). $$\frac{zPartdisj[Sx, Sy] + clstat_0[z, Sx, Sy] = 0}{(zPartSx \times zPartSy \times A_2[z, disj[Sx, Sy]])] = 0}.$$

This is part of lemma (ii), Ch. 1, 'Two well-formed formulae fail to overlap'. First some definitions:

$$\left.\begin{array}{rll} Eqx = 0 & \text{for} & x \text{ is } g.n. \text{ of an equation} \\ Ineqx = 0 & \text{for} & x \text{ is } g.n. \text{ of an inequation,} \\ Disjx = 0 & \text{for} & x \text{ is } g.n. \text{ of a disjunction,} \\ Conjx = 0 & \text{for} & x \text{ is } g.n. \text{ of a conjunction,} \\ Ex = 0 & \text{for} & x \text{ is } g.n. \text{ of an existential statement,} \end{array}\right\} = 1, \text{ otherwise.}$$

Then we have

$$\frac{clstatx = 0}{Eqx \times Ineqx \times Disjx \times Conjx \times Ex = 0} \quad \text{by definition,}$$

and

$$\frac{clstatx = 0}{(I \times D \times C \times E) + (Eq \times D \times C \times E) + (Eq \times I \times C \times E) + (Eq \times I \times D \times E) + (Eq \times I \times C \times D) = 1},$$

with obvious abbreviations, is easily **E**-proved.

$NLPx = \sum_{1 \leqslant x' \leqslant \varpi x} B_2[8, x^{\infty}_{x'}]$, the number of left parentheses in '$x$',

$NRPx = \sum_{1 \leqslant x' \leqslant \varpi x} B_2[9, x^{\infty}_{x'}]$ the number of right parentheses in '$x$'.

We now obtain:

$$\frac{Eqx = 0}{\overline{A_2[NLPx, NRPx] = 0}}, \ldots, \frac{Ex = 0}{\overline{A_2[NLPx, NRPx] = 0}}, \qquad \text{(vi)}$$

from this we easily obtain:

$$\frac{clstst_0 x = 0}{A_2[NLPx, NRPx] = 0}.$$

We first show that

$$\frac{tmx = 0}{A_2[NLPx, NRPx] = 0},$$

Now $tmx = A_2[x, 0] \times (A_2[x, \langle 8, 1\rangle^\frown x'^\frown\langle 9\rangle] + tmx')$

$$\times (A_2[x, \langle 8, 6, 11 + x''\rangle^\frown x'''^\frown\langle 9\rangle] + tmx''')$$

$$\times (A_2[x, \langle 8, 8, 8, 7, 8, 6, 10 + y, 8, 6, 11 + y + y'\rangle^\frown z^\frown\langle 9, 9\rangle^\frown z'^\frown\langle 9\rangle^\frown z''^\frown\langle 9\rangle] + tm[z, z', z'']). \quad \text{(vii)}$$

It suffices to show that

$$\frac{\Delta}{A_2[NLPx, NRPx]} = 0, \quad \text{(viii)}$$

where $\Delta$ is any one of the conjunctive terms in the definition of *tmx*, because if $tmx = 0$ then exactly one of the factors in the definition of *tm* is zero, this is easily **E**-proved. We first prove (viii) for $x = 0$ and then for $Sx$ assuming the result for $x' \leqslant x$. For $x = 0$ it is merely a matter of computation. Assuming the result for $x' \leqslant x$, take for instance the third factor,

$$A_2[x, \langle 8, 6, 11 + x''\rangle^\frown x'''^\frown\langle 9\rangle] + tmx'''$$

we have $A_2[NLPtmx''', NRPtmx'''] = 0$ by our induction hypothesis and the easily **E**-proved results,

$$\frac{\dfrac{\dfrac{A_2[Sx, \langle 8, 6, 10 + x''\rangle^\frown x'''^\frown\langle 9\rangle] + tmx'''}{x''' \leqslant x}}{NLPx''' = NRPx'''}}{NLPSx = NRPSx}$$

as desired. Now return to (vi), we have

$$\frac{\dfrac{Eqx = 0}{A_2[x, \langle 8, 8, 2\rangle^\frown x'^\frown\langle 9\rangle^\frown x''^\frown\langle 9\rangle] + tm[x', x''] = 0}}{NLPx = NRPx}$$

follows at once. Similarly for the other cases of (vi). To **E**-prove lemma (ii) is suffices to **E**-prove:

$$\frac{clstat_0 x}{NLPyprinsegx > NRPyprinsegx} \quad \frac{clstat_0 x}{NLPypresegx < NRPypresegx},$$

these are easily **E**-proved by cases and induction, supposing they hold for $x$ and then showing that they also hold for $Sx$. Here

*yprinsegx* for $y$ is a proper initial segment of $x$,

*ypresegx* for $y$ is a proper end segment of $x$.

We now have

$$\cfrac{\cfrac{\cfrac{zPartdisj[Sx, Sy] + clstat_0 z = 0}{zPart\langle 8,8,4\rangle^\frown Sx^\frown\langle 9\rangle^\frown Sy^\frown\langle 9\rangle + clstat_0 z = 0}}{zPart\langle 8,8,4\rangle^\frown Sx^\frown\langle 9\rangle^\frown Sy^\frown\langle 9\rangle + A_2[NLPz, NRPz] = 0}}{zPartSx \times zPartSy \times A_2[z, disj[Sx, Sy]] \times A_2[z, \langle 8,4\rangle^\frown Sx^\frown\langle 9\rangle \times A_0[z, \langle 4\rangle^\frown Sx]] = 0.}$$

But it is easily **E**-proved that $clstat_0[\langle 8,4\rangle^\frown Sx^\frown\langle 9\rangle] = clstat_0[\langle 4\rangle^\frown Sx] = 1$, and so lemma (ii) follows. Returning now to the case $\theta = 2$ and the snbcase of $Rl^{\mathrm{I}a}$, the conditions of case $\theta = 2$ are for this rule

$$\frac{\beta_2[x, x', x''] + Rl^{\mathrm{I}a}[disj[Sy, Sy'], disj[Sz, Sz']] = 0}{A_2[uPartSz, uPartdisj[Sy, Sy']] = 0} \tag{ix}$$

using $u$ as an extra variable of type $\iota$. But case $\beta_2$ gives

$$A_2[uPartSz, uPartdisj[Sy, Sy']] = 1,$$

if $u$ is an *f.a.*

Hence from (ix) we get

$$\frac{\beta_2[x, x', x''] = 0}{Rl^{\mathrm{I}a}[disj[Sy, Sy'], disj[Sz, Sz']] = 1}.$$

Thus we have $(\mathrm{iii})_2$ for the case when X is I$a$, because

$$Rl^{\mathrm{I}a}[\tau x^{\infty}_{x'}, \tau x^{\infty}_{x'}] = Rl^{\mathrm{I}a}[\tau' x^{\infty}_{x'}, \tau' x^{\infty}_{x'}] = 1.$$

This case can be completed for the remaining rules on the same lines. In the case of $Rl^{\mathrm{II}a}$, we find that this rule can be satisfied before in which case since II$a$ only introduces atoms the *m.f.* of the rule is the *f.a.* itself, so that after removal of *f.a.*'s we are left with a repetition. The remaining cases and rules are dealt with similarly. There are altogether 500 sub-cases. There are three two-premiss rules each of 125 cases, R 1, R 3, M.P.

For $\beta_7$ in the case of $Rl_0^{\mathrm{II}d}$ we added the remark that though the figure we got could be a case of rule II$d$ yet this would fail to arise in an $A_0$-proof of $\alpha = \beta \vee \omega$ with $\alpha = \beta$ $A_0$-false. This follows since a statement once governed by $E$ remains so throughout the $A_0$-proof, and so occurs governed by $E$ in the theorem, whereas in our case this situation fails to arise. In this case we want to **E**-prove

$$\langle 8,10,8,6,11+y\rangle^\frown z^\frown\langle 9,9\rangle\, Part\, \tau x^{\infty}_{x'} = 0,$$

$$\langle 8,10,8,6,11+y\rangle^\frown z^\frown\langle 9,9\rangle\, Part\, l.c.\, \tau x = 0$$

then since $\langle 8,10,8,6,11+y\rangle^\frown z^\frown\langle 9,9\rangle\, Part\, l.c.\, \tau x = 1$, we see that this case fails to arise.

It would seem that sufficient has been said so that the full proofs of all the 500 subcases can be written down. This then completes the case of an $\mathbf{A}_I$-proof of $\alpha = \beta \vee \omega\{\mathfrak{n}\}$ for all $\mathfrak{n}$, **E**-correct, with one level.

COR. $\mathbf{A}_{00}$-*cuts can be removed from an* $\mathbf{A}_{IC}$-*proof.*

Now suppose that we have demonstrated the result for $\mathbf{A}_{IC}$-proofs with $\lambda$ levels and consider an $\mathbf{A}_{IC}$-proof of $S\lambda$ levels. We may suppose that all $\mathbf{A}_{00}$-cuts have already been removed from the $\mathbf{A}_{IC}$-proofs of the production of level $\lambda$. Thus consider an $\mathbf{A}_{IC}$-proof with $S\lambda$ levels in which the cuts all occur in level $S\lambda$, and the cut formula is an equation.

We proceed exactly as in the case of level one $\mathbf{A}_{IC}$-proofs, the only difference is that we now have $ded_0$ instead of $prf_0$, and an hypothesis might contain an occurrence of a false ancestor. As in the case of level one the false ancestor may be removed in an hypothesis and the details of removing false ancestors in the body of the deduction is similar to what we have already done, so we omit further details.

## 10.14 *Cut removal with a weaker form of R* 3

Let $\mathscr{A}_{0\Phi}$ be the set of formulae which can be built up from closed $\mathbf{A}_0$-statements and **A**-true closed universal **A**-statements belonging to a bounded set $\Phi$ by the building rules of $\mathbf{A}_0$, i.e. by disjunction, conjunction and existential quantification acting on atomic formulae and members of $\Phi$.

LEMMA B. $\mathscr{A}_{0\Phi}$ *is complete with respect to* **A**-*truth.*

This means that if a closed **A**-statement $\psi$ of $\mathscr{A}_{0\Phi}$ is **A**-true then it can be $\mathbf{A}_0$-deduced from $\Phi$ as hypotheses. Where $\Phi$ is a set of **A**-true closed universal **A**-statements.

We use induction on the build up of $\psi$ from atomic formulae and the hypotheses.

$\psi$ is atomic and is **A**-true. Then $\psi$ is $\mathbf{A}_{00}$-true, hence is an $\mathbf{A}_{00}$-theorem, hence is an $\mathbf{A}_0$-theorem, the result follows.

$\psi$ is one of the hypotheses, then $\psi$ is **A**-true and it is immediately clear that $\psi$ is an $\mathbf{A}_0$-deductand from $\Phi$ as hypotheses.

$\psi$ is of the form $\psi' \vee \psi''$ and the result holds for $\psi'$ and for $\psi''$.

Since $\psi$ is **A**-true then either $\psi'$ or $\psi''$ is **A**-true, both belong to $\mathscr{A}_{0\Phi}$. Hence by induction hypothesis either $\psi'$ or $\psi''$ is an $\mathbf{A}_0$-deductand from

$\Phi$ as hypotheses, by II$a$ the same holds for $\psi$. Note that an hypothesis occurs either in $\psi'$ or in $\psi''$ or in both but fails to overlap $\psi'$ and $\psi''$.

$\psi$ is of the form $\psi'$ & $\psi''$ and the result holds for $\psi'$ and for $\psi''$.

Since $\psi$ is **A**-true then $\psi'$ and $\psi''$ are both **A**-true, and both belong to $\mathscr{A}_{0\Phi}$, hence by induction hypothesis both $\psi'$ and $\psi''$ are $\mathbf{A}_0$-deductands from $\Phi$ as hypotheses. By II$b'$ the same holds for $\psi$.

$\psi$ is of the form $(E\xi)\,\omega\{\xi\}$ and the result holds for $\omega\{\nu\}$ for all numerals $\nu$ and belongs to $\mathscr{A}_{0\Phi}$. Since $\psi$ is **A**-true then $\omega\{\kappa\}$ is **A**-true for some numeral $\kappa$ and it belongs to $\mathscr{A}_{0\Phi}$, by induction hypothesis $\omega\{\kappa\}$ is an $\mathbf{A}_0$-deductand from $\Phi$ as hypothesis for some numeral $\kappa$, by II$d$ the same holds for $\psi$.

These are all the cases that can occur because the build up of $\psi$ can only use the $\mathbf{A}_0$-building rules. Thus the case when $\psi$ is of the form $(\mathbf{A}\xi)\,\omega\{\xi\}$ is absent, unless $\psi$ is an hypothesis. This completes the demonstration of the lemma.

If the hypotheses are all **A**-true universal **A**-statements and if $\psi$ is obtained from the hypotheses by an $\mathbf{A}_{0C}$-deduction then $\psi$ belongs to $\mathscr{A}_{0\Phi}$ and is **A**-true hence it can be obtained from the hypotheses by an $\mathbf{A}_0$-deduction. Thus the cut can be eliminated from the system $\mathbf{A}_{I'}$, namely the system $\mathbf{A}_I$ with the restriction that R 3 is of the form:

$$\frac{\phi\{\mathfrak{n}\}}{(A\mathfrak{x})\,(\phi\{\mathfrak{x}\})}\quad \text{for all } \mathfrak{n}, \textbf{E}\text{-correct},$$

that is R 3 without subsidiary formula. We can then proceed as in the case of level 1.

The system $\mathbf{A}_{I'}$ is weaker than the system $\mathbf{A}_I$. Because if $(A\xi)\,\phi\{\xi\}$ can be $\mathbf{A}_{I'}$-proved then the same proof is an $\mathbf{A}_I$-proof. But T.N.D. fails in $\mathbf{A}_{I'}$. Consider

$$(Ax)\,G\{x\} \vee (Ex)\,N[\text{“}G\{x\}\text{”}], \tag{i}$$

where $(Ax)\,G\{x\}$ is an **A**-true irresolvable **A**-statement. (i) is an $\mathbf{A}_I$-theorem, but it fails to be an $\mathbf{A}_{I'}$-theorem. The only way of $\mathbf{A}_{I'}$-proving (i) is to prove $(Ax)\,G\{x\}$ and then dilute to obtain (i). This is absurd. (If $(Ax)\,G\{x\}$ were $\mathbf{A}_{I'}$-provable then it would be $\mathbf{A}_I$-provable.) But

$$(Ax)\,(G\{x\} \vee (Ex)\,N[\text{“}G\{x\}\text{”}]) \tag{ii}$$

is an $\mathbf{A}_{I'}$-theorem. Nevertheless (ii) is equivalent to (i), in a similar manner we see that the two systems $\mathbf{A}_I$ and $\mathbf{A}_{I'}$ produce equivalent theorems. Any theorem of the one system is equivalent to a theorem of the other system. It is merely that the scopes of quantifiers in $\mathbf{A}_{I'}$ are usually greater than in $\mathbf{A}_I$.

An $\mathbf{A}_I$-proof of $\phi$ can be converted into an $\mathbf{A}_{I'}$-proof of $\phi'$, where $\phi'$ is equivalent to $\phi$ and differs from it merely in that the quantifiers have larger scopes.

Now consider the case of $\mathbf{A}_{IC}$-proofs of $S\lambda$ levels and suppose that we have demonstrated cut-elimination for $\mathbf{A}_{IC}$-proofs with $\lambda$ levels. Then we may suppose that all the cuts in an $\mathbf{A}_{IC}$-proof with $S\lambda$ levels all occur in the $S\lambda$th level. The $S\lambda$th level consists of an $\mathbf{A}_0$-deduction from the production of level $\lambda$ as hypotheses or of a single application of R 3. The cut formula and its negation arise from the disjunctands of the hypotheses and atomic formulae by disjunction, conjunction and existential quantification only. Any quantifier in a cut owes its origin to an hypothesis because an existential quantifier in a cut formula corresponds to a universal quantifier in its negation and conversely, and a universal quantifier can only arise from an hypothesis.

Level $S\lambda$ can either be an $\mathbf{A}_0$-deduction from the production of level $\lambda$ as hypotheses or a single application of R 3. In the first case the cut may be removed by the method used in Prop. 4, Ch. 3. In this case we are without the necessity of proving **E**-correctness. Thus we need only consider the second case. We have $\phi\{\mathfrak{n}\} \vee \omega$ for all $\mathfrak{n}$ are $\mathbf{A}_{0C}$-deductands from the production of level $\lambda$ as hypotheses, **E**-correct.

We wish to remove the cuts, if any. We proceed as in Prop. 4, Ch. 3. This was by induction on the construction of the cut formula. Now in our present case we have

$$ded_{0C}[\rho\mathfrak{x}, \tau\mathfrak{x}, \sigma[\mathfrak{Num}\,\mathfrak{x}]] = 0,$$

where $\rho x$ is the *g.n.* of the hypotheses, $\tau x$ is equal to the *g.n.* of the $\mathbf{A}_{0C}$-deduction of $\phi\{\mathfrak{n}\} \vee \omega$ from the hypotheses and $\sigma[\mathfrak{Num}\,\mathfrak{n}]$ is equal to the *g.n.* of $\phi\{\mathfrak{n}\} \vee \omega$. Let us try to remove the highest cut first. The place where this occurs (if there is a cut) will in general vary with $\mathfrak{n}$, also the cut formula itself will in general vary with $\mathfrak{n}$. Thus a straightforward induction on the construction of the cut formula is impossible. But the demonstration of Prop. 4, Ch. 3 gives us a method of replacing a cut by a simpler cut, i.e. one with a simpler cut formula or by one higher up the tree. This gives us a relation which turns out to be primitive recursive between the *g.n.* of the original deduction and the *g.n.* of the deduction when we have made one application of the procedure in Prop. 4, Ch. 3.

Let the order of a cut formula be the number of logical signs used in its construction. If $f_\pi[\mu, \kappa]$ is the *g.n.* of the result of cut elimination in the $\mathbf{A}_{0C}$-deduction from hypotheses whose *g.n.* is $\pi$, then this enables us to get a relation between $f_\pi[\mu, S\kappa]$ and $f_{\pi'}[h\mu, \kappa]$ which turns out to be primitive recursive. We then merely want to check that if $ded_{0C}[z, x, u] = 0$ then so is $ded_0[z', f[x, y], u] = 0$ in the case when there is a single cut. Thus we can remove the highest cut, by repetition we can remove all the cuts.

We now proceed to the details.

## 10.15 *Cut removal in general*

PROP. 15. *The cut is redundant in* $\mathbf{A}_I$.

In other words the systems $\mathbf{A}_I$ and $\mathbf{A}_{IC}$ are equivalent. We proceed by induction on the number of levels in an $\mathbf{A}_{IC}$-proof. We take $\mathbf{A}_I$ in the equivalent form in which the primitive logical connectives are $\vee$, $N$ and $E$.

If there is only one level then the matter is settled by lemma A. If there are $S\lambda$ levels then we suppose that we have shown the result when there are $\lambda$ levels and then demonstrate it for $S\lambda$ levels. Thus suppose that we have an $\mathbf{A}_{IC}$-proof of $S\lambda$ levels, we may suppose that all cuts as far as the $\lambda$th level inclusive have been removed. Thus we may suppose that all the cuts in our $\mathbf{A}_{IC}$-proof occur in the $S\lambda$th level. We also suppose that the $S\lambda$th level consists of a single application of R 3, otherwise the result is trivial, by Prop. 4, Ch. 3. The $\mathbf{A}_{IC}$-theorem proved is $(A\mathfrak{x})\,\phi\{\mathfrak{x}\} \vee \omega$. Now we know that the cut can be removed from the $\mathbf{A}_{0C}$-deduction of $\phi\{\mathfrak{n}\} \vee \omega$. (The method of doing this is given in Prop. 4, Ch. 3.) Thus if $\tau\mathfrak{n}$ is equal to the *g.n.* of the deduction of $\phi\{\mathfrak{n}\} \vee \omega$ from the production of level $\lambda$ as hypotheses then $f\tau\mathfrak{n}$ for some function $f$ is equal to the *g.n.* of the $\mathbf{A}_I$-deduction without cut of the same formula from the same hypotheses. In the first case we have: ($\rho\mathfrak{n}$ is equal to the *g.n.* of the hypotheses)

$$ded_{0C}[\rho\mathfrak{x}, \tau\mathfrak{x}, \sigma[\mathfrak{Num}\,\mathfrak{x}]] = 0 \qquad \text{(i)}$$

is an **E**-theorem, we wish to show that:

$$ded_0[g\rho\mathfrak{x}, f\tau\mathfrak{x}, \sigma[\mathfrak{Num}\,\mathfrak{x}]] = 0 \qquad \text{(ii)}$$

is also an **E**-theorem, where $g\rho$ is some definite primitive recursive function defined from $\rho$. To do this implies that we must show that $f$ is a primitive recursive function.

We apply the process repeatedly. We wish to show:

$$\frac{ded_{OC}[g^{(\nu)}[\rho x], f^{(\nu)}[\tau x], \sigma[Num\, x]] = 0}{ded_{OC}[g^{(S\nu)}[\rho x], f^{(S\nu)}[\tau x], \sigma[Num\, x]] = 0} \text{ is an E-rule,} \qquad \text{(iii)}$$

then we shall have:

$$ded_{OC}[g^{(hx)}[\rho x], f^{(hx)}[\tau x], \sigma[Num\, x]] = 0 \text{ is an E-theorem.} \qquad \text{(iv)}$$

We now choose $hx$ so large that all cuts are removed, reduced to atomic cuts or removed to level $\lambda$.

We require a few definitions.

D 240 $ct[\alpha,\beta,\gamma]$ for $\dfrac{`\alpha\text{'} \quad `\beta\text{'}}{`\gamma\text{'}}$ is a cut.

D 241 $ct[1,\theta]$ for $\underset{u \leqslant \varpi\theta}{Min}\,[(Ex',x'')_{\varpi\theta}\,(x' < x'' < u\ \&\ ct[\theta^{\infty}_{x'}, \theta^{\infty}_{x''}, \theta^{\infty}_{u}])]$.

This gives the position of the highest cut if $\theta$ is $g.n.$ of a proof with cut.

$$ct[S\kappa,\theta] \text{ for } \underset{u \leqslant \varpi\theta}{Min}\,[(Ex',x'')_{\varpi\theta}\,(x' < x'' < u\ \&\ u > ct[\kappa,\theta]\ \&\ ct[\theta^{\infty}_{x'}, \theta^{\infty}_{x''}, \theta^{\infty}_{u}])].$$

This gives the position of the $S\kappa$th cut.

D 242 $D_1\theta$ for $\theta$ is the $g.n.$ of a disjunction and $D_1\theta$ is equal to the $g.n.$ of its first component.

D 243 $D_2\theta$ for $\theta$ is the $g.n.$ of a disjunction and $D_2\theta$ is equal to the $g.n.$ of the second component.

D 244 $ctfmla[1,\theta]$ for the $g.n.$ of the cut formula, if $\theta$ is the $g.n.$ of a cut.

D 245 $Ord\,\theta$ for the number of logical connectives $N, \vee, E$ in the formula whose $g.n.$ is $\theta$.

D 246 $\pi\, Comp\,\theta$ for all the components of the tuplet $\pi$ are components of the tuplet $\theta$, multiplicities included, and order is preserved.

D 247 $T_1[\theta]$ for $\theta$ is the $g.n.$ of an $\mathbf{A}_0$-deduction and the last rule used is a two-premiss rule and $T_1[\theta]$ is equal to the $g.n.$ of the $\mathbf{A}_0$-deduction of the left upper formula. This can be defined thus:

$$T_1[\theta] \text{ for } \underset{u \leqslant \theta}{Min}\,[u\, Comp\,\theta\ \&\ ded_I[\alpha, u, \kappa]] \ (\alpha\ g.n. \text{ of hypotheses}).$$

Here $\kappa$ is the $g.n.$ of the left upper formula.

D 248 $T_2[\theta]$ is equal to the *g.n.* of the $\mathbf{A}_0$-deduction of the right upper formula.

D 249 $C_1\theta$ for $\theta$ is the *g.n.* of a proof of $(\phi' \,\&\, \phi'') \vee \omega$ and $C_1\theta$ is equal to the *g.n.* of a proof of $\phi' \vee \omega$.

D 250 $C_2\theta$ for $\theta$ is the *g.n.* of $(\phi' \,\&\, \phi'') \vee \omega$ and $C_2\theta$ is equal to the *g.n.* of a proof of $\phi'' \vee \omega$.

D 251 $O\theta$ for $\theta$ is the *g.n.* of an $\mathbf{A}_{0C}$-deduction and the last rule used is a cut and this is the only cut in the deduction and

$$O\theta = \begin{cases} 0 \text{ if the cut formula is a disjunction statement} \\ 1 \quad \text{,,} \quad\quad \text{,,} \quad\quad \text{negation} \quad \text{,,} \\ 2 \quad \text{,,} \quad\quad \text{,,} \quad\quad \text{existential} \quad \text{,,} \\ 3 \quad \text{,,} \quad\quad \text{,,} \quad\quad \text{atomic} \quad \text{,,} \\ 4 \quad \text{,,} \quad\quad \text{,,} \quad\quad \text{removed to level } \lambda. \end{cases}$$

D 252 $\theta \upharpoonright \nu$ for the ordered tuplet consisting of the first $\nu$ members of the ordered tuplet $\theta$, order preserved.

D 253 $\theta \urcorner \nu$ for the ordered tuplet consisting of the last $\nu$ members of the ordered tuplet $\theta$, order preserved.

Clearly these are all primitive recursive functions.

If we follow the procedure of Prop. 4, Ch. 3 to an $\mathbf{A}_{0C}$-deduction from hypotheses $\phi', \ldots, \phi^{(\kappa)}$ then the formulae we arrive at as the result of such a deduction are built up from atomic formulae and the various disjunctands of the hypotheses by the building methods allowed in $\mathbf{A}_0$, that is to say by $\vee$, $N$ and $E$ only.

Now let

$$\frac{\omega \vee \psi \quad N\psi \vee \chi}{\omega \vee \chi}$$

be a cut. Apply the procedure of Prop. 4, Ch. 3. We go on until we reach a cut formula which is a disjunctand of an hypothesis, say $\psi_0$, or is an atomic formula.

In the course of the $\mathbf{A}_{0C}$-deduction the hypotheses can only enter the deduction *in toto*, but disjunctands of hypotheses can leave the deduction by the cut. Suppose $\psi_0$ is a disjunctand of the hypothesis $\phi_0$, so that $\phi_0$ is $\psi_0 \vee \phi_0^*$, except for permutations. Then if we add a new hypothesis

$\chi \vee \phi_0^*$ or $\omega \vee \phi_0^*$ according as whether $\psi_0$ is under the scope of an even or odd number of negations when we express the original cut in terms of $N \vee, E$ only, we shall arrive at $\omega \vee \chi$ without cut, by precisely the same build up.

Thus cuts may be removed from $\mathrm{A}_{0C}$-deductions provided we add some new hypotheses. Also it is an effective process to find these new hypotheses. In our case we shall deal with $\mathrm{A}_{0C}$-deductions which contain a single cut, and this cut is the last rule used, so that it is quite easy to find the new hypotheses to add. There may, of course, be several new hypotheses to add from one cut, because the cut formula may contain several disjunctands of hypotheses.

(i) $Octfmla[1, \theta] = 0$, the cut formula is a disjunction, $\phi' \vee \phi''$. From the procedure of Prop. 4, Ch. 3 we replace the cut by

$$\dfrac{\dfrac{\dfrac{\dfrac{\dfrac{\vdots}{\omega \vee (\phi' \vee \phi'')}}{(\omega \vee \phi') \vee \phi''}\;\mathrm{I}a \qquad \dfrac{\vdots}{N\phi'' \vee \chi}}{(\omega \vee \phi') \vee \chi}\;cut}{(\omega \vee \chi) \vee \phi'}\;\mathrm{I}a \qquad \dfrac{\vdots}{N\phi'' \vee \chi}}{\dfrac{(\omega \vee \chi) \vee \chi}{\omega \vee \chi}\;\mathrm{I}a, b.}\;cut \tag{o}$$

The original cut has been replaced by two cuts higher up the tree and with simpler cut formulae. The *g.n.* of the new deduction is

$$T_1(\theta \lceil ct[1, \theta])^\frown C_1 T_2(\theta \lceil ct[1, \theta])^\frown C_2 T_2(\theta \lceil ct[1, \theta])^\frown F\theta^\frown \theta \rceil (\varpi\theta \dot{-} ct[1, \theta]),$$

where $F\theta$ takes account of the applications of $\mathrm{I}a, b$ in the new deduction.

(ii) $Octfmla[1, \theta] = 1$, the cut formula is a negation. We have

$$\dfrac{\dfrac{\vdots}{\omega \vee N\phi} \qquad \dfrac{\vdots}{NN\phi \vee \chi}}{\omega \vee \chi}\;cut.$$

Using the reversibility of $\mathrm{II}c$ we replace this by:

$$\dfrac{\dfrac{\dfrac{\vdots}{\omega \vee N\phi}}{\chi \vee \phi}\;\mathrm{I}a \qquad \dfrac{\dfrac{\vdots}{\phi \vee \chi}}{N\phi \vee \omega}}{\dfrac{\chi \vee \omega}{\omega \vee \chi}\;\mathrm{I}a}\;cut \tag{i}$$

if $\omega$ is present, or by:

$$\dfrac{\dfrac{\dfrac{\underline{N\phi\ \mathrm{II}a}}{\vdots}\qquad}{\chi \vee N\phi \quad \phi \vee \chi}\ \mathrm{I}a}{\dfrac{\chi \vee \phi \quad N\phi \vee \chi}{\dfrac{\chi \vee \chi}{\chi}\ \mathrm{I}b}\ cut} \tag{i}$$

if $\omega$ is absent. In both cases the cut may remain at the same distance from the top of the deduction, but it is replaced by a simpler cut formula.

(iii) $Octfmla[1, \theta] = 2$, the cut formula is an existential statement, say $(E\xi)\,\phi\{\xi\}$, so that the cut is

$$\frac{\omega \vee (E\xi)\,\phi\{\xi\} \quad (A\xi)\,N\phi\{\xi\} \vee \chi}{\omega \vee \chi}\ cut. \tag{ii}$$

In Prop. 4, Ch. 3 we replaced this by

$$\frac{\omega \vee \Sigma(E\xi)\,\phi\{\xi\} \quad N\phi\{\alpha\} \vee \chi}{\omega \vee \chi}\ E\text{-}cut. \tag{iii}$$

This scheme we call an *E-cut*. The right upper formula comes from the reversibility of R 3, and the left upper formula is to take account of all the applications of I $b$ between the left upper formula of (ii) and the next building rule above it, so that the rule next above the left upper formula of (iii) is other than I $b$.

According to Prop. 4, Ch. 3 we use theorem induction on the left upper formula of (iii). That is to say we replace (iii) by

$$\frac{\omega' \vee \Sigma(E\xi)\,\phi\{\xi\} \quad N\phi\{\alpha\} \vee \chi}{\dfrac{\omega' \vee \chi}{\omega \vee \chi}\ \mathrm{R}\,x}\ E\text{-}cut, \tag{iv}$$

where R $x$ is the rule next above the left upper formula of (ii) if this is other than an introduction of $(E\xi)\,\phi\{\xi\}$ by II $d$.

If the rule is an introduction of $(E\xi)\,\phi\{\xi\}$ by II $d$ then we replace (iii) by

$$\frac{\omega \vee \phi\{\alpha\} \vee \Sigma(E\xi)\,\phi\{\xi\} \quad N\phi\{\alpha\} \vee \chi}{\dfrac{\dfrac{\omega \vee \phi\{\alpha\} \vee \chi}{} \ \mathrm{I}a \quad N\phi\{\alpha\} \vee \chi}{\dfrac{(\omega \vee \chi) \vee \chi}{\omega \vee \chi}\ \mathrm{I}a, b.}\ cut}\ E\text{-}cut \tag{v}$$

The new *E-cut* is at least one place higher up the tree, the second cut is where the first one was but has a simpler cut formula.

(iv) $Octfmla[1, \theta] = 3$, the cut formula is atomic and is removed by the method of false ancestors.

(v) $Octfmla[1, \theta] = 4$, the cut formula is removed to level $\lambda$. This can occur, as explained above, when $Octfmla[1, \theta] = 3$. But it can occur in other cases, namely when the cut formula is part of an hypothesis. Suppose in case (i) that $N(\phi' \vee \phi'')$ is part of an hypothesis. We require $N\phi' \vee \chi$ and $N\phi'' \vee \chi$, these will have to be new hypotheses. The cut still remains in level $S\lambda$ but the hypotheses have been enlarged. The old hypothesis can be discarded because it can be obtained from the new ones. Similarly in case (ii) we replace an hypothesis of the form $NN\phi \vee \chi$ by $\phi \vee \chi$, from this the original hypothesis can be obtained. In case (iii) the building rule next above the left upper formula might be R 3, in this case the cut is removed into level $\lambda$.

$Ord\,\theta$ is the number of $N$, $\vee$, $E$ in the statement '$\theta$'. The greatest order of a statement in an $\mathrm{A}_{0C}$-deduction of $g.n.\ \theta$ is

D 254 $\quad \Omega\theta$ for $\underset{1 \leqslant \xi \leqslant \varpi\theta}{Max} Ord\, \theta_\xi^\infty$.

Thus $\Omega\theta$ steps will suffice to reduce any cut to an atomic cut, remove it to level $\lambda$ or put it at the top of the deduction. Hence $\varpi\theta \times \Omega\theta$ applications of our cut removal process will suffice to reduce all cuts to atomic cuts, remove to level $\lambda$ or place at the top, except possibly $E$-cuts, since there are at most $\varpi\theta$ cuts. These are replaced by cuts higher up the tree, hence, at most $\varpi\theta$ applications of the cut removal process will bring any one of them to the top of the tree where it disappears into level $\lambda$. There are at most $\varpi\theta$ cuts, hence at most $(\varpi\theta)^2$ applications of the cut removal process will bring them all to the top of the tree. Altogether at most $(\varpi\theta)^2 + \varpi\theta \times \Omega\theta$ applications of the cut removal process are required to reduce all cuts to atomic cuts or to push them into level $\lambda$. This is a primitive recursive bound. Thus the function $h$ in (iv) is primitive recursive. We have shown that we have (iii). Thus we obtain (iv). It remains to consider the function $g$. The alterations we wanted in the hypotheses are to replace $N(\phi' \vee \phi'')$ by $N\phi'$ and by $N\phi''$, and to replace $NN\phi$ by $\phi$. The simplest thing is to replace every disjunctand of an hypothesis of either of these forms by those stated before we begin. Then the function $g$ is the identity function. This will require an addition to $\tau x$, namely the proofs of the original hypothesis from the new hypotheses also alterations in level $\lambda$ corresponding to the reversibility of II$b'$ and II$c$. To obtain (ii) from (i) we deal with the highest cut, the alterations to be made have been fully displayed, hence by Prop. 7 we obtain (ii) from (i) in each particular case. Thus

D 255
$$f\tau x = \begin{cases} \tau_0 x & \text{if } \mathit{Octfmla}[ct[1,\tau c]] = 0, \\ \tau_1 x & \text{,,} \qquad\qquad\qquad\quad \text{,,} \quad = 1, \\ \tau_2 x & \text{,,} \qquad\qquad\qquad\quad \text{,,} \quad = 2, \\ \tau_3 x & \text{,,} \qquad\qquad\qquad\quad \text{,,} \quad = 3, \\ \tau_4 x & \text{,,} \qquad\qquad\qquad\quad \text{,,} \quad = 4. \end{cases}$$

D 256 $O_i\tau xx$ is the characteristic function of the $i$th case.

The *g.n.* of the new deduction is

$$\tau_0 x O_0 \tau x + \ldots + \tau_4 x O_4 \tau x.$$

D 257 $g\rho x$ *for the result of making the above changes to* $\rho x$,
$\tau' x$ ,, ,, ,, $\tau x$.

We then require

$$ded_{OC}[g[\rho x], \tau_0' x O_0 \tau' x + \ldots + \tau_4' x O_4 \tau' x, \sigma[Num\, x]] = 0 \tag{vi}$$

we show

$$O_i \tau' x ded_{OC}[g[\rho x], \tau_0' x O_i \tau' x, \sigma[Num\, x]] = 0, \tag{vii}$$

from which by addition we obtain (ii).

We have

$$ded_{OC}[g\rho x, \tau' x \restriction ct[1, \tau' x], \tau' x^{\infty}_{ct[1,\tau' x]}] = 0 \text{ is an } \mathbf{E}\text{-theorem}$$

because it is a part summation of $ded_{OC}[g\rho x, \tau' x, \sigma[Num\, x]]$ which by an **E**-theorem is zero. Let $F\tau' x$ be equal to the *g.n.* of the alteration, then

$$ded_{OC}[\rho x, F\tau' x, \tau' x_{ct[1,\tau' x]}] = 0 \text{ is an } \mathbf{E}\text{-theorem}$$

because it is displayed, Prop 7, again

$$ded_{OC}[g\rho x, \tau' x \urcorner (\varpi\tau' x \dot{-} ct[1, \tau' x]), \sigma[Num\, x]] = 0 \text{ is an } \mathbf{E}\text{-theorem}$$

because it is a part summation of $ded_{0C}[g\rho x, \tau' x, \sigma[Num\, x]]$ which by an E-theorem is zero. Thus

D 258 $$f\tau x \quad \text{for} \quad \tau x \ulcorner ct[1, \tau x]^\frown F\tau x^\frown \tau x \urcorner (\varpi \tau x \dot{-} ct[1, \tau x]),$$

and $ded_{0C}[\rho x, f\tau x, \sigma[Num\, x]] = 0$ is an **E**-theorem, and we are finished. We alter $\rho x$ and $\tau x$ before we begin in the manner described to deal with alterations in the hypotheses, then $g$ is the identity function.

To summarize we have found that the *g.n.* of a cut-free $\mathbf{A}_0$-deduction is a primitive recursive function of the *g.n.* of the original $\mathbf{A}_{0C}$-deduction. This still holds when the original $\mathbf{A}_{0C}$-deduction has parameters. We have, so far, only demonstrated this for $\mathbf{A}_{0C}$-deductions ending in a single cut. But the method allows us to remove the cuts one after the other. In fact we have: If $f'[\pi, \kappa, 1]$ is equal to the *g.n.* of the result of cut-removal from an $\mathbf{A}_{0C}$-deduction, whose *g.n.* is $\kappa$ from hypotheses whose *g.n.* is $\pi$, ending in a single cut, and if $f'[\pi, \kappa, \nu]$ is the *g.n.* of the result of cut-removal from an $\mathbf{A}_{0C}$-deduction, whose *g.n.* is $\kappa$ from hypotheses whose *g.n.* is $\pi$, having $\nu$ cuts, then it is easily seen that if $f'[\pi, \kappa, \nu]$ is a primitive recursive function then so is $f'[\pi, \kappa, S\nu]$. Hence by substitution so is $f'[\pi, \kappa, h\kappa]$, where $h\kappa$ is the number of cuts in the original $\mathbf{A}_{0C}$-deduction whose *g.n.* is $\kappa$, here $h$ is a primitive recursive function and $h\kappa$ is the number of cuts in the $\mathbf{A}_{0C}$-deduction whose *g.n.* is $\kappa$. Now let

$\rho\nu$ be equal to the *g.n.* of the hypotheses,

$\tau\nu$ be equal to the *g.n.* of the $\mathbf{A}_{0C}$-deduction of the **A**-statement whose *g.n.* is $\sigma[Num\, \nu]$.

Suppose that we have

$$ded_{0C}[\rho x, \tau x, \sigma[Num\, x]] = 0 \quad \text{is an } \mathbf{E}\text{-theorem.} \qquad \text{(viii)}$$

We wish to show that

$$ded_0[\rho x, f'[\rho x, \tau x, hx], \sigma[Num\, x]] = 0 \quad \text{is an } \mathbf{E}\text{-theorem.} \qquad \text{(ix)}$$

Then we shall have removed the cut from an $\mathbf{A}_{IC}$-proof.

It remains then to **E**-deduce (ix) from (viii). This follows quite easily using Prop. 7. If we examine the scheme (o) for removal of a cut when the cut formula is a disjunction we see that if we have an $\mathbf{A}_{0C}$-deduction of $(N\phi' \,\&\, N\phi) \vee \chi$ for all n, **E**-correct then by Prop. 12 we have

$\mathbf{A}_{0C}$-deductions of $N\phi' \vee \chi$ and of $N\phi'' \vee \chi$ for all $\mathfrak{n}$, **E**-correct. Thus by Prop. 7 we have the scheme (o) for all $\mathfrak{n}$, **E**-correct. Similarly, if from the $\mathbf{A}_{0C}$-deduction of the lower formula of the first cut in the scheme (o) for all $\mathfrak{n}$, **E**-correct we can obtain a cut-free $\mathbf{A}_0$-deduction of its lower formula for all $\mathfrak{n}$, **E**-correct, and if the same applies to the second cut, then by Prop. 7 we have a cut-free $\mathbf{A}_0$-deduction of the final formula of the scheme (o) for all $\mathfrak{n}$, **E**-correct. But if we have just seen that from an $\mathbf{A}_{0C}$-deduction of $\psi$ for all $\mathfrak{n}$, **E**-correct we can obtain an $\mathbf{A}_0$-deduction of the same formula, for all $\mathfrak{n}$, **E**-correct. The same remarks apply to the scheme (v) for cut elimination in the case when the cut formula is an existential statement. Thus altogether from an $\mathbf{A}_{0C}$-deduction from hypotheses for all $\mathfrak{n}$, $E$-correct, we have obtained an $\mathbf{A}_0$-deduction of the same formula, but possibly from an enlarged set of hypotheses. This enlarged set of hypotheses can be effectively found. It still remains to show that the extra hypotheses are $\mathbf{A}_I$-theorems. The extra hypotheses arise by replacing a disjunctand $\phi$ of an hypothesis by $\chi$. This comes about by the process in Prop. 4, Ch. 3 of cut removal. In this process a cut is reduced in that it is replaced by cuts of lower order or cuts higher up the deduction-tree. Thus we treat the cut until it is either atomic or a disjunctand of an hypothesis. If it is atomic it can be removed altogether, if it is a disjunctand of an hypothesis we can proceed in either of two ways, that already described or the following. We have in the course of the reduction

$$\frac{\omega \vee \phi \quad N\phi \vee \chi}{\omega \vee \chi} \quad \text{cut.} \tag{x}$$

Say the cut formula $\phi$ is a part disjunctand of the hypothesis

$$(A\xi)\,\psi\{\xi\} \vee \omega' \vee \phi,$$

the hypothesis must be of this form because it is a production of level $\lambda$. The cut amounts to the replacement of $\phi$ by $\chi$. We should arrive at the lower formula of the cut (x) if we had the hypothesis $(A\xi)\,\psi\{\xi\} \vee \omega' \vee \chi$ and this is done without a cut. The disjunctand of the hypothesis remains intact in the course of the deduction-tree until it disappears in the cut and is replaced by $\chi$, so if we make the substitution in the hypothesis we achieve the same lower formula of the cut without using the cut.

To obtain the new hypothesis we proceed as follows:
Suppose that the disjunction came from the hypothesis

$$(A\mathfrak{x})\,\psi\{\mathfrak{x}\}\vee\omega'\vee\phi.$$

In the course of the deduction-tree we had $N\phi\vee\chi$. Now consider the cut:

$$\frac{(A\mathfrak{x})\,\psi\{\mathfrak{x}\}\vee\omega'\vee\phi \qquad N\phi\vee\chi}{(A\mathfrak{x})\,\psi\{\mathfrak{x}\}\vee\omega'\vee\chi}\ \text{cut.}$$

This gives us the new hypothesis. Now this cut, which involves one of the productions of level $\lambda$ can be pushed into level $\lambda$, and by our induction on the number of levels can then disappear. The cut can be replaced by

$$\cfrac{\cfrac{\psi\{\mathfrak{n}\}\vee\omega'\vee\phi \qquad N\phi\vee\chi}{\psi\{\mathfrak{n}\}\vee\omega'\vee\chi}\ \text{cut, for all } \mathfrak{n}, \textbf{E}\text{-correct}}{(A\mathfrak{x})\,\psi\{\mathfrak{x}\}\vee\omega'\vee\chi}\ \text{R 3} \quad \text{for all } \mathfrak{n}, \textbf{E}\text{-correct}$$

as required.

There is however still one more adjustment to make. It could happen that we arrived at

$$(A\mathfrak{x})\,\psi\{\mathfrak{x}\}\vee\omega\vee\phi \qquad N\phi\vee\omega'\vee(A\mathfrak{y})\,\psi'\{\mathfrak{y}\}, \tag{xi}$$

i.e. the cut formula and its negation both are disjunctands of hypotheses. Applying the cut we arrive at

$$(A\mathfrak{x})\,\psi\{\mathfrak{x}\}\vee\omega\vee(A\mathfrak{y})\,\psi'\{\mathfrak{y}\}\vee\omega'. \tag{xii}$$

If we push the cut into level $\lambda$ by the same device as before we obtain instead of (xii)

$$\psi(\mathfrak{n}\}\vee\omega\vee\psi'\{\mathfrak{n}'\}\vee\omega', \quad \text{for all } \mathfrak{n}, \mathfrak{n}', \textbf{E}\text{-correct}, \tag{xiii}$$

without quantifiers. Now the present form of R 3 allows us to quantify the two sets of variables $\mathfrak{n}$ and $\mathfrak{n}'$ simultaneously giving

$$(A\mathfrak{x},\mathfrak{y})\,(\psi\{\mathfrak{x}\}\vee\psi'\{\mathfrak{y}\})\vee\omega\vee\omega', \tag{xiv}$$

but we wish to have

$$(A\mathfrak{x})\,\psi\{\mathfrak{x}\}\vee(A\mathfrak{y})\,\psi'\{\mathfrak{g}\}\vee\omega\vee\omega'. \tag{xv}$$

By Prop. 10 we can pass from (xiv) to (xv). Thus altogether we can get (xiv) without cut and then (xv) from (xiv) without cut, hence we can get (xv) without cut as desired. There are only a bounded number of cases of extra hypotheses, so we are without the necessity of proving **E**-correctness.

This completes the demonstration of Prop. 15.

## 10.16 *Further properties of the system* $\mathbf{A}_I$

PROP. 16

$$\frac{\phi\{0\}}{(A\xi)(\xi = 0 \to \phi\{\xi\})} \qquad \frac{(A\xi)(\xi = 0 \to \phi\{\xi\})}{\phi\{0\}}*$$

$$\frac{(E\xi)(\xi = 0 \,\&\, \phi\{\xi\})}{\phi\{0\}}* \qquad \frac{\phi\{0\}}{(E\xi)(\xi = 0 \,\&\, \phi\{\xi\})}.$$

We have

$$\frac{\phi\{0\}}{0 = 0 \to \phi\{0\}}\ \text{II}\,a \quad \text{and} \quad \frac{S\nu \neq 0}{S\nu = 0 \to \phi\{S\nu\}}\ \text{Ax. 2.2}\quad \text{II}\,a.$$

Here $\to$ is material implication. Hence we may apply R 3 and obtain

$$\frac{\phi\{0\}}{(A\xi)(\xi = 0 \to \phi\{\xi\})}.$$

From this, by the Deduction Theorem, we get

$$\phi\{0\} \to (A\xi)(\xi = 0 \to \phi\{\xi\}),$$

from this, using negations, we get

$$(E\xi)(\xi = 0 \,\&\, \phi\{\xi\}) \to \phi\{0\},$$

hence by the cut

$$\frac{(E\xi)(\xi = 0 \,\&\, \phi\{\xi\})}{\phi\{0\}}*.$$

Again we have

$$\cfrac{\cfrac{\phi\{0\} \quad 0 = 0}{\phi\{0\} \,\&\, 0 = 0}\ \text{II}\,b}{(E\xi)(\xi = 0 \,\&\, \phi\{\xi\})}\ \text{II}\,d$$

whence
$$\frac{\phi\{0\}}{\overline{(E\xi)\,(\xi = 0 \,\&\, \phi\{\xi\})}}.$$

From this by the Deduction Theorem we get

$$\phi\{0\} \to (E\xi)\,(\xi = 0 \,\&\, \phi\{\xi\}),$$

in this using negations and we get

$$(A\xi)\,(\xi = 0 \to \phi\{\xi\}) \to \phi\{0\},$$

whence by the cut
$$\frac{(A\xi)\,(\xi = 0 \to \phi\{\xi\})}{\phi\{0\}}\,*.$$

Cor. (i).
$$\frac{\phi\{\alpha\}}{\overline{(A\xi)\,(\xi = \alpha \to \phi\{\xi\})}} \qquad \frac{(A\xi)\,(\xi = \alpha \to \phi\{\xi\})}{\phi\{\alpha\}}\,*$$

*and similarly for the E-form.*

Cor. (ii).
$$\phi\{\alpha\} \leftrightarrow (A\xi)\,(\xi = \alpha \to \phi\{\xi\})$$

*and similarly for the E-form.*

Prop. 17. $\mathbf{A}_I$ *is regular.*

We have to show
$$\frac{\phi \leftrightarrow \psi \quad \chi\{\phi\}}{\chi\{\psi\}}\,*$$

We shall show
$$\frac{\phi \to \psi \quad \psi \to \phi}{\chi\{\phi\} \to \chi\{\psi\}}\,*$$

the result will then follow by the cut which can be eliminated.

We use induction on the construction of $\chi\{\Gamma_o\}$. We follow Prop. 5, Ch. 3 and note that it gives

if $\phi\{\nu\} \leftrightarrow \psi\{\nu\}$ for all $\nu$, **E**-correct

then $\chi\{\nu, \phi\{\nu\}\} \leftrightarrow \chi\{\nu, \psi\{\nu\}\}$ for all $\nu$, **E**-correct.

From this the case of a generalized statement is dealt with.

PROP. 18. *An existential* **A**-*statement can be an* $\mathbf{A}_I$-*theorem without an example being available.*

Consider the $\mathbf{A}_I$-theorem

$$(Ex'')\,(prf_I[x'', \nu] \vee (Ax')\,N[\text{“}prf_I[x', \nu]\text{”}]) \tag{i}$$

where $prf_I$ is the $\mathbf{A}_I$-proof-predicate and $(Ax')\,N[\text{“}prf_I[x', \nu]\text{”}]$ is the $\mathbf{A}_I$-irresolvable statement of Prop. 8. The $\mathbf{A}_I$-statement (i) is an $\mathbf{A}_I$-theorem from T.N.D. and (7) of Prop. 6, Ch. 3, this holds in $\mathbf{A}_I$ by Prop. 10.

Suppose

'if $(E\xi)\,\phi\{\xi\}$ is an $\mathbf{A}_I$-theorem then an example is available'

then we shall have

$$prf_I[\alpha, \nu] \vee (Ax')\,N[\text{“}prf_I[x', \nu]\text{”}] \tag{ii}$$

will be an $\mathbf{A}_I$-theorem for some numerical term $\alpha$. Now $prf_I[\xi, \eta]$ is of the form $(Ex'')\,(\rho[x'', \xi, \eta] = 0)$ where $\rho$ is a primitive recursive characteristic function. Using Prop. 6, Ch. 3 and Prop. 10 we see that:

$$(Ex'')\,((\rho[x'', \alpha, \nu] = 0) \vee (Ax')\,N[\text{“}prf_I[x', \nu]\text{”}])$$

is an $\mathbf{A}_I$-theorem. Hence by our hypothesis there will be a closed numerical term $\beta$ such that:

$$(\rho[\beta, \alpha, \nu] = 0) \vee (Ax')\,N[\text{“}prf_I[x', \nu]\text{”}] \tag{iii}$$

is an $\mathbf{A}_I$-theorem. Now $\alpha$ and $\beta$ are closed numerical terms and $\rho$ is a primitive recursive function whose value is either 0 or 1 also $(Ex'')\,(\rho[x'', \alpha, \nu] = 0)$ is $\mathbf{A}_0$-false by our choice of $\nu$, hence $\rho[\beta, \alpha, \nu] = 1$ is $\mathbf{A}_{00}$-provable. Thus from (iii) and R 1 we obtain

$$1 = 0 \vee (Ax')\,N[\text{“}prf_I[x', \nu]\text{”}]$$

is an $\mathbf{A}_I$-theorem. Now $1 \neq 0$ is an $\mathbf{A}_I$-theorem, whence by the cut which can be eliminated or by the process of eliminating false ancestors we see that $(Ax')\,N[\text{“}prf_I[x', \nu]\text{”}]$ is an $\mathbf{A}_I$-theorem. But this is our irresolvable statement. Thus the $\mathbf{A}_I$-statement (i) is without an example, that is to say, (ii) is **A**-true, but fails to be an $\mathbf{A}_I$-theorem for any numerical term $\alpha$.

## 10.17 *The consistency of* $\mathbf{A}_I$

We have come across several methods for showing that an **A**-statement of the form $(A\xi)\,\phi\{\xi\}$ is **A**-true. In one of the methods $(A\xi)\,\phi\{\xi\}$ turned out to be an $\mathbf{A}_I$-theorem and in another method it failed to be an $\mathbf{A}_I$-theorem, although **A**-true. These various methods together with one which will occur later are as follows: Suppose that $\phi\{\nu\}$ is an $\mathbf{A}_0$-statement,

(i) We are able to give a displayed proof-scheme for $\phi\{\nu\}$, so that we can write out the full $\mathbf{A}_0$-proofs $\phi\{0\}, \phi\{1\}, \dots$ as far as we like. This proof-scheme is in the form of proof by cases and deductions from hypotheses and is such that the proof of $\phi\{\nu\}$ is on a primitive recursive plan which we can show to be **E**-correct. Examples are T.N.D. laws of algebra, etc.

(ii) We are unable to give a displayed proof-scheme as in case (i) because the demonstration that $\phi\{\nu\}$ is $\mathbf{A}_0$-true depended on the consistency of **A** or $\mathbf{A}_I$ and these fail to be $\mathbf{A}_I$-theorems. An example is Gödel's irresolvable statement.

(iii) We are unable to give a displayed proof-scheme for $\phi\{\nu\}$ as in case (i). But we can, using abbreviations, describe a proof-scheme for $\phi\{\nu\}$ in **A** plus M.P., the abbreviations arise because the number of quantifications in the cut formulae are unbounded as $\nu$ increases, thus violating one of the conditions for being an $\mathbf{A}_I$-proof, namely that the number of inductions in the proof of $\phi\{\nu\}$ be bounded as $\nu$ increases. We shall come across a case of this in the next chapter.

(iv) We are unable to give a displayed proof-scheme for $\phi\{\nu\}$ as in case (i), but we can describe a general method for deciding whether $\phi\{\nu\}$ is $\mathbf{A}_0$-true or is $\mathbf{A}_0$-false, but we are without any means of telling if $\phi\{\nu\}$ is always $\mathbf{A}_0$-true. An example is $\rho[\nu] = 0$ where $\rho$ is a primitive recursive function. We have the standard proof of $\rho[\nu] = \kappa$, where $\kappa$ is the standard determination of $\rho[\nu]$, but we lack a method to show that $\kappa$ is always zero. A particular case is Goldbach's conjecture.

Similar remarks hold when $\phi\{\nu\}$ is any **A**-statement. We have come across several examples of case (i), and in most of those cases it was a complicated process to show that the proof-scheme was **E**-correct. Case (iii) will be dealt with in Ch. 11, where we shall be able to reduce the case to a proof with a bounded number of levels by using variables of higher type. Case (iv) occurred in Ch. 5 in our discussion of certain unsolvable problems in primitive recursive theory. We now turn to case (ii) and show:

PROP. 19. *If* $\mathbf{A}_I$ *is consistent then the consistency of* $\mathbf{A}_I$ *fails to be* $\mathbf{A}_I$*-provable.*

More precisely we shall demonstrate:

If $Con_{i1}\,\mathbf{A}_I$ is $\mathbf{A}$-true then it fails to be an $\mathbf{A}_I$-theorem.

We do this by showing:

If $Con_{i1}\,\mathbf{A}_I$ is $\mathbf{A}$-true and if it is an $\mathbf{A}_I$-theorem then we could $\mathbf{A}_I$-prove the $\mathbf{A}$-statement which we had shown was $\mathbf{A}_I$-irresolvable.

We do this by a deep analysis of Prop. 8. This began:

(i) Consider $(Ax')\,N[\text{``}(Ex'')\,P[x', s[x,x], x'']\text{''}]$ and let its *g.n.* be $\kappa$. Let $\nu$ be the *g.n.* of $(Ax')\,N[\text{``}(Ex'')\,P[x', s[\kappa,\kappa], x'']\text{''}]$, then $\nu = s[\kappa,\kappa]$ is an $\mathbf{A}_{00}$-theorem.

The demonstration went on:

(ii) If $(Ax')\,N[\text{``}(Ex'')\,P[x',\nu,x'']\text{''}]$ is an $\mathbf{A}_I$-theorem then by R 1 so is

$$(Ax')\,N[\text{``}(Ex'')\,P[x', s[\kappa,\kappa], x'']\text{''}].$$

The converse also holds. We can express the converse of (ii) in $\mathbf{A}_I$

$$(Ex')\,(Ex'')\,P[x',\nu,x''] \rightarrow (Ex')\,(Ex'')\,P[x',\lambda,x''], \qquad \text{I}$$

where $\lambda$ is the *g.n.* of

$$(Ax')\,N[\text{``}(Ex'')\,P[x',\nu,x'']\text{''}].$$

The demonstration continues:

(iii) If $(Ax')\,N[\text{``}(Ex'')\,P[x', s[\kappa,\kappa], x'']\text{''}]$ is an $\mathbf{A}_I$-theorem then by the properties of the proof-predicate, $(Ex'')\,P[\pi,\nu,x'']$, for some numeral $\pi$ is an $\mathbf{A}_0$-theorem.

We can express (iii) in $\mathbf{A}_I$ thus:

$$(Ex'')\,P[\pi,\nu,x''] \rightarrow (Ex')\,(Ex'')\,P[x', s[\theta,\pi], x''], \qquad \text{II}$$

where $\theta$ is the *g.n.* of $(Ex'')\,P[x,\nu,x'']$, more fully we should write $\theta[Num\,\nu]$. II says that if $(Ex'')\,P[\pi,\nu,x'']$ is $\mathbf{A}_0$-true then there is an $\mathbf{A}_I$-proof of it.

From II we get by II*d* and R 3, if II holds for all $\pi$, $\mathbf{E}$-correct:

$$(Ex')\,(Ex'')\,P[x',\nu,x''] \rightarrow (Ex')\,(Ex'')\,(Ex''')\,P[x', s[\theta, x'''], x'']. \qquad \text{II}'$$

The demonstration continues: if $(Ax')\,N[\text{``}(Ex'')\,P[x',\nu,x'']\text{''}]$ is an $\mathbf{A}_I$-theorem, then by the reversibility of R 3

$$N[\text{``}(Ex'')\,P[\pi,\nu,x'']\text{''}] \quad \text{for all } \pi, \ \mathbf{E}\text{-correct.}$$

Again we can express this in $\mathbf{A}_I$ thus:

$$(Ex')\,(Ex'')\,P[x',\lambda,x''] \rightarrow (Ax''')\,(Ex')\,(Ex'')\,P[x',Neg[s[\theta,x''']],x'']. \quad \text{III}$$

III says that if $(Ax')\,N[\text{``}(Ex'')\,P[x',\nu,x'']\text{''}]$ is an $\mathbf{A}_I$-theorem then so is $N[\text{``}(Ex'')\,P[\pi,\nu,x'']\text{''}]$ for each numeral $\pi$.

From I and III we get, using M.P. (which can be eliminated)

$$(Ex')\,(Ex'')\,P[x',\nu,x''] \rightarrow (Ax''')\,(Ex')\,(Ex'')\,P[x',Neg[s[\theta,x''']],x''],$$

whence from II′ and II$b'$

$$\begin{aligned}(Ex')\,(Ex'')\,P[x',\nu,x''] &\rightarrow (Ax''')\,(Ex')\,(Ex'')\,P[x',Neg[s[\theta,x''']],x'']\\ &\quad \& \ (Ex')\,(Ex'')\,(Ex''')\,P[x',s[\theta,x''],x''] \rightarrow N[\text{``}Con_{11}\mathbf{A}_I\text{''}].\end{aligned}$$

Thus if $Con_{11}\mathbf{A}_I$ were an $\mathbf{A}_I$-theorem then by M.P. (which can be eliminated) so is

$$(Ax')\,N[\text{``}(Ex'')\,P[x',\nu,x'']\text{''}],$$

but this is our $\mathbf{A}_I$-irresolvable statement. Hence if I, II, III are $\mathbf{A}_I$-theorems then $Con_{11}\mathbf{A}_I$ fails to be an $\mathbf{A}_I$-theorem.

It remains to show that I, II, III are $\mathbf{A}_I$-theorems, and that II holds for all $\nu$, E-correct. Or it remains to show I, II′ and III are $\mathbf{A}_I$-theorems.

I. Let $ded_I[\lambda,\mu,\nu]$ be the deduction predicate for $\mathbf{A}_I$. Then $ded_I[\lambda,\mu,\nu]$ is an $\mathbf{A}_0$-theorem if and only if $\mu$ is the *g.n.* of an $\mathbf{A}_I$-deduction of the closed A-statement whose *g.n.* is $\nu$ from the sequence of hypotheses whose *g.n.* is $\lambda$. We have

$$\begin{array}{ll}
N[\text{``}prf_I[\pi,\mu]\text{''}] \vee prf_I[\pi,\mu] \quad ded_I[\mu,\pi',\mu'] & \text{T.N.D.}\\
\hline\hline
 & \text{II}a;\ \text{II}b'\\
N[\text{``}prf_I[\pi,\mu]\text{''}] \vee (prf_I[\pi,\mu]\ \&\ ded_I[\mu,\pi',\mu']) & \\
\hline
 & \text{II}d\\
N[\text{``}prf_I[\pi,\mu]\text{''}] \vee prf_I[\pi^\frown\pi',\mu'] & \\
\hline
 & \text{II}d\\
N[\text{``}prf_I[\pi,\mu]\text{''}] \vee (Ex')\,prf_I[x',\mu'] & \\
\hline
 & \text{R 3}\\
(Ex')\,prf_I[x',\mu] \rightarrow (Ex')\,prf_I[x',\mu']. &
\end{array}$$

The application of R 3 is correct because the proof-scheme is displayed and we can use Prop. 6 and Prop. 9. Now take $\mu = \nu$ and $\mu' = \lambda$ then from (ii) $\pi'$ is easily found and $ded_I[\mu,\pi',\mu']$ is an $\mathbf{A}_0$-theorem. Thus I is an $\mathbf{A}_I$-theorem.

III. We have:

$$\cfrac{\cfrac{N[``\phi\{\kappa\}"] \vee \phi\{\kappa\}}{(E\xi)\, N[``\phi\{\xi\}"] \vee \phi\{\kappa\}}\ \text{II}d \quad \text{T.N.D.}}{(A\xi)\,\phi\{\xi\} \rightarrow \phi\{\kappa\}.}$$

If $\mu$ is the $g.n.$ of $(A\xi)\,\phi\{\xi\}$ and $\theta$ is the $g.n.$ of $\phi\{x\}$ then as in the demonstration of I

$$\frac{(Ex')\,prf_I[x', \mu] \rightarrow (Ex')\,prf_I[x', s[\theta, \kappa]]}{(Ex')\,prf_I[x', \mu] \rightarrow (Ax'')\,(Ex')\,prf_I[x', s[\theta, x'']]}\ \text{R}\,3,$$

the application of R 3 is correct because the proof-scheme has been displayed from what is given here and under I, so we can use Prop. 6 and Prop. 9. Taking $(Ax')\,N[``prf_I[x', \nu]"]$, whose $g.n.$ is $\lambda$, for $(A\xi)\,\phi\{\xi\}$, we see that III is an $\mathbf{A}_I$-theorem.

II. This is more troublesome. $prf_I[\pi, \nu]$ is of the form

$$(Ex'')\,(f[\pi, \nu, x''] = 0),$$

where $f$ is a primitive recursive function. II is

$$(Ex'')\,(f[\pi, \nu, x''] = 0) \rightarrow (Ex')\,(Ex'')\,P[x', s[\theta, \pi], x''], \qquad \text{(iv)}$$

where $\theta$ is the $g.n.$ of $\quad (Ex'')\,(f[x, \nu, x''] = 0).$

We shall show that (iv) is an $\mathbf{A}_I$-theorem for all $\pi, \nu$, **E**-correct, for each primitive recursive function $f$. We proceed by induction on the construction sequence for $f$.

We first obtain:

$$f[\pi, \nu, \mu] = 0 \rightarrow (Ex')\,prf_I[x', s[\theta', \mu]], \qquad \text{II}_1$$

where $\theta'$ is the $g.n.$ of $f[\pi, \nu, x] = 0$, for all $\pi$, $\nu$, $\mu$, **E**-correct, more fully $\theta'$ should be $\theta'[Num\,\pi, Num\,\nu]$. We do this by induction on the construction sequence for $f$. We have

$$f[\pi, \nu, \mu] = 0 \rightarrow (Ex)\,(f[\pi, \nu, x] = 0),$$

whence as in I

$$(Ex')\,prf_I[x', s[\theta', \mu]] \rightarrow (Ex')\,prf_I[x', \kappa] \quad \text{is an } \mathbf{A}_I\text{-theorem}, \qquad \text{(v)}$$

where $\kappa$ is the $g.n.$ of $(Ex)\,(f[\pi, \nu, x] = 0)$.

From $\text{II}_1$ and (v) we get by M.P. (which can be eliminated)

$$f[\pi, \nu, \mu] = 0 \rightarrow (Ex')\,prf_I[x', \kappa], \qquad \text{(vi)}$$

but (vi) holds for all $\mu$, **E**-correct, and $\kappa = s[\theta, \pi]$, so we can apply R 3 and obtain

$$(Ex)(f[\pi, \nu, x] = 0) \rightarrow (Ex')\, prf_I[x', s[\theta, \pi]],$$

but this is II as desired. It remains to give an $\mathbf{A}_I$-proof of $\mathrm{II}_1$, for all $\nu$, $\pi$, $\mu$, **E**-correct.

We shall suppose that $f[\pi, \nu, \mu]$ is without parts of the form $((\lambda\xi.\alpha\{\xi\})\beta)$, such a term will be called a normal term. A normal term is built up from zero by applications of the successor function $S$ and formation of $\mathscr{I}\rho\alpha\beta$, where $\rho$ is of the form $(\boldsymbol{\lambda}\xi\eta, \gamma\{\xi, \eta\})$ and $\alpha, \beta, \gamma\{\xi, \eta\}$ are normal terms, also $f[\pi, \nu, \mu]$ is closed.

Write

$$P(\text{‘}f[\mu, \pi] = 0\text{’}) \quad \text{for} \quad (Ex')\, prf_I[x', s[\theta, \mu]],$$

where $\theta$ is the *g.n.* of $f[x, \pi] = 0$. The inverted commas refer to the fact that $P(\text{‘}f[\mu, \pi] = 0\text{’})$ fails to contain $f[\mu, \pi] = 0$ but only its *g.n.* Hence if $f[\mu, \pi] = g[\mu, \pi]$ has been proved and if we have proved $P(\text{‘}f[\mu, \pi] = 0\text{’})$ then we are unable to obtain $P(\text{‘}g[\mu, \nu] = 0\text{’})$. Write

$$F(\text{‘}\phi\text{’}) \quad \text{for} \quad \phi \rightarrow P(\text{‘}\phi\text{’}).$$

We now give displayed $\mathbf{A}_I$-proofs of the following for all $\mu, \pi, \lambda$, **E**-correct.

(vii) $F(\text{‘}0 = \lambda\text{’})$,

(viii) if $F(\text{‘}f[\mu, \pi] = \lambda\text{’})$ then $F(\text{‘}Sf[\mu, \pi] = \lambda\text{’})$,

(ix) if $F(\text{‘}f[\mu', \pi] = \lambda\text{’})$ and $F(\text{‘}g[\mu, \pi] = \mu''\text{’})$ then $F(\text{‘}f[g[\mu, \pi], \pi] = \lambda\text{’})$,

(x) if $F(\text{‘}f[\pi] = \lambda\text{’})$ and $F(\text{‘}h[\mu, \kappa, \pi] = \lambda\text{’})$ then $F(\text{‘}k[\mu, \pi] = \lambda\text{’})$,

where $f$ and $h$ are primitive recursive functions and

$$k[0, \pi] = f[\pi], \quad k[S\mu, \pi] = h[\mu, k[\mu, \pi], \pi].$$

Case (x) occurs when $k[\mu, \pi] = \mathscr{I}hf[\pi]\mu, h$ for $(\boldsymbol{\lambda}\xi\eta\,.\,h[\xi, \eta])$.

(vii) $F(\text{‘}0 = \lambda\text{’})$ is $0 = \lambda \rightarrow (Ex')\, prf_I[x', \theta]$, where $\theta$ is the *g.n.* of $0 = \lambda$. Clearly $prf_I[\langle\theta\rangle, \theta]$ if $\lambda$ is zero, and the result follows by II*d*. If $\lambda$ is different from zero then the result follows from the axiom $0 \neq \lambda$ by dilution.

(viii) Write $\alpha$ for $f[\mu, \pi]$, we have $\alpha = \lambda \rightarrow P(\text{‘}\alpha = \lambda\text{’})$. Now

$$\alpha = \lambda \rightarrow S\alpha = S\lambda \quad \text{from R 2 and T.N.D.,}$$

hence

$$P(\text{‘}\alpha = \lambda\text{’}) \rightarrow P(\text{‘}S\alpha = S\lambda\text{’}),$$

and so

$$\alpha = \lambda \rightarrow P(\text{‘}S\alpha = S\lambda\text{’}). \tag{xi}$$

Also

$$\alpha = \lambda \,\&\, S\alpha = \kappa \rightarrow \kappa = S\lambda,$$

now $\kappa = S\lambda \to P(`\kappa = S\lambda')$ is of the form $\kappa = S\lambda \to P_i[\kappa, S\lambda]$, for some $\mathbf{A}_0$-predicate $P_i$, this is $\mathbf{A}_I$-provable from

$$(Ax)(x = S\lambda \to P_i[x, S\lambda]),$$

which is equivalent to $P_i[S\lambda, S\lambda]$, which is $P(`S\lambda = S\lambda')$, and this clearly is an $\mathbf{A}_I$-theorem. Hence

$$\kappa = S\lambda \to P(`\kappa = S\lambda').$$

Thus $$\alpha = \lambda \to .\, S\alpha = \kappa \to P(`\kappa = S\lambda'), \quad \text{(xii)}$$

but from $$\kappa = S\lambda \to .\, S\alpha = S\lambda \to S\alpha = \kappa$$

we get $$P(`\kappa = S\lambda') \to .\, P(`S\alpha = S\lambda') \to P(`S\alpha = \kappa'). \quad \text{(xiii)}$$

From (xi), (xii) and (xiii) we get by $\mathscr{P}_C$

$$\alpha = \lambda \to .\, S\alpha = \kappa \to P(`S\alpha = \kappa').$$

To this we may apply R 3 on $\lambda$ since the proof has been displayed and we can use Prop. 7, hence

$$(Ex)(\alpha = x) \to .\, S\alpha = \kappa \to P(`S\alpha = \kappa'),$$

whence by the cut (which can be eliminated) and using the $\mathbf{A}_I$-theorem $(Ex)(\alpha = x)$,

$$S\alpha = \kappa \to P(`S\alpha = \kappa') \quad \text{(xiv)}$$

as desired.

(ix) We have $$f[\mu', \pi] = \lambda \to P(`f[\mu', \pi] = \lambda')$$

and $$g[\mu, \pi] = \mu' \to P(`g[\mu, \pi] = \mu''). \quad \text{(xiv}')$$

Now $$g[\mu, \pi] = \mu' \to .\, f[g[\mu, \pi], \pi] = \lambda \to f[\mu', \pi] = \lambda$$

whence $$g[\mu, \pi] = \mu' \to .\, f[g[\mu, \pi], \pi] = \lambda \to P(`f[\mu', \pi] = \lambda'). \quad \text{(xv)}$$

Again from $$g[\mu, \pi] = \mu' \to .\, f[\mu', \pi] = \lambda \to f[g[\mu, \pi], \pi] = \lambda$$
we get

$$P(`g[\mu, \pi] = \mu'') \to .\, P(`f[\mu', \pi] = \lambda') \to P(`f[g[\mu, \pi], \pi] = \lambda').$$

From (xiv′) this gives

$$g[\mu, \pi] = \mu' \to .\, P(`f[\mu', \pi] = \lambda') \to P(`f[g[\mu, \pi], \pi] = \lambda') \quad \text{(xvi)}$$

from (xv) and (xvi)

$$g[\mu, \pi] = \mu' \to .\, f[g[\mu, \pi], \pi] = \lambda \to P(`f[g[\mu, \pi], \pi] = \lambda').$$

By R 3, which is applicable since the proof-scheme has been displayed,

$$(Ex)\,(g[\mu,\pi] = x) \rightarrow .\, f[g[\mu,\pi],\pi] = \lambda \rightarrow P(\text{`}f[g[\mu,\pi],\pi] = \lambda\text{'})$$

whence the result follows using the cut (which can be eliminated).

(x) We have

$$f[\pi] = \kappa \rightarrow P(\text{`}f[\pi] = \kappa\text{'}) \tag{xvii}$$

and

$$h[\mu,\kappa,\pi] = \lambda \rightarrow P(\text{`}h[\mu,\kappa,\pi] = \lambda\text{'}).$$

Now

$$k[\mu,\pi] = \lambda \rightarrow P(\text{`}k[\mu,\pi] = \lambda\text{'})$$

is of the form

$$k[\mu,\pi] = \lambda \rightarrow P_{ii}[\mu,\pi,\lambda],$$

where $P_{ii}$ is an $\mathbf{A}_0$-predicate. We require to $\mathbf{A}_I$-prove

$$k[0,\pi] = \kappa \rightarrow P_{ii}[0,\pi,\kappa]$$

and

$$\frac{k[\mu,\pi] = \kappa \rightarrow P_{ii}[\mu,\pi,\kappa]}{k[S\mu,\pi] = \kappa \rightarrow P_{ii}[S\mu,\pi,\kappa]}.$$

From

$$k[0,\pi] = \kappa \quad \text{and} \quad k[0,\pi] = f[\pi]$$

we get

$$f[\pi] = \kappa \rightarrow k[0,\pi] = \kappa, \tag{xviii}$$

$$k[0,\pi] = \kappa \rightarrow f[\pi] = \kappa. \tag{xix}$$

Hence, from (xvii) and (xix)

$$k[0,\pi] = \kappa \rightarrow P(\text{`}f[\pi] = \kappa\text{'}). \tag{xx}$$

From (xviii)

$$P(\text{`}f[\pi] = \kappa\text{'}) \rightarrow P(\text{`}k[0,\pi] = \kappa\text{'}). \tag{xxi}$$

From (xx), (xxi) and the cut (which can be eliminated)

$$k[0,\pi] = \kappa \rightarrow P(\text{`}k[0,\pi] = \kappa\text{'}).$$

This is the first of our results.

For the second result we have

$$k[S\mu,\pi] = h[\mu, k[\mu,\pi],\pi]$$

whence

$$k[\mu,\pi] = \kappa \rightarrow .\, k[S\mu,\pi] = \lambda \rightarrow h[\mu,\kappa,\pi] = \lambda \tag{xxii}$$

and

$$k[\mu,\pi] = \kappa \rightarrow .\, h[\mu,\kappa,\pi] = \lambda \rightarrow k[S\mu,\pi] = \lambda. \tag{xxiii}$$

We have

$$h[\mu,\kappa,\pi] = \lambda \rightarrow P(\text{`}h[\mu,\kappa,\pi] = \lambda\text{'})$$

from (xxii) this gives

$$k[\mu,\pi] = \kappa \rightarrow .\, k[S\mu,\pi] = \lambda \rightarrow P(\text{`}h[\mu,\kappa,\pi] = \lambda\text{'}). \tag{xxiv}$$

But from (xxiii)

$$P(\text{'}k[\mu,\pi]=\kappa\text{'})\rightarrow.P(\text{'}h[\mu,\kappa,\pi]=\lambda\text{'})\rightarrow P(\text{'}k[S\mu,\pi]=\lambda\text{'}). \quad \text{(xxv)}$$

From (xxiv), (xxv), the cut (which can be eliminated) and $\mathscr{F}_C$ we get

$$k[\mu,\pi]=\kappa\rightarrow:k[\mu,\pi]=\kappa\rightarrow P(\text{'}k[\mu,\pi]=\kappa\text{'}).\rightarrow.k[S\mu,\pi]=\lambda \rightarrow P(\text{'}k[S\mu,\pi]=\lambda\text{'})$$

to this we can apply II$d$ and get

$$k[\mu,\pi]=\kappa\rightarrow:(Ax)(k[\mu,\pi]=x\rightarrow P_{ii}[\mu,\pi,x])\rightarrow.k[S\mu,\pi]=\lambda \rightarrow P(\text{'}k[S\mu,\pi]=\lambda\text{'}),$$

the $\mathbf{A}_I$-proof of this has been displayed so by Prop. 7 we may apply R 3 to $\kappa$ and obtain

$$(Ex)(k[\mu,\pi]=x)\rightarrow:(Ax)(k[\mu,\pi]=x\rightarrow P_{ii}[\mu,\pi,x])\rightarrow.k[S\mu,\pi]=\lambda \rightarrow P(\text{'}k[S\mu,\pi]=\lambda\text{'})$$

but $(Ex)(k[\mu,\pi]=x)$ is an $\mathbf{A}_I$-theorem hence by the cut and R3 on $\lambda$ we get

$$(Ax)(k[\mu,\pi]=x\rightarrow P_{ii}[\mu,\pi,x])\rightarrow(Ax)(k[S\mu,\pi]=x\rightarrow P_{ii}[S\mu,\pi,x]). \quad \text{(xxvi)}$$

From the first part we get on applying R 3 which is possible since the proof-scheme has been displayed

$$(Ax)(k[0,\pi]=x\rightarrow P_{ii}[0,\pi,x]), \quad \text{(xxvii)}$$

and so from (xxvi) and (xxvii)

$$(Ax)(k[\mu,\pi]=x\rightarrow P_{ii}[\mu,\pi,x]).$$

Lastly by the reversibility of R 3 we have

$$k[\mu,\pi]=\lambda\rightarrow P(\text{'}k[\mu,\pi]=\lambda\text{'}),$$

as desired. Thus $\text{II}_1$ for all $\mu$, $\pi$, $\lambda$, **E**-correct.

This completes the demonstration of Prop. 19.

## Historical remarks to Chapter 10

Induction has been a philosophical and mathematical problem for many centuries. The first explicit statement of an induction axiom (or rather scheme or rule) is in Peano's (1891, 1897) axioms, these are fully discussed in Hilbert–Bernays (1934–6). The object of induction is to allow us to have a proof of $(Ax)\phi\{x\}$ when we can prove each of $\phi\{0\}, \phi\{1\}, \phi\{2\}, \ldots$ it, so to speak, telescopes an infinite number of separate proofs into one overriding or master proof. Our attitude is that our system must be kept formal (otherwise how can one check a proof?) and that proofs must be kept direct (so that $0 = 1$ is unprovable).

A system like the system **E**, an *equation calculus*, has been studied by Goldstein (1957), he has used a symbol like our iterator symbol for much the same purpose. Prop. 1 is based on Gödel's (1933, 1934) incompleteness theorem.

The rule of induction we use is often called the rule of infinite induction. It was apparently first proposed by Hilbert (1930) and more recently discussed by Schütte (1951), Lorenzen (1951) and others. We seem to go a bit further in requiring the proofs of $\phi\{\nu\}$ to be on a primitive recursive plan, **E**-correct. The system then becomes a formal system, otherwise we have a system often called a *semi-formal* system. Our system being formal it is possible to construct a proof-predicate for it and so it suffers the disadvantages of such systems in being incomplete, etc.

Prop. 9 states that T.N.D. holds in $\mathbf{A}_I$, so this is the first place apparently where we depart from the requirements of the intuitionists, though our axioms and rules seem innocent enough, but our definition of negation is classical rather than intuitionistic, so this is the place where we depart from the requirements of the intuitionists.

We have kept track of rule I*b* because it is intimately connected with undecidability. If it is indispensible then we have an undecidable system, but if it can be dispensed with then we have a decidable system, because any possible proof of a given statement without using I*b* can only be one of a bounded number of tree-like figures, whereas with I*b* an unbounded number of tree-like figures arise.

Props. 12–19 inclusive contain the characteristic difficulties of dealing with the system $\mathbf{A}_I$, these arise entirely from the requirements that the proofs of $\phi\{\nu\}$ be on a primitive recursive plan **E**-correct.

Prop. 19 is Gödel's (1933, 1934) second incompleteness theorem for $\mathbf{A}_I$, here we lean heavily on the account given by Bernays in *Hilbert–Bernays* (1934–6). Another account is given by Rogers–Shoenfield (1967).

It is fairly easy to see that if a universal **A**-statement is provable by ordinary Mathematical Induction then it is $\mathbf{A}_I$-provable, hence the system $\mathbf{A}_I$ is at least as strong as Peano Arithmetic. (Clearly all other Peano axioms and rules are $\mathbf{A}_I$-derivable.)

The advantage of direct proofs is that it is so easy to see that $0 = 1$ is impossible to prove so that the system is consistent in one sense.

## Examples 10

1. Prove the **E**-theorems:
   (i) $(x+x'') \dot{-} (x'+x'') = x \dot{-} x'$,
   (ii) $(x \dot{-} x') \dot{-} x'' = x \dot{-} (x'+x'')$,
   (iii) $(x \times x') \dot{-} (x \times x'') = x \times (x' \dot{-} x'')$,
   (iv) $Sx' \dot{-} x = \sum_{0 \leqslant x'' \leqslant x'} B_1[x \dot{-} x'']$,
   (v) $A_1[Sx' \dot{-} x] = B_1[x \dot{-} x']$,
   (vi) $(x \dot{-} x') \times A_1[x' \dot{-} x] = 0$,
   (vii) $A_1[x \dot{-} x'] \times A_1[x' \dot{-} x] = 0$,
   (viii) $x \times B_1[x' \dot{-} x] + x' \times A_1[x' \dot{-} x] = x' \times B_1[x \dot{-} x'] + x \times A_1[x \dot{-} x']$.
2. Define order in the system **E** as follows:

$$\alpha \leqslant \beta \quad \text{for} \quad \alpha + (\beta \dot{-} \alpha) = \beta, \qquad \alpha < \beta \quad \text{for} \quad S\alpha \leqslant \beta.$$

Prove the **E**-theorems or derived rules:
   (i) $x \leqslant x$,
   (ii) $x \dot{-} x' \leqslant x$,
   (iii) $\dfrac{\alpha \leqslant 0}{\alpha = 0}$,
   (iv) $\dfrac{\alpha < \beta \quad \beta < \gamma}{\alpha < \gamma}$,
   (v) $\dfrac{\alpha \leqslant \beta \quad \beta \leqslant \alpha}{\alpha = \beta}$,
   (vi) $0 \leqslant x$,
   (vii) $\dfrac{\alpha < \beta}{\alpha \times (S\gamma) < \beta \times (S\gamma)}$ and $\dfrac{\alpha \times (S\gamma) < \beta \times (S\gamma)}{\alpha < \beta}$,
   (viii) $A_2[x+x'', x'+x'''] \leqslant A_2[x, x'] + A_2[x'', x''']$.
3. Prove the following **E**-theorems or derived rules:
   (i) $x \times B_2[x, x'] = x' \times B_2[x, x']$,
   (ii) $\dfrac{\rho 0 = \sigma 0 \quad \rho(Sx) = \sigma(Sx)}{\rho x = \sigma x}$,

(iii) $x < 2^x$,

(iv) $\dfrac{\rho x < \sigma x}{\sum_{0 \leqslant x'' \leqslant x'} \rho x'' < \sum_{0 \leqslant x'' \leqslant x'} \sigma x''}$,

(v) $\dfrac{\rho[x, x'] = \rho[x, x' \dot{-} 1] + \rho[x \dot{-} 1, x']}{\rho[x, x'] = 0}$,

(vi) $\dfrac{\rho x < \rho(Sx)}{x \leqslant \rho x}$,

(vii) $Max[x, x'] \dot{-} x'' = Max[x \dot{-} x'', x' \dot{-} x'']$,

(viii) $A_2[2 \times x, 2 \times x' + 1] = 1$.

4. Call an **A**-statement decidable if either it or its negation is an $\mathbf{A}_I$-theorem. Show that if $\phi$ is decidable then we have a method for deciding its truth value.

5. Let $\phi$ be decidable. Investigate $F$('$\phi$').

6. Prove the **E**-theorem

$$x \dot{-} (x \dot{-} x') = x' \dot{-} (x' \dot{-} x).$$

7. Give the demonstration of Prop. 16 without using the cut.

## Chapter 11

# Extensions of the system $\mathbf{A}_I$

### 11.1 *The system* $\mathbf{A}'$

We have seen that the system $\mathbf{A}_I$ is incomplete and that any extension of it which remains a formal system will also be incomplete. We could add, as extra axioms, some $\mathbf{A}$-true but $\mathbf{A}_I$-unprovable statements so as to obtain more $\mathbf{A}$-true statements as theorems in the resulting extended system. But as long as we have a formal system it will still be incomplete and an irresolvable statement can be constructed on the same lines as before.

We can do this programme in a systematic manner as follows: We have an effective method for constructing an irresolvable $\mathscr{L}$-true $\mathscr{L}$-statement in a formal system $\mathscr{L}$ which contains recursive number theory and negation. Call this $\mathscr{L}$-statement $G\{\mathscr{L}\}$. We first form $G\{\mathbf{A}_I\}$ and then the system $G'$ which consists of the system $\mathbf{A}_I$ with the extra axiom $G\{\mathbf{A}_I\}$; having formed $G^{(\lambda)}$ we construct $G^{(S\lambda)}$ by adding the extra axiom $G\{G^{(\lambda)}\}$. Having formed the systems $\mathbf{A}_I, G', \ldots, G^{(\lambda)}, \ldots$ we then form the system $G^*$ as the union of all the systems $\mathbf{A}_I$, $G', \ldots$, i.e. the system $\mathbf{A}_I$ with all the extra axioms we added in forming the systems $G', \ldots$; this system again will be formal, hence we can form $G\{G^*\}$ and the system $G^{*\prime}$ which is the system $G^*$ plus the extra axiom $G\{G^*\}$. So we can proceed through the *constructive ordinals*. But we shall have to stop before we come to the end of the constructive ordinals, otherwise we shall cease to have a formal system.

Another way of extending the system $\mathbf{A}_I$ is to add *property variables*, i.e. variables of type $o\iota$. But we shall need care in doing this. A property is a term of type $o\iota$, for instance $\boldsymbol{\lambda}\xi\,.\,\phi\{\xi\}$.

Suppose that we have introduced property variables and quantifiers for them, then we can form the property $\boldsymbol{\lambda}\xi\,.\,(A\Xi)\,\phi\{\Xi, \xi\}$; this defines a property whose definition depends on all properties including the very one defined, such a property is called *impredicative*, and on the face of it seems to be objectionable, it seems to possess an unfortunate circular character. Secondly in order to be able to prove more **A**-true **A**-statements in the resulting system we shall have to have some method for ejecting property variables from a proof. If we continue to keep proofs direct so that anything which enters a proof will remain in that proof till the end, except that duplicates can be removed and equals replaced by equals and a term may get lost by introduction of an existential quantifier, then we shall have to have an extra rule which will enable us to *eject* a property variable. One such rule is Modus Ponens or the cut. In this rule the whole of the cut formula is ejected in one step of the proof. We have shown that the cut can be eliminated from the system $\mathbf{A}_I$, but we shall show that this is no longer possible when property variables are introduced. The cut thus provides a method by which property variables can be ejected from a proof.

We now construct a hierarchy of formal systems, each is a primary extension of its predecessor, and has direct proofs and predicative properties. $\mathbf{A}^\circ$ is $\mathbf{A}_I$ with closed properties $\boldsymbol{\lambda}\xi.\phi\{\xi\}$, $\boldsymbol{\lambda}\xi\xi'.\phi\{\xi,\xi'\}, \ldots$, these can only occur as $(\boldsymbol{\lambda}\xi.\phi\{\xi\})\,\alpha$, and so, can be replaced by $\phi\{\alpha\}$.

*The system* $\mathbf{A}'$

| Symbols | Type | Name |
|---|---|---|
| $0$ | $\iota$ | Zero |
| $S$ | $\iota\iota$ | Successor function |
| $x$ | $\iota$ | Variable for a natural number |
| $\mathscr{I}$ | $\iota\iota(\iota\iota)$ | Iterator operator |
| $=$ | $o\iota\iota$ | Equality predicate |
| $\sim$ | $oo$ | Negation |
| $\vee$ | $ooo$ | Disjunction |
| $A$ | $o(o\iota)$ | Universal quantifier for natural numbers |
| $X_{o\iota}$ | $o\iota$ | One-place property variable |
| $X_{o\iota\iota}, X_{o\iota\iota\iota}, \ldots$ | $o\iota\iota, o\iota\iota\iota, \ldots$ | Two-, three-, ... place property variables |
| $A_{o\iota}$ | $o(o(o\iota))$ | Universal quantifiers for one-place property variables |
| $A_{o\iota\iota}, A_{o\iota\iota\iota}, \ldots,$ | $o(o(o\iota\iota)),$ $o(o(o\iota\iota\iota)),$ ... | Universal quantifier for two- three-, ... place, property variables |
| $\boldsymbol{\lambda}$ | | Abstraction symbol |
| ( | | Left parenthesis |
| ) | | Right parenthesis |
| $'$ | | Generating symbol. |

We use $\Xi$ with or without superscripts or type subscripts for an undetermined property variable of the type shown by the subscript, (if a type subscript is absent then the context should indicate the type or the type is immaterial), we similarly use $\Delta$ for an undetermined property. Atomic statements are:

$$(\alpha = \beta), \quad (\Xi_\iota \alpha), \quad (\Xi_{\iota\iota} \alpha\beta), \quad \ldots, \quad (\Delta_\iota \alpha), \quad \ldots,$$

where $\alpha$, $\beta$ are numerical terms.

Properties are without bound property variables.

D 259 $\quad (\alpha \neq \beta) \quad$ for $\quad \sim(\alpha = \beta)$,

D 260 $\quad (\phi \,\&\, \psi) \quad$ for $\quad \sim(\sim\phi \vee \sim\psi)$,

D 261 $\quad (E\xi)\,\phi\{\xi\} \quad$ for $\quad \sim(A\xi) \sim \phi\{\xi\}$,

$\quad (E\Xi)\,\phi\{\Xi\} \quad$ for $\quad \sim(A\Xi) \sim \phi\{\Xi\}$.

We write $(A\Xi)\,\phi\{\Xi\}$ for $A(\lambda\Xi\,.\,\phi\{\Xi\})$ as before.

*Axioms*

$Ax\,1 \quad (\alpha = \alpha)$,

$Ax\,2.1 \quad S\alpha \neq 0, \qquad Ax\,2.2 \quad 0 \neq S\alpha$,

$Ax\,3.1 \quad \mathscr{I}\rho\alpha 0 = \alpha, \qquad Ax\,3.2 \quad \mathscr{I}\rho\alpha(S\beta) = \rho\beta(\mathscr{I}\rho\alpha\beta)$,

$Ax\,4.1 \quad (\boldsymbol{\lambda}\xi\,.\,\alpha\{\xi\})\,\beta\beta' \ldots \beta^{(\pi)} = \alpha\{\beta\}\beta' \ldots \beta^{(\pi)}$,

$Ax\,4.2 \quad (\boldsymbol{\lambda}\xi\,.\,\alpha\{\xi\})\,\beta\beta' \ldots \beta^{(\pi)} = (\boldsymbol{\lambda}\xi'\,.\,\alpha\{\xi'\})\,\beta\beta' \ldots \beta^{(\pi)}$,

here $\alpha, \alpha\{0\}, \beta, \beta', \ldots, \beta^{(\pi)}$ and $\rho$ are closed. $\xi$, $\xi'$ fail to be free in $\alpha\{\Gamma_\iota\}$.

*Rules*

Remodelling schemes.

$$\text{I}a \quad \frac{\omega' \vee \phi \vee \psi \vee \omega}{\omega' \vee \psi \vee \phi \vee \omega}; \qquad \text{I}b \quad \frac{\phi \vee \phi \vee \omega}{\phi \vee \omega}$$

permutation $\qquad$ cancellation

Building schemes.

$$\text{II}a \quad \frac{\chi}{\phi \vee \chi} \qquad \text{II}b \quad \frac{\sim\phi \vee \omega \quad \sim\psi \vee \omega}{\sim(\phi \vee \psi) \vee \omega} \qquad \text{II}c \quad \frac{\phi \vee \omega}{\sim\sim\phi \vee \omega}$$

dilution, $\phi$ atomic $\qquad$ composition $\qquad$ double negation

$$\text{II}d \quad \frac{\sim\phi\{\alpha\} \vee \omega}{\sim(A\xi)\,\phi\{\xi\} \vee \omega}, \text{ where } \alpha \text{ is free in } \phi\{\alpha\},$$

existential dilution for natural numbers,

$$\text{II}f \quad \frac{\sim\phi\{\Delta\}\vee\omega}{\sim(A\Xi)\,\phi\{\Xi\}\vee\omega}, \text{ where } \Delta \text{ is free in } \phi\{\Delta\},$$

existential dilution for properties.

Arithmetic schemes.

$$\text{R 1} \quad \frac{\phi\{\alpha\}\vee\omega \quad (\alpha=\beta)\vee\omega}{\phi\{\beta\}\vee\omega}, \qquad \text{R 2} \quad \frac{(\alpha\neq\beta)\vee\omega}{(S\alpha\neq S\beta)\vee\omega}$$

substitution — progression

$$\text{R 3} \quad \frac{\phi\{0\}\vee\omega \quad \phi\{1\}\vee\omega,\ldots,\phi\{\nu\}\vee\omega,\ldots}{(A\xi)\,\phi\{\alpha\{\xi\}\}\vee\omega} \quad \text{for all } \nu, \mathbf{E}\text{-correct},$$

induction

$$\text{R 3*} \quad \frac{\phi\{\Delta'\}\vee\omega, \phi\{\Delta''\}\vee\omega,\ldots}{(A\Xi)\,\phi\{\Delta\{\Xi\}\}\vee\omega} \quad \text{for all properties } \Delta', \Delta'', \ldots, \mathbf{E}\text{-correct},$$

property induction (we allow a substitution)

$$\text{R 4} \quad \frac{\phi\{(\boldsymbol{\lambda}\xi.\psi\{\xi\})\,\alpha\}}{\phi\{\psi\{\alpha\}\}}, \qquad \text{R 4}' \quad \frac{\phi\{\psi\{\alpha\}\}}{\phi\{(\boldsymbol{\lambda}\xi.\psi\{\xi\})\,\alpha\}}, \qquad \text{R 4}'' \quad \frac{\phi\{(\boldsymbol{\lambda}\xi.\psi\{\xi\})\,\alpha\}}{\phi\{(\boldsymbol{\lambda}\eta.\psi\{\eta\})\,\alpha\}}.$$

abstraction — alteration of variable

$\alpha$ free in $\psi\{\alpha\}$, $\xi$, $\eta$ fail to occur free in $\psi\{\Gamma_\iota\}$.

In the above $\omega$ is the subsidiary formula and can be absent, $\chi$ is a secondary formula and must be present, the others are main formulae and must be present. Properties are without bound property variables. All formulae in axioms or rules are to be closed.

*Property induction* goes as follows:

D 262 *prop* $\nu$ for $\nu$ is the *g.n.* of a property without bound property variables.

Let $\tau\nu$ be equal to the *g.n.* of an $\mathbf{A}'$-proof of $\phi\{\Delta\}\vee\omega$ when $\nu$ is the *g.n.* of an $\mathbf{A}'$-property $\Delta$, and let $\sigma\nu$ be equal to the *g.n.* of $\phi\{\Delta\}\vee\omega$, then we require $(1 \dot{-} propx)\times prf_{\mathbf{A}'}[\tau x, \sigma x]=0$ to be an $\mathbf{E}$-theorem. In R 3* $\Delta, \Delta', \ldots$ are closed properties which fail to contain any property variables at all. All $\mathbf{A}'$-theorems are closed.

The only practical way of using R 3* is as follows: let $\mathbf{A}'_v$ be the system $\mathbf{A}'$ when we allow free property variables and when we use a rule like II $e'$ for property variables instead of R 3*. Suppose that we can $\mathbf{A}'_v$-prove $\phi\{\Xi\}\vee\omega$, then from this $\mathbf{A}'_v$-proof we can obtain one of $\phi\{\Delta\}\vee\omega$ merely by everywhere replacing $\Xi$ by $\Delta$ and using II $a$ for any formula $\phi$. Let $p$

be the *g.n.* of an $\mathbf{A}'_v$-proof of $\phi\{\Xi\} \vee \omega$, and let $q$ be the *g.n.* of $\phi\{\Xi\} \vee \omega$, then if $\nu$ is the *g.n.* of $\Xi$ and $\kappa$ is the *g.n.* of $\Delta$, then $Subst\,[\kappa, \nu, p]$ is equal to the *g.n.* of an $\mathbf{A}'$-proof of $\phi\{\Delta\} \vee \omega$ and $Subst\,[\kappa, \nu, q]$ is equal to the *g.n.* of $\phi\{\Delta\} \vee \omega$. Thus to obtain R 3* we have to **E**-prove:

$$(1 \dot{-} propx) \times prf_{\mathbf{A}'}[Subst[x, \nu, p],\ Subst\,[x, \nu, q]] = 0. \tag{i}$$

Now $p = \langle p', p'', \ldots, p^{(\pi)} \rangle$ and we have $prf_{\mathbf{A}'}[p, q] = 0$, $p^{(\pi)} = q$. Now $prf_{\mathbf{A}'}[p, q] = 0$ splits up into a set of $\mathbf{A}_{00}$-provable clauses of the types:

$$Ax\,[p^{(\lambda)}] = 0,$$

$$Rl^1[p^{(\lambda')}, p^{(\lambda'')}] = 0,$$

$$Rl^2[p^{(\lambda_1)}, p^{(\lambda_2)}, p^{(\lambda_3)}] = 0,$$

for certain values of $\lambda$, $\lambda'$, etc., which can be found from $prf_{\mathbf{A}'}[p, q]$. So it suffices to **E**-prove a set of statements of the forms:

$$(1 \dot{-} propx) \times Ax\,[Subst[x, \nu, p^{(\lambda)}]] = 0,$$

$$(1 \dot{-} propx)\ \times Rl^1[Subst[x, \nu, p^{(\lambda')}], Subst\,[x, \nu, p^{(\lambda'')}]] = 0,$$

$$(1 \dot{-} propx) \times Rl^2[Subst[x, \nu, p^{(\lambda_1)}], Subst[x, \nu, p^{(\lambda_2)}], Subst[x, \nu, p^{(\lambda_3)}]] = 0$$

for the same values of $\lambda$, $\lambda'$, etc.

In the first of these there is nothing to do because axioms are without property variables in $\mathbf{A}'_v$. Thus $Subst[x, \nu, p^{(\lambda)}] = p^{(\lambda)}$, and since we have $Ax\,[p^{(\lambda)}] = 0$ then we also have $(1 \dot{-} propx) \times Ax\,[Subst[x, \nu, p^{(\lambda)}]] = 0$. Suppose we have

$$Rl^{\mathrm{II}a}[p^{(\lambda')}, p^{(\lambda'')}] = 0,$$

which is the same as $\quad A_2[p^{(\lambda'')}, disj\,[p^*, p^{(\lambda')}]] = 0,$

where $p^*$ can be explicitly defined in terms of $p^{(\lambda'')}$. Then this becomes

$$(1 \dot{-} propx) \times A_2[Subst[x, \nu, p^{(\lambda'')}], disj\,[Subst[x, \nu, p^*], Subst[x, \nu, p^{(\lambda')}]]] = 0$$

which follows from the **E**-theorem $A_2[x, x] = 0$ by substitution and $p^{(\lambda'')} = disj[p^*, p^{(\lambda)}]$. The other cases are dealt with similarly. Hence we can obtain (i). Thus we have a case of R 3*.

An $\mathbf{A}'$-proof, like an $\mathbf{A}_I$-proof, is divided up into a number of levels. Level one consists of a bounded set of single applications of R 3 and R 3* these are called the production of level one. When level $\lambda$ has been defined then level $S\lambda$ consists of a bounded set of single applications of R 3 and R 3* from the production of level $\lambda$ as hypotheses. The last level consists either of a deduction, without using R 3 or R 3*, from the production of the penultimate level as hypotheses, or of a single application of R 3 or R 3* from the production of the penultimate level as hypotheses. The full definitions are exactly as for $\mathbf{A}_I$, except that now we have a rule of generalization for property variables as well as for numerical variables.

## 11.2 *Remarks*

(*a*) We could dispense with many-place properties by using ordered sets. Thus we could use $\Delta\langle\alpha', \ldots, \alpha^{(\pi)}\rangle$ instead of $\Delta\alpha' \ldots \alpha^{(\pi)}$.

(*b*) We prove $(A\xi)\,\phi\{\xi\}$ by proving $\phi\{\nu\}$ for each numeral $\nu$, i.e. for each member of type $\iota$. If we allowed property variables to vary over all properties then we should be unable to define them all. We can enumerate all **A**-definable properties of type $o\iota$ by generating the natural numbers and when the natural number $\nu$ has been generated we test whether it is the *g.n.* of a property of type $o\iota$, if so we write it down in a list. This gives us: $\Delta, \Delta', \ldots$. Now consider the property $(\boldsymbol{\lambda}\xi\,.\sim\Delta^{(\xi)}\xi)$, if this is in our list then it is $\Delta^{(\kappa)}$ for some natural number $\kappa$, we shall then have

$$\Delta^{(\kappa)}\nu \leftrightarrow (\boldsymbol{\lambda}\xi\,.\sim\Delta^{(\xi)}\xi)\,\nu \quad \text{for all numerals } \nu,$$

taking $\kappa$ for $\nu$ and using the $\boldsymbol{\lambda}$-rule we obtain an absurdity.

We avoid this state of affairs by using a *hierarchy of properties*, each member of a hierarchy being definable in our system and those in each hierarchy being enumerable. Of course the properties in all these hierarchies still form an enumerable set, since the union of an enumerated set of enumerated sets is again enumerable, but we hope that we have a sufficient variety of properties to obtain interesting results, anyway we shall have all the properties that can be defined using our symbols.

(*c*) The rule R 4′ for properties corresponds to a class existence axiom. We have in $\mathbf{A}_I$: $(A\xi)\,(\phi\{\xi\}\leftrightarrow\phi\{\xi\})$, where $\xi$ fails to occur free in $\phi\{\Gamma_\iota\}$, whence by R 4′ we obtain: $(A\xi)\,(\phi\{\xi\}\leftrightarrow(\boldsymbol{\lambda}\eta\,.\,\phi\{\eta\})\,\xi)$ whence by II*f*

$$(EX)\,(A\xi)\,(\phi\{\xi\}\leftrightarrow X\xi).$$

This corresponds to the class existence (comprehension) axiom of set theory.

(*d*) We could have extended the system $\mathbf{A}_I$ by adding *class variables* and adding the *class membership symbol* $\epsilon$, or we could have added *variables for functions of natural numbers*, i.e. variables of types $\iota\iota$, $\iota\iota\iota$, ....

But a class may be represented by its characteristic function, and a function may be represented by a class of ordered pairs, such that when the first member is given then the second member is unique. Also the distinction between properties and classes is largely notational, e.g. write $\Delta\alpha$ as $(\alpha\epsilon\Delta)$ or $((\epsilon\,\Delta)\,\alpha)$, where $\epsilon$ is a symbol of type $o\iota(o\iota)$.

(*e*) If we add T.N.D. as an extra axiom and Modus Ponens as an extra rule then we can dispense with *Ax* 1 and rule R 1 if we define

$$(\alpha = \beta) \quad \text{for} \quad (AX)\,(X\alpha \to X\beta).$$

We have $(AX)\,({\sim}\, X\alpha \vee X\alpha)$; for any property $\Delta\,({\sim}\,\Delta\alpha \vee \Delta\alpha)$ is an axiom, since we are having T.N.D., now apply R 3*, (the required **E**-proof is easy), but this is *Ax* 1. For R 1 we have:

$$\dfrac{\text{R 4}\;\dfrac{\phi\{\alpha\}\vee\omega}{(\boldsymbol{\lambda}\xi.\phi\{\xi\})\,\alpha\vee\omega}\qquad \dfrac{\dfrac{(AX)\,(X\alpha\to X\beta)}{(\boldsymbol{\lambda}\xi.\phi\{\xi\}\alpha)\to(\boldsymbol{\lambda}\xi.\phi\{\xi\}\beta)}\,*}{(\boldsymbol{\lambda}\xi.\phi\{\xi\})\,\alpha\vee\omega\,.\!\to\!.\,(\boldsymbol{\lambda}\xi.\phi\{\xi\})\,\beta\vee\omega}}{\dfrac{(\boldsymbol{\lambda}\xi.\phi\{\xi\})\,\beta\vee\omega}{\phi\{\beta\}\vee\omega}\;\text{R 4}}\;\text{cut}$$

We have assumed the reversibility of R 3*!

(*f*) Remark (*e*) raises the question whether we can dispense with any of the other symbols. For instance in the system $\mathbf{A}_{00}$ we could dispense with the propositional connectives since:

$$\begin{array}{lcr}
\alpha = 0 \vee \beta = 0 & \text{may be replaced by} & \alpha\times\beta = 0,\\
\alpha = 0\,\&\,\beta = 0 & ,, & A_1[\alpha+\beta] = 0,\\
\alpha \neq 0 & ,, & 1 \dot{-} \alpha = 0,\\
\alpha = \beta & ,, & A_2[\alpha,\beta] = 0.
\end{array}$$

We have already used these in forming characteristic functions. We could also replace $(A\xi)\,\phi\{\xi\}$ by $Max[\alpha\{\xi\}] = 0$, where $\alpha\{\xi\}$ is the characteristic function of $\phi\{\xi\}$, similarly $(E\xi)\,\phi\{\xi\}$ may be replaced by

$$Min\,[\alpha\{\xi\}] = 0.$$

All our statements would then reduce to the form $\alpha = 0$, so we could omit '$=0$' and just transform numerical terms according to certain rules. This procedure would identify truth with zero and falsity with unity.

We prefer to have some duplication, it seems to facilitate reading, after all each numerical term has many others equal to it, and we should only get each numerical term uniquely represented if we discarded the iterator symbol and the abstraction symbol, so that the only terms of type $\iota$ were the numerals themselves. But then we would have taken the whole interest of the system away. Thus, however we go about it, some duplication is going to occur, in the sense that some concepts are going to be represented in many equivalent ways, it is a matter of taste how much duplication of this sort we allow.

## 11.3 *The hierarchy of systems* $\mathbf{A}^{(\nu)}$

We now define a hierarchy of formal systems starting with the systems $\mathbf{A}^{\circ}$, $\mathbf{A}'$ which have just been constructed. We add new symbols $X_{\kappa\iota}$ of type $(o\iota)$, $X_{\kappa\iota\iota}$ of type $(o\iota\iota)$, etc., called *property variables of order* $\kappa$. The system $\mathbf{A}^{(S\nu)}$ will contain these symbols for $\kappa = 0, 1, 2, \ldots, \nu$ and the only properties allowed in $\mathbf{A}^{(S\nu)}$ will be those in which $X_{\kappa\iota}$, $X_{\kappa\iota\iota}$, etc., occur, free or bound, for $\kappa = 0, 1, 2, \ldots, \nu$. The passage from $\mathbf{A}^{(\nu)}$ to $\mathbf{A}^{(S\nu)}$ then consists in adjoining the property variables $X_{\nu\iota}$ of types $(o\iota)$, $X_{\nu\iota\iota}$ of type $(o\iota\iota)$, etc., the property quantifiers of $\mathbf{A}'$ may be applied to them. The only way a property variable can enter an $\mathbf{A}^{(\nu)}$-proof is via II*a*, *f*, R 3*.

*The order of a statement or a property* in one of these systems is the greatest of the orders of the free property variables which it contains, zero if it contains none, and of the successors of the orders of the bound property variables which it contains, but zero if it contains none. Then the system $\mathbf{A}^{(S\nu)}$ contains properties of order $\nu$ and statements of order $S\nu$. The axioms of $\mathbf{A}^{(\nu)}$ are all closed and the rules preserve closure, hence $\mathbf{A}^{(\nu)}$-theorems are closed. In R 3* the properties run through all closed properties of order $\leqslant \kappa$ while the variable in the lower formula is of order $\kappa$, for $1 \leqslant \kappa \leqslant \nu$. In II*f* the property variable in the lower formula is of the same order as the property displayed in the upper formula. (So in $\mathbf{A}^{(\nu)}$ we must avoid properties with bound property variables of order $\nu$.)

## 11.4 *Properties of the systems* $\mathbf{A}^{(\nu)}$

PROP. 1. *If $\phi$ is a closed* **A**-*statement which is an* $\mathbf{A}^{(S\nu)}$-*theorem, then $\phi$ is an* $\mathbf{A}_I$-*theorem.*

In other words the systems $\mathbf{A}^{(S\nu)}$ are useless for proving **A**-true **A**-statements which are $\mathbf{A}_I$-unprovable.

Suppose that $\phi$ is an A-statement which is an $\mathbf{A}^{(S\nu)}$-theorem. The $\mathbf{A}^{(S\nu)}$-proof of $\phi$ will be entirely free of property variables, because once a property variable enters an $\mathbf{A}^{(S\nu)}$-proof then a property variable remains in that $\mathbf{A}^{(S\nu)}$-proof from that place to the end and so appears in the theorem proved. Property variables can only enter an $\mathbf{A}^{(S\nu)}$-proof at II*a*, *f* or R 3*, and once it has entered then we are without a method for removing it. At applications of II*f* and R* 3 any property variable in $\Delta$ is lost but then there is another to replace it in the lower formula. Again a property variable may disappear at an application of I*b* but then only one duplicate is removed so that a property variable is still left. Thus an $\mathbf{A}^{(S\nu)}$-proof of $\phi$, where $\phi$ is without property variables, is entirely without property variables, except that it may contain zero order properties outside the scope of a quantifier, if this is so then they can be removed by applying R 4, the resulting $\mathbf{A}^{(S\nu)}$-proof is then an $\mathbf{A}_I$-proof of $\phi$, $\phi$ itself is unchanged by these applications of R 4, since being an A-statement it is without occurrences of properties of any order other than zero order properties governed by quantifiers.

PROP. 2. *The systems* $\mathbf{A}^{(\nu)}$ *are consistent in the sense of Con* (i), *undecidable and incomplete.*

By Prop. 1 if we could $\mathbf{A}^{(\nu)}$-prove $0 = 1$ then we could $\mathbf{A}_I$-prove $0 = 1$, and hence we could $\mathbf{A}_{00}$-prove $0 = 1$, which is absurd. If $\mathbf{A}^{(\nu)}$ were decidable then so would be $\mathbf{A}_I$, which is absurd. If $\mathbf{A}^{(\nu)}$ were complete in the sense that for a closed $\mathbf{A}^{(\nu)}$-statement $\phi$ either $\phi$ or $\sim\phi$ is an $\mathbf{A}^{(\nu)}$-theorem, then by Prop. 1 the same would hold for $\mathbf{A}_I$ which again is absurd.

PROP. 3. *Rules* I*a*, I*b*, II*b*, *c*, R 2, 3, 3*, 4, 4′, 4″ *are reversible.*

The results for rules I*a*, *b*, R 2, 4, 4′, 4″ are trivial. The results for II*b*, *c*, R 3, 3* are dealt with as for the similar results for $\mathbf{A}_I$. For II*c* we omit all the ancestors of $\sim\sim$ in $\sim\sim\phi \vee \omega$, and then proceed on the lines of Prop. 12, Ch. 10. The details are left to the reader.

## 11.5 *The system* $\mathbf{A}^{(\nu)*}$

We are unable to show that T.N.D. holds in $\mathbf{A}^{(S\nu)}$, or that Modus Ponens can be eliminated. If we try to use formula induction in $\mathbf{A}^{(S\nu)}$ then when $\phi$ is atomic we have the cases

$$\alpha = \beta \quad \text{and} \quad \Delta_{\nu\iota}\alpha \quad \text{and} \quad \Delta_{\nu\iota\iota}\alpha\beta, \text{ etc.}$$

the latter cases are of the form

$$(\boldsymbol{\lambda}\xi\,.\,\psi\{\xi\})\,\alpha, \quad (\boldsymbol{\lambda}\xi\eta\,.\,\psi\{\xi,\eta\})\,\alpha\beta, \quad \text{etc.}$$

but these, if we apply R 4, can be any $\mathbf{A}^{(\nu)}$-statements. We can get over this difficulty by using formula induction on formulae to which R 4 is inapplicable. But then when we come to formulae of the form $(A\Xi)\,\phi\{\Xi\}$ in our formula induction then we would be referred to formulae of the type $\phi\{\Delta\}$ where $\Delta$ is any property of the same or less order than that of the variable $\Xi$. Now $\phi\{\Delta\}$ must be of the form $\phi\{(\boldsymbol{\lambda}\xi\,.\,\psi\{\xi\})\,\alpha\}$, and this by R 4 becomes $\phi\{\psi\{\alpha\}\}$, where $\psi\{\alpha\}$ is any formula form of the same or less order than the variable $\Xi$. Let $(A\Xi)\,\phi\{\Xi\}$ be closed and of order $S\nu$, then this is so either $(a)$ because $\Xi$ is of order $\nu$ or $(b)$ because $\phi\,\{\Gamma_{\nu\iota}\}$ contains an occurrence of a bound property variable of order $\nu$, or contains a free property variable of order $S\nu$. In case $(a)$, $\phi\{\psi\{\alpha\}\}$ contains one less variable of highest order, and if case $(b)$ fails, it is of lesser order, but in the case when $(b)$ holds and $(a)$ fails, $\phi\{\psi\{\alpha\}\}$ is of the same order and has the same number of variables of highest order, and also contains more logical symbols, so our formula induction falls to pieces. But these considerations are heuristic. We obtain something more rigorous when we show that the consistency of $\mathbf{A}_I$ can be proved in the system consisting of $\mathbf{A}'$ plus T.N.D. and Modus Ponens. The role of Modus Ponens in this proof is to eject property variables.

We take Modus Ponens in the form:

R 5
$$\frac{\omega \vee \phi \quad \sim \phi \vee \chi}{\omega \vee \chi},$$

$\phi$ is called the cut formula, $\omega$ is subsidiary and can be absent, $\chi$ is secondary and must be present. If both were absent we should end up with the null formula.

We take T.N.D. in the general form $\phi \vee \sim \phi$, where $\phi$ is any closed $\mathbf{A}^{(\nu)}$-statement. This is different from the other axioms for they are atomic statements. But, as discussed above, any attempt to deduce the general case from atomic cases seems doomed to failure.

We call the system $\mathbf{A}^{(\nu)}$ plus T.N.D. and Modus Ponens the system $\mathbf{A}^{(\nu)*}$.

PROP. 4. *If $\phi$ is a many-sorted $\mathscr{F}_{2C}$-theorem and if $\psi$ is obtained from $\phi$ by replacing many-sorted atomic predicates by atomic $\mathbf{A}^{(\nu)*}$-predicates with exactly the same number and sort of free variable and if the same atomic*

$\mathbf{A}^{(\nu)*}$ *predicate is used to replace a given atomic* $\mathscr{F}_{2C}$*-predicate at all its occurrences (except for change of free variables), then* $\psi$ *is an* $\mathbf{A}^{(\nu)*}$*-theorem.*

The $\mathscr{F}_{2C}$-proof of $\phi$ can be laid out in levels like an $\mathbf{A}^{(\nu)*}$-proof. The axioms in the $\mathscr{F}_{2C}$-proof are all cases of T.N.D. which translate into cases of T.N.D. in $\mathbf{A}^{(\nu)*}$, and these are $\mathbf{A}^{(\nu)*}$-axioms. All the $\mathscr{F}_{2C}$-rules except II*e* are $\mathbf{A}^{(\nu)*}$-rules. A case of II*e* can be replaced by a case of R 3 or R 3*, according to the sort of variable, as explained in Prop. 10, Ch. 10. The final result is an $\mathbf{A}^{(\nu)*}$-proof.

COR. (i). $\mathbf{A}^{(\nu)*}$*-tautologies are* $\mathbf{A}^{(\nu)*}$*-theorems.*

COR. (ii). $A^{(\nu)*}$ *is regular.*

COR. (iii). *The Deduction Theorem holds in* $\mathbf{A}^{(\nu)*}$.

## 11.6 *The definition of* A*-truth in* A′*

PROP. 5. A*-truth can be defined in the system* A′*.

We have to define an A′-statement $\chi\{\nu\}$, such that $\chi\{\nu\}$ is an A′*-theorem if and only if $\nu$ is the *g.n.* of an A-true A-statement. We defined A-truth by the following scheme (A) in Ch. 8;

$$\left.\begin{array}{l}(Ax)\,(P[x,0]\leftrightarrow Q\{x\})\\ (Ax,x')\,(P[x,Sx']\leftrightarrow R\{x,x',P[r'\{x,x',x''\},x'],\ldots,\\ \qquad\qquad\qquad\qquad\qquad P[r^{(\pi)}\{x,x',x''\},x']\}),\end{array}\right\}\quad(\mathrm{A})$$

where the variable $x''$ is bound by a quantifier in $R$. This scheme defines a property $P$. This scheme resembles the scheme of recursion with substitution in parameter, but differs in having bound variables. This is the cause of the undefinability of A-truth in A itself. We can cast scheme (A) into a scheme for defining the characteristic function of the property $P$, if we do this then the bound variable $x''$ will give rise to an infinite sum or infinite product or to an unlimited least number operation. In either case we have a case of definition by a form of induction quite different from any we have encountered before.

To put the scheme (A) into a scheme for definition of the characteristic function we proceed as in Prop. 3, Ch. 5, except that we add 'replace $(E\xi)\,\phi\{\xi\}$ by $\Pi\alpha\{\xi\}=0$, where $\alpha\{\xi\}$ is the characteristic function of $\phi\{\xi\}$', or we could replace it by $Min\,\alpha\{\xi\}=0$ or by $\alpha\{\mu_\xi[\alpha\{\xi\}=0]\}=0$;

having dealt with $E$ we can deal with $A$. This scheme of definition is more general than that of recursion and fails to be constructive, because we are in general unable to decide the unlimited least number operation.

Note that scheme (A) defines a sequence of properties

$$P_0[x], P_1[x], \ldots, P_\kappa[x], \ldots,$$

but the numeral $\kappa$ fails to occur in $P_\kappa[x]$, it only appears indirectly as the number of quantifiers in $P_\kappa[x]$, so that $P_{x'}[x]$ fails to be **A**-definable (it being impossible for an **A**-formula to have $x'$ quantifiers, cf. the discussion on *Val* in Ch. 5).

Instead of the general scheme (A) consider the simplified scheme

$$\left.\begin{array}{c}(Ax)\,(P[x,0] \leftrightarrow Q\{x\}) \\ (Ax,x')\,(P[x,Sx'] \leftrightarrow R\{x,x',P[r\{x,x',x''\},x']\}),\end{array}\right\} \qquad \text{(B)}$$

where the variable $x''$ is bound in $R$. We can put scheme (B) into the equivalent form:

$$(Ax,x')\,(P[x,x'] \leftrightarrow :x' = 0 \,\&\, Q\{x\}. \vee (Ex'')\,(x' = Sx'' \,\&\, R\{x,x'', P[r\{x,x'',x'''\},x'']\})), \qquad \text{(C)}$$

where $x'''$ is bound in $R$. We then wish to define a property $P$ which satisfies (C), i.e. such that (C) is an **A'***-theorem.

If (C) is an **A'***-theorem, then, by the reversibility of R 3 so is

$$P[\nu,q\nu] \leftrightarrow :q\nu = 0 \,\&\, Q\{\nu\}. \vee q\nu = S\nu' \,\&\, R\{\nu,\nu',P[r\{\nu,\nu',x'''\},\nu']\},$$

where $\nu$ is the *g.n.* of a closed **A**-statement and $q\nu$ is the number of quantifiers in the prenex normal form of '$\nu$'. Hence if $P[\nu,q\nu]$ is an **A'***-theorem, then:

if $q\nu = 0$, i.e. if '$\nu$' is an $\mathbf{A}_{00}$-statement, then $Q\{\nu\}$, i.e. '$\nu$' is an $\mathbf{A}_{00}$-theorem, (see scheme (A) in Ch. 8),

if $\qquad q\nu = S\nu' \quad$ then $\quad R\{\nu,\nu',P[r\{\nu,\nu',x'''\},\nu']\}$.

Now assume, as induction hypothesis, if $P[\nu,\nu'']$ is an **A'***-theorem and $\nu'' \leqslant \nu'$ then '$\nu$' is an **A**-statement with $\nu''$ quantifiers in its prefix when in prenex normal form then '$\nu$' is **A**-true, then if $q\nu = S\nu'$ and $P[\nu,q\nu]$ is an **A'***-theorem we shall have '$\nu$' is a closed **A**-statement and if $\nu$ is the *g.n.* of $(A\xi)\,\phi\{\xi\}$ then $\phi\{\pi\}$ is **A**-true for all numerals $\pi$, but if $\nu$ is the *g.n.* of $(E\xi)\,\phi\{\xi\}$ then $\phi\{\pi\}$ is **A**-true for some numeral $\pi$ (see scheme (A) in

Ch. 8). Thus the result holds for $S\nu'$ if it holds for $\nu'$. Thus it remains to **A**′*-prove scheme (C). The gist of the argument is as follows:

We wish to define a property $P$ such that

$$(Ax, x')\,(P[x, x'] \leftrightarrow H[x, x', P]),$$

for given $H$. We say that a property $\Delta$ satisfies condition C at $\nu$, $\nu'$ if

$$\Delta[\nu, \nu'] \leftrightarrow H[\nu, \nu', \Delta].$$

We want to define a property which satisfies condition C everywhere. Define

$$K\{x', X\} \quad \text{for} \quad (Ax, x'')\,(x'' \leqslant x' \rightarrow .\, X[x, x''] \leftrightarrow H[x, x'', X]),$$

where $X$ is of order zero. Then

$$K\{\nu', \Delta\} \quad \text{iff} \quad \Delta[\nu, \nu''] \leftrightarrow H[\nu, \nu'', \Delta] \quad \text{for} \quad \nu'' \leqslant \nu' \quad \text{and all} \quad \nu,$$

i.e. if and only if $\Delta$ satisfies condition C for $\nu'' \leqslant \nu'$ and any $\nu$. Thus if $K\{\nu', \Delta\}$ then $\Delta[\nu, \nu'']$ is the same as the property we wish to define up to $\nu'$ inclusive and all $\nu$. We first show that there are such properties and that they are unique up to $\nu'$ inclusive and that they can be **A**-defined.

Now define

$$P[x, x'] \quad \text{for} \quad (AX)\,(K\{x', X\} \rightarrow X[x, x']),$$

then $P$ will be a property of order one, and $P[\nu, \nu']$ will hold if

$$K\{\nu', \Delta\} \rightarrow \Delta[\nu, \nu']$$

for all properties $\Delta$ of order zero. In particular if $\Delta[\nu, \nu'']$ satisfies condition C for $\nu'' \leqslant \nu'$ and all $\nu$, then $K\{\nu', \Delta\}$ holds and $P[\nu, \nu']$ will be $\Delta[\nu, \nu']$ and so $P[\nu, \nu']$ will satisfy condition C, for all $\nu, \nu'$.

Note that $P[\nu, \nu']$ is the intersection of all zero order properties which satisfy condition C up to $\nu'$ inclusive and all $\nu$. The definition of the property $P$ requires a bound property variable of order zero, hence $P$ is a property of order one. Strictly we should speak of the statement $P[\nu, \nu']$.

LEMMA (i). *$(EX)\,K\{0, X\}$ is an* **A**′**-theorem.*

From T.N.D. and R 4 we obtain

$$(\boldsymbol{\lambda} xx'\,.\,Q\{x\})\,\nu 0 \rightarrow Q\{\nu\},$$

by Ax. 1 and dilution $\quad (\boldsymbol{\lambda} xx'\,.\,Q\{x\})\,\nu 0 \rightarrow 0 = 0,$

by II b′ $\quad (\boldsymbol{\lambda} xx'\,.\,Q\{x\})\,\nu 0 \rightarrow .\, 0 = 0 \,\&\, Q\{\nu\},$

by dilution, and writing $\Delta$ for $(\boldsymbol{\lambda} xx'.Q\{x\})$

$$\Delta[\nu,0] \rightarrow :0=0 \,\&\, Q\{\nu\}. \vee (Ex)(0=Sx \,\&\, G\{\nu,x,\Delta\}), \tag{1}$$

where $\qquad G\{x,x',X\}$ for $R\{x,x',X[r\{x,x',x'''\},x']\}$.

Again from T.N.D., R 4 and dilution,

$$0=0 \,\&\, Q\{\nu\}. \rightarrow \Delta[\nu,0].$$

From Ax. 2.2, dilution, R 3 and dilution again

$$(Ax)(0 \neq Sx \vee \sim G\{\nu,x,\Delta\}) \vee \Delta[\nu,0].$$

By II$b'$ $\quad 0=0 \,\&\, Q\{\nu\}. \vee (Ex)(0=Sx \,\&\, G\{\nu,x,\Delta\}) : \rightarrow \Delta[\nu,0]. \qquad (2)$

From (1), (2) by II$b'$

$$\Delta[\nu,0] \leftrightarrow :0=0 \,\&\, Q\{\nu\}. \vee (Ex)(0=Sx \,\&\, G\{\nu,x,\Delta\}).$$

From Prop. 16, Ch. 10, R 3 and II$f$ for property variables of order zero,

$$(EX)(Ax',x'')(x''=0 \rightarrow :.(X[x',x''] \leftrightarrow :x''=0 \,\&\, Q\{x'\}, \vee (Ex)(x''=Sx \,\&\, G\{x',x,X\}))),$$

whence, the result.

Note that $\Delta[\nu,0]$ is **A**-defined.

Lemma (ii). $(EX)K\{\nu,X\} \rightarrow (EX)K\{S\nu,X\}$ *is an* **A**$'$**-theorem.*

From R 4 $\qquad \Delta'[\theta,\kappa] \leftrightarrow H\{\theta,\kappa,X\}$, $\quad X$ of order zero, $\qquad (3)$

where we have written $\Delta'$ for $\lambda xx'.H\{x,x',X\}$, then $\Delta'$ is of order zero. We have the tautology:

$$\kappa \leqslant \nu \rightarrow .X[\theta,\kappa] \leftrightarrow H\{\theta,\kappa,X\} : \rightarrow :.\Delta'[\theta,\kappa] \leftrightarrow H\{\theta,\kappa,X\}. \rightarrow :\kappa \leqslant \nu \rightarrow .X[\theta,\kappa] \leftrightarrow \Delta'[\theta,\kappa],$$

where $X$ is a property variable of order zero, hence by II$d$, R 3

$$(Ax,x')(x \leqslant \nu \rightarrow .X[x',x] \leftrightarrow H\{x',x,X\}) \rightarrow .(Ax,x')(\Delta'[x',x] \leftrightarrow H\{x',x,X\}) \rightarrow (Ax,x')(x \leqslant \nu \rightarrow .X[x',x] \leftrightarrow \Delta'[x',x]).$$

From the derived rule $\qquad \dfrac{(A\xi)\phi\{\xi\} \vee \omega}{(A\xi)\phi\{\rho\{\xi\}\} \vee \omega}$ *

the last clause may be replaced by

$$(Ax,x',x'',x''')(x \leqslant S\nu \,\&\, x=Sx''. \rightarrow .X[\rho[x,x'',x'''],x''] \leftrightarrow \Delta'[\rho[x',x'',x'''],x''])$$

whence by regularity

$$(Ax, x')\,(x \leqslant S\nu \rightarrow .(Ex'')\,(x = Sx'' \,\&\, G\{x', x'', X\}) \leftrightarrow (Ex'')\,(x = Sx'' \,\&\, G\{x', x'', \Delta'\})),$$

whence by $\mathscr{F}_C$

$$(Ax, x')\,(x \leqslant S\nu \rightarrow :\Delta'[x', x] \leftrightarrow H\{x', x, X\}. \leftrightarrow .\Delta'[x', x] \leftrightarrow H\{x', x, \Delta'\}),$$

again by $\mathscr{F}_C$

$$(Ax, x')\,(\Delta'[x', x] \leftrightarrow H\{x', x, X\}) \rightarrow (Ax, x')\,(x \leqslant S\nu \rightarrow .\Delta'[x', x] \leftrightarrow H\{x', x, \Delta'\}),$$

Altogether so far we have

$$(Ax, x')\,(x \leqslant \nu \rightarrow .X[x', x] \leftrightarrow H\{x', x, X\}) \rightarrow .(Ax, x')\,(\Delta'[x', x] \leftrightarrow H\{x', x, X\}) \rightarrow (Ax, x')\,(x \leqslant S\nu \rightarrow .\Delta'[x', x] \leftrightarrow H\{x', x, \Delta'\})$$

using the cut to omit the **A**′*-theorem obtained from (3) by R 3 and then applying II*f* and R 3* the result follows. We have used the free property variable $X$ so that the application of R 3* is clear. Note that

$$\Delta[x, S\nu] \leftrightarrow H\{x, \nu, \Delta[x, \nu]\}$$

is **A**′*-defined.

LEMMA (iii). $(Ax)\,(EX)\,K\{x, X\}$ *is an* **A**′**-theorem.*

This follows from lemmas (i) and (ii) by R 3, the **A**′*-proofs have been written out in full, so that we can apply Prop. 5, Ch. 10 to $\tau x$ where $\tau\nu$ is the *g.n.* of an **A**′*-proof of $(EX)\,K\{\nu, X\}$.

LEMMA (iv). $(Ax, x')\,(x \leqslant 0 \,\&\, K\{0, X\} \,\&\, K\{0, X'\}. \rightarrow .X[x', x] \leftrightarrow X'[x', x])$ *is an* **A**′**-theorem.*

We should have added 'where $X$ and $X'$ are zero order properties. We have

$$K\{0, X\} \leftrightarrow (Ax, x')\,(x = 0 \rightarrow :.X[x', x] \leftrightarrow :x = 0 \,\&\, Q\{x'\}. \vee (Ex'')\,(x = Sx'' \,\&\, G\{x', x, X\}))$$

$$\rightarrow (Ax')\,(X[x', 0] \leftrightarrow :0 = 0 \,\&\, Q\{x'\}. \vee (Ex'')\,(0 = Sx'' \,\&\, G\{x', 0, X\}))$$ by Prop. 16, Ch. 10

$$\rightarrow (Ax')\,(X[x', 0] \leftrightarrow Q\{x'\})$$ by the cut, Ax. 2.2,

$$\rightarrow (Ax, x')\,(x \leqslant 0 \rightarrow .X[x', x] \leftrightarrow Q\{x'\}).$$

Hence by the predicate calculus

$$K\{0, X\} \,\&\, K\{0, X'\} . \rightarrow (Ax, x')\,(x \leqslant 0 \rightarrow : X[x', x] \leftrightarrow Q\{x'\} . \,\&\, . X'[x' . x] \leftrightarrow Q\{x'\}),$$

the result now follows by $\mathscr{F}_C$.

LEMMA (v). $\dfrac{(Ax, x')\,(x \leqslant \nu \,\&\, K\{\nu, X\} \,\&\, K\{\nu, X'\} . \rightarrow . X[x', x] \leftrightarrow X'[x', x])}{(Ax, x')\,(x \leqslant S\nu \,\&\, K\{S\nu, X\} \,\&\, K\{S\nu, X'\} . \rightarrow . X[x', x] \leftrightarrow X'[x', x])}$ *is an* $\mathbf{A}'^*$-*deduction* (see remark after lemma (iv)).

We have $\dfrac{K\{S\nu, X\}}{K\{\nu, X\}}$.

Hence $K\{S\nu, X\} \quad K\{S\nu, X'\}$

$K\{\nu, X\} \,\&\, K\{\nu, X'\} \quad Hyp.$

$(Ax, x')\,(x \leqslant \nu \rightarrow . X[x', x] \leftrightarrow X'[x', x])$ by regularity,

$(Ax, x')\,(x \leqslant \nu \rightarrow . G\{x', x, X\} \leftrightarrow G\{x', x, X'\})$

since $(Ax, x')\,(x \leqslant S\nu \,\&\, x = Sx'' . \leftrightarrow x'' \leqslant \nu)$,

then

$(Ax, x')\,(x \leqslant S\nu \rightarrow . (Ex'')\,(x = Sx'' \,\&\, G\{x', x'', X\}) \leftrightarrow (Ex'')\,(x = Sx'' \,\&\, G\{x', x'', X'\}))$ by $\mathscr{F}_C$,

$(Ax, x')\,(x \leqslant S\nu \rightarrow . H\{x', x, X\} \leftrightarrow H\{x', x, X'\})$ by regularity,

$K\{S\nu, X\} \,\&\, K\{S\nu, X'\} \,\&\, (Ax, x')\,(x \leqslant S\nu \rightarrow . H\{x', x, X\} \leftrightarrow H\{x', x, X'\})$ by $\mathscr{P}_C$,

$(Ax, x')\,(x \leqslant S\nu \rightarrow : X[x', x] \leftrightarrow H\{x', x, X\} . \,\&\, . X'[x', x] \leftrightarrow H\{x', x, X'\} . \,\&\, . H\{x', x, X\} \leftrightarrow H\{x', x, X'\})$ by $\mathscr{F}_C$,

$(Ax, x')\,(x \leqslant S\nu \rightarrow : X[x', x] \leftrightarrow X'[x', x])$.

The result now follows from the Deduction Theorem.

LEMMA (vi). $(Ax, x')\,(x < x'' \,\&\, K\{x'', X\} \,\&\, K\{x'', X'\} . \rightarrow . X[x', x] \leftrightarrow X'[x', x])$ *is an* $\mathbf{A}'^*$-*theorem.*

This follows from lemmas (iv), (v) by R 3; this can be applied since we have written out the full $\mathbf{A}'^*$-proof for each numeral $\nu$ on a primitive recursive plan, **E**-correct (see Prop. 5, Ch. 10).

LEMMA (vii). $(Ax, x')(K\{x, X\} \to . P\{x, x'\} \leftrightarrow X[x, x'])$ *is an* **A**′**-theorem* (see remark after lemma (iv)).

From lemma (vi) by $\mathscr{F}_C$: where $X$ and $X'$ are of zero order,

$$(Ax, x')(K\{x, X\} \,\&\, X[x', x] . \to . K\{x, X'\} \leftrightarrow X'[x', x]),$$

hence by the reversibility of R 3, then using R 3* and R 3 we get

$$(Ax, x')(K\{x, X\} \,\&\, X[x', x] . \to (AX')(K\{x, X'\} \to X'[x', x])).$$

Thus, by definition of $P$,

$$(Ax, x')(K\{x, X\} \to . X[x', x] \to P\{x', x\}). \tag{4}$$

Again, from $\mathscr{F}_C$ and lemma (vi)

$$(Ax, x')(K\{x, X\} \to : K\{x, X'\} \to X'[x', x] . \to X[x', x]),$$

and so by the reversibility of R 3, using II$f$, R 3 and the definition of $P$,

$$(Ax, x')(K\{x, X\} \to . (AX)(K\{x, X\} \to X[x', x]) \to X[x', x]) \\ \to . P\{x', x\} \to X[x', x]. \tag{5}$$

The lemma now follows from (4) and (5) by $\mathscr{F}_C$.

LEMMA (viii). $(Ax')(P\{x', 0\} \leftrightarrow Q\{x'\})$,

$$(Ax, x')(P\{x', Sx\} \leftrightarrow R\{x', x, P[r\{x', x, x''\}, x]\}),$$

*are* **A**′**-theorems.*

By definition we have

$$(Ax')(K\{0, X\} \to . X[x', 0] \leftrightarrow Q\{x'\}),$$

by lemma (vii) $\quad (Ax')(K\{0, X\} \to . P\{x', 0\} \leftrightarrow X[x', 0]),$

hence by $\mathscr{F}_C$ $\quad K\{0, X\} \to . (Ax')(P\{x', 0\} \leftrightarrow Q\{x'\}).$

By R 3*, lemma (i) and the cut, the first part of the lemma follows.

Again, by definition

$$(Ax, x')(K\{Sx, X\} \to . X[x', Sx] \leftrightarrow G\{x', x, X\}),$$

by lemma (vii)

$$(Ax, x')(K\{Sx, X\} \to . P\{x', Sx\} \leftrightarrow X[x', Sx]),$$

whence by $\mathscr{F}_C$

$$(Ax, x')\,(K\{Sx, X\} \rightarrow .\,P\{x', Sx\} \leftrightarrow G\{x', x, P\}) \rightarrow .\,P\{x', Sx\} \leftrightarrow R\{x', x, P[r\{x', x, x''\}, x]\}),$$

using the definition of $G$.

By the reversibility of R 3, II$f$, II$d$ and using R 3 we get

$$(Ax)\,(EX)\,K\{Sx, X\}. \rightarrow .\,(Axx')\,(P\{x', Sx\} \leftrightarrow R\{x', x, P[r\{x', x, x''\}, x]\}),$$

the lemma now follows from lemma (iii) using the cut.

This completes the demonstration of Prop. 6.

Cor. (i). *We can* $\mathbf{A}^{(S\kappa)}$*-define a property which satisfies scheme* (A) *when* $Q$, $R$ *contain bound property variables of order* $\kappa$.

The demonstration is similar. We just need to keep careful track of the orders of properties. For instance $\Delta$ in lemma (i) will now be of order $S\kappa$. $X$ throughout the other lemmas will have to be of order $S\kappa$. Finally $P$ will be of order $SS\kappa$, and all the lemmas are $\mathbf{A}^{S\kappa}*$-proved.

D 263 $q\nu$ for the number of quantifiers in the prefix of '$\nu$', when in prenex normal form.

Cor. (ii). $P\{\nu, q\nu\} \leftrightarrow \phi$ *is an* $\mathbf{A}'*$*-theorem, where* $\nu$ *is the g.n. of* $\phi$, *and* $\phi$ *is a closed* $\mathbf{A}$*-statement having* $q\nu$ *quantifiers in its prenex normal form.*

The corollary says that $P$ is an adequate truth-definition for $\mathbf{A}$ in $\mathbf{A}'*$. From Cor. (ii) we see that $P\{\nu, q\nu\}$ is an $\mathbf{A}'*$-theorem if and only if $\phi$ is an $\mathbf{A}'*$-theorem, or $\phi$ is an $\mathbf{A}'*$-theorem if and only if we can $\mathbf{A}'*$-prove that it is $\mathbf{A}$-true.

To demonstrate the corollary we use induction on the number of quantifiers in the prefix of $\phi$ when in prenex normal form. If $\phi$ is an $\mathbf{A}_{00}$-statement then $P\{\nu, 0\} \leftrightarrow Th_{00}[\nu]$, so we have to show that

$$Th_{00}[\nu] \leftrightarrow \phi.$$

Now this is $(Ex)\,prf_{00}[x, \nu] \leftrightarrow \phi$. It suffices to $\mathbf{A}_I$-prove:

$$\text{(i)} \qquad (Ex)\,prf_{00}[x, \nu] \rightarrow \phi \quad \text{and} \quad \phi \rightarrow (Ex)\,prf_{00}[x, \nu].$$

These follow easily if $\phi$ is $\mathbf{A}_{00}$-true in which case it is an $\mathbf{A}_{00}$-theorem and so $(Ex)\,prf_{00}[x, \nu]$ is an $\mathbf{A}_0$-theorem. Similarly if $\phi$ is $\mathbf{A}_{00}$-false, in which case $\sim \phi$ is an $\mathbf{A}_{00}$-theorem the second of (i) is easily proved.

There remains the first of (i) when $\phi$ is $\mathbf{A}_{oo}$-false. Then $\sim \phi$ is $\mathbf{A}_{oo}$-true and we have

$$\sim \phi \rightarrow (Ex)\, prf_{oo}[x, Neg\, \nu],$$

whence
$$\sim (Ex)\, prf_{oo}[x, Neg\, \nu] \rightarrow \phi.$$

Now
$$\dfrac{\dfrac{(Ex, x')\,(prf_{oo}[x, \nu]\, \&\, prf_{oo}[x', Neg\, \nu])}{(Ex, x')\, prf_{oo}[x^{\frown} x'^{\frown} \langle conj[\nu, Neg\, \nu]\rangle, conj[\nu, Neg\, \nu]]}}{(Ex)\, prf_{oo}[x, conj\, [\nu, Neg\, \nu]].}$$

By the Deduction Theorem in $\mathbf{A}_I$

$$(Ex, x')\,(prf_{oo}[x, \nu]\, \&\, prf_{oo}[x', Neg\, \nu]) \rightarrow (Ex)\, prf_{oo}[x, conj\, [\nu, Neg\, \nu]],$$

but
$$(\phi\, \&\, \sim \phi) \rightarrow 0 = 1,$$

whence

$$(Ex)\, prf_{oo}[x, conj[\nu, Neg\, \nu]] \rightarrow (Ex)\, prf_{oo}\,[x, g.n. \text{ of } (0 = 1)].$$

But$(Ax) \sim prf_{oo}[x, g.n. \text{ of } (0 = 1)]$ and so $\sim (Ex)\, prf_{oo}[x, conj[\nu, Neg\, \nu]]$. Thus $\sim (Ex, x')\,(prf_{oo}[x, \nu]\, \&\, prf_{oo}[x', Neg\, \nu])$ and so

$$(Ex)\, prf_{oo}[x, \nu] \rightarrow\, \sim (Ex)\, prf_{oo}[x', Neg\, \nu]$$

and so finally $(Ex)\, prf_{oo}\,[x, \nu] \rightarrow \phi$ as desired.

What we have done so far can be put down as a detailed $\mathbf{A}_I$-proof of

$$\frac{\phi}{P\{\nu, 0\} \leftrightarrow \phi} \quad \text{and} \quad \frac{\sim \phi}{P\{\nu, 0\} \leftrightarrow \phi}.$$

From these by Deduction Theorem and the cut we obtain $P\{\nu, 0\} \leftrightarrow \phi$. Now if in this we take $\phi\{\mathfrak{k}\}$ instead of $\phi$ and if $\sigma[\mathfrak{Num}\, \mathfrak{k}]$ is equal to the *g.n.* of $\phi\{\mathfrak{k}\}$ then we obtain a detailed $\mathbf{A}_I$-proof of $P\{\nu, 0\} \leftrightarrow \phi\{\mathfrak{k}\}$ for all $\mathfrak{k}$, **E**-correct. Now take as our induction hypothesis:

$$`P\{\sigma[\mathfrak{Num}\, \mathfrak{k}], q\sigma[\mathfrak{Num}\, \mathfrak{k}]\} \leftrightarrow \phi\{\mathfrak{k}\} \quad \text{for all } \mathfrak{k}, \mathbf{E}\text{-correct}'.$$

Then we have just shown that this holds for closed **A**-statements $\phi\{\mathfrak{k}\}$ free from quantifiers. Now assume it holds for closed **A**-statements with $\nu$ quantifiers in their prefix when in prenex normal form, and let $\phi\{\mathfrak{k}\}$ be a closed **A**-statement with $S\nu$ quantifiers in its prefix when in prenex normal form, and as first case suppose that the initial quantifier is universal. Then by our induction hypothesis we shall have:

(ii) $P\{st[Num\, \pi, g.n.\, x, m[pr\sigma[\mathfrak{Num}\, \mathfrak{k}]]], qm\sigma[\mathfrak{Num}\, \mathfrak{k}]\} \leftrightarrow \phi\{\mathfrak{k}, \pi\}$

for all $\mathfrak{k}, \pi$, **E**-correct, hence we get

$$P\{\sigma'[\mathfrak{Num}\,\mathfrak{k}], q\sigma'[\mathfrak{Num}\,\mathfrak{k}]\} \leftrightarrow (Ax)\,\phi\,\{\mathfrak{k}, x\} \quad \text{for all } \mathfrak{k}, \mathbf{E}\text{-correct},$$

here $\sigma'[\mathfrak{Num}\,\mathfrak{k}]$ is equal to the *g.n.* of $(Ax)\,\phi\{\mathfrak{k}\}$.

The case when the initial quantifier is an existential quantifier is dealt with similarly. This completes the demonstration of the corollary.

## 11.7 *Consistency of* $\mathbf{A}_I$

We had previously shown that the consistency of $\mathbf{A}_I$ failed to be $\mathbf{A}_I$-provable though it is $\mathbf{A}_I$-definable. We now show that the consistency of $\mathbf{A}_I$ can be $\mathbf{A}'^*$-proved. This will involve among other things that T.N.D. and Modus Ponens fail to be jointly eliminable from $\mathbf{A}'^*$. The role of Modus Ponens in this demonstration is to eject property variables from the $\mathbf{A}'^*$-proof of an **A**-statement.

**Prop** 6. *The consistency of* $\mathbf{A}_I$ *can be* $\mathbf{A}'^*$*-proved.*

We take the consistency of $\mathbf{A}_I$ in the form $Con_{(11)}\,\mathbf{A}_I$, namely

$$(Ax)\,(\sim (Ex')\,Prf_I[x', x] \vee \sim (Ex')\,Prf_I[x', Negx]). \tag{1}$$

$Prf_I\,[\kappa, \nu]$ is the proof predicate for $\mathbf{A}_I$. We shall take $\mathbf{A}_I$ in an equivalent form; namely that sub-system of $\mathbf{A}'$ which is obtained from $\mathbf{A}'$ by omitting all property variables and all rules in which they occur. This amounts to taking negation, disjunction and universal quantification as primitive and defining conjunction and existential quantification in terms of them. Let T be the truth-function for $\mathbf{A}_I$, then by Prop. 5 T can be $\mathbf{A}'^*$-defined.

In order to $\mathbf{A}'^*$-prove (1) we first give detailed $\mathbf{A}'^*$-proofs of

$$(Ex)\,Prf_I[x, \nu] \rightarrow \mathrm{T}[\nu], \tag{2}$$

and

$$\mathrm{T}[\nu] \rightarrow \sim \mathrm{T}[Neg\,\nu]. \tag{3}$$

From (2) we get

$$\sim \mathrm{T}[Neg\,\nu] \rightarrow \sim (Ex)\,Prf_I[x, Neg\,\nu]. \tag{4}$$

From (2) and (3) by the cut we get

$$\mathrm{T}[Neg\,\nu] \rightarrow \sim (Ex)\,Prf_I[x, \nu]. \tag{5}$$

From (4) and (5) by the cut we get

$$\sim (Ex)\,Prf_I\,[x, \nu] \vee \sim (Ex)\,Prf_I\,[x, Neg\,\nu]. \tag{6}$$

If we can display these **A**′*-proofs in detail then we can use R 3 and obtain (1), which is what we want to do, we can also get:

$$(Ax)((Ex')\,Prf_I\,[x',x] \rightarrow \mathrm{T}[x])$$

which is the consistency of $\mathbf{A}_I$ with respect to its truth-definition. Note that this **A**′*-proof of (1) contains property variables which occur in $T$ and are ejected by the cut.

Similarly from (3) we get:

$$(Ax)(\sim \mathrm{T}[x] \vee \sim \mathrm{T}[Negx])$$

which expresses the consistency of the **A**-truth definition.

In order to carry out this programme we shall have to formalize a great deal of what we have said informally about **A**-truth. We start with (3), we have

$$\mathrm{T}[\nu, 0] \leftrightarrow (Ex)\,prf_{00}\,[x, \nu],$$

$$\begin{aligned}\mathrm{T}[\nu, S\kappa] \leftrightarrow :\, & clstat\,\nu\,\&\,(pr\nu)_2^\infty = (g.n.\text{ of }A)\,\&\,(Ax')\,\mathrm{T}[st[m[pr\,\nu], (g.n.\text{ of }x),\\ & Numx'], \kappa].\vee.(pr\,\nu)_2^\infty = (g.n.\text{ of }E)\\ & \&\,(Ex')\,\mathrm{T}[st[m[pr\,\nu], (g.n.\text{ of }x), Numx'], \kappa].\end{aligned} \tag{D}$$

In this replace $\nu$ by $Neg\,\nu$ and note that

$$clstat\,\nu\,\&\,(pr\,\nu)_2^\infty = (g.n.\text{ of }A).\leftrightarrow.clstatNeg\,\nu\,\&\,(prNeg\,\nu)_2^\infty = (g.n.\text{ of }E), \tag{7}$$

and the same with $E$ and $A$ interchanged, (8)

and $$clstat\,\nu \leftrightarrow clstatNeg\,\nu. \tag{9}$$

Thus

$$\begin{aligned}\mathrm{T}[Neg\,\nu, S\kappa] \leftrightarrow\, & clstat\,\nu\,\&\,((pr\,\nu)_2^\infty = (g.n.\text{ of }A)\,\&\,(Ex')\,\mathrm{T}[st[m[pr[Neg\,\nu]],\\ & (g.n.\text{ of }x), Numx'], \kappa]).\vee.(pr\nu)_2^\infty = (g.n.\text{ of }E)\\ & \&\,(Ax')\,\mathrm{T}[st[m[pr[Neg\,\nu]], (g.n.\text{ of }x), Numx'], \kappa].\end{aligned} \tag{10}$$

Suppose that we have shown that

$$\mathrm{T}[Neg\,\nu, \kappa] \rightarrow \sim T[\nu, \kappa],$$

then since

$$clstat\,\nu \rightarrow st[m[pr[Neg\,\nu]], (g.n.\text{ of }x), Numx'] = Neg\,[st\,[m[pr\,\nu], (g.n.\text{ of }x), Numx']]$$

we get by $I\mathscr{F}_C$

$$\mathrm{T}[Neg\,\nu, S\kappa] \rightarrow clstat\,\nu\,\&\,((pr\,\nu)_2^\infty = (g.n.\text{ of } E)\,\&\sim(Ex')\,\mathrm{T}[st[m[pr\,\nu], (g.n.\text{ of } x), Numx'], \kappa]\,.\vee.(pr\,\nu)_2^\infty = (g.n.\text{ of } A)\,\&\sim(Ax')\,\mathrm{T}[st[m[pr\,\nu], (g.n.\text{ of } x), Numx'], \kappa]) \rightarrow\sim\mathrm{T}[\nu, S\kappa]. \tag{11}$$

Note that we have

$$\sim((pr\,\nu)_2^\infty = (g.n.\text{ of } E)\,\&\,(pr\,\nu)_2^\infty = (g.n.\text{ of } A))$$

and
$$(pr\,\nu)_2^\infty = (g.n.\text{ of } A) \vee (pr\,\nu)_2^\infty = (g.n.\text{ of } E).$$

The $\mathbf{A}_I$-proofs of (7), (8) and (9) can be written out in full and that of (11) from the hypothesis (10). We wish to do the same for

$$\mathrm{T}[Neg\,\nu, 0] \rightarrow \sim\mathrm{T}[\nu, 0]$$

and then we can apply R 3 and obtain

$$(Ax)(\sim Tx \vee \sim T[Neg\,x]),$$

the consistency of the **A**-truth definition.

Now $\mathrm{T}[\nu, 0]$ could have been defined as

$$clstat_{00}\,\nu\,\&\,\alpha\{\nu\} = 0,$$

where $\alpha\{\nu\}$ is the characteristic function of '$\nu$'. Then

$$\mathrm{T}[Neg\,\nu, 0] \leftrightarrow .\, clstat_{00}\,\nu\,\&\,\alpha\{\nu\} = 1.$$

Thus we have to show

$$\sim(clstat_{00}\nu\,\&\,\alpha\{\nu\} = 1) \vee \sim(clstat_{00}\,\nu\,\&\,\alpha\{\nu\} = 0), \tag{12}$$

i.e.
$$clstat_{00}\,\nu \rightarrow .\sim(\alpha\{\nu\} = 0\,\&\,\alpha\{\nu\} = 1). \tag{13}$$

But $\sim(\alpha\{\nu\} = 0\,\&\,\alpha\{\nu\} = 1)$ is an $\mathbf{A}_{00}$-theorem for all $\nu$, **E**-correct, hence so is (12) and hence so is

$$\mathrm{T}[Neg\,\nu, 0] \rightarrow \mathrm{T}[\nu, 0].$$

Thus we have (3) for all $\nu$, **E**-correct.

(13) is $Con_{(11)}\,\mathbf{A}_{00}$ which is thus $\mathbf{A}_I$-proved. We had already noted that it fails to be an **A**-theorem. It requires T.N.D. for the above proof, but T.N.D. can be eliminated from an $\mathbf{A}_I$-proof.

Now we want to give an **A**$'$*-proof of (2) for all $\nu$, **E**-correct. To do this we formalize the demonstration we gave in Prop. 6, Ch. 10. To do this it suffices to give **A**$'$*-proofs for all $\nu, \nu', \nu''$, **E**-correct of the following

$$Ax[\nu] \rightarrow \mathrm{T}\nu, \tag{14}$$

$$Rl^1[\nu', \nu] \,\&\, \mathrm{T}[\nu'] \,.\rightarrow \mathrm{T}\nu, \tag{15}$$

$$Rl^2[\nu', \nu'', \nu] \,\&\, \mathrm{T}[\nu'] \,\&\, \mathrm{T}[\nu''] \,.\rightarrow \mathrm{T}\nu, \tag{16}$$

$$clstat\,[Gen\,\nu] \,\&\, (Ax)\,\mathrm{T}[s[\nu, x]] \,.\rightarrow \mathrm{T}[Gen\,\nu], \tag{17}$$

where

D 264 $\qquad Gen\,\nu \quad \text{for} \quad (g.n. \text{ of } (Ax))^\frown \nu.$

Then

$$Prf[\kappa, \nu] \rightarrow \mathrm{T}[\nu] \tag{18}$$

has an **A**$'$*-proof for all $\nu$, **E**-correct.

For (14) we have

$$Ax[\nu] \rightarrow .\,\alpha\{\nu\} = 0 \,\&\, clstat\,[\alpha\{\nu\}]$$
$$\rightarrow \mathrm{T}[\nu, 0]$$
$$\rightarrow \mathrm{T}[\nu].$$

To deal with (15), (16) and (17) it suffices to give **A**$'$*-proofs for all $\nu, \nu', \nu''$, **E**-correct of

$$\sim \mathrm{T}[\nu] \leftrightarrow \mathrm{T}[Neg\,\nu], \tag{19}$$

$$\mathrm{T}[\nu'] \vee \mathrm{T}[\nu''] \,.\leftrightarrow \mathrm{T}[disj[\nu', \nu'']], \tag{20}$$

$$(Ax)\,\mathrm{T}[s[\nu, x]] \leftrightarrow \mathrm{T}[Gen\,\nu]. \tag{21}$$

We have already done this for (3), in a similar manner we do it for (19). For (21), in scheme (A) write $Gen\,\nu$ for $\nu$ then $st[m[pr\,\nu]] = \nu$ and

$$(pr\,\nu)_2^\infty = (\text{g}.n. \text{ of } A) \quad \text{and} \quad clstat\,\nu$$

so we get

$$\mathrm{T}[Gen\,\nu, S\kappa] \leftrightarrow (Ax')\,\mathrm{T}[s[\nu, x']],$$

whence

$$\mathrm{T}[Gen\,\nu] \leftrightarrow (Ax)\,\mathrm{T}[s[\nu, x]].$$

For (20) we have to formalize Prop. 3 of Ch. 4. We give **A**$'$*-proofs for all $\nu', \nu''$, **E**-correct of

$$\mathrm{T}[\nu', 0] \vee \mathrm{T}[\nu'', 0] \leftrightarrow \mathrm{T}[disj[\nu', \nu''], 0], \tag{22}$$

and if we have **A**$'$*-proofs for all $\nu', \nu''$, **E**-correct of

$$\mathrm{T}[\nu', \kappa] \vee \mathrm{T}[\nu'', 0] \leftrightarrow \mathrm{T}[disj[\nu', \nu''], \kappa] \tag{23}$$

then we show how to obtain **A′***-proofs for all $\nu', \nu''$, **E**-correct of

$$\mathrm{T}[\nu', S\kappa] \vee \mathrm{T}[\nu'', 0] \leftrightarrow \mathrm{T}[disj[\nu', \nu''], S\kappa], \tag{23'}$$

and if we have **A′***-proofs for all $\nu', \nu''$, **E**-correct of

$$\mathrm{T}[\nu', \kappa] \vee \mathrm{T}[\nu'', \kappa'] \leftrightarrow \mathrm{T}[disj[\nu', \nu''], \kappa + \kappa'] \tag{24}$$

then we show how to obtain **A′***-proofs for all $\nu', \nu''$, **E**-correct of

$$\mathrm{T}[\nu', S\kappa] \vee \mathrm{T}[\nu'', S\kappa'] \leftrightarrow \mathrm{T}[disj[\nu', \nu''], S\kappa + S\kappa'], \tag{24'}$$

(22) is

$$clstat\,[\nu', \nu''] \,\&\, (\alpha\{\nu'\} = 0 \vee \alpha\{\nu''\} = 0) \leftrightarrow clstat[disj[\nu', \nu'']] \\ \&\, \alpha\{\nu'\} \times \alpha\{\nu''\} = 0. \tag{25}$$

We can easily give **A′***-proofs for all $\nu', \nu''$, **E**-correct of

$$clstat\,[\nu', \nu''] \leftrightarrow clstat\,[disj[\nu', \nu'']]$$

and of $$\alpha\{\nu'\} = 0 \vee \alpha\{\nu''\} = 0 \leftrightarrow \alpha\{\nu'\} \times \alpha\{\nu''\} = 0,$$

whence (22) follows.

For (23′) we have

$$\mathrm{T}[\nu', S\kappa] \vee \mathrm{T}[\nu'', 0] \leftrightarrow R\{\nu', \kappa, \mathrm{T}[r[\nu', \kappa, x''], \kappa]\} \vee \mathrm{T}[\nu'', 0] \\ \leftrightarrow R\{\mathrm{disj}[\nu', \nu''], \kappa, T[r[disj[\nu', \nu''], \kappa, x''], \kappa]\}$$

by tautology, using the special construction of $R$ and the hypothesis (23) hence

$$\leftrightarrow \mathrm{T}[disj[\nu', \nu''], S\kappa]$$

as desired.

We proceed similarly for (24′). Thus we have obtained **A′***-proofs of (19) for all $\nu$, **E**-correct. Hence we can obtain **A′***-proofs of (14)–(17) inclusive for all $\nu$, **E**-correct, and hence of (18) for all $\nu$, **E**-correct. Thus we have **A′***-proofs of (2) for all $\nu$, **E**-correct, and hence similarly of (6). Thus, finally (1) is an **A′***-theorem, as desired. This completes the demonstration of Prop. 7.

Note that the property variables which are in T in (2) and in (3) get eliminated by Modus Ponens.

COR. (i). *It is impossible to eliminate Modus Ponens from* **A′***.

More precisely **A′** is strictly weaker, than **A′***. Suppose that we could eliminate Modus Ponens from **A′***, then we could do so in the **A′***-proof of $Con_{(11)}\mathbf{A}_I$ in Prop. 7. The result would be an $\mathbf{A}_I$-proof of $Con_{(11)}\mathbf{A}_I$, because it would be entirely without property variables, for once a property variable enters such a proof then a property variable remains in the proof from that place to the end, but $Con_{(11)}\mathbf{A}_I$ is without property variables.

The proof may use T.N.D., but cases of T.N.D. which are free of property variables can be $\mathbf{A}_I$-proved. Thus we could obtain an $\mathbf{A}_I$-proof of $Con_{(11)}\mathbf{A}_I$ which by Prop 19, Ch. 10 is absurd.

COR. (ii). *The $\mathbf{A}_I$ irresolvable statement can be $\mathbf{A}'^*$-proved.*

We had found in Ch.10 §17 that if we could $\mathbf{A}_I$-prove the consistency of $\mathbf{A}_I$ then we could $\mathbf{A}_I$-prove the $\mathbf{A}_I$-irresolvable statement. Thus if we can $\mathbf{A}'^*$-prove the consistency of $\mathbf{A}_I$ then we can $\mathbf{A}'^*$-prove the $\mathbf{A}_I$-irresolvable statement.

## 11.8 *Definition of $\mathbf{A}^{(\kappa)}$-truth*

Now let us consider a truth-definition for $\mathbf{A}^{(\kappa)*}$. This is the same as for $\mathbf{A}^{(\kappa)}$. We know from Prop. 14 and Prop. 11, Cor. (i), Ch. 8 that a formal system which contains negation and recursive function theory fails to contain its truth-definition, provided the system is consistent. The two propositions just referred to give two different ways of framing a truth-definition. Let $\nu$ be the *g.n.* of the closed $\mathscr{L}$-statement $\phi$. Then we frame the two truth-definitions as follows:

(*a*) T is a truth-definition of $\mathscr{L}$ in $\mathscr{L}'$ if $T\nu \leftrightarrow \phi$ is an $\mathscr{L}'$-theorem; here T is some one-place $\mathscr{L}'$-statement,

(*b*) T is a truth-definition of $\mathscr{L}$ in $\mathscr{L}'$ if $T\nu \leftrightarrow \phi$ is $\mathscr{L}'$-true. In (*b*) we are assuming that we have a definition of $\mathscr{L}'$-truth.

If $\mathscr{L}'$ is $\mathscr{L}$ then T becomes a truth-definition for $\mathscr{L}$ in itself, and T is a one-place $\mathscr{L}$-statement. Definition (*a*) is clearer than definition (*b*), because the method of testing is exactly stated, namely to be an $\mathscr{L}'$ theorem. In (*b*) if $\mathscr{L}'$ is $\mathscr{L}$ then it becomes; T is a truth definition for $\mathscr{L}$ in itself if $T\nu \leftrightarrow \phi$ is $\mathscr{L}$-true, i.e. if $T\mu$, where $\mu$ is the *g.n.* of $T\nu \leftrightarrow \phi$. Here 'if $T\mu$' can only mean 'if $T\mu$ is $\mathscr{L}$-true', i.e. 'if $T\lambda$' where $\lambda$ is the *g.n.* of $T\mu$, and so on. So let us adopt the definition (*a*). There are two things to do in defining a truth-definition according to (*a*).

First, we must produce a one-place closed statement T in $\mathscr{L}'$, secondly we must $\mathscr{L}'$-prove $T\nu \leftrightarrow \phi$. It might be possible to construct a one-place

$\mathscr{L}'$-statement T which intuitively possessed the required properties, but such that $T\nu \leftrightarrow \phi$ failed to be $\mathscr{L}'$-provable. On the other hand it might be impossible to construct a one-place closed $\mathscr{L}'$-statement T which even intuitively possessed the required properties, we would then, of course, be without anything to $\mathscr{L}'$-prove.

We have had various cases of truth-definitions before connected with the various formal systems we have constructed. $\mathbf{A}_{00}$ fails to contain its own truth-definition, because this is a general recursive property which fails to be primitive recursive, and $\mathbf{A}_{00}$ can only deal with primitive recursive properties. $\mathbf{A}_0$ contains its own truth-definition and is consistent. $\mathbf{A}_0$ is without a negation, so the contradiction which would be forthcoming in a formal system containing negation and recursive function just fails to arise.

$\mathbf{A}$, $\mathbf{A}_I$ both have the same truth-definition, but this fails to be expressible in them, because the set of $\mathbf{A}$-true $\mathbf{A}$-statements is a productive set and such a set fails to be arithmetically definable.

We shall find that $\mathbf{A}^{(\kappa)}$-truth can be $\mathbf{A}^{(S\kappa)}$-defined. This we will now discuss.

The definition of $\mathbf{A}^{(\kappa)}$-truth is by induction on the construction of the statement.

(i) the $\mathbf{A}^{(\kappa)}$-truth of $\phi\{(\boldsymbol{\lambda}\xi.\psi\{\xi\})\alpha\}$ is the same as that of $\phi\{\psi\{\alpha\}\}$, so we may suppose that rule R 4 has been applied as long as possible, we then say that the statement is in $\boldsymbol{\lambda}$-*normal form*.

(ii) closed $\mathbf{A}_{00}$-equations are $\mathbf{A}^{(\kappa)}$-true if and only if they are $\mathbf{A}_{00}$-true.

(iii) $\sim\phi$ is $\mathbf{A}^{(\kappa)}$-true if and only if $\phi$ fails to be $\mathbf{A}^{(\kappa)}$-true.

(iv) $\phi' \vee \phi''$ is $\mathbf{A}^{(\kappa)}$-true if and only if $\phi'$ is $\mathbf{A}^{(\kappa)}$-true or if $\phi''$ is $\mathbf{A}^{(\kappa)}$-true.

(v) $(A\xi)\,\phi\{\xi\}$ is $\mathbf{A}^{(\kappa)}$-true if and only if $\phi\{\nu\}$ is $\mathbf{A}^{(\kappa)}$-true for each numeral $\nu$.

(vi) $(A\Xi)\,\phi\{\Xi\}$ is $\mathbf{A}^{(\kappa)}$-true if and only if $\phi\{\Delta\}$ is $\mathbf{A}^{(\kappa)}$-true for each closed property $\Delta$ of order less than or equal to that of the variable $\Xi$.

In the last clause $\phi\{\Xi\}$ must be of the form

$$\phi\{A\Xi, \Xi\alpha', \ldots, \Xi\alpha^{(\pi)}\},$$

hence $\phi\{\Delta\}$ will be of the form

$$\phi\{A(\lambda\xi.\psi\{\xi\}), \psi\{\alpha'\}, \ldots, \psi\{\alpha^{(\pi)}\}\},$$

where $\psi\{\Gamma_\iota\}$ is a closed $\mathbf{A}^{(\kappa)}$-statement-form of order less than or equal to that of the variable $\Xi$, thus any property variables there may be in $\psi\{\Gamma_\iota\}$ are bound and of order at least one less than that of the variable $\Xi$.

All the clauses in the definition of $\mathbf{A}^{(\kappa)}$-truth are the same as in the definition of **A**-truth except the first and the last, the last clause is far more complicated than anything we have encountered so far. In it we pass from $(A\Xi)\,\phi\{A\Xi, \Xi\alpha', \ldots, \Xi\alpha^{(\pi)}\}$ to

$$\phi\{(A\xi)\,\psi\{\xi\}, \psi\{\alpha'\}, \ldots, \{\psi\alpha^{(\pi)}\}\}$$

which, in general, is a more complicated statement.

According to our rules of formation the following will be $\mathbf{A}^{(\kappa)}$-statements

$$AX_{o\iota}, A(X_{o\iota\iota}\,\xi), A_{o\iota}(\boldsymbol{\lambda} X_{o\iota}\,.\,AX_{o\iota}), \quad E_{o\iota}(\boldsymbol{\lambda} X_{o\iota}\,.\,(\sim AX_{o\iota}\,\&\,EX'_{o\iota}))\text{ etc.}$$

of these the third is closed, the others open. According to our $\mathbf{A}^{(\kappa)}$-truth definition we see that the third one is $\mathbf{A}^{(\kappa)}$-true if and only if $A(\boldsymbol{\lambda}\xi\,.\,\phi\{\xi\})$ is $\mathbf{A}^{(\kappa)}$-true for every $\phi$ such that $\phi\{\xi\}$ is of the same order as $\chi_{o\iota}$. That is to say if and only if the universal statement $(A\xi)\,\phi\{\xi\}$ is $\mathbf{A}^{(\kappa)}$-true for every $\phi$ of order at most that of $X_{o\iota}$. This obviously fails to be the case, hence $A_{o\iota}(\boldsymbol{\lambda} X_{o\iota}\,.\,AX_{o\iota})$ is $\mathbf{A}^{(\kappa)}$-false. $\mathbf{A}^{(\kappa)}$-statements of these types are harmless oddities, and so we allow them, though we could do without them, anyway they give us a wider range of statements.

Now let us see what happens when we try to find the $\mathbf{A}^{(\kappa)}$-truth of a closed $\mathbf{A}^{(\kappa)}$-statement according to the definition above. Let us take the closed $\mathbf{A}^{(\kappa)}$-statement in $\boldsymbol{\lambda}$-normal form and then put it into prenex normal form, called $\boldsymbol{\lambda}$-*prenex normal form* (this fails to alter the truth-value). The prefix will consist of quantifiers, general or restricted, numerical or property of orders less than $\kappa$. If the prefix is absent then we have an $\mathbf{A}_{00}$-statement for which we have a truth-definition. If there is a prefix then remove the first quantifier, if it is numerical then proceed as for $\mathbf{A}_I$, but if it is a property quantifier of order $\mu$, then we pass from $(Q\Xi)\,\phi\{\Xi\}$ to $\phi\{\Delta\}$ where $\Delta$ is a property of order less than or equal to $\mu$. Now $\phi\{\Delta\}$ must be of the form: $\phi\{A\Delta, \Delta\alpha', \ldots, \Delta\alpha^{(\pi)}\}$, where

$$\phi\{A\Gamma_{o\iota}, \Gamma_{o\iota}\alpha', \ldots, \Gamma_{o\iota}\alpha^{(\pi)}\}$$

is without free occurrences of $\Xi$. Thus $\phi\{\Delta\}$ is of the form

$$\phi\{(A\xi)\,\psi\{\xi\}, \psi\{\alpha'\}, \ldots, \psi\{\alpha^{(\pi)}\}\},$$

we then put this into $\boldsymbol{\lambda}$-prenex normal form. Suppose that the original $\boldsymbol{\lambda}$-prenex normal form was $(Q\Xi)\,(\mathfrak{Q}'\mathfrak{X}')\,\phi'\{\Xi, \mathfrak{X}'\}$, then the new one is:

$(\mathfrak{Q}'\mathfrak{X}')(\mathfrak{Q}''\mathfrak{X}'')\phi''\{\mathfrak{X}', \mathfrak{X}''\}$, where $(\mathfrak{Q}'\mathfrak{X}')$ is a prefix of quantifiers all of order less than $\kappa$ or numerical and $(\mathfrak{Q}''\mathfrak{X}'')$ is another prefix of quantifiers all of order at most that of the variable $\Xi$. The second prefix is, in general, longer than the first prefix, but, in a sense to be described, it is simpler. The prefix of $\phi$ is associated with an $S\kappa$-typlet $\{\theta, \theta', \theta'', \ldots, \theta^{(S\kappa)}\}$, where $\theta^{(S\lambda)}$ is the number of property quantifiers of order $\lambda$, and $\theta$ is the number of numerical quantifiers. When we remove the first quantifier in the prefix of $\phi$ then we pass from the $SS\kappa$-tuplet

$$\{\theta, \theta', \theta'', \ldots, \theta^{(S\kappa)}\}$$

to the $S\kappa$-tuplet

$$\{\pi, \pi', \pi'', \ldots, \pi^{(\mu)}, \theta^{(S\mu)} \dot{-} 1, \theta^{(SS\mu)}, \ldots, \theta^{(S\kappa)}\},$$

where $\mu$ is the greatest order of one of the new batch of quantifiers.

Now let us order $SS\kappa$-tuplets by *last differences*, so that

$$\{\theta, \theta', \theta'', \ldots, \theta^{(S\kappa)}\} < \{\pi, \pi', \pi'', \ldots, \pi^{(S\kappa)}\}$$

if and only if $\theta^{(\mu)} < \pi^{(\mu)}$ and $\theta^{(\lambda)} = \pi^{(\lambda)}$ for $\mu < \lambda \leqslant S\kappa$. We then say that the tuplet on the left is inferior to the tuplet on the right.

Thus in removing the initial quantifier of $\phi$ we pass from an ordered tuplet to another one which is inferior in the order of the removed quantifier. Thus the process will eventually stop. It is thus a process that we would expect an oracle to do for us.

## 11.9 *Scheme for an* $\mathbf{A}^{(\kappa)}$*-truth-definition*

We will now find a scheme satisfied by the $\mathbf{A}^{(\kappa)}$-truth-predicate. Let

$$\mathrm{T}[\pi^{(\kappa)}, \pi^{(\kappa \dot{-} 1)}, \pi^{(\kappa \dot{-} 2)}, \ldots, \pi', \pi; \nu]$$

hold if and only if $\nu$ is the *g.n.* of a closed $\mathbf{A}^{(\kappa)}$-statement in $\boldsymbol{\lambda}$-prenex normal form which is $\mathbf{A}^{(\kappa)}$-true and has $\pi^{(\kappa)}$ initial quantifiers up to and including the right-most one of order $\kappa$, the highest order allowed in $\mathbf{A}^{(\kappa)}$, and has $\pi^{(\kappa \dot{-} 1)}$, initial quantifiers after the right-most quantifier of order $\kappa$ and up to and including the right-most quantifier of order $(\kappa \dot{-} 1)$, and so on, till $\pi$ is the number of numerical quantifiers to the right of all the property quantifiers. Then we shall have

D 265 $LP^{\kappa}\,\nu$ for $\nu$ is the *g.n.* of a closed $\mathbf{A}^{(\kappa)}$-statement in $\boldsymbol{\lambda}$-prenex normal form.

D 266 *l.p.* $\nu$ for the *g.n.* of the $\boldsymbol{\lambda}$-prenex normal form of $\nu$.
m$\nu$ is defined in D 232.

$$\mathrm{T}[(\pi^{(\kappa)},\ldots,\pi^{(\kappa\dot-1)},\pi^{(\kappa\dot-2)},\ldots,\pi',\pi;\nu]\leftrightarrow LP^{\kappa}\,\nu\,\&\,(((\nu_2^{\infty}=(g.n.\text{ of }A_{\mu}))\,\&\,(Ax)\\ (prop^{\mu}x\rightarrow\mathrm{T}[\pi^{(\kappa\dot-1)},\ldots,\pi^{(\mu)},\pi^{(\mu\dot-1)}+h_1[x,\nu],\ldots,\pi+h_{\kappa}[x,\nu];\\ l.p.subst[x,\nu_5^{\infty},m\nu]])\vee((\nu_2^{\infty}=g.n.\text{ of }E_{\mu})\,\&\,(Ex)\,(prop^{\mu}x\\ \&\,\mathrm{T}[\pi^{(\kappa\dot-1)},\ldots,\pi^{(\mu)},\pi^{(\mu\dot-1)}+h_1[x,\nu],\ldots,\pi+h[x,\nu];\\ l.p.subst[x,\nu_5^{\infty},m\nu]]),$$

$\mu$ can be explicitely defined,

$$\mathrm{T}[1,\pi^{(\kappa\dot-1)},\ldots,\pi;\nu]\rightarrow LP^{\kappa}\,\nu\,\&\,((Ax)\,(\nu_2^{\infty}=g.n.\text{ of }A_{\kappa})\,\&\,(prop^{(\kappa)}x\\ \rightarrow\mathrm{T}[\pi^{(\kappa\dot+1)}+h_1[x,\nu],\ldots,\pi+h_{\kappa}[x,\nu];l.p.subst[x,\nu_5^{\infty},m\nu]])\\ \vee(((\nu_2^{\infty}=g.n.\text{ of }E_{\kappa})\,\&\,(Ex)\,\mathrm{T}[\pi^{(\kappa\dot-1)}+h_1[x,\nu],\ldots,\pi+h_{\kappa}[x,\nu];\\ l.p.subst\,[x,\nu_5^{\infty},m\nu]]))), \tag{E}$$

here $h_2[x,x'],\ldots,h_{\kappa}[x,x']$ are primitive recursive functions of $x,x'$. Scheme (E) is much more complicated than any scheme we have had so far but clearly we ultimately reduce

$$\mathrm{T}[\pi^{(\kappa)},\pi^{(\kappa\dot-1)},\ldots,\pi;\nu]\quad\text{to}\quad\mathrm{T}[\pi;\nu],$$

where $\mathrm{T}[\pi;\nu]$ is the truth definition for $\mathbf{A}_I$.

PROP. 7. *$\mathbf{A}^{(\kappa)}$-truth can be $\mathbf{A}^{(S\kappa)*}$-defined.*

We proceed by induction on $\kappa$. We have by Prop. 5 $\mathrm{T}[x;y]$ can be $\mathbf{A}'^{*}$-defined, we assume that $\mathbf{A}^{(\kappa)}$-truth, i.e. $\mathrm{T}[x^{(\kappa)},\ldots,x;y]$ can be $\mathbf{A}^{(S\kappa*)}$-defined and show that $\mathbf{A}^{(S\kappa)}$-truth can be $\mathbf{A}^{(SS\kappa)*}$-defined, i.e.

$$\mathrm{T}[x^{(S\kappa)},x^{(\kappa)},\quad x^{(\kappa\dot-1)},\ldots,x;y]$$

satisfying scheme (E), but with $S\kappa$ instead of $\kappa$, can be $\mathbf{A}^{(SS\kappa)*}$-defined. We proceed as in Prop. 5, the main difference is that here we have a set of parameters instead of just one, there we had $P[x,x']$ with the single parameter $x$, here we have $\mathrm{T}[x,x',\ldots,x^{(S\kappa)};y]$ with the parameters $x,x',\ldots,x^{(S\kappa)}$. We want to define a predicate $\mathrm{T}[x,x',\ldots,x^{(S\kappa)};y]$ which satisfies:

$$(Ax,x',\ldots,x^{(S\kappa)},y)\,(\mathrm{T}[x,x',\ldots,x^{(S\kappa)};y]\leftrightarrow H[x,x',\ldots,x^{(S\kappa)},y,T]),$$

where T is bound in $H$. As in Prop. 5 we define

$$K[x,x',\ldots,x^{(\kappa)},X]\quad\text{for}\quad(Ay,z)\,(z\leqslant x\rightarrow .X[z,x',\ldots,x^{(\kappa)},y]\\ \leftrightarrow H[z,x',\ldots,x^{(\kappa)},y,X]),$$

$$\mathrm{T}[x,x',\ldots,x^{(\kappa)},y]\quad\text{for}\quad(AX)\,(K[x,x',\ldots,x^{(\kappa)},X]\rightarrow X[x,x',\ldots,x^{(\kappa)},y]),$$

where $X$ is of order $S\kappa$. We then proceed to $\mathbf{A}^{(SS\kappa)*}$-prove 8 lemmas following those in Prop. 5, the final result is that T, as just defined, satisfies scheme (E) in $\mathbf{A}^{(SS\kappa)*}$. We use Cor. (ii), Prop. 5 to keep track of the orders of property variables. For instance, since $\mathrm{T}[x;y]$ is $\mathbf{A}'^*$-defined, then $\mathrm{T}[x,x';y]$ is $\mathbf{A}''^*$-defined, and so on.

As in Cor. (ii), Prop. 5, we can show that T as just defined is an adequate truth-predicate for $\mathbf{A}^{(\kappa)}$. In a similar manner we can define $Con_{(11)}\mathbf{A}^{(\kappa)}$ and show that this can be $\mathbf{A}^{(S\kappa)*}$-proved, following Prop. 6.

Now let us look at $\mathbf{A}^{(\omega)}$, the union of all the systems $\mathbf{A}^{(\kappa)}$ for $\kappa = 0, 1, 2, \ldots$. It will be convenient to alter the notation. Instead of

$$\mathrm{T}[x, x', \ldots, x^{(\kappa)}; y] \quad \text{write} \quad \mathrm{T}[\langle x, x', \ldots, x^{(\kappa)}\rangle, y],$$

and similarly for $K$. Note that the scheme that T is to satisfy is of the form ($h$ and $s$ are primitive recursive):

$$\mathrm{T}[\langle Sx\rangle^\frown x', y] \leftrightarrow H[y, T[\langle x\rangle^\frown h[x,y], s[x,y]]], \tag{i}$$

where $x$ is bound in $H$. Then we define

$$K[\langle x\rangle^\frown x', X] \leftrightarrow (Az, y)\,(z \leqslant x \rightarrow .\, X[\langle z\rangle^\frown x', y] \leftrightarrow H[y, X]), \tag{ii}$$

$$\mathrm{T}[\langle x\rangle^\frown x', y] \leftrightarrow (Ax)\,(K[\langle x\rangle^\frown x', X] \rightarrow X[\langle x\rangle^\frown x', y]). \tag{iii}$$

The orders of $X$ and $y$ are the same. Now an expression with a property variable of undetermined order is absent from our symbols. However we can always raise the order of a property variable in the definition of T provided $K[\langle x\rangle^\frown x', X]$ fails for all $X$ of order greater than $\operatorname{ord} y$. We could then replace the property variable $X$ in the above definitions of $K$ and T by $X_\omega$ the property variable introduced in forming $\mathbf{A}^{(\omega)'}$ or $\mathbf{A}^{(\omega+1)}$. But let us make the definitions of $K$ and T above but with the variable $X_\omega$ of order $\omega$, then we can carry through an $\mathbf{A}^{(\omega+1)*}$-proof that T as defined by (iii) satisfies its defining scheme (i). We can also $\mathbf{A}^{(\omega+1)*}$-prove that it is an adequate $\mathbf{A}^{(\omega)}$-truth-predicate. And so we can go on through the constructive ordinals.

## 11.10 *Truth-definitions in impredicative systems*

Now let us look at the situation in impredicative systems, that is systems in which we allow impredicative properties. Here, in the simplest case, we need only one kind of property variable, and a property can contain bound property variables, so that a property can depend, for its definition on all properties including itself, e.g. $\lambda\xi.(A\Xi)\,\phi\{\Xi\xi\}$. To define truth in an impredicative system we should require: $(A\Xi)\,\phi\{\Xi\alpha\}$ is true if and only if $\phi\{\Delta\alpha\}$ is true for all closed properties $\Delta$; but such a property could be $\lambda\xi.(A\Xi)\,\phi\{\Xi\xi\}$ itself.

Thus to decide the truth of $(A\Xi)\,\phi\{\Xi\alpha\}$ we should be referred to, among other things, $\phi\{(A\Xi)\,\phi\{\Xi\alpha\}\}$ and to decide the truth of this we should, in turn, be referred to $\phi\{\phi\{(A\Xi)\,\phi\{\Xi\alpha\}\}\}$, and so on without end. Even an oracle would fail at such a decision! If we put this process down as we did in the predicative case then the following arises; we start with an ordered set of zero and ones (these denote the sorts of quantifiers in the prefix when in **λ**-prenex normal form, zero for a numerical quantifier and one for a property quantifier) we then strike out the first quantifier and add at the end a bounded set of zeros and ones if we had struck out a one, but we leave the set unaltered if we had struck out a zero. This corresponds to passing from $(A\Xi)\,\phi\{\Xi\}$ to $\phi\{\Delta\}$ and then putting the result into **λ**-prenex normal form. We repeat the process. It is clear that eventually every zero and every one, whether an original one or an added one, will be struck out, but it is equally clear that at any stage there will be a sequence of zeros and ones, in general; in fact, in general, the sequence will grow in length. Thus, in general, the process will fail to terminate. These considerations give us cogent reasons for discarding impredicative systems. Nevertheless if we construct one, then, if it is consistent, it will, by Prop. 6, Ch. 12, have an $H$-model. The question is, can it be consistent? and if so, what is its truth-definition?

## 11.11 *Further extensions of the systems* $\mathbf{A}^{(\kappa)}$

Further extensions of the systems $\mathbf{A}^{(\kappa)}$ may be obtained by adjoining variables for properties of properties, that is variables of types $o(o\iota)$, $o(o\iota\iota)$, etc. We already have a constant of type $o(o\iota)$ namely $A$, the universal quantifier, so that $A\Delta_\iota$ is a statement in $\mathbf{A}^{(\kappa)}$. $A\Delta_\iota$ is read '$\Delta_\iota$ is a universal property', i.e. every natural number has property $\Delta_\iota$. The advantage of this type of extension is that we can then make statements about properties of a more general kind than we could before. Formulae of these types would be: $\boldsymbol{\lambda}X.\phi\{X\}$, $\boldsymbol{\lambda}XX'.\phi\{X,X'\}$, etc. The resulting system would allow us fuller use of the abstraction symbol. In fact, we could add variables for any type that we can get from the means at our disposal.

So far our systems have been limited in that we have only used a few of the types at our disposal. The difference between predicative and impredicative systems would still arise, and to avoid it we should require the concept of order for properties of properties. But one fails to reach finality. If we introduce an enumerated sequence of variables of every possible type available with the symbols at our disposal then we shall still only have an enumerable set of properties of type $o\iota$, these we can enumerate and by a diagonal process obtain a property of type $o\iota$ different from any in the enumerated list. It is our form of introducing universal quantifiers that restricts the possible things of a given type to being enumerable.

If we want to talk about all possible properties of type $o\iota$ whether we had a formula for them or were unable to find one, in a given system, then we should require a different method for introducing the universal quantifier, probably somewhat as in the system $A$. But it seems somewhat odd to want to talk about things which are unnameable in the system (in the language used).

This introduces us to the idea of starting with an unbounded set of symbols, perhaps one for each possible thing of a certain type. But this seems a bit absurd, because we are unable to use an unbounded set of symbols in any case, though we might be able to describe certain processes to do with them. But if we want to construct a language, we want to do things in that language itself and we want to avoid the use of another language to assist us to express what we want to express. Any extra equipment that we need in order to describe processes using an unbounded set of symbols should have been incorporated in the original construction of the system. Thus it would seem that formal systems with an unbounded set of symbols are impossible. Compare our discussion on Turing machines when we argued in favour of a displayed set of symbols, instructions, etc. *g.n.*'s are impossible for an uncountable set of symbols.

The hierarchy of systems $\mathbf{A}^{(\kappa)}$ for $\kappa = 0, 1, 2, \ldots$ can be continued into the constructible ordinals. Thus $\mathbf{A}^{(S\alpha)}$ is obtained from $\mathbf{A}^{(\alpha)}$ by adjoining new sorts of variables $X_{(\alpha\iota)}, X_{(\alpha\iota\iota)}, \ldots$ of type $(o\iota)$ and order $\alpha$, and quantifiers for them. But if $\alpha$ is a limit number then $\mathbf{A}^{(\alpha)}$ consists of the union of all the preceeding $\mathbf{A}^{(\beta)}$ for $\beta < \alpha$. When we adjoin a new sort of variable in these ways we also adjoin new rules II$d$, R 3* for their use.

Useful systems are $\mathbf{A}^{(\omega)}$ and $\mathbf{A}^{(\epsilon_0)}$, where $\epsilon_0$ is the first $\epsilon$-number.

## 11.12 *Incompleteness of extended systems*

To show that the extensions of $\mathbf{A}_I$ are incomplete and contain an irresolvable statement we require a slightly different argument to that we have used so far.

PROP. 8. *$A^{(\kappa)*}$ contains an irresolvable statement, if it is consistent.*

Let $Prf^{(\kappa)*}[\lambda, \mu]$ be the proof-predicate for the system $\mathbf{A}^{(\kappa)*}$. Then $Prf^{(\kappa)*}[\lambda, \mu]$ is a general recursive function of $\lambda$ and $\mu$. Consider:

(i) $(Ax')(\sim Prf^{(\kappa)*}[x', s[x,x]] \vee (Ex'')(x'' \leqslant x' \,\&\, Prf^{(\kappa)*}[x'', Negs[x,x]]))$, let its $g.n.$ be $\theta$, and let $\nu$ be the $g.n.$ of:

(ii) $(Ax')(\sim Prf^{(\kappa)*}[x', s[\theta,\theta]] \vee (Ex'')(x'' \leqslant x' \,\&\, Prfr^{(\kappa)*}$ $[x'', Negs[\theta,\theta]]))$;

(iii) (ii) is equivalent to

$$(Ax')(\sim Prf^{(\kappa)*}[x', \nu] \vee (Ex'')(x'' \leqslant x' \,\&\, Prf^{(\kappa)*}[x'', Neg\,\nu])),$$

since $\nu = s[\theta,\theta]$ is an $\mathbf{A}_{00}$-theorem.

(iii) says that if (ii) is an $\mathbf{A}^{(\kappa)*}$-theorem then there is an $\mathbf{A}^{(\kappa)*}$-proof of the negation of (ii) and this $\mathbf{A}^{(\kappa)*}$-proof has a lesser $g.n.$ than the $\mathbf{A}^{(\kappa)*}$-proof of (ii) itself. Hence if $\mathbf{A}^{(\kappa)*}$ is consistent with respect to negation (iii) fails to be an $\mathbf{A}^{(\kappa)*}$-theorem.

If (iii) is an $\mathbf{A}^{(\kappa)*}$-theorem then so is (ii), since $\nu = s[\theta,\theta]$ is an $\mathbf{A}_{00}$-theorem. Hence

(iv) $$Prf^{(\kappa)*}[\pi, \nu] \text{ for some numeral } \pi,$$

and since $\mathbf{A}^{(\kappa)*}$ is consistent

(v) $$(Ax'') \sim Prf^{(\kappa)*}[x'', Neg\,\nu].$$

Hence from (iii)

$$\sim Prf^{(\kappa)*}[\pi, \nu] \vee (Ex'')(x'' \leqslant \pi \,\&\, Prf^{(\kappa)*}[x'', Neg\,\nu]),$$

but this is the negation of ((iv) & (v)). This is absurd, thus if $\mathbf{A}^{(\kappa)*}$ is consistent then (iii) fails to be an $\mathbf{A}^{(\kappa)*}$-theorem.

Now suppose that the negation of (iii) is an $\mathbf{A}^{(\kappa)*}$-theorem. Then the negation of (ii) is an $\mathbf{A}^{(\kappa)*}$-theorem, thus:

(vi) $Prf^{(\kappa)*}[\pi, Neg\,\nu]$ for some numeral $\pi$. From this we get

$$(Ax')(x' \geqslant \pi \rightarrow (Ex'')(x'' \leqslant x' \,\&\, Prf^{(\kappa)*}[x'', Neg\,\nu])).$$

If $\mathbf{A}^{(\kappa)*}$ is consistent with respect to negation then

$$\sim Prf^{(\kappa)*}[0, \nu], \ldots, \sim Prf^{(\kappa)*}[\pi \dot{-} 1, \nu]$$

are $\mathbf{A}^{(\kappa)*}$-theorems, whence:

(vii) $$(Ax')(x' < \pi \rightarrow \sim Prf^{(\kappa)*}[x', \nu])$$

is an $\mathbf{A}^{(\kappa)*}$-theorem.

From (vi) and (vii) we get

$$(Ax')(\sim Prf^{(\kappa)*}[x', \nu] \vee (Ex'')(x'' \leqslant x' \,\&\, Prf[x'', Neg\,\nu])),$$

which is (iii). This again is absurd, hence the negation of (iii) also fails to be an $\mathbf{A}^{(\kappa)*}$-theorem. This completes the demonstration of the proposition.

This gives us a standard method for producing an irresolvable statement from a given consistent system.

## 11.13 *Real numbers*

The systems $\mathbf{A}^{(\kappa)}$ give rise to various orders of analysis. We have already explained how the rational numbers, positive and negative, can be represented by ordered triplets of natural numbers, and so can be discussed in $\mathbf{A}_{00}$. Properties could have been called classes of natural numbers, it is a matter of taste whether we use the word 'property' or the word 'class', thus we could consider classes of rational numbers, because in our construction of ordered triplets there is a (1–1)-correspondence between natural numbers and ordered triplets of natural numbers, thus a property could be considered as representing a class of rational numbers, or a class of natural numbers, the context should make it clear which was intended.

In the system $\mathbf{A}^{(\kappa)}$ we introduce real numbers thus:

D 267 $\quad Sect\,X \quad$ for $\quad (Ax, x')(Xx \,\&\, x' <_r x . \rightarrow Xx') \,\&\, (Ex)\,Xx$
$$\& \, (Ex) \sim Xx \,\&\, (Ax)(Xx \rightarrow (Ex')(x <_r x' \,\&\, Xx')),$$

this defines a Dedekind section; here $<_r$ is the relation of order between ordered triplets (see Ex. 7, Ch. 5). The last clause in the definition limits us to sections without last member, this is a matter of taste, it means that a section is uniquely represented. If *Sect* $\Delta$ then $\Delta$ is called a real number of order $\kappa$, where $\kappa$ is the order of $\Delta$.

We shall find that in a first course in analysis we can get on without using bound variables for sets of real numbers if we state our theorems as meta-theorems. We did this in $\mathbf{A}_{00}$ where we stated the commutative law of addition as : $\alpha+\beta = \beta+\alpha$, where $\alpha$ and $\beta$ are numerical terms. We can define a set of real numbers by abstraction thus: $\boldsymbol{\lambda}X.(\phi\{X\} \,\&\, \mathit{Sect}\,X)$, the set of sections which satisfy $\phi$.

The most important concept in analysis is the *l.u.b.* of a non-empty bounded set of real numbers. The *l.u.b.* of the set of real numbers $\boldsymbol{\lambda}X.(\phi\{X\} \,\&\, \mathit{Sect}\,X)$ is given by:

D 268 $\quad l.u.b.\boldsymbol{\lambda}X.(\phi\{X\} \,\&\, \mathit{Sect}\,X) \quad$ for $\quad \boldsymbol{\lambda}x.(EX')(X'x \,\&\, \phi\{X'\} \,\&\, \mathit{Sect}\,X').$

The union of all the sections which satisfy $\phi$. The first thing we notice is that the order of the *l.u.b.* of a set of real numbers is at least one greater than the orders of the members comprising the set, this is the order of the variable $X$ in $\phi\{X\}$. All members of a set of real numbers are of the same order. We can $\mathbf{A}^{(S\kappa)}$-prove $\mathit{Sect}\,l.u.b.\,\boldsymbol{\lambda}X.(\phi\{X\} \,\&\, \mathit{Sect}\,X)$, where $X$ is a variable of order $\kappa$ and $\phi\{\Gamma_{o\iota}\}$ is of order $\kappa$. We can also $\mathbf{A}^{(S\kappa)}$-prove:

(i) $(AX)(\phi\{X\} \,\&\, \mathit{Sect}\,X. \rightarrow .X \leqslant_s l.u.b.(\boldsymbol{\lambda}X.(\phi\{X\} \,\&\, \mathit{Sect}\,X)),$

(ii) $(AX)((X <_s l.u.b.(\boldsymbol{\lambda}X.\phi\{X\} \,\&\, \mathit{Sect}\,X)) \rightarrow (EX')(\mathit{Sect}\,X' \,\&\, \phi\{X'\} \,\&\, X <_s X')).$

The two fundamental properties of *l.u.b.*. $X <_s X'$ is the relation of order between real numbers, viz. $(Ex)(X'x \,\&\, \sim Xx)$ or $X \subset X'$ if $X$, $X'$ are sections. The order of the *l.u.b.* of a set of real numbers of order $\mu$ might be of any order greater than $\mu$, because there might be bound property variables in $\phi\{X\}$ of any order less than $\kappa$, so that in $\mathbf{A}^{(\kappa)}$ we could have a set of real numbers of order one whose *l.u.b.* was of order $\kappa$. But note that a real number can be the same, in the sense of consisting of exactly the same rationals, as other real numbers of any order, we need only use $\boldsymbol{\lambda}X.(\phi\{X\} \,\&\, \psi \,\&\, \mathit{Sect}\,X)$ instead of $\boldsymbol{\lambda}X.(\phi\{X\} \,\&\, \mathit{Sect}\,X)$, where

$\psi$ is a theorem of the required order and is without free occurrences of $X$. Thus the real numbers of order $\kappa$ are incomplete. This difficulty is overcome by using the system $\mathbf{A}^{(\omega)}$. A set of real numbers in $\mathbf{A}^{(\omega)}$ will be of some order given by a natural number, say $\mu$, the *l.u.b.* of the set (if bounded) will also be represented by some greater natural number $\lambda$ hence it will be in $\mathbf{A}^{(\lambda)}$ and so in $\mathbf{A}^{(\omega)}$. Thus $\mathbf{A}^{(\omega)}$ will be complete. In general, analysis in $\mathbf{A}^{(\alpha)}$, where $\alpha$ is a limit number, will be complete, but if $\alpha$ is a successor then $\mathbf{A}^{(\alpha)}$ will be incomplete.

Classical analysis is impredicative. There we have only one sort of variable for classes and so the class given by *l.u.b.* depends (on account of the bound property variable) on all classes including the very one defined. This is a blemish in classical analysis and is avoided in our case by having various orders of properties. The resulting analysis is naturally more complicated.

Having defined the *l.u.b.* of a non-null bounded set of real numbers we can then define $\overline{lim}$ and $\underline{lim}$ and so limits, but in general these will be of higher order than the members of the set or sequence. Weirstrass' theorem will follow in $\mathbf{A}^{(\omega)}$, but will fail in $\mathbf{A}^{(\kappa)}$ for certain sets.

Functions of real variables will be given by $\boldsymbol{\lambda}x.\phi\{x,X\}$, the domain of the function is given by the set of $X$ for which

$$Sect\, X \,\&\, Sect\, \boldsymbol{\lambda}x.\phi\{x,X\}.$$

A sequence of real numbers is represented by $\boldsymbol{\lambda}x.\phi\{x,\nu\}$, provided $Sect\,\boldsymbol{\lambda}x\,\phi\{x,\nu\}$. We can now prove the usual theorems on sequences and series, but we shall have to keep note of possible increase in order, this will be immaterial if we are dealing with $\mathbf{A}^{(\omega)}$.

The next type of theorem we come to concerns continuous functions, for this discussion we require the Heine–Borel Theorem. This we state in the form:

PROP. 9. *Heine–Borel Theorem.*

$$(AX_s)\,(0_s \leqslant_s X_s \leqslant_s 1_s . \rightarrow . \boldsymbol{\lambda}x.\phi\{x,X_s\} <_s X_s <_s \boldsymbol{\lambda}x.\psi\{x,X_s\})$$
$$\rightarrow (E\nu)\,(AX_s)\,(EY_s)\,(0_s \leqslant_s X_s \leqslant_s 1_s : \rightarrow . \boldsymbol{\lambda}x.\phi\{x,Y_s\}$$
$$<_s X_s <_s \boldsymbol{\lambda}x.\psi\{x,Y_s\} . \,\&\, \boldsymbol{\lambda}x.\phi\{x,Y_s\} +_s \frac{1}{\nu_s}{}_s <_s \boldsymbol{\lambda}x.\psi\{x,Y_s\}),$$

where $X_s$ is a variable for a real number of order $\kappa$; $0_s$, $1_s$ are the real numbers zero and one respectively $<_s, +_s, \frac{1_s}{\nu_s}{}_s$ are order, addition and reciprocal for real numbers, we are supposing that

$$(AX_s)\,(0_s \leqslant_s X_s \leqslant_s 1_s. \rightarrow .\, Sect\, \boldsymbol{\lambda} x.\phi\{x, X_s\} \,\&\, Sect\, \boldsymbol{\lambda} x.\psi\{x, X_s\}).$$

$(AX_s)\,\phi\{X_s\}$, $(EX_s)\,\phi\{X_s\}, \boldsymbol{\lambda} x.\phi\{X_s\}$ are the relativized quantifiers and properties, viz.:

$$(AX_s)\,\phi\{X_s\} \quad \text{for} \quad (AX)\,(Sect\, X \rightarrow \phi\{X\}),$$

$$(EX_s)\,\phi\{X_s\} \quad \text{for} \quad (EX)\,(Sect\, X \,\&\, \phi\{X\}),$$

$$\boldsymbol{\lambda} x.\phi\{x, X_s\} \quad \text{for} \quad \boldsymbol{\lambda} x.(\phi\{x, X\} \,\&\, Sect\, X).$$

Write

$$H_\nu[U_s, V_s] \quad \text{for} \quad (AX_s)\,(EY_s)\,(U_s \leqslant_s X_s \leqslant_s V_s. \rightarrow .\, \boldsymbol{\lambda} x.\phi\{x, Y_s\} <_s X_s <_s \boldsymbol{\lambda} x.\psi\{x, Y_s\} \,\&\, \boldsymbol{\lambda} x.\phi\{x, Y_s\} +_s \frac{1_s}{\nu_s}{}_s <_s \boldsymbol{\lambda} x.\psi\{x, Y_s\})$$

this is only wanted when; $0_s \leqslant_s U_s <_s V_s \leqslant_s 1_s$.

$$Hp \rightarrow (AX_s)\,(0_s \leqslant_s X_s <_s Min_s[1_s, \boldsymbol{\lambda} x.\psi\{x, 0_s\}]. \rightarrow .\, \boldsymbol{\lambda} x.\phi\{x, 0_s\} <_s X_s <_s \boldsymbol{\lambda} x.\psi\{x, 0_s\}),$$

hence

$$Hp \rightarrow (EV_s)\,(E\nu)\,H_\nu[0_s, V_s],$$

i.e. the set of real numbers $\boldsymbol{\lambda} x.(E\nu)\,H_\nu[0_s, V_s]$. If two overlapping intervals have property $H$ then so does their union, in fact if one has property $H$ for $\nu'$ and the other has property $H$ for $\nu''$ then their union has property $H$ for $Max\,[\nu', \nu'']$, thus:

$$0_s \leqslant_s U'_s <_s U''_s <_s V'_s <_s V''_s \leqslant_s 1_s. \rightarrow :(E\nu')\,H_{\nu'}[U'_s, V'_s] \,\&\, (E\nu'')\, H_{\nu''}[U''_s, V''_s] \rightarrow (E\nu)\,H_\nu[U'_s, V''_s].$$

Again $$Hp \rightarrow (AV_s)\left(0_s <_s V_s <_s 1_s \rightarrow (E\mu, \nu)\,H_\nu\left[V_s -_s \frac{1_s}{\mu_s}{}_s, V_s +_s \frac{1_s}{\mu_s}{}_s\right]\right),$$

$$Hp \rightarrow \left(AV_s)\,(0_s <_s V_s <_s 1_s. \,\&\, (E\nu)\,H_\nu[0, V_s] : \rightarrow (E\mu, \nu)\,H_\nu\left[0_s, V_s +_s \frac{1}{\mu_s}{}_s\right]\right)$$

$$\rightarrow (AV_s)\,(0_s <_s V_s <_s 1_s. \rightarrow V_s \neq l.u.b.\, \boldsymbol{\lambda} x.(E\nu)\,H_\nu[0_s, V_s])$$

$$\rightarrow l.u.b.\, \boldsymbol{\lambda} x.(E\nu)\,H_\nu[0_s, V_s] = 1_s.$$

Thus

$$Hp \rightarrow .(A\mu)(E\nu)H_\nu\left[0_S, 1_S -_S \frac{1_S}{2\mu}{}_S\right] \& (E\mu,\nu)H_\nu\left[1_S -_S \frac{1}{\mu_S}{}_S, 1_S\right]$$
$$\rightarrow (E\nu)H_\nu[0_S, 1_S]$$

as desired.

Note that the demonstration is in $\mathbf{A}^{(S\kappa)}$ on account of the use of *l.u.b.* The ordinary first course of analysis can now proceed without difficulty, series, limits, Riemann integral (best done by dividing interval of integration into equal parts), differential calculus, still without the need to introduce variables for properties of properties or any higher order variables, these would correspond to variables for classes of classes, etc., in our case sets of real numbers. This is because in the elementary parts of analysis we need only use free variables for sets of real numbers, so that our theorems can be stated as meta-theorems. We did this in $\mathbf{A}_{OO}$, as mentioned before. In the ordinary development of analysis, having developed the theory of series and limits, we go on to the Riemann integral and the differential calculus, and then to the Theory of Functions of a Complex Variable. This requires the construction of complex numbers. These are defined as ordered pairs of real numbers with special definitions of equality, addition and multiplication; an ordered pair of real numbers is represented by:

D 269 $\{\{X, X'\}\}$ for $\boldsymbol{\lambda}x.(X \cup X')$, i.e. $\boldsymbol{\lambda}x.(Xx \vee X'x)$,

this is only required when

$$Sect\, X \,\&\, Sect\, X'.\&.(Xx \rightarrow Evx).\&.(X'x \rightarrow Odx).$$

One property gives rise to two other properties, the one applying to the even numbers and the other applying to the odd numbers, similarly from two properties we can get one.

D 270 $Evx$ for $2/x$, i.e. 2 divides $x$,

D 271 $Odx$ for $2\not/x$, otherwise.

Having defined the complex numbers we can proceed to develop the theory of functions of a complex variable until we come to the definition of a simply-connected domain. This requires quantification over functions of real variables. The crucial clause is: 'any two points of the set can be joined by a continuous curve lying in the set'. This requires 'there is a function...' the function, of course, gives the curve, and to express the crucial clause we require quantification over function variables. Thus at this point we are forced to introduce variables of more complicated types.

A function of a real variable is of type $o\iota(o\iota)$. Particular functions are given by

$$\boldsymbol{\lambda} X.\boldsymbol{\lambda} x.(E!X')(\phi\{X,X'\}\,\&\,X'x) \tag{a}$$

D 272 $(E!X)\phi\{X\}$ for $(EX)(\phi\{X\}\,\&\,(AX')(\phi\{X'\}$
$\to (Ax)(X'x \leftrightarrow Xx)))$,

read 'there is exactly one $X$ such that $\phi\{X\}$'. The same with a suffix $s$ on all the $X'$s would read 'there is exactly one real number $X_s$ such that $\phi\{X_s\}$ & *Sect* $X_s$'. We want $(a)$ with the suffix $s$ on all the $X$'s.

As one develops analysis every now and again one will require to introduce variables and constants of more and more complicated types. One can try to keep the development predicative or one can take the easier course offered by the impredicative approach. But sooner or later one comes across something that can be done using *Mult. Ax.* and maybe fails without its use. In the predicative case, given, say a set of real numbers of order $\kappa$, we can always pick out a member of the set, though the choice fails to be defined in the system we are using. In the predicative case there is an enumerated set of real numbers of order $\kappa$, enumerated say by increasing *g.n.*, so we can always pick out the one in the set with least *g.n.* But this process fails to be definable in any of the systems $\mathbf{A}^{(\kappa)}$. Also there may be cases when we are unable to decide whether a given real number belongs to a given set of real numbers. Thus in any case we should require a function of type $\iota(o\iota)$ which associates a real number with each property, and an axiom for its use, e.g. $X(\mathscr{F}X)$, $\mathscr{F}X$ being the natural number picked from those having the property $X$, etc. for real numbers.

Having developed the theory of functions of a real variable to a certain extent we next require the Lebesgue Integral. In any of the systems $\mathbf{A}^{(\alpha)}$ there are only an enumerable set of real numbers, in fact we can enumerate them by increasing *g.n.* This is possible because a real number is a well-formed formula of type $o\iota$ and we have an effective method for testing this; the well-formed-formula (w.f.f.) must further satisfy the condition of being a section, this might be difficult to decide; but, if $\boldsymbol{\lambda} x.\phi\{x\}$ denotes a w.f.f. of type $o\iota$ then, provided $(Ex)\phi\{x\}$

$$\boldsymbol{\lambda} x.(Ex')(x <_r x'\,\&\,\phi\{x'\})$$

is a section, and by enumerating these we get the same set of real numbers as we get by enumerating those of $\lambda x\phi\{x\}$ which are real numbers. Thus the real numbers are enumerable, but this enumeration is outside $\mathbf{A}^{(\alpha)}$. There will be some sets of real numbers given by $\psi\{X\}$ which fail to be enumerable in $\mathbf{A}^{(\alpha)}$ simply because the system $\mathbf{A}^{(\alpha)}$ is too weak in modes of expression to be able to express a (1–1)-correspondence between the real numbers of the set and the natural numbers.

On the other hand in classical analysis we can only name an enumerable set of real numbers and we can enumerate them, say by increasing *g.n.* Yet, in classical analysis we talk about all real numbers whether we can name them or whether they can only be named in some more extended system. Surely it is senseless to talk about, or to imagine that one is talking about, things that are utterly without names in any formal system. In the systems $\mathbf{A}^{(\alpha)}$ all the real numbers dealt with have names. This is like the system $\mathbf{A}_{00}$ where all the natural numbers have names. Note in passing that the Roman system of naming the natural numbers gives out after a bit. Elementary analysis up to the Lebesgue integral goes quite easily in $\mathbf{A}^{(\alpha)}$, where $\alpha$ is a limit number. This is because it is largely about sequences of real numbers of *l.u.b.*'s of sets of real numbers such as

$$\overline{lim}_{X\to Y+}\, \boldsymbol{\lambda} x\,(\phi\{x,X\}\,\&\,Sect\,[X,\boldsymbol{\lambda} x\phi\{x,X\}]) \quad \text{for}$$

$$\underset{\nu\to\infty}{g.l.b.}\ \underset{Y<_s X<_s Y+_s \frac{1}{\nu^s}}{l.u.b.\,\boldsymbol{\lambda} x}\,(\phi\{x,X\}\,\&\,Sect\,[X,\boldsymbol{\lambda} x\phi\{x,X\}]),$$

and

$$\overline{S}\Big|_0^1\, \boldsymbol{\lambda} x(\phi\{x,X\}\,\&\,Sect\,[X,\boldsymbol{\lambda} x\phi\{x,X\}]) \quad \text{for}$$

$$\underset{\nu\to\infty}{g.l.b.}\ \sum_{\mu=0}^{\nu\dot{-}1}\ \underset{\frac{\mu}{\nu^s}<_s X<_s \frac{S\mu}{\nu^s}}{l.u.b.\,\boldsymbol{\lambda} x}\,(\phi\{x,X\}\,\&\,Sect\,[X,\boldsymbol{\lambda} x\phi\{x,X\}])\,\frac{1}{\nu_s},$$

and similarly for $\underline{S}\Big|_0^1$, the Riemann integral is then expressed by means of sequences.

When we come to the Lebesgue integral we want to define the inner and outer measures of arbitrary sets of real numbers (to take the one dimensional case). Now all the sets of real numbers that we can define are enumerable and as such should have measure zero. So it looks as if the whole theory of Lebesgue integration will be unexpressible in any $\mathbf{A}^{(\alpha)}$. Now a set of real numbers is of measure zero if and only if it can be covered with a set of intervals whose total length is arbitrarily small. We can cover up all the real numers in $\mathbf{A}^{(\alpha)}$ by the set of intervals

$$\left[\Delta_\mu -_s \frac{1_s}{\nu_s \mu_s^2}{}_s, \quad \Delta_\mu +_s \frac{1_s}{\nu_s \mu_s^2}{}_s\right]$$

where $\nu$ is any numeral and $\Delta_\mu$ for $\mu = 1, 2, \ldots$ runs through all real numbers. But this set fails to belong to any $\mathbf{A}^{(\alpha)}$ and so we are unable to use it. Similarly for the set of real numbers given by $\hat{X}_s\phi\{X_s\}$, in the case when we are unable to enumerate the sections $\Delta$ which satisfy $\phi\{\Delta\}$. The set $\hat{X}_s\phi\{X_s\}$ will then behave like a set of any cardinal $\leqslant c$, and maybe there will be sets of intervals expressible in $\mathbf{A}^{(\alpha)}$ and lying in $\hat{X}_s\phi\{X_s\}$ having a positive total length. It seems that this would be the case if all we could $\mathbf{A}^{(\alpha)}$-prove about real numbers of the set $\hat{X}_s\phi\{X_s\}$ was that they were in certain intervals, but we were unable to $\mathbf{A}^{(\alpha)}$-prove exactly where they were. This reminds one of the uncertainty principle of Quantum Mechanics!

From our definitions we have

$$\hat{X}_s\phi\{X_s\} \subseteq \hat{X}_s\chi\{X_s\} \leftrightarrow (AX_s)\,(\phi\{X_s\} \to \psi\{X_s\}).$$

Thus if all the real numbers in $\mathbf{A}^{(\alpha)}$ in a certain interval are also in another set of real numbers then the second set of real numbers contains the interval. The inner measure of a set of real numbers is the upper bound of the sum of the lengths of non-overlapping open intervals contained in the set.

D 273 $$\underline{m}\hat{X}_s\phi\{X_s\} \quad \text{for} \quad l.u.b. \sum_{\mu=1}^{\nu} \frac{\epsilon_\mu^\nu}{\nu},$$

where $$\epsilon_\mu^\nu = \begin{cases} 1 & \text{if} \quad \hat{X}_s\left(\frac{\mu_s}{\nu_s}{}_s <_s X_s <_s \frac{(\mu + 1_s)}{\nu_s}{}_s\right) \subseteq \hat{X}_s\phi\{X_s\}, \\ 0 & \text{otherwise.} \end{cases}$$

In order to keep within our formal system $\mathbf{A}^{(\alpha)}$ the numerical term must be explicitly defined in that system. Thus we shall have to define $\epsilon_\mu$ as follows

$$\epsilon_\mu^\nu = \underset{\kappa \to \infty}{l.u.b.}\,(1 \dot{-} prf[\kappa, \theta]),$$

where $\theta$ is the $g.n.$ of $\hat{X}_s\left(\frac{\mu_s}{\nu_s}{}_s <_s X_s <_s \frac{\mu_s + 1_s}{\nu_s}{}_s\right) \subseteq \hat{X}_s\phi\{X_s\}$, and $\kappa$ is the $g.n.$ of a $A^{(\kappa)}$ proof of '$\theta$'.

The outer measure is defined by

D 274 $$\overline{m}\hat{X}_s\phi\{X_s\} \quad \text{for} \quad 1_s \dot{-}_s \underline{m}([0_s, 1_s] - \hat{X}_s\phi\{X_s\}),$$

for a set in the interval $[0_s, 1_s]$, similarly for other intervals. Note that in the definition of $\epsilon_\mu^\nu$ we are unable to replace *prf* by truth because the truth-definition of $\mathbf{A}^{(\kappa)}$ is outside $\mathbf{A}^{(\kappa)}$.

As we pass through the hierarchy of systems $\mathbf{A}^{(\kappa)}$ so we obtain more and more real numbers. Thus an interval $\hat{X}_s[\Delta_s <_s X_s <_s \Delta_s']$ should be given an order, namely the order of the variable $X_s$. The order of the set $\hat{X}_s\phi\{X_s\}$ is the order of the variable $X_s$ even though the order of $\phi\{\Gamma_{o\iota}\}$ may be greater. Thus the same statement form $\phi\{\Gamma_{o\iota}\}$ can give rise to sets of all orders. Of course the set $\hat{X}_s\phi\{X_s\}$ can only be handled in the systems $\mathbf{A}^{(\kappa)}$ which contain $\phi\{X_s\}$. Thus we can have a set of order one which can only be handled in $\mathbf{A}''$ onwards, because of the orders of the bound variables in $\phi\{\Gamma_{o\iota}\}$.

Now suppose that we can $\mathbf{A}^{(S\kappa)}$-prove

$$(AX_s)(\Delta_s <_s X_s <_s \Delta_s' \rightarrow \phi\{X_s\}), \tag{a}$$

so that the open interval $(\Delta_s, \Delta_s')$ of order $\kappa$ lies in the set $\hat{X}_s\phi\{X_s\}$ of order $\kappa$, then this will still hold for orders higher than $\kappa$. This follows because the only way we have of $\mathbf{A}^{(S\kappa)}$-proving $(a)$ is by a free variable $\mathbf{A}^{(S\kappa)}$-proof of

$$\Delta_s <_s X_s <_s \Delta_s'. \rightarrow \phi\{X\}_s. \tag{b}$$

Now this $\mathbf{A}^{(S\kappa)}$ free variable proof (i.e. a proof in the system obtained from $A^{(S\kappa)}$ by allowing free variables) will be without bound variables of order $\kappa$ because $\hat{X}_s\phi\{X_s\}$ is of order $\kappa$. Thus we will still have a free variable $\mathbf{A}^{(\pi)}$-proof if we replace the free variable $X_s$ everywhere by one of higher order $\pi$. Thus if we can $\mathbf{A}^{(S\kappa)}$-prove $(a)$ then we can $\mathbf{A}^{(\pi)}$-prove the corresponding statement when the variable $X_s$ is replaced by one of higher order $\pi$. Similarly for any two ordinals (constructive, of course).

Thus if an interval of order $\kappa$ is contained in a set of order $\kappa$, then the corresponding interval of order $\pi$ greater than $\kappa$ lies in the corresponding set of that higher order. Where the set $\hat{X}_s\phi\{X_s\}$ with $X_s$ of order $\kappa$ corresponds to the set $Y_s\phi\{Y_s\}$ with $Y_s$ of order $\pi$. Thus the inner measure of a set is the same as the inner measure of all corresponding sets of higher order. Thus if an interval of order $\kappa$ is $\mathbf{A}^{(S\kappa)}$-proved to lie in a set of order $\kappa$, and if a real number of higher order can be proved in a suitable $\mathbf{A}^{(\pi)}$

to lie in the corresponding interval of that higher order then that real number also can be proved in a suitable order to lie in the corresponding set of higher order.

## 11.14 *The analytical hierarchy*

We have already noted that instead of property variables we could have used function variables. Set variables are much the same as property variables. In the theory of Turing oracle machines we used functions rather than sets. We could build up a similar theory using sets instead of functions, the oracle would then tell us, on demand, whether a natural number was in a certain set or otherwise. But the theory of Turing oracle machines seems to go nicer using functions rather than sets, and most of the literature uses functions rather than sets. In the same way the analytical hierarchy, about to be described, seems to go more easily using functions rather than sets.

In Ch. 9 we defined a hierarchy of arithmetically definable sets of natural numbers, and noted that we could define a similar hierarchy of $f$-definable sets of natural numbers. We wish to extend this hierarchy to sets whose definition involves bound set or function variables. In general these will fail to be arithmetically definable.

In our account of the extensions of the system $\mathbf{A}_I$ we introduced property variables of various orders. Similarly we shall consider functions of various orders. In our systems only the primitive recursive functions get explicit definition. General recursive functions were given an implicit definition as $\phi\{x,y\}$ represented $fx = y$, if and only if,

$$(Ax)\,(Ey)\,\phi\{x,y\}\,\&\,(Ax,y,z)\,(\phi\{x,y\}\,\&\,\phi\{x,z\} \rightarrow y = z),$$

is **A**-true. For the function $f$ we could take $\boldsymbol{\lambda} x \iota y \phi\{x,y\}$, where $\iota$ is the symbol introduced in our discussion of resolved $\mathscr{F}_C$ in Ch. 3.

A function of order 0 is one which is representable in the system **A**.

$$fx = y \leftrightarrow \phi\{x,y\},$$

where $\quad (Ax)\,(Ey)\,\phi\{x,y\}\,\&\,(Ax,y,z)\,(\phi\{x,y\}\,\&\,\phi\{x,z\}.\rightarrow y = z).$

Thus a function defines a set of ordered pairs. Similarly functions of order $S\kappa$ are defined where $\phi\{x, y\}$ contains bound $\kappa$-order functions. The function $f$ can be defined as $\boldsymbol{\lambda} x \iota y \phi\{x,y\}$ as before.

In the literature most of the work about the analytical hierarchy has been done using function variables rather than set variables (or property variables). We define systems $\mathbf{A}_f^{(\kappa)}$ as for the systems $\mathbf{A}^{(\kappa)}$ except that we use function variables instead of property variables. The sets of natural numbers that we are concerned with are given by statements in $\mathbf{A}_f^{(\kappa)}$ with $\pi$ free numerical variables and have all function variables bound. These statements can be put into prenex normal form. We consider first the system $\mathbf{A}_f'$, by using ordered tuplets we may consider that all functions are one-place functions. Write $A^1$, $E^1$ for function quantifiers and $A^0$, $E^0$ for numerical quantifiers. Then the prefix of an $\mathbf{A}_f'$-statement in prenex normal form will consist of a sequence of $A^1$'s, $E^1$'s, $A^0$'s and $E^0$'s, in some order. In this sequence omit all numerical quantifiers and the resulting sequence of function quantifiers is called the *reduced prefix*.

A $\Sigma_{S_\nu}'$*-prefix* is one whose reduced prefix contains $\nu$ alternations of quantifiers and begins with $E'$.

A $\Pi_{S_\nu}'$*-prefix* is one whose reduced prefix has $\nu$ alternations of quantifiers and begins with $A'$.

A $\Sigma_\nu'$*-statement* is one whose reduced prefix is a statement with a $\Sigma_\nu'$-prefix.

A $\Pi_\nu'$*-statement* is a statement with a $\Pi_\nu'$-prefix.

$\Sigma_\nu'$ is the class of all $\Sigma_\nu'$-statements,

$\Pi_\nu'$ is the class of all $\Pi_\nu'$-statements.

$\Sigma_0'$ and $\Pi_0'$ are statements with empty reduced prefix, thus they are $\mathbf{A}$-statements.

We now generalize what we did in Ch. 9

PROP. 10. $\Sigma_\nu' \cup \Pi_\nu' \subset \Sigma_{S_\nu}^{\nu} \cap \Pi_{S_\nu}'$.

The demonstration is as for Prop. 2, Ch. 9.

PROP. 11. *The following prefix transformations are permissible. I.e. in each case, for any statement with a prefix in the left column there is an equivalent statement with a prefix in the same row but in the right column, and there is an effective way of finding this statement.*

$$\text{(i)} \qquad \begin{cases} \ldots A^0A^0\ldots & \qquad \ldots A^0\ldots \\ \ldots E^0E^0\ldots & \qquad \ldots E^0\ldots \end{cases}$$

(ii) $\begin{cases} \ldots A^1A^1\ldots & \ldots A^1\ldots \\ \ldots E^1E^1\ldots & \ldots E^1\ldots \end{cases}$

(iii) $\begin{cases} \ldots A^0\ldots & \ldots A^1\ldots \\ \ldots E^0\ldots & \ldots E^1\ldots \end{cases}$

(iv) $\begin{cases} \ldots A^0E^1\ldots & \ldots E^1A^0\ldots \\ \ldots E^0A^1\ldots & \ldots A^1E^0\ldots \end{cases}$

(i) We have already done this in Ch. 9.

(ii) Replace $\ldots(Af)(Ag)\ldots\phi\{\ldots,f,g\ldots\}$ by $\ldots(Ah)\ldots\phi\{\ldots,h_1^1,h_2^1,\ldots\}$, where $h_1^1$ is $\boldsymbol{\lambda}x.(hx)_1^1$, etc. Similarly for $E$.

(iii) Replace $\ldots(A^0x)\ldots\phi\{\ldots,x,\ldots\}$ by $\ldots(A^1f)\ldots\phi\{\ldots,f0,\ldots\}$.

(iv) Replace $\ldots(A^0x)(E^1f)\ldots\phi\{\ldots,x,f,\ldots\}$ by

$$\ldots(E^1f)(A^0x)\ldots\phi\{\ldots,x,\boldsymbol{\lambda}yf\langle x,y\rangle\ldots\}.$$

Similarly for $E^0A^1$. Or take negations using $\sim\phi$ instead of $\phi$. Note that (iv) requires the axiom of choice to establish equivalence and that the equivalent statements are in $\mathbf{A}'_f$ and are of the same order as before. It can be shown that the converses of (iii) and (iv) fail.

Cor. *Any $\Sigma'_\nu$-statement is equivalent to one whose prefix consists of $S\nu$ alternating quantifiers of which the first $\nu$ are function quantifiers, and the first of these is $E^1$, and only the last one is a numerical quantifier. Similarly for $\Pi'_\nu$-statements except that the first quantifier is $A^1$.*

Using Prop. 11 all function quantifiers can be moved to the left of all numerical quantifiers. Then contract quantifiers, then replace all numerical quantifiers of the same kind (universal or existential) as the rightmost function quantifier by the corresponding function quantifier, move all these function quantifiers to the left and contract them into the original last function quantifier, finally contract all the numerical quantifiers, which are now all of the same kind, into one numerical quantifier.

Consider a $\Sigma'_\nu$-statement $(\mathfrak{Q}'\mathfrak{f})(Q^0x)\psi\{\mathfrak{f},x,\mathfrak{y}\}$ where all the function variables $\mathfrak{f}$ are quantified by the sequence of function quantifiers $\mathfrak{Q}'$. The truth value of $\psi\{\mathfrak{f},\nu,\mathfrak{m}\}$ is known if we know the values of $f'\lambda,\ldots,f^{(\pi)}\lambda$, where $\mathfrak{f}$ is $f',\ldots,f^{(\pi)}$, for a certain initial set of values of $\lambda$. Hence, if $\mathfrak{m}$ is $\mu',\ldots,\mu^{(\kappa)}$,

$$\psi\{\mathfrak{f},\nu,\mathfrak{m}\} \quad \text{is equivalent to} \quad (E\xi)(Un^\kappa_{1,\ldots,1}[\theta,\nu,\mathfrak{m},\mathfrak{f}\xi,\xi,0]=0)$$

for some $\theta$ (see Prop. 1, Ch. 6). Thus

PROP. 12. *A $\Sigma'_{2\nu}$-statement is equivalent to*

$$(\mathfrak{Q}'\mathfrak{f})\,(E^0x)\,(Un^{\kappa}_{1,\dots,1}[\theta, x^2_1, \mathfrak{y}, \mathfrak{f}\mathfrak{x}^2_2, x^2_2, 0] = 0),$$

*where $\mathfrak{Q}'$ begins with $E'$ and ends with $A'$, and consists of alternate kinds of quantifiers.*

*A $\Sigma'_{2\nu+1}$-statement is equivalent to*

$$(\mathfrak{Q}'\mathfrak{f})\,(A^0x)\,(Un^{\kappa}_{1,\dots,1}[\theta', x, \mathfrak{y}, \mathfrak{f}\mathfrak{x}'^2_1, x'^2_2, 0] \neq 0),$$

*where $\mathfrak{Q}'$ begins with $E'$ and ends with $E'$ and consists of alternate kinds of quantifiers.*

*Similarly for $\Pi'_\nu$-statements.*

The first part follows at once on contracting a couple of $E^0$'s. The second part follows since $\sim \psi\{\mathfrak{f}, \nu, \mathfrak{m}\}$ is recursive in $\mathfrak{f}$ thus

$$\sim (E^0x')\,(Un^{\kappa}_{1,\dots,1}[\theta, x, \mathfrak{y}, \mathfrak{f}\mathfrak{x}', x', 0] = 0)$$

is equivalent to

$$(E^0x')\,(Un^{\kappa}_{1,\dots1,}[\theta', x, \mathfrak{y}, \mathfrak{f}\mathfrak{x}', x', 0] = 0) \quad \text{for some } \theta'.$$

Using this we see that we get a prefix $(\mathfrak{Q}'f)\,(A^0x)\,(A^0x')\,(Un[\dots] \neq 0)$. Contracting the two $A^0$'s the result follows.

Prop. 12 allows us to enumerate $\Sigma^1_\nu$- and $\Pi^1_\nu$-statements. We denote the set in $\Sigma^1_{2\nu}$ given by

$$(\mathfrak{Q}\mathfrak{f})\,(E^0x)\,(Un^{\kappa}_{1,\dots,1}[\theta, \dots] = 0) \quad \text{by} \quad \Sigma^1_{2\nu}[\theta],$$

and the set in $\Sigma^1_{2\nu+1}$ given by

$$(\mathfrak{Q}'\mathfrak{f})\,(A^0x)\,(Un^{\kappa}_{1,\dots,1}[\theta', \dots] = 0) \quad \text{by} \quad \Sigma^1_{2\nu+1}[\theta'].$$

Similarly for $\Pi^1_\nu$-sets. The corresponding $\mathbf{A}'$-statements with free variables shown are denoted by

$$\Sigma^1_{2\nu}\,[\theta; \nu', \dots, \nu^{(\kappa)}], \quad \Sigma^1_{2\nu+1}[\theta'; \nu', \dots, \nu^{(\kappa)}].$$

Similarly for $\Pi^1_\nu$-statements.

PROP. 13. *The following prefix transformations are permissible in the sense of Prop.* 11:

$$\text{(i)} \qquad \begin{cases} \dots E' & \qquad \dots E^0 \\ \dots A' & \qquad \dots A^0 \end{cases}$$

$$\text{(ii)} \qquad \begin{cases} \dots(A'f)\,(E^0x)\,(E'g)\dots & \dots(E'g)\,(A'f)\,(E^0x)\dots \\ \dots(E'f)(A^0x)\,(E'g)\dots & \dots(A'g)\,(E'f)\,(A^0x)\dots \end{cases}$$

*where in each case* (i) *the quantifier changed is the final quantifier in the prefix, and in* (ii) *in each case the matrix of the statement in prenex normal form contains f only in the form* $f\mathfrak{x}$.

(i) Suppose we have

$$\ldots(E'f)\,(E^0x)\,(Un^{\kappa}_{1,\ldots,1}[\theta,\nu',\ldots,fx,x,0]=0)$$

as an equivalent statement, this in turn is equivalent to

$$\ldots(Ey)\,((E^0x)\;Un^{\kappa}_{1,\ldots,1}[\theta,\nu',\ldots,y,x,0]=0)\;\&\;\varpi y = x).$$

Now contract the two existential quantifiers. Similarly for $\ldots A'$.

(ii) Suppose we have

$$\ldots(A'f)\,(E^0x)\,(E'g)\ldots\psi\{.,f\mathfrak{x},.,x,.,g,.\},$$

where the only occurrences of $f$ are marked. This is equivalent to

$$\ldots(E'g)\,(A'f)\,(E^0x)\ldots\{.,f\mathfrak{x},.,x,.,\boldsymbol{\lambda} yg\langle f\mathfrak{x},y\rangle,.\}.$$

Similarly for the other case. Note Mult. Ax!

This proposition sometimes enables us to reduce the place in the hierarchy of given $\Sigma^1_\nu$- or $\Pi^1_\nu$-sets or statements.

PROP. 14. *There exists a hierarchy of* **A**′*-definable sets of lattice points in* $\mathcal{R}^{\kappa}$ *defined by* **A**′*-statements of the forms*:

$$\left.\begin{matrix}\Sigma^1_0(=\Pi^1_0)\,\Sigma^1_1[\theta]\,\Sigma^1_2[\theta]\ldots\\ \Pi^1_1[\theta]\,\Pi^1_2[\theta]\ldots\end{matrix}\right\}\theta=0,1,2,\ldots$$

*having* $\kappa$ *free numerical variables. Each type contains a set which fails to be equivalent to any set in any of the earlier types or to any set in the type in the same column if any.*

The demonstration is similar to that for Prop. 4, Ch. 9.

PROP. 15. $\Sigma^1_1\cap\Pi^1_1$ *contains a set outside the class of arithmetically definable sets.*

Consider the set consisting of the *g.n.*'s of true closed **A**-statements. This set is a $\Pi^1_1$-set by Prop. 5, Ch. 11. (This proposition was demonstrated using property variables, but could equally well have been done using function variables. ) If $T[\nu]$ is the $\mathbf{A}'_f$-statement which says that $\nu$ is the *g.n.* of an **A**-true closed **A**-statement, then $Clstat\,\nu\;\&\;\sim T[Neg\,\nu]$ is also a truth definition for the system **A**, but this is a $\Sigma^1_1$-statement. Thus the set of *g.n.*'s of closed **A**-true statements is a set in $\Sigma^1_1\cap\Pi^1_1$. But we had already found that this set is outside the class of arithmetically definable sets.

This is one point where the analogy with the $\mathscr{P}^{\kappa}_{\nu}$, $\mathscr{Q}^{\kappa}_{\nu}$ breaks down.

D 275 $$\Delta^1_\nu \quad \text{for} \quad \Sigma^1_\nu \cap \Pi^1_\nu.$$

Instead of using variables for functions or sets or properties we could have used variables for real numbers in [0, 1). Because the binary expansion of a real number in [0, 1) defines a set, namely the set whose characteristic function is $f\{\nu\}$ where the binary expansion of the real number is $.f\{1\}f\{2\}\ldots$. Because of this we use the term 'analytical', the final 'al' is added to distinguish between 'analytic' as used in mathematics.

## 11.15 *On the length of proofs*

**Prop.** 16. *For each primitive recursive function $f$, there exists an $\mathbf{A}_I$-theorem $\phi$ such that the minimal $\mathbf{A}_I$- and $\mathbf{A}'^*$-proofs of $\phi$ have g.n.'s $\pi_1$ and $\pi'^*$ respectively, and $\pi_I > f\pi'^*$.*

Note that an $\mathbf{A}_I$-theorem is also an $\mathbf{A}'^*$-theorem. Let $Prf_I[y, x]$ and $Prf'^*[y, x]$ be the proof-predicates for $\mathbf{A}_I$ and $\mathbf{A}'^*$ respectively. Define

$$Prf^M_I[y, x] \quad \text{for} \quad Prf_I[y, x] \,\&\, (Az)_y \sim Prf_I[z, x],$$

so that if $Prf^M_I[\theta, \pi]$ then $\theta$ is the least *g.n.* of an $\mathbf{A}_I$-proof of '$\pi$'. Similarly for $\mathbf{A}'^*$ and $Prf'^{*M}[y, x]$. Clearly

$$Prf^M_I[y, x] \,\&\, Prf^M_I[z, x] \leftrightarrow y = z. \tag{i}$$

Define $\quad Prf[y, x] \quad \text{for} \quad Prf'^{*M}[y, x] \,\&\, (Ez)_{fy+1} Prf^M_I[z, x].$

If the Prop. fails then for some primitive recursive function $f$ $Prf$ gives a proof-predicate for $\mathbf{A}_I$. Now consider $(Ay) \sim Prf[y, s[x, x]]$, let its *g.n.* be $\kappa$, let $\nu$ be the *g.n.* of $(Ay) \sim Prf[y, s[\kappa, \kappa]]$, then $\nu = s[\kappa, \kappa]$. As before '$\nu$' is an irresolvable $\mathbf{A}_I$-statement, but each of $\sim Prf[\pi, \nu]$ is an $\mathbf{A}_I$-theorem for $\pi = 0, 1, \ldots$. Also from Cor. (ii), Prop. 6 of this chapter this irresolvable statement is $\mathbf{A}'^*$-provable. Let its minimal $\mathbf{A}'^*$-proof have *g.n.* $\pi$. We have

$$\vdash Prf'^{*M}[\pi, \nu] \tag{ii}$$

$$\vdash_{\mathbf{A}_I} (Ay)\,(y < f\pi \rightarrow \sim Prf[y, \nu]). \tag{iii}$$

Also

$$\vdash_{\mathbf{A}_I} (Ax) \sim Prf[x, \nu] \leftrightarrow (Ax)\,(Prf'^*[x, \nu] \rightarrow (Ay)\,(y < fx \rightarrow \sim Prf[y, \nu]))$$
$$\leftrightarrow (Ay)\,(y < f\pi \rightarrow \sim Prf[y, \nu]).$$

The last equivalence follows from (i), (ii), and Prop. 16, Ch. 10.

Whence from (iii) $\vdash_{\mathbf{A}_I} (Ax)\,Prf[x,\nu]$, which is absurd, because it is our $\mathbf{A}_I$-irresolvable statement.

COR. *For each primitive recursive function f there exists an* $\mathbf{A}_I$ *theorem* $\phi$ *such that the* $\mathbf{A}_I$*- and* $\mathbf{A}'^*$*-proofs of* $\phi$ *of least length* $\lambda_I, \lambda'^*$ *respectively are such that* $\lambda_I > f\lambda'^*$.

We proceed as before except that we define

$$Prf_I^{M'}[y,x] \quad \text{for} \quad Prf_I[y,x] \,\&\, (Az)\,((\varpi z < \varpi y \vee (\varpi z = \varpi y \rightarrow z < y)) \rightarrow \sim Prf[z,x]).$$

## HISTORICAL REMARKS TO CHAPTER 11

The method of extending a formal system by continually adjoining a true irresolvable statement was first done by Turing (1939), he was trying to get a complete system.

The introduction of variables of higher type is due to Russell (1905), he introduced type theory as a method for eliminating the paradoxes which were appearing in Cantor's set theory. He also introduced the ramified theory of types, wherein each type was divided up into orders. Here we have only one type of property but an unending series of orders within the type. We close the book before introducing higher types, viz. properties of properties, etc.

Rule R 3* has been used by Lorenzen (1955) and others but without the proviso that the general proof be demonstrated correct in some other system.

Something similar to Prop. 1 has been found by Wang.

The form of definition by induction used in Prop. 5 to show that **A**-truth may be defined in a higher system is due to Wang (1953). The proof of the consistency of $\mathbf{A}_I$ in a higher system is due to Tarski (1933) and Bernays (see *H–B* 1934–6).

Predicative and impredicative systems were first discussed by Poincaré (1905), he strongly objected to impredicative systems. Russell tried to keep his work to the predicative case.

The formation of languages with an unbounded set of symbols was first considered by Gödel (1946) and Henkin (1949); see in this connection work on languages with statements of unbounded length by Karp (1964) and Barwise (1968), where further references are given.

The existence of an $\mathbf{A}^{(\kappa)*}$-statement, given in Prop. 9 is related to Rosser's version of Gödel's construction of an irresolvable statement.

Real numbers defined as sections were first thought of by Dedekind (1892, 1893), see also (1909) and Wang (1963) p. 73ff. Other constructions of real numbers as fundamental sequences, binary decimals, etc. are less suitable for our purposes. Various writers have written out real number theory analysis in full in the notation of some formal system. The first of these was Peano (1897), followed by Whitehead and Russell (1910–13), since then there have been several others notably Hilbert and Bernays (1934–6), Rosser (1953), Lorenzen (1955), and from the constructive point of view, Goodstein (1961) and Klaua (1961).

The development of real number analysis as we now know it is due to Cauchy, Weierstraus, Abel, Lebesgue, etc.

The analytical hierarchy has been studied by Kleene (1955, 1959, 1962, 1963), Addison (1955, 1959, 1962*a*, 1962*b*) and Spector (1958, 1959).

Prop. 16 on the length of proofs is due to Gödel (1936), we follow the account given by Mostowski (1952).

## EXAMPLES 11

1. Devise a system of *g.n.*'s for $\mathbf{A}^{(\kappa)}$ in such a way that there is a (1–1)-correspondence between symbols and *g.n.*'s. Do the same for $\mathbf{A}^{\omega}$.

2. Give the full definitions D 263–266 inclusive.

3. Obtain a proof-predicate for $\mathbf{A}^{(\kappa)*}$.

4. Discuss the consistency of $\mathbf{A}^{(\kappa)*}$ on the same lines as the consistency of $\mathbf{A}_I$.

5. State and investigate the analogue of Prop. 18, Ch. 10 for $\mathbf{A}^{(\kappa)*}$.

6. Find the truth-values of $(E_{o\iota}X_{o\iota})AX_{o\iota}$, and generally of $(PX_{o\iota})QX_{o\iota}$, where $P, Q$ are either $A$ or $E$ of appropriate types.

7. Find the truth values of

$$(P_{o\iota}X_{o\iota})(P'X_{o\iota} \vee P''X_{o\iota}), (P_{o\iota}X_{o\iota})(P'_{o\iota}X'_{o\iota})(Q'X_o \,\&\, QX'_o),$$

where $P', P'', Q', Q''$ are $A$ or $E$.

8. Prove $\quad Sect\, l.u.b.(\lambda X . \phi\{X\} \,\&\, Sect\, X)$.

9. Prove

(i) $\phi\{X\} \,\&\, Sect\, X . \rightarrow . X \leqslant_s l.u.b.(\lambda X . \phi\{X\} \,\&\, Sect\, X)$,

(ii) $X <_s l.u.b.(\lambda X . \phi\{X\} \,\&\, Sect\, X) . \rightarrow (EX')(Sect\, X' \,\&\, \phi\{X'\}$
$\qquad\qquad \& \, X <_s X')$,

where $X <_s X'$ for $(Ex)(X'x \,\&\, \sim Xx)$.

10. Express the axiom of choice in $\mathbf{A}^{(\omega)}$.

11. Define $Con_{(11)}\mathbf{A}^{(\kappa)}$ and show that it can be $\mathbf{A}^{(S\kappa)}$-proved.

12. Discuss $Con_{(11)}\mathbf{A}^{(\omega)}$.

13. Develop extension of $\mathbf{A}_{00}$ using variables for classes and the $\epsilon$-symbol. Discuss predicative and impredicative classes.

14. Develop extensions of $\mathbf{A}_I$ using function variables. Define predicative and impredicative extensions.

15. Complete the demonstration of Prop. 3.

16. Define

$$(X +_s X') \quad \text{for} \quad \lambda x.(Ex', x'')\,(x = x' + x'' \,\&\, Xx' \,\&\, X'x'').$$

Now define successively

$$|X|_s^2, \; -_s X, \; X -_s X', \; \frac{X}{2},$$

then
$$X \times_s X' \quad \text{for} \quad \frac{|X +_s X'|_s^2 -_s |X|_s^2 -_s |X'|_s^2}{2}.$$

Show that all these are sections, and prove the commutative, associative and distributive laws of addition and multiplication.

17. Prove Wierstrass' theorem that any bounded set containing an unbounded number of members has a limit point. If the set is of order $\kappa$ what is the order of the limit point?

18. Prove that a $\kappa$th order function which is continuous in a closed interval is uniformly continuous in that interval. In which $\mathbf{A}^{(\pi)}$ is this proved?

19. Prove that a $\kappa$th order function which is continuous in a closed interval is bounded in that interval, attains its bounds and also takes each value between its bounds. What restrictions, if any, are there on the order of the intermediate value? In which $\mathbf{A}^{(\pi)}$ does your proof take place?

20. Define $\lim\limits_{X_s \to \Delta} f\{X_s\}$. Here $f\{X_s\}$ is of the form

$$\lambda x'(Ex)\,(\Delta\{X_s\}x \,\&\, x' < x).$$

21. State and prove Rolles theorem for $\kappa$th order functions.

22. Define the upper and lower Riemann integral for a bounded function over a bounded interval by using subdivisions of the bounded interval into intervals of equal length only. This method of subdivision gives a sequence of upper sums and a sequence of lower sums, and is

logically easier to handle than the normal method of using all possible subdivisions. If we used any subdivision then we should require some variables of higher types.

23. Define the exponential function in the complex plane, and prove its functional equation.

24. A set of points in a plane can be defined as a set of complex numbers

$$\text{thus } \hat{X}.(EY, Y')(\phi\{Y, Y'\} \,\&\, X = \lambda y.(Yy \vee Y'y) \,\&\, (Ay)(Yy. \rightarrow Evy) \\ \&\, (Y'y. \rightarrow Ody)).$$

Using D 270, 271 define the closure of a set of points in a plane and prove that the closure is closed.

25. Define the interior, exterior and boundary of a set of points in a plane. If the set is of order $\kappa$ what are the orders of its interior, exterior and boundary?

26. Prove Wierstrass' theorem for a sequence of distinct points in a square.

27. Define the distance between two sets of points in a plane. Prove that the distance between two closed bounded sets of points in a plane is zero if and only if the sets have a point in common.

28. Show that a nested sequence of bounded closed sets of points in a plane have a point in common.

# Chapter 12

# Models

## 12.1 *Models and truth-definitions*

Models of formal systems are intimately connected with truth-definitions. They are mainly constructed when the system has variables and statements. A system with statements but without variables can usually act as its own model, so the construction of models for such a system is unnecessary. A truth-definition is connected with statements, so only applies to systems with statements, and can be attempted whether the system has variables or otherwise. In previous chapters we have stated exactly what we require of a truth-definition, now we must do the same for a model. We have omitted to discuss whether a given formal system has two or more distinct truth-definitions, but we shall, in this chapter, discuss whether a formal system can have more than one model, and from each model we can get a truth-definition. The truth-definitions we have given so far for various systems of arithmetic are what we shall call *standard truth-definitions.*

Our aim in constructing formal systems of arithmetic was to invent a language in which we could express thoughts about natural numbers and give some method of procedure whereby we could obtain those statements about natural numbers which are true in the standard truth-definition. Some statements about natural numbers are intuitively obvious (their formal proof in any formal system would be very easy) and we scarcely need to form a formal system if we are only going to use such statements, e.g. that part of arithmetic used in banking, cooking, etc.

Other statements about natural numbers are true in the standard sense but are far from obvious and we require some method of demonstrating their truth. This we do by the rules of procedure whereby from statements which are easily seen to be true we obtain another whose truth is easily seen to follow from that of the former. After a number of steps of this kind, each perfectly simple, we arrive at a statement which is far from obvious, but the rules of procedure convince us of its truth.

For instance Lagrange's theorem to the effect that every natural number is the sum of four squares, or Waring's theorem that every natural number is the sum of at most $G\{k\}$ $k$th powers are true but far from obvious, they require the rules of procedure to demonstrate without doubt that they are true. Again other statements about natural numbers, such as Goldbach's conjecture or the existence of an unbounded set of prime twins are such that it seems unlikely that they will ever be proved, they seem impossible to handle, and their truth seems impossible to decide. Again we found an irresolvable arithmetic statement which it was fairly easy to see is true.

We now want to see if these systems have other interpretations than the one which gives rise to the standard truth-definition. Whether they are *categorical*, i.e. whether they characterize the natural numbers uniquely. One of our objects in forming a formal system of arithmetic was to construct a complete categorical system. We have shown that it is impossible to construct a formal system of arithmetic in which we can prove exactly all the true statements about natural numbers. We now want to show that any such system will have an unbounded set of distinct models or interpretations. In other words, in such a system we shall be talking about an unbounded set of different things instead of about one particular thing, as we wanted to. To do this we want to make precise our intuitive notion of model or interpretation.

A *model of a formal system* $\mathscr{S}$ is another formal system $\mathscr{M}$ which satisfies:

(i) $\mathscr{M}$ has all the constants of $\mathscr{S}$ and of the same type as in $\mathscr{S}$.

(ii) If $\mathscr{S}$ has variables of type $\alpha$ then $\mathscr{M}$ has constants of type $\alpha$.

(iii) $\mathscr{M}$ contains constants of type $o$, there are at least two of them, they are of two kinds called *designated* and *undesignated*, there is to be at least one constant of each kind, if there is exactly one of each kind then the model is called *two-valued*.

(iv) $\mathscr{M}$ is without axioms.

(v) The rules of procedure of $\mathscr{M}$ enable us to replace $\mathscr{S}$-theorems uniquely by designated $\mathscr{M}$-constants of type $o$. Thus the $\mathscr{M}$-rules enable us to replace the $\mathscr{S}$-axioms by designated $\mathscr{M}$-constants of type $o$, and if the upper formulae of an $\mathscr{S}$-rule have been replaced by designated $\mathcal{M}$-constants of type $o$ then the $\mathcal{M}$-rules allow us to replace the lower formula by a designated $\mathcal{M}$-constant of type $o$.

(vi) If $\tau_{\alpha\beta}$ and $\sigma_\beta$ are $\mathcal{M}$-constants of the types shown in the suffixes then there is a unique $\mathcal{M}$-constant of type $\alpha$, say $\gamma_\alpha$, such that the $\mathcal{M}$-rules allow us to replace $\tau_{\alpha\beta}\sigma_\beta$ by $\gamma_\alpha$.

An $\mathcal{S}$-statement which can be replaced by a designated $\mathcal{M}$-constant of type $o$ is uniquely so replaceable and is called an *$\mathcal{M}$-true $\mathcal{S}$-statement.* This definition of a model for a formal system corresponds very closely to the definition of truth we gave for the systems $A_{oo}$, $A_o$ and A. There we had a two-valued model with elements ***t*** and ***f*** of which ***t*** is designated and ***f*** is undesignated, and the $\mathcal{M}$-constants of type $\iota$ are exactly the natural numbers. This we call the *standard model* for arithmetic.

If the value of an $\mathcal{M}$-term of the type $o$ is always ***t*** or always ***f*** then the corresponding model is called trivial, such models are without interest or use, and a system whose only models are trivial will itself be called trivial. Again such a system is without interest.

We are mainly interested in two-valued systems, but many-valued systems are of considerable interest for many diverse purposes.

## 12.2 *Models for* $A_{oo}$

The system $A_{oo}$ has the following constants:

| constant | $\vee$ | $\&$ | $=$ | $\neq$ | $S$ | $0$ | $\mathcal{I}$ |
|---|---|---|---|---|---|---|---|
| type | $ooo$ | $ooo$ | $o\iota\iota$ | $o\iota\iota$ | $\iota\iota$ | $\iota$ | $\iota\iota(\iota\iota)$ |

it also has the variables $x, x', x'', \ldots$ used only in the formation of functors. We want to find tables for these constants in a two-valued model, call it $\mathcal{M}_{oo}$. By condition (i) 0 is a constant of the model. By condition (ii) $\mathcal{M}_{oo}$ has constants of type $\iota$. By condition (vi) one of these constants can replace $S0$, another can replace $SS0$, and so on. We take equality in $\mathcal{M}_{oo}$ to be identity and $\neq \alpha\beta$ to have opposite value to $= \alpha\beta$. By $Ax_{oo}$ 2.1, 2.2 and R 2, and $Ax_{oo}$ 1 we see that $0, S0, SS0, \ldots$ are all replaced by distinct constants of $\mathcal{M}_{oo}$. From results demonstrated for $A_{oo}$ we know that a closed numerical term $\alpha$ determines a unique numeral $\nu$ such that $\alpha = \nu$ is an $A_{oo}$-theorem. Hence the term $\alpha$ which in $\mathcal{M}_{oo}$ is replaced by an $\mathcal{M}_{oo}$ constant must in $\mathcal{M}_{oo}$ be replaced by the $\mathcal{M}_{oo}$ constant which replaces the numeral $\nu$. Thus the only constants in $\mathcal{M}_{oo}$ of type $\iota$ are the $\mathcal{M}_{oo}$-numerals, i.e. the $\mathcal{M}_{oo}$-constants which replace the $A_{oo}$-numerals. We could add, if we like, more constants to $\mathcal{M}_{00}$ but they would be redundant, never used, and useless.

From II$a$ we see that if the last argument of the $\mathcal{M}_{00}$-constant $\vee$ is $\boldsymbol{t}$ then the values of $\vee\boldsymbol{tt}$ and $\vee\boldsymbol{ft}$ are both $\boldsymbol{t}$, from I$a$ we see that $\vee\boldsymbol{tf}$ and $\vee\boldsymbol{ft}$ have the same values. If $\vee\boldsymbol{ff}$ is $\boldsymbol{t}$ then every disjunction is $\boldsymbol{t}$ and it is pointless to have that constant. Thus take $\vee\boldsymbol{ff}$ to be $\boldsymbol{f}$.

From II$b'$ we see that $\&\boldsymbol{tt}$ is $\boldsymbol{t}$, and that $\&\boldsymbol{tf}$ and $\&\boldsymbol{ft}$ have the same value, if this is $\boldsymbol{t}$ then $\&\boldsymbol{ff}$ must be $\boldsymbol{f}$, otherwise every conjunction is $\boldsymbol{t}$, and it would be pointless to have such a constant. But if $\&\boldsymbol{ff}$ is $\boldsymbol{f}$ while $\&\boldsymbol{tf}$ and $\&\boldsymbol{ft}$ are $\boldsymbol{t}$ then $\&$ becomes the same as $\vee$ and II$b'$ becomes derivable from II$a$ and I$a$, so this possibility can be discarded. The remaining alternative is when $\&\boldsymbol{tf}$ and $\&\boldsymbol{ft}$ are both $\boldsymbol{f}$.

Thus a non-trivial model for $\mathbf{A}_{00}$ consists of representatives of the numerals, identity for equality, inequality having the opposite value to equality, and conjunction and disjunction having the familiar two-valued truth-tables.

## 12.3 *Models for* $\mathbf{A}_0$

The system $\mathbf{A}_0$ has the same constants as $\mathbf{A}_{00}$ together with the constant $E$ of type $o(o\iota)$, it contains variables of type $\iota$ which are only used to form functors and properties.

Thus a model for $\mathbf{A}_0$ will contain the same constants as a model for $\mathbf{A}_{00}$ together with a new constant $E$ of type $o(o\iota)$. The constants of $\mathbf{A}_0$ other than $E$ will be the same as in the model for $\mathbf{A}_{00}$. As regards the constant $E$, from II$d$ we see that if $\phi\{\alpha\}$ is $\boldsymbol{t}$ for some $\alpha$ then $(E\xi)\,\phi\{\xi\}$ must also be $\boldsymbol{t}$, if $(E\xi)\,\phi\{\xi\}$ is $\boldsymbol{t}$ when $\phi\{\alpha\}$ is always $\boldsymbol{f}$ then $(E\xi)\,\phi\{\xi\}$ is always $\boldsymbol{t}$ and the symbol is useless. Thus we shall have in the model $\mathcal{M}_0$ for $\mathbf{A}_0$ $(E\xi)\,\phi\{\xi\}$ is $\boldsymbol{t}$ if and only if $\phi\{\nu\}$ is $\boldsymbol{t}$ for some numeral $\nu$. This assumes that we have used the same set of constants of type $o$ in $\mathcal{M}_0$ as before. That is it assumes that we are dealing with the standard model. We could, however, have some further constants of type $\iota$ in the model, say $\alpha$ is one of them, then we could have a model in which $(E\xi)\,\phi\{\xi\}$ was $\boldsymbol{t}$ even though $\phi\{\nu\}$ was $\boldsymbol{f}$ for all numerals $\nu$, but $\phi\{\alpha\}$ was $\boldsymbol{t}$. We can form a new system $\mathbf{A}_{0\alpha}$ which is the same as $\mathbf{A}_0$ except that it has $\phi\{\alpha\}$ as an extra axiom and $\alpha$ as an extra symbol. The resulting system would be consistent in the sense that $0 = 1$ would fail to be a theorem, because if the axiom $\phi\{\alpha\}$ was used in an $\mathbf{A}_0$-proof of $0 = 1$ then either $\alpha$ or $E$ would appear in the theorem proved, and this is absurd if the theorem is $0 = 1$. But an $A_0$-proof is without occurrences of $\alpha$ because $\alpha$ is foreign to the $A_0$-symbols. The model just described is called a *non-standard model*, there are many of them. If $\alpha$ is a *non-standard constant* of type $\iota$ then $f\alpha$ is to be replaceable in the model by another non-standard constant of type $\iota$ for each term $f$ of type $\iota\iota$ in $A_0$.

## 12.4 *Models for* $A_I$, $A^{(\kappa)}$

In a standard model for $A_I$ we have representatives of the natural numbers as constants of type $\iota$ and $(A\xi)\,\phi\{\xi\}$ will be $\boldsymbol{t}$ if and only if $\phi\{\nu\}$ is $\boldsymbol{t}$ for each numeral $\nu$. We can get non-standard models as for $A_0$ by taking an irresolvable statement $G$ and adding an extra non-standard constant which is such that whichever of $G$ or $\sim G$ is $\boldsymbol{f}$ in the standard model has the value $\boldsymbol{t}$ in the non-standard model.

The situation is similar in models for $A^{(\kappa)}$, except that here we have another type of variable. We shall have more to say about non-standard models in the sequel.

## 12.5 *General models*

If a formal system has variables of type $(\alpha\beta)$ then a model will require to have constants of type $(\alpha\beta)$. These are functions from a domain of constants of type $\beta$ to a domain of constants of type $\alpha$. A model will be called *general* if the constants of type $(\alpha\beta)$ is some set of functions of type $(\alpha\beta)$ which can be a proper subset of all such functions.

A statement of a formal system is *valid over a general model* if it always reduces to a designated value no matter how the free variables are replaced by constants of that general model and of the same type. A statement of a formal system is said to be *satisfiable over a general model* if it is possible to replace the variables of the statement by constants of that general model of the same type in such a manner that the statement reduces to a designated value.

If a model contains all functions of type $(\alpha\beta)$ for which there are variables of this type in the formal system then the model is called a *full model*. Thus a full model is a general model, and to be valid over every general model is a stronger condition than to be valid over a full model.

## 12.6 *Satisfaction*

We go into the notion of satisfaction a little more fully. We suppose that the statement is in **λ**-normal form.

A statement in a formal system is built up from atomic statements by the connectives of the system. In the cases we are interested in the atomic statements are equations or statements of the form $\Xi\alpha$ where $\Xi$ is a property variable and $\alpha$ is a numerical term, and the connectives are negation, disjunction and universal quantification. (Or the system can be put in an equivalent form where this is so.) Thus associated with each statement $\phi$ in a formal system there corresponds a *construction sequence* or *C-sequence* $\phi', \phi'', \ldots, \phi^{(S\pi)}$, where $\phi^{(S\pi)}$ is $\phi$ and $\phi^{(S\pi')}$ is for $\pi' < \pi$, either an atomic statement, or is:

(i) $\sim\phi^{(S\pi'')}$ for $\pi'' < \pi'$,

(ii) $\phi^{(S\pi'')} \vee \phi^{(S\pi''')}$ for $\pi'', \pi''' < \pi'$,

(iii) $(A\Xi)\,\phi^{(S\pi'')}$ for $\pi'' < \pi'$, where $\Xi$ is a variable of the system.

To each member of a $C$-sequence for statement $\phi$ we attach a set of functions, giving a sequence of sets of functions, called an *S-sequence*, this is done as follows:

Let $\sigma^{(S\theta)}$ be the set of affixes of the free numerical variables in $\phi^{(S\theta)}$, and let $\tau_\beta^{(S\theta)}$ be the set of affixes of the free variables of sort $\beta$ in $\phi^{(S\theta)}$.

(i) $\phi^{(S\theta)}$ is an equation $\alpha = \beta$, let $\mathscr{F}^{(S\theta)}$ be the set of functions over $\sigma^{(S\theta)}$ whose values satisfy $\alpha = \beta$. By this we mean, if $\rho$ is one of these functions and if $\sigma^{(S\theta)}$ consists of $\kappa', \ldots, \kappa^{(\mu)}$ so that $\alpha = \beta$ is of the form:

$$\alpha\{x^{(\kappa')}, \ldots, x^{(\kappa^{(\mu)})}\} = \beta\{x^{(\kappa')}, \ldots, x^{(\kappa^{(\mu)})}\},$$

then $\alpha\{\rho[\kappa'], \ldots, \rho[\kappa^{(\mu)}]\}$ and $\beta\{\rho[\kappa'], \ldots, \rho[\kappa^{(\mu)}]\}$ both determine the the same numeral $\nu$ when $\rho[\kappa'], \ldots, \rho[\kappa^{(\mu)}]$ are replaced by the numerals which they determine. As a simple example consider the equation $x'' + x''' = x$, here $\mathscr{F}$ is the class of functions over $[0, 2, 3]$ with values $\nu' + \nu''$, $\nu'$, $\nu''$ at those places. Thus $\rho[0] = \nu' + \nu''$, $\rho[2] = \nu'$ and $\rho[3] = \nu''$. We denote this function by

$$\begin{matrix}[\nu' + \nu'', & \nu', & \nu'']. \\ [\;\;0\;\;\;, & 2, & 3\;]\end{matrix}$$

(ii) $\phi^{(S\theta)}$ is $X_\beta^{(\kappa)}\alpha$, let $\sigma^{(S\theta)}$ consist of $[\kappa', \ldots, \kappa^{(\mu)}]$, so that $X_\beta^{(\kappa)}\alpha$ is of the form $X_\beta^{(\kappa)}\alpha\{x^{(\kappa')}, \ldots, x^{(\kappa^{(\mu)})}\}$, now replace the numerical variables by $\rho[\kappa'], \ldots, \rho[\kappa^{(\mu)}]$ and then replace $\alpha\{\rho[\kappa'], \ldots, \rho[\kappa^{(\mu)}]\}$ by the numeral it determines, say $\nu$, when $\rho[\kappa'], \ldots, \rho[\kappa^{(\mu)}]$ have been replaced by the numerals which they determine. Lastly replace $X_\beta^{(\kappa)}$ by any property $\Delta_\beta^{(\nu)}$ of sort $\beta$ which the numeral $\nu$ possesses. Then $\mathscr{F}^{(S\theta)}$ is the set of functions over $[\kappa; \kappa', \ldots, \kappa^{(\mu)}]$ whose value at that place is given by $[\Delta_\beta^{(\nu)}; \rho[\kappa'], \ldots, \rho[\kappa^{(\mu)}]]$ we denote this function by

$$\begin{matrix}[\Delta_\beta^{(\nu)}; \nu', \ldots, \nu^{(\mu)}] \\ [\quad \kappa; \kappa', \ldots, \kappa^{(\mu)}]\end{matrix}$$

where $\nu', \ldots, \nu^{(\mu)}$ are the values of $\rho[\kappa'], \ldots, \rho[\kappa^{(\mu)}]$ respectively, and $\Delta_\beta^{(\nu)}$ is a property which $\nu$ possesses.

(iii) $\phi^{(S\theta)}$ is $\sim \phi^{(S\theta')}$ where $\theta' < \theta$. $\mathscr{F}^{(S\theta)}$ is the set of functions over $[\tau^{(S\theta)}; \sigma^{(S\theta)}]$ which fail to satisfy $\mathscr{F}^{(S\theta')}$, thus the set of functions over the same domain which fail to be in the set $\mathscr{F}^{(S\theta')}$.

(iv) $\phi^{(S\theta)}$ is $\phi^{(S\theta')} \vee \phi^{(S\theta'')}$ where $\theta', \theta'' < \theta$. $\mathscr{F}^{(S\theta)}$ is the set of functions over $[\tau^{(S\theta')} \cup \tau^{(S\theta'')}; \sigma^{(S\theta')} \cup \sigma^{(S\theta'')}]$ which confined to $[\tau^{(S\theta')}; \sigma^{(S\theta')}]$ belong to $\mathscr{F}^{(S\theta')}$ and confined to $[\tau^{(S\theta'')}; \sigma^{(S\theta'')}]$ belong to $\mathscr{F}^{(S\theta'')}$.

(v) $\phi^{(S\theta)}$ is $(Ax^{(\kappa)})\, \phi^{(S\theta')}\{x^{(\kappa)}\}$, $\theta' < \theta$. $\mathscr{F}^{(S\theta)}$ is the set of functions over $[\tau^{(S\theta)}; \sigma^{(S\theta)}]$ where $\kappa$ is outside $\sigma^{(S\theta)}$, and such that for each numeral $\nu$ any function of $\mathscr{F}^{(S\theta)}$ extended so as to take the value $\nu$ at $\kappa$ in the second set of arguments belongs to $\mathscr{F}^{(S\theta')}$.

(vi) $\phi^{(S\theta)}$ is $(AX_\beta^{(\kappa)})\, \phi^{(S\theta')}\{X_\beta^{(\kappa)}\}$, $\theta' < \theta$. $\mathscr{F}^{(S\theta)}$ is the set of functions over $[\tau^{(S\theta)}; \sigma^{(S\theta)}]$, where $\kappa$ is outside $\tau^{(S\theta)}$, and such that if $\Delta$ is a property of sort $\beta$, then any function of $\mathscr{F}^{(S\theta)}$ extended so as to take the value $\Delta$ at $\kappa$ in the first group of arguments belongs to $\mathscr{F}^{(S\theta')}$.

In the last two cases the domain of the functions of $\mathscr{F}^{(S\theta')}$ contains exactly one more argument place than the domain of $\mathscr{F}^{(S\theta)}$.

Thus we have defined a $C$-sequence and an $S$-sequence for any statement of the system. Denote the $S$-sequence by $\mathscr{F}', \ldots, \mathscr{F}^{(S\pi)}$. We say that a function $\rho$ *satisfies* $\phi$ if there is an $S$-sequence $\{\mathscr{F}\}$ such that $\rho$ belongs to $\mathscr{F}^{(S\pi)}$.

If $\phi$ is a closed statement of the system then the only function which could possibly belong to $\mathscr{F}^{(S\pi)}$ is the void function (the function whose domain is empty). Call this function [ ]. If $\phi$ is closed and if $\mathscr{F}^{(S\theta)}$ consists of [ ] alone then we say that $\phi$ is $\mathscr{A}$-true, where $\mathscr{A}$ is the formal system, otherwise that $\phi$ is $\mathscr{A}$-false. If $\phi$ is open we say that $\phi$ is $\mathscr{A}$-valid if its closure is $\mathscr{A}$-true, otherwise $\phi$ is $\mathscr{A}$-invalid. We illustrate with a few examples. Note that our construction demands an unbounded domain.

**12.7** *Examples*

(*a*) $(Ax)(Ex')(x \neq x')$, i.e. $(Ax) \sim (Ax')(x = x')$.

The $C$-sequence is:

$$x = x', \quad (Ax')(x = x'), \quad \sim (Ax')(x = x'), \quad (Ax) \sim (Ax')(x = x').$$

The $S$-sequence is:

$\dfrac{[\nu, \nu]}{[0, 1]}$, null set of functions over 0, set of all functions over 0, [ ].

In the first of these functions the values at 0 and 1 are the same, but otherwise can be any numeral.

(*b*) $(Ex)(Ax')(x \neq x')$, i.e. $\sim (Ax) \sim (Ax') \sim (x = x')$.

The $C$-sequence is:

$$x = x', \quad \sim (x = x'), \quad (Ax') \sim (x = x'), \quad \sim (Ax') \sim (x = x'),$$
$$(Ax) \sim (Ax') \sim (x = x'), \quad \sim (Ax) \sim (Ax') \sim (x = x').$$

The $S$-sequence is:

$\dfrac{[\nu, \nu]}{[0, 1]}$, $\dfrac{[\nu, \kappa]}{[0, 1]}$ $(\nu \neq \kappa)$, null set of functions over 0, all functions over 0, [ ], null set of void functions.

Clearly (*a*) is **A**-true and (*b*) is **A**-false. (*a*) is false in a one-element model, but our construction forbids this.

(*c*) $(EX)(Ax)Xx$, i.e. $\sim (AX) \sim (Ax)Xx$.

The $C$-sequence is:

$$Xx, \quad (Ax)Xx, \quad \sim (Ax)Xx, \quad (AX) \sim (Ax)Xx, \quad \sim (AX) \sim (Ax)Xx.$$

The $S$-sequence is:

$\dfrac{[\Delta;\ \nu]}{[0;\ 0]}$ $\Delta\nu$ is $\mathbf{A}^{(\kappa)}$-true, $\dfrac{[\Delta;\ ]}{[0;\ ]}$ $\Delta\nu$ is $\mathbf{A}^{(\kappa)}$-true for all numerals $\nu$,

$\dfrac{[\Delta;]}{[0;]}$ where $\sim \Delta\nu$ is $\mathbf{A}^{(\kappa)}$-true for some numeral $\nu$, null set of void functions, the void function.

(*d*) $(Ex)(AX)Xx$, i.e. $\sim (Ax) \sim (AX)Xx$.

The $C$-sequence is:

$$Xx, \quad (AX)Xx, \quad \sim (AX)Xx, \quad (Ax) \sim (AX)Xx, \quad \sim (Ax) \sim (AX)Xx.$$

The $S$-sequence is:

$\frac{[\Delta;\ \nu]}{[0;\ 0]}$ where $\Delta\nu$ is $\mathbf{A}^{(\kappa)}$-true, null set of functions over [; 0], the set of all functions over [; 0], [ ], null set of void functions.

(*e*) $$(AX)(Ax) \sim (Ax') \sim (Xx \vee \sim Xx').$$

The $C$-sequence is:

$$Xx,\ Xx',\ \sim Xx',\ Xx \vee \sim Xx',\ \sim (Xx \vee \sim Xx'),\ (Ax') \sim (Xx \vee \sim Xx'),$$
$$\sim (Ax') \sim (Xx \vee \sim Xx'),\quad (Ax) \sim (Ax') \sim (Xx \vee \sim Xx'),$$
$$(AX)(Ax) \sim (Ax') \sim (Xx \vee \sim Xx').$$

The $S$-sequence is:

$\frac{[\Delta;\ \nu]}{[0;\ 0]}$, where $\Delta\nu$ is $\mathbf{A}^{(\kappa)}$-true,

$\frac{[\Delta';\ \kappa]}{[0\ ;\ 1]}$, where $\Delta'\kappa$ is $\mathbf{A}^{(\kappa)}$-true,

$\frac{[\Delta'';\ \kappa]}{[0\ ;\ 1]}$, where $\Delta''\kappa$ is $\mathbf{A}^{(\kappa)}$-false,

$\frac{[\Delta \cup \Delta';\ \nu\ \kappa]}{[\ \ 0\ \ ;\ 0\ 1]}$, where $\Delta\nu$ is $\mathbf{A}^{(\kappa)}$-true or $\Delta'\kappa$ is $\mathbf{A}^{(\kappa)}$-false,

$\frac{[\bar{\Delta} \cap \bar{\Delta}';\ \nu\ \kappa]}{[\ \ 0\ \ ;\ 0\ 1]}$, where $\bar{\Delta}\nu$ is $\mathbf{A}^{(\kappa)}$-false and $\bar{\Delta}'\kappa$ is $\mathbf{A}^{(\kappa)}$-true,

The null set of functions over [0; 0],
All functions over [0 ; 0],
All functions over [0 ;],
The void function.
Thus (*c*) and (*e*) are $\mathbf{A}^{(\kappa)}$-true while (*d*) is $\mathbf{A}^{(\kappa)}$-false.

## 12.8 *Non-standard models*

We now want to find out something about non-standard models whose constants of type $\iota$ contain members distinct from the numerals. Every theorem of the system will be satisfied in the model. Thus if $\alpha$ and $\alpha'$ are two distinct constants of the model of type $\iota$ then exactly one of $\alpha < \alpha'$, $\alpha > \alpha'$, $\alpha = \alpha'$ will hold in the model. If $\alpha$ is a non-standard constant of type $\iota$ then it is distinct from 0 and so $\alpha \dot{-} 1$ is another constant of the model, or rather is replaceable by a constant of the model; this constant must be non-standard, because if $\alpha \dot{-} 1$ were replaceable by the numeral $\nu$ then $\alpha$ would be replaceable by the numeral $S\nu$.

Similarly each of $\alpha \dot{-} 2$, $\alpha \dot{-} 3, \ldots$ will all be replaceable by non-standard constants of the model. Similarly, of course, $\alpha+1, \alpha+2, \ldots$ are all replaceable by non-standard constants of the model. Thus associated with each non-standard constant of the model there is a *batch* of non-standard constants of the model in the order type $\omega^*+\omega$, where $\omega$ is the order type of the natural numbers and $\omega^*$ is the order type of the negative integers, the reverse order type, and the addition sign indicates that we have the order type $\omega^*$ followed by the order type $\omega$, so the final order type is that of the positive and negative integers. Any member of this batch can be considered as occupying the central position.

If $\alpha$ and $\beta$ are non-standard constants of the model which belong to distinct batches then if they are both of the same parity $\left[\frac{\alpha+\beta}{2}\right]$ will be replaceable by a non-standard constant of the model and so gives rise to another batch of non-standard constants of the model. This batch will lie between the two batches which give rise to it. Similarly if one is even and the other is odd then $\left[\frac{\alpha+\beta+1}{2}\right]$ gives rise to another batch of non-standard constants of the model lying between the two batches which give rise to it.

Thus the non-standard constants of the model has a part of the order type $\omega+\eta\times(\omega^*+\omega)$, where $\eta$ is the order type of the rational numbers. All the non-standard constants of the model are greater than all the numerals. The product of the order types $\eta$ and $(\omega^*+\omega)$ is the order type of a set obtained from the rational numbers by replacing each rational by a set of the order type of the positive and negative integers.

Note that since $\dfrac{(Sx+x')\dot{-}x'' = Sx}{x'' = x'}$ is **A**-valid, then if $(S\nu+\alpha)\dot{-}\nu'' = S\nu$, we should have $\alpha = \nu''$, i.e. $\alpha$ would be standard, hence any non-standard constant of the model is greater than every standard constant of the model. Also if $\alpha$, $\beta$ are of the same parity and if $\left[\frac{\alpha+\beta}{2}\right] = \alpha\dot{-}\nu$, so that they belong to the same batch, then $\beta = \alpha\dot{-}2\nu$, and so $\alpha$ and $\beta$ would belong to the same batch. Thus if $\alpha$ and $\beta$ belong to the different batches then $\left[\dfrac{\alpha+\beta}{2}\right]$ belongs to another batch distinct from both of them. Similarly in the case of different parities.

We can realize this order type by functions of natural numbers ordered in a certain manner.

LEMMA 1. *Any enumerable set of one-place functions of natural numbers can be linearly ordered.*

Let the enumerable set of one-place functions of natural numbers be: $f, f', f'', \ldots$. Order distinct pairs of natural numbers with the first member less than the second member, and let the $\nu$th pair be $\{\nu_1, \nu_2\}$, where $\nu_1 < \nu_2$. Generate the natural numbers and when the natural number $\nu$ has been generated consider the $\nu$th pair $f^{(\nu_1)}, f^{(\nu_2)}$ of functions of our set. Consider first the pair $f, f'$. We shall have:

$$\left.\begin{aligned} & f\kappa > f'\kappa, \\ \text{or}\quad & f\kappa = f'\kappa, \\ \text{or}\quad & f\kappa < f'\kappa, \end{aligned}\right\}\quad \text{for an unbounded set of natural numbers.}$$

We take the first case which is satisfied. Say it is the first in the above list, then we shall write: $f \gtrdot f'$. Let $g0, g1, g2, \ldots$ be equal to the natural numbers in order of magnitude for which the first alternative holds. Now consider the $\nu$th pair $f^{(\nu_1)}$ and $f^{(\nu_2)}$, if the functions $g, g', g'', \ldots, g^{(\nu \dot{-} 2)}$ have been defined then consider the sequences formed by composition with association to the right:

$$f^{(\nu_1)}gg'g'' \ldots g^{(\nu \dot{-} 2)}\kappa \quad f^{(\nu_2)}gg'g'' \ldots g^{(\nu \dot{-} 2)}\kappa.$$

Then we shall have:

$$\left.\begin{aligned} & f^{(\nu_1)}gg'g'' \ldots g^{(\nu \dot{-} 2)}\kappa > f^{(\nu_2)}gg'g'' \ldots g^{(\nu \dot{-} 2)}\kappa, \\ \text{or}\quad & f^{(\nu_1)}gg'g'' \ldots g^{(\nu \dot{-} 2)}\kappa = f^{(\nu_2)}gg'g'' \ldots g^{(\nu \dot{-} 2)}\kappa, \\ \text{or}\quad & f^{(\nu_1)}gg'g'' \ldots g^{(\nu \dot{-} 2)}\kappa < f^{(\nu_2)}gg'g'' \ldots g^{(\nu \dot{-} 2)}\kappa, \end{aligned}\right\}\quad \begin{array}{r}\text{for an unbounded set}\\ \text{of numerals.}\end{array}$$

Take the first case which is satisfied and let $g^{(\nu \dot{-} 1)}0, g^{(\nu \dot{-} 1)}1, \ldots$ be the natural numbers in order of magnitude for which this case is satisfied. Suppose that it is the second case that is first satisfied, then we shall write $f^{(\nu_1)} \doteq f^{(\nu_2)}$. Let $g^*\nu$ be $gg'g'' \ldots g^{(\nu \dot{-} 1)}\nu$, then for sufficiently large $\kappa$ we shall have exactly one of:

$$f^{(\nu_1)}g^*\kappa \gtrdot f^{(\nu_2)}g^*\kappa \quad \text{or} \quad f^{(\nu_1)}g^*\kappa \doteq f^{(\nu_2)}g^*\kappa \quad \text{or} \quad f^{(\nu_1)}g^*\kappa \lessdot f^{(\nu_2)}g^*\kappa.$$

Thus the functions of our enumerable set of functions have been linearly ordered as desired. Note that the ordering will depend on the initial enumeration and on the chosen ordering of distinct pairs of natural numbers.

LEMMA 2. *The closure of an enumerable set of one-place functions under composition by an enumerable set of functions is an enumerable set of one-place functions.*

Let the enumerable set of one-place functions be $f, f', f'', \ldots$ and let the enumerable set of compounding functions be $g, g', \ldots$, let $g^{(\nu)}$ have $k\nu$ places, where $k$ is a primitive recursive function. Consider $g^{(\nu)}$ to have any number of places greater than $k\nu$. This amounts to taking

$$g^{(\nu)}\theta'\theta'' \ldots \theta^{(k\nu)} = g^{(\nu)}\theta'\theta'' \ldots \theta^{(k\nu)} \ldots \theta^{(\pi)} \quad \text{for} \quad k\nu < \pi.$$

Now form a sequence of functions as follows:

(1) $f; \underset{k0 \text{ places}}{g\{f, \ldots, f\}}; f'$; followed by $g$ and $g'$ with $Max[k0, k1]$ places filled by all possible permutations of preceding functions; $f''$; followed by $g$, $g'$ and $g''$ with $Max[k0, k1, k2]$ places filled with all possible permutations of preceding functions; $f'''$;, etc. This sequence gives an enumeration of all functions obtained from the initial set by composition with the functions of the compounding set. Note that the enumeration will contain the functions of the first set. There will be repetitions. To obtain the required enumeration we proceed as follows:

Let the number of functions which precede the function $f^{(\nu)}$ in the order (1) be $\sigma[\nu]$. Then

$$\sigma[0] = 0, \quad \sigma[S\nu] = (S\nu) \times (\sigma[\nu] + 1)^{\tau[\nu]} + \sigma[\nu] + 1,$$

where $\tau[\nu] = \underset{0 \leqslant \xi \leqslant \nu}{Max}[k[\xi]]$.

$\sigma$ is a primitive recursive function because $k$ is a primitive recursive function. The place number of $f^{(\nu)}$ in the list (1) is $\sigma[\nu] + 1$. Let $h, h', \ldots$ be the successive functions in the list (1). Then $h$ is $f$, $h^{(S\nu)}$ is $f^{(\kappa)}$ if $\nu = \sigma[\kappa]$. If $\sigma[\kappa] < S\nu < \sigma[S\kappa]$ then $\nu = \sigma[\kappa] + (S\kappa) \times \pi + \pi'$, where $\pi' < S\kappa$ and $\pi = \theta' + \theta'' \times (S\sigma[\kappa]) + \ldots + \theta^{(\tau[\kappa])} \times (S\sigma[\kappa])^{\tau[\kappa] \dot{-} 1}$ with $\theta', \ldots, \theta^{(\tau[\kappa])} < S\sigma[\kappa]$, then

$$h^{(S\nu)} = g^{(\pi')}[h^{(\theta')}, \ldots, h^{(\theta^{(\tau[\kappa])})}].$$

This gives the required enumeration. By lemma (i) the closure of an enumerated set of functions by composition with an enumerated set of functions can be put into linear order. This order will start with the constant functions if the constant function zero is in the first set and the successor function is in the second set. The constant functions will be followed by unbounded functions.

If the set of compounding functions contains the functions $S$ and $P$, in particular if the set of compounding functions contains all primitive recursive functions, then associated with an unbounded function there will be a batch of functions $\ldots PPf, Pf, f, Sf, SSf, \ldots$ of order type $\omega^* + \omega$. There will fail to be any function of the set formed by composition between any two consecutive functions of a batch. If $f$ and $g$ are two functions of the set formed by composition then if the compounding functions contain addition and the function $\left[\frac{\nu}{2}\right]$, we shall have an unbounded function in a new batch lying between them. Thus the set of functions formed by composition with primitive recursive functions from an enumerated set of functions which contains some unbounded functions will contain a part of the order type $\omega + \eta \times (\omega^* + \omega)$.

## 12.9 *A non-standard model for* $\mathbf{A}_I$

PROP. 1. *The system* $\mathbf{A}_I$ *fails to characterize the natural numbers.*

Take a set $\mathscr{F}$ of unbounded one-place functions of natural numbers and form their closure with respect to $S$, $\mathscr{I}$ and the resolving functions of the $\mathbf{A}_I$-theorems. A resolving function is obtained as follows: Let $\phi$ be an $\mathbf{A}_I$-theorem, put it into prenex normal form, remove the prefix and replace the restricted variables by such functions of the superior general variables that when we replace the general variables by numerals in any manner the resulting $\mathbf{A}_{00}$-statement is $\mathbf{A}_{00}$-true. These functions can be uniquely defined by using the least natural number which satisfies as the value of the function. Details are left to the reader. Call the set of resolving functions $\mathscr{R}$. Then we form the closure of $\mathscr{F}$ with respect to $S$, $\mathscr{I}$ and the functions of $\mathscr{R}$, and put the resulting set of functions $\mathscr{C}$ into linear order as described in lemma (i), (ii). This set of functions is in a different order to that of the natural numbers. We now show that $\mathscr{C}$ is a set of constants of type $\iota$ in a non-standard model $\mathscr{M}$ for $\mathbf{A}_I$. It is non-standard because it has a different order type to that of the natural numbers.

The constants of type $o$ in the model $\mathscr{M}$ are $\boldsymbol{t}$ and $\boldsymbol{f}$ and we evaluate statements as in the two-valued case. To show that $\mathscr{M}$ is a model for $\mathbf{A}_I$ we have to show that the $\mathbf{A}_I$-axioms are $\mathscr{M}$-true and that the $\mathbf{A}_I$-rules preserve $\mathscr{M}$-truth, so that $\mathbf{A}_I$-theorems are $\mathscr{M}$-true. To determine $\mathscr{M}$-truth we have to replace the $\mathbf{A}_I$-constants by the corresponding $\mathscr{M}$-constants. Numerical terms are replaced by functions which have a constant value for sufficiently large argument and this constant value is that of the numerical term replaced. We then immediately see that the $\mathbf{A}_I$-axioms are $\mathscr{M}$-true. Further we have to replace the general variables by one-place functions of $\mathscr{M}$ and restricted variables by certain one-place functions of $\mathscr{M}$, namely the member of $\mathscr{M}$ which is obtained by the closure operation when we replace the general variables as above and apply the appropriate resolving function. We then give all these one-place functions the same argument $\kappa$, then if $\kappa$ is sufficiently large these functions are in linear order. If the upper formulae of an $\mathbf{A}_0$-rule other than R 1 is $\mathbf{A}$-true then it is $\mathscr{M}$-true because we are using the same two-valued tables to evaluate them and the lower formula is likewise $\mathscr{M}$-true. Note that the constant $\alpha$ in II $d$

$$\frac{\phi\{\alpha\} \vee \omega}{(E\xi)\,\phi\{\xi\} \vee \omega}$$

has to be a numerical term, these are the only constants allowed in $\mathbf{A}_I$. With regard to R 1, 3 if the upper formulae are $\mathbf{A}_I$-theorems then the resolving functions they provide are in $\mathscr{R}$ and are used in forming the closure of $\mathscr{F}$. For instance a case of R 3 is

$$\frac{(E\eta)\,\phi\{\eta, \nu\}}{(A\xi)\,(E\eta)\,\phi\{\eta, \alpha\{\xi\}\}} \quad \text{for all } \nu,\ E\text{-correct,}$$

the upper formula gives the resolving function $f$, say, suppose that $f\nu = \mu$ then to prove $(E\eta)\,\phi\{\eta, \nu\}$ we need only prove $\phi\{\mu, \nu\}$, the function $f$ fails to appear, but in the lower formula the function $f$ is required to get $\mathscr{M}$-truth. We first form $\phi\{f\{\alpha\{g\}\}, \alpha\{g\}\}$ $\alpha\{g\}$ is a constant of $\mathscr{M}$ of type $\iota$ and so is $f\{\alpha\{g\}\}$, both by our closure operation, we now give an argument $\kappa$ to $g$ and get $\phi\{f\{\alpha\{g\kappa\}\}, \alpha\{g\kappa\}\}$, for sufficiently large $\kappa$ this is of the form $\phi\{f\mu, \mu\}$ and this is $\mathbf{A}$-true by our definition of resolving function. Similarly for R 1, these two rules are different from the other rules in that they contain undetermined statements in the main formulae. Thus $\mathbf{A}_I$-theorems are $\mathscr{M}$-true and so $\mathscr{M}$ is a model for $\mathbf{A}_I$. Note that there are an unbounded set of such models varying with the choice of the initial set of unbounded functions $\mathscr{F}$.

We express this state of affairs by saying that $\mathbf{A}_I$-fails to characterize the natural numbers.

The constants of the non-standard model $\mathscr{M}$ satisfy all the $\mathbf{A}_I$-theorems. Thus each constant is either even or is odd; it is uniquely, except for order, factorizable into prime factors. Consider the constant of $\mathscr{M}$ corresponding to the function $!x$ It is divisible by every standard prime, yet all its prime divisors are bounded, being less than some non-standard constant, $!x$ itself will do. Again the non-standard constant corresponding to the function $!x+1$ is without standard primes, yet it is either itself a prime or is the product of non-standard primes.

$\mathbf{A}_I$ contains an irresolvable statement, $G$ say; we can add either $G$ or $\sim G$ as an extra axiom and the resulting system will be consistent in the sense that there will be unprovable statements. Suppose we add $G$ as an extra axiom and that the system is inconsistent in the above sense, then we can prove $\sim G$; by the Deduction Theorem we could $\mathbf{A}_I$-prove $\sim G \vee \sim G$ whence by I $b$ we could $\mathbf{A}_I$-prove $\sim G$, which is absurd. We shall show later in this chapter that a consistent system has a model. Suppose that $G$ is of the form $(A\xi)(E\eta)\,\phi\{\xi,\eta\}$ and is $\mathbf{A}$-true, then $\phi\{\nu, f\nu\}$ is $\mathbf{A}$-true for some function $f$ and all numerals $\nu$. In the non-standard model $\mathscr{M}$ the $\mathscr{M}$-truth of $(A\xi)(E\eta)\,\phi\{\xi,\eta\}$ requires the $\mathbf{A}$-truth of $\phi\{h\nu, fh\nu\}$ for all sufficiently large $\nu$ (depending on $h$) and all functions $h$ in $\mathscr{M}$. If the function $f$ is absent from $\mathscr{M}$ then $(A\xi)$ $(E\eta)$ $\phi\{\xi,\eta\}$ will be $\mathscr{M}$-false, thus we shall have a non-standard model $\mathscr{M}$ for $\mathbf{A}_I$ in which $\sim G$ is $\mathbf{A}$-true but $\mathscr{M}$-false.

## 12.10 *Induction*

The rule of induction R 3 is different from the other arithmetic axioms and rules in that if we allowed free individual and function variables and a substitution rule then the arithmetic axioms and rules could, except for the subsidiary formula, be put down as single cases, e.g. axiom 1 as $x = x$. R 2 as $\dfrac{x \neq 0 \vee \omega}{Sx \neq 0 \vee \omega}$. But to do anything similar for the induction rule we should require a new type of variable namely a property variable, and this would fundamentally alter the character of $\mathbf{A}_I$. The other changes mentioned such as allowing free variables and having a substitution rule is more or less a matter of taste. We now show, by means of a non-standard model that the induction rule fails to be equivalent to a bounded number of particular cases.

PROP. 2. *The system* $\mathbf{A}_I$ *restricted to a displayed set of cases of the induction rule is strictly weaker than the system* $\mathbf{A}_I$.

Let $(A\xi)\,\phi\{\xi\}, \ldots, (A\xi^{(\kappa)})\,\phi^{(\kappa)}\{\xi^{(\kappa)}\}$ be the main formulae in the displayed set of cases of R 3. Put those of them that are **A**-true into resolved form and let the functions thus required be: $f', \ldots, f^{(\lambda)}$, where $f^{(\theta)}$ has $h\theta$ places. Let $\alpha$ be a non-standard constant of type $\iota$, then we can form a non-standard model $\mathscr{M}_\alpha$ by taking the closure of applying $S, \mathscr{I}, f', \ldots, f^{(\lambda)}$ to $\alpha$. This non-standard model will satisfy the system $\mathbf{A}_{I,\kappa}$ which is the system $\mathbf{A}_I$ restricted to the cases of R 3 which have the above listed main formulae. We will now construct a case of R 3 for which the upper formulae are $\mathbf{A}_{I,\kappa}$-provable, **E**-correct, but for which the lower formula fails to be $\mathscr{M}_\alpha$-true. The argument will depend on the fact, already noted, that each non-standard constant of type $\iota$ is greater than each numeral.

The *order* of a constant of type $\iota$ of $\mathscr{M}_\alpha$ with respect to the constant $\alpha$ is equal to the number of $S, \mathscr{I}, f', \ldots, f^{(\lambda)}$ used in its construction starting with $\alpha$. Thus associated with the non-standard constant $\bar{\alpha}$ of $\mathscr{M}_\alpha$ of type $\iota$ there is a construction sequence $\alpha, \alpha', \ldots, \bar{\alpha}$, where each member of the sequence is obtained from its predecessors by applying $S, \mathscr{I}, f', \ldots, f^{(\lambda)}$. Each member of this construction sequence will be given an order as defined above. We can express this in **A** as follows:

D 276 $\quad U[\alpha, \bar{\alpha}, \nu] \quad$ for $\quad (E\xi, \eta)$

$$((\xi_1^\infty = \alpha \,\&\, \eta_1^\infty = 0 \,\&\, l.c.\,\xi = \bar{\alpha} \,\&\, l.c.\,\eta = \nu \,\&\, \varpi\xi = \varpi\eta)$$

$$\&\, (A\xi')_{\varpi\xi \dot{-} 1}(E\eta')_{\xi'}((\xi^\infty_{S\xi'} = S\xi^\infty_{\eta'} \,\&\, \eta^\infty_{S\xi'} = S\eta^\infty_{S\eta'})$$

$$\vee\, (E\eta', \eta'', \eta''')_{\xi'}(\xi^\infty_{S\xi'} = \mathscr{I}(\boldsymbol{\lambda}\zeta'\zeta''.\,\xi^\infty_{\eta'})\,\xi^\infty_{\eta''}\,\xi^\infty_{\eta'''} \,\&\, \eta^\infty_{S\xi'} = \eta^\infty_{\eta'} + \eta^\infty_{\eta''} + \eta^\infty_{\eta'''} + 1) \tag{1}$$

$$\vee\, \sum_{\zeta=1}^{\lambda}(E\eta', \ldots, \eta^{(h\zeta)})_{\xi'}\,(\xi^\infty_{S\xi'} = f^{(\zeta)}\xi^\infty_{\eta'} \ldots \xi^\infty_{\eta(h\zeta)} \,\&\, \eta^\infty_{S\xi'} = S\sum_{\zeta'=1}^{h\zeta}\eta^\infty_{\eta(\zeta')})),$$

read as: '$\bar{\alpha}$ *is a* $\nu$*th descendant of* $\alpha$'. Thus if $\bar{\alpha}$ is a $\nu$th descendant of $\alpha$ it is formed from $\alpha$ using exactly $\nu$ $S$'s, $\mathscr{I}$'s, and function signs $f', \ldots, f^{(\lambda)}$.

The constants of the model of type $\iota$ which are at most $\nu$th descendents of $\alpha$ form a displayable set, they are then bounded, in fact their sum would do as a bound.

D 277 $$V[\alpha, \bar{\alpha}, \nu] \quad \text{for} \quad \sum_{\zeta=0}^{\nu} U[\alpha, \bar{\alpha}, \zeta],$$

read '$\bar{\alpha}$ is an at most $\nu$th descendant of $\alpha$'.

The at most $\nu$th descendants of $\alpha$ are bounded is expressed by

$$(A\xi, \eta)(Ev)(A\zeta)(V[\xi, \zeta, \eta] \rightarrow \zeta \leqslant v).$$

Now consider
$$(A\xi)(Ev)(A\zeta)(V[\xi, \zeta, \nu] \rightarrow \zeta \leqslant v)$$
call it $P[\nu]$.

LEMMA (i). *$P[0]$ is an $\mathbf{A}_{I,\kappa}$-theorem.*

Let (1) shorn of its unbounded quantifiers be denoted by $U[\xi, \eta, \alpha, \bar{\alpha}, \nu]$. We have $V[\alpha, \bar{\alpha}, 0] = U[\alpha, \bar{\alpha}, 0]$. If $U[\xi, \eta, \alpha, \bar{\alpha}, 0]$ then *l.c.* $\eta = 0$. If $\varpi\eta > 1$ then *l.c.* $\eta > 1$ hence $\varpi\eta = 1$ and so the 2nd, 3rd and 4th clauses fail, thus $\varpi\xi = 1$ and $\bar{\alpha} = \alpha$. Hence

$$V[\alpha, \bar{\alpha}, 0] \rightarrow \bar{\alpha} \leqslant \alpha,$$

and so $P[0]$ is an $\mathbf{A}_{I,\kappa}$-theorem.

LEMMA (ii). *$P[\nu]$ is an $\mathbf{A}_{I,\kappa}$-theorem for each $\nu$, $\mathbf{E}$-correct.*

We say that $g$ is a majorizing function for functions $f', \ldots, f^{(\kappa)}$ of $h1, \ldots, h\kappa$ places respectively, if:

$$f^{(\lambda)}\nu' \ldots \nu^{(h\lambda)} \leqslant g\nu \quad \text{for} \quad \nu', \ldots, \nu^{(h\lambda)} \leqslant \nu \quad (1 \leqslant \lambda \leqslant \kappa).$$

If $\mathscr{M}$ is a non-standard model consisting of a set of functions arranged in linear order as in lemma 1, then any bounded set of these functions has a majorizing function in the model, because the model is closed with respect to primitive recursive functions, thus their sum will do for the majorizing function.

The 0th descendants of a non-standard constant $\alpha$ are bounded, because $\alpha$ is the only such descendant. Suppose that the $\pi$th descendants of $\alpha$ are bounded by the non-standard constant represented by the function $f$, i.e. $g\nu' \ldots \nu^{(\theta)} < f\nu$ for $\nu', \ldots, \nu^{(\theta)} \leqslant \nu$, where the function $g$

represents a non-standard constant which is a $\pi$th descendant of $\alpha$ and has $\theta$ places. Then the $S\pi$th descendants of $\alpha$ are of one of the forms:

$f^{(\lambda)}$ with one place containing $g\nu' \dots \nu^{(\theta)}$,

$g$ with one place filled with $f^{(\lambda)}\nu' \dots \nu^{(h\lambda)}$, $Sg\nu' \dots \nu^{(\theta)}$,

$g$ with $\mu$th place filled with $S\nu^{(\mu)}$,

$\mathscr{I}(\lambda\zeta'\zeta''.f_1\nu'\nu'' \dots \nu^{(\theta')})f_2\nu^{(S\theta')} \dots \nu^{(\theta'')}f_3\nu^{(S\theta'')} \dots \nu^{(\theta''')}$,

$f^{(\lambda)}$ with all places filled with earlier descendants of $\alpha$,

where $f_1, f_2, f_3$ are at most $\pi$th descendants of $\alpha$. Let $h\nu' \dots \nu^{(\theta)}$ be at most an $S\pi$th descendant of $\alpha$, then $h\nu' \dots \nu^{(\theta)} \leqslant ff\nu$ altogether $h\nu' \dots \nu^{(\theta)} \leqslant F\nu$, where $F\nu = ff\nu^{\pi}$. Thus the $S\pi$th descendants of $\alpha$ are bounded.

We have used the result:

If $f_1, f_2, f_3$ are majorized by $f$ then

$$\mathscr{I}(\lambda\zeta'\zeta''.f_1\zeta'\zeta''\nu' \dots \nu^{(\theta')})f_2\nu^{(S\theta')} \dots \nu^{(\theta'')}f_3\nu^{(S\theta'')} \dots \nu^{(\theta''')} \leqslant ff\nu$$

where $\nu', \dots, \nu^{(\theta''')} \leqslant \nu$.

This demonstration that $\dfrac{P[\nu]}{P[S\nu]}$ can be fully displayed, hence we can get $P[\nu]$ for each $\nu$, **E**-correct.

LEMMA (iii). $(Ax)P[x]$ *is* $\mathscr{M}$*-false.*

If $(Ax)P[x]$ is $\mathscr{M}_\alpha$-true then for a non-standard constant $\beta$ we shall have $P[\beta]$. Now $\nu < \beta$ for each numeral $\nu$ and each non-standard constant $\beta$. Again $V[\alpha, \bar{\alpha}, \beta]$ for every constant $\alpha$ in $\mathscr{M}_\alpha$, thus if $P[\beta]$ then $\bar{\alpha} \leqslant \gamma$ for some constant $\gamma$ in $\mathscr{M}_\alpha$. But $\bar{\alpha}$ can be any constant in $\mathscr{M}_\alpha$, say $S\gamma$, but this is absurd. This completes the demonstration of the proposition.

## 12.11 *S-models*

Given models $\mathscr{M}', \mathscr{M}''$, ... of a formal system $\mathscr{L}$ we can form another model $\mathscr{M}$ using the direct product of the models $\mathscr{M}', \mathscr{M}''$, .... We do this as follows: the constants of type $\iota$ of the model are functions over $\mathscr{N}$, the natural numbers, such that $f\nu$ is in $\mathscr{M}^{(\nu)}$. An atomic statement $\phi\{f\}$ is $\mathscr{M}$-true if and only if $\phi\{f\nu\}$ is $\mathscr{M}^{(\nu)}$-true for each $\nu$. Now $\sim\phi\{f\}$ is $\mathscr{M}$-true if and only if $\sim\phi\{f\nu\}$ is $\mathscr{M}^{(\nu)}$-true for each $\nu$, i.e. if $\phi\{f\nu\}$ is $\mathscr{M}^{(\nu)}$-false for each $\nu$. Thus we require $\phi\{f\nu\}$ to be $\mathscr{M}^{(\nu)}$-true for each $\nu$ or to be $\mathscr{M}^{(\nu)}$-false for each $\nu$. We must somehow avoid the possibility of $\phi\{f\nu\}$

being sometimes $\mathscr{M}^{(\nu)}$-true and sometimes being $\mathscr{M}^{(\nu)}$-false. The case when the models are all the same is of particular interest.

If $g$ is a $\kappa$-place functor in $\mathscr{L}$ then the corresponding functor in $\mathscr{M}$ is the function over $\mathscr{N}$ $g\{f'\nu, f''\nu, \ldots, f_\nu^{(\kappa)}\}$ whose value is in $\mathscr{M}^{(\nu)}$. In order to have closed atomic statements either $\mathscr{M}$-true or $\mathscr{M}$-false we proceed as follows: we arrange the atomic statements in linear order, say $\phi', \phi'', \ldots$ then

$$\phi'\{f1\} \text{ is } \mathscr{M}'\text{-true}, \quad \phi'\{f2\} \text{ is } \mathscr{M}''\text{-true}, \quad \ldots$$

will either be correct for an unbounded set of numerals or will be incorrect for an unbounded set of numerals. Consider the case which first arises and let $g'1, g'2, \ldots$ be in increasing order the numerals for which this is so. Now consider

$$\phi''\{f\{g'1\}\}, \quad \phi''\{f\{g'2\}\}, \quad \ldots$$

and let $g''1, g''2, \ldots$ be in increasing order the numerals for which the first of the two above cases holds.

So we continue, having found $g^{(\nu)}$ we consider

$$\phi^{(S\nu)}\{fg'g'' \ldots g^{(\nu)}1\}, \quad \phi^{(S\nu)}\{fg'g'' \ldots g^{(\nu)}2\}, \quad \ldots$$

and let $g^{(S\nu)}1, g^{(S\nu)}2, \ldots$ be in increasing order the numerals for which the first of the two above mentioned cases holds. Now let $g\nu$ stand for $g'g'' \ldots g^{(\nu)}\nu$ then $\phi^{(\kappa)}\{fg\nu\}$ is either $\mathscr{M}^{(\nu)}$-true for all sufficiently large $\nu$ or is $\mathscr{M}^{(\nu)}$-false for all sufficiently large $\nu$. Thus we can associate $\boldsymbol{t}$ or $\boldsymbol{f}$ with $\phi^{(\kappa)}\{f\}$. Similarly we deal with $\phi^{(\kappa)}\{f'\}$. Here we start off with the sequence $\phi'\{f'g\nu\}$ and repeating the process we arrive at a function $g^*$ such that $\phi^{(\kappa)}\{f'gg^*\nu\}$ is either $\mathscr{M}^{(\nu)}$-true for all sufficiently large $\nu$ or is $\mathscr{M}^{(\nu)}$-false for all sufficiently large $\nu$. Now consider $\phi'\{f''gg^*\nu\}$, and so on. Finally we form the function $G\nu$ for $gg^* \ldots g^{*\cdots*}\nu$, where the last $g$ has $\nu$ asterisks. Then $\phi^{(\kappa)}\{f^{(\mu)}G\nu\}$ will either be $\mathscr{M}^{(\nu)}$-true for all sufficiently large $\nu$ or will be $\mathscr{M}^{(\nu)}$-false for all sufficiently large $\nu$. So we have associated $\boldsymbol{t}$ or $\boldsymbol{f}$ with each atomic formula $\phi^{(\kappa)}\{f^{(\mu)}\}$, as desired. By applying the familiar two-valued truth tables we then uniquely associate $\boldsymbol{t}$ or $\boldsymbol{f}$ with each compound formula. This is easily seen to give us a model consisting of the functions $f^{(\mu)}$ such that $f^{(\mu)}\nu$ is in $\mathscr{M}^{(\nu)}$.

## 12.12 *Ultraproducts*

Another way of obtaining new models from old ones is to use *ultraproducts*. Let $\mathscr{M}^{(\nu)}$ for $\nu = 1, 2, \ldots$ be an enumerable set of models for a formal system $\mathscr{L}$, in particular they could all be the same. Let $\mathscr{Q}$ denote this enumerable set of models. Let $\mathscr{D}$ be an ultrafilter in $\mathscr{N}$ (maximal dual ideal in the Boolean algebra of subsets of $\mathscr{N}$). That is $\mathscr{D}$ satisfies the following:

(i) The null set is absent from $\mathscr{D}$.

(ii) If $\mathscr{A}$ and $\mathscr{B}$ belong to $\mathscr{D}$ then so does their intersection.

(iii) If $\mathscr{A}$ belongs to $\mathscr{D}$ and is a subset of $\mathscr{B}$ then $\mathscr{B}$ belongs to $\mathscr{D}$.

(iv) If $\mathscr{A}$ is a set of natural numbers then either $\mathscr{A}$ or its complement $\bar{\mathscr{A}}$ belongs to $\mathscr{D}$ (by (i) and (ii) exactly one of $\mathscr{A}$ and $\bar{\mathscr{A}}$ belongs to $\mathscr{D}$).

We now define another model for $\mathscr{L}$ denoted by $\mathscr{Q}_{\mathscr{D}}$, and called an ultraproduct, as follows:

The set of constants of type $\iota$ is a set of functions defined over the natural numbers and closed with respect to composition with the functions of $\mathscr{L}$, and such that $f\nu$ belongs to $\mathscr{M}^{(\nu)}$ for each natural number $\nu$. Now let $R[x', \ldots, x^{(\kappa)}]$ be an atomic $\mathscr{L}$-statement then we say that $R[f' \ldots f^{(\kappa)}]$ is $\mathscr{Q}_{\mathscr{D}}$-true if and only if the set of natural numbers $\nu$ such that $R[f'\nu, \ldots, f^{(\kappa)}\nu]$ is $\mathscr{M}^{(\nu)}$-true belongs to the ultrafilter $\mathscr{D}$.

If $\rho$ is a function in $\mathscr{L}$ then $\boldsymbol{\lambda} x . \rho f x, f\nu$ in $\mathscr{M}^{(\nu)}$, is the corresponding function in $\mathscr{Q}_{\mathscr{D}}$. If $P$ is an $S\kappa$-place predicate in $\mathscr{L}$ then $\boldsymbol{\lambda} x . P[fx, f'x, \ldots, f^{(\kappa)}x]$, where $P[f'\nu, f''\nu, \ldots, f^{(\kappa)}\nu]$ is evaluated in $\mathscr{M}^{(\nu)}$, is the corresponding predicate in $\mathscr{Q}_{\mathscr{D}}$.

For compound $\mathscr{L}$-statements the $\mathscr{Q}_{\mathscr{D}}$-truth value is found in the usual way from the two-valued truth tables using the $\mathscr{Q}_{\mathscr{D}}$-truth values of its atomic components. An $\mathscr{L}$-statement corresponds to a formula of the form $\psi\nu$ when we replace the variable $x^{(\theta)}$ by $f^{(\theta)}\nu$; now let $\mathscr{A}$ stand for the set of natural numbers $\nu$ such that $\psi\nu$ is $\mathscr{M}^{(\nu)}$-true. Then an $\mathscr{L}$-statement corresponds to a set of natural numbers, called its *corresponding set.*

LEMMA. *An $\mathscr{L}$-statement is $\mathscr{Q}_{\mathscr{D}}$-true if and only if its corresponding set is in $\mathscr{D}$.*

We use formula induction:

(*a*) If $\phi$ is an atomic $\mathscr{L}$-statement then by construction of $\mathscr{Q}_{\mathscr{D}}$ it is $\mathscr{Q}_{\mathscr{D}}$-true if and only if it corresponds to a set in $\mathscr{D}$.

(*b*) If $\phi$ is an $\mathscr{L}$-disjunction $\phi' \vee \phi''$ and if $\phi'$ and $\phi''$ correspond to $\mathscr{A}'$ and $\mathscr{A}''$ respectively then $\phi$ corresponds to $\mathscr{A}' \cup \mathscr{A}''$. If one of $\mathscr{A}'$ or $\mathscr{A}''$ is in $\mathscr{D}$ then so is $\mathscr{A}' \cup \mathscr{A}''$ by (iii), so $\phi$ is $\mathscr{Q}_{\mathscr{D}}$-true if one of $\mathscr{A}'$, $\mathscr{A}''$ is

$\mathcal{Q}_{\mathcal{D}}$-true. If both of $\mathscr{A}'$ and $\mathscr{A}''$ are outside $\mathscr{D}$ then $\bar{\mathscr{A}}'$ and $\bar{\mathscr{A}}''$ are both in $\mathscr{D}$ by (iv), also $\phi$ corresponds to $\mathscr{A}' \cup \mathscr{A}''$, now $\overline{\mathscr{A}' \cup \mathscr{A}''} = \bar{\mathscr{A}}' \cap \bar{\mathscr{A}}''$ is in $\mathscr{D}$, thus $\mathscr{A}' \cup \mathscr{A}''$ is outside $\mathscr{D}$. Thus $\phi' \vee \phi''$ is $\mathcal{Q}_{\mathcal{D}}$-true if and only if one of $\phi'$ and $\phi''$ is $\mathcal{Q}_{\mathcal{D}}$-true.

(*c*) If $\phi$ is $\sim\phi'$ and if $\phi'$ corresponds to $\mathscr{A}$ then $\sim\phi$ corresponds to $\bar{\mathscr{A}}$. If $\mathscr{A}$ is in $\mathscr{D}$ then $\bar{\mathscr{A}}$ is outside $\mathscr{D}$ and conversely. Thus the lemma holds in this case.

(*d*) If $\phi$ is an existential statement $(E\xi)\,\phi'\{\xi\}$, then by our induction hypothesis $\phi'\{f\}$ is $\mathcal{Q}_{\mathcal{D}}$-true if and only if $\mathscr{A}_f$ is in $\mathscr{D}$, where $\mathscr{A}_f$ is the set corresponding to $\phi'\{f\}$. If one of $\mathscr{A}_f$ is in $\mathscr{D}$ then $\bigcup_f \mathscr{A}_f$ is in $\mathscr{D}$ by (iii) and $\phi'\{f\}$ is $\mathcal{Q}_{\mathcal{D}}$-true by induction hypothesis, so that $(E\xi)\,\phi'\{\xi\}$ is $\mathcal{Q}_{\mathcal{D}}$-true and its corresponding set $\bigcup_f \mathscr{A}_f$ is in $\mathscr{D}$. If all $\bar{\mathscr{A}}_f$ are in $\mathscr{D}$ then $\bigcap_f \bar{\mathscr{A}}_f$ is in $\mathscr{D}$. Suppose $\bigcap_f \bar{\mathscr{A}}_f$ is outside $\mathscr{D}$ while all intersections of bounded sets of $\bar{\mathscr{A}}_f$ are in $\mathscr{D}$. Consider the sets $\mathscr{B}$ such that $\mathscr{B}$ is outside $\mathscr{D}$, by Zorn's Lemma there will be a greatest such set, say $\mathscr{B}_0$, then $\mathscr{B}_0$ is in $\bar{\mathscr{D}}$ while $\mathscr{B}_0 \cup \{\nu\}$ is in $\mathscr{D}$ for every natural number $\nu$. Thus since $\bar{\mathscr{B}}_0$ is in $\mathscr{D}$ then $(\bar{\mathscr{B}}_0 \cup \{\nu\}) \cap \bar{\mathscr{B}}_0$ is in $\mathscr{D}$ by (ii), i.e. $\{\nu\} \cap \bar{\mathscr{B}}_0$ is in $\mathscr{D}$, and so by (i) is different from the null set. Hence $\nu \in \bar{\mathscr{B}}_0$ and so $(\bar{\mathscr{B}}_0 \cap \{\nu\}) = \{\nu\}$, thus $\{\nu\} \in \mathscr{D}$ but this holds for every natural number $\nu$, but this violates (i) and (ii), hence $\bigcap_f \bar{\mathscr{A}}_f$ is in $\mathscr{D}$. Thus $\bigcap_f \bar{\mathscr{A}}_f$ is in $\mathscr{D}$ and so by (iv) $\overline{\bigcap_f \bar{\mathscr{A}}_f}$ is outside $\mathscr{D}$, i.e. $\bigcup_f \mathscr{A}_f$ is outside $\mathscr{D}$. Thus altogether $(E\xi)\,\phi\{\xi\}$ is $\mathcal{Q}_{\mathcal{D}}$-true if and only if its corresponding set is in $\mathscr{D}$.

The model we first constructed over the direct product of models reduces to an ultraproduct. We have an enumeration of $\mathscr{L}$-statements $\phi'\nu, \phi''\nu, \ldots$ with values (designated or undesignated) in $\mathscr{M}_\nu$. Let $\mathscr{S}'$ be the set of natural numbers for which $\phi'\nu$ has a designated value provided that this set is unbounded, otherwise let $\mathscr{S}'$ be the set of natural numbers for which $\phi'\nu$ has an undesignated value. Then $\mathscr{S}'$ is unbounded. Let $\mathscr{S}''$ be the set of natural numbers for which $\phi''\nu$ has a designated value provided that the intersection of $\mathscr{S}'$ and $\mathscr{S}''$ is unbounded, otherwise let $\mathscr{S}''$ be the set of natural numbers for which $\phi''\nu$ has an undesignated value.

Generally when we have defined $\mathscr{S}', \mathscr{S}'', \ldots, \mathscr{S}^{(\lambda)}$ then $\mathscr{S}^{(S\lambda)}$ is defined as the set of natural numbers for which $\phi^{(S\lambda)}\nu$ has a designated value provided that the intersection of $\mathscr{S}^{(S\lambda)}$ and $\prod_{\theta=1}^{\lambda} \mathscr{S}^{(\theta)}$ is unbounded, otherwise let $\mathscr{S}^{(S\lambda)}$ be the set of natural numbers for which $\phi^{(S\lambda)}\nu$ has an undesignated value. Then all the sets $\mathscr{S}', \mathscr{S}'', \ldots$ are unbounded and the intersection of any two is also unbounded. Let $\Sigma$ be the set of all $\mathscr{S}^{(\lambda)}$ closed with respect to intersection. Then $\Sigma$ satisfies conditions (i) and (ii) for being an ultrafilter. Consider the extensions of $\Sigma$ which also satisfy (i) and (ii) of the conditions for being an ultrafilter, by Zorn's Lemma among these there will be a maximal such extension. Let $\mathscr{D}$ be one of these extensions, we show that $\mathscr{D}$ satisfies (iii) and (iv) of the conditions for being an ultrafilter.

Let $\mathscr{D}'$ be the set of subsets of $\mathscr{N}$ which contain a member of $\mathscr{D}$, $\hat{\mathscr{X}}[\mathscr{A} \subseteq \mathscr{X} \,\&\, \mathscr{A} \epsilon \mathscr{D}]$. Then $\mathscr{D} \subseteq \mathscr{D}'$. If $\mathscr{B} \epsilon \mathscr{D}'$ then

$$\mathscr{B} \supseteq \mathscr{A} \,\&\, \mathscr{A} \epsilon \mathscr{D} \,\&\, \mathscr{A} \neq \varnothing,$$

hence $\mathscr{B} \neq \varnothing$. If $\mathscr{B}, \mathscr{B}' \epsilon \mathscr{D}$ then $\mathscr{B} \supseteq \mathscr{A} \,\&\, \mathscr{B}' \supseteq \mathscr{A}' \,\&\, \mathscr{A}, \mathscr{A}' \epsilon \mathscr{D}$, whence $\mathscr{B} \cap \mathscr{B}' \supseteq \mathscr{A} \cap \mathscr{A}' \,\&\, \mathscr{A} \cap \mathscr{A}' \epsilon \mathscr{D}$ hence $\mathscr{B} \cap \mathscr{B}' \epsilon \mathscr{D}'$, thus $\mathscr{D}'$ satisfies (i) and (ii) of the conditions for being an ultrafilter. But $\mathscr{D}$ is maximal, hence $\mathscr{D} = \mathscr{D}'$. Thus $\mathscr{D}$ satisfies condition (iii) for being an ultrafilter. We now show that $\mathscr{D}$ satisfies conditions (iv) for being an ultrafilter as well.

Let $\mathscr{A}$ be a set of natural numbers, then of $\mathscr{A}$ and $\bar{\mathscr{A}}$ one fails to belong to $\mathscr{D}$ since their intersection is null. Suppose that for some set $\mathscr{A}$ both $\mathscr{A}$ and $\bar{\mathscr{A}}$ fail to belong to $\mathscr{D}$, then let $\mathscr{D}_1$ be the set of subsets of $\mathscr{N}$ $\hat{\mathscr{X}}[(E\mathscr{Y}, \mathscr{Z})(\mathscr{X} = \mathscr{Y} \cap \mathscr{Z} \,\&\, \mathscr{A} \subseteq \mathscr{Y} \,\&\, \mathscr{Z} \epsilon \mathscr{D})]$, then $\mathscr{D} \subseteq \mathscr{D}_1$, for if $\mathscr{Y} = \mathscr{N}$ then $\mathscr{X} = \mathscr{Z} \epsilon \mathscr{D}$, if $\mathscr{X}, \mathscr{X}' \epsilon \mathscr{D}_1$ then $\mathscr{X} = \mathscr{Y} \cap \mathscr{Z}$, $\mathscr{X}' = \mathscr{Y}' \cap \mathscr{Z}'$, hence $\mathscr{X} \cap \mathscr{X}' = (\mathscr{Y} \cap \mathscr{Y}') \cap (\mathscr{Z} \cap \mathscr{Z}')$, but $\mathscr{A} \subseteq \mathscr{Y} \cap \mathscr{Y}'$ and $\mathscr{Z} \cap \mathscr{Z}' \epsilon \mathscr{D}$, hence $\mathscr{X} \cap \mathscr{X}' \epsilon \mathscr{D}_1$.

Also $\mathscr{A} \epsilon \mathscr{D}_1$, for take $\mathscr{Y} = \mathscr{A}$, $\mathscr{Z} = \mathscr{N}$ then $\mathscr{A} \epsilon \mathscr{D}_1$. Thus $\mathscr{D}_1$ is a proper extension of $\mathscr{D}$ and $\mathscr{D}_1$ satisfies condition (ii) for being an ultrafilter, hence $\mathscr{D}_1$ must fail to satisfy condition (i) for being an ultrafilter, thus $\varnothing$ is a member of $\mathscr{D}_1$, thus there is a $\mathscr{B}_1 \epsilon \mathscr{D}$ such that $\mathscr{B}_1 \cap \mathscr{A} = \varnothing$. Similarly there is a $\mathscr{B}_2 \epsilon \mathscr{D}$ such that $\mathscr{B}_2 \cap \bar{\mathscr{A}} = \varnothing$. Now

$$\begin{aligned}\mathscr{B}_1 \cap \mathscr{B}_2 = \mathscr{N} \cap \mathscr{B}_1 \cap \mathscr{B}_2 &= (\mathscr{A} \cup \bar{\mathscr{A}}) \cap \mathscr{B}_1 \cap \mathscr{B}_2 \\ &= (\mathscr{A} \cap \mathscr{B}_1 \cap \mathscr{B}_2) \cup (\bar{\mathscr{A}} \cap \mathscr{B}_1 \cap \mathscr{B}_2) \\ &\subseteq (\mathscr{A} \cap \mathscr{B}_1) \cup (\bar{\mathscr{A}} \cap \mathscr{B}_2) \\ &= \varnothing,\end{aligned}$$

but this is absurd. Hence $\mathscr{D}$ satisfies condition (iv) for being an ultrafilter.

It remains to show that $\mathscr{Q}_{\mathscr{D}}$ is a model for $\mathscr{L}$. Let $R[x, \rho x, x', \sigma[x, x']]$ be the resolved form of an $\mathscr{L}$-axiom, then if $f, f' \in \mathscr{Q}_{\mathscr{D}}$ so are $\rho f$ and $\sigma[f, f']$. Now the $\mathscr{Q}_{\mathscr{D}}$-truth of an $\mathscr{L}$-statement depends on whether the intersection of the set

$$\hat{\nu}[R[f\nu, \rho f\nu, f'\nu, \sigma[f\nu, f'\nu]] \text{ is } \mathscr{M}_\nu\text{-true}] \tag{i}$$

with the sets associated with all the atomic components of $R$ is in $\mathscr{D}$ or is in $\overline{\mathscr{D}}$. Now let the atomic components of $R$ be $R^{(\theta)}[x, x']$, $1 \leqslant \theta \leqslant \pi$, and let

$$\mathscr{A}^{(\theta)} = \hat{\nu}[R^{(\theta)}[f\nu, f'\nu] \text{ is } \mathscr{M}_\nu\text{-true}].$$

Let
$$\mathscr{A}_1^{(\theta)} = \begin{cases} \mathscr{A}^{(\theta)} & \text{if } \mathscr{A}^{(\theta)} \text{ is in } \mathscr{D}, \\ \overline{\mathscr{A}}^{(\theta)} & \text{if } \mathscr{A}^{(\theta)} \text{ is in } \overline{\mathscr{D}}. \end{cases}$$

Then $\mathscr{A} = \mathscr{A}_1' \cap \ldots \cap \mathscr{A}_1^{(\pi)}$ is in $\mathscr{D}$, thus the set of natural numbers associated with the axiom $R$ contains $\mathscr{A}$ and so is in $\mathscr{D}$. Thus the $\mathscr{L}$-axioms are $\mathscr{Q}_{\mathscr{D}}$-true. The $\mathscr{L}$-rules preserve $\mathscr{Q}_{\mathscr{D}}$-truth because if the premisses correspond to sets $\mathscr{A}', \ldots, \mathscr{A}^{(\kappa)}$ in $\mathscr{D}$ then the conclusion corresponds to a set which contains $\mathscr{A}' \cap \ldots \cap \mathscr{A}^{(\kappa)}$ and this is in $\mathscr{D}$.

## 12.13 *H-models*

Prop. 3. *If every bounded subset of $\Lambda$ is an $\mathbf{A}^{(\kappa)}*$-consistent set of closed $\mathbf{A}^{(\kappa)}$-statements then there is a two-valued model in which every domain is denumerable and with respect to which $\Lambda$ is satisfiable.*

Note that it can happen that $\Lambda$ is inconsistent with respect to negation while each bounded subset of $\Lambda$ is consistent with respect to negation. For let $(A\xi)\,\phi\{\xi\}$ be irresolvable, while each of $\phi\{0\}, \phi\{1\}, \ldots$ are $\mathbf{A}^{(\kappa)}*$-theorems. Then the $\mathbf{A}^{(\kappa)}*$-proofs of $\phi\{\nu\}$ for $\nu = 0, 1, \ldots$ will either fail to be on a primitive recursive plan or if so this will fail to be $\mathbf{E}$-correct, otherwise $(A\xi)\,\phi\{\xi\}$ would be an $\mathbf{A}^{(\kappa)}*$-theorem. Now if $\Lambda$ contains all of $\phi\{\nu\}$ for $= 0, 1, \ldots$ and also contains $(E\xi) \sim \phi\{\xi\}$ then we can deduce $(A\xi)\,\phi\{\xi\}$ because $\phi\{\nu\}$ now becomes its own deduction from the hypotheses $\Lambda$. Thus from $\Lambda$ we shall be able to deduce $(A\xi)\,\phi\{\xi\}$ and its negation $(E\xi) \sim \phi\{\xi\}$. To avoid this we use bounded subsets of $\Lambda$ as stated in the proposition.

We have to show that there is a general model, that is we have to exhibit such a model, in which each domain $\mathscr{D}_0, \mathscr{D}^{(\alpha)}, \ldots, \mathscr{D}^{(\beta)}$ is denumerable and such that each $\mathbf{A}^{(\kappa)}$-axiom and each $\mathbf{A}^{(\kappa)}$-statement belonging to $\Lambda$ takes the value $\boldsymbol{t}$.

Let $u', u'', \ldots$ be a sequence of new constants of type $\iota$ and let $U', U'', \ldots$ be a sequence of new constants of type $(o\iota)$ and order one. Let $U_1', U_1'', \ldots$ be a sequence of new constants of type $(o\iota\iota)$ and order one, etc. for the other types and orders. This gives a denumerable sequence of denumerable sequences and so their union is denumerable.

Let every bounded subset of $\Lambda$ be an $\mathbf{A}^{(\kappa)*}$-consistent set of $\mathbf{A}^{(\kappa)}$-statements. Let $\mathbf{A}U^{(\kappa)*}$- be the system $\mathbf{A}^{(\kappa)*}$ plus all the new constants and with all of $Uu \vee \sim Uu$, etc. as new axioms. Suppose the closed $\mathbf{A}U^{(\kappa)*}$-statements arranged in some order, say by increasing *g.n.* We keep this order fixed and call it the *standard order.* From this standard order we can obtain a standard ordering of those closed $\mathbf{A}^{(\kappa)*}$-statements of the form $(E\Sigma)\,\phi\{\Sigma\}$, where $\Sigma$ is a variable of some sort, let this order be $(E\Sigma_1)\,A_1\{\Sigma_1\}, (E\Sigma_2)\,A_2\{\Sigma_2\}, \ldots$.

We now define a sequence $j_1, j_2, \ldots$ as follows:

$j_1 = \underset{j}{Min}\,[V_j$ fails to occur in $(E\Sigma_1)\,A_1\{\Sigma_1\}]$, where $V_j$ is $u^{(j)}$ if $\Sigma_1$ is of type $\iota$, is $U^{(j)}$ if $\Sigma_1$ is of type $(o\iota)$, etc.

$j_{\nu+1} = \underset{j}{Min}\,[V_j$ fails to occur in $A_1\{V_{j_1}\}, \ldots, A_{j_\nu}\{V_{j_\nu}\}, (E\Sigma_{\nu+1})\,A_{\nu+1}\{\Sigma_{\nu+1}\}]$, where $V_j$ is as before.

Clearly $j_\nu$ is a primitive recursive function of $\nu$.

Let $\mathscr{R}$ be the class of all statements of the form

$$(E\Sigma_\theta)\,A_\theta\{\Sigma_\theta\} \to A_\theta\{V_{j_\theta}\}, \ldots \quad \text{for} \quad \theta = 1, 2, \ldots,$$

where $V_{j_\theta}$ is as before.

Lemma (i). *If every bounded subset of* $\Lambda$ *is* $\mathbf{A}U^{(\kappa)*}$*-consistent then every bounded subset of* $\Lambda \cup \mathscr{R}$ *is* $\mathbf{A}U^{(\kappa)*}$*-consistent.*

Denote $\Lambda \cup \mathscr{R}$ by $\Lambda^*$. Write $f$ for $0 = 1$. If a bounded subset of $\Lambda^*$ is $\mathbf{A}U^{(\kappa)*}$-inconsistent then we shall have $\Lambda_\nu^* \vdash_{\mathbf{A}U^{(\kappa)*}} f$, for some bounded subset $\Lambda_\nu^*$ of $\Lambda^*$. But this deduction of $f$ from $\Lambda_\nu^*$ as hypotheses uses only a bounded set of members of $\mathscr{R}$, say

$$(E\Sigma_{h_\theta})\,A_{h_\theta}\{\Sigma_{h_\theta}\} \to A_{h_\theta}\{V_{h_\theta}\}, \quad \text{for} \quad 1 \leqslant 0 \leqslant \mu.$$

Call these $B_1, \ldots, B_\mu$, or more fully, $B_1\{V_{j_{h_1}}\}, \ldots, B_\mu\{V_{j_{h_\mu}}\}$.

Then $V_{j_{h_\mu}}$ occurs only in $B_\mu$ and there only in the place shown; this follows from the definition of $j_\nu$. From the Deduction Theorem we obtain:

$$\Lambda_\nu, B_1, \ldots, B_{\mu \dot{-} 1} \vdash_{\mathbf{A}U^{(\kappa)*}} B_\mu \to f,$$

where $\Lambda_\nu$ is a bounded subset of $\Lambda$.

This $\mathbf{A}U^{(\kappa)}$*-deduction remains correct if we replace $V_{j_{h_\mu}}$ by a new free variable of the same type and order, because $V_{j_{h_\mu}}$ fails to occur in the hypotheses. Thus we obtain, by an easy induction:

$$\Lambda_\nu, B_1, \ldots, B_{\mu \dot{-} 1} \vdash_{\mathbf{A}U^{(\kappa)*}} (E\Sigma_\mu)\, B_\mu\{\Sigma_\mu\} \rightarrow f.$$

But $(E\Sigma_\mu)\, B_\mu\{\Sigma_\mu\}$ is an $\mathbf{A}U^{(\kappa)}$*-theorem, because it is

$$(E\Sigma'_{h_\mu})\,((E\Sigma_{h_\mu})\, A_{h_\mu}\{\Sigma_{h_\mu}\} \rightarrow A_{h_\mu}\{\Sigma'_{h_\mu}\}),$$

and this is equivalent to

$$(E\Sigma_{h_\mu})\, A_{h_\mu}\{\Sigma_{h_\mu}\} \rightarrow (E\Sigma'_{h_\mu})\, A_{h_\mu}\{\Sigma'_{h_\mu}\},$$

since $\Sigma'_{h_\mu}$ occurs only in the last place as shown. Thus using the cut we obtain

$$\Lambda_\nu, B_1, \ldots, B_{\mu \dot{-} 1} \vdash_{\mathbf{A}U^{(\kappa)*}} f.$$

Similarly we can dispense with all of $B_1, \ldots, B_\mu$ and obtain

$$\Lambda_\nu \vdash_{\mathbf{A}U^{(\kappa)*}} f \quad \text{and hence} \quad \Lambda_\nu \vdash_{\mathbf{A}^{(\kappa)*}} f,$$

(because any $u$'s or $U$'s left in the deduction may be replaced by closed $\mathbf{A}^{(\kappa)}$ terms or properties), this is absurd. This completes the demonstration of the lemma.

We now form a maximal consistent set $\Gamma^*$ which contains $\Lambda^*$ and is such that every bounded subset of $\Gamma^*$ is $\mathbf{A}U^{(\kappa)}$*-consistent. Let $C_1$ be the first closed $\mathbf{A}U^{(\kappa)}$-statement in the standard ordering such that every bounded subset of $\{\Lambda^*, C_1\}$ is $\mathbf{A}U^{(\kappa)}$*-consistent. When $C_1, \ldots, C_\theta$ have been determined let $C_{S\theta}$ be the first closed $\mathbf{A}U^{(\kappa)}$-statement which comes later than $C_\theta$ in the standard ordering and is such that every bounded subset of $\{\Lambda^*, C_1, \ldots, C_{S\theta}\}$ is $\mathbf{A}U^{(\kappa)}$*-consistent. Then $\Gamma^*$ is the set $\{\Lambda^*, C_1, \ldots\}$. Clearly every bounded subset of $\Gamma^*$ is $\mathbf{A}U^{(\kappa)}$*-consistent.

LEMMA (ii). *For any closed $\mathbf{A}U^{(\kappa)}$*-statement $\phi$ we have $\Gamma^*_\nu \vdash_{\mathbf{A}U^{(\kappa)*}} \phi$ or $\Gamma^*_{\nu'} \vdash_{\mathbf{A}U^{(\kappa)*}} \sim \phi$ for some bounded subsets $\Gamma^*_\nu$, $\Gamma^*_{\nu'}$ of $\Gamma^*$. In the first case $\phi$ is in $\Gamma^*$, in the second case $\sim \phi$ is in $\Gamma^*$.*

If every bounded subset of $\{\Gamma^*, \phi\}$ is $\mathbf{A}U^{(\kappa)}$*-consistent let $\phi$ come after $C_\theta$ and before or at $C_{S\theta}$ in the standard ordering of closed $\mathbf{A}U^{(\kappa)}$-statements. Then every bounded subset of $\{\Gamma^*, C_1, \ldots, C_\theta, \phi\}$ is $\mathbf{A}U^{(\kappa)}$*-consistent, whence, by construction of $\Gamma^*$ $\phi$ is $C_{S\theta}$ and so $\phi$ is in $\Gamma^*$ and hence $\Gamma^*_\nu \vdash_{\mathbf{A}U^{(\kappa)*}} \phi$ for some bounded subset $\Gamma^*_\nu$ of $\Gamma^*$. Conversely, if $\Gamma^*_\nu \vdash_{\mathbf{A}U^{(\kappa)*}} \phi$ for some bounded subset $\Gamma^*_\nu$ of $\Gamma^*$ then $\{\Gamma^*_\nu, \phi\}$ is $\mathbf{A}U^{(\kappa)}$*-consistent and so $\phi$ is in $\Gamma^*$. If $\{\Gamma^*_\nu, \phi\}$ is $\mathbf{A}U^{(\kappa)}$*-inconsistent for some bounded subset $\Gamma_\nu^*$ of $\Gamma^*$ then $\Gamma_\nu^*, \phi \vdash_{\mathbf{A}U^{(\kappa)*}} f$, whence by the Deduction Theorem $\Gamma_\nu^* \vdash_{\mathbf{A}U^{(\kappa)*}} \phi \rightarrow f$, and so $\Gamma_\nu^* \vdash_{\mathbf{A}U^{(\kappa)*}} \sim \phi$, hence by what we have already shown $\sim \phi$ is in $\Gamma^*$. This completes the demonstration of the lemma.

$\Gamma^*$ is closed with respect to rejection. Thus if $\Gamma^*$ contains $(E\Sigma_\theta)\, A_\theta\{\Sigma_\theta\}$ then it will contain $A_\theta\{V_{j\theta}\}$ as well, because $\Gamma^*$ contains

$$(E\Sigma_\theta)\, A_\theta\{\Sigma_\theta\} \rightarrow A_\theta\{V_{j\theta}\}.$$

Thus $\Gamma^*$ will contain an example for each existential statement which it contains.

We now form a model for $\Lambda^*$ over the domains formed by the new contants, $u', u'', \ldots, U', U'', \ldots$, etc. We have to assign $\boldsymbol{t}$ or $\boldsymbol{f}$ to each $\mathbf{A}U^{(\kappa)}$-statement $(\alpha = \beta)$, where $\alpha$ and $\beta$ are closed $\mathbf{A}U^{(\kappa)}$-terms of type $\iota$, and to each $\mathbf{A}U^{(\kappa)}$-atomic statement $\Delta\alpha, \Delta'\alpha\alpha', \ldots$ where $\Delta$ and $\Delta'$ are closed $\mathbf{A}U^{(\kappa)}$-properties and $\alpha$, $\alpha'$ are closed $\mathbf{A}U^{(\kappa)}$-terms of type $\iota$.

We assign $\boldsymbol{t}$ or $\boldsymbol{f}$ to a closed atomic $\mathbf{A}U^{(\kappa)}$-statement according as it is in $\Gamma^*$ or otherwise, compound $\mathbf{A}U^{(\kappa)}$-statements are then given $\boldsymbol{t}$ or $\boldsymbol{f}$ by the usual truth-tables.

LEMMA (iii). *For each closed $\mathbf{A}U^{(\kappa)}$-statement $\phi$ the assigned value is $\boldsymbol{t}$ or $\boldsymbol{f}$ according as $\Gamma_\nu^* \vdash_{\mathbf{A}U^{(\kappa)*}} \phi$ or $\Gamma_{\nu'}^* \vdash_{\mathbf{A}U^{(\kappa)*}} \sim \phi$, for some bounded subsets $\Gamma_\nu^*, \Gamma_{\nu'}^*$ of $\Gamma^*$, that is according as $\phi$ is in $\Gamma^*$ or otherwise.*

Clearly if $\phi$ is in $\Gamma^*$ then $\Gamma_\nu^* \vdash_{\mathbf{A}U^{(\kappa)*}} \phi$ for some bounded subset $\Gamma_\nu^*$ of $\Gamma^*$, any subset containing $\phi$ would do.

We demonstrate the proposition by formula induction. If $\phi$ is atomic then the lemma holds for $\phi$ by definition of the assignment.

If $\phi$ is of the form $\sim \psi$ and if the lemma holds for $\psi$, then $\psi$ is assigned $\boldsymbol{t}$ or $\boldsymbol{f}$ according as $\psi$ is in $\Gamma^*$ or $\sim \psi$ is in $\Gamma^*$, thus $\sim \psi$ is assigned $\boldsymbol{t}$ or $\boldsymbol{f}$ according as $\sim \psi$ in is $\Gamma^*$ or $\psi$, i.e. $\sim \sim \psi$, is in $\Gamma^*$, and the lemma is correct for $\sim \psi$.

If $\phi$ is of the form $\psi' \vee \psi''$ and the lemma is correct for $\psi'$ and for $\psi''$; then, if $\psi'$ is assigned $\boldsymbol{t}$ so is $\psi' \vee \psi''$ by the truth-tables and by induction hypothesis $\Gamma_\nu^* \vdash_{\mathbf{A}U^{(\kappa)*}} \psi'$, whence $\Gamma_\nu^* \vdash_{\mathbf{A}U^{(\kappa)*}} \psi' \vee \psi''$ by II*a*, by lemma (ii) $\psi' \vee \psi''$ is in $\Gamma^*$. Similarly if $\psi''$ is assigned $\boldsymbol{t}$. If both $\psi'$ and $\psi''$ are assigned $\boldsymbol{f}$, then by induction hypothesis $\Gamma_\nu^* \vdash_{\mathbf{A}U^{(\kappa)*}} \sim \psi'$ and $\Gamma_{\nu'}^* \vdash_{\mathbf{A}U^{(\kappa)*}} \sim \psi''$ whence by II$b'$ $\Gamma_{\nu''}^* \vdash_{\mathbf{A}U^{(\kappa)*}} (\sim \psi' \,\&\, \sim \psi'')$ so $\Gamma_{\nu''}^* \vdash_{\mathbf{A}U^{(\kappa)*}} \sim (\psi' \vee \psi'')$ where $\Gamma_{\nu''}^*$ is $\Gamma_\nu^* \cup \Gamma_{\nu'}^*$, where $\Gamma_\nu^*$, $\Gamma_{\nu'}^*$ and hence

$\Gamma^*_{\nu''}$ are bounded subsets of $\Gamma^*$. But $\psi' \vee \psi''$ is assigned $\boldsymbol{f}$ by truth-tables and so the lemma is correct in this case.

If $\psi$ is of the form $(E\Sigma)\,\psi\{\Sigma\}$. If $\Gamma^*_\nu \vdash_{\mathbf{A}U^{(\kappa)*}} (E\Sigma)\,\psi\{\Sigma\}$ for some bounded subset $\Gamma^*_\nu$ of $\Gamma^*$, then as we have already remarked $\Gamma^*_\nu \vdash_{\mathbf{A}U^{(\kappa)*}} \psi\{V\}$ for some $V$. By induction hypothesis the lemma is correct for $\psi\{V\}$, hence $\psi\{V\}$ is assigned $\boldsymbol{t}$ and so $(E\Sigma)\,\psi\{\Sigma\}$ is assigned $\boldsymbol{t}$ by truth-tables and the lemma is correct for $(E\Sigma)\,\psi\{\Sigma\}$ in this case. If $\Gamma^*_\nu \vdash_{\mathbf{A}U^{(\kappa)*}} \sim (E\Sigma)\,\psi\{\Sigma\}$ for some bounded subset $\Gamma^*_\nu$ of $\Gamma^*$, then $\Gamma^*_\nu \vdash_{\mathbf{A}U^{(\kappa)*}} (A\Sigma) \sim \psi\{\Sigma\}$, whence $\Gamma^*_\nu \vdash_{\mathbf{A}U^{(\kappa)*}} \sim \psi\{\Delta\}$ for every property $\Delta$ of the appropriate kind, using the reversibility of R 3. By induction hypothesis the lemma is correct for $\psi\{\Delta\}$ for all properties $\Delta$ of the appropriate kind, hence by what has been shown above the lemma is correct for $\sim \psi\{\Delta\}$ for all appropriate $\Delta$, thus $\psi\{\Delta\}$ is assigned $\boldsymbol{f}$ for each property $\Delta$ of the appropriate kind and so $(E\Sigma)\,\psi\{\Sigma\}$ is assigned $\boldsymbol{f}$ by truth-tables, and the lemma is again correct in this case. This completes the demonstration of the lemma.

We can now demonstrate the proposition. If $\phi$ is in $\Lambda$ then $\phi$ is in $\Gamma^*$ and so $\phi$ is assigned $\boldsymbol{t}$ in the model just described. Thus the set $\Lambda$ of closed $\mathbf{A}^{(\kappa)}$-statements are simultaneously satisfied over a denumerable domain.

Note that the elements of type $\iota$ in the model of Prop. 3 are just $u', u'', \ldots$, hence among these there must be representatives of the numerals. These will come about as follows: $(Ex)\,(x = 0)$ will be one of the statements in the list $(E\Sigma)\,A\psi\{\Sigma\}$, hence in $\mathscr{R}$ we shall have: $(Ex)\,(x = 0) \rightarrow u^{(\lambda)} = 0$. As $(Ex)\,(x = 0)$ is an $\mathbf{A}U^{(\kappa)*}$-theorem then so is $u^{(\lambda)} = 0$. So $u^{(\lambda)}$ will be a representative of 0, etc.

Cor. (i). *If the set $\Lambda$ of closed $\mathbf{A}^{(\kappa)}$-statements fails to be simultaneously satisfiable over a denumerable domain then some bounded subset of $\Lambda$ is $\mathbf{A}U^{(\kappa)*}$-inconsistent.*

Cor. (ii). *If the system $\mathscr{L}$ is based on $\mathscr{F}_C$ and if every bounded subset of the $\mathscr{L}$-axioms is consistent, then $\mathscr{L}$ has a two-valued denumerable model.*

We repeat the demonstration of Prop. 3.

Order in H-models is given by

$$\text{if} \quad (Ex)\,(Sx + u = u') \quad \text{is in } \Gamma^* \text{ then} \quad u < u'.$$

If $u \neq 0, u \neq 1, \ldots$ are all in $\Gamma^*$ then $u$ is non-standard. We could take all these as axioms of a system $\mathscr{L}$ as in Cor. (ii) above and so obtain a consistent system with a non-standard model. If

$$(Ex)\left(x = \left[\frac{u+u'}{2}\right]\right)$$

is in $\Gamma^*$ then
$$\left[\frac{u+u'}{2}\right]$$

is equal to a non-standard element between $u$ and $u'$. If

$$(Ex)\,(x = \nu \,\&\, u+x = u')$$

is in $\Gamma^*$ then $u = u' \dot{-} \nu$, and so on.

PROP. 4. *A closed* $\mathbf{A}^{(\kappa)}$*-statement is an* $\mathbf{A}^{(\kappa)}$*-theorem if and only if it is satisfiable over every denumerable general model.*

First, by definition of model, the $\mathbf{A}^{(\kappa)}$-axioms are satisfied and the $\mathbf{A}^{(\kappa)}$-rules preserve satisfaction over any model, thus the $\mathbf{A}^{(\kappa)}$-theorems are satisfied over any model. Secondly, if a closed $\mathbf{A}^{(\kappa)}$-statement $\phi$ is satisfied over every two-valued general model then $\phi \rightarrow f$ has the value $f$ over every two-valued general model, hence by Cor. (i) above the set consisting of $\phi \rightarrow f$ alone is inconsistent, thus $\phi \rightarrow f \vdash_{\mathbf{A}^{(\kappa)*}} f$. By the Deduction Theorem $\vdash_{\mathbf{A}^{(\kappa)*}} \phi \rightarrow f. \rightarrow f$, whence by the cut and the $\mathbf{A}^{(\kappa)*}$-theorem $\sim f$ we see that $\phi$ is an $\mathbf{A}^{(\kappa)*}$-theorem.

As already remarked the condition of being satisfied over every two-valued general model is a stronger condition than being satisfied over the full two-valued standard model. Thus a closed $\mathbf{A}^{(\kappa)}$-statement fails to be an $\mathbf{A}^{(\kappa)*}$-theorem if and only if its negation can be satisfied over some two-valued general model.

## 12.14 *Satisfactions by $\Delta_2$-predicates*

PROP. 5. *If* $\Lambda$ *is a recursive set of closed* $\mathbf{A}^{(\kappa)}$*-statements and if every bounded subset of* $\Lambda$ *is* $\mathbf{A}^{(\kappa)*}$*-consistent, then* $\Lambda$ *can be simultaneously satisfied by predicates in* $\mathscr{P}_2 \cap \mathscr{Q}_2$.

We have to show that the properties $U$ can be replaced by $\mathscr{P}_2 \cap \mathscr{Q}_2$ properties and that the equality predicate can be replaced by a $\mathscr{P}_2 \cap \mathscr{Q}_2$ predicate.

The closed $\mathbf{A}U^{(\kappa)}$-statements and the set $\mathscr{R}$ are both primitive recursive sets, that is their *g.n.*'s are primitive recursive sets of natural numbers. The standard order of $\mathbf{A}U^{(\kappa)}$-statements is when they are arranged in order of increasing *g.n.* Also $j_\nu$ is a primitive recursive function of $\nu$.

The following are primitive recursive:

D 278 $M[\nu, \pi]$: $\nu$ is the *g.n.* of a closed $\mathbf{A}U^{(\kappa)}$-statement which is a member of a bounded sequence of closed $\mathbf{A}U^{(\kappa)}$-statements in standard order and the *g.n.* of the sequence is $\pi$.

D 279 $L[\nu, \pi]$: $\nu$ is the *g.n.* of the last member of an initial segment, whose *g.n.* is $\pi$, of closed $\mathbf{A}U^{(\kappa)}$-statements in standard order. $\nu$ is thus the $\varpi\pi$th closed $\mathbf{A}U^{(\kappa)}$-statement in the standard order.

D 280 $P[\theta, \nu, \pi]$: $\theta$ is the *g.n.* of an $\mathbf{A}U^{(\kappa)*}$-deduction of a closed $\mathbf{A}U^{(\kappa)}$-statement whose *g.n.* is $\nu$ from a bounded sequence of premisses whose *g.n.* is $\pi$, and each premiss is used in the $\mathbf{A}U^{(\kappa)*}$-deduction. If $\pi$ is zero the sequence is void.

We easily see that if $M[\nu, \pi]$ then $\nu \leqslant \pi$ and if $P[\theta, \nu, \pi]$ then $\pi \leqslant \theta$.

Let $\Lambda$ be a primitive recursive and consistent set of $\mathbf{A}^{(\kappa)}$-statements. Let $\Lambda^*$ be $\Lambda \cup \mathscr{R}$ as before, then $\Lambda^*$ is a primitive recursive set. The following are primitive recursive:

D 281 $\theta \epsilon \Lambda^*$: $\theta$ is the *g.n.* of a closed $\mathbf{A}U^{(\kappa)}$-statement of $\Lambda^*$.

D 282 $P^{\Lambda^*}[\theta, \nu]$ for $(Ex)_\theta(P[\theta, \nu, x] \mathbin{\&} (Ax')_x(M[x', x] \rightarrow x' \in \Lambda^*))$.

$P^{\Lambda^*}$ is a primitive recursive predicate because the quantifiers are bounded. Thus if $P^{\Lambda^*}[\theta, \nu]$ then $\theta$ is the *g.n.* of an $\mathbf{A}U^{(\kappa)*}$-deduction from the premisses $\Lambda^*$ of a closed $\mathbf{A}U^{(\kappa)}$-statement whose *g.n.* is $\nu$.

D 283 $f[\lambda]$: the *g.n.* of the negation of the conjunction of the members of a bounded sequence of $\mathbf{A}U^{(\kappa)}$-statements in standard order whose *g.n.* is $\lambda$. If $\lambda$ fails to be the *g.n.* of such a sequence then $f[\lambda] = 0$.

D 284 $g[\lambda]$: $\lambda$ is the *g.n.* of a bounded sequence of closed $\mathbf{A}U^{(\kappa)}$-statements in standard order, $\lambda = \langle \nu^{(\lambda')}, \ldots, \nu^{(\lambda^{(\mu)})} \rangle$ where $\nu^{(\lambda')} < \ldots < \nu^{(\lambda^{(\mu)})}$, if the standard order of $\mathbf{A}U^{(\kappa)}$-statements is;

$\Phi', \Phi'', \ldots$ and their *g.n.*'s are $\lambda', \lambda'', \ldots$ then $g[\lambda]$ is the *g.n.* of the conjunction of

$$\sim (\Phi' \vee \ldots \vee \Phi^{(\lambda' \dot{-} 1)}),$$
$$(\Phi^{(\lambda')} \rightarrow \sim (\Phi^{(\lambda'+1)} \vee \ldots \vee \Phi^{(\lambda'' \dot{-} 1)}),$$
$$(\Phi^{(\lambda')} \,\&\, \Phi^{(\lambda'')}) \rightarrow \sim (\Phi^{(\lambda''+1)} \vee \ldots \vee \Phi^{(\lambda''' \dot{-} 1)}),$$
$$\vdots$$
$$(\Phi^{(\lambda')} \,\&\, \ldots \,\&\, \Phi^{\lambda^{(\mu \dot{-} 1)}}) \rightarrow \sim (\Phi^{(\lambda^{(\mu \dot{-})}\ 1)} \vee \ldots \vee \Phi^{(\lambda^{(\mu)} \dot{-} 1)}),$$

if $\lambda$ fails to be the *g.n.* of such a sequence then $g[\lambda]$ is zero.

Then $g$ is a primitive recursive function.

Now $\lambda$ is the *g.n.* of an initial segment of $\Gamma^*$ in standard order if and only if

$$(Ex)\, P^{\Lambda^*}[x, g[\lambda]] \,\&\, \sim (Ex)\, P^{\Lambda^*}[x, f[\lambda]].$$

Also $\nu$ is the *g.n.* of a closed $\mathbf{A}U^{(\kappa)}$-statement of $\Gamma^*$ if and only if

$$(Ex)\, (L[\nu, x] \,\&\, (Ex')\, P^{\Lambda^*}[x', g[x]]) \,\&\, \sim (Ex'')\, P^{\Lambda^*}[x, f[\nu]].$$

In prenex normal form this has a prefix $(Ex, x')\,(Ax'')$ and a primitive recursive matrix. By contraction of quantifiers the prefix becomes $(Ex)\,(Ax')$ with a primitive recursive matrix. Thus $\nu$ is the *g.n.* of a closed $\mathbf{A}U^{(\kappa)}$-statement in $\Gamma^*$ if and only if $(Ex)\,(Ax')\,R[\nu, x, x']$ is $\mathbf{A}$-true, where $R$ is a primitive recursive predicate.

Note that, if $\Phi^{(\lambda')}, \ldots, \Phi^{(\lambda^{(\mu)})}$ is an initial segment of $\Gamma^*$ in standard order then

$$\Lambda^* \vdash_{\mathbf{A}U^{(\kappa)*}} \sim \Phi', \ldots, \Lambda^* \vdash_{\mathbf{A}U^{(\kappa)*}} \sim \Phi^{(\lambda \dot{-} 1)} \quad \text{empty if } \lambda' = 0,$$

whence $$\Lambda^* \vdash_{\mathbf{A}U^{(\kappa)*}} \sim (\Phi' \vee \ldots \vee \Phi^{(\lambda' \dot{-} 1)}).$$

Also $$\Lambda^*, \Phi^{(\lambda')} \vdash_{\mathbf{A}U^{(\kappa)*}} \sim \Phi^{(\lambda'-1)}, \ldots, \Lambda^*, \Phi^{(\lambda')} \vdash_{\mathbf{A}U^{(\kappa)*}} \sim \Phi^{(\lambda'' \dot{-} 1)},$$

whence $$\Lambda^* \vdash_{\mathbf{A}U^{(\kappa)*}} \Phi^{(\lambda')} \rightarrow \sim (\Phi^{(\lambda'-1)} \vee \ldots \vee \Phi^{(\lambda'' \dot{-} 1)});$$
and so on.

On the other hand we can express the condition that $\nu$ is the *g.n.* of a closed $\mathbf{A}U^{(\kappa)}$-statement in $\Gamma^*$ by a predicate of the form

$$(Ax)\,(Ex')\,R'[\nu, x, x']$$

with $R'$ primitive recursive.

D 285 $C[\nu]$: $\nu$ is the *g.n.* of a closed $\mathbf{A}U^{(\kappa)}$-statement.

Then $C$ is primitive recursive. Now a closed $\mathbf{A}U^{(\kappa)}$-statement $\phi$ is in $\Gamma^*$

if and only if $\sim\phi$ is outside $\Gamma^*$. Thus $\nu$ is the *g.n.* of a closed $\mathbf{A}U^{(\kappa)}$-statement in $\Gamma^*$ if and only if

$$C[\nu] \mathbin{\&} \sim (Ex)(Ax')\,R[Neg\,[\nu], x, x'],$$

in prenex normal form this is $(Ax)(Ex')\,R'[\nu, x, x']$, where $R'$ is primitive recursive.

Denote either of the conditions

$$(Ex)(Ax')\,R[\nu, x, x'], \quad (Ax)(Ex')\,R'[\nu, x, x']$$

by $H[\nu]$. Let $k[\lambda]$ be the *g.n.* of $u^{(\lambda)}$, then $k$ is primitive recursive (we can certainly arrange the Gödel numbering so that this is so). Let ${}_{S\mu}U^{S\lambda}_{S\nu}$ be the new property constants we use in forming the denumerable model, where $S\mu$ is the order, $S\nu$ is the number of argument places and the superfix denotes the number of superscripted primes. Let:

D 286 ${}_{S\mu}G^{S\lambda}_{S\nu}[k[\theta'], \ldots, k[\theta^{(S\nu)}]]$ be equal to the *g.n.* of ${}_{S\mu}U^{S\lambda}_{S\nu}\, u^{(\theta')} \ldots u^{(\theta^{(S\nu)})}$, and let $G^0_2[k[\theta'], k[\theta'']]$ be equal to the *g.n.* of $(u^{(\theta')} = u^{(\theta'')})$.

Then the $G$s are primitive recursive functions.

D 287 ${}_{S\mu}\mathscr{G}^{S\lambda}_{S\nu}[\theta', \ldots, \theta^{(S\nu)}]$ for $H[{}_{S\mu}G^{S\lambda}_{S\nu}[k[\theta'], \ldots, k[\theta^{(S\nu)}]]]$.

In the model the property ${}_{S\mu}U^{S\lambda}_{S\nu}$ is replaced by the predicate ${}_{S\mu}\mathscr{G}^{S\lambda}_{S\nu}$ in $\mathscr{P}_2 \cap \mathscr{Q}_2$. This demonstrates the proposition.

## 12.15 *H$\epsilon$-models*

We now have two different methods for forming models, viz.: first, the method of ultraproducts, these will be called S-models, they depend on the ultrafilter used and on the sequence $\mathscr{Q}$ of models, usually these are all the same, namely the standard model; secondly, the method of finding a maximal consistent set as in Prop. 3, these will be called H-models, they depend on the ordering of the statements of the system for which we require a model. In either case the method yields a variety of different models and this variety is clearly unbounded.

However the two methods are distinct in that there are H-models for consistent systems which fail to have S-models. To show this consider a closed **A**-statement $(Ax)(\alpha\{x\} \neq 0)$ which is **A**-true but $\mathbf{A}_I$-irresolvable. There fails to be an S-model for which $(Ex)(\alpha\{x\} = 0)$ is satisfied, for this would involve that for some function $f$ we have $\alpha\{f\nu\} = 0$ for $\nu$ in $d$, where $d$ is a member of an ultrafilter $\mathscr{D}$. Now $\alpha\{\nu\} \neq 0$ for all natural numbers $\nu$, hence $\alpha\{f\nu\} \neq 0$ for all functions $f$ and numerals $\nu$. But we can obtain a consistent set by adding $(Ex)(\alpha\{x\} = 0)$ as an extra axiom and so as in Prop. 3 we can obtain an H-model.

The members of the S- and of the H-models are of the same order types, namely $\omega + \eta(\omega^* + \omega)$, where $\eta$ is the order type of the rationals. One can obtain such an order type from elements of the form

$$i^{\pi}(\nu'/\mu') \pm i^{\pi \dot{-} 1}(\nu''/\mu'') \pm \ldots \pm i(\nu^{(\pi)}/\mu^{(\pi)}) \pm \lambda \quad (\mu', \ldots, \mu^{(\pi)} > 0),$$

where addition and multiplication are carried out according to the usual rules ($i$ commutes with the rationals) and where $i > (\nu/\mu)$ for all $\nu$ and $\mu$, and $i^2 > i$, etc. These elements are easily defined formally.

The elements in an H-model are merely a new set of symbols, but the elements in an S-model are functions (in the usual case from $\mathscr{N}$ to $\mathscr{N}$). Can we find out what the nature of the elements of an H-model could be? We now show that the elements of an H-model can be taken to be the $\epsilon$-terms that we introduced in Ch. 3 in connection with the systems $E\mathscr{F}_C$ and $EI\mathscr{F}_C$. These terms are $\epsilon(\lambda x.\phi\{x\})$ where $\epsilon$ is of type $\iota(o\iota)$. In these systems we defined $E$ in terms of $\epsilon$, namely $(Ex)\phi\{x\}$ for $\phi\{\epsilon_x\phi\{x\}\}$. Now in Prop. 3 instead of $(Ex)\phi\{x\} \to \phi\{u_j\}$ we will take $(Ex)\phi\{x\} \to \phi\{\epsilon_x\phi\{x\}\}$, which by the definition of $E$ is a case of T.N.D., and so is a theorem. Similarly for $(EX)\phi\{X\}$, etc. Then lemma (i) of Prop. 3 holds so does the construction of $\Gamma^*$. The rest of Prop. 3 goes through as before. Thus we have a model whose elements are exactly the $\epsilon$-terms. We call these models H$\epsilon$-models. There is an unbounded variety of them, this comes about from different orderings of the closed statements of the system. If $\phi$ is irresolvable then we can either take $\phi$ or $\sim\phi$ as true in the model.

It would seem that there is considerable advantage in using the $\epsilon$-terms, it provides for quantifiers, a choice function and supplies models for every consistent system. But there is a considerable disadvantage, namely the complications one soon gets into on account of the internesting of $\epsilon$-terms.

## 12.16 *Completeness of higher order Predicate Calculi*

The Completeness Theorem for the first order Predicate Calculus $\mathscr{F}_C$, Prop. 12, Ch. 3, says that a closed $\mathscr{F}_C$-statement is an $\mathscr{F}_C$-theorem if and only if it is generally valid over the natural numbers; i.e. the domain $\mathscr{D}_\iota$ is enumerable while the domains $\mathscr{D}_{o\iota}, \mathscr{D}_{o\iota\iota}, \ldots$ are full. The Completeness Theorem for $\mathbf{A}^{(\kappa)*}$ that we demonstrated in Prop. 4 can be modified to give a similar result for the second order Predicate Calculus $\mathscr{F}^2_C$, namely: A closed $\mathscr{F}^2_C$-statement is an $\mathscr{F}^2_C$-theorem if and only if it is satisfiable over every denumerable model. Here all the domains are denumerable. This is a stronger condition than being valid over the full model with a denumerable individual domain. The reason for a stronger condition is seen from the following:

Prop. 6. *If we add a constant predicate of type $o\iota$ to $\mathscr{F}^2_C$ then we can represent primitive recursive number theory in the resulting system by adding a bounded list of axioms.*

Let $S$ be the constant predicate of type $o\iota$.

D 288 $Z_0 x$ for $(Ax') \sim S[x, x']$, read '$x = 0$',

$Z_1 x$ for $(Ex')(Z_0 x' \mathbin{\&} S[x, x'])$, read '$x = 1$', etc.

Read $S[x, x']$ as '$x$ is a successor of $x'$'.

The axioms are:

$Ax\,1$ $(Ex')\,S[x', x]$,

$Ax\,2$ $S[x', x] \rightarrow . S[x'', x] \rightarrow x' = x''$,

i.e. each $x$ has a unique successor.

$Ax\,3$. $S[x, x'] \rightarrow . S[x, x''] \rightarrow x' = x''$,

i.e. if $x$ has a predecessor then it is unique,

$Ax\,4$. $(Ex)\,Z_0 x$,

$Ax\,5$. $(Z_0 x \rightarrow Xx) \mathbin{\&} (Ax')(Xx' \rightarrow (Ax'')(S[x'', x'] \rightarrow Xx'')) . \rightarrow Xx'$,

the principle of mathematical induction.

We also want Modus Ponens and a substitution rule. Call this system $\mathbf{A}_2$.

To show how primitive recursive number theory can be represented in $\mathbf{A}_2$ we first show how addition and multiplication can be represented and then how any primitive recursive function, following its construction sequence, can be represented in terms of the representations of addition and multiplication.

D 289 $\Sigma[x, x', x'']$ for $(AX)((Ay)(Z_0 y \rightarrow X[y, x])$

$\mathbin{\&} (Ay, z)(X[y, z] \rightarrow (Au)(S[y, u] \rightarrow (Av)(S[z, v] \rightarrow X[u, v])))) \rightarrow X[x', x'']$.

This says that $x''$ is the sum of $x$ and $x'$ if for every binary predicate $X$, if $X[0,x]$ and $X[y,z] \to X[Sy,Sz]$ then $X[x',x'']$, where $Sx$ denotes the successor of $x$. We can eliminate the function $S$ in favour of the predicate $S$. Similarly for multiplication we define:

D 290 $\Pi[x,x',x'']$ for $(AX)((Ay)(Z_0 y \to (Az)(Z_0 z \to X[y,z]))$
$\& (Ay,z)(X[y,z] \to (Au)(S[u,y] \to (Av)(\Sigma[x,z,v] \to X[u,v])))) \to X[x',x'']$.

Clearly we can continue this type of definition so as to have a representation for each primitive recursive function. But we can also proceed as follows:

D 291 $R[x,x',x'']$ for $(Ey,z)(\Pi[y,x',z] \& \Sigma[z,x'',x]$
$\& (Eu,v)(S[v,u] \& \Sigma[x'',v,x']))$,

i.e. $x''$ is the remainder on dividing $x$ by $x'$.

Let $\pi$ be a numeral which is divisible by each numeral from 1 to $\nu$ inclusive. Let $\theta', \theta'', \ldots, \theta^{(S\nu)}$ be a given sequence of numerals. Choose $\pi$ greater than the greatest of $\theta', \ldots, \theta^{(S\nu)}$. Then we can solve the congruences

$$\kappa \equiv \theta^{(\lambda)}\, Mod\, [(\lambda \times \pi)+1] \quad \text{for} \quad \lambda = 1, 2, \ldots, S\nu.$$

Now suppose that we have representations for the functions $f$ and $g$, one-place and three-place respectively:

$$F[x,x'] \quad \text{represents} \quad fx = x',$$

$$G[x,x',x'',y] \quad \text{represents} \quad g[x,x',x''] = y.$$

Let $\theta', \ldots, \theta^{(S\nu)}$ satisfy

$$\theta' = f\mu, \quad \text{i.e.} \quad F[\mu,\theta'],$$

$$\theta^{(S\lambda)} = g[\theta^{(\lambda)}, \lambda, \mu], \quad \text{i.e.} \quad G[\theta^{(\lambda)}, \lambda, \mu, \theta^{(S\lambda)}].$$

The existence of $\kappa$ is stated by:

(i) $(Ex,y)(Rem\,[y,Sx] = f\mu \,\&\, (Az)(z < \nu \to Rem\,[y,(Sz \times x)+1]$
$= g[Rem\,[y,(z \times x)+1], z, \mu]))$.

Denote this by $(Ex,y)\,L[x,y,\nu,\mu]$ and label it D 292.

D 293 $M[\nu,\mu]$ for $\lambda x(Ey)\,L[x,y,\nu,\mu]$,

D 294 $N[\nu,\mu]$ for $\lambda y L[M[\nu,\mu],y,\nu,\mu]$,

D 295 $\rho[\nu,\mu]$ for $Rem\,[N[\nu,\mu], \nu \times M[\nu,\mu]+1]$.

These can be represented without the least number symbol and using the predicate $S$ instead of the function $S$ and the predicate $R$ instead of the function *Rem*. Then we can demonstrate the following:

(ii) $$\rho[1,\mu] = f\mu, \quad \rho[S\nu,\mu] = g[\rho[\nu,\mu],\nu,\mu],$$

or rather their representations by means of predicates. The full demonstration of (i) and (ii) is long and tedious.

Having defined, or rather represented, primitive recursive functions, we can repeat the demonstration that the system $\mathbf{A}_2$, if consistent, contains an irresolvable statement $G$. Let $K$ be the conjunction of the axioms and suppose that $G$ is true when we interpret $S$ as the successor function over the natural numbers, then $K \to G$ will be generally valid over the natural numbers, but will fail to be an $\mathscr{F}^2_C$-theorem for if it were then $G$ would be an $\mathbf{A}_2$-theorem which is absurd. Thus the reason for a stronger condition for completeness.

The above considerations show that we can dispense with all our arithmetic symbols except $S$ if we allow quantification over predicate variables. Thus we could modify the system $\mathbf{A}'$ in this way, and similarly for succeeding systems.

Lastly we show that we can dispense with even $S$ if we adjoin variables for properties of properties and quantification over these variables. The resulting system is $\mathscr{F}^4_C$. The new variables are of types $o(o\iota)$, $o(o\iota)$, etc. But to get the full force of these definitions we have to use $\mathscr{F}^6_C$ by adjoining properties of properties of properties and quantification over variables of this type.

PROP. 7. *The successor function can be defined in* $\mathscr{F}^4_C$.

First we require some definitions.

D 296 $(1\text{–}1)r$ for $(Ax,y,z)(r[x,y] \,\&\, r[x,z] . \to y = z)$
$\&\, (Ax,y,z)(r[x,z] \,\&\, r[y,z] . \to x = y)$,

read '$r$ is a one–one relation'.

D 297 $Sim[p,q]$ for $(Er)((Ax)(px \to (Ey)(r[x,y] \,\&\, qy)$
$\&\, (Ay)(qy \to (Ex)(r[x,y] \,\&\, px)) \,\&\, (1\text{–}1)r)$,

read '$p$ and $q$ are similar' or there is a one–one correspondence between the things which have the property $p$ and those which have the property $q$.

D 298 $Card$ for $\boldsymbol{\lambda}P(Ap,q)(Pp \,\&\, Pq . \to . Sim[p,q] : \& : Pp$
$\&\, Sim[p,q] \to Pq)$,

read *Card P* as '*P* is a cardinal number', by this definition a cardinal number is a class of similar classes. It might happen that one of the classes failed to be bounded, we should then have what we call a *transfinite cardinal number*. *Num* is of type $o(o(o\iota)$, numbers are of type $o(o\iota)$.

D 299 0 for $\lambda p . \sim (Ex)px$, the natural number zero,

D 300 1 for $\lambda p.(Ex)(px \,\&\, (Ay)(py \rightarrow x = y))$, the natural number one,

D 301 2 for $\lambda p.(Ex, y)(x \neq y \,\&\, px \,\&\, py \,\&\, (Az)(pz \rightarrow . z = x \vee z = y))$, the natural number two, etc.

D 302 $S$ for $\lambda P \lambda q\, Sim[\lambda x(Ep, y)(Pp \,\&\, px \,\&\, \sim py . \vee y = x), q]$, the successor function.

D 303 *Nat Num* for $\lambda Q(A\mathscr{P})(\mathscr{P}0 \,\&\, (A\mathscr{P})(\mathscr{P}P \rightarrow \mathscr{P}SP) . \rightarrow \mathscr{P}Q)$.

Here $\mathscr{P}$ is of type $o(o(o\iota))$. This separates out the unbounded from the bounded numbers. Using relativized quantification we can now speak about all natural numbers.

There is however a difficulty. It concerns Ax. 2.1, 2.2. Apparently the set-up we have so far is incapable of distinguishing between the case when all properties are bounded and the case when this fails. If the only model that satisfies has to have a bounded individual domain then the successor of any bound will be the null set, i.e. 0, because there will fail to be a property applicable to this number of individuals. Thus in order to have properties applicable to any number of individuals so that each natural number, as defined, will have a successor different from zero it is necessary to have an axiom to ensure that this is so. This axiom is called the *axiom of infinity*, because it ensures that there are an unbounded number of individuals. Apart from this axiom the development is based on 'pure logic', i.e. $\mathscr{F}^6_C$, a system which could be used as basic to any 'universe of discourse'. So it is in the nature of a blemish on what is otherwise a purely logical development of number. One form of the axiom of infinity is:

D 304 $Ax\infty$ for $(Ep, q)((Ax)(qx \rightarrow px) \,\&\, (Ey)(py \,\&\, \sim qy) \,\&\, Sim[p, q])$,

i.e. there is a set which is similar to a proper part of itself. This is a distinctive feature of an unbounded set.

PROP. 8. *If we add* $Ax\infty$ *to* $\mathscr{F}^6_C$ *then we can develop primitive recursive arithmetic in the resulting system.*

The natural numbers satisfy *Nat Num* and are called the *inductive numbers*, because then they satisfy the principle of mathematical induction.

We have given a selection of systems in which primitive recursive function theory can be developed. It is matter of taste which one prefers. Each has advantages and disadvantages. Set theory as developed in Ch. 3 is a very powerful system for practically all mathematics, it is such a wonderful medium for model making, but it has the disadvantage of lacking a standard truth definition. But it has the advantage that all one needs is the null set and what one can construct from it by the methods available in set theory, but we have to have an axiom of infinity, and so the resulting blemish. In the beginning was the null set.

## 12.17 *Independence proofs*

In this final section we will give a very brief sketch of a remarkable breakthrough which occurred a few years ago in independence proofs in set theory, and invented by P. Cohen.

It seems almost natural that a consistent formal system should have a countable model, as first discovered by Skolem, when one considers that a formal system contains names for only a countable number of things. Thus a formal system for arithmetic contains a name for each natural number, but a formal system for analysis is without names for all real numbers so that quantifiers refer to a ghostly host of nameless entities besides the entities of that type that have names and so can be referred to directly.

Thus a countable model of a consistent system will contain elements corresponding to all the named elements in the system but the host of unnameable elements may be unrepresented, just absent as if they had never been. Thus the countable model only has elements corresponding to part of the potential universe in cases where this can be uncountable. But we must be careful with this kind of talk because the concept of countable and of cardinal number are relative to the formal system used. The countable model for set theory is obtained thus: let $\{\phi_\nu\}$ be an enumeration of the axioms of set theory. The set $\{\phi_\nu\}$, if consistent, will have a countable model. Now set theory contains sets that are uncountable, yet we have a countable model.

The solution to this apparent paradox is that we are using the term 'countable' in two different senses, once in set theory, and secondly in the language we are using for the model. This resolves the paradox, it is a confusion of languages, a common fault of unscrupulous persons.

Now suppose that we have a natural model of set theory. By this we mean that the membership relation in the set theory translates into a membership relation in the model and that the equality relation in set theory translates into the identity relation in the model. Thus if $(\alpha\epsilon\beta)$ is proved in set theory and if $\alpha, \epsilon, \beta$ translate respectively into $\dot{\alpha}, \dot{\epsilon}, \dot{\beta}$ in the model then if $\mathrm{e}$ is the membership relation in the model then we shall have in the model $(\dot{\alpha}\dot{\epsilon}\dot{\beta})$, by definition of model, but we shall also have $(\dot{\alpha}\mathrm{e}\dot{\beta})$ and $\dot{\alpha}, \dot{\beta}$ will be collections of which $\dot{\alpha}$ is a member of the collection $\dot{\beta}$. Similarly for equality.

Now take a consistent set theory and a countable model for it, $\mathscr{M}$. The model $\mathscr{M}$ being countable will fail to have elements corresponding to certain possible sets (unnameable in the set theory), for instance certain sets of natural numbers. The gist of the breakthrough is to add certain new sets of natural numbers to the model $\mathscr{M}$ and so obtain a new model $(\mathscr{M}, \alpha)$. Of course we must also add all the sets in the model that we can obtain from the new set $\alpha$ by the methods allowed for the construction of new sets from already existing sets, e.g. union, complementation, etc. and then one has to verify that $(\mathscr{M}, \alpha)$ actually is a model. One also wants to know how ordinals and cardinals behave in the model.

This operation of adjunction is similar to the adjunction of a root of an irreducible equation to an algebraic field. Gödel obtained his inner model of set theory by starting with the null set, which of course is named in set theory, and then constructing, by transfinite induction, all the sets that can be obtained from it by applying the methods allowed in set theory for the construction of new sets from old ones. By the very nature of his construction every set so constructed was associated with an ordinal, and so $L$, the class of constructible sets, was well-ordered and hence the axiom of choice A.C. holds in $L$. It was then shown that $L$ is a model of set theory, that is to say that the axioms of set theory restricted to $L$ are theorems of set theory. This showed that A.C. is consistent with the axiom of set theory, provided that these are themselves consistent. It is a relative consistency proof.

The logical position of A.C. has been in question for many years. Is it independent of the other axioms of set theory? Does it contradict them, or can it be proved from them? The postition of A.C. was thus similar to the position of Euclid's parallel axiom until it was shown to be independent of the other axioms of Euclidean Geometry. The proof of independence of the axiom of parallels was obtained by finding a model of Non-Euclidean Geometry in the Euclidean plane. So that non-Euclidean Geometry is consistent if Euclidean Geometry is.

In Gödel's contructible model we have $V = L$, i.e. the universe consists of just the constructible sets. Now take a countable model $\mathscr{M}$ of set theory, and, as explained above, adjoin a new set $a$ of natural numbers to it and obtain a set $(\mathscr{M}, a)$ consisting of all the elements that can be contructed from $\mathscr{M}$ and $a$ by the methods allowed in set theory for the construction of new sets from old sets. Then there seems a good chance that if $(\mathscr{M}, a)$ turns out to be a model then $L = V$ will fail. As yet we have left the set $a$ undefined. Obviously the set $a$ must be a set of a very special kind. Every sentence in the extended set theory, i.e. the original set theory plus the set $a$, is to have a truth-value and this must in some way depend upon the set $a$. Of course in order to have a model the truth-value of the axioms must be truth.

The method of defining the set $a$ is called the '*forcing*' method, and the sets which so arise are called '*generic*' sets. The sentences of extended set theory are those of the original set theory plus the set $a$. The method of forcing consists of considering bounded sets of '*conditions*'

$$\mu' \epsilon a, \ldots, \mu^{(\kappa)} \epsilon a, \quad \lambda' \bar{\epsilon} a, \ldots, \lambda^{(\pi)} \bar{\epsilon} a,$$

where the natural numbers $\mu', \ldots, \mu^{(\kappa)}$ and $\lambda', \ldots, \lambda^{(\pi)}$ are distinct. We denote a condition by $P$ or $Q$ with or without superscripts or subscripts. We say $P$ forces $\phi$ if $\phi$ takes the value truth $\boldsymbol{t}$ in consequence of $P$. In symbols $P \models \phi$. Thus the knowledge whether a bounded set of natural numbers belongs to $a$ or to $\bar{a}$ is to be sufficient to determine the truth-value of $\phi$. Evidently for this to be possible the set $a$ must be a very special kind of set, and one might wonder whether any such sets can be found at all.

We state the *forcing rules*:

(i) $P$ forces $(E\xi)\,\phi\{\xi\}$ if $P$ forces $\phi\{\beta\}$ for some term $\beta$ which may contain $a$, i.e. a term in $(\mathscr{M}, a)$,

(ii) $P$ forces $(A\xi)\,\phi\{\xi\}$ if for all conditions $Q \supset P$ and all terms $\beta$ of $(\mathscr{M}, a)$ $Q$ fails to force $\sim\phi\{\beta\}$,

(iii) $P$ forces $\sim\phi$ if for all conditions $Q \supset P$, $Q$ fails to force $\phi$,

(iv) $P$ forces $\phi$ & $\psi$ if $P$ forces $\phi$ and $P$ forces $\psi$,

(v) $P$ forces $\phi \rightarrow \psi$ if $P$ forces $\psi$ or $P$ forces $\sim\phi$,

(vi) $P$ forces $\phi \leftrightarrow \psi$ if $P$ forces $\phi \rightarrow \psi$ and $P$ forces $\psi \rightarrow \phi$,

(vii) $P$ forces $\alpha = \beta$ if $P$ forces $(A\xi)(\xi\epsilon\alpha \leftrightarrow \xi\epsilon\beta)$,

(viii) $P$ forces $\alpha\epsilon\hat{\xi}\phi\{\xi\}$ if $P$ forces $\phi\{\alpha\}$,

(ix) $P$ forces $\alpha\epsilon a$ if $P$ forces $\alpha = \nu$ & $\nu\epsilon a$, for some numeral $\nu$,

(x) $P$ forces $\nu\epsilon a$ if this is one of the equations in the condition $P$, similarly for $\nu\bar{\epsilon}a$.

The peculiar position of negation is to be noted. From these rules we easily obtain the following main properties of forcing:

I. For all conditions $P$ and statements $\phi$ it is impossible to have $P$ forces $\phi$ and $P$ forces $\sim\phi$,

II. If $P$ forces $\phi$ and $Q \supset P$, then $Q$ forces $\phi$,

III. For all conditions $P$ and statements $\phi$ there is a condition $Q$, $Q \supset P$ such that $Q$ forces $\phi$ or $Q$ forces $\sim\phi$.

This is the surprising thing about forcing; namely every statement of extended set theory is decided by a bounded set of statements of the form $\nu\epsilon a$ or $\nu\bar{\epsilon}a$.

A complete set of forcing conditions $\{P_\nu\}$, $\nu = 0, 1, 2, \ldots$, is obtained thus: enumerate the statements of extended set theory $\{\phi_\nu\}$. Let $P_0$ be $\varnothing$ let $P_{S\nu} \supset P_\nu$ be such that $P_{S\nu}$ forces $\phi_{S\nu}$ or forces $\sim\phi_{S\nu}$. The forcing conditions are partially ordered so we could make the definition of $P_{S\nu}$ definite. The complete sequence of forcing conditions determines the generic set $a$.

It seems almost intuitively clear that $a$ is a non-constructible set in the sense of Gödel. Thus we have obtained a model for set theory in which $V = L$ fails. This is the simplest example of the forcing method.

To show the independence of the axiom of choice, A.C., we have to construct things so that there are certain symmetries in the model $\mathscr{M}$. Symmetry is expressed by automorphisms. This time we extend our countable model by adjoining sets $a_\nu$ for $\nu = 1, 2, \ldots$. Conditions will then consist of a bounded set of atomic conditions of the forms $\mu\epsilon a_\nu$, $\mu\bar{\epsilon}a_\nu$ provided the set is consistent. Consider the group of bounded permutations of the set of natural numbers, i.e. only a bounded set of natural numbers are affected by the permutation. We want to extend the permutation to act on the conditions and on the statements of extended set theory. We do this in such a way that if $P$ forces $\phi$ then $\rho P$ forces $\rho\phi$, where $\rho$ is one of the permutations. Also we lay down if $\mu \epsilon a_\nu$ is in $P$ then $\mu \epsilon a_{\rho\nu}$ is in $\rho P$. We extend the permutations to the statements of extended set theory as follows:

(i) $\rho(\hat{\xi}\phi\{\xi\})$ is $\hat{\eta}(\eta = \{\rho\xi\} \,\&\, \phi\{\xi\})$, i.e. $\rho``\hat{\xi}\phi\{\xi\}$, if we have some ordinal ranking so that members of a set have lower rank than the set itself then this gives us a *well-founded* situation provided membership is a well-founded relation.

(ii) $\rho\xi$ is $\xi$,

(iii) $\rho(\alpha\epsilon\beta)$ is $\rho(\alpha)\,\epsilon\rho(\beta)$,

(iv) $\rho(\alpha = \beta)$ is $\rho(\alpha) = \rho(\beta)$,

(v) $(\phi \,\&\, \psi)$ is $\rho(\phi) \,\&\, \rho(\psi)$,

(vi) $\rho(A\xi)\,\phi\{\xi\}$ is $(A\xi)\,\rho(\phi\{\xi\})$,

(vii) $\rho({\sim}\phi)$ is ${\sim}\rho(\phi)$,

(viii) $\rho(a_\pi)$ is $a_{\rho\pi}$.

Then $\rho(\phi)$ is defined for every statement $\phi$ of extended set theory and $\rho(\beta)$ is defined for every term $\beta$ of extended set theory.

To show that A.C. fails in such a model we need only show that the following fails:

D.E. Each Dedekind-finite set is finite.

A set is *finite* if it can be put into (1–1)-correspondence with a natural number (remember that in set theory a natural number is the set of all smaller natural numbers), and a set is *Dedekind-finite* if it is impossible to put it into (1–1)-correspondence with a proper subset of itself. Using A.C. we can prove D.E. Thus to show that A.C. fails we need only show that D.E. fails. To do this it suffices to find a countable set $A$ which is Dedekind-finite in the extended model $(\mathcal{M}, \{a_\nu\})$. The set $A$ will be the set $\{a_\nu\}$, $\nu = 1, 2, \ldots$. All the adjoined elements are different, for two of them, say $a_\nu$ and $a_\mu$ can only be equal if they are forced to be equal by some condition $P$, but since $P$ is a bounded set of atomic conditions we can find a larger condition $Q$ which contains $\mu\epsilon a_\nu$ and $\mu\bar{\epsilon}a_\pi$, now if $P$ forces $a_\nu = a_\pi$ then every larger condition also forces $a_\nu = a_\pi$, but $Q$ forces $a_\nu \neq a_\pi$, hence every condition fails to force $a_\nu = a_\pi$, and so all the adjoined elements are different. Clearly they form an infinite set because each condition fails to make the set finite. This would cause two of them to be equal, which is impossible.

Now suppose that some condition $P_0$ forces

$$\mathscr{F}_n f \,\&\, \mathscr{F}_n f^{-1} \,\&\, \mathscr{D}f = A \,\&\, \mathscr{R}f \subset A.$$

We show that $\mathscr{R}f = A$ so that the set $A$ is Dedekind finite (in the sense of forcing) though infinite (also in the sense of forcing). Let $\mu$ be greater than all $\mu'$ which occur in $P_0$. There must be a $P'$ such that $P'$ forces $fa_\kappa = a_\pi$ for some $\kappa$, $\pi$, where $\pi > \mu$, since $f$ takes infinitely many distinct values. Let $\lambda > \mu$ be such that $\lambda$ is greater than any $\lambda'$ occurring in $P'$. Let $\rho$ be the permutation which interchanges $\lambda$ and $\pi$ and leaves every other natural number unaltered. If $P'' = \rho P'$, then $P''$ forces $fa_\kappa = a_\lambda$, because $\rho(fa_\kappa) = (\rho f)(\rho a_\kappa) = (\rho f)\, a_\kappa = \rho a_\pi = a_\lambda$. Also $P'$ and $P''$ are compatible, i.e. $Q = P' \cup P''$ is a forcing condition, since $P'$ and $P''$ are identical except for atomic conditions involving $a_\lambda$ and $a_\pi$, but $P'$ is without $a_\lambda$ and $P''$ is without $a_\pi$. Thus $Q$ forces $fa_\kappa = a_\pi$ and $fa_\kappa = a_\lambda$, which is impossible.

To show the independence of the continuum hypothesis, C.H., it is first necessary to say something about cardinal numbers. A cardinal number is often defined as the class of all classes similar to a given class. Two classes are similar if they can be put into (1–1)-correspondence. Another way of defining a class of cardinals, called *alephs*, is a follows: a cardinal number is an ordinal number such that it is impossible to put it into (1–1)-correspondence with a smaller ordinal. This arranges the alephs in order and this order is isomorphic to the order of the ordinals themselves. The alephs are denoted by $\aleph_\alpha$ where $\alpha$ is an ordinal. The smallest aleph is $\aleph_0$ which is $\omega$, the set of all the natural numbers. The next greatest aleph is $\aleph_1$, and so on. We require A.C. to show that every cardinal is an aleph. $\mathscr{A}^{\mathscr{B}}$, where $\mathscr{A}$ and $\mathscr{B}$ are classes is the class of functions whose domain is $\mathscr{B}$ and range is $\mathscr{A}$. Thus $2^\omega = 2^{\aleph_0}$ is the class of all subsets of $\omega$, a set being given by its characteristic function. The Continuum Hypothesis C.H. is $2^{\aleph_0} = \aleph_1$, i.e. $2^{\aleph_0}$ is the cardinal next greater than $\aleph_0$. The logical status of C.H. has been in doubt for some time. In Gödel's constructible model C.H. holds, and so does G.C.H., the general continuum hypothesis, $2^{\aleph_\alpha} = \aleph_{S_\alpha}$. Thus C.H. and G.C.H. are consistent with the other axioms of set theory, provided these are themselves consistent. To show that C.H. and G.C.H. are independent of the axioms of set theory we adjoin $a_\alpha$ new elements for ordinals $\alpha$ $1 \leqslant \alpha < \aleph_\beta$, $\beta \geqslant 2$, as before all these are different and are (in the sense of forcing of course) subsets of $\omega$. This seems impossible, but remember we are using two senses of being true. This model, in the sense of extended set theory satisfies $2^{\aleph_0} \geqslant \aleph_\beta$, where $\beta$ is any cardinal. Similarly we can find a model for $2^{\aleph_\nu} > \aleph_{\nu+\lambda}$.

The following are theorems of set theory:

$$V = L \rightarrow \text{G.C.H.}, \quad \text{G.C.H.} \rightarrow \text{A.C.}$$

We have briefly shown that this is as much as we can have.

Many other models have been constructed in this manner to show the independence of a variety of statements about cardinals. The basic ideas when they first appeared were couched in such strange language that they were understood only by the most resolute of professional logicians, but now they are appearing as quite simple, like Gödel's famous work on irresoluble statements, at first found difficult but now quite easy to understand.

So this theory stood for a few years until it was noticed that the main feature of forcing was the partial order of the forcing conditions. Now a partially ordered set gives rise to a Boolean Algebra, and a Boolean Algebra is easily made *complete*, i.e. such that every set of elments has a *l.u.b.* and a *g.l.b.* This then gave rise to the idea of having Boolean-valued truth-values, i.e. instead of having just truth and falsehood we have other truth-values ordered among themselves in the form of a Boolean Algebra.

This then is a very brief sketch of the forcing method and the generic sets which emerge from it, it is just the bare bones of the original idea. The working out in detail would necessitate another long chapter and the development of set theory, ordinals, cardinals, Boolean Algebra, etc., beyond what we have already done. It would also involve the careful construction of the models on one or other of the two lines in which the constructible set $L$ can be constructed, and the verification that the axioms of set theory are satisfied in the model, definition by transfinite induction of well-founded entities, the behaviour of ordinals and cardinals in the model, and many other things. All this would be best done gradually throughout the book, but the plan of the book was formed before forcing was known.

## Historical remarks to Chapter 12

The concepts of model or interpretation and of truth go back a long way, though originally most writers were mainly concerned with formalizing a given structure. That is to say they started with a model and tailor-made a formal system to fit it. The systematic definition of model is more recent, see for instance Kemeny (1948) and the symposium edited by J. W. Addison, L. Henkin, A. Tarski (1965). The term categorical is due to Huntington (1902) and Veblen (1904). The definition of a general model is due to Henkin (1950). The account given of satisfaction is due to Tarski (1933).

Non-standard models were first invented by Skolem (1934), he first showed that most systems of arithmetic failed to characterize the natural numbers. Prop. 2 to the effect that the induction rule limited to a bounded number of cases is weaker than the full rule is due to Ryll–Nardewski (1952).

The method of obtaining new models by using ultraproducts is due to Loś (1955), but see also A. Robinson (1966), where interesting applications of non-standard models are given, in particular with regard to making the use of infinitesimals rigorous, see also Machover (1969).

H-models are so called after Henkin (1949, 1950), to him is due Prop. 3 about consistent sets of statements being simultaneously satisfied over denumerable domains, and Prop. 4 N.S.C. for being an $\mathbf{A}^{(\kappa)}$-theorem, expressed in terms of satisfiability over general models. Prop. 5 to the effect that the satisfying predicates can be taken to be in $P_2 \cap Q_2$ is due to Mostowski.

The $\epsilon$-symbol, as already remarked, is due to Hilbert, see *H–B* (1936). The representation of recursive number theory in $\mathscr{F}^2_C$ plus a constant predicate of type $o\iota$ is also due to Hilbert–Bernays (1936), vol. 2, Suppl. IV G., so also is the definition of the successor function in $\mathscr{F}^4_C$, see Church (1956), Ch. V. The axiom of infinity goes back to Russell (1919), Ch. 13. It can be taken in many different forms, see Church (1956).

Inner models are due to Gödel (1940) who found one for set theory. His inner model is well-ordered and this easily shows that the Axiom of Choice is consistent with the other axioms of set theory provided that these are themselves consistent. He was also able to show that the same inner model satisfied the General Continuum Hypothesis, thus showing that this hypothesis is also consistent with the other axioms of set theory provided that these are themselves consistent. This part of Gödel's work proved the greatest stumbling block to would-be readers. Further work on inner models has been done by Sheperdson (1951–3) who showed that Gödel's work goes as far as possible. These consistency results raised the question of independence of the Axiom of Choice and the General Continuum Hypothesis. Recently an affirmative answer has been given by Cohen (1963–4), who has a vigorous, almost physical, intuition about set theory. He based his work on Gödel's inner model and the fact that we can have countable models. Starting with an undefined set $\mathscr{A}$ he gradually '*forced*' members into it until it was completely defined. In this way he was able to show that $V = L$ (viz. the universe consists of constructable sets), is independent of the other axioms of set theory, provided these are themselves consistent. The independence of the Axiom of Choice and the General Continuum Hypothesis are shown in a similar, but very much more complicated method, see Cohen (1966).

More recently Scott (not yet published) has noticed that the partial ordering of forcing conditions is one of their chief characteristics. Accordingly he has developed a theory of models for set theory in which the individuals are Boolean functions over Boolean functions. Rosser (1969) has published an account of these researches, see also Leven (1967).

Finally we must mention that new methods are required to elucidate the structure of the lattice of degrees of unsolvability. The most that has been done as yet in this direction is a proof by Sacks (1963) using both the priority method and the 'forcing' method of Cohen.

Recently much research has been devoted to model theory and a number of results have been collected in Crossley (1967). Finally we should mention articles by Mostowski (1965) and Wang (1958, 1963) on the recent history of foundational studies.

## Examples 12

1. Find $C$-sequences and $S$-sequences for:

(*a*) $(EX)((Ax)(X'x \vee Xx\,.\,\&\,.\,X''x \vee \sim Xx) \leftrightarrow (Ax)(Xx \vee X'x)$,

(b) $(EX)((Ax)(Ex')Xxx' \,\&\, (Axx')(Xxx' \rightarrow (Ax'')(Xxx'' \rightarrow x' = x''))$

$\&\, (Axx')(Xx'x \rightarrow (Ax'')(Xx''x \rightarrow x' = x'')) \,\&\, (Ex)(Ax') \sim Xx'x)$

(c) $(EX)((Ax)(Ex')Xxx' \,\&\, (Axx')(Xx'x \rightarrow$

$(Ax'')(Xx''x \rightarrow x' = x'')) \,\&\, (Ex)(Ax') \sim Xx'x).$

2. Show that every model of the following axiom system based on $I\mathscr{F}_C$ is bounded, $S$ is a constant predicate,

$$Sxx' \rightarrow .\, Sxx'' \rightarrow x' = x'',$$

$$(Ax)(px \rightarrow (Ax')(px' \rightarrow (Ax'')(Sx'x'' \rightarrow px''))) \rightarrow (Ax')px'.$$

# Epilogue

Anyone writing a book like this one is naturally drawn into some examination of the history of the subject. Logic is different from other subjects in that it requires no apparatus or external objects of any kind whatsoever, except pencil and paper, and even these could be dispensed with by someone with a very retentive mind. All other subjects rely to some extent on external objects of some kind or other. A few poems might require no external objects, but one cannot write an ode to a daffodil without daffodils. Parts of theology might get on with no external objects but most theologies require the founder, even Old Testament Jewish theology required the physical presence of the deity. Philosophy might claim that it requires no external objects, but I am one of those who think it is largely about pseudo-problems.

This being so, the question presents itself: 'Why weren't the main features of modern symbolic logic invented thousands of years ago?' 'What were the hurdles that held things up?' One obvious hurdle was a notation of the natural numbers. The Greeks had practically nothing and the Roman system was very unsatisfactory. The Arabs however invented a satisfactory notation still used. Euclid was familiar with the axiomatic method, but said very little or nothing about rules of formation or consequence. The idea of truth was known but it doesn't seem too clear that truth is a property of sentences. Aristotle missed the relation, and by his amazing influence stifled everything for 2,000 years or more. The relation was discovered by de Morgan and Peirce about a century ago. If Aristotle had only analysed a few lines of Homer or another writer he would have discovered the relation. As mentioned in this book a single binary relation suffices for the whole of mathematics. Algorithms were known to the Arabs but they did not get very far with them. Neither the Greeks, Romans, Arabs, Indians nor Chinese needed to do much computation because their science and technology were very immature, so their applied mathematics was non-existent. Hence there was no incentive to any of them to reduce computation to a series of atomic acts as did Turing.

Aristotle did not continue his logical studies deeply enough to get the idea of a formal system or language, still less did anyone think of metalanguage; that came in with the Vienna school. Hence Gödel's arithmetization of logic could not take place.

No one would have expected the Greeks or any other early civilization to have discovered the periodic table of the elements, because this requires millions of experiments and chance discovery of minerals and their analysis, apart from developments in electricity and other things. But the matter developed in this book requires absolutely nothing from outside, it does not even depend on the art of printing or of making pencil and paper.

The question discussed in this epilogue was crystallized in my mind by a remark someone made to the effect 'why wasn't Cohen's idea thought of twenty years ago?' I thought why not thousands of years ago?

Logic was never considered heresy and logicians were never persecuted, they were never thought harmful to Church or State, so there was no obstacle in their way in that direction. The Greeks and Romans had very fine languages, still much taught today, yet they did not analyse it. I am not blaming the ancient civilizations for failing to discover such a thing as a simple set or a hypersimple set, these arose from the peculiar workings of Post's mind, but it is a little difficult to understand why they did not have something like a propositional calculus and a predicate calculus. The next step is to consider equality and the one after that to consider the verb 'to be' giving set theory. These they all missed.

# Glossary of special symbols

Special symbols in order of their first appearance:

| Symbol | Page |
|---|---|
| T.N.D. | 9 |
| $\mathscr{L}$ | 10 |
| $'$ | 11 |
| $^{\cap}$ | 11 |
| $(\nu)$ | 11 |
| $(S\nu)$ | 11 |
| $\Phi, \Psi, \Xi$ | 12 |
| $\boldsymbol{\lambda}$ | 13 |
| w.f.f. | 13 |
| ( | 13 |
| ) | 13 |
| $o$ | 14 |
| $\iota$ | 14 |
| —— | 20 |
| —— * | 21 |
| ═══ | 22 |
| $\Gamma$ | 22 |
| { } | 22 |
| [“ ”] | 24 |
| $\lambda\xi\eta\,.\,\phi\{\xi, \eta\}$ | 27 |
| $\mathscr{S}, \mathscr{F}, \mathscr{A}, \mathscr{P}$ | 27 |
| $\mathscr{T}_{\mathscr{L}}$ | 29 |
| $\mathscr{F}_{\mathscr{L}}$ | 29 |
| $\mathscr{M}$ | 34 |
| $N_{oo}$ | 42 |
| $D_{ooo}$ | 42 |
| $\mathscr{P}_C$ | 42 |
| $p_o$ | 42 |
| I$a$ | 42 |
| II$a$, $v$, $c$ | 42 |
| I$b$ | 44 |
| $K$ | 45 |
| II$b'$ | 45 |
| $C$ | 48 |
| $B$ | 51 |
| $S$ | 55 |
| $S'$ | 55 |
| $\mathscr{M}_C$ | 55 |
| $\boldsymbol{t}$ | 56 |
| $\boldsymbol{f}$ | 56 |
| $\mathscr{P}_1$ | 60 |
| $=$ | 61 |
| $\neq$ | 61 |
| 0 | 61 |
| 1 | 61 |
| $\cup$ | 61 |
| $\cap$ | 61 |
| — | 63 |
| $<$ | 63 |
| $\leqslant$ | 63 |
| $\to$ | 63 |
| $\leftrightarrow$ | 63 |
| $\mathscr{P}_2$ | 69 |
| $*$ | 70 |
| $\mathscr{F}, \mathscr{F}'', \mathscr{F}'''$ | 73 |
| $\iota', \iota'',$ | 75 |
| $\mathscr{F}_C$ | 78 |
| $x_\iota$ | 78 |
| $p_{o\iota}, p_{o\iota\iota}, \ldots$ | 78 |
| $E$ | 78 |



20-2

| | |
|---|---|
| $\mathrm{II}d, e$ | 78 |
| $(E\xi, \xi')$ | 78 |
| $\vdash_{\mathscr{F}}$ | 80 |
| $\Pi$ | 81 |
| $\Sigma$ | 81 |
| $\overline{\phi}$ | 84 |
| $A$ | 89 |
| $\mathscr{F}''_C$ | 90 |
| $\mathrm{II}d', e'$ | 90 |
| $\mathscr{F}'_C$ | 92 |
| $\mathrm{II}d^*, d'^*$ | 93 |
| $O$ | 93 |
| $(Q\xi)$ | 93 |
| $H$-disjunction | 99 |
| $H$-scheme | 100 |
| $\mathscr{N}, \mathscr{N}_\kappa$ | 110 |
| $\mathscr{F}$ | 114 |
| $x_\iota(\kappa)$ | 115 |
| $\mathscr{F}^\kappa_C$ | 116 |
| $\mathscr{T}_\kappa$ | 117 |
| $I$ | 119 |
| $Ii, Iii$ | 119 |
| $I\mathscr{F}_C$ | 120 |
| $E!$ | 120 |
| $I\mathscr{F}^f_C$ | 123 |
| $\mathscr{F}_f, \mathscr{T}_f$ | 124 |
| $J$ | 127 |
| $s$ | 127 |
| $\epsilon$ | 127 |
| $-$ | 127 |
| $V$ | 127 |
| $\times$ | 127 |
| $\cup, \cap$ | 127 |
| $\mathscr{D}$ | 127 |
| $Cnv_1$ | 127 |
| $Cnv_2$ | 127 |
| $Id$ | 127 |
| $[T/\kappa\kappa']$ | 134 |
| $[T_1, T_2]$ | 135 |
| $\mathscr{P}, \mathscr{N}, \mathscr{T}, \mathscr{U}$ | 143 |
| $f$-column | 150 |
| $t$-column | 150 |
| $H$ | 160 |
| $\boldsymbol{\eta}$ | 160 |
| $\boldsymbol{\iota}$ | 161 |
| $J, J'$ | 161 |
| $\boldsymbol{\epsilon}$ | 161 |
| $E, E'$ | 162 |
| $A$ | 162 |
| $EI\mathscr{F}_C$ | 163 |
| $\mathscr{B}\mathscr{F}_C$ | 166 |
| $=, \neq$ | 167 |
| $\subseteq, \subset$ | 167 |
| $\mathscr{E}m$ | 167 |
| $\wedge$ | 167 |
| $\cup, \cap, -$ | 167 |
| $\mathfrak{S}$ | 169 |
| $\mathfrak{P}$ | 169 |
| $\{X, Y\}$ | 169 |
| $\Sigma$ | 169 |
| $P$ | 169 |
| $\{X\}$ | 169 |
| $\langle X, Y\rangle, \langle X, Y, Z\rangle, \ldots$ | 169 |
| $V$ | 169 |
| $\Lambda$ | 169 |
| $\hat{x}\phi\{x\}$ | 169 |
| $(UXV)$ | 170 |
| $\times$ | 170 |
| $\mathscr{D}, \mathscr{R}$ | 170 |
| $\breve{X}, Cnv_1$ | 170 |
| $Rel$ | 170 |
| “ | 170 |
| $Un$ | 170 |
| $\mathscr{E}$ | 170 |
| $I$ | 170 |

| | |
|---|---|
| $\phi, \psi, \chi, \omega$ | 214 |
| $\rho, \sigma, \tau$ | 214 |
| $\mathscr{T}_{00}$ | 218 |
| $\mathscr{F}_{00}$ | 218 |
| $N[``\phi"]$ | 227 |
| $[``\phi \rightarrow \psi"]$ | 227 |
| $[``\phi \leftarrow \psi"]$ | 227 |
| $\mathbf{B}_{00}$ | 228 |
| $O_{(S\pi)}$ | 233 |
| $U^{(\theta)}_{(S\pi)}$ | 233 |
| $(\sigma/[\tau', \ldots, \tau^{(S\theta)}])$ | 233 |
| $\widehat{\tau\sigma}(\pi)$ | 234 |
| $1, 2, 3, \ldots, 9$ | 236 |
| $(\alpha + \beta)$ | 236 |
| $(\alpha \times \beta)$ | 237 |
| $\alpha^{\beta}$, $\alpha \, exp \, \beta$ | 237 |
| $!\alpha$ | 237 |
| $(\alpha \dot{-} \beta)$ | 237 |
| $(\alpha < \beta), (\alpha > \beta)$ | 237 |
| $A_1, B_1$ | 237 |
| $A_2, B_2$ | 237 |
| $(\Sigma\tau), (\Pi\tau)$ | 238 |
| $Max[\nu, \kappa]$, $Min[\nu, \kappa]$ | 239 |
| $(Max\rho)\,\nu$, $(Min\rho)\,\nu$ | 239 |
| $\underset{0 \leqslant \xi \leqslant \nu}{Max}(\alpha\{\xi\} = 0)$ | 239 |
| $\underset{0 \leqslant \xi \leqslant \nu}{Min}(\alpha\{\xi\} = 0)$ | 239 |
| $\underset{0 \leqslant \xi \leqslant \nu}{Max}(\alpha\{\xi\} < \kappa)$ | 239 |
| $\underset{0 \leqslant \xi \leqslant \nu}{Min}(\alpha\{\xi\} < \kappa)$ | 239 |
| $[\kappa/\nu]$ | 239 |
| $D[\nu, \kappa]$ | 240 |
| $Rem[\nu, \kappa]$ | 240 |
| $\Delta[\nu, \kappa]$ | 240 |
| $\sigma_2[\nu, \kappa]$ | 240 |
| $N[\nu, \kappa, \pi]$ | 240 |
| $\sigma_3[\nu, \kappa, \pi]$ | 240 |

| | |
|---|---|
| $pt[\nu, \kappa, \pi]$ | 241 |
| $Pt[\nu, \pi]$ | 241 |
| $\underset{1 \leqslant \xi \leqslant \kappa}{\{\rho\xi\}}$ | 241 |
| $\underset{1 \leqslant \xi \leqslant \kappa}{\langle\rho\xi\rangle}$ | 241 |
| $(\nu^{\frown}\kappa)$ | 241 |
| $\underset{1 \leqslant \xi \leqslant \kappa}{\widehat{\rho\xi}}$ | 241 |
| $(E\xi)_\nu \phi\{\xi\}$ | 242 |
| $(A\xi)_\nu \phi\{\xi\}$ | 242 |
| $(E\xi, \eta)_\kappa$ | 244 |
| $(\nu ISeg \kappa)$ | 244 |
| $(\nu ESeg \kappa)$ | 244 |
| $(\nu Part \kappa)$ | 244 |
| $(\mu\xi)_\nu [``\phi\{\xi\}"]$ | 245 |
| $\equiv$ | 245 |
| $Rt(\nu)$ | 245 |
| $E(\nu)$ | 254 |
| $J[\nu, \kappa]$ | 254 |
| $U[\nu]$ | 254 |
| $Q[\nu]$ | 257 |
| $F[\nu]$ | 257 |
| $G[\nu]$ | 258 |
| $\Delta_\nu[\kappa, \pi]$ | 259 |
| $g.n.$ | 260 |
| $\rho_{(\kappa)}$ | 262 |
| $t_1[\nu, \kappa]$ | 262 |
| $t_{11}[\nu]$ | 263 |
| $S_1[\nu]$ | 263 |
| $k[\nu]$ | 263 |
| $f[\nu]$ | 263 |
| $Num[\nu]$ | 263 |
| $Val$ | 264 |
| $Var[\nu]$ | 265 |
| $Eq[\kappa, \nu]$ | 265 |
| $Ineq[\kappa, \nu]$ | 265 |
| $Disj[\kappa, \nu]$ | 265 |
| $Conj[\kappa, \nu]$ | 265 |

| | |
|---|---|
| $\mathscr{A} \equiv \mathscr{B}$ | 329 |
| $\Phi[\kappa; x', \ldots, x^{(\theta)}]$ | 330 |
| $1-1)$-*reducible* | 331 |
| $\mathscr{A} \leqslant_1 \mathscr{B}$ | 331 |
| $Un^{(\theta)}_{\theta',\ldots,\theta^{(\pi)}}[\kappa, \nu', \ldots, \nu^{(\theta)}, \mathfrak{f}\xi, \ldots, \mathfrak{f}^{(\pi)}\xi, \xi, \nu]$ | 337 |
| $\mathfrak{f}\xi$ | 338 |
| $q^{(\lambda)}$ | 338 |
| $J_{\theta',\ldots,\theta^{(\pi)}}(\pi)[\kappa, \nu, \mathfrak{f}'\alpha, \ldots, \mathfrak{f}^{(\pi)}_{\alpha}]$ | 340 |
| $\Theta[\kappa, \nu, \nu', \ldots, \nu^{(\pi)}, \lambda]$ | 340 |
| $S^{(\lambda')}_{\lambda,\theta',\ldots,\theta^{(\pi)}}[\eta, \eta', \ldots, \eta^{(\lambda')}]$ | 342 |
| $\mathfrak{f}$-*recursive* | 342 |
| $\mathscr{X}^{f}_{\kappa}, \mathscr{X}^{\mathscr{A}}_{\kappa}$ | 343 |
| $\Phi^{f}_{\kappa}, \Phi^{\mathscr{A}}_{\kappa}, \Phi^{f}_{\kappa}\theta$ | 343 |
| $\Phi^{[f]}_{\kappa}, \Phi^{[f]}_{\kappa}\theta$ | 343 |
| $f \equiv_T g$ | 343 |
| $\mathbf{a} < \mathbf{b}, \mathbf{a} > \mathbf{b}$ | 344 |
| $\mathbf{a} \leqslant \mathbf{b}, \mathbf{a} \nleqslant \mathbf{b}$ | 344 |
| $\mathbf{a}/\boldsymbol{b}$ | 344 |
| $\mathbf{a} = \mathbf{b}, \mathbf{a} \neq \mathbf{b}$ | 344 |
| $\mathbf{O}$ | 344 |
| *l.u.b.* | 344 |
| $\mathbf{a} \cup \mathbf{b}$ | 344 |
| $*_{\kappa}$ | 345 |
| $\mathbf{a}'$ | 345 |
| $\leqslant_T$ | 352 |
| $\rightarrow$ | 352 |
| $\leftrightarrow$ | 352 |
| $<_R$ | 362 |
| $[\nu, \mu]$ | 363 |
| $+_r, \times_r, -_r, =_r, <_r, 0_r, 1_r, \frac{1_r}{\nu_r}{}_r$ | 375 |
| $\mathscr{E}$ | 376 |
| $+_s, \times_s$ | 376 |
| $\mathscr{E}(\iota)$ | 377 |
| $+_c, \times_c, -_c, =_c, 0_c, 1_c, \|\|_c$ | 377 |
| $Max_s$ | 377 |
| $(m-1)$-*reducible* | 380 |

| | |
|---|---|
| $\mathbf{A}$ | 387 |
| $IId'$ | 387 |
| $A$ | 389 |
| $Prf_{\mathbf{A}}$ | 390 |
| $s[\xi, \eta]$ | 390 |
| *Fals* | 394 |
| $ClStat_{\mathbf{A}}$ | 395 |
| *Con*(i) | 399 |
| *Con*(ii) | 399 |
| *Con*(iii) | 399 |
| *Con*(iv) | 399 |
| *Neg* | 400 |
| *pr* | 402 |
| $st[\nu, \nu', \nu'']$ | 402 |
| $m$ | 402 |
| $T$ | 402 |
| $T[x, x']$ | 402 |
| $T^*$ | 403 |
| $\mathbf{A}_{ng}$ | 406 |
| $\mathbf{B}$ | 408 |
| $\mathscr{P}^{\kappa}_{\pi}$ | 409 |
| $\mathscr{Q}^{\kappa}_{\pi}$ | 409 |
| $\Delta^{\kappa}_{\pi}$ | 413 |
| $\mathscr{P}^{\kappa,\mathscr{A}}_{\pi}$ | 417 |
| $\mathscr{Q}^{\kappa,\mathscr{A}}_{\pi}$ | 417 |
| $\mathscr{P}^{\kappa,f}_{\pi}$ | 417 |
| $\mathscr{Q}^{\kappa,f}_{\pi}$ | 417 |
| $\varnothing, \varnothing', \varnothing'', \ldots$ | 419 |
| $\mathbf{O}, \mathbf{O}', \mathbf{O}'', \ldots$ | 419 |
| R 3′, R 3″, R 3″, R $3^{(\mathrm{iv})}$ | 426 |
| R $3^{(\mathrm{v})}$ | 427 |
| R $3^{(\mathrm{vi})}$ | 428 |
| $\mathbf{E}$ | 430 |
| R 3 | 438 |
| $\mathbf{A}_I$ | 438 |
| **E**-*correct* | 438 |
| $ded_0[\lambda, \mu, \nu]$ | 441 |
| $\langle \nu \rangle_{\kappa}$ | 446 |

| | |
|---|---|
| $\kappa_\theta^\pi$ | 449 |
| $\kappa_\theta^\infty$ | 449 |
| *l.c.* | 449 |
| $\Theta\kappa$ | 449 |
| $\Xi\delta$ | 449 |
| $\Delta\delta$ | 449 |
| $H[\pi, \pi']$ | 450 |
| $\mathscr{G}$ | 455 |
| $\mathbf{A}_{IC}$ | 470 |
| $\mathbf{A}_{0C}$ | 471 |
| *f.a.* | 476 |
| *u.f.*, *l.f.* | 476 |
| *m.f.*, *s.f.* | 476 |
| *n.d.* | 476 |
| *r.s.f.* | 476 |
| *Eqx*, *Ineqx* | 481 |
| *Disjx*, *Conjx* | 481 |
| *Ex* | 481 |
| *NLPx*, *NRPx* | 481 |
| $\mathbf{A}_{0,\Phi}$ | 484 |
| $\mathbf{A}_{I'}$ | 485 |
| $ct[1, \theta]$ | 488 |
| $ct[\alpha, \beta, \gamma]$ | 488 |
| $D_1\theta$ | 488 |
| $D_2\theta$ | 488 |
| $ctfmla[\kappa, \theta]$ | 488 |
| $Ord\,\theta$ | 488 |
| $T_1\theta, T_2\theta$ | 488 |
| $\pi Comp\theta$ | 488 |
| $C_1\theta, C_2\theta$ | 488 |
| $O\theta$ | 489 |
| $\theta \restriction \nu$ | 489 |
| $\theta \rceil \nu$ | 489 |
| *E-cut* | 491 |
| $P[`f[\mu, \pi] = 0']$ | 504 |
| $F[`\phi']$ | 504 |
| $\mathbf{A}'$ | 511 |
| $\mathscr{G}\{\mathscr{S}\}$ | 511 |
| $G^*$ | 511 |
| $\sim$ | 512 |
| $X_{o\iota}, X_{o\iota\iota}, \ldots$ | 512 |
| $A_{o\iota}, A_{o\iota\iota}, \ldots$ | 512 |
| $\mp$ | 513 |
| R 3* | 514 |
| R 4, R 4′, R 4″ | 514 |
| *prop* | 514 |
| $\epsilon$ | 516 |
| $X_{\kappa\iota}, X_{\kappa\iota\iota}, \ldots$ | 518 |
| $\mathbf{A}^{(\nu)}$ | 518 |
| $\mathbf{A}^{(\nu)*}$ | 519 |
| $Gen\nu$ | 533 |
| $LP_\nu^\kappa$ | 539 |
| *l.p.* | 539 |
| *SectX* | 544 |
| *l.u.b.* | 545 |
| $\{\{X, X'\}\}$ | 548 |
| *EvX* | 548 |
| *OdX* | 548 |
| $(E\,!\,X)$ | 549 |
| $\overline{lim}$, $\underline{lim}$ | 550 |
| $\bar{S}\Big\vert_0^1, \underline{S}\Big\vert_0^1$ | 550 |
| $\underline{m}\hat{X}_s\phi\{X_s\}$ | 551 |
| $\overline{m}\hat{X}_s\phi\{X_s\}$ | 552 |
| $\Sigma'_{S\nu}$-*prefix* | 554 |
| $\Pi'_{S\nu}$-*prefix* | 554 |
| $\Sigma'_\nu$-*statement* | 554 |
| $\Pi'_\nu$-*statement* | 554 |
| $\Delta'_\nu$ | 558 |
| *C-sequence* | 570 |
| *S-sequence* | 570 |
| $[\nu, \kappa]$ | 570 |
| $[0, 1]$ | 570 |
| $[\Delta; \nu]$, $[0; 0]$ | 570 |
| $\gtrdot$ | 573 |

| | |
|---|---|
| $\dot{=}$ | 573 |
| $\lessdot$ | 573 |
| $U[\alpha, \bar{\alpha}, \nu]$ | 578 |
| $V[\alpha, \bar{\alpha}, \nu]$ | 579 |
| *S-model* | 580 |
| *H-model* | 585 |
| $\mathbf{A}U^{(\kappa)*}$ | 586 |
| $M[\nu, \pi]$ | 591 |
| $L[\nu, \pi]$ | 591 |
| $P[\theta, \nu, \pi]$ | 591 |
| $\theta \epsilon \Lambda^*$ | 591 |
| $P^{\Lambda^*}[\theta, \nu]$ | 591 |
| $f[\lambda]$ | 591 |
| $g[\lambda]$ | 591 |
| $C[\nu]$ | 592 |
| ${}_{\mathbf{S}\mu}\mathscr{G}^{\mathbf{S}\lambda}_{\mathbf{S}\nu}[\theta', \ldots, \theta^{(\mathbf{S}\nu)}]$ | 593 |
| *H-model* | 593 |
| $\mathbf{Z}_0 x$ | 595 |
| $\mathbf{Z}_1 x$ | 596 |
| $S[x, x']$ | 596 |
| $\Sigma[x, x', x'']$ | 596 |
| $\Pi[x, x', x'']$ | 596 |
| $R[x, x', x'']$ | 596 |
| $M[\nu, \mu]$ | 596 |
| $N[\nu, \mu]$ | 596 |
| $\rho[\nu, \mu]$ | 596 |
| $\mathscr{F}^2_C$ | 597 |
| $\mathscr{F}^4_C, \mathscr{F}^6_C$ | 597 |
| $(1\text{–}1)r$ | 597 |
| $Sim[p, q]$ | 597 |
| *Card* | 597 |
| $0, 1, 2, \ldots$ | 598 |
| $S$ | 598 |
| *NatNum* | 598 |
| $Ax\infty$ | 598 |

Several symbols occur more than once owing to their use in different circumstances.

# Note on references

This bibliography only contains references to works that I am conscious of having used or which are mentioned for historical reasons even though they may not have been used. Thus many important works may not be listed in this bibliography, simply because I have not used them or because I may have forgotten that they were used.

Another reason for keeping a bibliography as short as possible is that it adds considerably to the cost of the book. Now-a-days the printer's time is as valuable as that of the reader.

Nevertheless Symbolic Logic is very fortunate in having a very complete bibliography, arranged by author and arranged by subject. Thus anyone wishing to find works on a given subject which is dealt with in symbolic logic can find papers which discuss it even though the title of the paper may bear little or no relation to the given topic. This bibliography is due to Church. It begins in *J.S.L.*, vol. 1, pp. 121–218, with additions and corrections in *J.S.L.*, vol. 3, pp. 178–212. Thereafter it is kept up to date at two-yearly intervals. However vol. 26 (1961) is entirely devoted to collecting together all these additions. It contains references to all papers printed in *J.S.L.* up to 1961 and separately references to all papers reviewed in *J.S.L.* up to 1961, listed by author and by subject. Thereafter additions are added at two-yearly intervals.

Thus if one wants to find all references to Three-valued logic one need only look up the entry 'Three-valued logic' in *J.S.L.*, vol. 26, p. 233, where one will find a host of references. To be quite up to date (as far as the refereeing has been done) one must also consult the same entry in vol. 28, 30 and then go through the reviews and papers in vol. 31 in order to have all the available references up to the end of 1966.

Every subject has furry edges. Thus the line between symbolic logic and pure mathematics on the one hand and between symbolic logic and philosophy on the other hand are both difficult or impossible to draw, so that a bibliographer has to make a judgement whether to include a given work or not. Church expands on this topic in his introduction to the bibliography in *J.S.L.*, vol. 1. These remarks apply to papers on computers, among other things.

In forming a bibliography one thing that strikes one is the small number of writers to whom most of the basic parts of the subject are due (such as Russell, Tarski, Hilbert, Gödel, Turing, to mention only a few of the greatest) compared with the vast number of writers altogether.

# References

ACKERMANN, Wilhelm, (1928). Zum Hilbertshen Aufbau der reellen Zahlen *Math. Ann.* **99**, 118–33.
(154). *Solvable cases of the decision problem.* North-Holland Publ. Co., vi + 114.
ADDISON, John W., (1955 *a*). Analogies in the Borel, Lusin and Kleene hierarchies, I. *Bull. Am. math. Soc.* **61**, 75.
155*b*). Separation principles in the hierarchies of classical and effective (descriptive set theory. *Fund. Math.* **46**, 123–35.
(1962*a*). Some Problems of hierarchy Theory. *Proc. Symposia pure Math.* V, 123–30. Amer. Math. Soc., Providence, R.I.
(1962*b*). The Theory of hierarchies. *Logic, Methodology and Phil.* Stanford Univ. Press.
ADDISON, J. W., HENKIN, L., TARSKI, A. (Eds.), (1963). *The theory of models.* North-Holland Publ. Co. *Proc.* 1963 *Internat. Symp. at Berkeley.* xv + 494.
ANDERSON, J. M. & JOHNSTONE, H. W. Jr., (1962). *Natural Deduction.* Wadsworth Publ. Co. Inc., Belmont, Calif. 418.
BACHMANN, H., (1955). *Transfinite Zahlen.* Springer Verlag. vii + 204.
BARWISE, J., (1959). *The Foundations of Mathematics.* North-Holland Publ. Co. xxvi + 741.
(1968). The Syntax and semantics of Infinitary Languages. *Lecture Notes on Mathematics*, No. 72. iv + 268. Springer-Verlag.
BERNSTEIN, F., (1905). Untersuchungen aus der Mengenlehre. *Math. Ann.* **61**, 117–55.
BOCHENSKI, L. M., (1961). *A History of formal Logic.* Trans. by Ivo Thomas. Univ. Notre Dame Press.
BOOLE, George, (1847). *The Mathematical Analysis of Logic, being an essay toward a calculus of deductive reasoning.* London and Cambridge. Reprinted in *George Boole's collected logical works.* Ed. P. E. B. Jourdain. **1**. Chicago and London (1916).
(1848). *The calculus of Logic.* Cambridge and Dublin Math. Jour. **3**, 183–98. Reprinted in *George Boole's collected logical works.*
(1854). *An Investigation of the Laws of Thought, on which are founded the mathematical theories of logic and probabilities.* London. v + iv + 424. Reprinted in *George Boole's collected logical works.* **2**.
BROUWER, L. E. J., (1908). De onbetrouwbaarheid der logische principes. (The untrustworthiness of the principles of logic.) *Tijds, v. wijsbegeerte.* **2**, 152–8. Reprinted in *Wiskunde. Waarheid, werkelijkheid.* Gröningen. (1919).
(1947). *Cambridge Lectures.* (Not yet published.)
CANTOR, Georg, (1932). *Gesammelte Abhandlungen mathematischen und philosophischen Intalts.* Ed. by E. Zermelo. Berlin.
*Transfinite numbers.* Dover Publ. Trans. by P. E. B. Jourdain.
CARNAP, Rudulf, (1937). *The logical syntax of language.* London. Kegan, Paul, Trench, Trubner and Co. Ltd. xvi + 352.

CHURCH, Alonzo, (1932). A set of postulates for the foundation of logic. *Ann. Math.* 2 s. **33**, 346–66.

(1936). A note on the Entscheidungsproblem. *J.S.L.* **1**, 40–1.

Correction, *ibid.* **1**, 101–2. Reprinted in M. Davis *The Undecidable* (1967). 440.

(1936). *Mathematical Logic.* Mimeographed lectures at Princeton. iii + 113.

(1941). *The Calculi of lambda-conversion.* Princeton Univ. Press. ii + 77.

(1956). *Introduction to Mathematical Logic I.* Princeton Univ. Press. x + 376.

CHWISTEK, Leon, (1921). Antynomje logiki formalnej. (Antinomies of formal logic.) *Przegl. fil.* **24**, 164–71.

(1925). The theory of constructive types. *Rocznik Polskiego Towarzystwa Mat.* **3**, 92–141.

COHEN, Paul J., (1963). The independence of the continuum Hypothesis. *Proc. Nat. Acad. Sci. U.S.A.* **50**, 1143–8.

(1964). The independence of the continuum Hypothesis II. *Proc. Nat. Acad. Sci. U.S.A.* **51**, 105–10.

(1965). *Independence results in set theory. The theory of Models.* Symposium at Berkley. North-Holalnd Publ. Co. 39–54.

(1967). *Set Theory and the Continuum Hypothesis.* W. A. Benjamin. Inc. N.Y. 154.

CRAIG, William, (1953). On axiomatizability within a system. *J.S.L.* **18**, 30–2.

CROSSLEY, J. N., (Ed.). (1967). *Sets, Models and Recursion Theory.* North-Holland Publ. Co. 331.

CURRY, H. B., (1929). An Analysis of logical Substitution. *Am. J. Math.* **51**, 363–84.

(1941). A Formalization of recursive Arithmetic. *Am. J. Math.* **63**, 263–82.

CURRY, H. B. & FEYS, R., (1958). *Combinatory Logic I.* North-Holland Publ. Co. xvi + 417.

DAVIS, Martin (Ed.), (1958). *Computability and Unsolvability.* McGraw Hill. xxv + 210.

(1965). *The Undecidable. Basic papers on undecidable propositions, unsolvable problems and computable functions.* Raven Press. 440.

DEDEKIND, Richard, (1888). *Was sind und was sollen die Zahlen?* Braunschweig. xv + 58. 6th Ed. (1930). Also in *Dedekind's Werke* **3**, Braunschweig (1932) 335–91. Also trans. by W. W. Bemen. *Essays on Number.* Open Court. Chicago (1909). 44–115.

(1892). *Stetigkeit und irrationale Zahlen.* 2nd ed. Braunschweig. Also trans. by W. W. Bemen. *Essays on Number.* Open Court Publ. Co. Chicago. (1909). 1–43.

DEKKER, James E., (1955). Productive sets. *Trans. Am. math. Soc.* **78**, 129–49.

DEKKER, James E. & MYHILL, J., (1960). *Recursive equivalence types.* Univ. Calif. Publs Math. **3**, No. **3**, 67–214.

DOPP, J., (1962). *Logiques construits par une méthode de déduction naturelle.* Paris. Gauthier-Villars. 191.

DURR, K., (1951). *The propositional Calculus of Boethius.* North-Holland Publ. Co. x + 79.

EASTON, W. B., (1964). *Powers of Regular Cardinals.* Ph.D. dissertation. Princeton.

FEFERMAN, Solomon, (1957). Degrees of unsolvability associated with classes of formalised theories. *J.S.L.* **22**, 161–75.

FITCH, F. B., (1942). A basic logic. *J.S.L.* **7**, 105–14.

FRAENKEL, A. A. & BAR-HILLEL, Y., (1958). *Foundations of Set Theory.* North-Holland Publ. Co. x + 415.

(1946). *Einleitung in die Mengenlehre.* 3rd ed. Dover Publ. xiv + 424.

FREGE, Gottlob, (1879). *Begriffsschrift, eine der arithmetischen nachgebildete Formelsprache des reinen Denkens*. Halle. viii + 88. Reprinted in English in: *Translations from the philosophical writings of Gottlob Frege*. Basil Blackwell. Oxford (1952). x + 244.

(1891). *Function und Begriff*. Jena. 31. Reprinted in *Translations from the writings of G. Frege.*

(1892). Über Sinn und Bedeutung. *Zeit. Phil. n.s.* **100**, 25–50. Reprinted in *Translations from the writings of G. Frege.*

(1893). *Grundgesetze der Arithmetic, begriffsschriftlich abgeleitet.* I. Jena. xxxii + 245. II, Jena (1903). xv + 265. Reprinted (1962) by George Ohms Hildesheim. xxxii + 254. xvi + 266.

(1884). *Die Grundlagen der Arithmetic, eine logisch-mathematische Untersuchungen über der Begriff der Zahlen.* Breslau. xix + 119. Reprinted Breslau (1934). Reprinted with English translation on facing page by J. L. Austin. Basil Blackwell. Oxford (1950). xii + 119 bis.

FREUDENTAL, Hans, (1960). *Lincos. Design for a language for cosmic intercourse.* North-Holland Publ. Co. 224.

FRIEDBERG, Richard, (1957). Two recursively enumerable sets of incomparable degrees of unsolvability. (Solution of Post's problem 1944.) *Proc. Nat. Acad. Sci.* **43**, 236–8.

(1957). A criterion for completeness of degrees of unsolvability. *J.S.L.* **22**, 159–60.

(1958). Three theorems on recursive enumeration. I. Decomposition, II. Maximal set, III. Enumeration. *J.S.L.* **23**, 309–16.

GENTZEN, Gerhard, (1934). Untersuchungen über des logische Schliessen. *Math. Z.* **39**, 176–210.

(1955). *Recherches sur la déduction logique.* Presses universitaires de France. xi + 167. Trans. by R. Feys and J. Ladiere.

GÖDEL, Kurt, (1930). Die Vollständigkeit der Axiome des Logischen Funktionenkalküls. *Monatsch. Math. Phys.* **37**, 349–60.

(1930). Über formal unentscheidbare Sätze der Principia Mathematica und verwandter System. *Monatsch. Math. Phys.* **37**, 349–60. Reprinted in M. Davis, *The Undecidable.* 4–38.

(1933). Zum Entscheidungsproblem des logischen Funktionenkalkuls. *Monatsch Math. Phys.* **40**, 433–43.

(1940). Consistency of the Continuum Hypothesis. *Annals of Math. Studies*, No. 3. Princeton. 2nd printing (1951). 69.

(1934). *On undecidable Propositions of formal mathematical Systems.* Mimeographed notes of lectures given by K. Gödel at Inst. for Advanced Study. Reprinted in *The Undecidable.* Ed. M. Davis. Raven Press, Hewlett, N.Y. (1965).

(1965). *Remarks before the Princeton bicentennial conference on problems in Mathematics* 1946. *The Undecidable.* M. Davis. Raven Press Books. 84–8.

(1936). Über die Läuge der Beweise. *Ergebnisse eines mathematischen Kolloquiums.* **7**, 23–4.

GOODSTEIN, R. L., (1957). *Recursive number theory.* North-Holland Publ. Co. xii + 190.

(1961). *Recursive Analysis.* North-Holland Publ. Co. viii + 138.

HALMOS, P. R., (1963). *Lectures on Boolean Algebras.* Van Nostrand. Princeton.

(1960). *Naive set theory.* Van Nostrand. vii + 104.

HENKIN, Leon, (1949). Completeness of the first order functional calculus. *J.S.L.* **14**, 159–66.

(1950). Completeness in the theory of types. *J.S.L.* **15**, 81–91.

HERBRAND, Jacques, (1930). Recherches sur la théorie de la démonstration. *Travuax d.l. Soc. d. Sci. e.d. Lettres d. Varsovie.* Cl. III. No. 33, 128.

(1931–2). Sur la non-contradiction de l'arithmétique. *Jour. r. angew. Math.* **166**, 1–8.

HERMES, H. & SCHOLZ, H., (1952). Mathematische Logik. *Encyk. d. Math.* Wissen. Bd. I 1, Heft. 1, Teil, 1, 82.

HERMES, Hans, (1961). *Aufzahlbarkeit, Entscheidbarkeit, Berechenbarkeit.* Springer-Verlag. Berlin. ix + 246.

(1965). Eine Termlogik mit Auswahloperator. *Lecture Notes on Mathematics.* No. 6, 42. Springer-Verleg.

HEYTING, Arend, (1930). Die formal Regeln der intuitionistischen Logik. *Sitz. d. Preus. Akad. d. Wiss. Ph-Math.* Kl. 42–56.

(1930). Die formal Regeln der intuitionistischen Mathematik. *Ibid.* 57–71, 158–169.

(1934). *Malhematische Grundlagenforschung intuitionismus Beweistheorie.* J. Springer, Berlin. iv + 73.

(1955). *Les fondements des mathematiques intuitionisme theorie de la demonstration.* Paris. Gauthier-Villars. 91.

(1956). *Intuition. An Introduction.* North-Holland Publ. Co. vi + 133.

HILBERT, David, (1904). Über die Grundlagen der logik und der Arithmetik. *Verhand. d. Dritten Internat. Math. Kong. in Heidelberg.* 174–85. Reprinted in *Grundlagen der Geometrie*, 4th ed. Leipzig (1913), 243–58. Also in Werke.

(1918). Axiomatische Denken. *Math. Ann.* **78**, 405–15. Reprinted in *Gesammelte Abhandlungen.* 3. Berlin (1935), 146–56.

(1922). Neubegrundung der Mathematik. *Abh. Math. Sem.* Hamburg Univ. **1**, 157–77.

(1922). *Grundlagen der Geometrie.* Teubner, Leipzig. 5th Ed. vi + 264.

(1926). Über das unendliche. *Math. Ann.* **95**, 161–90.

HILBERT, David & AKERMANN, W., (1949). *Grundzüge der theoretischen Logik.* Springer-Verlag. Berlin. 3rd ed. viii + 155.

(1930–1). Die Grundlagen der elementaren Zahlenlehre. *Math. Ann.* **104**, 485–94.

HILBERT, David & BERNAYS, P., (1934–6). *Grundlagen der Mathematik.* 2 Vols. J. Springer. Berlin.

HINTIKKA, K. Jaakko, (1953). A new approach to sentential logic. *Soc. Sci. Fenn. Comment. phys-math.* **17**, No. 2, 14.

(1955). Form and content in Quantification theory. *Acta phil. Fenn.* No. 8. Helsinki, 7–55.

(1955). Notes on quantification theory. *Soc. Sci. Fenn. Co. Comment. phys-math.* **17**, no. 12, 13.

JENSEN, R. B., (1967). *Modelle der Mengenlehre.* Springer-Verlag, Berlin. 176.

JORDAN, Z., (1945). *Polish science and learning. The development of Mathematical Logic and of logical Positivism in Poland between the two wars.* Oxford Univ. Press. 47.

KALMÁR, László, (1935). Über die axiomatisierbarkeit der Aussagenkalküls. *Scient. Math.* **7**, f 4. 222–43.

(1933). Über die Erfüllbarkeit der jenigen Zählausdruck welche in die Normalform zwei benachbarte Allzeichen enthalten. *Math. Ann.* **108**, 466–84.

(1940). On the possibility of definition by recursion. *Acta Sci. math. Szeged*, **9**, 227–32.

Kalmár, László & Suranyi, J., (1947). On the reduction of the decision problem. Second paper. Gödel prefix, a single binary predicate. *J.S.L.* **12**, 65–73.

Kamke, E., (1950). *Theory of Sets*. 2nd Trans. from German. Dover Publ. vi + 144.

Karp, C. R., (1964). *Languages with expressions of infinite length*. North-Holland Publ. Co. 183.

Kemeny, J. G., (1948). Models of logical systems. *J.S.L.* **13**, 16–30.

Klausa, Dieter, (1961). *Konstructive Analysis*. Deutsche der Wissenschaften. vii + 160.

Kleene, Stephen Cole, (1935). A theory of positive integers. *Am. J. Math.* **57**, 153–73, 219–44.

(1938). On notation for ordinal numbers. *J.S.L.* **3**, 150–5.

(1943). Recursive predicates and quantifiers. *Trans. Am. Math. Soc.* **53**, 41–73.

(1952*a*). Two papers on the predicate calculus. *Mem. Am. math. Soc.* No. 10. Providence. 27–66.

(1952*b*). *Introduction to Metamathematics*. North-Holland Publ. Co. x + 550.

(1955). Arithmetical predicates and function Quantifiers. *Trans. Am. Soc.* **79**, 312–40.

(1955). Hierarchies of number-theoretic predicates. *Bull. Am. math. Soc.* **61**, 193–213.

(1959*a*). Recursive Functionals and quantifiers of finite types, I. *Trans. Am math. Soc.* **91**, 1–52.

(1959*b*). Quantification of number theoretic functions. *Comp. Math.* **14**, 23–40.

(1962). Turing machine computable functions of finite types, I. *Logic, Methodology and Phil. of Sci. Proc.* 1960 *internat. Cong.* Stanford Univ. Press. 38–45.

(1963). Recursive functionals of finite type, II. *Trans. Am. math. Soc.* ser. 2. **59**, 106–407.

(1967). Mathematical Logic. John Wiley & Sons. x + 398.

Kleene, Stephen Cole & Post, E. L., (1954). The upper semi-lattice of degrees of recursive unsolvability. *Ann. Math.* **59**, 379–407.

Kronecker, L., (1887). Über den Zahlbegriff. *J. d. Math.* **101**, 337–55. Also in *Kronecker's Collected Works.* **III** 1, 249–74. Leipzig. (1899).

Landau, Edmund, (1930). *Grundlagen der Analysis*. Leipzig.

Leibnitz, G. W. V., (1936). See references in Bibliography. *J.S.L.* **1**, 23.

Leven, F. J., (1967). Modelle der Mengenlehre. *Lecture Notes on Mathematics*. No. 37. Springer-Verlag.

Lewis, C. I., (1918). *A survey of symbolic logic*. Berkley, Calif. vi + 406.

(1920). Strict implication – an emendation. *Jour. phil.* **17**, 300–2.

Lewis, C. I. & Langford, C. H., (1932). *Symbolic logic*. New York. iv + 506.

Littlewood, J. E., (1926). *The elements of the theory of real numbers*. Heffer, Cambridge. vii + 60.

Löb, M. H., (1953). Concatenation as a basis for a complete arithmetic. *J.S.L.* **18**, 1–6.

(1968). (Ed.) Proceedings of the Summer School in Logic, Leeds 1967. *Lecture Notes in Mathematics*, No. 70. 331. Springer-Verlag.

LORENZEN, Paul, (1938–9). Die definition durch vollständige Induction. *Monatsch. f. Math. u. Phys.* **47**, 356–8.

(1951). Algebraische und Logische Untersuchungen über freie Verbände. *J.S.L.* **16**, 81–106.

(1955). *Einführing in die operative Logik und Mathematik.* Springer-Verlag. Berlin. 298.

ŁOS, J., (1949). *O Matryeach Logiczynch.* Wrocław. 42.

(1955). Quelques remarques, théorèmes et problèmes sur les classes définissables d'algèbres. Mathematical Interpretation of formal Systems. (*Symposium Amsterdam* 1954) *Studies in Logic and the foundation of mathematics.* Amsterdam. 98–113.

LÖWENHEIM, Leopold, (1915). Über Möglichkeiten im Relationkalkül. *Math. Ann.* **76**, 447–70.

ŁUKASIEWICZ, Jan, (1920). O logice trójwartościowij. (On three-valued logic.) *Ruch. fil. Lwow.* **5**, 169–71.

ŁUKASIEWICZ, Jan & TARSKI, A., (1930). Untersuchungen über Aussagenkalkül. *C. R. d. l. Soc. d. Sci. et d. Lett. d. Varsovie.* Cl. III **23**, 1–21.

ŁUKASIEWICZ, Jan., (1941). *Die Logik und die Grundlagenproblem.* Entretiens de Zürich. Dec. 1938. S. A. Leeman frères & Cie. Zürich. 82–108.

(1951). *Aristotle's Syllogistic.* Oxford, Clarendon Press. x + 141.

MACHOVER, M., (1969). Lecture Notes on Non-standard Analysis. *Lecture notes on Mathematics*, No. 94. vi + 79. Springer-Verlag.

MARKOW, A. A., (1954). Theory of Algorithms. *Acad. Sci. U.S.S.* **v.** *Works of the Math. Inst.* **42.** Trans. J. J. Schon-Kon and P.S.T. staff. Publ. by Nat. Sci. Found. Washington, D.C. (1961), **444.**

MACCOLL, Hugh, (1877). Calculus of equivalent statements and integration limits. *Proc. Lond. math. Soc.* **9**, 9–20.

MCCALL, Storrs (Ed.), (1969). *Polish Logic*, 1920–1939. Introduction by Tadeusz Kotarbinski. Oxford Univ. Press. 406.

MOODY, E. A., (1953). *Truth and Consequence in Mediaeval Logic.* North-Holland Publ. Co. viii + 113.

MORGAN, Augustus de, (1847). *Formal Logic: or the calculus of reference necessary and probable.* London. xvi + 336.

(1864). On the syllogism, No. IV, and on the logic of relations. *Trans. Camb. Phil. Soc.* **10**, 331–58.

MOSTOWSKI, Andrzej, (1952). *Sentences undecidable in formalized Arithmetic.* North-Holland Publ. Co. viii + 117.

(1954). On definable sets of positive integers. *Fund. Math.* **34**, 81–112.

(1955). The present state of investigations on the foundations of mathematics. *Pol. Akad. Nauk Inst. Mat. Roz. Mat.* no. 9. Warszawa. 47.

MUČNIK, A. A., (1956). Nérazréšmosť prolémy svodimosti téorii algoritmov. (Negative answer to the problem of reducibility of the theory of algorithms.) *Dok. Acad. Nauk. S.S.S.R.* **108**, 194–7.

MYHILL, John R., (1950). A complete theory of natural, rational and real numbers. *J.S.L.* **15**, 185–96.

(1955). Creative sets. *Zeit. f. Math. Log. u. Grund. d. Math.* **1**, 97–108.

NEUMANN, John v., (1925). Eine Axiomatizierung der Mengenlehre. *Jour.r. angew. Math.* **154**, 219–40.

(1927). Zur Hilbertschen Beweistheorie. *Math. Zeit.* **26**, 1–46.

NICOD, Jean, (1916). A reduction in the number of primitive propositions of logic. *Proc. Camb. Phil. Soc.* **19**, 32–41.

OGLESBY, F. C., (1962). An examination of a Decision Procedure. *Memoir of the Am. Math. Soc.* **44**, 148.

PEANO, Giuseppe, (1936). See bibliography of symbolic logic. *J.S.L.* **1**, 139.

PEIRCE, C. S., (1933). *Collected works of Charles Sanders Pierce.* Ed. by C. H. Hartshorne and P. Weiss. Cambridge, Mass. *Original papers* (1867–1905). *Vol.* 2, *Elements of Logic. Vol.* 3, *Exact Logic. Vol.* 4, *The simplest Mathematics.*

PÉTER, Rózsa, (1951). *Rekursive funktionen.* Akad. Kiado. Budapest. 206.

POINCARÉ, Henri, (1905). Les mathématiques et la logique. *Rev. metaph. mor.* **13**, 815–35. Reprinted in *Science et Methode.* Paris (1908), 311.

POST, Emil L., (1921). Introduction to a general theory of elementary propositions. *Am. J. Math.* **43**, 163–85.

(1936). Finite combinatory processes – Formulation I. *J.S.L.* **1**, 103–5.

(1943). Formal reductions of the general combinatorial decision problem. *Am. J. Math.* **65**, 197–215.

(1944). Recursively enumerable sets of positive integers and their decision problem. *Bull. Am. math. Soc.* **50**, 284–316. Reprinted in M. Davis. *The Undecidable.* (1967).

(1965). *Absolutely unsolvable problems and relatively undecidable propositions. Account of an anticipation. The Undecidable.* M. Davis. Raven Press Books Ltd. 338–433.

QUINE, Willard van Oman, (1951). *Mathematical Logic.* Harvard Univ. Press. Revised ed.

RAMSEY, Frank Plumpton, (1920). The foundations of mathematics. *Proc. Lond. Math. Soc.* 2s. **25**, 338–84. Reprinted in *The foundations of mathematics and other logical essays.* Kegan, Paul, Trench, Trubner and Co. Ltd. xxviii + 292.

(1928). On a Problem of Formal Logic. *Proc. Lond. Math. Soc.* Ser. 2. **30**, 338–84. Reprinted in *The foundations of mathematics and other logical essays.*

RICE, H. G., (1953). Classes of recursively enumerable sets and their decision problems. *Trans. Am. math. Soc.* **74**, 358–86.

(1954). Recursive real numbers. *Proc. Am. math. Soc.* **5**, 784–91.

RIEGER, Ladislav, (1969). *Algebraic Methods of Mathematical Logic.* Academic Press. 210.

ROBINSON, Abraham, (1966). *Non-standard Analysis.* North-Holland Publ. Co. xi + 293.

ROBINSON, Raphael M., (1947). Primitive recursive functions. *Bull. Amer. Math. Soc.* **53**, 925–942.

ROGERS, Hartly Jr., (1967). *Theory of recursive functions and effective computability.* McGraw Hill. xix + 482.

ROSE, G. F. & ULLIAN, J. S., (1963). Approximations of functions on the integers. *Pac. J. Math.* **13**, 693–701.

ROSENBLOOM, Paul, (1945). An elementary constructive proof of the fundamental theorem of algebra. *Amer. math. Monthly*, **52**, 562–70.

(1949). *The elements of mathematical logic.* Dover Press. iv + 214.

ROSSER, J. Barkley, (1936). Extensions of some theorems of Gödel and Church. *J.S.L.* **1**, 87–91.

(1963). *Logic for mathematicians.* McGraw Hill. xiv + 530.

ROSSER, J. Barkley, (1969). *Boolean-valued models of set theory*. Academic Press. 210.

RUSSELL, Bertrand, (1903). *The Principles of mathematics, Vol. I.* (All published.) Cambridge University Press. xxix + 534.

(1905). On denoting. *Mind. n.s.* **14**, 479–93. Reprinted in *Bertrand Russell. Logic and Knowledge*. George Unwin. Ed. R. C. Marsh. x + 382.

(1908). Mathematical logic as based on the theory of types. *Amer. jour. Math.* **30**, 222–62. Reprinted in *Logic and Knowledge.*

(1906). Les paradoxes de la logique. *Rev. metaph. mor.* **14**, 627–50.

(1919). *Introduction to Mathematical Philosophy*. George Allen and Unwin. vii + 208.

*See also* Whitehead, A. N. and Russell.

RYLL-NARDZEWSKI, C., (1952). The role of the axiom of induction in elementary arithmetic. *Fund math.* **39**, 239–63.

SACKS, Gerald, (1963). Degrees of unsolvability. *Ann. Math. Studies*, no. 55. 174.

SCHMIDT, Arnold, (1938). Über deduktive Theorien mit mehreren Sorten von Grunddingen. *Math. Ann.* **115**, 485–506.

SCHOLZ, Heinrich, (1931). *Geschichte der Logik*. Berlin. vii + 78.

SCHÖNFINKEL, Moses, (1924). Über der baustein der Mathematischen Logik. *Math. Ann.* **92**, 305–16.

SCHRÖDER, Ernst, (1890). *Vorlesungen über die Algebra der Logik*. Vol. I, xii + 717. Vol. II, pt 1, xiii + 400 (1891). Vol. III, pt 1, (1895), viii + 649. Vol. II, pt 2, xxix + 205, with Life of Schröder (1905). Reprinted by Chelsea Publ. Co. (1966).

SCHÜTTE, Kurt, (1934–5). Über der erfüllbarkeit einer Klasse von logischen Formeln. *Math. Ann.* **110**, 161–94.

(1950). Schlussweisen-Kalküle der prädikatenlogik. *Math. Ann.* **122**, 47–65.

(1960). *Beweistheorie*. J. Springer. Berlin. x + 356.

SCOTT, Dana, (1966). *Boolean Valued Models for Higher Order Logic* (hectographed). Stanford.

(1966). *A proof of the independence of the Continuum Hypothesis* (mimeographed). Stanford.

SCOTT, Dana & SOLOVAY, R., (1970). *Boolean Valued Models for Set Theory*. Proc. Amer. Math. Soc. Summer Inst. Axiomatic Set Theory, 1967. University of California, Los Angeles. *Proc. Symposia pure Math.* **13** (1969).

SHAPIRO, N., (1956). Degrees of Computability. *Trans. Amer. Math. Soc.* **82**, 281–99.

SHEFFER, H. M., (1913). A set of five independent postulates for Boolean Algebras, with application to logical constants. *Trans. Amer. Math. Soc.* **14**, 481–8.

SHEPHERDSON, J. C., (1951). Inner models for set theory. *J.S.L.* **16**, 161–90; II, *J.S.L.* **17** (1952), 225–237; III, *J.S.L.* **18** (1953), 145–67.

SHOENFIELD, J. R., (1967). *Mathematical Logic*. Addison Wesley Publ. Co. vii + 344.

SIERPINSKI, W., (1928). *Nombres Transfinis*. Paris. Gauthier-Villais. vi + 240.

(1934). Hypothese du Continue. *Warsaw Monografie Mat.* **4**, v + 192.

(1958). Cardinal and ordinal Numbers. *Polska Akad. Nauk.* **34**, 487.

SIKORSKI, Roman, (1960). *Boolean Algebras*. Springer Verlag. ix + 176.

SKOLEM, Thoraf, (1919). Untersuchungen über die Axioms des Klassenkalkuls und über Produktations–und Summationsprobleme, welche gewisse Klassen von Aussagen betreffen. *Skrifter utgit av Vidensk. i. Kristiania. I Nat-Mat. Kl.* no. **3**, 37.

(1920). Logische-kombinatorische Untersuchungen über die Erfüllbarkeit oder Beweisbarkeit mathematische Sätz nebst einem Theorem über dichte Mengen. *Vid. Schr. I. Mat-Nat. Kl.* No. **4**, 36.

(1923). Begründung der elementaren Arithmetik dirch die rekurrierende Denkweise ohne Anwendung scheinbarer veränderlichen mit unendlichen Ausdehnungsbereich. *Vid. Skr. I. Mat-Nat. Kl.* no. **6**, 38.

(1934). Über die Nicht-charakterisierbarkeit der Zahlenreihe mittels endliche oder abzählbar unendliche vieler Aussagen mit ausschliesslic Zahlenvariablen. *Fund. math.* **23**, 150–61.

(1950). Schlussweisen-Kalküle der Pradikatenlogik. *Math. Ann.* **122**, 47.

(1959). Reduction of axiom systems with axiom schemes to systems with only simple axioms. *Logica Studia. Paul Bernays dedicata.* Griffon. Neuchatel. Suisse. 239–46.

(1962). *Abstract Set Theory.* Notre Dame, Indiana. 70.

SKOLIMOWSKI, Henryk, (1969). *Polish Analytical Philosophy.* Routledge and Kegan Paul. 275.

SMULLYAN, R. M., (1961). Theory of Formal Systems. Princeton. *Annals of math. Studies*, no. **47**, x + 142.

SPECKER, E., (1949). Nicht Konstructive beweisbare Sätze der Analysis. *J.S.L.* **55**, 145–58.

SPECTOR, Clifford, (1956). On degrees of recursive unsolvability. *Ann. Math.* **64**, 581–92.

(1958). Strongly invariant hierarchies. *Notices Am. math. Soc.* **5**, 851.

(1959). Hyperarithmetical quantifiers. *Fund. math.* **48**, 313–20.

SUPPES, P., (1960). *Axiomatic Set Theory.* van Nostrand. xii + 265.

SURANYI, Janos, (1943). A logikao függvénykakkulus eldöntésproblémájának nedukciójáról (on the reduction problem of the logical function calculus). *Különlenyomat a Mat. e. Fiz. Lapok.* **50.** kötenböl. Budapest. 51–74.

(1959). Reduktionsthéorie des Entscheidungsproblems im Prädikatenkalkül der ersten Stufe. *Hung. Akad. d. Wiss.* Budapest, 216.

TARSKI, Alfred, (1956). *The concept of truth in formalised languages.* Trans. by J. H. Woodger in *Logic, Semantics, metamathematics. Papers from* 1923–1938 *by Alfred Tarski.* Oxford, Clarendon Press. xiv + 471.

(1935). Grundzüge des Systemenkalkül I. *Fund. math.* **25**, 283–301; II, **26** (1936), 283–301. Trans. in Logic, Semantics, Metamathematics.

(1949). On essential undecidability. *J.S.L.* **14**, 75–6.

(1956). *Fundamental Concepts of the methodology of the deductive Sciences. Logic, Semantics, Metamathematics.* Oxford. 60–109. First publ. (1930). *Monatsch f. Math. u. Phys.* 361–404.

TARSKI, Alfred, MOSTOWSKI, A. & ROBINSON, R. M., (1953). *Undecidable Theories.* North-Holland Publ. Co. xi + 98.

TURING, ALAN M., (1936). On computable numbers, with an application to the Entscheidungsproblem. *Proc. Lond. math. Soc.*, ser. 2. **42**, 230–65; **43**, 544–6. Reprinted in M. Davis *The Undecidable.*

(1939). Systems of logics based on ordinals. *Proc. Lond. math. Soc.* ser. 2. **45**, 161–228. Reprinted in M. Davis *The Undecidable.*

VEBLEN, O., (1904). *Trans. Am. math. Soc.* **5**, 346–47.

WANG, Hao, (1952). Logic of many-sorted theories. *J.S.L.* **17**, 105–116. Reprinted in *A Survey of Mathematical Logic*. North-Holland Publ. Co (1963). 322–33.

(1953). Certain predicates defined by induction schemata. *J.S.L.* **18**, 49–59.

(1958). Eighty years of Foundational Research. *Bernays Festschrift, Dialectica* **12**, 465–97. Reprinted in *A Survey of Mathematical Logic*.

(1963). *A Survey of Mathematical Logic*. Science Press, Peking. x + 651.

WEYL, Hermann, (1918). *Das Kontinuum*. Leipzig. iv + 83. Reprinted, Chelsea Publ. Co.

WHITEHEAD, Alfred North & RUSSELL, Bertrand, (1925–7). *Principia Mathematica* 3 vols. 2nd ed. Cambridge University Press.

WIENER, Norbert, (1912–14). A simplification in the logic of relations. *Proc. Camb. Phil. Soc.* **17**, 387–90.

WITTGENSTEIN, Ludwig, (1922). *Tractatus Logico-Philosophicus*. Kegan, Paul, Trench, Trubner & Co., London. German with English translation on facing page. 189 bis.

YATES, C. E. M., (1965). Three theorems on the degree of recursively enumerable sets. *Duke math. J.* **32**, 461–8.

ZERMELO, Ernst, (1904). Beweis das jede Menge wohlgeordnet werden kann. *Math. Ann.* **59**, 514–16.

(1908). Untersuchungen über die Grundlagen der Mengenlehre. I. *Math. Ann.* **65**, 261–81.

# Index